Advances in Neurobiology

Volume 23

Series Editor
Arne Schousboe

More information about this series at http://www.springer.com/series/8787

Mauro DiNuzzo • Arne Schousboe
Editors

Brain Glycogen Metabolism

Springer

Editors
Mauro DiNuzzo
Center for Basic and Translational Neuroscience
University of Copenhagen
Copenhagen, Denmark

Arne Schousboe
Faculty of Health and Medical Sciences
University of Copenhagen
Copenhagen, Denmark

ISSN 2190-5215 ISSN 2190-5223 (electronic)
Advances in Neurobiology
ISBN 978-3-030-27479-5 ISBN 978-3-030-27480-1 (eBook)
https://doi.org/10.1007/978-3-030-27480-1

This Springer imprint is published by the registered company Springer Nature Switzerland AG
The registered company address is: Gewerbestrasse 11, 6330 Cham, Switzerland

Preface

Glycogen is the sole carbohydrate reserve of the brain. Glycogen granules are located in astrocytic processes surrounding neuronal elements, where they can be readily mobilized at times of increased energy need (e.g., during sensory stimulation) or energy failure (e.g., when blood glucose supply is inadequate). Remarkably, the long-held concept that brain glycogen serves solely as an emergency depot has been recently displaced by numerous, unequivocal, and fundamental observations that related its utilization to the support of cognitive functions. Any interference with brain glycogen metabolism affects neurophysiology at multiple hierarchical scales, including cellular (e.g., astrocyte-neuron interactions), network (e.g., neuronal excitability), and system (e.g., learning and memory) levels. Not surprisingly, specific forms of glycogen storage disease are associated with mental retardation, and glycogen has been implicated in aging and several pathological conditions, such as epilepsy, diabetes, and dementia. Glycogen turnover is under dynamic control by neurotransmitters, neuromodulators, and extracellular potassium, or put simply, brain activity.

It goes without saying that this seemed to us a particularly good time to have a book about brain glycogen. While many books have been published about energy metabolism in the central nervous system, none has focused specifically on brain glycogen. Cerebral glycogen metabolism is only briefly touched upon in many neuroscience books, and, with the exception of a couple of special issues, there is a substantial lack of journal titles centered on brain glycogen. It is our opinion that brain glycogen research is more than mature to have a dedicated book. Our idea of such a book has been that of providing an unbiased state-of-the-art summary about the current knowledge of brain glycogen metabolism. The underlying principal aim has been to allow the reader appreciating how the relatively small cerebral glycogen content is so crucial for normal brain function, thereby, fundamentally revisiting the vestigial nature commonly associated with the polysaccharide in the brain. As such, the book fills a gap in current literature. Leading international experts have contributed up-to-date accounts of cerebral glycogen metabolism, covering the most important progresses obtained in the last 25 years. Our intent has been to be as inclusive as possible, especially for the topics that are still without any apparent consensus. Indeed, while there is no doubt that glycogen is essential to brain function, there are ongoing debates about the relevant biochemical mechanisms. Thus, opposite views normally found as sealed-off compartments in scientific publications finally could coexist in one place. The book chapters accordingly discuss different potential fates for glycogen-derived metabolites and related metabolic pathways that might be implicated in supporting the energetics of neurons and glial cells. Some chapters review the metabolic and functional features of brain glycogen as well as of the brain-specific enzyme isoforms that synthetize and degrade the polysaccharide, which are exquisitely sensitive to small and physiological energy fluctuations. Other chapters cover the localization of glycogen in different brain structures and the relevant metabolic aspects in either white or gray matter. Several chapters describe how physiological and pathological conditions affect glycogen metabolism in the brain. The book also includes historical perspectives that outline how brain glycogen has progressively gained attention as an essential player in cerebral energy metabolism. We believe that the material contained in the book constitutes a good reference about brain glycogen metabolism, which holds a great potential for stimulating further research.

We express our sincere gratitude to all the authors of the book, and we are more than confident that many others will join us, including undergraduate and graduate students as well as researchers and professionals working with or interested in brain glycogen metabolism. All these people will benefit from having the current knowledge of the field at hand. We also thank Springer Science for the continuous support and helpful insights throughout all the stages of putting this work together. Finally, we wish to acknowledge the invited researchers who could not participate to the present endeavor, as many among them have had a pivotal role in developing new ideas and performing critical experiments that advanced the field of brain glycogen metabolism. We hope to have given their work the appropriate credit within the various chapters of the book. One of these researchers is our friend Leif Hertz (1930–2018), to whose memory this book is dedicated.

Copenhagen, Denmark Mauro DiNuzzo
Arne Schousboe

Contents

Major Advances in Brain Glycogen Research: Understanding of the Roles of Glycogen Have Evolved from Emergency Fuel Reserve to Dynamic, Regulated Participant in Diverse Brain Functions

Gerald A. Dienel and Gerald M. Carlson

Abstract Brain glycogen is extremely difficult to study because it is very labile to physiological status and postmortem autolysis, and glycogen degradative enzymes are rapidly activated by metabolites and signaling molecules. Glycogen is predominantly located within astrocytes in adult brain, and abnormal glycogen metabolism in neurons has lethal consequences. Diverse distribution of glycogen among subcellular compartments suggests local regulation and different functional roles, and recent studies have revealed critically important roles for glycogen in normal brain function and Lafora disease. This brief overview highlights some of the major advances in elucidation of glycogen's roles in astrocytic functions and neurotransmission and the severe consequences of aberrant neuronal glycogen metabolism.

Keywords Astrocyte · Brain · Glycogen · Glycogen debranching enzyme · Glycogen granule · Glycogen phosphorylase · Phosphorylase kinase · Lafora disease (Lafora progressive myoclonic epilepsy, MELF) · Neuron · Neurotransmission

---

G. A. Dienel (✉)
Department of Neurology, University of Arkansas for Medical Sciences, Little Rock, AR, USA

Department of Cell Biology and Physiology, University of New Mexico, Albuquerque, NM, USA
e-mail: gadienel@uams.edu

G. M. Carlson
Department of Biochemistry and Molecular Biology, University of Kansas Medical Center, Kansas City, KS, USA
e-mail: gcarlson@kumc.edu

M. DiNuzzo, A. Schousboe (eds.), *Brain Glycogen Metabolism*, Advances in Neurobiology 23, https://doi.org/10.1007/978-3-030-27480-1_1

Abbreviations

cAMP	cyclic AMP
CMR_{glc}	Cerebral metabolic rate for glucose
CMR_{O2}	Cerebral metabolic rate for oxygen
Glc	Glucose
Glc-6-P	Glucose-6-phosphate
KO	Knockout

1 Introduction

Studies on glycogen metabolism in liver, where it contributes to maintenance of blood glucose, and in muscle, where it supplies energy to sustain contraction, have a long and illustrious history, and many luminaries of science have worked on this polyglucan. Glycogen was discovered in the nineteenth century (1857) by the great physiologist Claude Bernard, who for decades investigated mammalian sugar metabolism (Young 1957). The following year, August Kekulé, a key figure in developing a structural theory of organic chemistry, accurately determined glycogen's empirical formula to be $C_6H_{10}O_5$ (Young 1957), consistent with a polymer of glucose linked via glycosidic bonds. In the twentieth century, work on glycogen metabolism continued, and directly resulted in six Nobel Prizes. At the level of enzymes that act directly on glycogen, Carl and Gerty Cori were awarded (1947) the honor for their work on glycogen phosphorylase, whereas Luis Leloir received the award (1970) for his studies on glycogen synthase. For regulatory aspects of the cascade activation of glycogenolysis, Earl Sutherland was honored (1971) for identifying cAMP as the second messenger linking epinephrine to the activation of glycogen phosphorylase. Finally, their discovery of the stimulation of the catalytic activity of glycogen phosphorylase and phosphorylase kinase through post-translational phosphorylation resulted in the award being presented (1992) to Edmond Fischer and Edwin Krebs. Studies of glycogen metabolism in the twenty first century have been steadily expanding, however, from muscle and liver into the brain, where glycogen seems to play different and sometimes surprising roles.

As an important energy reserve in brain, glycogen is predominantly localized in astrocytes, with very small amounts in neurons. Regulation of astrocytic glycogenolysis by neurotransmitters integrates astrocytic metabolism with neuronal signaling, and dysregulation of glycogen turnover causes severe epilepsy and, ultimately, death of patients with Lafora disease. Glycogen fuels K^+ uptake into cultured astrocytes in the presence of adequate levels of glucose, revealing a role in the energetics of control of the extracellular ionic milieu during neurotransmission. Astrocytic glycogen mobilization during brain activation can also spare blood-borne glucose for use by neurons as energy demand is increased. Glycogenolysis is required for learning and memory consolidation, but the mechanisms remain to be established. Putative functions (that are not mutually exclusive) include fuel for astrocytes, precursor for astrocytic neurotransmitter glutamate synthesis, fuel for neurons via shuttling of glycogen-derived lactate, and release of lactate to act as a signaling molecule that can regulate neuronal activity (Bergersen 2015; Bozzo et al. 2013; Tang et al. 2014). Our understanding of functions of brain glycogen, its metabolism, and its roles in pathophysiology is far from complete, thus brain glycogen is an important topic for further study.

Characterization of the levels, turnover, location, and functions of brain glycogen actually has a long history marked by major advances in methodology followed by slower but steady progress, with technical difficulties being a persistent problem. Initially, global methods were used in conjunction with chemical analyses, but these procedures were vulnerable to lability of glycogen and impurities in the isolated fractions. Development of microanalytical enzymatic methods and specialized tissue freezing and enzyme inactivation techniques enabled detailed quantification of regional distributions of glycogen and enzymes of glycogen metabolism in relation to those of the glycolytic and oxidative pathways. Use of knockout (KO) mice as models for Lafora disease provided remarkable insights into the roles of glycogen in epilepsy. In spite of many constraints, recent studies provide fascinating perspectives of the participation and regulation of glycogen in normal brain functions and disorders.

This prologue highlights selected discoveries that have had a high impact on brain glycogen research that are discussed in more detail in the chapters in the present volume. Interested readers are also referred to a recent minireview series on brain glycogen in the *Journal of Biological Chemistry* (Prats et al. 2018; Nadeau et al. 2018; Bak et al. 2018; Gentry et al. 2018; Carlson et al. 2018) that also address key aspects of brain glycogen biochemistry and neurobiology.

2 Brain Glycogen as an Emergency Energy Reserve

Glycogen is a macromolecular storage form of glucose, the obligatory fuel for brain, and along with glucose, ATP, and phosphocreatine, glycogen is one of the brain's energy stores. The presence of glycogen in brain was reported in the 1880s, but it was not until the 1930s that brain glycogen was carefully characterized, when classical work was carried out by Stanley E. Kerr and colleagues at the American University of Beirut and published in the *Journal of Biological Chemistry*. Kerr established the requirement for rapid *in situ* freezing of brain to obtain valid values for concentrations of phosphocreatine and lactic acid (Kerr 1935; Avery et al. 1935) and used these procedures to minimize postmortem metabolism of glycogen. He then devised methods to minimize interference from cerebrosides and other contaminates, and reported brain glycogen concentrations (in glucosyl units) of 3.9–5.5 μmol/g (Kerr 1936). Kerr proved that the polysaccharide he purified from brain was, in fact, glycogen, based on elemental composition, optical rotation, quantity of glucose produced by acid hydrolysis, reaction with iodine, melting point of its osazone, and other properties (Kerr 1938). Postmortem metabolism of brain glucose was found to be more rapid than glycogen, and the concomitant rise in lactate level corresponded to the loss of glucose and glycogen, indicating both to be precursors of lactate (Kerr and Ghantus 1937). Insulin-induced hypoglycemia depleted glycogen in brain, muscle, and liver (Kerr and Ghantus 1936), whereas epinephrine injections, which reduced liver and skeletal muscle glycogen levels, did not affect brain glycogen content (Kerr et al. 1937), nor did short-term exposure to various anesthetics or several convulsion-inducing drugs (Kerr and Antaki 1937). Kerr's pioneering work was the first major technological advance in studies of brain glycogen that had a high impact on the field for more than 40 years. Unfortunately, the necessity to use special procedures to preserve labile metabolites is still not implemented by all researchers.

Rapid depletion of brain glycogen after death led to studies of synthesis and turnover of glycogen, activation of glycogen phosphorylase, and responses of brain glycogen level to various treatments. Re-synthesis of glycogen in brain slices after postmortem autolysis was slow and it could be influenced by experimental conditions, including inhibition by high [K^+] (LeBaron 1955; Kleinzeller and Rybová 1957). Potential physiological roles for glycogen as an *active participant* in brain metabolism were discussed in the early literature, which also reported regional and age-related differences in its concentration (Merrick 1961) and in the levels of enzymes of glycogen metabolism (Breckenridge and Crawford 1961). Studies of glycogen levels and turnover that used acid hydrolysis to degrade polysaccharides and anthrone reagent to assay monosaccharides were sometimes compromised by the presence of non-glucose carbohydrates that turned over more slowly than glycogen (Strang and Bachelard 1971). Metabolic labeling with [^{14}C]glucose has provided considerable information, but rapid turnover of the outer tiers of glycogen influences interpretation of label incorporation or loss in relation to glycogen concentration because degradation coupled with re-synthesis can remove recently-incorporated label and stabilize glycogen concentration. At the local and cellular level in brain, we need to learn considerably more about the control of the number and size of glycogen particles and the regulation of the enzymes involved in glycogenolysis (Nadeau et al. 2018).

A second technological milestone was the enormous improvement in the specificity and sensitivity of quantitative glycogen assays on impure preparations. The Lowry-Passonneau group developed enzymatic methods using phosphorylase (Lowry et al. 1967; Passonneau et al. 1967) or amyloglucosidase (Passonneau and Lauderdale 1974; Lust et al. 1975) to degrade the glycogen, followed by fluorescent-enzymatic assay of glucose-1-phosphate or glucose (Lowry and Passonneau 1972). Their microanalytical analyses of 0.1–1 μg dry-weight samples revealed 300-fold concentration differences in glycogen levels among nine layers of rabbit retina, a much larger range than those of eight other metabolites (Matschinsky et al. 1968). On the other hand, glycogen levels were similar in five layers of mouse cerebral cortex, but higher in subcortical white matter (Folbergrova et al. 1970). After labeling with [^{14}C]glucose, use of phosphorylase, with or without glycogen debranching enzyme, enabled calculation of a 4.4-h turnover time in vivo for total glycogen in normal mouse brain, compared with 0.86 h for the outer tiers and 8.6 h for limit dextrin (Watanabe and Passonneau 1973). For comparison, the half-life for glucose in non-stimulated rat brain is ~1.5 min (Savaki et al. 1980). Enzymatic assays of glycogen opened the field to a larger group of neuroscientists. Faster turnover of outer tiers raises the possibility that these structural elements are most likely to be involved in moment-to-moment brain functions, and subcellular compartmentation of glycogen in endfeet, perisynaptic structures, and cytosol (see below) suggests localized roles and regulation, but lack of quantitative assays with high temporal-spatial resolution hinders progress.

A third major methodological advance was preservation of labile brain metabolites by new approaches [funnel-freezing (Ponten et al. 1973a; Ponten et al. 1973b), freeze-blowing (Veech et al. 1973), freeze-clamping (Quistorff 1975), and microwave fixation (Medina et al. 1975)], with each procedure having advantages and limitations, including use of anesthesia, stress, and ability to assay regions of interest. It is extremely difficult to fully prevent peri-mortem and postmortem changes in glycogen levels and label content because subject handling, environmental stimuli, and fixation procedures can involve stress- and sensory-related responses that cause glycogenolysis secondary to actions of neurotransmitters, especially norepinephrine released throughout the brain from the locus coeruleus. However, delays in postmortem freezing were useful in revealing the rapid activation of glycogen phosphorylase. For example, 18% of mouse brain glycogen phosphorylase was in the active form (i.e., activity assayed without added AMP as percent of that with 1 mM AMP) in cerebral cortex after in situ freezing through the skull, whereas 70% was in the active form when freezing was delayed for ~1 min (Breckenridge and Norman 1962). Freeze-blowing or freezing through the intact skull also showed ~20% in the active form, whereas it was only 8% after freezing through the intact dura, and it was two-to three-fold higher (~40–60%) after immersion of rats into liquid N_2 or decapitation into liquid N_2 (Lust et al. 1973; Folbergrova et al. 1978). These findings emphasize the significance of Kerr's work and the need for greater understanding of the regulation of brain glycogenolysis.

A fourth methodological advance was the use of ^{13}C NMR to detect glycogen non-invasively, which was first performed in muscle, where the high level of glycogen allows detection of its natural abundance C-1 resonance (Shulman et al. 1990; Jue et al. 1989a, 1989b) Because of the low level of glycogen in the brain this signal is too weak for reliable quantification. Consequently, the glycogen must be isotopically enriched through administration of a labeled glycogen precursor, such as [1-^{13}C]glucose. The tiered structure of glycogen makes labeling of its inner chains somewhat problematic, causing this technique to be better suited for metabolic studies than analysis of concentration. Nevertheless, *in vivo* measurements on rat brain have been performed, and the glycogen reported to be as high as 5.1 μmol/g (Choi et al. 1999), equivalent to the concentrations cited above. A recent study in human brains that utilized prolonged administration of [1-^{13}C]glucose in an attempt to achieve steady-state labeling of glycogen coupled with mathematical modeling based on turnover resulted in an even higher value of 7.8 μmol/g (Öz et al. 2015; DiNuzzo 2013).

Studies in many laboratories over several decades showed: (i) the resting brain glycogen level was relatively stable within the range of ~2–3 μmol/g, i.e., approximating or exceeding brain glucose concentration, with subsequent higher levels of 4–8 μmol/g determined with enzymatic assays in extracts of microwave-fixed brain and *in vivo* ^{13}C NMR assays, (ii) glycogen was consumed during pathophysiological conditions, e.g., ischemia, spreading cortical depression, seizures, and hypoglycemia, and (iii) glycogen levels increased and turnover slowed during prolonged, deep anesthesia. Of note, glycogen alone can support normal brain activity for only a few minutes: the resting rate of glucose (Glc) utilization (CMR_{glc}, Cerebral Metabolic Rate for the substrate designated in the subscript) in rat cerebral cortex is ~1 μmol/g/min (Sokoloff et al. 1977), and glycogen and glucose are depleted in <1 min during ischemia (Lowry et al. 1964). Considered together, the slow turnover under resting conditions, the relative stability to various normal physiological conditions, the rapid activation of phosphorylase, and the fast disappearance of glycogen during energy failure led to the long-held concept that brain glycogen was mainly an *emergency fuel reservoir.*

3 Cellular Localization of Glycogen in Brain

Astrocytes are the predominant, but not exclusive, brain cell type containing glycogen. Early histochemical assays in rapidly-fixed tissue to minimize autolysis (Shimizu and Kumamoto 1952), and early light and electron microscopic studies identified glycogen granules throughout the astrocytic cytoplasm, including their endfeet that surround the vasculature and in perisynaptic regions. Assays also demonstrated that astrocytic glycogen levels increased after exposure to ionizing radiation (Wolfe et al. 1962; Maxwell and Kruger 1965), prolonged barbiturate anesthesia (Phelps 1972), and treatment with the glutamine synthetase inhibitor methionine sulphoximine (Phelps 1975). Ultrastructural studies of glycogen phosphorylase and phosphorylase kinase demonstrated their predominant distribution throughout astrocytic cytosol, endfeet, and perisynaptic processes, with both brain and muscle isoforms of phosphorylase being co-expressed in the same astrocytes, and the brain isoform being expressed in very few neurons (Richter et al. 1996; Psarra et al. 1998; Pfeiffer-Guglielmi et al. 2000, 2003). Neuronal glycogen was usually undetectable, but with a few exceptions, e.g., neuronal dendrites in the vestibular nucleus were packed with mitochondria and glycogen granules (Sotelo and Palay 1968). More recently, through use of very sensitive assays, low levels of glycogen, glycogen phosphorylase, and glycogen synthase were detected in neurons (Saez et al. 2014). *In vitro* studies in brain slices and cultured astrocytes have been particularly important for studies of regulation and function that cannot be carried out *in vivo*, and they identified many compounds that influenced glycogen metabolism, e.g., β-adrenergic blocking drugs, amphetamine-like compounds, norepinephrine, serotonin, histamine, adenosine, vasoactive intestinal peptide, K^+, Ca^{2+}, and various receptor agonists and antagonists e.g., (Quach et al. 1988; Magistretti et al. 1986; Quach et al.

1978, 1980, 1981, 1982; Hof et al. 1988; Magistretti et al. 1981; Subbarao et al. 1995; Zhang et al. 1993; Subbarao and Hertz 1990). These findings led to linkage of astrocytic glycogen metabolism with neuronal signaling (e.g., glycogenolysis triggered by neurotransmitters) and control of the extracellular milieu (via astrocytic K^+ uptake fueled by glycogen). In contrast, functional studies of normal neuronal glycogen metabolism have barely scratched the surface.

A fifth methodological landmark in studies of glycogen was the use of molecular biological techniques to modify levels of glycogen and enzymes of its metabolism. For example, KO of glycogen synthase caused substantial deficits in cognitive ability, clearly demonstrating a requirement for glycogenolysis in learning and memory (Duran et al. 2013, 2019). Glycogen synthesis in neurons is strongly suppressed, but over-expression of protein targeting to glycogen led to neuronal glycogen accumulation and apoptosis (Vilchez et al. 2007). Animal models for Lafora disease have been created using different approaches: (i) a single KO of malin (an E3 ubiquitin ligase) or laforin (a glycogen phosphatase), (ii) a double KO of both malin and laforin, and (iii) a KO of either malin or laforin in conjunction with elimination of either protein targeting to glycogen or of glycogen synthase. These models have been instrumental in major advances to elucidate roles of malin, laforin, and glycogen in severe epilepsy and death (Gentry et al. 2018).

4 A Major Paradigm-Shift: Active Participation of Glycogen in Normal Brain Functions

The first report of a dynamic relationship between sensory stimulation and glycogenolysis in awake rats came from Swanson and colleagues (Swanson et al. 1992). These researchers showed that label is released from pre-labeled glycogen in two specific brain regions in which tactile whisker movement is known to increase neuronal activity: the whisker-barrel field of somatosensory cortex and the ventromedial thalamus. Glycogenolysis also occurs in the cerebral cortex during and after alerting and generalized sensory stimulation (Cruz and Dienel 2002) concomitant with the poorly-understood phenomenon called 'aerobic glycolysis' [i.e., the disproportionate increase in glucose utilization compared with oxygen consumption during brain activation (Dienel and Cruz 2016)]. In addition, the resting brain glycogen levels in ethanol extracts (to prevent autolysis that can occur during routine acid extraction) of funnel-frozen brain from very carefully-handled rats were much higher and more variable than generally reported [in the range of 12 μmol/g with coefficients of variation of 30–40% (Cruz and Dienel 2002)]. Similarly-high levels were also detected in microwave-fixed brain (Kong et al. 2002), with considerable regional and subcellular heterogeneity (Oe et al. 2016). If astrocytes contain most of this glycogen and account for ~20–25% of brain volume (Hertz 2011), their glycogen levels may be on the order of ~48–60 μmol/g, close to that in resting gastrocnemius muscle, ~70 μmol/g (Price et al. 1994). Note that this astrocytic volume fraction is probably overestimated, and may be in the range of 5–10% (Gundersen et al. 2015), with astrocytic glycogen levels being two- to tour-fold higher than calculated above. Brain glycogen levels are strongly influenced by animal handling, stress, and physiological activity, and label release from glycogen during activation of awake rats exceeded the percent decrease in concentration of unlabeled glycogen, suggesting that turnover of the outer tiers is enhanced during functional activity (Dienel and Cruz 2015; Dienel et al. 2007). Furthermore, inhibition of glycogenolysis caused large, regionally-heterogeneous increases in the local CMR_{glc} during sensory stimulation but not during rest, suggesting compensatory responses to lack of glycogen availability (Dienel et al. 2007) and major, unidentified roles for glycogen during brain activation. These findings clearly demonstrate that glycogen utilization in the presence of an adequate glucose supply occurs in living brain and raise the likelihood of functional compartmentation of astrocytic glycolytic pathways. Cultured astrocytes also consume glycogen in the presence of an adequate supply of glucose, e.g., during uptake of K^+ (Subbarao et al. 1995) and in response to oxidative stress (Rahman et al. 2000). Thus, glycogen is not just an emergency energy reserve; it is a dynamic, regulated participant in brain activity.

Liver glycogen has a key role in whole-body glucose homeostasis, and glucose-6-phosphate (Glc-6-P) derived from glycogen is, in part, dephosphorylated by Glc-6-phosphatase to generate glucose that is released to blood (Nordlie and Foster 2010). Muscle has ~1/40th the Glc-6-phosphatase activity of liver (Shieh et al. 2004) and may produce small amounts of glucose from glycogen (beyond the ~10% produced by the debranching enzyme), but most Glc-6-P is probably further metabolized by glycolysis. There are also reports that brain glycogen is degraded to produce glucose, but painstaking analysis of Glc-6-phosphatase activity in living rat brain (Dienel et al. 1988; Nelson et al. 1985, 1986) and in cultured astrocytes (Gotoh et al. 2000) revealed that its activity is negligible, even though the enzyme protein is detectable. In cultured astrocytes glycogenolysis is accompanied by lactate release to the medium (Dringen et al. 1993). Roles of glycogen in brain, muscle, and liver probably overlap somewhat, but are distinctly different with respect to other brain-specific functions and glucose vs. lactate production.

The fates of glycogen-derived Glc-6-P and pyruvate/lactate in brain *in vivo* under different conditions are still important, unresolved issues. When oxidative stress increases, glycogen is consumed and more Glc-6-P enters the pentose phosphate shunt pathway (Rahman et al. 2000). Metabolic modeling of brain activation led to the hypothesis that elevated levels of glycogen-derived Glc-6-P in astrocytes increase feedback inhibition of astrocytic hexokinase, so more blood-borne glucose is available for neurons (DiNuzzo et al. 2010, 2011). Glycolytic metabolism of glycogen-derived Glc-6-P provides 50% more ATP than glucose because the hexokinase step is by-passed, and rapid glycogenolytic generation of ATP may be important for the energetics of membrane transport processes. The pyruvate can be used by astrocytes as oxidative fuel, as an anaplerotic substrate to synthesize glutamate, glutamine, and aspartate, and/or converted to lactate. When glycogen-derived pyruvate is not oxidized, lactate must be released from these astrocytes to regenerate NAD^+ to sustain elevated glycolytic rates. Some lactate may be metabolized by other brain cells as fuel, but its oxidation is restricted by the accompanying smaller rise in oxygen consumption (CMR_{O2}) compared with blood-borne glucose utilization (and glycogen consumption that is usually not measured) during activation (Dienel and Cruz 2016). In fact, when the ratio of oxygen-to-glucose utilization (CMR_{O2}/CMR_{glc}) is calculated from arteriovenous differences for oxygen, glucose, and lactate, the ratio falls during activation, and if the glycogen consumed is included in the carbohydrate metabolized, the ratio drops further, suggesting that the metabolites of glycogen are not substantially oxidized. Glycogenolysis is claimed to involve astrocyte-neuron lactate shuttling during memory-evoking events and severe hypoglycemia or aglycemia. However, direct proof and quantification of astrocyte-neuron lactate transfer *in vivo* are lacking and the quantitative contribution of shuttled lactate to neuronal metabolism (versus glucose oxidation) is unknown but probably small (Dienel 2017a, b; 2019a; Dienel and Cruz 2016; Diaz-Garcia et al. 2017; Yellen 2018; Diaz-Garcia and Yellen 2019). The available data indicate that most of the lactate derived from the glucose and glycogen consumed in excess of oxygen is released from brain. Efflux of lactate from brain can occur by release to blood (Cruz et al. 1999) and to perivascular fluid (Ball et al. 2010). It has been proposed that brain lactate can be delivered to lymph nodes by perivascular fluid flow (Lundgaard et al. 2017); however, this conclusion is questioned because appropriate steps required to prevent postmortem metabolism (e.g., microwave fixation, freeze-blowing, or funnel freezing) were not carried out. The animals were decapitated and lymph nodes dissected out so that any glucose and glycogen present in the lymph nodes would be converted to lactate during the postmortem ischemic interval prior to tissue freezing. Furthermore, small molecules (molecular weight <500) in perivascular fluid are predominantly released to blood when the cerebral blood vessels traverse through the cribriform plate and the vessels become fenestrated instead of having tight junctions as in the brain (Bradbury and Westrop 1983). These findings indicate that the lactate recovered from the lymph nodes was predominantly from glycolytic metabolism in the node, not trafficking of lactate from the brain to the nodes. In terms of lactate shuttling within brain, if lactate were quantitatively oxidized, there would be stoichiometric equivalence of glucose plus glycogen utilization and oxygen consumption, which is not the case (Dienel 2017a, 2019a, b). Thus, lactate contributes to the manifestation of aerobic glycolysis when incoming blood-borne glucose cycles through glycogen to lactate that is released from brain. Re-synthesis of glycogen also contributes to aerobic glycolysis because glucose is consumed in excess of oxygen.

Lactate trafficking within brain after generation from glycogen and/or glucose also has potentially important consequences. For example, lactate signaling can have metabolic and transcriptional effects arising from shifts in intracellular pH, due to its co-transport with H^+ (the activity of phosphofructokinase is very pH sensitive (Halperin et al. 1969; Trivedi and Danforth 1966) and from changes in the $NADH/NAD^+$ ratio (Winkler and Hirrlinger 2015; Garriga-Canut et al. 2006; Kumar et al. 2002) via the lactate dehydrogenase reaction. Receptor-mediated signaling actions can modulate excitatory and inhibitory signaling, cAMP levels, and norepinephrine release (Bozzo et al. 2013; Lauritzen et al. 2014; Mosienko et al.

2015; Tang et al. 2014), thereby governing neuronal activity directly and indirectly. Lactate stimulates cerebral blood flow (Choi et al. 2012; Gordon et al. 2008; Hein et al. 2006; Yamanishi et al. 2006), and its release from astrocytic endfeet secondary to glycogenolysis would help increase fuel delivery to activated brain. These are exciting areas of on-going research that open new vistas to functions of glycogen in brain.

5 Concluding Remarks

There is a strong need for improved understanding of glycogen neurobiology: its composition and life cycle in neurons and astrocytes, differential regulation of enzymes involved in glycogenolysis in both major brain cell types, regulation of glycogen utilization during brain activity via intra- and inter-cellular signaling pathways and regulatory molecules, and consequences of inherited defects in neuronal glycogen metabolism. The reviews in this volume constitute a broad knowledge base from which our understanding of the complexity of glycogen turnover in brain activities and diseases can be extended. They should serve to stimulate advancement in this field by increasing research interest in brain glycogen and attracting application of new methodology and novel experimental approaches to study the normal and pathophysiological roles of glycogen in brain.

Acknowledgments *Conflict of Interest*: The authors declare that they have no conflicts of interest.

References

Avery BF, Kerr SE, Ghantus M (1935) The lactic acid content of mammalian brain. J Biol Chem 110(3):637–642

Bak LK, Walls AB, Schousboe A, Waagepetersen HS (2018) Astrocytic glycogen metabolism in the healthy and diseased brain. J Biol Chem 293(19):7108–7116. https://doi.org/10.1074/jbc.R117.803239

Ball KK, Cruz NF, Mrak RE, Dienel GA (2010) Trafficking of glucose, lactate, and amyloid-beta from the inferior colliculus through perivascular routes. J Cereb Blood Flow Metab 30(1):162–176. https://doi.org/10.1038/jcbfm.2009.206

Bergersen LH (2015) Lactate transport and signaling in the brain: potential therapeutic targets and roles in body-brain interaction. J Cereb Blood Flow Metab 35(2):176–185. https://doi.org/10.1038/jcbfm.2014.206

Bozzo L, Puyal J, Chatton JY (2013) Lactate modulates the activity of primary cortical neurons through a receptor-mediated pathway. PLoS One 8(8):e71721. https://doi.org/10.1371/journal.pone.0071721

Bradbury MW, Westrop RJ (1983) Factors influencing exit of substances from cerebrospinal fluid into deep cervical lymph of the rabbit. J Physiol 339:519–534

Breckenridge BM, Crawford EJ (1961) The quantitative histochemistry of the brain enzymes of glycogen metabolism. J Neurochem 7(3):234–240. https://doi.org/10.1111/j.1471-4159.1961.tb13508.x

Breckenridge BM, Norman JH (1962) Glycogen phosphorylase in brain. J Neurochem 9:383–392

Carlson GM, Dienel GA, Colbran RJ (2018) Introduction to the thematic minireview series: brain glycogen metabolism. J Biol Chem 293(19):7087–7088. https://doi.org/10.1074/jbc.TM118.00264

Choi IY, Tkac I, Ugurbil K, Gruetter R (1999) Noninvasive measurements of [1-(13)C]glycogen concentrations and metabolism in rat brain in vivo. J Neurochem 73(3):1300–1308

Choi HB, Gordon GR, Zhou N, Tai C, Rungta RL, Martinez J, Milner TA, Ryu JK, McLarnon JG, Tresguerres M, Levin LR, Buck J, Macvicar BA (2012) Metabolic communication between astrocytes and neurons via bicarbonate-responsive soluble adenylyl cyclase. Neuron 75(6):1094–1104. https://doi.org/10.1016/j.neuron.2012.08.032

Cruz NF, Dienel GA (2002) High glycogen levels in brains of rats with minimal environmental stimuli: implications for metabolic contributions of working astrocytes. J Cereb Blood Flow Metab 22(12):1476–1489. https://doi.org/10.1097/00004647-200212000-00008

Cruz NF, Adachi K, Dienel GA (1999) Rapid efflux of lactate from cerebral cortex during K+ – induced spreading cortical depression. J Cereb Blood Flow Metab 19(4):380–392. https://doi.org/10.1097/00004647-199904000-00004

Diaz-Garcia CM, Yellen G (2019) Neurons rely on glucose rather than astrocytic lactate during stimulation. J Neurosci Res 97(8):883–889. https://doi.org/10.1002/jnr.24374

Diaz-Garcia CM, Mongeon R, Lahmann C, Koveal D, Zucker H, Yellen G (2017) Neuronal stimulation triggers neuronal glycolysis and not lactate uptake. Cell Metab 26(2):361–374 e364. https://doi.org/10.1016/j.cmet.2017.06.021

Dienel GA (2017a) Lack of appropriate stoichiometry: strong evidence against an energetically-important astrocyte-neuron lactate shuttle in brain. J Neurosci Res 95(11):2103–2125. https://doi.org/10.1002/jnr.24015

Dienel GA (2017b) The metabolic trinity, glucose–glycogen–lactate, links astrocytes and neurons in brain energetics, signaling, memory, and gene expression. Neurosci Lett 637:18–25. https://doi.org/10.1016/j.neulet.2015.02.052

Dienel GA (2019a) Brain glucose metabolism: integration of energetics with function. Physiol Rev 99(1):949–1045. https://doi.org/10.1152/physrev.00062.2017

Dienel GA (2019b) Does shuttling of glycogen-derived lactate from astrocytes to neurons take place during neurotransmission and memory consolidation? J Neuro Res 97(8):863–882. https://doi.org/10.1002/jnr.24387

Dienel GA, Cruz NF (2015) Contributions of glycogen to astrocytic energetics during brain activation. Metab Brain Dis 30(1):281–298. https://doi.org/10.1007/s11011-014-9493-8

Dienel GA, Cruz NF (2016) Aerobic glycolysis during brain activation: adrenergic regulation and influence of norepinephrine on astrocytic metabolism. J Neurochem 138(1):14–52. https://doi.org/10.1111/jnc.13630

Dienel GA, Nelson T, Cruz NF, Jay T, Crane AM, Sokoloff L (1988) Over-estimation of glucose-6-phosphatase activity in brain in vivo. Apparent difference in rates of [2-^{3}H]glucose and [U-^{14}C]glucose utilization is due to contamination of precursor pool with ^{14}C-labeled products and incomplete recovery of ^{14}C-labeled metabolites. J Biol Chem 263(36):19697–19708

Dienel GA, Ball KK, Cruz NF (2007) A glycogen phosphorylase inhibitor selectively enhances local rates of glucose utilization in brain during sensory stimulation of conscious rats: implications for glycogen turnover. J Neurochem 102(2):466–478. https://doi.org/10.1111/j.1471-4159.2007.04595.x

DiNuzzo M (2013) Kinetic analysis of glycogen turnover: relevance to human brain ^{13}C-NMR spectroscopy. J Cereb Blood Flow Metab 33(10):1540–1548. https://doi.org/10.1038/jcbfm.2013.98

DiNuzzo M, Mangia S, Maraviglia B, Giove F (2010) Glycogenolysis in astrocytes supports blood-borne glucose channeling not glycogen-derived lactate shuttling to neurons: evidence from mathematical modeling. J Cereb Blood Flow Metab 30(12):1895–1904. https://doi.org/10.1038/jcbfm.2010.151

DiNuzzo M, Maraviglia B, Giove F (2011) Why does the brain (not) have glycogen? BioEssays 33(5):319–326. https://doi.org/10.1002/bies.201000151

Dringen R, Gebhardt R, Hamprecht B (1993) Glycogen in astrocytes: possible function as lactate supply for neighboring cells. Brain Res 623(2):208–214

Duran J, Gruart A, Varea O, López-Soldado I, Delgado-García JM, Guinovart (2019) Lack of neuronal glycogen impairs memory formation and learning-dependent synaptic plasticity in mice. Front in Cell Neurosci 13

Duran J, Saez I, Gruart A, Guinovart JJ, Delgado-Garcia JM (2013) Impairment in long-term memory formation and learning-dependent synaptic plasticity in mice lacking glycogen synthase in the brain. J Cereb Blood Flow Metab 33(4):550–556. https://doi.org/10.1038/jcbfm.2012.200

Folbergrova J, Lowry OH, Passonneau JV (1970) Changes in metabolites of the energy reserves in individual layers of mouse cerebral cortex and subjacent white matter during ischaemia and anaesthesia. J Neurochem 17(8):1155–1162

Folbergrova J, Nordström CH, Siesjo BK (1978) Labile metabolites and phosphorylase a in rapidly frozen rat cerebral cortex. J Neurochem 30(2):493–495

Garriga-Canut M, Schoenike B, Qazi R, Bergendahl K, Daley TJ, Pfender RM, Morrison JF, Ockuly J, Stafstrom C, Sutula T, Roopra A (2006) 2-Deoxy-D-glucose reduces epilepsy progression by NRSF-CtBP-dependent metabolic regulation of chromatin structure. Nat Neurosci 9(11):1382–1387. https://doi.org/10.1038/nn1791

Gentry MS, Guinovart JJ, Minassian BA, Roach PJ, Serratosa J (2018) Lafora disease offers a unique window into neuronal glycogen. J Biol Chem 293(19):7117–7125. https://doi.org/10.1074/jbc.R117.803064

Gordon GR, Choi HB, Rungta RL, Ellis-Davies GC, MacVicar BA (2008) Brain metabolism dictates the polarity of astrocyte control over arterioles. Nature 456(7223):745–749. https://doi.org/10.1038/nature07525

Gotoh J, Itoh Y, Kuang TY, Cook M, Law MJ, Sokoloff L (2000) Negligible glucose-6-phosphatase activity in cultured astroglia. J Neurochem 74(4):1400–1408

Gundersen V, Storm-Mathisen J, Bergersen LH (2015) Neuroglial Transmission. Physiol Rev 95(3):695–726. https://doi.org/10.1152/physrev.00024.2014

Halperin ML, Connors HP, Relman AS, Karnovsky ML (1969) Factors that control the effect of pH on glycolysis in leukocytes. J Biol Chem 244(2):384–390

Hein TW, Xu W, Kuo L (2006) Dilation of retinal arterioles in response to lactate: role of nitric oxide, guanylyl cyclase, and ATP-sensitive potassium channels. Invest Ophthalmol Vis Sci 47(2):693–699. https://doi.org/10.1167/iovs.05-1224

Hertz L (2011) Astrocytic energy metabolism and glutamate formation--relevance for ^{13}C-NMR spectroscopy and importance of cytosolic/mitochondrial trafficking. Magn Reson Imaging 29(10):1319–1329. https://doi.org/10.1016/j.mri.2011.04.013

Hof PR, Pascale E, Magistretti PJ (1988) K^+ at concentrations reached in the extracellular space during neuronal activity promotes a Ca^{2+}-dependent glycogen hydrolysis in mouse cerebral cortex. J Neurosci 8(6):1922–1928

Jue T, Rothman DL, Shulman GI, Tavitian BA, DeFronzo RA, Shulman RG (1989a) Direct observation of glycogen synthesis in human muscle with ^{13}C NMR. Proc Natl Acad Sci U S A 86(12):4489–4491

Jue T, Rothman DL, Tavitian BA, Shulman RG (1989b) Natural-abundance ^{13}C NMR study of glycogen repletion in human liver and muscle. Proc Natl Acad Sci U S A 86(5):1439–1442

Kerr SE (1935) Studies on the phosphous compounds of brain: I. Phosphocreatine. J Biol Chem 110(3):625–635

Kerr SE (1936) The carbohydrate metabolism of brain: 1. The determination of glycogen in nerve tissue. J Biol Chem 116:1):1–1):7

Kerr SE (1938) The carbohydrate metabolism of brain: VI. Isolation of glycogen. J Biol Chem 123(2):443–449

Kerr SE, Antaki A (1937) The carbohydrate metabolism of brain: V. The effect of certain narcotics and convulsant drugs upon the carbohydrate and phosphocreatine content of rabbit brain. J Biol Chem 122(1):49–52

Kerr SE, Ghantus M (1936) The carbohydrate metabolism of brain: II. The effect of varying the carbohydrate and insulin supply on the glycogen, free sugar, and lactic acid in mammalian brain. J Biol Chem 116(1):9–20

Kerr SE, Ghantus M (1937) The carbohydrate metabolism of brain: III. On the origin of lactic acid. J Biol Chem 117(1):217–225

Kerr SE, Hampel CW, Ghantus M (1937) The carbohydrate metabolism of brain: IV. Brain glycogen, free sugar, and lactic acid as affected by insulin in normal and adrenal-inactivated cats, and by epinephrine in normal rabbits. J Biol Chem 119(2):405–421

Kleinzeller A, Rybová R (1957) Glycogen synthesis in brain cortex slices and some factors affecting it. J Neurochem 2(1):45–57. https://doi.org/10.1111/j.1471-4159.1957.tb12352.x

Kong J, Shepel PN, Holden CP, Mackiewicz M, Pack AI, Geiger JD (2002) Brain glycogen decreases with increased periods of wakefulness: implications for homeostatic drive to sleep. J Neurosci 22(13):5581–5587

Kumar V, Carlson JE, Ohgi KA, Edwards TA, Rose DW, Escalante CR, Rosenfeld MG, Aggarwal AK (2002) Transcription corepressor CtBP is an NAD(+)-regulated dehydrogenase. Mol Cell 10(4):857–869

Lauritzen KH, Morland C, Puchades M, Holm-Hansen S, Hagelin EM, Lauritzen F, Attramadal H, Storm-Mathisen J, Gjedde A, Bergersen LH (2014) Lactate receptor sites link neurotransmission, neurovascular coupling, and brain energy metabolism. Cereb Cortex 24(10):2784–2795. https://doi.org/10.1093/cercor/bht136

LeBaron FN (1955) The resynthesis of glycogen by guinea pig cerebral-cortex slices. Biochem J 61(1):80–85

Lowry OH, Passonneau JV (1972) A flexible system of enzymatic analysis. Academic Press, New York

Lowry OH, Passonneau JV, Hasselberger FX, Schulz DW (1964) Effect of ischemia on known substrates and cofactors of the glycolytic pathway in brain. J Biol Chem 239:18–30

Lowry OH, Schulz DW, Passonneau JV (1967) The kinetics of glycogen phosphorylases from brain and muscle. J Biol Chem 242(2):271–280

Lundgaard I, Lu ML, Yang E, Peng W, Mestre H, Hitomi E, Deane R, Nedergaard M (2017) Glymphatic clearance controls state-dependent changes in brain lactate concentration. J Cereb Blood Flow Metab 37(6):2112–2124. https://doi.org/10.1177/0271678x16661202

Lust WD, Passonneau JV, Veech RL (1973) Cyclic adenosine monphosphate, metabolites, and phosphorylase in neural tissue: a comparison a methods of fixation. Science 181(4096):280–282

Lust WD, Passonneau JV, Crites SK (1975) The measurement of glycogen in tissues by amylo-alpha-1,4-alpha-1,6-glucosidase after the destruction of preexisting glucose. Anal Biochem 68(1):328–331

Magistretti PJ, Morrison JH, Shoemaker WJ, Sapin V, Bloom FE (1981) Vasoactive intestinal polypeptide induces glycogenolysis in mouse cortical slices: a possible regulatory mechanism for the local control of energy metabolism. Proc Natl Acad Sci U S A 78(10):6535–6539

Magistretti PJ, Hof PR, Martin JL (1986) Adenosine stimulates glycogenolysis in mouse cerebral cortex: a possible coupling mechanism between neuronal activity and energy metabolism. J Neurosci Off J Soc Neurosci 6(9):2558–2562

Matschinsky FM, Passonneau JV, Lowry OH (1968) Quantitative histochemical analysis of glycolytic intermediates and cofactors with an oil well technique. J Histochem Cytochem 16(1):29–39

Maxwell DS, Kruger L (1965) The fine structure of astrocytes in the cerebral cortex and their response to focal injury produced by heavy ionizing particles. J Cell Biol 25(2):141–157

Medina MA, Jones DJ, Stavinoha WB, Ross DH (1975) The levels of labile intermediary metabolites in mouse brain following rapid tissue fixation with microwave irradiation. J Neurochem 24(2):223–227

Merrick AW (1961) Encephalic glycogen differences in young and adult rats. J Physiol 158(3):476–485. https://doi.org/10.1113/jphysiol.1961.sp006781
Mosienko V, Teschemacher AG, Kasparov S (2015) Is L-lactate a novel signaling molecule in the brain? J Cereb Blood Flow Metab 35(7):1069–1075. https://doi.org/10.1038/jcbfm.2015.77
Nadeau OW, Fontes JD, Carlson GM (2018) The regulation of glycogenolysis in the brain. J Biol Chem 293(19):7087–7088. https://doi.org/10.1074/jbc.R117.803023
Nelson T, Lucignani G, Atlas S, Crane AM, Dienel GA, Sokoloff L (1985) Reexamination of glucose-6-phosphatase activity in the brain in vivo: no evidence for a futile cycle. Science 229(4708):60–62
Nelson T, Lucignani G, Goochee J, Crane AM, Sokoloff L (1986) Invalidity of criticisms of the deoxyglucose method based on alleged glucose-6-phosphatase activity in brain. J Neurochem 46(3):905–919
Nordlie RC, Foster JD (2010) A retrospective review of the roles of multifunctional glucose-6-phosphatase in blood glucose homeostasis: genesis of the tuning/retuning hypothesis. Life Sci 87(11–12):339–349. https://doi.org/10.1016/j.lfs.2010.06.021
Oe Y, Baba O, Ashida H, Nakamura KC, Hirase H (2016) Glycogen distribution in the microwave-fixed mouse brain reveals heterogeneous astrocytic patterns. Glia 64(9):1532–1545. https://doi.org/10.1002/glia.23020
Öz G, DiNuzzo M, Kumar A, Moheet A, Seaquist ER (2015) Revisiting glycogen content in the human brain. Neurochem Res 40(12):2473–2481. https://doi.org/10.1007/s11064-015-1664-4
Passonneau JV, Lauderdale VR (1974) A comparison of three methods of glycogen measurement in tissues. Anal Biochem 60(2):405–412
Passonneau JV, Gatfield PD, Schulz DW, Lowry OH (1967) An enzymic method for measurement of glycogen. Anal Biochem 19(2):315–326
Pfeiffer-Guglielmi B, Bröer S, Bröer A, Hamprecht B (2000) Isozyme pattern of glycogen phosphorylase in the rat nervous system and rat astroglia-rich primary cultures: electrophoretic and polymerase chain reaction studies. Neurochem Res 25(11):1485–1491
Pfeiffer-Guglielmi B, Fleckenstein B, Jung G, Hamprecht B (2003) Immunocytochemical localization of glycogen phosphorylase isozymes in rat nervous tissues by using isozyme-specific antibodies. J Neurochem 85(1):73–81
Phelps CH (1972) Barbiturate-induced glycogen accumulation in brain. An electron microscopic study. Brain Res 39(1):225–234
Phelps CH (1975) An ultrastructural study of methionine sulphoximine-induced glycogen accumulation in astrocytes of the mouse cerebral cortex. J Neurocytol 4(4):479–490
Ponten U, Ratcheson RA, Salford LG, Siesjo BK (1973a) Optimal freezing conditions for cerebral metabolites in rats. J Neurochem 21(5):1127–1138
Ponten U, Ratcheson RA, Siesjo BK (1973b) Metabolic changes in the brains of mice frozen in liquid nitrogen. J Neurochem 21(5):1211–1126
Prats C, Graham TE, Shearer J (2018) The dynamic life of the glycogen granule. J Biol Chem 293(19):7089–7098. https://doi.org/10.1074/jbc.R117.802843
Price TB, Taylor R, Mason GF, Rothman DL, Shulman GI, Shulman RG (1994) Turnover of human muscle glycogen with low-intensity exercise. Med Sci Sports Exerc 26(8):983–991
Psarra AM, Pfeiffer B, Giannakopoulou M, Sotiroudis TG, Stylianopoulou F, Hamprecht B (1998) Immunocytochemical localization of glycogen phosphorylase kinase in rat brain sections and in glial and neuronal primary cultures. J Neurocytol 27(11):779–790
Quach TT, Rose C, Schwartz JC (1978) [3H]Glycogen hydrolysis in brain slices: responses to neurotransmitters and modulation of noradrenaline receptors. J Neurochem 30(6):1335–1341
Quach TT, Duchemin AM, Rose C, Schwartz JC (1980) 3H-Glycogen hydrolysis elicited by histamine in mouse brain slices: selective involvement of H1 receptors. Mol Pharmacol 17(3):301–308
Quach TT, Duchemin AM, Rose C, Schwartz JC (1981) Specific desensitization of histamine H1 receptor-mediated [3H]glycogen hydrolysis in brain slices. Mol Pharmacol 20(2):331–338

Quach TT, Rose C, Duchemin AM, Schwartz JC (1982) Glycogenolysis induced by serotonin in brain: identification of a new class of receptor. Nature 298(5872):373–375

Quach TT, Duchemin AM, Rose C, Schwartz JC (1988) [^{3}H]glycogenolysis in brain slices mediated by beta-adrenoceptors: comparison of physiological response and [^{3}H]dihydroalprenolol binding parameters. Neuropharmacology 27(6):629–635

Quistorff B (1975) A mechanical device for the rapid removal and freezing of liver or brain tissue from unanaesthetized and nonparalyzed rats. Anal Biochem 68(1):102–118

Rahman B, Kussmaul L, Hamprecht B, Dringen R (2000) Glycogen is mobilized during the disposal of peroxides by cultured astroglial cells from rat brain. Neurosci Lett 290(3):169–172

Richter K, Hamprecht B, Scheich H (1996) Ultrastructural localization of glycogen phosphorylase predominantly in astrocytes of the gerbil brain. Glia 17(4):263–273. https://doi.org/10.1002/(SICI)1098-1136(199608)17:4<263::AID-GLIA1>3.0.CO;2-0

Saez I, Duran J, Sinadinos C, Beltran A, Yanes O, Tevy MF, Martínez-Pons C, Milán M, Guinovart JJ (2014) Neurons have an active glycogen metabolism that contributes to tolerance to hypoxia. J Cereb Blood Flow Metab 34(6):945–955. https://doi.org/10.1038/jcbfm.2014.33

Savaki HE, Davidsen L, Smith C, Sokoloff L (1980) Measurement of free glucose turnover in brain. J Neurochem 35(2):495–502

Shieh JJ, Pan CJ, Mansfield BC, Chou JY (2004) A potential new role for muscle in blood glucose homeostasis. J Biol Chem 279(25):26215–26219. https://doi.org/10.1074/jbc.M402036200

Shimizu N, Kumamoto T (1952) Histochemical studies on the glycogen of the mammalian brain. Anat Rec 114(3):479–497

Shulman GI, Rothman DL, Jue T, Stein P, DeFronzo RA, Shulman RG (1990) Quantitation of muscle glycogen synthesis in normal subjects and subjects with non-insulin-dependent diabetes by ^{13}C nuclear magnetic resonance spectroscopy. N Engl J Med 322(4):223–228. https://doi.org/10.1056/NEJM199001253220403

Sokoloff L, Reivich M, Kennedy C, Des Rosiers MH, Patlak CS, Pettigrew KD, Sakurada O, Shinohara M (1977) The [^{14}C]deoxyglucose method for the measurement of local cerebral glucose utilization: theory, procedure, and normal values in the conscious and anesthetized albino rat. J Neurochem 28(5):897–916

Sotelo C, Palay SL (1968) The fine structure of the lateral vestibular nucleus in the rat. I. Neurons and neuroglial cells. J Cell Biol 36(1):151–179

Strang RH, Bachelard HS (1971) Extraction, purification and turnover of rat brain glycogen. J Neurochem 18(6):1067–1076

Subbarao KV, Hertz L (1990) Effect of adrenergic agonists on glycogenolysis in primary cultures of astrocytes. Brain Res 536(1–2):220–226

Subbarao KV, Stolzenburg JU, Hertz L (1995) Pharmacological characteristics of potassium-induced, glycogenolysis in astrocytes. Neurosci Lett 196(1–2):45–48

Swanson RA, Morton MM, Sagar SM, Sharp FR (1992) Sensory stimulation induces local cerebral glycogenolysis: demonstration by autoradiography. Neurosci 51(2):451–461

Tang F, Lane S, Korsak A, Paton JF, Gourine AV, Kasparov S, Teschemacher AG (2014) Lactate-mediated glia-neuronal signalling in the mammalian brain. Nat Commun 5:3284. https://doi.org/10.1038/ncomms4284

Trivedi B, Danforth WH (1966) Effect of pH on the kinetics of frog muscle phosphofructokinase. J Biol Chem 241(17):4110–4112

Veech RL, Harris RL, Veloso D, Veech EH (1973) Freeze-blowing: a new technique for the study of brain in vivo. J Neurochem 20(1):183–188

Vilchez D, Ros S, Cifuentes D, Pujadas L, Valles J, Garcia-Fojeda B, Criado-Garcia O, Fernandez-Sanchez E, Medrano-Fernandez I, Dominguez J, Garcia-Rocha M, Soriano E, Rodriguez de Cordoba S, Guinovart JJ (2007) Mechanism suppressing glycogen synthesis in neurons and its demise in progressive myoclonus epilepsy. Nat Neurosci 10(11):1407–1413. https://doi.org/10.1038/nn1998

Watanabe H, Passonneau JV (1973) Factors affecting the turnover of cerebral glycogen and limit dextrin in vivo. J Neurochem 20(6):1543–1554

Winkler U, Hirrlinger J (2015) Crosstalk of signaling and metabolism mediated by the NAD(+)/NADH redox state in brain cells. Neurochem Res 40(12):2394–2401. https://doi.org/10.1007/s11064-015-1526-0

Wolfe LS, Klatzo I, Miquel J, Tobias C, Haymaker W (1962) Effect of alpha-particle irradiation on brain glycogen in the rat. J Neurochem 9(2):213–218. https://doi.org/10.1111/j.1471-4159.1962.tb11862.x

Yamanishi S, Katsumura K, Kobayashi T, Puro DG (2006) Extracellular lactate as a dynamic vasoactive signal in the rat retinal microvasculature. Am J Physiol Heart Circ Physiol 290(3):H925–H934. https://doi.org/10.1152/ajpheart.01012.2005

Yellen G (2018) Fueling thought: management of glycolysis and oxidative phosphorylation in neuronal metabolism. J Cell Biol 217(7):2235–2246. https://doi.org/10.1083/jcb.201803152

Young FG (1957) Claude Bernard and the discovery of glycogen; a century of retrospect. Br Med J 1(5033):1431–1437

Zhang X, Peng L, Chen Y, Hertz L (1993) Stimulation of glycogenolysis in astrocytes by fluoxetine, an antidepressant acting like 5-HT. Neuroreport 4(11):1235–1238

Brain Glycogen Structure and Its Associated Proteins: Past, Present and Future

M. Kathryn Brewer and Matthew S. Gentry

Abstract This chapter reviews the history of glycogen-related research and discusses in detail the structure, regulation, chemical properties and subcellular distribution of glycogen and its associated proteins, with particular focus on these aspects in brain tissue.

Keywords Glycogen structure · Glycogen architecture · Polyglucosan · Starch · Amylopectin · Pompe disease · Lafora disease · Glycophagy

Abbreviations

AMP	Adenosine monophosphate
AMPK	AMP-activated protein kinase
Atg	Autophagy-related
ATP	Adenosine triphosphate
cAMP	Cyclic adenosine monophosphate
CBM	Carbohydrate binding module
CLD	Chain length distribution
CNS	Central nervous system
DP	Degree of polymerization
EM	Electron microscopy
ER	Endoplasmic reticulum
GAA	Acid α-glucosidase
GABARAPL1	GABA type A receptor associated protein like 1
GBE	Glycogen branching enzyme
GDE	Glycogen debranching enzyme
GH	Glycosyl hydrolase

M. K. Brewer · M. S. Gentry (✉)
Department of Molecular and Cellular Biochemistry, Epilepsy and Brain Metabolism Center, Lafora Epilepsy Cure Initiative, and Center for Structural Biology, University of Kentucky College of Medicine, Lexington, KY, USA
e-mail: matthew.gentry@uky.edu

M. DiNuzzo, A. Schousboe (eds.), *Brain Glycogen Metabolism*, Advances in Neurobiology 23, https://doi.org/10.1007/978-3-030-27480-1_2

G_L	Liver-specific glycogen-targeting regulatory subunit of PP1
GlcNAc	N-acetyl-glucosamine
G_M	Muscle-specific glycogen-targeting regulatory subunit of PP1
GNIP	Glycogenin interacting protein
GPCPD1	Glycerophosphocholine phosphodiesterase 1
GSD	Glycogen storage disease
GSK3	Glycogen synthase kinase 3
GYG	Glycogenin
GYS	Glycogen synthase
HPAEC	High performance anion exchange chromatography
kDa	Kilodaltons
LAMP1	Lysosomal-associated membrane protein 1
LB	Lafora body
LC3	Microtubule associated light chain 3
LD	Lafora disease
LSF2	Like Sex Four 2
NADH	Nicotinamide adenine dinucleotide (reduced form)
NMR	Nuclear magnetic resonance
PAS	Periodic acid-Schiff
PASD	Periodic acid-Schiff with diastase
PhK	Phosphorylase kinase
PP1	Protein phosphatase 1
PTG	Protein targeting to glycogen (also known as R5)
PYG	Glycogen phosphorylase
R5	Ubiquitous glycogen-targeting regulatory subunit of PP1 (also known as PTG)
R6	Ubiquitous glycogen-targeting regulatory subunit of PP1
SEX4	Starch Excess 4
Stbd1	Starch binding domain-containing protein 1
TCA	Trichloroacetic acid
UDP	Uridine diphosphate

1 Introduction

In the 1840s, the French physiologist Claude Bernard made an astonishing observation: that sugar could be synthesized *de novo* by the liver of a dog that was exclusively fed meat (Bernard 1850). In 1857, he described the isolation of this sugar-forming substance, which he aptly called *la matière glycogène* (Bernard 1857). Soon afterward, glycogen in muscle was reported, shown to diminish with exercise, and discovered to "ferment" into lactic acid. Over the century that followed, the structure and regulation of glycogen were extensively characterized in the context of liver and muscle, leading to multiple Nobel prizes throughout the twentieth century. Although the mechanisms governing glycogen metabolism have been deeply studied, the work is not yet complete. Meticulous investigations of Nobel laureates Otto Meyerhof, Carl and Gerty Cori, Luis Leloir, Edwin Krebs and Edmond Fischer, and countless others have paved the way for an even more profound understanding of glycogen metabolism, even in organs where glycogen constitutes a smaller fraction of total weight, such as the brain. Despite glycogen metabolism being a mainstay in textbooks, new aspects of glycogen structure, metabolism and tissue-specific roles are still being defined, aided by modern technologies and the mechanistic investigations of glycogen storage diseases.

This chapter discusses the historical progression of glycogen research and its current condition with particular emphasis on how it has benefitted the overarching field of biochemistry, why the study of brain glycogen has lagged, and what gaps remain in our understanding of glycogen metabolism. There are aspects of glycogen structure and metabolism that are applicable to all types of glycogen, but some aspects are specific to organisms, tissues, and/or nutritional state, and insights can be gleaned through study of the other types of carbohydrates, particularly plant starch. Given the similarities and connections, it is useful to review the literature on glycogen (and starch) to understand what can be extrapolated to brain glycogen. This chapter provides a detailed overview of our current understanding of glycogen structure, architecture, associated proteins, tissue-specific properties and subcellular distribution. Later chapters of this book are devoted to the structure and regulation of glycogen synthase and phosphorylase (Chaps. 3 and 4), so they will not be discussed in depth here. Glycogen storage diseases inform our studies of glycogen in ways that would not otherwise be possible; insights into glycogen metabolism gained from these diseases are discussed in this chapter. The entire disease class has been recently reviewed elsewhere (Adeva-Andany et al. 2016).

2 Discovery of Glycogen and Its Associated Proteins

2.1 Claude Bernard and Early Reports of Brain Glycogen

Glycogen constitutes up to 8% of liver volume and up to 2% of muscle volume, and its content is higher in these tissues than in any other tissue of the adult mammal (Brown 2004). It is not surprising that the fundamental study of its structure and metabolism began in liver and muscle. Claude Bernard, through a series of fundamental experiments, discovered the primary role of the liver in supplying glucose to other organs through the bloodstream (reviewed by Young 1937, 1957; Grmek 1968). In 1848, Bernard noticed that after death, a large amount of glucose was released from the hepatic veins of a dog, even when the dog had not been fed any carbohydrates (Grmek 1968). He concluded that the liver could produce glucose, and by 1857, reported that he could extract glycogen (a term that literally means "sugar-former") by plunging the tissue in boiling water shortly after death and precipitating the material with alcohol (Bernard 1857) (Fig. 1). Through fermentation with yeast and basic chemical tests, he confirmed that its main constituent was glucose (Young 1957). He also noticed that the glycogen yielded a red wine color in the presence of iodine, which facilitated its

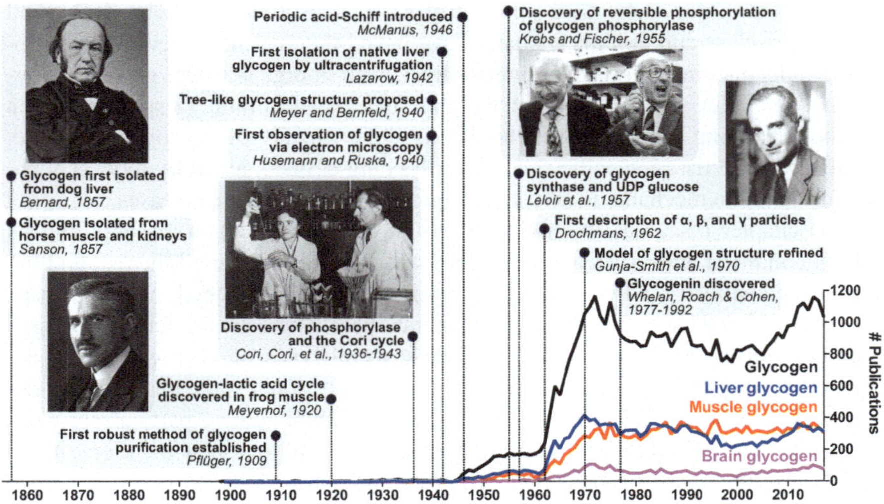

Fig. 1 Timeline of landmark discoveries in the history of glycogen-related research. Notable discoveries and papers are shown. Claude Bernard, the physiologist who discovered glycogen and its function in liver, and Nobel laureates are pictured. PubMed (https://www.ncbi.nlm.nih.gov/pubmed/) searches were performed using the keywords shown (glycogen, liver glycogen, muscle glycogen, or brain glycogen), and the number of publications per year for each set of keywords are shown. Key terms related to glycogen synthase kinase (GSK, GSK3, GSK3 beta) were excluded from search results since much of the GSK-related literature is not directly relevant to glycogen metabolism.

histological observation in other tissues (Foster 1899; Larner 1967). Very shortly afterward, Sanson reported the isolation of "a glycogen-like substance analogous to dextrin" from the spleen, muscle and kidneys of a horse (translated, Sanson 1857) (Fig. 1). By 1875, Bernard and others reported glycogen in mammalian muscle, embryos, and trace amounts in other tissues, but it was difficult to know for certain that those trace amounts were indeed glycogen (Carpenter et al. 1876).

Soon after Bernard's discovery of glycogen, it was noted that glycogen in muscle diminished upon contraction, and Bernard reported that "muscle glycogen always undergoes a lactic acid fermentation, and this is the only change that muscle glycogen ever undergoes, either in the living animal or after death" (Carpenter et al. 1876; Young 1957). In the early twentieth century, a landmark paper from Fletcher and Hopkins demonstrated that lactic acid was formed in muscle under anaerobic conditions (Fletcher and Hopkins 1907). Then, a German physician named Otto Meyerhof published a series of experiments demonstrating that the lactic acid produced by frog muscle was derived from glycogen, and that in the presence of oxygen, it could be either oxidized, or reconverted to glycogen (Meyerhof 1920a, b, c) (Fig. 1). "Meyerhof's brilliant analysis of the glycogen-lactic acid cycle and its relation to respiration explained the course of the heat production and, for the first time, established the cyclic character of energy transformations in the living cell" (Nachmanson et al. 1960). For this discovery and his seminal work on glycolysis, Meyerhof and his colleague A.V. Hill were awarded the 1922 Nobel Prize in Physiology or Medicine.

The German physiologist Eduard Pflüger is credited as the first to establish a method for obtaining pure fractions of glycogen (Fig. 1), and modifications of his method are still used today (Pflüger 1909; Young 1937; Passonneau et al. 1967). While Pflüger, Bernard and most others were unable to detect glycogen in the nervous system, traces of brain glycogen in the normal and diabetic human brain were reported by Cramer in 1880 and Pavy in 1881 (Gage 1917). Similar findings were published in an 1890 edition of *The Lancet*: "Dr. Fütterer has examined various organs of a diabetic person, finding glycogen in the medulla oblongata, spinal cord, and kidney in large quantities, and a little in the liver. A careful examination of the cerebral cortex showed that the vessels were full of glycogen. He concluded that extensive disturbances of nutrition were bound to result from this" (The Lancet 1890).

Although nearly all studies of glycogen in the early 1900s were in the liver and muscle of various organisms, as well as in yeast, a few groups reported glycogen in the central nervous system (CNS) of animals by staining tissues with an iodine solution. Glycogen was observed in the CNS, often in the retina, of the lancelet, lamprey, frog, pigeon, rabbit, and dog (Gage 1917; Holmes and Holmes 1926). Gage asserted its presence in the CNS of lower organisms, in the dorsal root ganglia of the pig, and in the choroid plexus of human embryos (Fig. 2). He validated the identity of the "mahogany-red substance" by incubating the sections with saliva, which con-

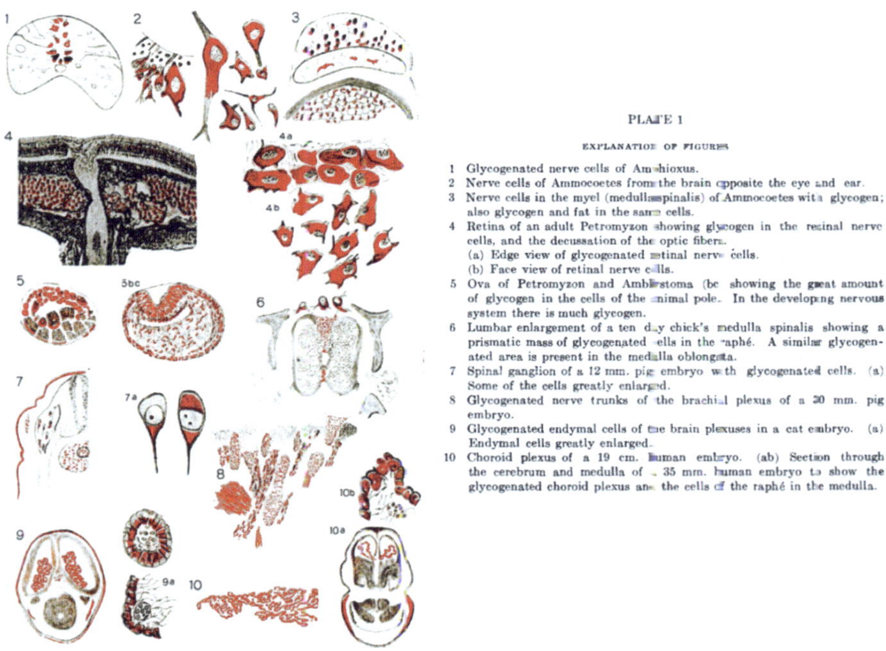

Fig. 2 Glycogen in the CNS of invertebrates and vertebrates. Drawings illustrating glycogen in the CNS of multiple organisms detected via iodine staining (Gage 1917).

tained amylases that digested the glycogen so that it no longer stained. "With the higher vertebrates, glycogen in demonstrable amount is not found in the nervous system after the embryonic period, the liver and muscles then assuming the main glycogenic function." Gage was hopeful that with the proper techniques and material, glycogen would also be found in the adult human CNS (Gage 1917).

With the relationship between glycogen and lactic acid demonstrated in muscle by Meyerhof, a few groups began investigating this relationship in brain. However, brain glycogen and lactic acid remained difficult to isolate and detect, and results were variable. Some groups reported a reduction in glycogen with insulin-stimulated convulsions or anesthetics, while others reported a rise in glycogen after convulsions produced by methylguanidine; Holmes and Holmes in reviewing these studies argued that the low values reported for brain glycogen were within the normal range of variation, and that brain glycogen is unlikely to produce lactic acid under normal conditions (Holmes and Holmes 1926). They astutely commented that "it is possible that the low values which we, in common with other workers, find for the glycogen content of the brain, may be due, not to a slow or inadequate breakdown mechanism, but to an extremely rapid one." This was indeed the case, and methods for capturing brain glycogen improved over the years; however, it was not until the end of the twentieth century that the most sophisticated methods for measuring brain glycogen emerged. "It seemed to us, therefore, that it is quite possible that glycogen plays a comparatively unimportant (or at least quite obscure) part in the carbohydrate metabolism of mammalian brain" (Holmes and Holmes 1926). This sentiment prevailed for many years, and it did not become clear until the following century that the role of glycogen in brain was more obscure than it was unimportant. Meanwhile, studies on glycogen enzymology had just begun, which led to a series of fundamental discoveries with major impacts on the wider field of biochemistry.

2.2 The Cori Years: Advances in Glycogen Enzymology and Glycogen Structure

Soon after Meyerhof's Nobel prize was awarded, Carl and Gerty Cori, Czech-born medical doctors who had emigrated to the United States, started working on the effects of insulin and epinephrine on liver glycogen. They reported that the hyperglycemia observed in the blood after epinephrine injections could not be accounted for solely by liver glycogen, and that lactic acid, which would be derived from muscle, could give rise to newly formed liver glycogen (Cori and Cori 1929). They theorized that there was a "cycle of carbohydrates" linking liver glycogen and muscle glycogen, with the intermediates between the two being glucose and lactic acid (Fig. 3a). Their subsequent work demonstrated this theory to be true, and it became known as the Cori cycle.

The Coris made a series of major discoveries in the 1930s (Fig. 1). In 1936 they isolated a novel ester, glucose-1-phosphate, from a mixture of glycogen, inorganic phosphate, and aqueous muscle extract (Cori and Cori 1936). So significant was this

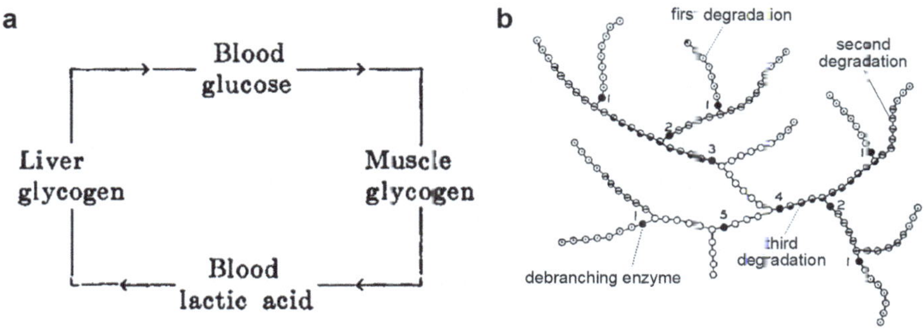

Fig. 3 The Cori cycle and the tree-like structure of glycogen. (**a**) The cycle of carbohydrates between liver and muscle proposed by Cori and Cori (1929) when they observed that ingestion or subcutaneous injection of sodium *d*-lactate led to glycogen deposition in the liver. Their subsequent work validated and elaborated this model, which became known as the Cori cycle. (**b**) A segment of the tree-like structure of glycogen based on Meyer's model and the results of the Cori group. Glucose residues are represented by circles; dotted, bisected, and half-filled circles correspond to glucose residues released by the first, second, and third rounds of degradation with phosphorylase, as indicated. Filled circles represent glucose residues released by α-1,6 glucosidase (i.e. debranching enzyme). Tiers are numbered. Modified from Larner et al. 1952. Copyright © American Society for Biochemistry and Molecular Biology. Used with permission

discovery that the compound became known as the Cori ester (Larner 1967). In subsequent years, the Coris showed that a phosphorylating enzyme found in muscle, heart, liver, brain, and yeast extracts catalyzed the release of glucosyl moieties from glycogen by esterifying them with inorganic phosphate. The resulting glucose-1-phosphate was converted to glucose-6-phosphate by an enzyme they named phosphoglucomutase (Cori et al. 1937, 1938). In a 1939 article in Science, they reported that the enzyme producing the Cori ester (which they aptly named phosphorylase) could also catalyze the reverse reaction, in order to synthesize polysaccharide (Cori et al. 1939). Almost simultaneously, phosphorolytic enzymes catalyzing polysaccharide synthesis were being described in the extracts of peas and potatoes and in yeast (Kiessling 1939; Hanes 1940). In 1943, Cori, Cori and Green crystallized glycogen phosphorylase and published a series of seminal papers on its properties (Cori et al. 1943; Cori and Cori 1943; Cori and Green 1943; Green and Cori 1943). In reward for their pioneering work, the Coris shared the 1946 Nobel Prize in Physiology or Medicine with Bernardo Houssay, who also made important contributions to the field of carbohydrate metabolism through his study of the role of the pituitary gland in hormonal regulation.

The discovery of phosphorylase in mammals, plants and yeast was a major breakthrough in the overarching field of biochemistry (Larner 1967). Phosphorylase was the first enzyme demonstrated to synthesize polysaccharide, and glycogen was the first macromolecule to be synthesized *in vitro* (Manners 1963; Whelan 2007). However, while mammals and yeast utilize glycogen as an energy source, plants synthesize starch, a glucose polymer with very different properties. Investigations of plant starch had been underway years before Bernard's discovery of glycogen and were progressing in parallel with those of glycogen. By the 1940s it had become clear that starch was composed of two distinct polymeric fractions: amylose, which was composed of long linear chains of glucose and gave an intense blue-black color with iodine, and amylopectin, which was a branched macromolecule and gave a purplish to reddish color (Bates et al. 1943). While the Coris were working out glycogen enzymology, others were studying the polymeric nature of both glycogen and starch. Haworth and Percival established a method for identifying the chemical linkages in glucose polymers through methylation of the free hydroxyls of the glucose polymer prior to hydrolysis with acid (Haworth and Percival 1932). They found that since most of the product had unmethylated hydroxyls at the 1- and 4- positions, the majority of the glucose units in glycogen must be linked by α-1,4 glycosidic bonds. A small proportion of glucose was methylated at the 4-hydroxyl position, which would represent what is called the non-reducing end of a glucose chain (i.e. the end lacking an aldehyde group). Based on the proportion of the methylated products, the average linear chain in glycogen was proposed to contain 12 glucose units. Similar work was done on several varieties of plant starch, and the average chain length in starch was estimated to be 24–30 units (Bawn et al. 1940). In 1947 and 1949, the α-1,6-linked disaccharide isomaltose was isolated from hydrolyzed starch and glycogen, providing definitive evidence for the chemical identity of the branch points (Montgomery et al. 1949; Wolfrom et al. 1951).

For many years it was believed that phosphorylase catalyzed both the synthesis and degradation of glycogen and starch, although synthesis by phosphorylase always required a primer, i.e. the addition of a small amount of carbohydrate upon which phosphorylase could act (Larner et al. 1952; Larner 1967). Curiously, phosphorylase from brain, heart and liver extracts produced branched polysaccharides that yielded a similar color to glycogen when stained with iodine, while the enzyme from potato and muscle extracts synthesized a linear, unbranched polysaccharide resembling amylose (Cori and Cori 1943). It was hypothesized that the former preparations contained a contaminating enzyme that was capable of introducing the α-1,6-linked branch points. A branching enzyme had been identified in potato, and in 1953, Larner characterized the branching enzyme from rat liver and muscle extracts. Using isotopic labeling, he showed that potato, liver and muscle branching enzymes transferred a 1,4-linked chain of 6–11 glucose moieties to create the 1,6-linked branch. Thus, branching enzyme is considered a transglucosidase (Larner 1953). His approach was novel in that he utilized radioactive isotopes rather than measuring branching by a shift in iodine color, susceptibility to phosphorylase, or change in end-group. With the combination of phosphorylase and branching enzyme, glycogen could be synthesized *in vitro*, and for the most part, it resembled the glycogen purified from tissues (Parodi et al. 1969).

Similarly, it was observed that while muscle phosphorylase could completely digest glycogen, recrystallized preparations of muscle phosphorylase and potato phosphorylase degraded glycogen only partially, stopping at the branch points, and leaving what was called a limit dextrin (Cori and Larner

1951). A second enzyme must be present in the crude extracts that could remove the branch points. In 1951, Cori and Larner discovered the contaminating enzyme, showing that it cleaved α-1,6 linkages and released free glucose (Cori and Larner 1951). Walker and Whelan later found that the phosphorylase limit dextrin contained not one, but four glucose units attached to branch points (Walker and Whelan 1960). It was then discovered that the enzyme they had identified, which they called amylo-1,6-glucosidase, also had transglucosidase activity, favoring the transfer of three glucose units from one chain to another (Brown and Illingworth 1962). It is now well established that when phosphorolysis terminates four units away from a branch point, the glycogen debranching enzyme, which possesses both transferase and glucosidase activities, moves three glucose units to another chain and then releases the single branched glucosyl moiety (Huijing 1975).

It appeared at that time that all of the enzymes required for glycogen synthesis and degradation had been identified: glycogen phosphorylase, branching enzyme, and debranching enzyme. Enzymes that were either analogous or distinct to these had also been identified in plants. It became possible to utilize these enzymes to more precisely understand the structure of the glycogen and amylopectin molecules. With such useful tools and insights in common, the fields of starch and glycogen metabolism continued to progress together and overlapped considerably. In 1940 a tree-like structure for glycogen and amylopectin was proposed by Meyer and Bernfeld based on serial degradation with various enzymes (Meyer and Bernfeld 1940) (Fig. 1). This model varied from the prior structures proposed by Staudinger and Haworth, which were based solely on data obtained by chemical methods. The Cori group used stepwise enzymatic degradation of various glycogens and amylopectins with phosphorylase and debranching enzyme to test the various models. It was evident that glycogen was composed of "tiers," and each successive tier could only be accessed by phosphorylase after removal of the branch points by the debranching enzyme; degradation of multiple tiers required alternating treatments of the two enzymes. "The Staudinger and the Haworth models would yield a constant percentage of the total branch points in each tier during successive enzymatic degradation, while the Meyer model would yield a diminishing percentage as one progresses from the outer to the inner tiers.... Only one kind of model could be made to fit this arrangement of branch points; namely, one which represents the polysaccharides as multibranched, tree-like structures" (Larner et al. 1952) (Fig. 3b). It was also noted in this paper that the average chain length of liver and muscle glycogen (15 glucose units) were shorter than those of wheat and corn amylopectin (18 and 24 glucose units, respectively). We now know that chain length is one of the major features that distinguish glycogen from amylopectin. Meyer's model was the accepted model for glycogen and amylopectin until experiments from Whelan's group in 1970 showed the absence of very long chains and refined the model (Fig. 1) (Gunja-Smith et al. 1970, 1971). The currently accepted model for glycogen is based on Whelan's work, but recent data indicates this model requires further refinement (see Sect. 3.1).

2.3 Emerging Technologies with Staying Power: Periodic Acid-Schiff, the Ultracentrifuge, and the Electron Microscope

When the Coris began their work, the most common procedure for purifying glycogen was based on the 1909 Pflüger method. The protocol is similar to what Bernard used and involves boiling tissue in hot concentrated alkali and then precipitating the glycogen with alcohol (Pflüger 1909). Modifications of this procedure, especially those of Somogyi (Somogyi 1934), were used by the Coris and are still the most commonly utilized protocols for glycogen extraction. Another frequently used method, introduced in 1934 (Willstätter and Rohdewald 1934), is extraction with cold trichloroacetic acid (TCA); however, it has been observed that not all glycogen can be extracted by this method (see Sect. 6.2), and a slightly different molecular weight is observed for the purified glycogen (Lazarow 1942; Calder 1991). In 1936, the first reliable procedure for extracting brain glycogen was established utilizing Somogyi's method (Kerr 1938). Kerr found that he obtained the highest yields when the brains of dogs were rapidly excised under amytal anesthesia and crushed directly into 60% potassium hydroxide. He also obtained high yields when the brains were frozen *in situ* in liquid air. Importantly, he showed that the purified brain glycogen "is indistinguishable from liver glycogen prepared by Somogyi's method. It dissolves readily in water to form a transparent solution, which is opalescent in reflected light. This when treated with iodine solution gives a Burgundy red color identical with that obtained with a solution of liver glycogen." The similar chemical properties of brain and liver glycogen were important to establish, and made it reasonable to extend the principles of glycogen structure and regulation that were being so thoroughly defined in liver and muscle to brain glycogen.

Quantifications of brain glycogen were now reproducible. Kerr and other groups showed that brain glycogen levels dropped during hypoglycemia in dogs, cats and rabbits (Carter and Stone 1961). Once a robust method for quantifying low levels of brain lactate was also established (Carr 1947), informative experiments on the effects of various stimuli on brain glycogen and lactate could begin. Using a circular saw to rapidly excise the brains of mice, Chance and Yaxley showed that lactate was elevated with both subconvulsive and convulsive levels of various seizure-inducing stimulants, while glycogen was only elevated when a convulsion was reached (Chance and Yaxley 1950). They also carefully defined the rapid decrease of glycogen and lactate levels after death in both normal and convulsed mice, demonstrating that most of the glycogen was lost after 5 min (Fig. 4a). Based on their own measurements and those of their contemporaries, they estimated that "a large number of vertebrates normally possess between 10 and 30 mg of glycogen per 100 g of brain tissue" and that glycogen distribution within the brain was heterogeneous. These values are in agreement with current estimates (Brown 2004). A few groups also began studying glycogen metabolism in isolated cerebral tissues and the effects of sleep on brain glycogen (LeBaron 1955; McIlwain and Tresize 1956; Svorad 1958).

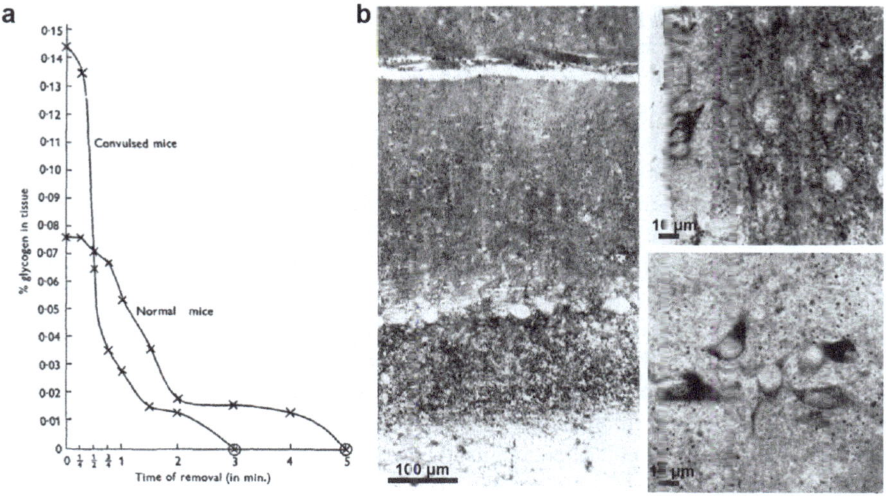

Fig. 4 Early studies of brain glycogen. (**a**) Rapid glycogen loss following extraction from brain of normal and convulsed mice; glycogen was measured using a modification of the method by Kerr (1938) (from Chance and Yaxley 1950). Copyright © Company of Biologists, Ltd. Used with permission. (**b**) Staining of glycogen (dark stain) in the perfused rabbit brain using the lead-tetra-acetate-Schiff method, a modification of PAS (Shimizu and Kumamoto 1952). The identity of stained regions as glycogen was confirmed by salivary digestion of comparable sections. Left: glycogen is abundant in the granular and molecular layers of the cerebellum, but Purkinje neurons lack glycogen. Upper right: ependymal cells of the hypothalamus lining the third ventricle show intense staining for glycogen, and granules are abundant in the neuropil, but nerve cells lack glycogen in this region. Lower right: some small nerve cells in the lateral hypothalamic nucleus appear to contain glycogen. Scale bars have been approximated based on magnification. Copyright © 2004 John Wiley and Sons, Inc. Used with permission

As glycogen quantification methods improved, so did microscopic techniques. Glycogen had long been visualized with iodine using the light microscope, but the stain was not sensitive enough to detect the small amounts found in the normal adult brain. Additionally, alcoholic fixation was also critical, as glycogen would dissolve in aqueous solutions (Gage 1917). Some attempts were made to visualize glycogen with Best's carnitine or the Bauer reaction, but these methods were also criticized for their lack of sensitivity (Kerr 1938; Lillie 1950; Mowry and Bangle 1951). In 1946, McManus introduced the periodic acid-Schiff (PAS) technique as a delicate and convenient way to stain carbohydrates, including glycogen, in tissue sections: the periodic acid reacts with the 1,2 glycol linkage of carbohydrates, producing an aldehyde that can be colored with Schiff reagent (McManus 1946, 1948) (Fig. 1). Although PAS stained most polysaccharides, including glycoproteins and mucins, its specificity for glycogen could be tested by incubation with diastase (a general term for amylase) or saliva (which contains amylase) to digest the glycogen and distinguish it from other types of polysaccharides, which were left intact after digestion (Mowry and Bangle 1951). Using PAS and a modification of this stain (lead-tetra-acetate-Schiff), Shimizu and Kumamoto published a series of papers in the

1950s on glycogen deposition in the normal brain and with pathological insults (Shimizu and Kumamoto 1952; Shimizu 1955; Shimizu and Kubo 1957; Shimizu and Hamuro 1958). Glycogen was detected in the area postrema (an area populated primarily by glial cells) in mammalian brains, particularly around blood vessels, and in ependymal cells of the supraoptic crest in rodents; it could also be detected in nerve cells of the hypothalamic nucleus and other regions when the brain was perfused with appropriate fixatives (Shimizu and Kumamoto 1952) (Fig. 4b). The PAS technique, usually in combination with diastase (PASD), is now the most widely used histochemical stain for glycogen detection in tissues (Bancroft and Gamble 2008).

Another extremely useful technology in visualizing glycogen structure and distribution was born with the invention of the electron microscope in 1931 by Ernst Ruska (Ruska and Knoll 1931). The electron microscope achieved superior resolution to that of the light microscope. In 1934, Marton developed a histological technique that allowed biological specimens to be visualized by the electron microscope without destruction (Marton 1934). With this technique, the fine structure of subcellular organelles could be visualized. The first report of glycogen using electron microscopy (EM) was from Husemann and Helmut Ruska, Ernst Ruska's brother (Husemann and Ruska 1940) (Fig. 1). The spherical particles, which were derivatized to produce better scattering of the x-ray beam, were about 15–30 nm in diameter. But visualization of glycogen was difficult for a number of reasons. Firstly, polysaccharides contain primarily light atoms (carbon, oxygen and hydrogen) and are intrinsically not very electron dense. Secondly, after the identification of ribonucleoprotein particles by EM, particulate components with a diameter of ~150 Å free in the cytoplasm or associated with the endoplasmic reticulum were typically interpreted as ribonucleoprotein (Revel et al. 1960; Revel 1964). Unknown tissue components were sometimes identified by comparing the thin EM sections with thicker, stained sections viewed under the light microscope, but this too had its limits.

The ultracentrifuge, invented by Theodor Svedburg in 1925, became a very powerful tool in the field of biochemistry and allowed glycogen to be isolated its native form. Lazarow showed in 1942 a species which he called "particulate glycogen" could be isolated from liver via ultracentrifugation without the use of harsh chemicals (Lazarow 1942) (Fig. 1). Its sedimentation constant suggested a molecular weight much larger than previous reports for Pflüger-purified glycogen, and the particle, which he showed was made of pure glycogen, could be dispersed by heating, trichloroacetic acid (TCA), or potassium hydroxide treatment. Lazarow concluded that "clearly, then, particulate glycogen is an aggregate of smaller glycogen units." In 1962, Drochmans, combining these two state-of-the art technologies and utilizing a negative staining technique with phosphotungstate, elegantly described the ultrastructure of glycogen particles purified by differential ultracentrifugation (Drochmans 1962) (Fig. 1). He defined three levels of glycogen structure in the rat liver visible by EM: the largest structure was the α particle, which was 60–200 nm in diameter and had a rosette like appearance; the β particle, 20–40 nm in diameter, which associated to form α particles; and the γ particle, which referred to the 3 nm fine filaments comprising the β particle (Fig. 5a, b). The α particles could be

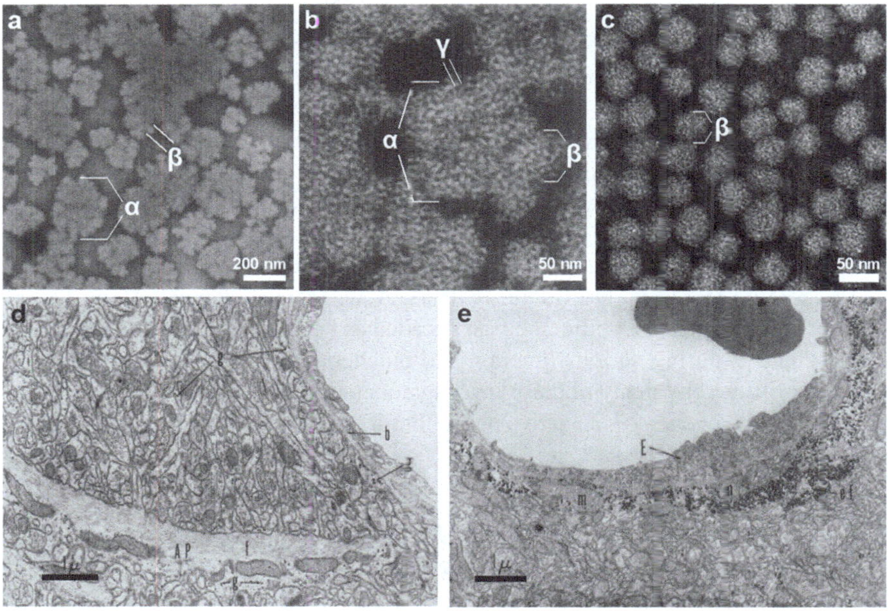

Fig. 5 Glycogen ultrastructure as demonstrated by electron microscopy. (**a**, **b**) Negative staining of natively purified rat liver glycogen from reveals the presence of α, β and γ particles (Drochmans 1962). Copyright © 1962 Elsevier. Used with permission. (**c**) Typically, purified muscle glycogen exists only as β particles (Wanson and Drochmans 1968). Scale bars have been approximated based on magnification. Copyright © Rockerfeller University Press. Used with permission. (**d**, **e**) Glycogen in the rat cerebral cortex visualized by EM (Maxwell and Kruger 1965). (**d**) In sections of normal cortex, glycogen β particles (*g*) can be found in astrocyte processes (*AP*) and astrocytic end feet applied to the basement membrane (*b*). An astrocytic bundle of fibrils (*f*) is labeled. (**e**) One day after irradiation with ionizing particles, end feet (*ef*) are enlarged and contain numerous glycogen granules and mitochondria (*m*). An endothelial cell lining the capillary is also labeled (*E*). Copyright © Rockerfeller University Press. Used with permission

dispersed into β particles in low pH. It became clear that Drochmans's α particle was equivalent to Lazarow's particulate glycogen, and that these particles represent the primary native form of glycogen in the liver. This helped to explain the discrepancies in molecular weights that others had observed for glycogen purified by different methods. In 1968, Wanson and Drochmans also visualized differentially centrifuged glycogen from rabbit skeletal muscle, showing the presence of solely β particles (Wanson and Drochmans 1968) (Fig. 5c). Dozens of EM studies on glycogen in various tissues and organisms followed.

The use of EM on sections of the mammalian brain in the 1960s and later years made it clear that the vast majority of brain glycogen was present in the form of β particles in astrocytic processes (Maxwell and Kruger 1965; Cataldo and Broadwell 1986) (Fig. 5d). Multiple groups showed striking accumulations of glycogen in reactive astrocytes in response to various traumas (Shimizu and Kubo 1957; Shimizu and Hamuro 1958; Maxwell and Kruger 1965) (Fig. 5e). Most groups reported that in the adult mammalian brain, neurons and microglia contained virtually no glycogen except under certain pathological conditions; however, in a few studies of normal tissue, glycogen was found in nerve cells (reviewed by Koizumi 1974). During this time, numerous groups began using PAS and EM to visualize the microscopic characteristics of polyglucosan bodies, abnormal carbohydrate structures found in a variety of tissues. Polyglucosan bodies were identified in both astrocytes and neurons, in the context of glycogen storage diseases, epilepsy, neurodegenerative disorders, and aging (Cavanagh 1999; Duran and Guinovart 2015). Although glycogen still received little attention in the brain compared to other tissues, its significance was becoming increasingly evident. In 1961, A.W. Merrick stated, "A substantial amount of Russian work within the past decade has been directed toward brain glycogen changes during various function and biochemical states of the animal. The opinion is offered by many of these investigators (several of whom are cited in this paper) that glycogen takes an active part in brain metabolism" (Merrick 1961).

2.4 Sugar Nucleotides, Reversible Phosphorylation, and a Primer for Glycogen Synthesis

In 1950, the Argentinian chemist Luis Leloir discovered the first sugar nucleotide: uridine-diphosphate (UDP) glucose (Caputto et al. 1950). It soon became apparent that the sugar nucleotides were found throughout nature and were essential for the interconversion of carbohydrates, occupying "a central position in carbohydrate metabolism" (Hassid et al. 1959). In 1957, Leloir made the shocking discovery that the true catalyst of glycogen synthesis was not glycogen phosphorylase, but a glucosyl transferase utilizing UDP-glucose as a glucosyl donor, which became known as glycogen synthase (Leloir and Cardini 1957) (Fig. 1). In the words of Larner, "Nature seems to have found a new more powerful glucose donor by making a pyrophosphate derivative of the Cori ester" (Larner 1967). Leloir's group went on to show that glycogen synthesized *in vitro* using glycogen synthase in combination with branching enzyme was identical to natively purified glycogen β particles from muscle; in contrast, glycogen synthesized by phosphorylase and branching enzyme was not exactly equivalent to native glycogen: it differed in its stability and response to various types of degradative treatments (Mordoh et al. 1966; Parodi et al. 1967). These results were confirmed and expanded upon by numerous groups, and Leloir was awarded the 1970 Nobel Prize in Chemistry for his discovery of the sugar nucleotides.

It became apparent that both glycogen phosphorylase and synthase are peculiar enzymes. The Coris observed that glycogen phosphorylase existed in two forms, an active *a* form, and inactive *b* form, which were differentially activated by adenylic acid (Cori and Cori 1946). The Coris had identified another enzyme capable of converting the *a* form to the *b* form, calling this the "prosthetic-group-removing" enzyme (Cori and Green 1943). Kinase activity, the covalent addition of phosphate to one protein by another, was first described in 1954 (Burnett and Kennedy 1954), and in 1955, Fisher and Krebs showed that phosphorylase *a* could be converted to phosphorylase *b* in the presence of adenosine triphosphate (ATP) by an enzyme from muscle that became known as phosphorylase kinase (PhK) (Fischer and Krebs 1955). Concurrently, Wosilait and Sutherland also demonstrated the presence of an equivalent converting enzyme in liver (Sutherland and Wosilait 1955). These were the first ever demonstrations of reversible phosphorylation, a momentous discovery that has reverberated throughout the life sciences. Fischer and Krebs published a series of papers on the details of this novel mechanism that earned them the 1970 Nobel Prize in Physiology or Medicine, and Sutherland received the 1971 Nobel Prize for his discovery of cyclic adenosine monophosphate (cAMP) and its role in hormonal regulation (Cohen 2002). A few years later, Larner and others showed that glycogen synthase is also subject to allosteric control, that it exists in two forms, and that the interconversion of these forms requires phosphorylation by another kinase (Friedman and Larner 1963). In fact, unlike phosphorylase, which has one phosphosite and one kinase, synthase is hierarchically phosphorylated at multiple sites by multiple kinases, inextricably linking glycogen metabolism to hormonal regulation, energy status, and a milieu of intracellular signals (see Chap. 3) (Roach 1990). One of the kinases discovered to phosphorylate synthase, glycogen synthase kinase 3 (GSK3), was later found to regulate a much greater array of cellular processes and play an important role in many pathologies including cancer, Alzheimer's disease, Parkinson's disease and diabetes (Cohen and Frame 2001; Jope and Johnson 2004). Many of the roles of GSK3 are quite unrelated to glycogen metabolism, leading to an unfortunately misleading keyword in literature searches.

While more pieces in the puzzle of glycogen regulation had fallen into place, other pieces were not even known to be missing. One enzyme must be mentioned at this point that has received insufficient attention despite its discovery in 1963. It was discovered as the deficient enzyme in Pompe disease, one of the glycogen storage diseases (GSDs). GSDs, also known as glycogenoses, are a group of diverse pathologies characterized by glycogen accumulation in various tissues. GSDs were first described in the early 1900s, and over the decades, the enzyme deficiencies causing many of these disorders were identified through biochemical analyses of patient tissues. Typically, patients were lacking one of the basic enzymes involved in glycogen metabolism (reviewed by Huijing 1975). However, in one of the most severe GSDs, Pompe disease, all of the known glycogen-related enzymes had normal activity. Lysosomes had been recently discovered by de Duve with the use of the ultracentrifuge (De Duve et al. 1955), and in 1963, Hers demonstrated that Pompe patients were lacking an enzyme called acid α-glucosidase or maltase, which was found in lysosomes and converts glycogen or maltose to glucose (Hers 1963). He identified Pompe disease as the first of the lysosomal storage diseases, now known to have a combined incidence of 1 in 5000–10,000 (Fuller et al. 2006; Raben et al. 2012). Brown, Brown and Jeffrey demonstrated that the lysosomal enzyme discovered by Hers could cleave both 1,4 and 1,6 glycosidic linkages (Brown et al. 1970; Jeffrey et al. 1970a, 1970b). The role of this enzyme in the degradation of glycogen was unclear, but presumably it was important since its absence resulted in a fatal condition. It is now well established, but not broadly known, that glycogen can be degraded within lysosomes, but the details of this pathway are still being defined (see Sect. 5).

The discoveries of Leloir, Krebs and Fischer spurred a major wave of glycogen-related research in the decade that followed (Fig. 1). By 1970, these fervent investigations were waning, as it seemed to most that the work on glycogen was virtually complete. In 1971, Ryman and Whelan synthesized an exhaustive review of glycogen metabolism, which spanned 158 pages and cited nearly 900 publications. They stated in their introduction:

> "The field of glycogen metabolism is one that to the outside observer has long seemed in a settled condition, with new discoveries being only likely to add gloss to existing facets. This is because it has been possible since the early 1940s to draw metabolic maps that seem to explain the process fully and satisfactorily, these maps being based on highly satisfactory *in vitro* experiments carried out with purified or semi-purified enzymes. This has been the polymer *par excellence* as far as *in vitro* work is concerned.
>
> This apparently settled condition is, in fact, illusory, and this has probably worked to the detriment of progress, since potentially interested investigators have almost certainly turned to other pursuits, feeling that no more major advances would be forthcoming. The true situation is the exact opposite, and the ferment of activity now going on testifies to the growing realization that studies of glycogen metabolism have thrown up key discoveries that have the widest implications throughout biochemistry" (Ryman and Whelan 1971).

One aspect that remained unclear was how glycogen synthesis was initiated *in vivo*. Both the Coris and Leloir observed that *in vitro* glucose polymerization by glycogen phosphorylase or glycogen synthase required the addition of a pre-formed carbohydrate primer (Swanson and Cori 1948; Leloir and Cardini 1957). Some investigators believed that glycogen synthesis could be initiated *in vivo* by glycogen synthase (Salsas and Larner 1975); others reported that a protein acted as the priming factor (Krisman and Barengo 1975). In 1977, Whelan's group discovered a protein covalently bound to glycogen in liver, and later showed that the protein, which became known as glycogenin, was linked to glycogen via a novel tyrosine-glucose linkage (Butler et al. 1977; Rodriguez and Whelan 1985) (Fig. 1). At first, Whelan's discovery was doubted, but his subsequent work and work from the laboratories of Cohen and Roach provided very strong support for this protein acting as the true primer of glycogen biogenesis. These groups established that glycogenin was a glucosyltransferase using UDP-glucose, and that after glycosylating itself, it could also extend the glucose chain to ~8 glucose residues, which would then be acted upon by glycogen synthase and branching enzyme (Lomako et al. 1988, 1990; Pitcher et al. 1988; Viskupic et al. 1992). The initiation of glycogen synthesis by glycogenin is now widely accepted and has been extensively characterized, and similar protein primers for starch synthesis have also been described (reviewed by Roach 2002; D'Hulst and Mérida 2010; Roach et al. 2012).

2.5 Reviving Interest in Brain Glycogen Metabolism

In recent years there has been a growing interest in studying brain metabolism, in part due to some important technological advancements. Firstly, in the 1970s, neurochemists began using focused microwave irradiation to rapidly and irreversibly inactivate enzymes in rodent brains in order to more accurately measure labile metabolites (reviewed by Schneider et al. 1981; Marani 1998). This technique was elegantly applied to determine the regional distribution of glycogen in the rat brain in the 1980s (Sagar et al. 1987; Swanson et al. 1989) and more recently in the mouse brain (Oe et al. 2016). Secondly, after it was suggested that elevated glycogen in a GSD patient could be observed by ^{1}H NMR, Choi et al. introduced the use of NMR spectroscopy to noninvansively measure the turnover of ^{13}C-labelled glycogen *in vivo* in rodents (Salvan et al. 1997; Choi et al. 1999). Öz and colleagues applied this technique to humans, publishing a series of studies on the role of glycogen in normal human brain metabolism and during hypoglycemia (Öz et al. 2003, 2007, 2009).

Another reason for the growing interest in brain glycogen is the emerging theme of metabolic coupling of astrocytes and neurons, particularly involving the transfer of lactate. These interactions are fascinating, complex, and highly debated. They have been recently discussed by multiple excellent reviews (Barros 2013; Dienel and Cruz 2016; Alberini et al. 2018; Bak et al. 2018; Magistretti and Allaman 2018) and in the subsequent chapters of this book. The coupling between astrocytes and neurons is reminiscent of the Cori cycle and the shuttling of lactate between liver and muscle; however, these interactions are more complicated and difficult to study. Glycogen metabolism is intimately linked to granule size, architecture, and its various associated proteins, which are still being investigated. Such topics are not frequently discussed in the context of brain glycogen metabolism. These nuanced aspects of glycogen regulation, in addition to its surprisingly dynamic nature, subcellular distribution, and alternative degradation pathways, are introduced in the next sections.

3 Glycogen Structure

3.1 Basic Structure of Glycogen and Other Glucose Polymers

Glycogen is one of many types of polysaccharides that exist in biology. A polysaccharide refers to any large polymer comprised of many covalently linked monosaccharides; the bonds between them are known as glycosidic linkages. Glycogen and the two primary constituents of plant starch, amylopectin and amylose, are all homopolymers of glucose designed for energy storage in various organisms. We and others refer to these glucose homopolymers as polyglucans. Glycogen and amylopectin contain branches, while amylose is almost exclusively composed of linear chains. The linear chains of glucose are connected by α-1,4 glycosidic linkages, and the branch points are comprised of α-1,6 glycosidic linkages (Fig. 6a). Because of the tree-like arrangement of branching, glycogen and amylopectin contain fewer reducing ends (i.e. ends with a free aldehyde group) than nonreducing ends, which are oriented outward. The α configuration of the glycosidic linkage causes a linear α-1,4 linked chain to twist, and when the chains are long enough, and uninhibited by branch points, they can form single or double helices (Gessler et al. 1999) (Fig. 6b). An alternative β-linked configuration of glucose polymers favors very

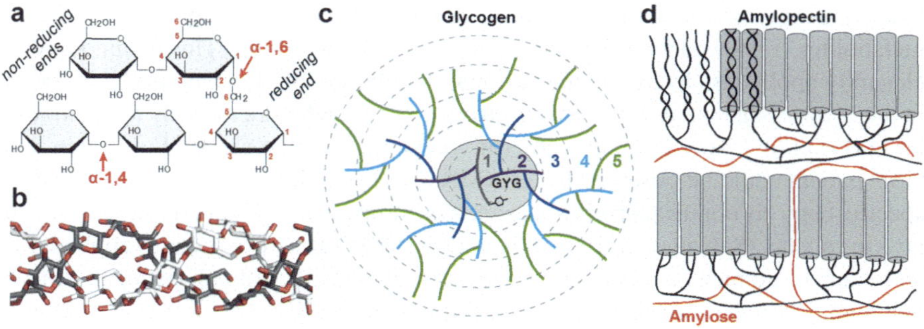

Fig. 6 Structure and synthesis of polyglucans. (**a**) Starch and glycogen contain linear chains of glucose joined by α-1,4 glycosidic linkages, and α-1,6 glycosidic linkages constitute the branch points. The reducing ends (containing an aldehyde group) are oriented near the interior of the polysaccharide molecule, while metabolic enzymes work on the nonreducing ends, which are oriented outward. (**b**) The α conformation of α-1,4 linked polyglucans gives them a propensity to twist, and long unbranched chains can form single or double helices; a model of a double helix is shown. (**c**) A model of the first five tiers of the glycogen particle according to the Whelan model. A tyrosine residue of glycogenin (GYG) is shown, which is covalently linked to the glucan chain making up the first tier. (**d**) A recent model for starch structure according to (Bertoft 2017). In amylopectin, the branched component of starch, branching is clustered, and the long linear chains form double helices (grey cylinders) that make up the crystalline regions of starch. Amylose, the unbranched component of starch, is believed to occupy areas around the amylopectin molecules.

straight chains that can associate into tight fibrils; this is characteristic of structural polyglucans such as cellulose. Mammals do not synthesize or hydrolyze β-glycosidic linkages, so they cannot digest cellulose, but they are well equipped to degrade glycogen and starch (Berg et al. 2002).

Although the models of glycogen and amylopectin structure are both based on the tree-like arrangement originally proposed by Meyer and Bernfeld, the models were individually refined as it became increasingly evident that the arrangement and length of the branches of these glucose polymers were quite different. Glycogen contains a high degree of branching and short linear chains. By enzymatic analyses, there are, on average, 13 glucose units per α-1,4-linked linear chain, and 8% of the total glycosidic linkages are α-1,6 branch points (Cori 1952; Illingworth et al. 1952; Melendez-Hevia et al. 1993; Melendez et al. 1997; Roach et al. 2012) (Fig. 6c). Branching in glycogen is believed to be continuous, meaning the branch points are evenly distributed within the molecule. The presently accepted model for glycogen is based on the revisions of Whelan's group (Calder 1991), but high resolution studies of chain length distribution (CLD) suggests this model may still not be completely accurate (discussed below). Typically, a single glycogen is depicted as originating from a single chain covalently attached to a glycogenin molecule, which gives rise to all subsequent chains (Roach et al. 2012; Prats et al. 2018) (Fig. 6c). Enzymatic experiments indicate 3 to 4 glucose units between each branch point, so each linear chain would yield two branches (Illingworth et al. 1952; Calder 1991). Each new set of branches is referred to as a tier (Fig. 6c), and since the number of linear chains doubles with each successive tier, the outermost tier would theoretically contain about one-third of the total glucose in the molecule, which is consistent with empirical observations (Larner et al. 1952; Melendez et al. 1997). Mathematical modeling has demonstrated that due to physical constraints, glycogen molecules can theoretically contain up to 12 tiers, corresponding to ~55,000 glucose molecules; beyond this size, the outer chains would become so crowded they would be inaccessible to enzymes (Melendez-Hevia et al. 1993). Indeed, the empirically observed upper limit of glycogen β-particles (44 nm) is consistent with the theoretical maximal diameter of a 12-tiered glycogen molecule (42 nm) (Shearer and Graham 2004; Prats et al. 2018). The continuous branching within glycogen prevents, or dramatically limits, the formation of double helices, which would lead to insolubility of glycogen and inaccessibility of the glucan chains to enzymatic degradation (Emanuelle et al. 2016). Mathematical modeling studies have also demonstrated that the empirically observed average chain length and branching degree are optimal for maintaining solubility and structural homogeneity (Melendez et al. 1998).

Amylopectin is also a branched polyglucan, but it contains longer chains of 20–25 glucose units per chain with infrequent and clustered branching (Manners 1989; Buleon et al. 1998) (Fig. 6d). The clustering of branch points means there are regions of long, unbranched chains that associate to form crystalline, water-excluding double helices (Fig. 6b). How these crystalline regions are arranged is still debated, but the most recent data suggests they are arranged along a polyglucan "backbone" (Bertoft 2017). The alternating crystalline and amorphous layers within starch are advantageous as it allows for very dense glucose packing within starch granules, which can be as large as 100 μm in diameter (Emanuelle et al. 2016). Amylose is believed to be interspersed among amylopectin molecules (Fig. 6d). Glycogen, amylopectin, and amylose have diverse structural properties due to differences in chain length and degree of branching. Glycogen has the shortest average chain length and the highest degree of branching, usually 8%. The degree of branching in amylopectin is nearly half that of glycogen, only 4–6% (Manners 1991; Buleon et al. 1998). Amylose is generally considered to be entirely composed of very long, linear chains, although infrequent branching has been reported (Manners 1989).

3.2 Glycogen Architecture

It is well established that amylopectin and glycogen differ in chain length and degree of branching. But a diverse array of polyglucans exist in nature with structures that vary beyond these two simple parameters. It is becoming increasingly apparent that the distribution of chain lengths and the arrangement of branch points within a polyglucan are variable, producing distinct biological and physiochemical properties. Furthermore, not only do amylopectin and glycogen differ significantly from each other, but each polyglucan comes in a variety of shapes and properties based on species, tissue of origin, and even the nutritional state of the tissue. Polyglucan diversity in the context of starch has been exhaustively studied and thoroughly parameterized (Pérez and Bertoft 2010; Nakamura 2015). We will use the term "architecture" to refer to the unique chain length distribution, branching frequency and arrangement, and quantity and distribution of non-glucose moieties (notably phosphate) characteristic of a certain type of glycogen or amylopectin.

Since the days of Bernard, investigators have utilized iodine to study polyglucan architecture. Bernard observed in 1877 that newly synthesized glycogen in rabbit muscle following exhaustive exercise gave a bluish coloration with iodine, in contrast to the reddish color of glycogen from rested muscle or well-fed liver (Young 1937). This colorimetric test can be made quantitative by an absorbance spectral scan: the wavelength of maximal absorption (λ_{max}) increases with longer chains and less branching, reflected by the staining color (Swanson 1948; Krisman 1962; Banks et al. 1971). Glycogen from different sources yields a range of colors from yellow to reddish brown with iodine and produces a λ_{max} of 420–490 nm; amylopectin gives a more intense color ranging from red to lavender with λ_{max} of 490–570 nm; and amylose yields a blue to green color with a λ_{max} of 580–640 nm (Archibald et al. 1961; Bailey and Whelan 1961; Krisman and Alfredo 1991). The iodine method is not perfectly quantitative, due to the various effects of the iodine and salt concentrations, temperature, and polysaccharide structure and source (Morris 1946; Archibald et al. 1961). Additionally, although chain length and degree of branching describe different aspects of glycogen architecture, both parameters are related to λ_{max} (Archibald et al. 1961; Bailey and Whelan 1961; Hirai et al. 1994). It is difficult to deconvolve the various architectural parameters of polyglucans using just iodine, but it remains a convenient technique that yields useful information.

High-performance anion exchange chromatography (HPAEC) was introduced in the 1990s to determine the chain length distribution (CLD) of enzymatically debranched polysaccharides with very high resolution (Rani et al. 1992). The CLD for various glycogens is asymmetrically unimodal: in bovine and rabbit liver glycogens, chains ranged from 3 to 35 degrees of polymerization (DP), peaking at DP 6–10 or 7–15, respectively (Rani et al. 1992; Matsui et al. 1993) (Fig. 7a). This range of chain lengths suggests the architecture of glycogen may be intermediary between the Meyer and Whelan models. Amylopectin CLD profiles display a strikingly different bimodal CLD, with distinct groupings of short and long chains, further evidence for amylopectin chains being organized very differently than glycogen (Fig. 7b) (Hanashiro et al. 1996). CLD profiles for skeletal muscle and brain glycogen from mice have also been described recently. The Minassian group showed that the CLD profiles of mouse skeletal muscle and brain glycogen are nearly identical, ranging from DP 2–35 or more, with most chains being DP 4–15 (Nitschke et al. 2013, 2017). Additionally, gradual isoamylolysis of muscle glycogen suggests that the internal chains are longer than the external chains (Nitschke et al. 2013). The Roach group showed a CLD for mouse skeletal muscle glycogen ranging from DP 3–45, with most chains being 7–15 units long (Irimia et al. 2015). After exercise and a 1 h recovery, the CLD shifted toward long chains

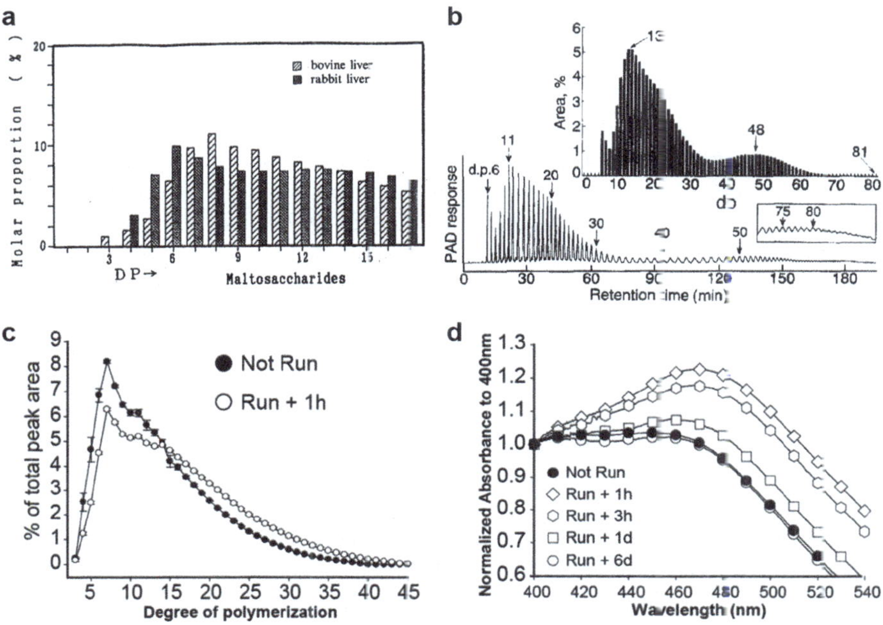

Fig. 7 Chain length distribution (CLD) of glycogen and amylopectin demonstrated by HPAEC. (**a**) CLD of bovine liver and rabbit liver glycogen determined by HPAEC (Matsui et al. 1993). Copyright © 1993 Japan Society for Bioscience and Agrochemistry, reprinted by permission of Taylor & Francis, Ltd. (**b**) HPAEC profile and CLD of potato amylopectin (Hanashiro et al. 1996). Copyright © 1996 Elsevier. Used with permission. (**c**) CLD of skeletal muscle glycogen from rested mice (not run) and 1 h following exhaustive exercise (Irimia et al. 2015). (**d**) Iodine spectra of skeletal muscle glycogen from rested mice and 1 h, 3 h, 1 day, and 6 days post-exercise (Irimia et al. 2015). Copyright © 2015 American Society for Biochemistry and Molecular Biology. Used with permission

(Fig. 7c). Similarly, muscle glycogen had an increased λ_{max} after exercise and recovery, which was gradually restored over the course of 6 days. These data suggest that newly synthesized glycogen is less branched with longer chains that are remodeled over time (Fig. 7d).

It appears that glycogen molecule is not as homogenously structured as the Whelan model suggests, and chain length and branching appear to fluctuate based on nutritional state. Gerty Cori stated in a 1952 lecture that "the relationship of structure to nutritional state has not yet been fully explored, but it seems that glycogen freshly deposited after a fasting period is least branched and has long outer chains whereas the opposite is true of 'old' glycogens" (Cori 1952). Additionally, mammalian and non-mammalian glycogens with similar average chain lengths produce differences in iodine spectra, suggesting that mammalian glycogens contain long chains in the interior of the granule that are not present in non-mammalian glycogens (Manners et al. 1983). Whole glycogen particles also display a continuum of sizes, and their size and ultrastructure vary with tissue type (see Sect. 6.1).

3.3 Glycogen Structure and Phosphate: Insights from Lafora Disease

Fontana first demonstrated that [^{32}P]-labeled glycogen could be purified from the livers of rats after the administration of radioactive phosphoric acid (Fontana 1980). Whelan's group subsequently reported that mammalian muscle glycogen contains about 0.064% phosphorus by weight (Lomako et al. 1993b, 1994). However, much like Whelan's reports of a proteinaceous component of glycogen that were later substantiated with the discovery of glycogenin, phosphate was initially treated as a contaminant. The physiological relevance of glycogen phosphate became evident through basic scientific investigations of Lafora disease (LD), a fatal, inherited childhood epilepsy and non-classical glycogen storage disease. LD is characterized by accumulations of abnormal polysaccharides known as Lafora bodies (LBs) that cause neurodegeneration (recently reviewed by Gentry et al. 2018) (Fig. 8a). In the 1960s and 1970s, it was demonstrated that LBs contain high levels of phosphorus and histologically resemble plant amylopectin (Yokoi et al. 1968; Sakai et al. 1970). In 1998 and 2003 the genes associated with the disease were identified, and it is now well established that virtually all LD patients carry recessive mutations in either the *EPM2A* or *EPM2B* gene (Minassian et al. 1998; Serratosa et al. 1999; Chan et al. 2003). *EPM2A* encodes laforin, a glycogen phosphatase (Worby et al. 2006; Tagliabracci et al. 2007), and *EPM2B* encodes malin, an E3 ubiquitin ligase (Gentry et al. 2005; Lohi et al. 2005). *Epm2a−/−* and *Epm2b−/−* mouse models recapitulate the disease with respect to LB formation and neurodegeneration (Ganesh et al. 2002; DePaoli-Roach et al. 2010; Valles-Ortega et al. 2011; Criado et al. 2012; Tiberia et al.

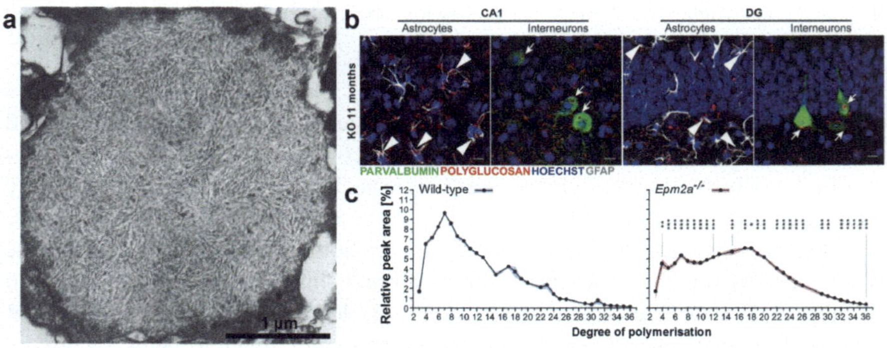

Fig. 8 Insights from Lafora disease, a nonclassical glycogen storage disease. (**a**) A typical LB visualized by electron microscopy in the human retina (Berard-Badier et al. 1980). Scale bar has been approximated based on magnification. Copyright © 1980 Springer Nature. Used with permission. (**b**) Neuronal (arrows) and astrocytic (arrowheads) polyglucosan accumulations visualized by immunostaining in the dentate gyrus (DG) and CA1 of the hippocampus in malin knockout (KO) mice (Valles-Ortega et al. 2011). Scale bars = 10 μm. (**c**) Normal and abnormal CLD of purified polysaccharides from WT and LD (*Epm2a−/−*) mice determined by HPAEC (Nitschke et al. 2017).

2012; Duran et al. 2014) (Fig. 8b). Purified polysaccharides from LD mice have a shifted λ_{max} reminiscent of amylopectin (Valles-Ortega et al. 2011; Brewer et al. 2019), and their CLD profile shows a higher proportion of long chains than wild-type mice, most prominently in brain tissue (Nitschke et al. 2017) (Fig. 8c). Unlike glycogen, purified LBs are quite large (>1 μm in diameter) and insoluble; they are presumed to be aggregates of a polysaccharide with abnormal architecture resulting from aberrant glycogen metabolism (Gentry et al. 2018; Brewer et al. 2019).

Laforin and malin regulate two aspects of glycogen architecture that may be intertwined: chain length and phosphate level. Through studies of its role in LD, laforin was discovered to be the founding member of a class of enzymes known as glucan phosphatases that directly dephosphorylate carbohydrate substrates (Worby et al. 2006; Gentry et al. 2007; Gentry and Pace 2009; Vander Kooi et al. 2010; Meekins et al. 2013, 2015). Laforin is the only known glucan phosphatase in mammals, and decreased laforin activity results in glycogen hyperphosphorylation. Malin ubiquitinates enzymes involved in glycogen metabolism, but the effects of ubiquitination are not clear (see Sect. 5.2). Surprisingly, the absence of malin also leads to hyperphosphorylation, and the absence of either enzyme leads to the accumulation of glycogen with abnormally long chains, which is puzzling (Sullivan et al. 2017). Recent results from mouse models overexpressing a catalytically inactive laforin demonstrate that the inactive laforin rescues the LD phenotype in the *Epm2a*-deficient mouse model (Gayarre et al. 2014; Nitschke et al. 2017). As a result, the relevance of laforin's catalytic activity has been questioned, despite the very high conservation of its catalytic residues and the presence of glucan phosphatases across multiple kingdoms (Gentry et al. 2007; Raththagala et al. 2015). It has also been suggested that laforin and malin form a ubiquitination-targeting complex involved in the disposal of glycogen molecules with aberrantly long chains that could crystallize and cause the molecule to precipitate (Sullivan et al. 2017). A growing body of evidence suggest that laforin and malin are an integral part of an alternative route for glycogen degradation known as glycophagy, involving members of the autophagic pathway and lysosomal α-glucosidase (see Sect. 5.2). However, glycophagy is probably not reserved only for aberrantly structured glycogen molecules, since the accumulated lysosomal glycogen in acid α-glucosidase deficiency (i.e. Pompe disease) is of normal structure based on iodine staining (Levin et al. 1968; Mahler 1969). Additional evidence for a physiological role of glycogen phosphate comes from the Roach lab and a substantial body of work from the field of starch metabolism.

The Roach lab analyzed glycogen CLD and phosphate levels of laforin-deficient mice pre- and post-exercise. When muscle glycogen phosphate was depleted, phosphate levels remained suppressed even after total glycogen and CLD returned to normal (Irimia et al. 2015). Laforin-deficient mice displayed exercise-induced glycogen depletion identical to wild-type mice, but phosphate levels remained elevated and iodine spectra suggested a delay in glycogen remodeling post-recovery. This study demonstrated that laforin dephosphorylates glycogen during exercise-induced cytosolic glycogeolysis, providing strong evidence for its physiological role as a glycogen phosphatase. Whelan suggested in 1994 that phosphate was a marker for the age of a glycogen molecule; the suppression of glycogen phosphate after exhaustive exercise in wild-type mice does indeed suggest that phosphate accumulates with age, possibly from multiple cycles of glycogen degradation and re-synthesis (Lomako et al. 1994; Irimia et al. 2015). The role of phosphate in glycogen metabolism is not clear, but studies from the starch field demonstrate that phosphate plays an important physiochemical and biological role in starch architecture and metabolism (Hejazi et al. 2008; Blennow and Engelsen 2010; Kotting et al. 2010).

In plants, phosphorylation is necessary for the proper synthesis of starch, and reversible phosphorylation is an integral part of starch breakdown. Only amylopectin contains significant amounts of phosphate, which is enriched in the amorphous regions (Takeda and Hizukuri 1982; Blennow et al. 2000). Root and tuber starches, which exhibit less densely packed crystalline helices than other types of starches, have high phosphate levels (Lim et al. 1994; Blennow et al. 1998). Starch phosphate is covalently linked to the C3 and C6 hydroxyls of glucose moieties; about 70–80% of the phosphate is esterified to C6 (Ritte et al. 2006). Phosphate, particularly at the C3 position, disrupts the crystalline helices in amylopectin by introducing steric hindrance, promoting their solubilization (Hansen et al. 2009). Starch degradation is believed to be a cyclic process involving reversible phosphorylation and the concerted action of multiple enzymes: glucan dikinases, amylases, and glucan phosphatases (Emanuelle et al. 2016). Two dikinases that respectively phosphorylate the C6 and C3 position in that order are glucan, water dikinase and phosphoglucan, water dikinase. Phosphorylation promotes solubilization of the glucan chains and facilitates their access by plant amylases. However, the amylases cannot proceed past a phosphate, so phosphate removal is achieved by the glucan phosphatases Starch Excess 4 (SEX4) and Like Sex Four 2 (LSF2), named after the plant phenotype that results from their deficiency and part of the same family as the glycogen phosphatase laforin (Edner et al. 2007; Kotting et al. 2009; Santelia et al. 2011; Gentry et al. 2016; Meekins et al. 2016). Crystal structures and structure-function studies of SEX4 and LSF2 preceded those of laforin (Vander Kooi et al. 2010; Meekins et al. 2013, 2014; Raththagala et al. 2015) and the cooperative study of both systems has led to a wealth of insight about polyglucan architecture and phosphorylation (Gentry et al. 2009, 2016; Emanuelle et al. 2016).

Through studies of LD, the Roach and Minassian groups determined that normal glycogen contains about 1 phosphate moiety per 600–2500 glucose units, depending on species and tissue type (Tagliabracci et al. 2007, 2008; Turnbull et al. 2010; Tiberia et al. 2012; DePaoli-Roach et al. 2015). Phosphates are present as monoesters linked to the C2, C3 and C6 hydroxyls of glucose moieties, in approximately equal quantities (Tagliabracci et al. 2011; Nitschke et al. 2013; DePaoli-Roach et al. 2015). Data from multiple groups using various approaches strongly suggest the phosphate is concentrated at the interior of glycogen molecules (Tagliabracci et al. 2007; Nitschke et al. 2013; Irimia et al. 2015). Liver glycogen contains less phosphate than muscle glycogen, while C6 phosphate has been detected in brain glycogen at similar levels to C6 phosphate in muscle glycogen (Lomako et al. 1994; Tagliabracci et al. 2007; Turnbull et al. 2010; Nitschke et al. 2017). The source of glycogen phosphate remains a mystery: although it has been reported that glycogen synthase can incorporate the β-phosphate of UDP-glucose into glycogen in a rare side reaction, this mechanism could only account for C2 and C3 phosphate, and the rate of incorporation is very low (1 in 10,000 catalytic cycles) compared to the actual levels observed in glycogen (Tagliabracci et al. 2011; Contreras et al. 2016). Although no glucan dikinase has been identified in mammals, the physiological role of glycogen phosphate will eventually be defined, likely through the continued investigations of starch metabolism and the molecular mechanisms of LD.

4 Glucosamine

Glycogen also contains small amounts of covalent glucosamine. In the 1960s and 70s, Maley et al. reported the incorporation of glycosidically linked [^{14}C]glucosamine into glycogen from [^{14}C]galactosamine, and showed that glycogen synthase could catalyze glucosamine incorporation *in vitro* by using the substrate UDP-glucosamine instead of UDP-glucose (Maley et al. 1966; Tarentino and Maley 1976). Whelan also reported that intraperitoneal injection of [^{14}C]galactosamine led to the incorporation of [^{14}C]glucosamine into glycogen, replacing as much as 10% of the glucose residues (Romero et al. 1980). Covalently linked glucosamine in glycogen apparently would not block phosphorolysis, since it could be released as glucosamine-1-phosphate by phosphorylase. Whelan reported that normal pig and rabbit liver contained 86–266 nmol glucosamine per g glycogen, which was randomly distributed throughout the molecule (Kirkman and Whelan 1986). It is worth noting that an earlier study also demonstrating the incorporation of [^{14}C] into glycogen from injected [^{14}C]glucosamine showed that glucosyl residues were labeled, indicating [^{14}C]glucosamine was converted to [^{14}C]glucose through a deamination reaction prior to glycogen incorporation (Khac et al. 1972).

Additional links between glycogen and glucosamine have been reported. Glycogen synthase can be modified by N-acetyl-glucosamine (GlcNAc), and it has been suggested that glycogenin could also be modified by GlcNAc (Parker et al. 2003; Tavridou and Agius 2003). Furthermore, studies have suggested that glycogen could be a carbohydrate source for protein glycosylation, for which glucosamine is a major precursor, and hypoglycosylation has been reported in GSDs (McMahon and Frost 1996; Hayee et al. 2011; Tegtmeyer et al. 2014; Ondruskova et al. 2018). A physiological role for glycogen in protein glycosylation and the relevance of covalently linked glucosamine await further investigation.

5 Cytosolic Glycogen Synthesis and Degradation

5.1 Glycogenin

Glycogen synthesis begins with glycogenin [UDP-α-glucose:glycogenin α-glucosyltransferase, EC 2.4.1.186], a member of glucosyltransferase family 8 (Campbell et al. 1997) (www.cazy.org). The enzyme possesses two distinct enzymatic activities: self-glucosylation and chain elongation, both requiring UDP-glucose as a glucosyl donor and both occurring at the same active site, though the chemistries of the reactions are different (Alonso et al. 1995; Lomako et al. 2004). For this reason the two activities were initially believed to belong to separate enzymes, until Whelan and Cohen discovered that both were accomplished by glycogenin (Lomako et al. 1988; Pitcher et al. 1988). Crystal structures and biochemical studies show that glycogenin functions as an obligate dimer and requires Mn^{2+} for activity (Gibbons et al. 2002; Chaikuad et al. 2011). Autoglucosylation on tyrosine creates a glucose-1-O-tyrosyl linkage, a chemical linkage rarely found in nature, then glycogenin synthesizes a short oligosaccharide primer of at least 7–8 glucose units (Smythe and Cohen 1991; Alonso et al. 1995). Even if this tyrosine is mutated, glycogenin is still capable of glucoslyation, just not of itself (Cao et al. 1993a; Alonso et al. 1994). Whether glucosyl additions occur via intra- or intersubunit reactions has been debated, but recent studies suggest that the first four glucosyl units may be added via an intrasubunit reaction, and longer chains require intersubunit interaction (Chaikuad et al. 2011). It has been suggested that although glycogenin functions as a dimer, at low enough concentrations the two subunits would dissociate after chain initiation, giving rise to two separate glycogen molecules (Lin et al. 1999). The presence of a single chain giving rise to an entire β-particle is supported by structural studies of glycogen (Calder 1991).

In most mammals, there is only one isoform of glycogenin that is widely expressed, but humans have two isoforms: glycogenin-1 (*GYG1*), expressed in all tissues, and glycogenin-2 (*GYG2*) which is primarily expressed only in liver, with minor expression in heart, pancreas, and adipose tissue (Mu et al. 1997; Roach 2002; www.proteinatlas.org). There is some evidence that glycogenin is phosphorylated, but it has not been corroborated (Lomako et al. 2004). Glycogenin can interact directly with glycogen synthase, and a crystal structure of the interaction provides evidence for their cooperation in the initiation of the glycogen granule (Skurat et al. 2006; Zeqiraj et al. 2014). Additionally, insufficiently glucosylated glycogenin does not serve as an efficient primer for glycogen synthase, and phosphorylase can reduce the glucosylation state of glycogenin, making it a less effective substrate for synthase (Cao et al. 1993b; Skurat et al. 1993). Phosphorylase induces the dissociation of glycogen synthase from a proteoglycan fraction in hepatocytes, presumably containing glycogenin (Tavridou and Agius 2003). Thus, priming through glycogenin may be regulated indirectly by the signals that govern glycogen synthase and phosphorylase activity (see Chaps. 3 and 4), rather than by direct regulation of glycogenin. Glycogenin interacting proteins (GNIPs) have been suggested to stimulate the activity of glycogenin (Graham et al. 2010). Paradoxically, glycogenin mutations in humans and mice lead to glycogen accumulation in some tissues and muscle weakness, indicating that without glycogenin, glycogen synthesis can still occur, but in a dysregulated and pathogenic manner (Malfatti et al. 2014; Testoni et al. 2017).

5.2 Glycogen Synthase and Phosphorylase

Glycogen synthase [UDPglucose:glycogen α-4-glucosyltransferase, EC 2.4.1.11] is a member of the glycosyltransferase family 3 that catalyzes the addition of alpha-1,4-linked glucosyl units to a glycogen chain, using UDP-glucose as the donor and releasing UDP as product (www.cazy.com). Two isoforms of synthase exist in mammals: *GYS1*, encoding muscle glycogen synthase, highly expressed in all tissues including brain, and *GYS2*, restricted to liver (Roach et al. 2012) (www.proteinatlas.org). Muscle glycogen synthase has nine phosphorylation sites and was one of the first examples of a hierarchically phosphorylated protein (Roach 1990). Although liver glycogen synthase is also multiply phosphorylated, its activity appears to be regulated by only one phospho-site (Ros et al. 2009). Phosphorylation inhibits synthase activity, although activity can be fully restored in the presence of the potent allosteric activator glucose-6-phosphate. Numerous kinases are responsible for phosphorylating synthase, including protein kinase A, protein kinase C, AMP-activated protein kinase (AMPK), caseine kinase 2, and glycogen synthase kinase 3 (GSK3). Thus, control of glycogen metabolism is inextricably linked to intracellular energy status, glucose homeostasis, insulin signaling, and other external signals through diverse signaling cascades (Roach 1990; Lawrence and Roach 1997; Roach et al. 2012). The structure and regulation of glycogen synthase warrant much attention and are discussed in Chap. 3.

Glycogen phosphorylase [1,4-α-glucan:orthophosphate α-glycosyltransferase, EC 2.4.1.1] is a member of the glycosyltransferase family 35, and catalyzes the transfer of glucose moieties from the glycogen molecule to inorganic phosphate, releasing the product glucose-1-phospate (www.cazy.com). Glucose-1-phosphate is converted to glucose-6-phosphate by phosphoglucomutase, facilitating its entry into glycolysis (see Chap. 6). In the presence of excess glucose-1-phosphate, phosphorylase can catalyze the reverse reaction (Cori and Cori 1943). There are three isoforms of phosphorylase corresponding to the tissues in which they are enriched, but not restricted: muscle (*PYGM*), liver (*PYGL*) and brain (*PYGB*) (www.proteinatlas.org). Only the brain isoform is expressed in fetal tissues, and all are expressed to some extent in adult brain (Pfeiffer-Guglielmi et al. 2000). A single site on phosphorylase is phosphorylated by only one kinase: phosphorylase kinase (PhK), a multisubunit enzyme with muscle, liver and brain isoforms. Both phosphorylase and PhK are activated by phosphorylation and allosteric modulators, and an excellent review on their regulation in brain has been published recently (Nadeau et al. 2018). Crystal structures of all phosphorylase isoforms have been determined (Rath et al. 2000; Lukacs et al. 2006; Mathieu et al. 2016). The structure and regulation of brain phosphorylase are also given a comprehensive review in Chap. 4.

5.3 Glycogen Branching and Debranching Enzymes

Like glycogenin, and unlike synthase and phosphorylase, glycogen branching enzyme (GBE) and debranching enzyme (GDE) do not appear to be regulated by phosphorylation or allosteric modulation. There is only one isoform of each, and both are highly expressed ubiquitously (www.proteinatlas.org). Due to their abundance, these enzymes are not considered to be rate-limiting under normal circumstances (Geddes 1986; Melendez et al. 1997; Roach et al. 2012). However, they are clearly essential for maintaining proper glycogen structure, as deficiencies in either enzyme result in GSDs characterized by distinctly abnormal glycogen deposits with pathological consequences (Adeva-Andany et al. 2016). Crystal structures of human GBE and yeast GDE have been determined and the effects of GSD-causing mutations defined, providing molecular level insights into how they affect glycogen structure (Froese et al. 2015; Zhai et al. 2016).

GBE (α-1,4-glucan:α-1,4-glucan 6-glycosyltransferase, EC 2.4.1.18) belongs to the subfamily 8 of the GH13 family of glucosyl hydrolases. Like other members of this subfamily, it contains a carbohydrate binding module, CBM48 (Janecek et al. 2011; www.cazy.com). Two reaction steps occur successively in its central catalytic core: hydrolysis and transglucosylation (Froese et al. 2015). In the first step, the enzyme cleaves an α-1,4 linkage on a glucan chain, forming a covalent enzyme-glycosyl intermediate; in the second step, an α-1,6 linkage is formed from the same chain or one nearby. The minimum length transferred is 6 or 7 glucosyl units (Brown and Brown 1966; Verhue and Hers 1966), and the average distance between chains is 3 or 4 glucose units (Gibson et al. 1971; Calder 1991). This chain length requirement is supported by the crystal structure of GBE in complex with a heptasaccharide (Froese et al. 2015).

Mammalian GBE can convert amylose or amylopectin to a glycogen-like polysaccharide. Krisman showed that incubation of amylopectin or amylose with GBE purified from rat liver led to a displacement of λ_{max} to a value that coincided with that of glycogen (Krisman 1962). However, GBE did not change the λ_{max} of rabbit liver glycogen, suggesting the molecule already had been maximally branched. Indeed, comparative studies on mammalian and plant branching enzymes indicate that the branching degree in polysaccharides is determined by intrinsic properties of the branching enzyme used to synthesize them (Tolmasky and Krisman 1987; Kuriki et al. 1997). In other words, the optimal structure of the mammalian glycogen molecule may be primarily attributed to the inherent "equilibrium" of the mammalian branching enzyme. Not surprisingly, synthetic glycogens produced *in vitro* by branching enzymes from different organisms have similar, but not identical, structural properties to natural glycogens (Kajiura et al. 2008, 2010; Takata et al. 2009). Perhaps these observations explain the delay in glycogen remodeling reported by the Roach lab after exercise (Irimia et al. 2015): glycogen super-compensation, characterized by high glycogen synthase activity, results in longer chains because branching enzyme has not yet introduced the appropriate branch points. Only over time does branching enzyme act so that glycogen structure returns to an equilibrium, with CLD and λ_{max} at basal levels. Indeed, it has been suggested that a chronic imbalance of glycogen synthase and GBE activities leads to abnormal polysaccharide formation in pathological conditions (Raben et al. 2001; Tagliabracci et al. 2008; Kakhlon et al. 2013). Also, the ubiquitous expression of a single GBE isoform may help to explain why brain glycogen has a similar CLD to muscle glycogen (Nitschke et al. 2017). Presumably all healthy tissues in a single species would reach the same CLD at "equilibrium."

GDE is required for cytosolic glycogen degradation since phosphorylase terminates four glucosyl residues away from a branch point. Mammalian GDE has two catalytic activities: 4-α-glucanotransferase activity (EC 2.4.1 25) involves transfer of a chain of three glucosyl units from the branch to a nearby nonreducing end, leaving a single α-1,6 glucosyl unit. Its amylo-α-1,6-glucosidase activity (EC 3.2.1.33) follows, and the α-1,6 linkage is hydrolyzed to release free glucose (Ryman and Whelan 1971). The release of free glucose instead of glucose-1-phosphate makes it very convenient to compare the levels of α-1,6 and α-1,4 linkages in polysaccharides. GDE is considered an indirect debranching enzyme; direct debranching enzymes such as isoamylase exclusively cleave α-1,6-linkages and release intact oligosaccharide chains. Direct debranching enzymes occur in plants, bacteria and yeast but not in mammals (Ryman and Whelan 1971). These enzymes are very useful in polysaccharide analysis, especially in determining CLD.

GDE catalytic mutants display a loss of either glucosidase or transferase activity, but not both, strongly suggesting disparate catalytic sites (Nakayama et al. 2001). This was confirmed when the crystal structure of yeast GDE was determined, revealing an elongated structure with catalytic domains at either end of the molecule (Zhai et al. 2016). Structures of both the free wild-type enzyme and a transferase-deficient mutant in complex with maltopentaose were reported. Maltopentaose molecules were bound at both catalytic domains and other sites, illustrating the specificity of GDE transferase activity for a glucan of five units or less and the necessity of non-catalytic binding sites for glycogen association. The data also suggested that the product of the transferase reaction, a chain containing a single α-1,6-linked glucosyl unit, completely dissociates from GDE prior to recruitment to the glucosidase catalytic site. Multiple binding sites are a recurring theme among glycogen- and starch-associated proteins (Baskaran et al. 2010; Meekins et al. 2013; Froese et al. 2015; Raththagala et al. 2015; Zhai et al. 2016). In observing that oligosaccharides were bound both to catalytic and non-catalytic sites in the crystal structure of GBE, Froese et al. stated: "[Non-catalytic binding sites] may provide GBEs the capability to anchor a complex glycogen granule and, as proposed previously, determine the chain length specificity for the branching reaction as a 'molecular ruler.' This agrees with the emerging concept of glycogen serving not only as the substrate and product of its metabolism but also as a scaffold for all acting enzymes" (2015) (See Sect. 7).

6 Glycophagy

6.1 *Lessons from Pompe Disease*

Phosphorolysis is not the sole means of glycogen degradation in cells. This became evident through studies of Pompe disease, a lysosomal storage disorder characterized by the buildup of vacuole-bound glycogen in virtually all tissues (Hers 1963; Raben et al. 2012). The deficient enzyme in Pompe disease is lysosomal acid α-glucosidase (called GAA or acid maltase) [EC 3.2.1.3] which hydrolyzes both α-1,4 and α-1,6 linkages, optimally at low pH (Hermans et al. 1991). It is a widely expressed enzyme, with a promoter characteristic of a housekeeping gene (Hoefsloot et al. 1990). GAA is synthesized as a precursor protein in the endoplasmic reticulum and undergoes extensive processing before it achieves optimal catalytic activity and reaches its lysosomal destination (Hermans et al. 1991; Wisselaar et al. 1993). It can fully hydrolyze glycogen as well as short oligosaccharides, maltose and other polyglucans to glucose (Hers 1963; Palmer 1971; Matsui et al. 1984). GAA is exo-acting, cleaving single glucosyl units successively from the non-reducing ends of glycogen. However, rather than releasing glucose-6-phosphate like phosphorylase, GAA liberates free glucose, which can translocate into the cytosol via glucose transporters in the lysosomal membrane (Mancini et al. 1990).

How glycogen gets into the lysosome is not well defined, but recent work demonstrates that it involves the autophagic machinery. Autophagy, or 'self-eating,' refers to multiple cellular pathways converging on the lysosomal breakdown of cellular components such as long-lived proteins and organelles (reviewed by Cuervo et al. 2005; Mizushima and Komatsu 2011; Schneider and Cuervo 2014). Autophagy is critical both for general cellular maintenance and for nutrient release in times of energy stress. It is especially critical for turning over worn or damaged cellular components in neurons and other post-mitotic cells (Cuervo et al. 2005; Komatsu et al. 2006; Mizushima and Komatsu 2011). The best characterized autophagic pathway is macroautophagy, activated in response to starvation; 'autophagy' and 'macroautophagy' are frequently used interchangeably. Macroautophagy is a highly conserved pathway in eukaryotes involving a family of autophagy-related (Atg) proteins and the formation of a structure with a characteristic double membrane known as the autophagosome, which engulfs substrates and fuses with the lysosome. Microautophagy and chaperone-mediated autophagy are alternative pathways in which cellular components are targeted directly to lysosomes in the absence of autophagosome formation. In recent years, numerous terms for the selective degradation of specific substrates (via any of the autophagic pathways) have been introduced; e.g. mitophagy (of mitochondria), pexophagy (of perioxisomes), nucleophagy (of nuclear components), reticulophagy (of ER), and xenophagy (of pathogens) (Mizushima and Komatsu 2011; Zhao et al. 2018). In 2011, the Roach group introduced the term 'glycophagy' to refer to the selective autophagic degradation of glycogen (Jiang et al. 2011). Ironically, the term 'glycogenosome' had already been used by some investigators to refer to the glycogen-laden lysosomes found in newborn rat liver, aged neural tissue, and glycogen storage diseases (Iwamasa et al. 1980, 1983; Cavanagh and Jones 2000), but the term did not gain traction.

In addition to glycogenosomes, which have a single membrane characteristic of a lysosome, Pompe (*GAA−/−*) tissues also display accumulated autophagosomes filled with ubiquitinated proteins and cellular debris. In order to study the effect of autophagy on lysosomal glycogen, Raben and colleagues generated a muscle-specific knockout of a critical autophagosome gene, *Atg5*, on the *GAA−/−* background. These mice had just as many glycogenosomes as the *GAA−/−* mice, but they had a reduced number of autophagosomes and no improvement in pathology. Ubiquitinated proteins were increased, but they were not surrounded by the double membrane as in the *GAA−/−* mice (Raben et al. 2008; Lim et al. 2017). The study suggested that the autophagy defect was caused by inefficient fusion of the autophagosomes with the glycogen-laden lysosomes, which formed by an independent pathway. In a later study, the group selectively inactivated a closely related gene, *Atg7*, in fast-twitch *GAA−/−* muscle (Raben et al. 2010). No autophagosomes were formed, glycogenosomes were still present, and there was no change in pathology; however, the glycogen level was reduced. These two studies demonstrate that macroautophagy contributes to some, but not all, of the delivery of glycogen to the lysosome (Raben et al. 2010). The group also illustrated that the muscle pathology of Pompe disease resulted largely from defective autophagic flux. The observation that Atg7, but not Atg5, abrogated glycogen accumulation may be explained by the fact that there are two primary conjugation systems in autophagy, and while Atg5 is only involved in one, Atg7 is required for both (Nakatogawa et al. 2009).

In 2002, Janeček identified a putative CBM20 in a protein of unknown function called starch binding domain-containing protein 1 (Stbd1, also called genethonin 1) (Janeček 2002). The only other proteins in mammals with a CBM20 are laforin and a glycosyltransferase (glycerophosphocholine phosphodiesterase 1, GPCPD1). It was known that the laforin CBM facilitates laforin binding to glycogen (Wang et al. 2002). The Roach group later showed that Stbd1 is highly expressed in glycogen-rich tissues such as muscle, liver, and heart, with trace amounts in brain, kidney and pancreas, and that Stbd1 levels mirrored glycogen levels in fed, fasted and transgenic mice (Jiang et al. 2010). Stbd1 binds glycogen *in vitro* and co-localizes with the lysosomal-associated membrane protein LAMP1 and the Atg8 homolog GABARAPL1 in cell culture, but not with the microtubule-associated light chain 3 (LC3), a marker for auotophagosomes (Jiang et al. 2010, 2011). Demetriadou et al. later showed that when overexpressed in HeLa cells, Stbd1 was N-myristoylated, and this modification facilitates its recruitment with glycogen to subcellular domains associated with autophagy (Demetriadou et al. 2017). The Sun group showed that Stbd1 is elevated in muscle tissue of *GAA−/−* mice, but not in liver or heart, but shRNA-mediated knockdown of Stbd1 did not alter glycogen levels in any tissues (Yi et al. 2013). A full knockout of Stbd1 in the Pompe mice (*Stbd1−/− GAA−/−*) did not change lysosomal accumulation of glycogen in muscle or heart tissue, but reduced liver glycogen by 73% (Sun et al. 2016). Glycogen accumulation could be restored by exogenous expression of human Stbd1, but not a mutant lacking the CBM20. Their study demonstrated that Stbd1 may significantly participate in glycophagy in the liver but not in the heart or skeletal muscle, and that the CBM20 is essential for its function.

6.2 Lessons from Lafora Disease

Another link between glycogen and autophagy emerged through studies of LD. Laforin- and malin-deficient mouse models were generated and analyzed by six different groups. Most observed autophagy defects in the brain of these mice (Criado et al. 2012; Puri et al. 2012; Duran et al. 2014; Gayarre et al. 2014). Some groups observed reduced conversion of LC3 from its cytosolic form (LC3-I) to lipidated form (LC3-II), which is associated with its recruitment to autophagosomal membranes (Criado et al. 2012; Gayarre et al. 2014). Both mouse models also displayed an increase in p62 (Criado et al. 2012; Puri et al. 2012; Duran et al. 2014), a protein that normally binds to both ubiquitinated cargo and LC3, is incorporated into the autophagosome, and degraded (Mizushima and Komatsu 2011). Reduced LC3-II and increased p62 levels strongly suggested defects in autophagic flux. Similar defects were also observed in liver of laforin-deficient mice, but not in muscle (Aguado et al. 2010; Irimia et al. 2015). In contrast, the Minassian group did not observe autophagy defects in muscle or brain in either mouse model (Wang et al. 2016; Nitschke et al. 2017).

The Roach, Minassian and Guinovart groups generated LD mouse models with genetically reduced glycogen synthesis and found that LB formation was reduced and neuropathology was abrogated (Turnbull et al. 2011a, 2014; Pederson et al. 2013; Duran et al. 2014). The Guinovart group elegantly demonstrated that the autophagy impairment was secondary to polyglucosan accumulation by crossing malin-deficient mice with mice specifically lacking muscle glycogen synthase in the brain, which eliminated cerebral glycogen synthesis, LB formation, autophagy defects, gliosis, and susceptibility to epilepsy (Duran et al. 2014). These studies indicate that polyglucosan accumulation drives the autophagy impairment and neurological phenotype in LD. Although some groups have argued that laforin and malin regulate autophagy independent of glycogen, this is unlikely to be the case *in vivo* (Gentry et al. 2018). However, as was demonstrated with the Pompe mice, insufficient clearance of glycogen via glycophagy may indeed provoke pathological damage by obstructing overall autophagic flux. This is a very prominent theme among lysosome storage disorders, which frequently have severe neurological phenotypes; this disease class is reviewed in depth elsewhere (Lieberman et al. 2012; Seranova et al. 2017). Corroborating this, recent studies from the Sanz group showed that in LD, mitophagy is impaired due to a global autophagic defect, and endocytic recycling of the astrocytic glutamate transporter GLT1 is altered (Munoz-Ballester et al. 2016; Garcia-Gimeno et al. 2018), likely also resulting from impaired autophagy.

The possibility that laforin and malin are involved in glycophagy builds on work from many labs showing that malin, typically in concert with laforin, ubiquitinates glycogen-associated substrates. Malin is an E3 ubiquitin ligase that was initially demonstrated to bind and ubiquitinate laforin in cell culture (Gentry et al. 2005; Lohi et al. 2005). Malin also promotes the ubiquitination of multiple proteins that bind to glycogen molecules: glycogen synthase, protein targeting to glycogen

(PTG), R6, AMPK and GDE (see Sect. 7). In most of these studies, ubiquitination was shown to require laforin, which is believed to act as a scaffold for malin interaction with its substrate (Cheng et al. 2007; Vilchez et al. 2007; Worby et al. 2008; Moreno et al. 2010; Rubio-Villena et al. 2013). Although ubiquitination led to the degradation of these substrates in cell culture, PTG, R6 and AMPK levels were unchanged in LD mouse models; in contrast, glycogen synthase accumulated in the insoluble glycogen fraction, as did laforin in malin-deficient mice (Wang et al. 2007; DePaoli-Roach et al. 2010; Valles-Ortega et al. 2011; Tiberia et al. 2012). Importantly, whenever the ubiquitin linkages incorporated by malin are defined, they are K63-linked (Moreno et al. 2010; Roma-Mateo et al. 2011a; Rubio-Villena et al. 2013; Sanchez-Martin et al. 2015). K63-linked ubiquitination is a substrate for p62 binding and specifically linked to selective autophagy (Tan et al. 2008; Shaid et al. 2013). The presence of a laforin-malin ubiquitination complex has been highly debated (Gentry et al. 2013, 2018; Sullivan et al. 2017).

Interestingly, the Roach group reported that there was no change in the levels of GABARAPL1 (a member of the Atg8 family and a homolog of LC3; see Lee and Lee 2016) and LAMP1 in laforin-deficient fibroblasts, but these proteins were decreased in malin-deficient fibroblasts and in fibroblasts with both laforin and malin knocked out (Garyali et al. 2014). These results suggest laforin is upstream of malin in the path from glycogen to the lysosome. The Sanz group showed that laforin, malin and p62 form a complex, and that malin mediates the interaction between laforin and p62, likely through ubiquitination (Sanchez-Martin et al. 2015). Additionally, malin and p62 do not co-localize in cells without the expression of laforin. It is well known that laforin binds to and co-localizes with glycogen *in vitro*, in cell culture, and *in vivo*.

A model to explain these studies that is compatible with available data posits: laforin binds to glycogen and recruits malin; the laforin-malin complex promotes K63-linked ubiquitination of various glycogen-bound proteins; p62 binds to K63-linked ubiquitin moieties, recruiting LC3 (or its homolog GABARAPL1) and the autophagosome machinery, which engulfs the glycogen particle and fuses with the lysosome. Since glycogen synthase and laforin expression parallel glycogen levels and these proteins accumulate with glycogen in LD mouse models, they may also be turned over inside the lysosome. The fact that a catalytically inactive form of laforin rescues laforin-deficient mouse models supports this scaffolding role for laforin (Gayarre et al. 2014; Nitschke et al. 2017), but does not exclude the possibility that glycogen dephosphorylation by laforin is still relevant, as during cytosolic glycogenolysis (Irimia et al. 2015). Also, phosphorylation of laforin by AMPK has been shown to enhance the interaction of laforin with malin (Roma-Mateo et al. 2011b); this regulation event could be one way to stimulate glycophagy in times of energy stress, when AMPK becomes activated. Finally, it is well established that laforin preferentially binds to polysaccharides with long chains and localizes to LBs in malin-deficient mice (Chan et al. 2004; Criado et al. 2012; Raththagala et al. 2015). Thus, laforin would selectively recruit the glycophagy machinery to precipitation-prone glycogen molecules with precariously long chains, as has been proposed (Sullivan et al. 2017).

It is worth noting that delivery of glycogen to lysosomes may involve multiple pathways, possibly with cell-specific relevance. Perhaps Stbd1 and laforin, each containing a CBM20, function as distinct receptors for glycophagy. It is interesting that although LBs occupy most tissues in LD, they are enriched in certain cell types within these tissues. For example, in LD mice, 97% of LBs in skeletal muscle are found in fast-twitch type IIb muscle fibers and 2% in slow-twitch type I fibers (Turnbull et al. 2011b). In human skin biopsy, LBs are only found in duct cells, but not secretory cells (Andrade et al. 2003). In human liver, LBs are enriched in the periportal regions, but not the perivenous regions (Carpenter et al. 1974). All of these cell types appear to make glycogen (Montagna et al. 1951; Vøllestad et al. 1984; Quistorff 1985), but the differences in whether or not they make LBs suggest the glycogen is handled differently. Cell-specific glycogen metabolism and polyglucosan body formation are of particular interest in the brain. LBs were first reported in neurons (Lafora 1911), and historically were believed to occupy this cell type almost exclusively (Minassian 2001). However, this observation has been perplexing since the vast majority of brain glycogen is stored in astrocytes. Recent work has demonstrated that neurons possess an active glycogen metabolism (Vilchez et al. 2007; Saez et al. 2014), and LD mouse models accumulate both LBs in neurons and a distinct polyglucosan body known as corpora amylacea in astrocytes (Augé et al. 2018; Rubio-Villena et al. 2018). Early literature on human patients corroborate these findings. Corpora amylacea in addition to LBs have been reported in multiple studies (Van Heycop Ten Ham 1975). Small inclusions, which may also correspond to corpora amylacea, have been identified in astrocytes and oligodendrocytes (Schwarz and Yanoff 1965). Recently, it has been asserted that the striking appearance of neuronal LBs left glial LBs largely overlooked over the years; upon careful review of the literature, small LBs that were assumed to be in neuronal processes are likely to actually be astrocytic (Minassian, personal communication). This topic is further discussed in Chap. 10.

6.3 *The benefits of Glycophagy*

The lysosomal glycogen pool plays an important role in circumstances when a burst of glucose is needed, such as during the neonatal starvation period (Kotoulas et al. 2004, 2006; Schiaffino et al. 2008). Glycogen-rich autophagosomes have been observed in the neonatal liver, muscle and heart, but they are depleted within hours after birth. A unique role for glycophagy in the heart is also emerging (Reichelt et al. 2013; Delbridge et al. 2015). While liver and muscle normally deplete their glycogen stores with fasting, cardiac tissues accumulate glycogen and upregulate glycogen autophagy under nutrient stress (Reichelt et al. 2013). An advantage of lysosomal glycogen is that it can be mobilized independently or in parallel with cytosolic glycogen by a distinct set of cellular stimuli. For example, glycophagy could be specifically regulated by lysosomal uptake of Ca^{2+}, which enhances GAA activity (Kotoulas et al. 2006; Delbridge et al. 2015). Very little is known about the role of glycophagy in brain tissue, besides the report of glycogenosomes in the aging brain (Cavanagh and Jones 2000). It has been suggested that glycogen of large molecular weight (i.e. α particles) is specifically degraded in lysosomes and proceeds by a random, rather than an ordered degradation mechanism (Geddes and Stratton 1977a; Devos and Hers 1980). Based on glycogen degradation patterns it has been speculated that brain glycogen is degraded primarily by phosphorolysis and not by lysosomal hydrolysis (Chee et al. 1983).

7 Glycogen Ultrastructure and Subcellular Distribution

In over a century of glycogen-related research many researchers have debated the existence of multiple forms of glycogen that differ in size, structure, associated proteins, and metabolic activity. The distinction between glycogen α and β particles came into focus with studies using the electron microscope and ultracentrifugation, and some progress has been made on defining the nature of these particles. Acid-soluble and acid-insoluble fractions of glycogen have also been observed, but the physiological relevance of these distinct pools has been highly debated. These topics, as well as a possible role for glycogen metabolic enzymes in the nucleus, are discussed in the next sections.

7.1 *Glycogen α, β, and γ Particles*

Drochmans and colleagues were the first to describe the various levels of glycogen structure in native glycogen isolated from liver and muscle, and their observations were corroborated by others (Drochmans 1962; Revel 1964; Wanson and Drochmans 1968; Ryu et al. 2009). The largest glycogen particle, named the α particle, is a rosette-shaped association of several smaller units, known as β particles, that can reach up to 300 nm in diameter and over 10^8 Daltons in molecular weight (Gilbert and Sullivan 2014; Prats et al. 2018). Glycogen β particles typically range in size from 10 nm to 44 nm and have a molecular weight of 10^6–10^7 Daltons (Shearer and Graham 2002; Sullivan et al. 2012). It is generally accepted that a β particle corresponds to a single molecular unit of glycogen that can reach up to 12 tiers as modeled by Whelan and others (See Sect. 3.1) (Fig. 6c). Drochmans designated the 3 nm filaments making up the β particles as γ particles (Drochmans 1962). However, 'γ particle' has also been used to refer to protein-rich subunits on the glycogen β particles (see Sect. 7) (Prats et al. 2018). Generally speaking, glycogen is primarily stored as α particles (or a mixture of α and β particles) in the liver and as β particles in muscle, brain, and other tissues (reviewed below). Recent work has demonstrated a mixture of α and β particles in cardiac tissue (Besford et al. 2012). It is apparent that these norms become disrupted in pathological conditions.

It is estimated that liver α particles contain, on average, 20 to 40 β particle subunits (Devos et al. 1983). Liver glycogen is typically cytoplasmic and often reported to closely associate with smooth endoplasmic reticulum (Bruni and Porter 1965; Cardell and Cardell 1990). In muscle, β particles are associated with sarcoplasmic reticulum, concentrated either in subsarcolemmal regions, between myofibrils, or within the myofibrils; these pools respond differently to metabolic cues (Graham et al. 2010). Aggregates up to 60 nm in diameter, apparently composed of 4–6 β particles, were also observed via EM by Wanson and Drochmans in muscle, and others have reported α particles in rat muscle and insect flight muscle (Wanson and Drochmans 1968; Calder 1991). In skin and adipose tissue, β particles appear as cytoplasmic granules interspersed between ribonucleoprotein particles (Napolitano and Fawcett 1958; Hashimoto and DiBella 1967). In adipose tissue, they are only prominent after refeeding following a fast (Eichner 1984). In leukocytes and thymus, β particles are scattered throughout the cytoplasm and are fairly homogenous in size (Scott and Still 1968; Salmoral et al. 1990). In the retina, cytoplasmic β particles and occasional small α particles (60 nm) have been observed in the cone, but not rod, photoreceptor cells, and in Müller cells (Okubo et al. 1998).

In the brain, glycogen exists primary as cytoplasmic β particles in the cell bodies and processes of astrocytes (Maxwell and Kruger 1965; Sotelo and Palay 1968; Phelps 1975; Cataldo and Broadwell 1986). Normally, they are sparse to nonexistent in neurons, microglia, or oligodendrocytes, but β particles have appeared in neuronal axons and dendrites following fasting or trauma (Phelps 1975; Cataldo and Broadwell 1986). Astroglial β particles become apparent in early embryos and increase in both abundance and diameter with development (Gadisseux and Evrard 1985). Small α particles (80–100 nm in diameter, composed of 4–8 β particle subunits by EM) have also been reported in rat astrocytes (Sotelo and Palay 1968). Biochemical studies of glycogen isolated via a mild cold-water extraction method show significant heterogeneity in molecular weight: a fraction of particles had molecular weights suggestive of small α-particles composed of 3–6 β particle subunits (Chee et al. 1983). In several cases of human glycogenoses, small (60–120 nm in diameter) and large (150–350 nm in diameter) α particles have been reported in astrocytes or neurons (Résibois-Grégoire and Dourov 1966; Kornfeld and LeBaron 1984; Towfighi et al. 1989). Aggregates of α or β particles have also been reported in Alzheimer's disease and the aging brain (Gertz et al. 1985; Mann et al. 1987), which are distinct from polyglucosan bodies (Cavanagh 1999). Reactive astrocytes are known to accumulate glycogen β particles in response to various pathological stimuli, including: brain infarction (Kajihara et al. 2001), methionine sulfoxide (Phelps 1975; Delorme and Hevor 1985), X-ray irradiation (Maxwell and Kruger 1965), and barbiturates (Phelps 1975). Occasionally, intramitochondrial glycogen has also been observed in rat retinal cells (Ishikawa and Pei 1965), in the human and canine myocardium (Buja et al. 1972; Maron and Ferrans 1975), and in astrocytes (Sotelo and Palay 1968; Alaraj et al. 2004).

β particle size exists around a normal distribution, with most granules being 20–30 nm (Calder 1991; Marchand et al. 2002) (Fig. 9a). Since each additional tier is estimated to contribute 3.8 nm to the diameter (Melendez-Hevia et al. 1993;

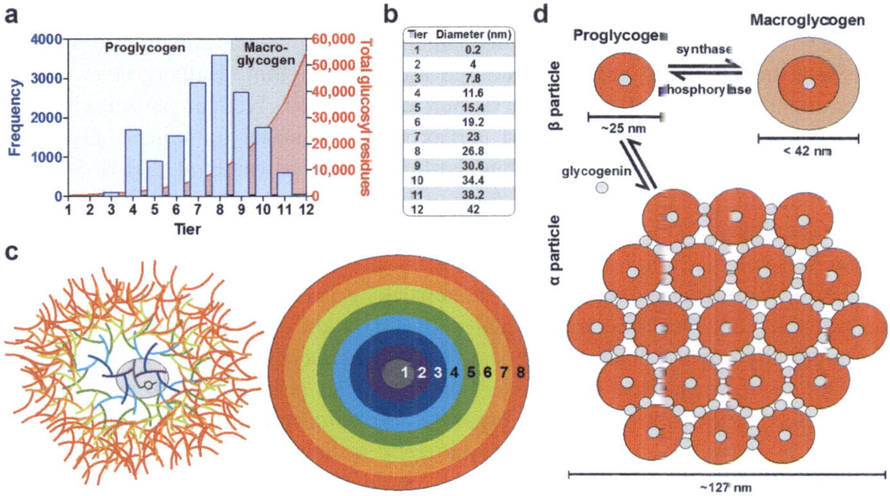

Fig. 9 Size and structure of glycogen α and β particles, proglycogen and macroglycogen. (**a**) Distribution of β glycogen particle size in muscle (Marchand et al. 2002) compared to the estimated number of glucosyl residues per glycogen molecule (Shearer and Graham 2002). The theoretical threshold between proglycogen and macroglycogen is after tier 8. (**b**) Estimated diameter of the tiers of glycogen particles. (**c**) Schematic diagrams of an 8-tiered glycogen molecule based on the studies of Whelan and mathematical modeling of (Melendez-Hevia et al. 1993). (**d**) A possible relationship between proglycogen and macroglycogen and α and β particles and approximated sizes. Grey circles represent glycogenin molecules. Diagram of α particle was adapted from (Tan et al. 2018).

Marchand et al. 2002), the observed size distribution implies that most β particles have only about 7–9 tiers (Fig. 9a, b, c). It is still not clear why most β particles do not reach the maximum diameter, since such large granules can store a significantly greater amount of glucose (Graham et al. 2010). A growing body of evidence suggests that glycogenin regulates β particle size. In cell culture and in tissue, glycogenin does not appear to exist in a free form; it is always attached to glycogen (Alonso et al. 1995; Skurat et al. 1997; Testoni et al. 2017). Overexpression of glycogenin in cell culture did not change total glycogen levels, but it increased the total number of glycogen molecules, which were of a smaller size (Skurat et al. 1997). The Guinovart group showed that mice lacking glycogenin display reduced exercise endurance and an accumulation of glycogen particles in muscle that were larger than those in wild-type mice (Testoni et al. 2017). Also, crystal structures and data from the Sicheri lab suggest glycogen particle size is regulated by the length of the linker between the glycogenin catalytic domain and the region that binds to glycogen synthase; this region is alternatively spliced and varies in length across species (Zeqiraj et al. 2014; Zeqiraj and Sicheri 2015).

In healthy organisms, α particles seem to be a unique feature of the liver. It is well established that total liver glycogen depletes with fasting and exceeds basal levels upon refeeding, a phenomenon known as glycogen super-compensation that is also observed in muscle and brain (Nilsson and Hultman 1973; Terjung et al. 1974; Matsui et al. 2012). Additionally, hepatic glycogen levels and the activities of glycogen metabolizing enzymes undergo a diurnal rhythm (Halberg et al. 1960; Roesler and Khandelwal 1985). The Gilbert lab analyzed glycogen particles by size exclusion chromatography and EM and proposed a model for diurnal glycogen cycling in liver based on their data and earlier work (Sullivan et al. 2014). Starved livers maintain low glycogen in the form of small β particles. Upon refeeding, glycogen is quickly synthesized on these preexisting β particles, which provide optimal surface-to-volume ratio for rapid resynthesis. Once glycogen levels peak, the β particles are assembled into α particles. The α particles persist as glycogen levels begin to gradually fall, but they are gradually disassembled into β particles, which are preferentially degraded. Thus it appears that the primarily role of α particles is to facilitate a slower release of glucose than β particles (Chandramouli et al. 2015). This indeed seems optimal for the organ whose primary role is to maintain blood glucose homeostasis. Additionally, α particles in diabetic livers are more fragile than those of healthy mice, suggesting they could more easily be broken into β particles and contribute to hyperglycemia (Deng et al. 2015; Hu et al. 2018).

The nature of α particle assembly has remained elusive for decades. Drochmans observed that purified α particles could be dissociated into β particles under acidic conditions (Drochmans 1962). Lazarow, Orrell and Bueding previously showed using ultracentrifugation that acid, alkali and heat drastically changed the sedimentation of liver glycogen, suggesting these treatments disrupt α particle structure (Lazarow 1942; Orrell and Bueding 1964). The Gilbert lab has devoted a significant body of work to elucidating the nature of the α particle bond due to its relevance to diabetes (Sullivan et al. 2010, 2012; Deng et al. 2015, 2016; Tan et al. 2018). The lack of significant dissociation of normal α particles in dimethyl sulfoxide, 2-mercaptoethanol, or sodium dodecyl sulfate suggested the association is not through hydrogen or disulfide bonds or weak protein-protein interactions (Sullivan et al. 2010, 2012). It was reported that unexpectedly high levels of glycogenin were found in a proteomic study of liver glycogen, suggesting that glycogenin was present on the granule surface in addition to the granule core (Stapleton et al. 2010). The Gilbert lab then used proteomics to identify glycogenin as the molecular "glue" holding α particles together (Tan et al. 2018) (Fig. 9d).

7.2 *Proglycogen and Macroglycogen*

There is a lot of controversy surrounding proglycogen and macroglycogen, which have gone by different terms over the years and have been discussed in great detail by others (Lomako et al. 1993a; Alonso et al. 1995; Rybicka 1996; Shearer and Graham 2002). Although the precise nature of these pools is debated, there does appear to be sufficient evidence for their distinct physiological roles in glycogen turnover (Shearer and Graham 2002; Graham et al. 2010). The relationship between proglycogen/macroglycogen and α/β particles is not completely clear as these distictions are discussed seperately. However, since they are relevant to tissues generally lacking α particles, proglycogen and macroglycogen likely represent two states of β particle synthesis (Alonso et al. 1995).

When the TCA extraction method was first introduced for the purification of glycogen, it was observed that a portion of the glycogen was resistant to extraction by cold TCA (Willstätter and Rohdewald 1934). It could not be released unless the tissue was treated with heat, alkali, or protease, so it was concluded that this resistant fraction was attached to protein, and the two fractions were called lyo- ("free") and desmo- ("fixed") glycogen (Willstätter and Rohdewald 1934; Van Heijningen and Kemp 1955). Such observations also gave rise to the terms acid-soluble, i.e. extractable, glycogen, and acid-insoluble, i.e. residual, glycogen (Russell and Bloom 1955; Stetten Jr and Stetten 1960). In a 1960 literature review, Stetten and Stetten noted that while the quantity of free glycogen was widely variable with nutritional state, the quantity of fixed glycogen remained constant. However, experiments suggested the latter was more metabolically active, since injected [^{14}C]glucose was more readily incorporated into fixed glycogen (Stetten Jr and Stetten 1960). For decades, researchers continued to debate whether such observations were artefactual, merely reflecting the association of known metabolic enzymes with glycogen (Whelan 1976). Whelan revisited the subject in the late twentieth century while investigating the origin of glycogen synthesis. Using cultured primary astrocytes, his group showed that fixed glycogen, which was re-named proglycogen, contained 10% protein by weight and was rapidly resynthesized upon refeeding after glucose starvation, reaching a size of 400,000 Da. The proglycogen was later converted to macroglycogen, as demonstrated by pulse-chase labeling (Lomako et al. 1993a). The group proposed that proglycogen was an intermediate between glycogenin and the mature glycogen granule (Lomako et al. 1993a; Alonso et al. 1995). Resynthesis of proglycogen prior to macroglycogen following exhaustive exercise was also confirmed *in vivo* in humans (Adamo 1998).

It has been suggested that the upper limit of proglycogen size corresponds to a diameter of ~30 nm, a molecule of approximately 8 tiers; anything larger would be considered macroglycogen (Melendez et al. 1997; Marchand et al. 2002; Shearer and Graham 2002) (Fig. 9a, d). However, Whelan's group estimated that the upper limit of proglycogen was 400 kDa (Lomako et al. 1993a; Alonso et al. 1995), which corresponds to a glycogen molecule with ~2500 glucose residues, i.e. approximately 5 tiers. Whelan's group also suggested that proglycogen and macroglycogen are synthesized by different enzymes, however it may be that the different activities associated with the pro- and macroglycogen fractions both correspond to glycogen synthase. Two independent groups have demonstrated that glycogen synthase in the proglycogen fraction has a higher affinity for UDP-glucose than synthase in the macroglycogen fraction (Curtino and Lacoste 2000; Tavridou and Agius 2003). Curtino and Lacoste proposed that the insolubility of proglycogen in TCA was due to its heavy association with glycogen synthase, not with glycogenin, as Whelan had proposed (Curtino and Lacoste 2000). Tavridu and Agnius showed that the hormone insulin increased the association of glycogen synthase with the proglycogen fraction, and glucagon had the opposite effect (Tavridou and Agius 2003). Insulin is known to stimulate dephosphorylation (and activation) of glycogen synthase, and the effects of both hormones are well known to regulate glycogen synthesis through signaling cascades, so the difference in synthase activity may be due to its phosphorylation state. Indeed, phosphorylation-dependent stimulation of glycogen synthase and phosphorylase has been shown to induce translocation of these proteins to actin-rich structures proximal to sarcoplasmic reticulum in skeletal muscle (Prats et al. 2005, 2009). In liver, newly synthesized glycogen has also been reported to be associated with a protein backbone (Geddes and Stratton 1977b) and smooth endoplasmic reticulum (Cardell and Cardell 1990). Additionally, glycogenin has been shown to bind to actin (Baqué et al. 1997). *In vitro* studies suggest that glycogen synthase is more active when it is bound to glycogenin (Pitcher et al. 1987, 1988). When glycogen content decreases, glycogen synthase has been shown to translocate from a glycogen-enriched membrane fraction to the cytoskeleton, concurrent with a decrease in activity (Nielsen et al. 2001). In isolated hepatocytes, glucose stimulated the cytoskeleton-dependent translocation of synthase from the cytosol to the plasma membrane, and newly synthesized glycogen molecules always appeared first at the cell membrane, but moved inward as they were replaced by newer molecules (García-Rocha et al. 2001; Fernández-Novell et al. 2002; Ferrer et al. 2003).

Proglycogen may represent a membrane-associated and/or actin-bound form of glycogen that is primed and ready for active anabolism/catabolism. When it grows to a certain size, it becomes macroglycogen, synthesis levels off, and it is released into the cytosol. A similar model has been proposed recently (Prats et al. 2018) and prompts further discussion of glycogen and its associated proteins as a dynamic subcellular entity rather than a static molecule (See Sect. 7) (Rybicka 1996). It is still not clear whether synthesis and degradation can occur simultaneously on the same granule, but proximity of these activities is suggested (Prats et al. 2005). The spatiotemporal regulation of glycogen metabolism is an interesting area of research requiring further elucidation.

7.3 Nuclear Glycogen

Dozens of microscopy studies have reported intranuclear glycogen deposits, most often in pathological conditions of the liver (Chipps and Duff 1942; Sheldon et al. 1962; Tanikawa 1965). In a 1975 case report, Ferrans et al. beautifully reviewed 24 prior studies on intranuclear glycogen either in the context of diabetes, glycogen storage diseases, or as an incidental feature in healthy humans and animals (Ferrans et al. 1975). In these reports, nuclear glycogen was usually in the form of β particles, sometimes dispersed, other times clustered; occasionally a few α particles were reported. The findings of Ferrans et al. were unusual in that nuclear glycogen was observed in the diseased myocardium. Intranuclear glycogen was only found in 6 out of 90 patients with cardiac diseases, and only in a few nuclei per patient. Most of the glycogen was present as β particles, often aggregated and filling the entire nucleoplasm. The presence of nuclear glycogen was not associated with any cellular damage. The authors proposed three mechanisms for the formation of nuclear glycogen, but shrewdly concluded that only one was possible: (1) there was no ultrastructural evidence supporting phagocytosis or endocytosis of glycogen particles by the nuclear membrane; (2) glycogen particles were too large to pass through the nuclear pore complex; (3) the glycogen must be synthesized *in situ,* which had been previously demonstrated in normal Müller cells of the rat retina (Amemiya 1970) and in Novikoff hepatoma cells (Karasaki 1971). Ferrans et al. proposed that conditions favoring increased permeability of the nuclear membrane facilitated the entry of the metabolic enzymes into the nucleus, but that they became trapped when permeability returned to normal. This implies that atypical trapping of glycogenic enzymes into the nucleus may result in excessive levels of glycogen. Nuclear α and β particles have also been reported in the context of various cancers: human gastric adenocarcinoma (Ohyumi and Takano 1977), sublines of the Ehrlich-Lettré mouse ascites tumor (Kopun et al. 1989), chicken sarcoma (Binggeli 1959) and human and mouse hepatoma (Leduc and Wilson 1959; Ghandially and Parry 1966). Nuclear glycogen synthase activity was also detected in the Ehrlich-Lettré subline (Granzow et al. 1981; Kopun et al. 1982). In the brain, intranuclear glycogen deposits have been observed in the abnormal astrocytes of Alzheimer's disease (Horita et al. 1981; Miyakawa et al. 1982) and in the pituicytes of aged rats (Lafarga et al. 1991). Nuclear polyglucosan was also reported in a neuronal cell culture model of GBE deficiency (Kakhlon et al. 2013).

In more recent years, Guinovart, Gentry and others have shown that nearly all the central glycogen metabolic enzymes can be found in the nucleus: glycogenin (Miozzo et al. 1996; Baqué et al. 1997), glycogen synthase (Ferrer et al. 1997; Cid et al. 2005; Vilchez et al. 2007) glycogen phosphorylase (Sun et al. 2019), and debranching enzyme (Cheng et al. 2007). Guinovart's group published a series of studies demonstrating the translocation of the muscle isoform of glycogen synthase from the cytosol to the nucleus in conditions of glucose deprivation, which required glycogen binding (Ferrer et al. 1997; Cid et al. 2005; Díaz et al. 2011). They also showed that neurons display primarily nuclear localization of glycogen synthase, characteristic of cells lacking glycogen, in contrast to glycogen-rich astrocytes with cytosolic synthase (Vilchez et al. 2007). Additionally, GDE and glucokinase, a key glucose sensing enzyme, also shuttle to the nucleus upon glucose deprivation (Fernández-Novell et al. 1999; Cheng et al. 2007). In contrast, glucose starvation has an opposite effect on the gluconeogenic enzyme fructose bisphosphatase, inducing its translocation to the cytosol (Yáñez et al. 2004). It has been suggested that the nuclear sequestration of these enzymes is yet another way to regulate their activity, preventing them from acting inappropriately in times of energy stress (Jurczak et al. 2008). Laforin and AMPK, which both contain CBMs and bind to glycogen *in vitro* and in cell culture, also translocate to the nucleus with glucose deprivation and glycogen depletion (Cheng et al. 2007; Singh et al. 2012) (Fig. 10a, b).

Nuclear-cytoplasmic shuttling has also been observed for yeast glycogen synthase. In a review discussing these data, it was postulated that "the freeing of glycogen synthase from its cytoplasmic tether to the glycogen particle as these stores reduced would therefore be a signal that carbon and energy reserves were low. The uptake of glycogen synthase into the nucleus might therefore represent a form of molecular 'fuel gauge.' It is possible that glycogen synthase could regulate transcription in response to energy availability by some as yet undetermined means" (Wilson et al. 2010). 'Moonlighting' of metabolic enzymes in the nucleus is not a new theme. Moonlighting proteins perform multiple unrelated functions, expanding the functional repertoire of the cell without expanding the number of genes, and the first to be described was the lens structural protein ε-crystallin, also known as lactate dehydrogenase (Hendriks et al. 1988). All essential glycolytic enzymes have been observed in the nucleus, as well as some mitochondrial proteins and enzymes involved in methylation and acetylation (Kim and Dang 2005; Boukouris et al. 2016). These energy-sensing proteins have been shown to regulate transcription by diverse means in a "metabolism-epigenetic axis" that is critical for cellular homeostasis (Boukouris et al. 2016). Although the epigenetic role of glycolytic proteins is well described, it is still not clear whether glycolysis occurs in the nucleus, although its products ATP and NADH could certainly be useful. Glycogen-associated enzymes may also have both an epigenetic and energy producing role, and perhaps excessive glycogen deposits result from misregulated nuclear sequestration. AMP-activated protein kinase (AMPK), a master regulator of cellular energy known to bind to and sense glycogen levels (McBride et al. 2009) (see Sect. 7), also translocates to the nucleus and regulates transcription and the circadian clock through phosphorylation (Leff 2003; Lamia et al. 2009).

8 The Glycogen Granule and Its Associated Proteins

In 1968, Scott and Still recognized the dynamic nature of glycogen and introduced the term "glycosome" to refer to the molecule and its associated proteins, stating that the nature of glycogen as a cellular organelle had not been appreciated (Scott and Still 1968). Rybicka reintroduced the term in a detailed historical review of the evidence for such a classification (Rybicka 1996). Unfortunately, the term "glycosome" has been used far more frequently to refer to the glycolytic enzyme-containing peroxisomes of trypanosomatids, a family of unicellular parasites (Michels et al. 2006), which creates confusion. Also, in the strict definition, organelles are membrane-bound entities (Klausner et al. 1992; Mellman and Warren 2000), and glycogen is primarily cytosolic. Perhaps it is safest to simply recognize that a host of proteins are associated with the glycogen granule, its structure is complex and highly regulated, and it is a dynamic participant in cellular metabolism.

The γ particle named by Drochmans was a description of the ramifying fine fibers (3 nm in diameter) comprising β particle structure, likely corresponding to glucan chains (Drochmans 1962). Later authors came to identify protein-rich subdomains on the glycogen molecule as γ particles, which may be a misinterpretation (Prats et al. 2018). Protein is more electron dense than carbohydrate, so with lead and uranyl acetate staining, protein-rich regions appear as darkly stained clusters

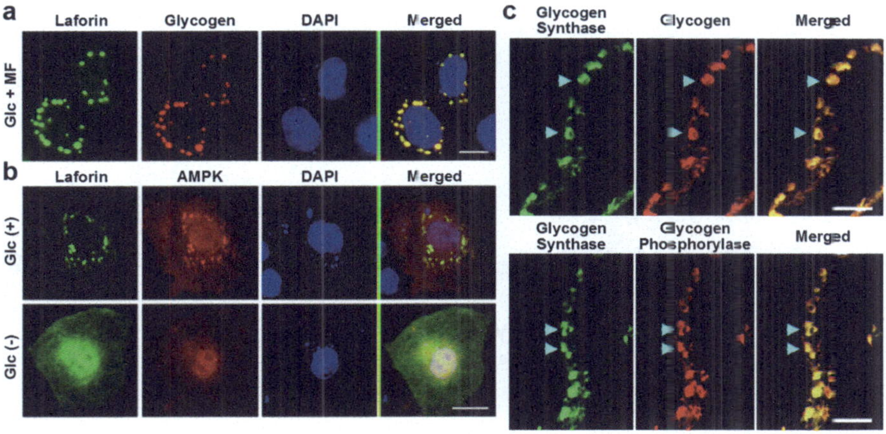

Fig. 10 Nuclear-cytoplasmic shuttling and colocalization of glycogen associated enzymes. (**a**) In media containing glucose and metformin, laforin colocalizes with glycogen in Neuro2A cells (Singh et al. 2012). (**b**) When expressed in COS-7 cells, laforin and AMPK colocalize (with glycogen) in the cytosol in the presence of glucose, and translocate to the nucleus upon glucose starvation (Singh et al. 2012). For (**a**, **b**), scale bars = 10 μm. Copyright © 2012 American Society for Microbiology. Used with permission. (**c**) Immunolabeling of glycogen synthase, phosphorylase, and glycogen itself in tilapia gill sections (Chang et al. 2007). Scale bars = 20 μm. Copyright © Company of Biologists. Used with permission

about 2–3 nm in diameter; alternatively, staining with the histochemical method of Thiéry using silver proteinate results in electron-dense clusters of the same size corresponding to vicinal glycols (Ferrans et al. 1973; Rybicka 1995; Prats et al. 2018). Rybicka refers to the electrodense vicinal glycols as γ particles, which is likely consistent with the meaning originally intended by Drochmans. Due to their very small size, both γ particles and protein-rich subdomains are only visible by EM. It is not yet clear why they exist as discrete, regular entities and which proteins correspond to the protein subdomains. However, it is well established that all of the primary glycogen metabolic enzymes (glycogenin, synthase, phosphorylase, GDE and GBE) bind to and colocalize with glycogen granules in cell culture and *in vivo* (Chang et al. 2007; Cheng et al. 2007; Vilchez et al. 2007; Stapleton et al. 2013; Zhai et al. 2016) (Figs. 10c and 11). In fact, due to its very tight association, glycogen synthase is frequently used as a marker for glycogen and polyglucosan bodies (Criado et al. 2012; Augé et al. 2018). There are a few antibodies that directly bind to the polysaccharide chains of glycogen, but they are not yet commercially available and are still being characterized (Nakamura-Tsuruta et al. 2012; Oe et al. 2016). Other proteins known to closely associate with the glycogen molecule all contain a glycogen-binding CBM: laforin and Stbd1 (Sect. 5), the β-subunit of AMPK, and the glycogen-targeting subunits of protein phosphatase 1 (discussed below) (Fig. 10b, c). Phosphorylase kinase also binds to glycogen (Chebotareva et al. 2009; Nadeau et al. 2018).

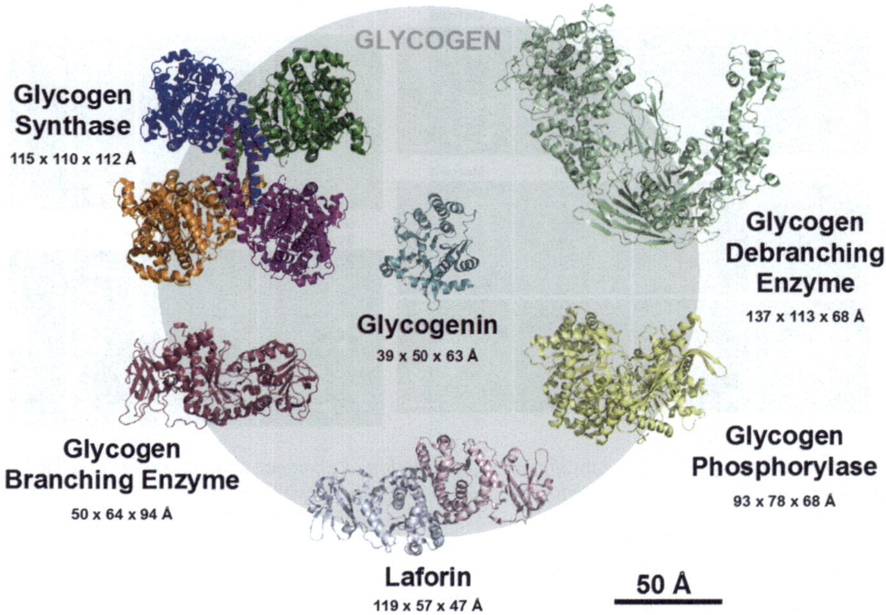

Fig. 11 Crystal structures and relative sizes of glycogen-associated proteins. Crystal structures of yeast glycogen synthase (PDB: 3NAZ), yeast glycogen debranching enzyme (PDB: 5D06), human branching enzyme (PDB: 4BZY), human brain glycogen phosphorylase (PDB: 5IKO), rabbit muscle glycogenin (PDB: 1LL2), and human laforin (PDB: 4RKK). Structures were superimposed manually in PyMol and are shown at equivalent zoom. The size of an average glycogen β particle (diameter = 250 Å) is shown for scale to illustrate the relative sizes of the proteins and glycogen particle. Crystal structure dimensions were calculated based on an inertia axis aligned bounding box in PyMol. In multimeric structures, individual subunits are colored separately

Protein phosphatase 1 (PP1) catalyzes a vast array of dephosphorylation events in cells, regulation of which is achieved by diverse targeting subunits that bring the catalytic subunit (PP1c) into the vicinity of its target (Bollen et al. 2010; Heroes et al. 2013). Seven glycogen-targeting subunits of PP1 have been identified (*PPP1R3A-G*), but four have been most extensively characterized as primary regulators of glycogen synthesis (Newgard et al. 2000; Roach et al. 2012). G_M (also known as R_{GL}, encoded by *PPP1R3A*) expression is restricted to heart and skeletal muscle; G_L (encoded by *PPP1R3B*) is primarily expressed in liver; and Protein Targeting to Glycogen (PTG, also known as R5, encoded by *PPP1R3C*) and R6 (encoded by *PPP1R3D*) are more ubiquitously expressed, with enrichment in glycogen-rich tissues (Roach et al. 2012; www.proteinatlase.org). All these subunits possess both a PP1-binding domain and a putative CBM21 and recruit PP1 to glycogen granules (Newgard et al. 2000; Christiansen et al. 2009; www.cazy.com). Dephosphorylation by PP1 has opposing effects on synthase (activating) and phosphorylase (inhibitory), leading to net glycogen synthesis (Newgard et al. 2000; Greenberg et al. 2006; Vilchez et al. 2007). Overexpression of PTG, G_M, G_L, and R6 leads to glycogen accumulation in cell culture and mouse models (Printen et al. 1997; Gasa et al. 2000; Greenberg et al. 2006; Worby et al. 2008; Montori-Grau et al. 2011; Duran et al. 2014).

AMPK, a cellular energy sensor with multiple intracellular targets, is also considered an important glycogen sensor (Hardie 2014). AMPK is activated by elevated levels of AMP relative to ATP reflecting energy deficits. It is a heterotrimeric complex composed of a catalytic α subunit and regulatory β and γ subunits; the β subunit contains a CBM48 and targets AMPK to glycogen granules (Polekhina et al. 2003; McBride et al. 2009; Janecek et al. 2011). AMPK phosphorylates both isoforms of glycogen synthase, inhibiting its activity (Bultot et al. 2012; Jørgensen et al. 2004). It has been speculated that recruitment of activated AMPK to glycogen granules would lead to decreased glycogen synthesis. There is also significant evidence for regulation of AMPK activity by glycogen via the β subunit (Hardie and Sakamoto 2006). AMPK is particularly inhibited by the presence of a single α-1,6 linked glucose unit or by a glycogen molecule that has been partially degraded by phosphorylase (McBride et al. 2009; Koay et al. 2010). A single glucose unit at a branch point could be exposed during debranching, since GDE is likely to dissociate from glycogen between its transferase and glucosidase activities (Zhai et al. 2016). Thus, glycogen, especially when being actively degraded, is a potent inhibitor of AMPK, and there is likely to be reciprocal regulation between glycogen and AMPK.

Proteomic studies have been performed on glycogen isolated from mouse and rat liver and the mouse adipocyte cell line 3 T3-L1 (Stapleton et al. 2010, 2013). Synthase, phosphorylase, GBE, GDE, and glycogenin were all abundant, as well as PP1 regulatory subunits. Stbd1, laforin, and proteins associated with endoplasmic reticulum were only found in the hepatic proteome; lysosomal α amylase was identified only in the 3 T3-L1 proteome, indicative of cell-specific differences in glycogen metabolism. The lack of AMPK or PhK binding was consistent with earlier studies and attributed to the stringency of the isolation procedure and the nutritional state of the liver (Parker et al. 2007). A number of proteins associated with mitochondria, ribosomes and nuclei were also identified in lower abundance, suggestive of both a heterogeneous subcellular distribution and the link between glycogen and a variety of cellular processes, as has been discussed in previous sections.

It should be noted that many structures of glycogen-associated enzymes have been determined, and they are quite large relative to the size of the glycogen granule (Fig. 11). The dimeric structure of rabbit muscle glycogenin is ~80 kDa in solution and the crystal structure of the monomer is 63 Å at its widest point (Gibbons et al. 2002). This corresponds to one fourth of the diameter of the average glycogen β particle (25 nm, i.e. 250 Å). Yeast glycogen synthase is tetrameric, measuring 115 Å at its widest point, approximately half the diameter of a glycogen granule; each subunit is about 90 Å long (Baskaran et al. 2010). Monomeric human brain phosphorylase is 78 Å long, but it also is known to function as a dimer (Mathieu et al. 2016). Yeast GDE and human GBE monomers are 137 Å and 94 Å long, respectively (Froese et al. 2015; Zhai et al. 2016). The elongated human laforin dimer measures 119 Å in length both in crystals and in solution (Raththagala et al. 2015). PhK and AMPK are both very large multisubunit complexes (Hardie 2014;

Nadeau et al. 2018). Although a complete structure of AMPK is not yet available, structures of the PhK holoenzyme and PhK in complex with phosphorylase determined by cryo-EM reveal that these complexes are 270 Å by 225 Å and 310 Å by 250 Å, respectively, roughly the same dimensions as glycogen β particles (Vénien-Bryan et al. 2009). With this scale in mind, one can begin to envision that the surface of glycogen molecules is likely to be very crowded, and enzymes may compete for access to glucan chains. Molecular crowding has been shown to influence the interaction of PhK with phosphorylase and with the glycogen molecule (Chebotareva et al. 2009). Glycogen and its associated proteins may function as molecular scaffolds, binding to one another and recruiting a variety of enzymes that respond to glycogen levels and regulate energy homeostasis.

9 The future of Brain Glycogen Research

The groundbreaking discoveries of the twentieth century greatly expanded our understanding of glycogen metabolism and monumentally impacted the broader field of biochemistry. As a result, there is a widespread belief that glycogen metabolism is an antiquated area of research that requires no further investigation. However, there is a milieu of interesting questions that remain unresolved regarding glycogen, particularly on its complex yet elusive role in the mammalian brain. Defects in glycogen metabolism are a hallmark of many diseases, including neurodegenerative diseases. The advent of new technologies makes it possible to better study brain glycogen and will most certainly lead to a better understanding of its architecture and metabolism in a variety of contexts.

References

Adeva-Andany MM, González-Lucán M, Donapetry-García C, Fernández-Fernández C, Ameneiros-Rodríguez E (2016) Glycogen metabolism in humans. BBA Clin 5:85–100

K. B. Adamo, M. A. Tarnopolsky, T. E. Graham, (1998) Dietary carbohydrate and postexercise synthesis of proglycogen and macroglycogen in human skeletal muscle. American Journal of Physiology-Endocrinology and Metabolism 275 (2):E229-E234

Aguado C, Sarkar S, Korolchuk VI, Criado O, Vernia S, Boya P et al (2010) Laforin, the most common protein mutated in Lafora disease, regulates autophagy. Hum Mol Genet 19(14):2867–2876

Alaraj M, Gajkowska B, Cholewinski M, Lazarewicz JW (2004) Hyperglycaemia and intramitochondrial glycogen granules in the brain of mice. Ultrastructural study. Folia Neuropathol 42(2):113–118

Alberini CM, Cruz E, Descalzi G, Bessières B, Gao V (2018) Astrocyte glycogen and lactate: New insights into learning and memory mechanisms. Glia 66(6):1244–1262

Alonso MD, Lomako J, Lomako WM, Whelan WJ (1994) Tyrosine-194 of glycogenin undergoes autocatalytic glucosylation but is not essential for catalytic function and activity. FEBS Lett 342(1):38–42

Alonso MD, Lomako J, Lomako WM, Whelan WJ (1995) A new look at the biogenesis of glycogen. FASEB J 9(12):1126–1137

Amemiya T (1970) Intranuclear polysaccharide synthesis in müller's cells of the normal rat retina. Acta Histochem Cytochem 3(2):41–51

Andrade DM, Ackerley CA, Minett TSC, Teive HAG, Bohlega S, Scherer SW et al (2003) Skin biopsy in Lafora disease: genotype–phenotype correlations and diagnostic pitfalls. Neurology 61(11):1611–1614

Archibald AR, Fleming ID, Liddle AM, Manners DJ, Mercer GA, Wright A (1961) 232. α-1, 4-Glucosans. Part XI. The absorption spectra of glycogen–and amylopectin–iodine complexes. J Chem Soc:1183–1190

Augé E, Pelegrí C, Manich G, Cabezón I, Guinovart JJ, Duran J et al (2018) Astrocytes and neurons produce distinct types of polyglucosan bodies in Lafora Disease. Glia 66(10):2094–2107

Bailey JM, Whelan WJ (1961) Physical properties of starch I. Relationship between iodine stain and chain length. J Biol Chem 236(4):969–973

Bak LK, Walls AB, Schousboe A, Waagepetersen HS (2018) Astrocytic glycogen metabolism in the healthy and diseased brain. J Biol Chem 293(19):7108–7116

Bancroft JD, Gamble M (2008) Theory and practice of histological techniques. Churchill Livingstone, London

Banks W, Greenwood CT, Khan KM (1971) The interaction of linear, amylose oligomers with iodine. Carbohydr Res 17(1):25–33

Baqué S, Guinovart JJ, Ferrer JC (1997) Glycogenin, the primer of glycogen synthesis, binds to actin. FEBS Lett 417(3):355–359

Barros LF (2013) Metabolic signaling by lactate in the brain. Trends Neurosci 36(7):396–404

Baskaran S, Roach PJ, DePaoli-Roach AA, Hurley TD (2010) Structural basis for glucose-6-phosphate activation of glycogen synthase. Proc Natl Acad Sci 107(41):17563–17568

Bates FL, French D, Rundle RE (1943) Amylose and amylopectin content of starches determined by their iodine complex formation. J Am Chem Soc 65(2):142–148

Bawn CEH, Hirst EL, Young GT (1940) The nature of the bonds in starch. Trans Faraday Soc 36:880–885

Berg JM, Tymoczko JL, Stryer L (2002) Biochemistry. W.H. Freeman, New York

Bernard C (1850) Sur une nouvelle fonction du foie chez l'homme et les animaux. Institut de France, Paris

Bernard C (1857) Sur le mécanisme physiologique de la formation du sucre dans le foie. C R Acad Sci (Paris) 44:578–586

Bertoft E (2017) Understanding starch structure: recent progress. Agronomy 7(3):–56

Besford QA, Sullivan MA, Zheng L, Gilbert RG, Stapleton D, Gray-Weale A (2012) The structure of cardiac glycogen in healthy mice. Int J Biol Macromol 51(5):887–891

Binggeli MF (1959) Abnormal intranuclear and cytoplasmic formations associated with a chemically induced, transplantable chicken sarcoma. J Biophys Biochem Cytol 5(1):143

Blennow A, Engelsen SB (2010) Helix-breaking news: fighting crystalline starch energy deposits in the cell. Trends Plant Sci 15(4):236–240

Blennow A, Bay-Smidt AM, Olsen CE, Møller BL (1998) Analysis of starch-bound glucose 3-phosphate and glucose 6-phosphate using controlled acid treatment combined with high-performance anion-exchange chromatography. J Chromatogr A 829(1–2):385–391

Blennow A, Bay-Smidt AM, Olsen CE, Møller BL (2000) The distribution of covalently bound phosphate in the starch granule in relation to starch crystallinity. Int J Biol Macromol 27(3):211–218

Bollen M, Peti W, Ragusa MJ, Beullens M (2010) The extended PP1 toolkit: designed to create specificity. Trends Biochem Sci 35(8):450–458

Boukouris AE, Zervopoulos SD, Michelakis ED (2016) Metabolic enzymes moonlighting in the nucleus: metabolic regulation of gene transcription. Trends Biochem Sci 41(8):712–730

Brown AM (2004) Brain glycogen re-awakened. J Neurochem 89(3):537–552

Brown DH, Brown BI (1966) Action of a muscle branching enzyme on polysaccharides enlarged from UDP [14C] glucose. Biochimica et Biophysica Acta (BBA)-General Subjects 130(1):263–266

Brown DH, Illingworth B (1962) The properties of an oligo-1, 4→ 1, 4-glucantransferase from animal tissues. Proc Natl Acad Sci 48(10):1783–1787
Brown BI, Brown DH, Jeffrey PL (1970) Simultaneous absence of α-1, 4-glucosidase and α-1, 6-glucosidase activities (pH 4) in tissues of children with type II glycogen storage disease. Biochemistry 9(6):1423–1428
Bruni C, Porter KR (1965) The fine structure of the parenchymal cell of the normal rat liver: I. General observations. Am J Pathol 46(5):691
Buja LM, Ferrans VJ, Levitsky S (1972) Occurrence of intramitochondrial glycogen in canine myocardium after prolonged anoxic cardiac arrest. J Mol Cell Cardiol 4(3):237–242
Buleon A, Colonna P, Planchot V, Ball S (1998) Starch granules: structure and biosynthesis. Int J Biol Macromol 23(2):85–112
Bultot L, Guigas B, Wilamowitz-Moellendorff AV, Maisin L, Vertommen D, Hussain N, Beullens M, Guinovart JJ, Foretz M, Viollet B, Sakamoto K, Hue L, Rider MH (2012) AMP-activated protein kinase phosphorylates and inactivates liver glycogen synthase. Biochemical Journal 443(1):193–203
Burnett G, Kennedy EP (1954) The enzymatic phosphorylation of proteins. J Biol Chem 211(2):969–980
Butler NA, Lee EYC, Whelan WJ (1977) A protein-bound glycogen component of rat liver. Carbohydr Res 55(1):73–82
Calder PC (1991) Glycogen structure and biogenesis. Int J Biochem 23(12):1335–1352
Campbell JA, Davies GJ, Bulone V, Henrissat B (1997) A classification of nucleotide-diphospho-sugar glycosyltransferases based on amino acid sequence similarities. Biochem J 326(Pt 3):929
Cao Y, Mahrenholz AM, DePaoli-Roach AA, Roach PJ (1993a) Characterization of rabbit skeletal muscle glycogenin. Tyrosine 194 is essential for function. J Biol Chem 268(20):14687–14693
Cao Y, Skurat AV, DePaoli-Roach AA, Roach PJ (1993b) Initiation of glycogen synthesis. Control of glycogenin by glycogen phosphorylase. J Biol Chem 268(29):21717–21721
Caputto R, Leloir LF, Cardini CE, Paladini AC (1950) Isolation of the coenzyme of the galactose phosphate-glucose phosphate transformation. J Biol Chem 184(1):333–350
Cardell RR, Cardell EL (1990) Heterogeneity of glycogen distribution in hepatocytes. J Electron Microsc Tech 14(2):126–139
Carpenter WB, Smith FG, Meneses HP (1876) Principles of Human Physiology. Henry C. Lea, Philadelphia, PA
Carpenter S, Karpati G, Andermann F, Jacob JC, Andermann E (1974) Lafora's disease: peroxisomal storage in skeletal muscle. Neurology 24(6):531–538
Carr C (1947) Practical Physiological Chemistry. By Philip B. Hawk, Bernard L. Oser, and William H. Summerson. J Phys Colloid Chem 51(5):1214–1215
Carter SH, Stone WE (1961) Effect of convulsants on brain glycogen in the mouse. J Neurochem 7(1):16–19
Cataldo AM, Broadwell RD (1986) Cytochemical identification of cerebral glycogen and glucose-6-phosphatase activity under normal and experimental conditions: I. Neurons and glia. J Electron Microsc Tech 3(4):413–437
Cavanagh JB (1999) Corpora-amylacea and the family of polyglucosan diseases. Brain Res Brain Res Rev 29(2–3):265–295
Cavanagh JB, Jones HB (2000) Glycogenosomes in the aging rat brain: their occurrence in the visual pathways. Acta Neuropathol 99(5):496–502
Chaikuad A, Froese DS, Berridge G, von Delft F, Oppermann U, Yue WW (2011) Conformational plasticity of glycogenin and its maltosaccharide substrate during glycogen biogenesis. Proc Natl Acad Sci U S A 108(52):21028–21033
Chan EM, Young EJ, Ianzano L, Munteanu I, Zhao X, Christopoulos CC et al (2003) Mutations in NHLRC1 cause progressive myoclonus epilepsy. Nat Genet 35(2):125–127
Chan EM, Ackerley CA, Lohi H, Ianzano L, Cortez MA, Shannon P et al (2004) Laforin preferentially binds the neurotoxic starch-like polyglucosans, which form in its absence in progressive myoclonus epilepsy. Hum Mol Genet 13(11):1117–1129

Chance MRA, Yaxley DC (1950) Central nervous function and changes in brain metabolite concentration: I. glycogen and lactate in convulsing mice. J Exp Biol 27(3):311–323
Chandramouli C, Varma U, Stevens EM, Xiao RP, Stapleton DI, Mellor KM et al (2015) Myocardial glycogen dynamics: new perspectives on disease mechanisms. Clin Exp Pharmacol Physiol 42(4):415–425
Chang JC-H, Wu S-M, Tseng Y-C, Lee Y-C, Baba O, Hwang P-P (2007) Regulation of glycogen metabolism in gills and liver of the euryhaline tilapia (Oreochromis mossambicus) during acclimation to seawater. J Exp Biol 210(19):3494–3504
Chebotareva NA, Meremyanin AV, Makeeva VF, Eronina TB, Kurganov BI (2009) Glycogen phosphorylase b and phosphorylase kinase binding to glycogen under molecular crowding conditions. Inhibitory effect of FAD. Biochemistry (Mosc) 74(5):562–568
Chee NP, Geddes R, Wills PR (1983) Metabolic heterogeneity in rabbit brain glycogen. Biochim Biophys Acta 756(1):9–12
Cheng A, Zhang M, Gentry MS, Worby CA, Dixon JE, Saltiel AR (2007) A role for AGL ubiquitination in the glycogen storage disorders of Lafora and Cori's disease. Genes Dev 21(19):2399–2409
Chipps HD, Duff GL (1942) Glycogen infiltration of the liver cell nuclei. Am J Pathol 18(4):645
Choi IY, Tkáčc I, Ugurbil K, Gruetter R (1999) Noninvasive measurements of [1-13C] glycogen concentrations and metabolism in rat brain in vivo. J Neurochem 73(3):1300–1308
Christiansen C, Abou Hachem M, Janecek S, Vikso-Nielsen A, Blennow A, Svensson B (2009) The carbohydrate-binding module family 20--diversity, structure, and function. FEBS J 276(18):5006–5029
Cid E, Cifuentes D, Baqué S, Ferrer JC, Guinovart JJ (2005) Determinants of the nucleocytoplasmic shuttling of muscle glycogen synthase. FEBS J 272(12):3197–3213
Cohen P (2002) The origins of protein phosphorylation. Nat Cell Biol 4(5):E127
Cohen P, Frame S (2001) The renaissance of GSK3. Nat Rev Mol Cell Biol 2(10):769
Contreras CJ, Segvich DM, Mahalingan K, Chikwana VM, Kirley TL, Hurley TD et al (2016) Incorporation of phosphate into glycogen by glycogen synthase. Arch Biochem Biophys 597:21–29
Cori GT (1952) Glycogen structure and enzyme deficiencies in glycogen storage disease. Harvey Lect 48:145
Cori CF, Cori GT (1929) Glycogen formation in the liver from d-and l-lactic acid. J Biol Chem 81(2):389–403
Cori CF, Cori GT (1936) Mechanism of formation of hexosemonophosphate in muscle and isolation of a new phosphate ester. Proc Soc Exp Biol Med 34(5):702–705
Cori GT, Cori CF (1943) Crystalline muscle phosphorylase IV. Formation of glycogen. J Biol Chem 151(1):57–63
Cori CF, Cori GT (1946) Carbohydrate metabolism. Annu Rev Biochem 15(1):193–218
Cori GT, Green AA (1943) Crystalline muscle phosphorylase II. Prosthetic group. J Biol Chem 151(1):31–38
Cori GT, Larner J (1951) Action of amylo-1, 6-glucosidase and phosphorylase on glycogen and amylopectin. J Biol Chem 188(1):17–29
Cori CF, Colowick SP, Cori GT (1937) The isolation and synthesis of glucose-1-phosphoric acid. J Biol Chem 121:465–477
Cori GT, Colowick SP, Cori CF (1938) The enzymatic conversion of glucose-1-phosphoric ester to 6-ester in tissue extracts. J Biol Chem 124(2):543–555
Cori CF, Schmidt G, Cori GT (1939) The synthesis of a polysaccharide from glucose-1-phosphate in muscle extract. Science 89(2316):464–465
Cori CF, Cori GT, Green AA (1943) Crystalline muscle phosphorylase III. Kinetics. J Biol Chem 151(1):39–55
Criado O, Aguado C, Gayarre J, Duran-Trio L, Garcia-Cabrero AM, Vernia S et al (2012) Lafora bodies and neurological defects in malin-deficient mice correlate with impaired autophagy. Hum Mol Genet 21(7):1521–1533

Cuervo AM, Bergamini E, Brunk UT, Dröge W, Ffrench M, Terman A (2005) Autophagy and aging: the importance of maintaining "clean" cells. Autophagy 1(3):131–140
Curtino JA, Lacoste ER (2000) Two glycogen synthase activities associated with proteoglycogen in retina. Neurochem Res 25(1):129–132
D'Hulst C, Mérida Á (2010) The priming of storage glucan synthesis from bacteria to plants: current knowledge and new developments. New Phytol 188(1):13–21
De Duve C, Pressman B, Gianetto R, Wattiaux R, Appelmans F (1955) Tissue fractionation studies. 6. Intracellular distribution patterns of enzymes in rat-liver tissue. Biochem J 60(4):604
Delbridge LMD, Mellor KM, Taylor DJR, Gottlieb RA (2015) Myocardial autophagic energy stress responses—macroautophagy, mitophagy, and glycophagy. Am J Physiol Heart Circ Physiol 308(10):H1194–HH204
Delorme P, Hevor TK (1985) Glycogen particles in methionine sulfoximine epileptogenic rodent brain and liver after the administration of methionine and actinomycin D. Neuropathol Appl Neurobiol 11(2):117–128
Demetriadou A, Morales-Sanfrutos J, Nearchou M, Baba O, Kyriacou K, Tate EW et al (2017) Mouse Stbd1 is N-myristoylated and affects ER-mitochondria association and mitochondrial morphology. J Cell Sci 130(5):903–915
Deng B, Sullivan MA, Li J, Tan X, Zhu C, Schulz BL et al (2015) Molecular structure of glycogen in diabetic liver. Glycoconj J 32(3–4):113–118
Deng B, Sullivan MA, Chen C, Li J, Powell PO, Hu Z et al (2016) Molecular structure of human-liver glycogen. PLoS One 11(3):e0150540
DePaoli-Roach AA, Tagliabracci VS, Segvich DM, Meyer CM, Irimia JM, Roach PJ (2010) Genetic depletion of the malin E3 ubiquitin ligase in mice leads to lafora bodies and the accumulation of insoluble laforin. J Biol Chem 285(33):25372–25381
DePaoli-Roach AA, Contreras CJ, Segvich DM, Heiss C, Ishihara M, Azadi P et al (2015) Glycogen phosphomonoester distribution in mouse models of the progressive myoclonic epilepsy, Lafora disease. J Biol Chem 290(2):841–850
Devos P, Hers H-G (1980) Glycogen in rat adipose tissue: sequential synthesis and random degradation. Biochem Biophys Res Commun 95(3):1031–1036
Devos P, Baudhuin P, Van Hoof F, Hers H-G (1983) The alpha particulate liver glycogen. A morphometric approach to the kinetics of its synthesis and degradation. Biochem J 209(1):159–165
Díaz A, Martínez-Pons C, Fita I, Ferrer JC, Guinovart JJ (2011) Processivity and subcellular localization of glycogen synthase depend on a non-catalytic high-affinity glycogen-binding site. J Biol Chem 286(21):18505–18514
Dienel GA, Cruz NF (2016) Aerobic glycolysis during brain activation: adrenergic regulation and influence of norepinephrine on astrocytic metabolism. J Neurochem 138(1):14–52
Drochmans P (1962) Morphologie du glycogène: Etude au microscope électronique de colorations négatives du glycogène particulaire. J Ultrastruct Res 6(2):141–163
Duran J, Guinovart JJ (2015) Brain glycogen in health and disease. Mol Asp Med 46:70–77
Duran J, Gruart A, Garcia-Rocha M, Delgado-Garcia JM, Guinovart JJ (2014) Glycogen accumulation underlies neurodegeneration and autophagy impairment in Lafora disease. Hum Mol Genet 23(12):3147–3156
Edner C, Li J, Albrecht T, Mahlow S, Hejazi M, Hussain H et al (2007) Glucan, water dikinase activity stimulates breakdown of starch granules by plastidial beta-amylases. Plant Physiol 145(1):17–28
Eichner RD (1984) Adipose tissue glycogen synthesis. Int J Biochem 16(3):257–261
Emanuelle S, Brewer MK, Meekins DA, Gentry MS (2016) Unique carbohydrate binding platforms employed by the glucan phosphatases. Cell Mol Life Sci 73(14):2765–2778
Fernández-Novell JM, Castel S, Bellido D, Ferrer JC, Vilaró S, Guinovart JJ (1999) Intracellular distribution of hepatic glucokinase and glucokinase regulatory protein during the fasted to refed transition in rats. FEBS Lett 459(2):211–214
Fernández-Novell JM, López-Iglesias C, Ferrer JC, Guinovart JJ (2002) Zonal distribution of glycogen synthesis in isolated rat hepatocytes. FEBS Lett 531(2):222–228

Ferrans VJ, Buja LM, Jones M (1973) Ultrastructure and cytochemistry of glycogen in cardiac diseases. Recent Adv Stud Cardiac Struct Metab 3:97
Ferrans VJ, Maron BJ, Buja LM, Ali N, Roberts WC (1975) Intranuclear glycogen deposits in human cardiac muscle cells: ultrastructure and cytochemistry. J Mol Cell Cardiol 7(6):381–386
Ferrer JC, Baqué S, Guinovart JJ (1997) Muscle glycogen synthase translocates from the cell nucleus to the cytosol in response to glucose. FEBS Lett 415(3):249–252
Ferrer JC, Favre C, Gomis RR, Fernández-Novell JM, García-Rocha M, de la Iglesia N et al (2003) Control of glycogen deposition. FEBS Lett 546(1):127–132
Fischer EH, Krebs EG (1955) Conversion of phosphorylase b to phosphorylase a in muscle extracts. J Biol Chem 216(1):121–132
Fletcher WM, Hopkins FG (1907) Lactic acid in amphibian muscle 1. J Physiol 35(4):247–309
Fontana JD (1980) The presence of phosphate in glycogen. FEBS Lett 109(1):85–92
Foster M (1899) Claude Bernard. Unwin, London
Friedman DL, Larner J (1963) Studies on UDPG-α-glucan transglucosylase. III. Interconversion of two forms of muscle UDPG-α-glucan transglucosylase by a phosphorylation-dephosphorylation reaction sequence. Biochemistry 2(4):669–675
Froese DS, Michaeli A, McCorvie TJ, Krojer T, Sasi M, Melaev E et al (2015) Structural basis of glycogen branching enzyme deficiency and pharmacologic rescue by rational peptide design. Hum Mol Genet 24(20):5667–5676
Fuller M, Meikle PJ, Hopwood JJ (2006) Epidemiology of lysosomal storage diseases: an overview
Gadisseux J-F, Evrard P (1985) Glial-neuronal relationship in the developing central nervous system. Dev Neurosci 7(1):12–32
Gage SH (1917) Glycogen in the nervous system of vertebrates. J Comp Neurol 27(3):451–465
Ganesh S, Delgado-Escueta AV, Sakamoto T, Avila MR, Machado-Salas J, Hoshii Y et al (2002) Targeted disruption of the Epm2a gene causes formation of Lafora inclusion bodies, neurodegeneration, ataxia, myoclonus epilepsy and impaired behavioral response in mice. Hum Mol Genet 11(11):1251–1262
Garcia-Gimeno MA, Knecht E, Sanz P (2018) Lafora disease: a ubiquitination-related pathology. Cell 7(8):E87
García-Rocha M, Angela R, de la Iglesia N, Otto B, Fernández-Novell JM, Ferrer JC et al (2001) Intracellular distribution of glycogen synthase and glycogen in primary cultured rat hepatocytes. Biochem J 357(1):17–24
Garyali P, Segvich DM, DePaoli-Roach AA, Roach PJ (2014) Protein degradation and quality control in cells from laforin and malin knockout mice. J Biol Chem 289(30):20606–20614
Gasa R, Jensen PB, Berman HK, Brady MJ, DePaoli-Roach AA, Newgard CB (2000) Distinctive regulatory and metabolic properties of glycogen targeting subunits of protein phosphatase-1 (PTG, GL, GM/RGl) expressed in hepatocytes. J Biol Chem 275(34):26396–26403
Gayarre J, Duran-Trio L, Criado Garcia O, Aguado C, Juana-Lopez L, Crespo I et al (2014) The phosphatase activity of laforin is dispensable to rescue Epm2a−/− mice from Lafora disease. Brain 137(Pt 3):806–818
Geddes R (1986) Glycogen: a metabolic viewpoint. Biosci Rep 6(5):415–428
Geddes R, Stratton GC (1977a) The influence of lysosomes on glycogen metabolism. Biochem J 163(2):193–200
Geddes R, Stratton GC (1977b) Molecular and metabolic heterogeneity of liver glycogen. Carbohydr Res 57:291–299
Gentry MS, Pace RM (2009) Conservation of the glucan phosphatase laforin is linked to rates of molecular evolution and the glycogen metabolism of the organism. BMC Evol Biol 9(1):138
Gentry MS, Worby CA, Dixon JE (2005) Insights into Lafora disease: malin is an E3 ubiquitin ligase that ubiquitinates and promotes the degradation of laforin. Proc Natl Acad Sci U S A 102(24):8501–8506
Gentry MS, Dowen RH 3rd, Worby CA, Mattoo S, Ecker JR, Dixon JE (2007) The phosphatase laforin crosses evolutionary boundaries and links carbohydrate metabolism to neuronal disease. J Cell Biol 178(3):477–488

Gentry MS, Dixon JE, Worby CA (2009) Lafora disease: insights into neurodegeneration from plant metabolism. Trends Biochem Sci 34(12):628–639
Gentry MS, Roma-Mateo C, Sanz P (2013) Laforin, a protein with many faces: glucan phosphatase, adapter protein, et alii. FEBS J 280(2):525–537
Gentry MS, Brewer MK, Vander Kooi CW (2016) Structural biology of glucan phosphatases from humans to plants. Curr Opin Struct Biol 40:62–69
Gentry MS, Guinovart JJ, Minassian BA, Roach PJ, Serratosa JM (2018) Lafora disease offers a unique window into neuronal glycogen metabolism. J Biol Chem 293(19):7117–7125
Gertz HJ, Cervos-Navarro J, Frydl V, Schultz F (1985) Glycogen accumulation of the aging human brain. Mech Ageing Dev 31(1):25–35
Gessler K, Uson I, Takaha T, Krauss N, Smith SM, Okada S et al (1999) V-Amylose at atomic resolution: X-ray structure of a cycloamylose with 26 glucose residues (cyclomaltohexaicosaose). Proc Natl Acad Sci U S A 96(8):4246–4251
Ghandially FN, Parry EW (1966) Ultrastructure of a human hepatocellular carcinoma and surrounding non-neoplastic liver. Cancer 19(12):1989–2004
Gibbons BJ, Roach PJ, Hurley TD (2002) Crystal structure of the autocatalytic initiator of glycogen biosynthesis, glycogenin. J Mol Biol 319(2):463–477
Gibson WB, Brown BI, Brown DH (1971) Glycogen branching enzyme. Preparation and properties of α-1, 4-glucan-α-1, 4-glucan 6-glycosyltransferase and its action on the characteristic polysaccharide of the liver of children with type IV glycogen storage disease. Biochemistry 10(23):4253–4262
Gilbert RG, Sullivan MA (2014) The molecular size distribution of glycogen and its relevance to diabetes. Austral J Chem 67(4):538–543
Graham TE, Yuan Z, Hill AK, Wilson RJ (2010) The regulation of muscle glycogen: the granule and its proteins. Acta Physiol (Oxf) 199(4):489–498
Granzow C, Kopun M, Zimmermann H-P (1981) Role of nuclear glycogen synthase and cytoplasmic UDP glucose pyrophosphorylase in the biosynthesis of nuclear glycogen in HD33 Ehrlich-Lettré ascites tumor cells. J Cell Biol 89(3):475–484
Green AA, Cori GT (1943) Crystalline muscle phosphorylase I. Preparation, properties, and molecular weight. J Biol Chem 151(1):21–29
Greenberg CC, Danos AM, Brady MJ (2006) Central role for protein targeting to glycogen in the maintenance of cellular glycogen stores in 3T3-L1 adipocytes. Mol Cell Biol 26(1):334–342
Grmek MD (1968) First steps in Claude Bernard's discovery of the glycogenic function of the liver. J Hist Biol 1(1):141–154
Gunja-Smith Z, Marshall JJ, Mercier C, Smith EE, Whelan WJ (1970) A revision of the Meyer-Bernfeld model of glycogen and amylopectin. FEBS Lett 12(2):101–104
Gunja-Smith Z, Marshall JJ, Smith EE (1971) Enzymatic determination of the unit chain length of glycogen and related polysaccharides. FEBS Lett 13(5):309–311
Halberg F, Albrecht PG, Barnum CP Jr (1960) Phase shifting of liver-glycogen rhythm in intact mice. Am J Phys 199(3):400–402
Hanashiro I, Abe J-I, Hizukuri S (1996) A periodic distribution of the chain length of amylopectin as revealed by high-performance anion-exchange chromatography. Carbohydr Res 283:151–159
Hanes CS (1940) The reversible formation of starch from glucose-1-phosphate catalysed by potato phosphorylase. Proc R Soc Lond B 129(855):174–208
Hansen PI, Spraul M, Dvortsak P, Larsen FH, Blennow A, Motawia MS et al (2009) Starch phosphorylation—Maltosidic restrains upon 3′-and 6′-phosphorylation investigated by chemical synthesis, molecular dynamics and NMR spectroscopy. Biopolymers 91(3):179–193
Hardie DG, Sakamoto K (2006) AMPK: A Key Sensor of Fuel and Energy Status in Skeletal Muscle. Physiology 21(1):48–60
Hardie DG (2014) AMPK—sensing energy while talking to other signaling pathways. Cell Metab 20(6):939–952

Hashimoto K, DiBella RJ (1967) Electron microscopic studies of normal and abnormal elastic fibers of the skin. J Invest Dermatol 48(5):405–423
Hassid WZ, Neufeld EF, Feingold DS (1959) Sugar nucleotides in the interconversion of carbohydrates in higher plants. Proc Natl Acad Sci 45(7):905–915
Haworth WN, Percival EGV (1932) 324. Polysaccharides. Part XI. Molecular structure of glycogen. J Chem Soc:2277–2282
Hayee BH, Antonopoulos A, Murphy EJ, Rahman FZ, Sewell G, Smith BN et al (2011) G6PC3 mutations are associated with a major defect of glycosylation: a novel mechanism for neutrophil dysfunction. Glycobiology 21(7):914–924
Hejazi M, Fettke J, Haebel S, Edner C, Paris O, Frohberg C et al (2008) Glucan, water dikinase phosphorylates crystalline maltodextrins and thereby initiates solubilization. Plant J 55(2):323–334
Hendriks W, Mulders JW, Bibby MA, Slingsby C, Bloemendal H, De Jong WW (1988) Duck lens epsilon-crystallin and lactate dehydrogenase B4 are identical: a single-copy gene product with two distinct functions. Proc Natl Acad Sci 85(19):7114–7118
Hermans MM, Kroos MA, Van Beeumen J, Oostra BA, Reuser AJ (1991) Human lysosomal alpha-glucosidase. Characterization of the catalytic site. J Biol Chem 266(21):13507–13512
Heroes E, Lesage B, Görnemann J, Beullens M, Van Meervelt L, Bollen M (2013) The PP1 binding code: a molecular-lego strategy that governs specificity. FEBS J 280(2):584–595
Hers HG (1963) α-Glucosidase deficiency in generalized glycogen-storage disease (Pompe's disease). Biochem J 86(1):11
Hirai M, Hirai T, Ueki T (1994) Effect of branching of amylopectin on complexation with iodine as steric hindrance. Polymer 35(10):2222–2225
Hoefsloot LH, Hoogeveen-Westerveld M, Reuser AJJ, Oostra BA (1990) Characterization of the human lysosomal α-glucosidase gene. Biochem J 272(2):493–497
Holmes EG, Holmes BE (1926) Contributions to the study of brain metabolism: Carbohydrate metabolism relationship of glycogen and lactic acid. Biochem J 20(6):1196
Horita N, Matsushita M, Ishii T, Oyanagi S, Sakamoto K (1981) Ultrastructure of Alzheimer type II glia in hepatocerebral disease. Neuropathol Appl Neurobiol 7(2):97–102
Hu Z, Deng B, Tan X, Gan H, Li C, Nada SS et al (2018) Diurnal changes of glycogen molecular structure in healthy and diabetic mice. Carbohydr Polym 185:145–152
Huijing F (1975) Glycogen metabolism and glycogen-storage diseases. Physiol Rev 55(4):609–658
Husemann E, Ruska H (1940) Versuche zur Sichtbarmachung von Glykogenmolekülen. J für Praktische Chemie 156(1–3):1–10
Illingworth B, Larner J, Cori GT (1952) Structure of glycogens and amylopectins. I. Enzymatic determination of chain length. J Biol Chem 199(2):631–640
Irimia JM, Tagliabracci VS, Meyer CM, Segvich DM, DePaoli-Roach AA, Roach PJ (2015) Muscle glycogen remodeling and glycogen phosphate metabolism following exhaustive exercise of wild type and laforin knockout mice. J Biol Chem 290(37):22686–22698
Ishikawa T, Pei YF (1965) Intramitochondrial glycogen particles in rat retinal receptor cells. J Cell Biol 25(2):402
Iwamasa T, Tsuru T, Hamada T, Takeuchi T (1980) Physicochemical and ultrastructural studies on glycogenosomes in newborn rat hepatocytes. Pathol Res Pract 167(2–4):363–373
Iwamasa T, Fukuda S, Tokumitsu S, Ninomiya N, Matsuda I, Osame M (1983) Myopathy due to glycogen storage disease: pathological and biochemical studies in relation to glycogenosome formation. Exp Mol Pathol 38(3):405–420
Janeček Š (2002) A motif of a microbial starch-binding domain found in human genethonin. Bioinformatics 18(11):1534–1537
Janecek S, Svensson B, MacGregor EA (2011) Structural and evolutionary aspects of two families of non-catalytic domains present in starch and glycogen binding proteins from microbes, plants and animals. Enzym Microb Technol 49(5):429–440
Jeffrey PL, Brown DH, Brown BI (1970a) Lysosomal α-glucosidase. I. Purification and properties of the rat liver enzyme. Biochemistry 9(6):1403–1415

Jeffrey PL, Brown DH, Brown BI (1970b) Lysosomal α-glucosidase. II. Kinetics of action of the rat liver enzyme. Biochemistry 9(6):1416–1422
Jiang S, Heller B, Tagliabracci VS, Zhai L, Irimia JM, DePaoli-Roach AA et al (2010) Starch binding domain containing protein 1/genethonin 1 is a novel participant in glycogen metabolism. J Biol Chem 285(45):34960–34971
Jiang S, Wells CD, Roach PJ (2011) Starch-binding domain-containing protein 1 (Stbd1) and glycogen metabolism: Identification of the Atg8 family interacting motif (AIM) in Stbd1 required for interaction with GABARAPL1. Biochem Biophys Res Commun 413(3):420–425
Jope RS, Johnson GVW (2004) The glamour and gloom of glycogen synthase kinase-3. Trends Biochem Sci 29(2):95–102
Jorgensen SB, Nielsen JN, Birk JB, Olsen GS, Viollet B, Andreelli F, Schjerling P, VaulontS, Hardie DG, Hansen BF, Richter EA, Wojtaszewski J, FP (2004) The 2-5'AMP-Activated Protein Kinase Is a Site 2 Glycogen Synthase Kinase in Skeletal Muscle and Is Responsive to Glucose Loading. Diabetes 53(12):3074–3081
Jurczak MJ, Danos AM, Rehrmann VR, Brady MJ (2008) The role of protein translocation in the regulation of glycogen metabolism. J Cell Biochem 104(2):435–443
Kajihara H, Tsutsumi E, Kinoshita A, Nakano J, Takagi K, Takeo S (2001) Activated astrocytes with glycogen accumulation in ischemic penumbra during the early stage of brain infarction: immunohistochemical and electron microscopic studies. Brain Res 909(1–2):92–101
Kajiura H, Kakutani R, Akiyama T, Takata H, Kuriki T (2008) A novel enzymatic process for glycogen production. Biocatal Biotransformation 26(1–2):133–140
Kajiura H, Takata H, Kuriki T, Kitamura S (2010) Structure and solution properties of enzymatically synthesized glycogen. Carbohydr Res 345(6):817–824
Kakhlon O, Glickstein H, Feinstein N, Liu Y, Baba O, Terashima T et al (2013) Polyglucosan neurotoxicity caused by glycogen branching enzyme deficiency can be reversed by inhibition of glycogen synthase. J Neurochem 127(1):101–113
Karasaki S (1971) Cytoplasmic and nuclear glycogen synthesis in Novikoff ascites hepatoma cells. J Ultrastruct Res 35(1–2):181–196
Kerr SE (1938) The carbohydrate metabolism of brain VI. Isolation of glycogen. J Biol Chem 123(2):443–449
Khac LD, Eboué-bonis D, Chambaut AM, Clauser H (1972) The use of glucosamine as a metabolic probe in the rat diaphragm: the effect of insulin on its metabolic fate. Eur J Biochem 31(1):86–94
Kiessling W (1939) Über ein neues Fermentprotein der Hefe und eine reversible enzymatische Synthese des Glykogens. Naturwissenschaften 27(8):129–130
Kim J-W, Dang CV (2005) Multifaceted roles of glycolytic enzymes. Trends Biochem Sci 30(3):142–150
Kirkman BR, Whelan WJ (1986) Glucosamine is a normal component of liver glycogen. FEBS Lett 194(1):6–11
Klausner RD, Donaldson JG, Lippincott-Schwartz J (1992) Brefeldin A: insights into the control of membrane traffic and organelle structure. J Cell Biol 116(5):1071–1080
Koay A, Woodcroft B, Petrie EJ, Yue H, Emanuelle S, Bieri M et al (2010) AMPK beta subunits display isoform specific affinities for carbohydrates. FEBS Lett 584(15):3499–3503
Koizumi J (1974) Glycogen in the central nervous system. Prog Histochem Cytochem 6(4):III–I35
Komatsu M, Waguri S, Chiba T, Murata S, Iwata J-I, Tanida I et al (2006) Loss of autophagy in the central nervous system causes neurodegeneration in mice. Nature 441(7095):880
Kopun M, Spring H, Granzow C (1982) Nuclear glycogen synthase—fact or artifact? FEBS Lett 147(2):207–210
Kopun M, Granzow C, Krisman CR (1989) Comparative study of nuclear and cytoplasmic glycogen isolated from mutant HD33 ascites cells. J Cell Biochem 39(2):185–195
Kornfeld M, LeBaron M (1984) Glycogenosis type VIII. J Neuropathol Exp Neurol 43(6):568–579
Kotoulas OB, Kalamidas SA, Kondomerkos DJ (2004) Glycogen autophagy. Microsc Res Tech 64(1):10–20

Kotoulas OB, Kalamidas SA, Kondomerkos DJ (2006) Glycogen autophagy in glucose homeostasis. Pathol Res Pract 202(9):631–638

Kotting O, Santelia D, Edner C, Eicke S, Marthaler T, Gentry MS et al (2009) STARCH-EXCESS4 Is a Laforin-like phosphoglucan phosphatase required for starch degradation in arabidopsis thaliana. Plant Cell 21(1):334–346

Kotting O, Kossmann J, Zeeman SC, Lloyd JR (2010) Regulation of starch metabolism: the age of enlightenment? Curr Opin Plant Biol 13(3):321–329

Krisman CR (1962) A method for the colorimetric estimation of glycogen with iodine. Anal Biochem 4:17–23

Krisman CR, Alfredo CJ (1991) Corn starch (α1, 4-α1, 6) glucopolysaccharides-correlation between amylose: amylopectin ratios and physical properties of the grains. Starch Stärke 43(8):291–294

Krisman CR, Barengo R (1975) A precursor of glycogen biosynthesis: α-1, 4-glucan-protein. Eur J Biochem 52(1):117–123

Kuriki T, Stewart DC, Preiss J (1997) Construction of chimeric enzymes out of maize endosperm branching enzymes I and II: activity and properties. J Biol Chem 272(46):28999–29004

M. Kathryn Brewer, Annette Uittenbogaard, Grant L. Austin, Dyann M. Segvich, Anna DePaoli-Roach, Peter J. Roach, John J. McCarthy, Zoe R. Simmons, Jason A. Brandon, Zhengqiu Zhou, Jill Zeller, Lyndsay E.A. Young, Ramon C. Sun, James R. Pauly, Nadine M. Aziz, Bradley L. Hodges, Tracy R. McKnight, Dustin D. Armstrong, Matthew S Gentry, (2019) Targeting Pathogenic Lafora Bodies in Lafora Disease Using an Antibody-Enzyme Fusion. Cell Metabolism S1550-4131(19)30375-4

Lafarga M, Berciano MT, Pérez-Figares JM, Andrés MA, Maquiera E (1991) Influence of age on nuclear bodies and nuclear volume in pituicytes of the rat neurohypophysis. Anat Rec 230(3):319–324

Lafora GR (1911) Uber des Vorkommen amyloider KJrperchen im innern der Ganglienzellen. Virchows Arch f Path Anat 205:295

Lamia KA, Sachdeva UM, DiTacchio L, Williams EC, Alvarez JG, Egan DF et al (2009) AMPK regulates the circadian clock by cryptochrome phosphorylation and degradation. Science 326(5951):437–440

Larner J (1953) The action of branching enzymes on outer chains of glycogen. J Biol Chem 202(2):491–503

Larner J (1967) The discovery of glycogen and glycogen today. Claude Bernard and Experimental Medicine, Schenkman Publ, Cambridge, MA, p 135

Larner J, Illingworth B, Cori GT, Cori CF (1952) Structure of glycogens and amylopectins II. Analysis by stepwise enzymatic degradation. J Biol Chem 199(2):641–651

Lawrence JC, Roach PJ (1997) New insights into the role and mechanism of glycogen synthase activation by insulin. Diabetes 46(4):541–547

Lazarow A (1942) Particulate glycogen: A submicroscopic component of the guinea pig liver cell; its significance in glycogen storage and the regulation of blood sugar. Anat Rec 84(1):31–50

LeBaron FN (1955) The resynthesis of glycogen by guinea pig cerebral-cortex slices. Biochem J 61(1):80

Leduc EH, Wilson JW (1959) An electron microscope study of intranuclear inclusions in mouse liver and hepatoma. J Cell Biol 6(3):427–430

Lee Y-K, Lee J-A (2016) Role of the mammalian ATG8/LC3 family in autophagy: differential and compensatory roles in the spatiotemporal regulation of autophagy. BMB Rep 49(8):424

Leff T (2003) AMP-activated protein kinase regulates gene expression by direct phosphorylation of nuclear proteins. Portland Press Limited: Biochemical Society Transactions. https://doi.org/10.1042/bst0310224

Leloir LF, Cardini CE (1957) Biosynthesis of glycogen from uridine diphosphate glucose1. J Am Chem Soc 79(23):6340–6341

Levin B, Burgess EA, Mortimer PE (1968) Glycogen storage disease type IV, amylopectinosis. Arch Dis Child 43(231):548

Lieberman AP, Puertollano R, Raben N, Slaugenhaupt S, Walkley SU, Ballabio A (2012) Autophagy in lysosomal storage disorders. Autophagy 8(5):719–730
Lillie RD (1950) Further exploration of the HIO4-Schiff reaction with remarks on its significance. Anat Rec 108(2):239–253
Lim ST, Kasemsuwan T, Jane J-L (1994) Characterization of phosphorus in starch by 31P-nuclear magnetic resonance spectroscopy. Cereal Chem 71:488–493
Lim J-A, Zare H, Puertollano R, Raben N (2017) Atg5flox-derived autophagy-deficient model of Pompe disease: does it tell the whole story? Mol Ther Methods Clin Dev 7:11–14
Lin A, Mu J, Yang J, Roach PJ (1999) Self-glucosylation of glycogenin, the initiator of glycogen biosynthesis, involves an inter-subunit reaction. Arch Biochem Biophys 363(1):163–170
Lohi H, Ianzano L, Zhao X-C, Chan EM, Turnbull J, Scherer SW et al (2005) Novel glycogen synthase kinase 3 and ubiquitination pathways in progressive myoclonus epilepsy. Hum Mol Genet 14(18):2727–2736
Lomako J, Lomako WM, Whelan WJ (1988) A self-glucosylating protein is the primer for rabbit muscle glycogen biosynthesis. FASEB J 2(15):3097–3103
Lomako J, Lomako WM, Whelan WJ (1990) The biogenesis of glycogen: nature of the carbohydrate in the protein primer. Biochem Int 21(2):251–260
Lomako J, Lomako WM, Whelan WJ, Dombro RS, Neary JT, Norenberg MD (1993a) Glycogen synthesis in the astrocyte: from glycogenin to proglycogen to glycogen. FASEB J 7(14):1386–1393
Lomako J, Lomako WM, Whelan WJ, Marchase RB (1993b) Glycogen contains phosphodiester groups that can be introduced by UDPglucose: glycogen glucose 1-phosphotransferase. FEBS Lett 329(3):263–267
Lomako J, Lomako WM, Kirkman BR, Whelan WJ (1994) The role of phosphate in muscle glycogen. Biofactors 4(3–4):167–171
Lomako J, Lomako WM, Whelan WJ (2004) Glycogenin: the primer for mammalian and yeast glycogen synthesis. Biochim Biophys Acta 1673(1–2):45–55
Lukacs CM, Oikonomakos NG, Crowther RL, Hong LN, Kammlott RU, Levin W et al (2006) The crystal structure of human muscle glycogen phosphorylase a with bound glucose and AMP: an intermediate conformation with T-state and R-state features. Proteins 63(4):1123–1126
Magistretti PJ, Allaman I (2018) Lactate in the brain: from metabolic end-product to signalling molecule. Nat Rev Neurosci 19(4):235–249
Mahler R (1969) Glycogen storage diseases. J Clin Pathol Suppl (Ass Clin Path) 2:32
Maley F, McGarrahan JF, DelGiacco R (1966) Galactosamine: a precursor of glycogen glucosamine. Biochem Biophys Res Commun 23(1):85–91
Malfatti E, Nilsson J, Hedberg-Oldfors C, Hernandez-Lain A, Michel F, Dominguez-Gonzalez C et al (2014) A new muscle glycogen storage disease associated with glycogenin-1 deficiency. Ann Neurol 76(6):891–898
Mancini GM, Beerens CE, Verheijen FW (1990) Glucose transport in lysosomal membrane vesicles. Kinetic demonstration of a carrier for neutral hexoses. J Biol Chem 265(21):12380–12387
Mann DMA, Sumpter PQ, Davies CA, Yates PO (1987) Glycogen accumulations in the cerebral cortex in Alzheimer's disease. Acta Neuropathol 73(2):181–184
Manners DJ (1963) Enzymic synthesis and degradation of starch and glycogen. Advances in carbohydrate chemistry: Elsevier. p. 371–430 https://doi.org/10.1016/S0096-5332(08)60139-3
Manners DJ (1989) Recent developments in our understanding of amylopectin structure. Carbohydr Polym 11(2):87–112
Manners DJ (1991) Recent developments in our understanding of glycogen structure. Carbohydr Polym 16(1):37–82
Manners DJ, Stark JR, Thambyrajah V (1983) Iodine staining properties of branched α-(14)-D-glucanst. J Jap Soc Starch Sci 30(1):13–18
Marani E (1998) Microwave applications in neuromorphology and neurochemistry: safety precautions and techniques. Methods 15(2):87–99

Marchand I, Chorneyko K, Tarnopolsky M, Hamilton S, Shearer J, Potvin J et al (2002) Quantification of subcellular glycogen in resting human muscle: granule size, number, and location. J Appl Physiol 93(5):1598–1607

Maron BJ, Ferrans VJ (1975) Intramitochondrial glycogen deposits in hypertrophied human myocardium. J Mol Cell Cardiol 7(9):697–698

Marton L (1934) Electron microscopy of biological objects. Phys Rev 46(6):527

Mathieu C, De La Sierra-Gallay IL, Duval R, Xu X, Cocaign A, Léger T et al (2016) Insights into brain glycogen metabolism: the structure of human brain glycogen phosphorylase. J Biol Chem 291(35):18072–18083

Matsui H, Sasaki M, Takemasa E, Kaneta T, Chiba S (1984) Kinetic studies on the substrate specificity and active site of rabbit muscle acid α-glucosidase. J Biochem 96(4):993–1004

Matsui M, Kakuta M, Misaki A (1993) Comparison of the unit-chain distributions of glycogens from different biological sources, revealed by anion exchange chromatography. Biosci Biotechnol Biochem 57(4):623–627

Matsui T, Ishikawa T, Ito H, Okamoto M, Inoue K, Lee MC et al (2012) Brain glycogen supercompensation following exhaustive exercise. J Physiol 590(3):607–616

Maxwell DS, Kruger L (1965) The fine structure of astrocytes in the cerebral cortex and their response to focal injury produced by heavy ionizing particles. J Cell Biol 25(2):141–157

McBride A, Ghilagaber S, Nikolaev A, Hardie DG (2009) The glycogen-binding domain on the AMPK β subunit allows the kinase to act as a glycogen sensor. Cell Metab 9(1):23–34

McIlwain H, Tresize MA (1956) The glucose, glycogen and aerobic glycolysis of isolated cerebral tissues. Biochem J 63(2):250

McMahon RJ, Frost SC (1996) Glycogen: a carbohydrate source for GLUT-1 glycosylation during glucose deprivation of 3T3-L1 adipocytes. Am J Physiol Endocrinol Metabol 270(4):E640–E645

McManus JFA (1946) Histological demonstration of mucin after periodic acid. Nature 158(4006):202

McManus JFA (1948) Histological and histochemical uses of periodic acid. Stain Technol 23(3):99–108

Meekins DA, Guo HF, Husodo S, Paasch BC, Bridges TM, Santelia D et al (2013) Structure of the Arabidopsis glucan phosphatase like sex four2 reveals a unique mechanism for starch dephosphorylation. Plant Cell 25(6):2302–2314

Meekins DA, Raththagala M, Husodo S, White CJ, Guo HF, Kotting O (2014) Phosphoglucan-bound structure of starch phosphatase Starch Excess4 reveals the mechanism for C6 specificity. Proc Natl Acad Sci U S A 111(20):7272–7277

Meekins DA, Raththagala M, Auger KD, Turner BD, Santelia D, Kotting O et al (2015) Mechanistic insights into glucan phosphatase activity against polyglucan substrates. J Biol Chem 290(38):23361–23370

Meekins DA, Vander Kooi CW, Gentry MS (2016) Structural mechanisms of plant glucan phosphatases in starch metabolism. FEBS J 283(13):2427–2447

Melendez R, Melendez-Hevia E, Cascante M (1997) How did glycogen structure evolve to satisfy the requirement for rapid mobilization of glucose? A problem of physical constraints in structure building. J Mol Evol 45(4):446–455

Melendez R, Melendez-Hevia E, Mas F, Mach J, Cascante M (1998) Physical constraints in the synthesis of glycogen that influence its structural homogeneity: a two-dimensional approach. Biophys J 75(1):106–114

Melendez-Hevia E, Waddell TG, Shelton ED (1993) Optimization of molecular design in the evolution of metabolism: the glycogen molecule. Biochem J 295(Pt 2):477–483

Mellman I, Warren G (2000) The road taken: past and future foundations of membrane traffic. Cell 100(1):99–112

Merrick AW (1961) Encephalic glycogen differences in young and adult rats. J Physiol 158(3):476–485

Meyer KH, Bernfeld P (1940) Recherches sur l'amidon V. L'amylopectine Helv chim Acta 23:875

Meyerhof O (1920a) Die Energieumwandlungen im Muskel. Pflugers Arch Gesamte Physiol Menschen Tiere 182(1):232–283
Meyerhof O (1920b) II. Das Schicksal der Milchsäure in der Erholungsperiode des Muskels. Pflugers Arch 182:284
Meyerhof O (1920c) Kohlenhydrat und Milchsäureumsatz im Froschmuskel. Pflüg Arch ges Physiol 185:11–20
Michels PAM, Bringaud F, Herman M, Hannaert V (2006) Metabolic functions of glycosomes in trypanosomatids. Biochim Biophys Acta 1763(12):1463–1477
Minassian BA (2001) Lafora's disease: towards a clinical, pathologic, and molecular synthesis. Pediatr Neurol 25(1):21–29
Minassian BA, Lee JR, Herbrick JA, Huizenga J, Soder S, Mungall AJ et al (1998) Mutations in a gene encoding a novel protein tyrosine phosphatase cause progressive myoclonus epilepsy. Nat Genet 20(2):171–174
Miozzo MC, Maldonado C, Curtino JA (1996) Cellular and subcellular localization of glycogenin in chicken retina. IUBMB Life 40(1):173–180
Miyakawa T, Kuramoto R, Shimoji A, Higuchi Y (1982) Fine structure of inclusion body in the nucleus of Alzheimer glia type II in the brain of hepatocerebral degeneration. Acta Neuropathol 56(4):315–319
Mizushima N, Komatsu M (2011) Autophagy: renovation of cells and tissues. Cell 147(4):728–741
Montagna W, Chase HB, Hamilton JB (1951) The distribution of glycogen and lipids in human skin. J Invest Dermatol 17(3):147–157
Montgomery EM, Weakley FB, Hilbert GE (1949) Isolation of 6-[α-D-glucopyranosyl]-D-glucose (isomaltose) from enzymic hydrolyzates of starch2. J Am Chem Soc 71(5):1682–1687
Montori-Grau M, Guitart M, García-Martínez C, Orozco A, Gómez-Foix AM (2011) Differential pattern of glycogen accumulation after protein phosphatase 1 glycogen-targeting subunit PPP1R6 overexpression, compared to PPP1R3C and PPP1R3A, in skeletal muscle cells. BMC Biochem 12(1):57
Mordoh J, Krisman CR, Leloir LF (1966) Further studies on high molecular weight liver glycogen. Arch Biochem Biophys 113(2):265–272
Moreno D, Towler MC, Hardie DG, Knecht E, Sanz P (2010) The laforin-malin complex, involved in Lafora disease, promotes the incorporation of K63-linked ubiquitin chains into AMP-activated protein kinase beta subunits. Mol Biol Cell 21(15):2578–2588
Morris DL (1946) Colorimetric determination of glycogen. Disadvantages of the iodine method. J Biol Chem 166:199–203
Mowry RW, Bangle R Jr (1951) Histochemically demonstrable glycogen in the human heart: with special reference to glycogen storage disease and diabetes mellitus. Am J Pathol 27(4):611
Mu J, Skurat AV, Roach PJ (1997) Glycogenin-2, a novel self-glucosylating protein involved in liver glycogen biosynthesis. J Biol Chem 272(44):27589–27597
Munoz-Ballester C, Berthier A, Viana R, Sanz P (2016) Homeostasis of the astrocytic glutamate transporter GLT-1 is altered in mouse models of Lafora disease. Biochim Biophys Acta 1862(6):1074–1083
Nachmanson D, Ochoa S, Lipmann FA (1960) Otto Meyerhof-1884-1951: a biographical memoir. National Academy of Sciences, Washington, DC
Nadeau OW, Fontes JD, Carlson GM (2018) The regulation of glycogenolysis in the brain. J Biol Chem 293(19):7099–7107
Nakamura Y (ed) (2015) Starch: metabolism and structure. Springer, Akita, Japan
Nakamura-Tsuruta S, Yasuda M, Nakamura T, Shinoda E, Furuyashiki T, Kakutani R et al (2012) Comparative analysis of carbohydrate-binding specificities of two anti-glycogen monoclonal antibodies using ELISA and surface plasmon resonance. Carbohydr Res 350:49–54
Nakatogawa H, Suzuki K, Kamada Y, Ohsumi Y (2009) Dynamics and diversity in autophagy mechanisms: lessons from yeast. Nat Rev Mol Cell Biol 10(7):458
Nakayama A, Yamamoto K, Tabata S (2001) Identification of the catalytic residues of bifunctional glycogen debranching enzyme. J Biol Chem 276(31):28824–28828

Napolitano L, Fawcett D (1958) The fine structure of brown adipose tissue in the newborn mouse and rat. J Cell Biol 4(6):685–692
Newgard CB, Brady MJ, O'Doherty RM, Saltiel AR (2000) Organizing glucose disposal: emerging roles of the glycogen targeting subunits of protein phosphatase-1. Diabetes 49(12):1967–1977
Nielsen JN, Derave W, Kristiansen S, Ralston E, Ploug T, Richter EA (2001) Glycogen synthase localization and activity in rat skeletal muscle is strongly dependent on glycogen content. J Physiol 531(3):757–769
Nilsson LH, Hultman E (1973) Liver glycogen in man–the effect of total starvation or a carbohydrate-poor diet followed by carbohydrate refeeding. Scand J Clin Lab Invest 32(4):325–330
Nitschke F, Wang P, Schmieder P, Girard JM, Awrey DE, Wang T et al (2013) Hyperphosphorylation of glucosyl c6 carbons and altered structure of glycogen in the neurodegenerative epilepsy lafora disease. Cell Metab 17(5):756–767
Nitschke F, Sullivan MA, Wang P, Zhao X, Chown EE, Perri AM et al (2017) Abnormal glycogen chain length pattern, not hyperphosphorylation, is critical in Lafora disease. EMBO Mol Med 9(7):906–917
Oe Y, Baba O, Ashida H, Nakamura KC, Hirase H (2016) Glycogen distribution in the microwave-fixed mouse brain reveals heterogeneous astrocytic patterns. Glia 64(9):1532–1545
Ohyumi M, Takano S (1977) Intranuclear synthesized and native glycogen particles in human gastric cancer: ultrastructure and histochemistry. Histochemistry 50(3):239–250
Okubo A, Sameshima M, Unoki K, Uehara F, Ohba N (1998) Ultracytochemical demonstration of glycogen in cone, but not in rod, photoreceptor cells in the rat retina. Ann Anat 180(4):307–314
Ondruskova N, Honzik T, Kolarova H, Pakanova Z, Mucha J, Zeman J et al (2018) Aberrant apolipoprotein C-III glycosylation in glycogen storage disease type III and IX. Metabolism 82:135–141
Orrell SA, Bueding E (1964) A comparison of products obtained by various procedures used for the extraction of glycogen. J Biol Chem 239(12):4021–4026
Öz G, Henry P-G, Seaquist ER, Gruetter R (2003) Direct, noninvasive measurement of brain glycogen metabolism in humans. Neurochem Int 43(4–5):323–329
Öz G, Seaquist ER, Kumar A, Criego AB, Benedict LE, Rao JP et al (2007) Human brain glycogen content and metabolism: implications on its role in brain energy metabolism. Am J Physiol Endocrinol Metab 292(3):E946–EE51
Öz G, Kumar A, Rao JP, Kodl CT, Chow L, Eberly LE et al (2009) Human brain glycogen metabolism during and following hypoglycemia. Diabetes 58(9):1978–1985
Palmer TN (1971) The substrate specificity of acid α-glucosidase from rabbit muscle. Biochem J 124(4):701–711
Parker GJ, Lund KC, Taylor RP, McClain DA (2003) Insulin resistance of glycogen synthase mediated by O-linked N-acetylglucosamine. J Biol Chem 278(12):10022–10027
Parker GJ, Koay A, Gilbert-Wilson R, Waddington LJ, Stapleton D (2007) AMP-activated protein kinase does not associate with glycogen α-particles from rat liver. Biochem Biophys Res Commun 362(4):811–815
Parodi AJ, Krisman CR, Leloir LF, Mordoh J (1967) Properties of synthetic and native liver glycogen. Arch Biochem Biophys 121(3):769–778
Parodi AJ, Mordoh J, Krisman CR, Leloir LF (1969) In vitro synthesis of particulate glycogen from uridine diphosphate glucose. Arch Biochem Biophys 132(1):111–117
Passonneau JV, Gatfield PD, Schulz DW, Lowry OH (1967) An enzymic method for measurement of glycogen. Anal Biochem 19(2):315–326
Pederson BA, Turnbull J, Epp JR, Weaver SA, Zhao X, Pencea N et al (2013) Inhibiting glycogen synthesis prevents Lafora disease in a mouse model. Ann Neurol 74(2):297–300
Pérez S, Bertoft E (2010) The molecular structures of starch components and their contribution to the architecture of starch granules: A comprehensive review. Starch Stärke 62(8):389–420
Pfeiffer-Guglielmi B, Bröer S, Bröer A, Hamprecht B (2000) Isozyme pattern of glycogen phosphorylase in the rat nervous system and rat astroglia-rich primary cultures: electrophoretic and polymerase chain reaction studies. Neurochem Res 25(11):1485–1491

Pflüger E (1909) Meine Methode der quantitativen Analyse des Glykogenes und die Arteigenthümlichkeit der Substanzen des Thierleibes. Archiv für die gesamte Physiologie des Menschen und der Tiere 129(6–7):362–378

Phelps CH (1975) An ultrastructural study of methionine sulphoximine-induced glycogen accumulation in astrocytes of the mouse cerebral cortex. J Neurocytol 4(4):479–490

Pitcher J, Smythe C, Campbell DG, Cohen P (1987) Identification of the 38-kDa subunit of rabbit skeletal muscle glycogen synthase as glycogenin. Eur J Biochem 169(3):497–502

Pitcher J, Smythe C, Cohen P (1988) Glycogenin is the priming glucosyltransferase required for the initiation of glycogen biogenesis in rabbit skeletal muscle. Eur J Biochem 176(2):391–395

Polekhina G, Gupta A, Michell BJ, van Denderen B, Murthy S, Feil SC et al (2003) AMPK β subunit targets metabolic stress sensing to glycogen. Curr Biol 13(10):867–871

Prats C, Cadefau JA, Cussó R, Qvortrup K, Nielsen JN, Wojtaszewki JFP et al (2005) Phosphorylation-dependent translocation of glycogen synthase to a novel structure during glycogen re-synthesis. J Biol Chem 280(24):23165–23172

Prats C, Helge JW, Nordby P, Qvortrup K, Ploug T, Dela F et al (2009) Dual regulation of muscle glycogen synthase during exercise by activation and compartmentalization. J Biol Chem 284(23):15692–15700

Prats C, Graham TE, Shearer J (2018) The dynamic life of the glycogen granule. J Biol Chem 293(19):7089–7098

Printen JA, Brady MJ, Saltiel AR (1997) PTG, a protein phosphatase 1-binding protein with a role in glycogen metabolism. Science 275(5305):1475–1478

Puri R, Suzuki T, Yamakawa K, Ganesh S (2012) Dysfunctions in endosomal-lysosomal and autophagy pathways underlie neuropathology in a mouse model for Lafora disease. Hum Mol Genet 21(1):175–184

Quistorff B (1985) Gluconeogenesis in periportal and perivenous hepatocytes of rat liver, isolated by a new high-yield digitonin/collagenase perfusion technique. Biochem J 229(1):221–226

Raben N, Danon M, Lu N, Lee E, Shliselfeld L, Skurat AV et al (2001) Surprises of genetic engineering: a possible model of polyglucosan body disease. Neurology 56(12):1739–1745

Raben N, Hill V, Shea L, Takikita S, Baum R, Mizushima N et al (2008) Suppression of autophagy in skeletal muscle uncovers the accumulation of ubiquitinated proteins and their potential role in muscle damage in Pompe disease. Hum Mol Genet 17(24):3897–3908

Raben N, Schreiner C, Baum R, Takikita S, Xu S, Xie T et al (2010) Suppression of autophagy permits successful enzyme replacement therapy in a lysosomal storage disorder—murine Pompe disease. Autophagy 6(8):1078–1089

Raben N, Wong A, Ralston E, Myerowitz R (2012) Autophagy and mitochondria in Pompe disease: nothing is so new as what has long been forgotten. Am J Med Gen C Semin Med Genet 160C(1):13–21

Ramon C. Sun, Vikas V. Dukhande, Zhengqiu Zhou, Lyndsay E.A. Young, Shane Emanuelle, Christine Fillmore Brainson, and Matthew S. Gentry, (2019) Nuclear Glycogenolysis Modulates Histone Acetylation in Human Non-Small Cell Lung Cancers. Cell Metabolism (In Press)

Rani MRS, Shibanuma K, Hizukuri S (1992) The fine structure of oyster glucogen. Carbohydr Res 227:183–194

Rath VL, Ammirati M, LeMotte PK, Fennell KF, Mansour MN, Danley DE et al (2000) Activation of human liver glycogen phosphorylase by alteration of the secondary structure and packing of the catalytic core. Mol Cell 6(1):139–148

Raththagala M, Brewer MK, Parker MW, Sherwood AR, Wong BK, Hsu S et al (2015) Structural mechanism of laforin function in glycogen dephosphorylation and lafora disease. Mol Cell 57(2):261–272

Reichelt ME, Mellor KM, Curl CL, Stapleton D, Delbridge LMD (2013) Myocardial glycophagy—a specific glycogen handling response to metabolic stress is accentuated in the female heart. J Mol Cell Cardiol 65:67–75

Résibois-Grégoire A, Dourov N (1966) Electron microscopic study of a case of cerebral glycogenosis. Acta Neuropathol 6(1):70–79

Revel JP (1964) Electron microscopy of glycogen. J Histochem Cytochem 12(2):104–114
Revel JP, Napolitano L, Fawcett DW (1960) Identification of glycogen in electron micrographs of thin tissue sections. J Cell Biol 8(3):575–589
Ritte G, Heydenreich M, Mahlow S, Haebel S, Kotting O, Steup M (2006) Phosphorylation of C6- and C3-positions of glucosyl residues in starch is catalysed by distinct dikinases. FEBS Lett 580(20):4872–4876
Roach PJ (1990) Control of glycogen synthase by hierarchal protein phosphorylation. FASEB J 4(12):2961–2968
Roach PJ (2002) Glycogen and its metabolism. Curr Mol Med 2:101–120
Roach PJ, Depaoli-Roach AA, Hurley TD, Tagliabracci VS (2012) Glycogen and its metabolism: some new developments and old themes. Biochem J 441(3):763–787
Rodriguez IR, Whelan WJ (1985) A novel glycosyl-amino acid linkage: rabbit-muscle glycogen is covalently linked to a protein via tyrosine. Biochem Biophys Res Commun 132(2):829–836
Roesler WJ, Khandelwal RL (1985) Diurnal variations in the activities of the glycogen metabolizing enzymes in mouse liver. Int J Biochem 17(1):81–85
Roma-Mateo C, Moreno D, Vernia S, Rubio T, Bridges TM, Gentry MS et al (2011a) Lafora disease E3-ubiquitin ligase malin is related to TRIM32 at both the phylogenetic and functional level. BMC Evol Biol 11:225
Roma-Mateo C, Solaz-Fuster Mdel C, Gimeno-Alcaniz JV, Dukhande VV, Donderis J, Worby CA et al (2011b) Laforin, a dual-specificity phosphatase involved in Lafora disease, is phosphorylated at Ser25 by AMP-activated protein kinase. Biochem J 439(2):265–275
Romero PA, Smith EE, Whelan WJ (1980) Glucosamine as a substitute for glucose in glycogen metabolism. Biochem Int 1(1):1–9
Ros S, García-Rocha M, Domínguez J, Ferrer JC, Guinovart JJ (2009) Control of liver glycogen synthase activity and intracellular distribution by phosphorylation. J Biol Chem 284(10):6370–6378
Rubio-Villena C, Garcia-Gimeno MA, Sanz P (2013) Glycogenic activity of R6, a protein phosphatase 1 regulatory subunit, is modulated by the laforin-malin complex. Int J Biochem Cell Biol 45(7):1479–1488
Rubio-Villena C, Viana R, Bonet J, Garcia-Gimeno MA, Casado M, Heredia M et al (2018) Astrocytes: new players in progressive myoclonus epilepsy of Lafora type. Hum Mol Genet 27(7):1290–1300
Ruska E, Knoll M (1931) Die magnetische Sammelspule für schnelle Elektronenstrahlen. The magnetic concentrating coil for fast electron beams. Z techn Physik 12:389–400
Russell JA, Bloom WL (1955) Extractable and residual glycogen in tissues of the rat. Am J Phys 183(3):345–355
Rybicka KK (1996) Glycosomes—the organelles of glycogen metabolism. Tissue Cell 28(3):253–265
Ryman BE, Whelan WJ (1971) New aspects of glycogen metabolism. Adv Enzymol Relat Areas Mol Biol 34:285–443
Ryu JH, Drain J, Kim JH, McGee S, Gray-Weale A, Waddington L et al (2009) Comparative structural analyses of purified glycogen particles from rat liver, human skeletal muscle and commercial preparations. Int J Biol Macromol 45(5):478–482
Saez I, Duran J, Sinadinos C, Beltran A, Yanes O, Tevy MF et al (2014) Neurons have an active glycogen metabolism that contributes to tolerance to hypoxia. J Cereb Blood Flow Metab 34(6):945–955
Sagar SM, Sharp FR, Swanson RA (1987) The regional distribution of glycogen in rat brain fixed by microwave irradiation. Brain Res 417(1):172–174
Sakai M, Austin J, Witmer F, Trueb L (1970) Studies in myoclonus epilepsy (Lafora body form). II. Polyglucosans in the systemic deposits of myoclonus epilepsy and in corpora amylacea. Neurology 20(2):160–176
Salmoral EM, Tolmasky DS, Krisman CR (1990) Evidence for the presence of glycogen in rat thymus. Cell Mol Biol 36(2):163–174

Salsas E, Larner J (1975) Glycogen synthase can use glucose as an acceptor. J Biol Chem 250(5):1833–1837
Salvan AM, Vion-Dury J, Confort-Gouny S, Dano P, Cozzone PJ (1997) Increased cerebral glycogen detected by localized 1H-magnetic resonance spectroscopy in a patient with suspected McArdle's disease. Eur Neurol 37(4):251–253
Sanchez-Martin P, Roma-Mateo C, Viana R, Sanz P (2015) Ubiquitin conjugating enzyme E2-N and sequestosome-1 (p62) are components of the ubiquitination process mediated by the malin-laforin E3-ubiquitin ligase complex. Int J Biochem Cell Biol 69:204–214
Sanson MA (1857) Note sur la formation physiologique du sucre dans l'eçonomie animale. Comptes rendus des seances de l'Acad des Sci 44:1323–1325
Santelia D, Kotting O, Seung D, Schubert M, Thalmann M, Bischof S et al (2011) The phosphoglucan phosphatase like sex Four2 dephosphorylates starch at the C3-position in Arabidopsis. Plant Cell 23(11):4096–4111
Schiaffino S, Mammucari C, Sandri M (2008) The role of autophagy in neonatal tissues: just a response to amino acid starvation? Autophagy 4(5):727–730
Schneider JL, Cuervo AM (2014) Autophagy and human disease: emerging themes. Curr Opin Genet Dev 26:16–23
Schneider DR, Felt BT, Goldman H (1981) Microwave radiation energy: a probe for the neurobiologist. Life Sci 29(7):643–653
Schwarz GA, Yanoff M (1965) Lafora's disease. Distinct clinico-pathologic form of Unverricht's syndrome. Arch Neurol 12:172–188
Scott RB, Still WJS (1968) Glycogen in human peripheral blood leukocytes: II. The macromolecular state of leukocyte glycogen. J Clin Invest 47(2):353–359
Seranova E, Connolly KJ, Zatyka M, Rosenstock TR, Barrett T, Tuxworth RI et al (2017) Dysregulation of autophagy as a common mechanism in lysosomal storage diseases. Essays Biochem 61(6):733–749
Serratosa JM, Gomez-Garre P, Gallardo ME, Anta B, de Bernabe DB, Lindhout D et al (1999) A novel protein tyrosine phosphatase gene is mutated in progressive myoclonus epilepsy of the Lafora type (EPM2). Hum Mol Genet 8(2):345–352
Shaid S, Brandts CH, Serve H, Dikic I (2013) Ubiquitination and selective autophagy. Cell Death Differ 20(1):21
Shearer J, Graham TE (2002) New perspectives on the storage and organization of muscle glycogen. Can J Appl Physiol 27(2):179–203
Shearer J, Graham TE (2004) Novel aspects of skeletal muscle glycogen and its regulation during rest and exercise. Exerc Sport Sci Rev 32(3):120–126
Sheldon H, Silverberg M, Kerner I (1962) On the differing appearance of intranuclear and cytoplasmic glycogen in liver cells in glycogen storage disease. J Cell Biol 13(3):468
Shimizu N (1955) Histochemical studies of glycogen of the area postrema and the allied structures of the mammalian brain. J Comp Neurol 102(2):323–339
Shimizu N, Hamuro Y (1958) Deposition of glycogen and changes in some enzymes in brain wounds. Nature 181(4611):781
Shimizu N, Kubo Z (1957) Histochemical studies on brain glycogen of the guinea pig and its alteration following electric shock. J Neuropathol Exp Neurol 16(1):40–47
Shimizu N, Kumamoto T (1952) Histochemical studies on the glycogen of the mammalian brain. Anat Rec 114(3):479–497
Singh PK, Singh S, Ganesh S (2012) The laforin-malin complex negatively regulates glycogen synthesis by modulating cellular glucose uptake via glucose transporters. Mol Cell Biol 32(3):652–663
Skurat AV, Cao Y, Roach PJ (1993) Glucose control of rabbit skeletal muscle glycogenin expressed in COS cells. J Biol Chem 268(20):14701–14707
Skurat AV, Lim SS, Roach PJ (1997) Glycogen biogenesis in rat 1 fibroblasts expressing rabbit muscle glycogenin. Eur J Biochem 245(1):147–155
Skurat AV, Dietrich AD, Roach PJ (2006) Interaction between glycogenin and glycogen synthase. Arch Biochem Biophys 456(1):93–97

Smythe C, Cohen P (1991) The discovery of glycogenin and the priming mechanism for glycogen biogenesis. Eur J Biochem 200(3):625–631
Somogyi M (1934) The solubility and preparation of phosphorus-and nitrogen-free glycogen. J Biol Chem 104(2):245–253
Sotelo C, Palay SL (1968) The fine structure of the lateral vestibular nucleus in the rat: I. Neurons and neuroglial cells. J Cell Biol 36(1):151–179
Stapleton D, Nelson C, Parsawar K, McClain D, Gilbert-Wilson R, Barker E et al (2010) Analysis of hepatic glycogen-associated proteins. Proteomics 10(12):2320–2329
Stapleton D, Nelson C, Parsawar K, Flores-Opazo M, McClain D, Parker G (2013) The 3T3-L1 adipocyte glycogen proteome. Proteome Sci 11(1):11
Stetten D Jr, Stetten MR (1960) Glycogen metabolism. Physiol Rev 40(3):505–537
Sullivan MA, Vilaplana F, Cave RA, Stapleton D, Gray-Weale AA, Gilbert RG (2010) Nature of α and β particles in glycogen using molecular size distributions. Biomacromolecules 11(4):1094–1100
Sullivan MA, O'Connor MJ, Umana F, Roura E, Jack K, Stapleton DI et al (2012) Molecular insights into glycogen α-particle formation. Biomacromolecules 13(11):3805–3813
Sullivan MA, Aroney ST, Li S, Warren FJ, Joo JS, Mak KS et al (2014) Changes in glycogen structure over feeding cycle sheds new light on blood-glucose control. Biomacromolecules 15(2):660–665
Sullivan MA, Nitschke S, Steup M, Minassian BA, Nitschke F (2017) Pathogenesis of Lafora disease: transition of soluble glycogen to insoluble polyglucosan. Int J Mol Sci 18(8):E1743
Sun T, Yi H, Yang C, Kishnani PS, Sun B (2016) Starch binding domain-containing protein 1 plays a dominant role in glycogen transport to lysosomes in liver. J Biol Chem 291(32):16479–16484
Sun RC, Dukhande VV, Zhou Z, Young LEA, Emanuelle S, Brainson CF, Gentry MS (2019) Nuclear Glycogenolysis Modulates Histone Acetylation in Human Non-Small Cell Lung Cancer. Cell Metabolism. https://doi.org/10.1016/j.cmet.2019.08.014
Sutherland EW, Wosilait WD (1955) Inactivation and activation of liver phosphorylase. Nature 175(4447):169
Svorad D (1958) Diurnal changes in the brain glycogen. Experientia 14(12):452
Swanson MA (1948) Studies on the structure of polysaccharides; relation of the iodine color to the structure. J Biol Chem 172(2):825–837
Swanson MA, Cori CF (1948) Studies on the structure of polysaccharides III. Relation of structure to activation of phosphorylases. J Biol Chem 172(2):815–824
Swanson RA, Sagar SM, Sharp FR (1989) Regional brain glycogen stores and metabolism during complete global ischaemia. Neurol Res 11(1):24–28
Tagliabracci VS, Turnbull J, Wang W, Girard JM, Zhao X, Skurat AV et al (2007) Laforin is a glycogen phosphatase, deficiency of which leads to elevated phosphorylation of glycogen in vivo. Proc Natl Acad Sci U S A 104(49):19262–19266
Tagliabracci VS, Girard JM, Segvich D, Meyer C, Turnbull J, Zhao X et al (2008) Abnormal metabolism of glycogen phosphate as a cause for Lafora disease. J Biol Chem 283(49):33816–33825
Tagliabracci VS, Heiss C, Karthik C, Contreras CJ, Glushka J, Ishihara M et al (2011) Phosphate incorporation during glycogen synthesis and Lafora disease. Cell Metab 13(3):274–282
Takata H, Kajiura H, Furuyashiki T, Kakutani R, Kuriki T (2009) Fine structural properties of natural and synthetic glycogens. Carbohydr Res 344(5):654–659
Takeda Y, Hizukuri S (1982) Location of phosphate groups in potato amylopectin. Carbohydr Res 102(1):321–327
Tan JMM, Wong ESP, Dawson VL, Dawson T, Lim K-L (2008) Lysine 63-linked polyubiquitin potentially partners with p62 to promote the clearance of protein inclusions by autophagy. Autophagy 4(2):251–253
Tan X, Sullivan MA, Nada SS, Deng B, Schulz BL, Gilbert RG (2018) Proteomic investigation of the binding agent between liver glycogen β particles. Acs Omega 3(4):3640–3645
Tanikawa K (1965) Electron microscopic observation of the intranuclear glycogen of the liver in diabetes mellitus. Kurume Med J 11(4):177–181

Tarentino AL, Maley F (1976) Direct evidence that D-galactosamine incorporation into glycogen occurs via UDP-glucosamine. FEBS Lett 69(1–2):175–178

Tavridou A, Agius L (2003) Phosphorylase regulates the association of glycogen synthase with a proteoglycogen substrate in hepatocytes. FEBS Lett 551(1–3):87–91

Tegtmeyer LC, Rust S, van Scherpenzeel M, Ng BG, Losfeld M-E, Timal S et al (2014) Multiple phenotypes in phosphoglucomutase 1 deficiency. N Engl J Med 370(6):533–542

Terjung RL, Baldwin KM, Winder WW, Holloszy JO (1974) Glycogen repletion in different types of muscle and in liver after exhausting exercise. Am J Phys 226(6):1387–1391

Testoni G, Duran J, García-Rocha M, Vilaplana F, Serrano AL, Sebastián D et al (2017) Lack of glycogenin causes glycogen accumulation and muscle function impairment. Cell Metab 26(1):256–266. e4

The Lancet. J. Onwhyn (1890) https://doi.org10.1016/S0140-6736(02)15107-5; link to the publisher site: https://www.sciencedirect.com/science/article/pii/S0140673602151075?via%3Dihub

Tiberia E, Turnbull J, Wang T, Ruggieri A, Zhao XC, Pencea N et al (2012) Increased laforin and laforin binding to glycogen underlie Lafora body formation in malin-deficient Lafora disease. J Biol Chem 287(30):25650–25659

Tolmasky DS, Krisman CR (1987) The degree of branching in (α1, 4)-(α1, 6)-linked glucopolysaccharides is dependent on intrinsic properties of the branching enzymes. Eur J Biochem 168(2):393–397

Towfighi J, Yoss BS, Wasiewski WW, Vannucci RC, Bentz MS, Mamourian A (1989) Cerebral glycogenosis, alpha particle type: morphologic and biochemical observations in an infant. Hum Pathol 20(12):1210–1215

Turnbull J, Wang P, Girard JM, Ruggieri A, Wang TJ, Draginov AG et al (2010) Glycogen hyperphosphorylation underlies lafora body formation. Ann Neurol 68(6):925–933

Turnbull J, Depaoli-Roach AA, Zhao X, Cortez MA, Pencea N, Tiberia E et al (2011a) PTG depletion removes lafora bodies and rescues the fatal epilepsy of Lafora disease. PLoS Genet 7(4):e1002037

Turnbull J, Girard JM, Pencea N, Zhao X, Graham TE, Wang P et al (2011b) Lafora bodies in skeletal muscle are fiber type specific. Neurology 76(19):1674–1676

Turnbull J, Epp JR, Goldsmith D, Zhao X, Pencea N, Wang P et al (2014) PTG protein depletion rescues malin-deficient Lafora disease in mouse. Ann Neurol 75(3):442–446

Valles-Ortega J, Duran J, Garcia-Rocha M, Bosch C, Saez I, Pujadas L et al (2011) Neurodegeneration and functional impairments associated with glycogen synthase accumulation in a mouse model of Lafora disease. EMBO Mol Med 3(11):667–681

Van Heijningen AJMK, Kemp A (1955) Free and fixed glycogen in rat muscle. Biochem J 59(3):487

Van Heycop Ten Ham MW (1975) Lafora disease, a form of progressive myoclonus epilepsy. In: Vinken PJ, Bruyn GW (eds) Handbook of clinical neurology. North Holland Publishing Company, Holland, Amsterdam, pp 382–422

Vander Kooi CW, Taylor AO, Pace RM, Meekins DA, Guo HF, Kim Y et al (2010) Structural basis for the glucan phosphatase activity of Starch Excess4. Proc Natl Acad Sci U S A 107(35):15379–15384

Vénien-Bryan C, Jonic S, Skamnaki V, Brown N, Bischler N, Oikonomakos NG et al (2009) The structure of phosphorylase kinase holoenzyme at 9.9 Å resolution and location of the catalytic subunit and the substrate glycogen phosphorylase. Structure 17(1):117–127

Verhue W, Hers HG (1966) A study of the reaction catalysed by the liver branching enzyme. Biochem J 99(1):222

Vilchez D, Ros S, Cifuentes D, Pujadas L, Valles J, Garcia-Fojeda B et al (2007) Mechanism suppressing glycogen synthesis in neurons and its demise in progressive myoclonus epilepsy. Nat Neurosci 10(11):1407–1413

Viskupic E, Cao Y, Zhang W, Cheng C, DePaoli-Roach AA, Roach PJ (1992) Rabbit skeletal muscle glycogenin. Molecular cloning and production of fully functional protein in Escherichia coli. J Biol Chem 267(36):25759–25763

Vøllestad MK, Vaage ODD, Hermansen L (1984) Muscle glycogen depletion patterns in type I and subgroups of type II fibres during prolonged severe exercise in man: Glycogen depletion in muscle fibres during exercise. Acta Physiol Scand 122(4):433–441
Walker GJ, Whelan WJ (1960) The mechanism of carbohydrase action. 8. Structures of the muscle-phosphorylase limit dextrins of glycogen and amylopectin. Biochem J 76(2):264
Wang J, Stuckey JA, Wishart MJ, Dixon JE (2002) A unique carbohydrate binding domain targets the lafora disease phosphatase to glycogen. J Biol Chem 277(4):2377–2380
Wang W, Lohi H, Skurat AV, Depaoli-Roach AA, Minassian BA, Roach PJ (2007) Glycogen metabolism in tissues from a mouse model of Lafora disease. Arch Biochem Biophys 457(2):264–269
Wang P, Israelian L, Xue Y, Song S, Attisano L, Minassian BA (2016) SGK1 (glucose transport), dishevelled2 (wnt signaling), LC3/p62 (autophagy) and p53 (apoptosis) proteins are unaltered in Lafora disease. All Results J Biol 7(3):28
Wanson J-C, Drochmans P (1968) Rabbit skeletal muscle glycogen: a morphological and biochemical study of glycogen ß-particles isolated by the precipitation-centrifugation method. J Cell Biol 38(1):130–150
Whelan WJ (1976) On the origin of primer for glycogen synthesis. Trends Biochem Sci 1(1) 13–15
Whelan WJ (2007) Why the linkage of glycogen to glycogenin was so difficult to determine. Biochem Mol Biol Educ 35(5):313–315
Willstätter R, Rohdewald M (1934) Über den Zustand des Glykogens in der Leber, im Muskel und in Leukocyten.(Zur Kenntnis der Proteinbindung physiologisch wichtiger Stoffe.). Hoppe-Seyler´ s Zeitschrift für physiologische Chemie 225(2–3):103–124
Wilson WA, Roach PJ, Montero M, Baroja-Fernández E, Muñoz FJ, Eydallin G et al (2010) Regulation of glycogen metabolism in yeast and bacteria. FEMS Microbiol Rev 34(6):952–985
Wisselaar HA, Kroos MA, Hermans MM, Van Beeumen J, Reuser AJ (1993) Structural and functional changes of lysosomal acid alpha-glucosidase during intracellular transport and maturation. J Biol Chem 268(3):2223–2231
Wolfrom ML, Lassettre EN, O'Neill AN (1951) Degradation of glycogen to isomaltose1. J Am Chem Soc 73(2):595–599
Worby CA, Gentry MS, Dixon JE (2006) Laforin: a dual specificity phosphatase that dephosphorylates complex carbohydrates. J Biol Chem 281(41):30412–30418
Worby CA, Gentry MS, Dixon JE (2008) Malin decreases glycogen accumulation by promoting the degradation of protein targeting to glycogen (PTG). J Biol Chem 283(7):4069–4076
Yáñez AJ, Garcia-Rocha M, Bertinat R, Droppelmann C, Concha II, Guinovart JJ et al (2004) Subcellular localization of liver FBPase is modulated by metabolic conditions. FEBS Lett 577(1–2):154–158
Yi H, Fredrickson KB, Das S, Kishnani PS, Sun B (2013) Stbd1 is highly elevated in skeletal muscle of Pompe disease mice but suppression of its expression does not affect lysosomal glycogen accumulation. Mol Genet Metab 109(3):312–314
Yokoi S, Austin J, Witmer F, Sakai M (1968) Studies in myoclonus epilepsy (Lafora body form). I. Isolation and preliminary characterization of Lafora bodies in two cases. Arch Neurol 19(1):15–33
Young FG (1937) Claude Bernard and the theory of the glycogenic function of the liver. Ann Sci 2(1):47–83
Young FG (1957) Claude Bernard and the discovery of glycogen. Br Med J 1(5033):1431
Zeqiraj E, Sicheri F (2015) Getting a handle on glycogen synthase–Its interaction with glycogenin. Mol Asp Med 46:63–69
Zeqiraj E, Tang X, Hunter RW, García-Rocha M, Judd A, Deak M et al (2014) Structural basis for the recruitment of glycogen synthase by glycogenin. Proc Natl Acad Sci U S A 201402926
Zhai L, Feng L, Xia L, Yin H, Xiang S (2016) Crystal structure of glycogen debranching enzyme and insights into its catalysis and disease-causing mutations. Nat Commun 7:11229
Zhao H, Tang M, Liu M, Chen L (2018) Glycophagy: An emerging target in pathology. Clin Chim Acta 484:298–303

Structure and Regulation of Glycogen Synthase in the Brain

Bartholomew A. Pederson

Abstract Brain glycogen synthesis is a regulated, multi-step process that begins with glucose transport across the blood brain barrier and culminates with the actions of glycogen synthase and the glycogen branching enzyme to elongate glucose chains and introduce branch points in a growing glycogen molecule. This review focuses on the synthesis of glycogen in the brain, with an emphasis on glycogen synthase, but draws on salient studies in mammalian muscle and liver as well as baker's yeast, with the goal of providing a more comprehensive view of glycogen synthesis and highlighting potential areas for further study in the brain. In addition, deficiencies in the glycogen biosynthetic enzymes which lead to glycogen storage diseases in humans are discussed, highlighting effects on the brain and discussing findings in genetically modified animal models that recapitulate these diseases. Finally, implications of glycogen synthesis in neurodegenerative and other diseases that impact the brain are presented.

Keywords Glucose metabolism · Glycogen metabolism · Glycogen storage disease · Knockout mice · Transgenic mice

Abbreviations

AD	Alzheimer's disease
Akt/PKB	Protein kinase B
AMPK	Adenosine monophosphate-activated protein kinase
Ap-2	Activator protein 2
ATP	Adenosine triphosphate
bp	Base pairs

B. A. Pederson (✉)
Indiana University School of Medicine-Muncie, Ball State University, Muncie, IN, USA
e-mail: bapederson@bsu.edu

M. DiNuzzo, A. Schousboe (eds.), *Brain Glycogen Metabolism*, Advances in Neurobiology 23, https://doi.org/10.1007/978-3-030-27480-1_3

CA	Corpora amylacea
CAMKII	Ca + 2/calmodulin-dependent protein kinase II
cAMP	Cyclic adenosine monophosphate
CEBP	CAAT/enhancer binding protein
CK1	Casein kinase 1
CK2	Casein kinase 2
CRE	cAMP responsive element
dpc	Days post coitum
DYRK	Dual-specificity tyrosine-phosphorylated and regulated kinase
EDL	Extensor digitorum longus
EEG	Electroencephalogram
FISH	Fluorescence in situ hybridization
GFP	Green fluorescent protein
Glucose-1-P	Glucose-1-phosphate
Glucose-6-P	Glucose-6-phosphate
GLUT	Glucose transporter
GN	Glycogenin
GS	Glycogen synthase
GSD	Glycogen storage disease
GSK3	Glycogen synthase kinase
GYG1	Glycogenin 1
GYG2	Glycogenin 2
GYS1	Muscle isoform of glycogen synthase
GYS2	Liver isoform of glycogen synthase
HK	Hexokinase
HPLC	High performance liquid chromatography
HRE	Hypoxia response element
kD	Kilodaltons
MEF2	Myocyte enhancer factor 2
MGSKO	Muscle glycogen synthase knockout
MGSKO/GSL30	Muscle glycogen synthase knockout/glycogen synthase line 30
MINI	Muscle initiator
NA	Noradrenaline
Oct1	Octamer binding protein 1
p38beta	Stress-activated protein kinase 2b
PAS	Periodic acid Schiff
PASK	Per/Arnt/Som domain-containing protein kinase
PEPCK	Phosphoenolpyruvate carboxykinase
PGM1	Phosphoglucomutase 1
PhK	Phosphorylase kinase
PI3K	Phosphoinositide 3-kinase
PKA	Proteins kinase A
PKC	Proteins kinase C
PML	Promyelocytic leukemia
PP1c	Protein phosphatase 1

PSSM	Polysaccharide storage myopathy
PTG	Protein targeting to glycogen
SDS-PAGE	Sodium dodecyl sulfate-polyacrylamide gel electrophoresis
SRE	Sterol responsive elements
SSRI	Serotonin reuptake inhibitor
TCA	Tricarboxylic acid
TEF-1	Transcriptional enhancer factor 1
UDP	Uridine diphosphate
UMP	Uridine monophosphate
UTP	Uridine triphosphate
VIP	Vasoactive intestinal peptide

1 Brain Glycogen Measurement

Before delving into the synthesis of glycogen, it is important to discuss the lability of glycogen in the brain. The glycogen concentration in a tissue reflects the net of synthesis of the glucose polymer by glycogen synthase and the degradation by glycogen phosphorylase. Thus, it is important to effectively preserve glycogen in samples. While true for all tissues, brain glycogen in particular is rapidly degraded post-mortem. In adult mouse, brain glycogen levels were reduced ~50% 1 min following decapitation, as compared to mice sacrificed by whole body immersion in liquid nitrogen (Hutchins and Rogers 1970). By 5 min after decapitation, only ~20% of the brain glycogen remained (Hutchins and Rogers 1970). This glycogen remained until 20 min post-decapitation, the last time it was measured. Lowry et al. (1964) reported an even greater loss in brain glycogen (~85%) at 1 min after decapitation in a study where heads were frozen in dichlorodifluoromethane.

Numerous techniques have been used to minimize brain glycogen degradation, including whole body immersion into liquid nitrogen, decapitation into liquid nitrogen or dichlorodifluoromethane, funnel freezing, and head-focused microwave. Decapitation followed by flash freezing of mouse heads in liquid nitrogen preserved brain glycogen to the same extent as when mice were immersed into liquid nitrogen (Hutchins and Rogers 1970). Measuring cooling rates inside mouse brain revealed that it takes 6 s to cool the cortex to 0 °C and 12 s to cool to −30 °C when decapitated heads were immersed in liquid nitrogen (Swaab 1971). In our hands (Canada et al. 2011), when mouse heads were immediately frozen in liquid nitrogen after decapitation or cervical dislocation brain glycogen levels were ~2.5 μmol/g, almost identical to that reported by others (Hutchins and Rogers 1970; Lowry et al. 1964). Two other techniques provide optimal preservation of brain glycogen; freeze-blowing and head-focused microwave. The former technique uses high pressure to expel the brain from the skull cavity onto a metal plate cooled in liquid nitrogen (Ghajar et al. 1982; Veech et al. 1973). This process removes and freezes the brain in less than 1 s. A limitation of this technique is that it is only available for rats and requires that the animal is anesthetized. The latter technique is amenable to both mice and rats and does not require anesthesia, which increases brain glycogen levels (Morgenthaler et al. 2006; Nelson et al. 1968; Nordstrom and Siesjo 1978). Two limitations of this method are cost of the instrument and the inability to measure enzyme activity in the brain tissue. Head-focused microwave heats brains to ~80 °C in less than 1 s resulting in rapid inactivation of brain enzymes and thus preservation of glycogen and other metabolites. To reach ~85 °C in rat brain, Kong et al. (2002) used 10 kW for 1.2 s. With these parameters, they measured a glycogen concentration of 8 μmol/g in cortex. In comparison, the concentration was ~5 μmol/g and <0.5 μmol/g at 6 kW and 3.5 kW respectively. Subjecting anesthetized C57Bl/6 J mice to a 3.5 kW microwave, which heated the brain to 85 °C in 0.5 s, resulted in brain glycogen levels of ~3 μmol/g (Franken et al. 2003). Higher glycogen levels (>10 μmol/g) were reported in mouse brain when microwave (5Kw) was focused on the brain for ~1 s (Oe et al. 2016). In the author's hands, similar brain glycogen concentrations (~2.5 μmol/g tissue) are obtained when comparing mice sacrificed by microwave (5Kw) with mice sacrificed by cervical dislocation or decapitation when heads are immediately frozen in liquid nitrogen (author's unpublished observations). When studying rodent models, the choice of rodent, the amount of stress the animal experiences, and the age of the animal are important considerations. Rats have the disadvantages of longer times to freeze the brain, due to size, and the elevated glycogen phosphorylase activity in rats (60% active; Lust et al. 1973) as compared to mice (30% active; Lust and Passonneau 1976). Various types of stress and sensory activation are reported to impact brain glycogen levels (Cruz and Dienel 2002). In contrast to the rapid brain glycogen degradation observed in adult rodents, in 10 day old mouse pups, brain glycogen levels were relatively stable for 2 min post-mortem, but after an additional 8 min had dropped to near zero (Lowry et al. 1964).

2 Glycogen Synthetic Pathway

Glucose transport. The pathway for forming glycogen begins with glucose moving from the blood through the blood brain barrier and into astrocytes and neurons (reviewed in McEwen and Reagan 2004; Vannucci et al. 1997) (Fig. 1). This movement of glucose is facilitated by the glucose transporter (GLUT) proteins. While several members of the GLUT family (GLUT1,2,3,4,5,8) are expressed in brain, GLUT1 and GLUT3 appear to be responsible for the majority of glucose uptake. GLUT1 (55kD isoform), which is expressed in the blood-brain barrier (Farrell and Pardridge 1991), serves to transport glucose across this barrier while the 45kD isoform of GLUT1 is expressed in astrocytes (Leino et al. 1997). GLUT3 transports glucose into neurons (Leino et al. 1997). The K_m for glucose of GLUT1 and GLUT3 are ~8.5 mM and 1.4 mM, respectively, giving neurons preference to glucose (Jensen et al. 2014). However, astrocytic end feet completely cover blood microvessels (Mathiisen et al. 2010). The result is that the majority of glucose, and other metabolites, entering the brain will pass through these astrocytic end feet

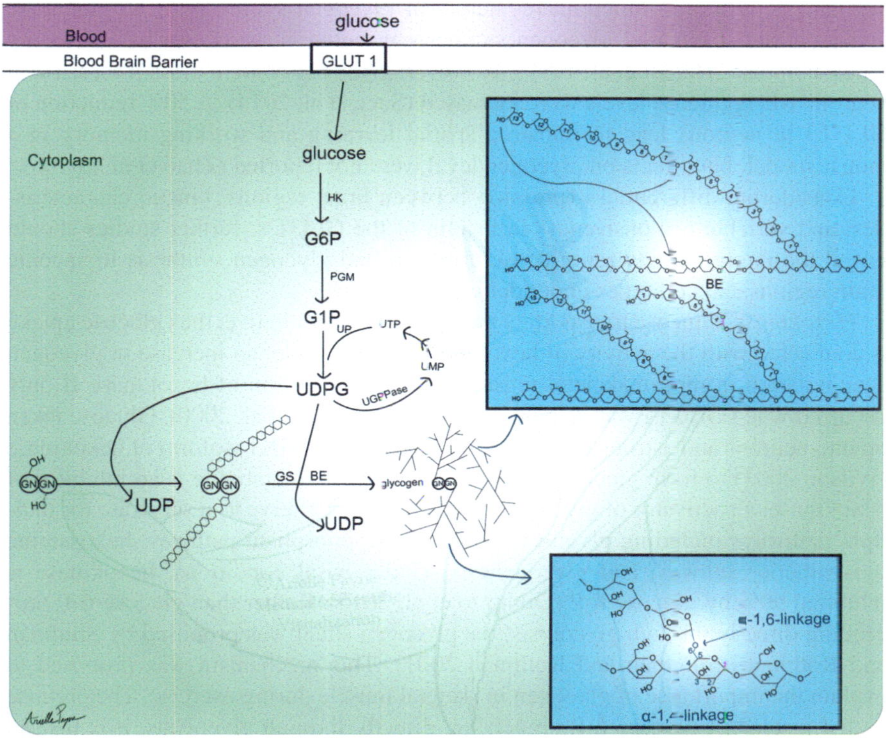

Fig. 1 Glycogen synthetic pathway in astrocytes. *GLUT* glucose transporter, *HK* hexokinase, *G6P* glucose-6-phosphate, *G1P* glucose-1-phosphate, *PGM* phosphoglucomutase, *UDPG* uridine diphosphate glucose, *UTP* uridine triphosphate, *UMP* uridine monophosphate, *UP* UDP-glucose pyrophosphorylase, *UGPPase* UDP-glucose pyrophosphophosphatase, *GS* glycogen synthase, *BE* branching enzyme, *GN* glycogenin. Hexagons depict glucose molecules. Illustration by Arielle Payne

(Teschemacher et al. 2015). The significance of the other glucose transporters expressed in the brain is still not completely clear, with some of the transporters being expressed in discrete brain regions (reviewed in Choeiri et al. 2002). Although the brain has generally been considered to be insulin-independent, the insulin sensitive glucose transporters GLUT4 and GLUT8 along with the insulin receptor are expressed (reviewed in McEwen and Reagan 2004).

Blood glucose levels tend to correlate with brain glycogen levels (Poitry-Yamate et al. 2009) and incubation of cultured astrocytes in high concentrations of glucose leads to increased glycogen accumulation (Magistretti et al. 1993). Overexpression of GLUT1 in mouse skeletal muscle increased glycogen concentration several fold without effects on glycogen synthase activity, suggesting that glucose transport is limiting in glycogen synthesis, at least in muscle (Ren et al. 1993). However, overexpression of glycogen synthase in muscle also increases glycogen accumulation (Manchester et al. 1996) suggesting that both glucose transport and glycogen synthase activity are important regulators of glycogen synthesis. Both glutamate and aspartate stimulated glycogen synthesis in cultured astrocytes by increasing glucose uptake (Hamai et al. 1999). However, glycogen content did not increase in neurons when glucose levels were increased (Saez et al. 2014). A 50% reduction of GLUT3 in neurons lead to impaired spatial learning and working memory in a mouse model, but effects on glycogen level were not reported (Zhao et al. 2010b).

Considering differential expression between brain regions, kinetic characteristics, and variable responsiveness to insulin of the GLUT's, further studies should reveal the interplay between glucose transport and glycogen synthesis in specific brain regions and cell types under defined conditions.

Hexokinase. Glucose transporters in the brain are facilitative, thus glucose uptake is connected with the activity of hexokinase. For example, an increase in glycogen accumulation in mouse skeletal muscle by over expression of hexokinase II only occurred when GLUT1 was also overexpressed (Hansen et al. 2000). Glucose taken up into neurons and astrocytes is phosphorylated by the HK1 isoform of hexokinase (Wilson 2003) to form glucose-6-P. This branch point metabolite is an inhibitor of hexokinase, an activator of glycogen synthase, and can serve as a substrate for multiple pathways including glycolysis, the pentose phosphate pathway, hexosamine biosynthetic pathway, and glycogenesis. Under basal conditions, hexokinase is inhibited 90% by glucose-6-P (DiNuzzo et al. 2015). Rather than glucose-6-P proceeding directly through glycolysis, the glycogen shunt was proposed by Shulman and Rothman (Shulman and Rothman 2001). This mechanism was proposed to explain the importance of glycogen in skeletal muscle during exercise. The premise is that instead of glucose-6-P proceeding directly through glycolysis, it is instead used to synthesize glycogen, which is then metabolized to generate ATP for muscle contraction. In support of this concept is the exercise intolerance observed in patients with McArdle disease, caused by a deficiency in muscle glycogen phosphorylase. Also, mice in which the muscle isoform of glycogen synthase (GYS1) is disrupted specifically in the skeletal muscle of adult mice have impaired exercise endurance capacity (Xirouchaki et al. 2016). In contrast, mice with global GYS1 disruption had normal exercise capacity (Pederson et al. 2005a). However, this latter animal model has metabolic changes in the muscle which enhance oxidative capacity. Thus, these adaptations perhaps overcome a normal reliance on the glycogen shunt. To examine the significance of the glycogen shunt in the brain, Walls et al. (Walls et al. 2009) manipulated glycogen metabolism in astrocytes and monitored the TCA cycle and glycolysis. This study indicated a significant role for the glycogen shunt in this cell type. Further studies in the brain will be needed to determine the importance of the glycogen shunt in physiological and pathological conditions.

Phosphoglucomutase. Phosphoglucomutase converts glucose-6-P to glucose-1-P. PGM1 is the main isoform in most tissues and deficiency results in a glycogen storage disease (GSDXIV). The first patient identified had abnormal accumulations of muscle glycogen (Stojkovic et al. 2009). In another study of 19 patients, fibroblast glycogen levels were normal (Tegtmeyer et al. 2014). Apart from one patient with mild intellectual disability (Ondruskova et al. 2014), patients with mutations in

PGM1 do not have adversely affected brains (Tegtmeyer et al. 2014), possibly due to the expression of other isozymes in the brain.

UDP-glucose pyrophosphorylase. Glucose-1-P reacts with UTP producing UDP-glucose through the action of UDP-glucose pyrophosphorylase. In Saccharomyces cerevisiae, disruption of UDP-glucose pyrophosphorylase indicated that this enzyme is essential for viability (Daran et al. 1995). Overexpression of this enzyme in mouse skeletal muscle increased levels of UDP-glucose threefold, but did not impact glycogen levels in muscle, suggesting that it is not rate limiting for glycogen synthesis (Reynolds et al. 2005). Whether this enzyme plays a regulatory role in the brain is not clear. Besides serving as the substrate for glycogen synthase, the UDP-glucose produced by UDP-glucose pyrophosphorylase can be converted to glucose-1-P and UMP through the action of UDP-glucose pyrophosphatase. This protein is expressed in multiple tissues including brain (Heyen et al. 2009) and could serve as a regulator of glycogen synthesis. Though no regulation has yet been demonstrated, the protein ran on SDS-PAGE as a doublet suggesting some form of post-translational modification.

Glycogenin. The glycogenin protein is found as a dimer at the core of glycogen molecules (Fig. 1). This protein is unique in that it serves in the roles of enzyme, substrate, and product. Glycogenin initially self glucosylates, adding a glucose molecule to a tyrosine residue (Tyr195), and then extends this glucose chain to 10–20 residues in length (Roach and Skurat 1997; Smythe and Cohen 1991). After reaching this length, the protein-glucose complex serves as a scaffold for glycogen synthase to continue elongation of the glucose chains. The crystal structure for rabbit muscle glycogenin was solved (Gibbons et al. 2002). It is a member of the GT8 glycosyltransferase family with a single Rossman-fold domain. UDP-glucose binds in a metal-dependent manner in the central beta-sheet structure.

There are two genes, GYG1 and GYG2, encoding glycogenin in humans. GYG2 is expressed in liver, heart, and pancreas (Mu et al. 1997), while GYG1 is more widely expressed, with highest expression in cardiac and skeletal muscle (Barbetti et al. 1996). Imagawa et al. (2014) reported that GYG2, but not GYG1, is expressed in human brain. Consistent with this expression pattern, a patient with an inactivating GYG1 mutation (Thr83Met) was found to have depleted glycogen stores in skeletal muscle, and a switch to slow-twitch oxidative muscle fibers (Moslemi et al. 2010), but no cognitive abnormalities were observed. This mutation prevents the formation of the glucose-O-tyrosine linkage at Tyr195 (Nilsson et al. 2012). In another study, 7 patients with mutations in GYG1 were found to have either depleted glycogenin protein, or in 1 case impaired interaction with glycogen synthase, varying levels of normal glycogen, and an accumulation of polyglucosan in skeletal muscle (Malfatti et al. 2014). This observation challenges the notion that glycogenin is required for the synthesis of glycogen. In contrast, two male siblings with Leigh syndrome, a progressive neurodegenerative disorder, were found to have a hemizygous missense mutation in GYG2 (Imagawa et al. 2014). Both brothers suffered from seizures and neurodegeneration (Imagawa et al. 2014). According to the study's authors, it remains to be established whether the GYG2 mutation is causative of Leigh syndrome in these patients.

In contrast to humans, non-primates appear to only have one gene for glycogenin, GYG. Glycogenin activity was demonstrated in rat brain (Tolmasky et al. 1991). In cultured astrocytes starved of glucose, free glycogenin was present (Lomako et al. 1993). When fed with glucose, a low-molecular weight form of glycogen termed "pro glycogen", was made. However, another study suggested that this pro glycogen may be an artifact of the extraction procedures used (James et al. 2008). Lomako et al. (1993) additionally proposed a "proglycogen synthase" separate from glycogen synthase, but to date this enzyme has not been identified.

In the mouse, glycogenin was identified as a lethal gene (Dickinson et al. 2016). Homozygous null embryos died perinatally. Characteristics included abnormal morphology of brain at E18.5, consistent with neural degeneration. At E15.5, embryos had a flattened forebrain with reduced lateral ventricles. In another study, when GYG was disrupted, approximately 85% of pups lacking glycogenin died shortly after birth, due to cardiopulmonary defects (Testoni et al. 2017). Fetal glycogen levels were not reported, but perhaps the lethality of the disruption is due to the inability to synthesize glycogen, as we observed with the disruption of the GYS1 isoform of glycogen synthase (Pederson et al. 2004a). Surprisingly, surviving GYG knockout mice had normal levels of brain glycogen, despite no detectable expression of GYG in this tissue (Testoni et al. 2017). These findings indicate the importance of glycogenin for survival but challenge the idea that glycogenin is required for the synthesis of glycogen.

Glycogenin associates with glycogen synthase via two mechanisms (Skurat et al. 2006). Namely, glycogen synthase associates with the glucose-primer chain on glycogenin. In addition, there is interaction of glycogen synthase with the C-terminus of glycogenin; the C-terminal 33 amino acids purify glycogen synthase from tissue extracts. Zeqiraj et al. (Zeqiraj and Sicheri 2015; Zeqiraj et al. 2014) solved the structure of *C. elegans* glycogen synthase in complex with a minimal targeting region of glycogenin, finding that glycogenin binds to the first of two Rossman fold domains of glycogen synthase in a region of the protein not involved in tetramer interactions or binding to glucose-6-P or UDP-glucose. When glycogen synthase or glycogenin was mutated in a manner that prevented interaction between the two expressed proteins, glycogen was not synthesized in *S. cerevisiae*, demonstrating the importance of this interaction. The region that separates the glycogen synthase binding domain and the catalytic domain is variable in length between isoforms and species and is a site of alternative splicing in humans (Zhai et al. 2000). Zeqiraj et al. (Zeqiraj and Sicheri 2015; Zeqiraj et al. 2014) suggests that the length of this linker region may control the size of glycogen particles. Glycogenin mRNA levels were increased in cerebral cortex of mice that underwent gentle sleep deprivation (Petit et al. 2010). The significance of glycogenin regulation will have to be borne out by further studies.

Glycogen synthase. The bulk synthesis of glycogen is catalyzed by glycogen synthase (EC2.4.1.11) which forms alpha-1,4 glycosidic linkages using UDP-glucose as the glucosyl donor (Leloir et al. 1959; Villar-Palasi and Larner 1958) (Fig. 1). There are two genes encoding this enzyme in mammals (GYS1 and GYS2). In rat, expression of GYS2 appears to be limited to liver (Kaslow et al. 1985), while

GYS1 is expressed in muscle, heart, fat, kidney, and brain (Kaslow and Lesikar 1984).

Glycogen branching enzyme. The glycogen branching enzyme (amylo-(1,4 to 1,6) transglycosylase) introduces branch points in the glycogen molecule producing chains of glucose that range from 1 to more than 30 glucose residues in both muscle and brain (Nitschke et al. 2017) (Fig. 1). An average chain length of ~13 glucose residues is found in multiple tissues and organisms. Branching increases the solubility of glycogen and provides multiple points of attack for glycogen phosphorylase (Melendez-Hevia et al. 1993). Demonstrating the importance of proper branching, mice that overexpress constitutively active GYS1 in skeletal muscle overaccumulate poorly branched glycogen, i.e., polyglucosan bodies (Pederson et al. 2003). Interestingly, the over expression of glycogen synthase results in increased expression of the glycogen branching enzyme, though not to the extent that it completely compensates for the increase in glycogen synthesis. Complete loss of branching enzyme activity is lethal in utero in a mouse model (Lee et al. 2011) replicating the severe form of glycogen storage disease type IV. There has been limited study of the glycogen branching enzyme in brain, but Tomalsky et al. (1998) characterized the mechanism of action in rat. Mice that are homozygous for a hypomorphic allele of the branching enzyme (GBE1) have a 90% reduction in branching enzyme activity and accumulate polyglucosan bodies in several tissues including brain (Akman et al. 2011). In another mouse model of the most common human GBE1 mutation, Y329S, adult mice had progressive muscular dysfunction, and died prematurely (Orhan Akman et al. 2015). The phenotype of this model was similar to human adult polyglucosan body disease. Polyglucosan accumulates in several tissues in newborn mice, but not in cerebral cortex. As mice age, polyglucosan accumulates in neurons and their axons causing neuropathy. With our collaborators, we found that crossing mice globally heterozygous for GYS1 with hypomorphic GBE1 mutant mice resulted in reduced formation of polyglucosan bodies, improved mobility, and extended lifespan (Chown 2018).

3 Glycogen Synthase Structure

Gene structure. Zhang et al. (1989) determined the sequence of rabbit skeletal muscle glycogen synthase from cDNA clones. A 3.6 kb mRNA was identified that coded for a protein of 734 amino acids and a molecular weight of 83,480. The 5′ untranslated and coding regions were 79% and 90% identical to human muscle glycogen synthase. The 3′ untranslated region was less similar. cDNA for human muscle glycogen synthase was cloned and sequenced by Browner et al. (1989), who found that it encodes a protein of 737 amino acids with a predicted molecular weight of 83,645 daltons. The 3′ untranslated region of the glycogen synthase cDNA is 1.2 kb. Using this cDNA as a probe, mRNA of ~3.5 kb was identified in human fetal and adult heart and skeletal muscle. In rabbit liver and brain, a single hybridizable mRNA of ~2.4 kb was identified, leading Browner et al. to suggest that the liver isoform of glycogen synthase would be expressed in the brain. The cDNA for mouse glycogen synthase was isolated from a cortical astrocyte cDNA library (Pellegri et al. 1996). The 3534 bp cDNA encoded a protein of 737 amino acids with 87% nucleotide identity and 96% amino acid identity with the human and rabbit muscle glycogen synthase and a 67% nucleotide and 71% amino acid identity with liver glycogen synthase. The cDNA has a 64 bp 5'untranslated region upstream of the first in-frame ATG and a 1259 bp 3'untranslated region. The 2211 bp opens reading frame codes for a protein with a calculated molecular weight of 83,823. These results suggested that the muscle isoform of glycogen synthase (GYS1) is expressed in the brain, and that the small sequence difference between the mouse brain cDNA and human could be a species difference, which indeed appears to the be case (https://www.uniprot.org/uniprot/Q9Z1E4). In humans, the GYS1 gene is located on chromosome 19 (19q13.3) (Groop et al. 1993; Lehto et al. 1993) and spans 32,229 bp with 16 exons. Promoter analysis indicated a number of putative transcription factor binding sites (Fredriksson et al. 2004) (Fig. 2). These included sites for Sp1, activator protein 2 (Ap-2), octamer binding protein 1 (Oct1), myocyte enhancer factor 2

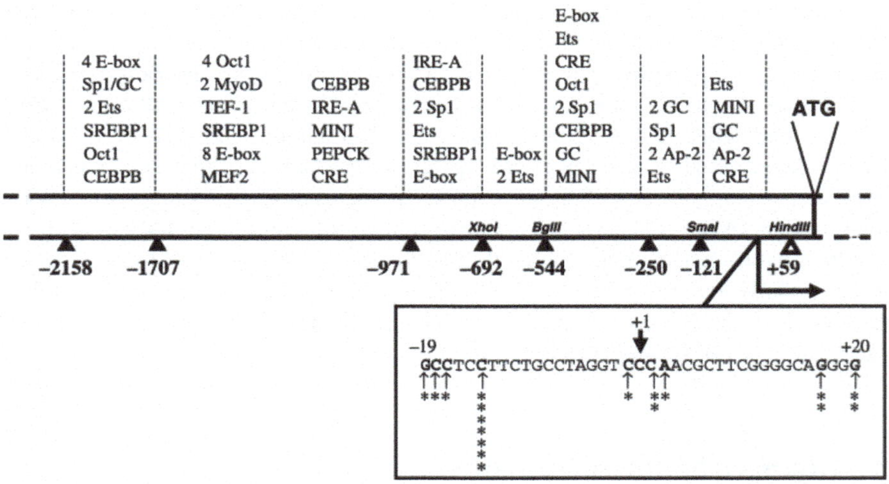

Fig. 2 Glycogen synthase gene (GYS1) promoter constructs and putative transcription factor-binding sites within the constructs. Filled triangles indicate 5′-ends of the different constructs. The −250, −971, −1707 and −2158 constructs were made by introducing a MluI site in a linker primer, whereas the −692, −544, and −121 constructs were created using internal restriction sites. All constructs contained 59 bp of the 5′-untranslated region (open triangles) numbered according to the previously determined transcription initiation site (Bjorbaek et al. 1994; Orho et al. 1995). The major transcriptional initiation site, proposed by seven different clones in the present study, is indicated at position −14 relative to the previously determined single initiation site. Four sites, recognized by one clone each, are not shown in the figure. These sites were located at nt +92, +108, +125 and +168. Ap-2, activator protein 2; *CEBPB* CAAT/enhancer-binding protein beta, *CRE* cAMP responsive element, *IRE-A* insulin response element A, *MEF2* myocyte enhancer factor, *MINI* muscle initiator, *PEPCK* phosphoenolpyruvate carboxykinase, *SREBP* sterol responsive element binding protein, *TEF-1* transcriptional enhancer factor 1. Used with permission from (Fredriksson et al. 2004)

(MEF2), CAAT/enhancer binding protein (CEBP), and enhancer factor-1. In addition, cAMP responsive elements (CRE), muscle initiator sequences (MINI), sterol responsive elements (SRE), E-boxes, Ets-like motifs and phosphoenolpyruvate carboxykinase (PEPCK)-like motifs were found. No TATA or CAAT boxes were identified. The first 250 bp of the 5′ flanking region were determined to account for basal promoter activity and contain elements characteristic of housekeeping genes. More recently, a hypoxia response element (HRE) was identified at position −314 in GYS1 (Pescador et al. 2010).

Protein structure. The human glycogen synthase protein (GYS1) contains 737 amino acids. The molecular weight of the GYS1 holoenzyme, as judged by HPLC gel filtration was 225 kD from brain and 274 kD from muscle, while the molecular weight of the subunit was 88 kD and 87 kD from brain and muscle respectively (Inoue et al. 1987). The reason for these differences in GYS1 between the brain and muscle is not clear. Besides catalytic residues, there are multiple phosphorylation sites in the protein and a region that confers sensitivity to glucose-6-P (Fig. 3). The crystal structure of yeast glycogen synthase, Gsy2p, was solved and found to be a tetramer (Baskaran et al. 2010) (Figs. 4 and 5). Two arginine residues in the C-terminus confer sensitivity to glucose-6-P, which when bound causes translations and rotations among the subunits. This rearrangement permits easier substrate access. Further studies showed that Arg589 and Arg592 are key in allowing the conversion to and from the T state of the enzyme (Mahalingan et al. 2017). Four additional arginine residues in this same helix are involved in the enzyme's response to phosphorylation (Baskaran et al. 2010). The control of the enzyme by glucose-6-P and phosphorylation can be described by a three-state model (Fig. 6) (Baskaran et al. 2010; Pederson et al. 2000). Unphosphorylated glycogen synthase is in an intermediate state (I state or state II). Binding of glucose-6-P converts the enzyme to a high activity state (R state or state III), while phosphorylation converts the

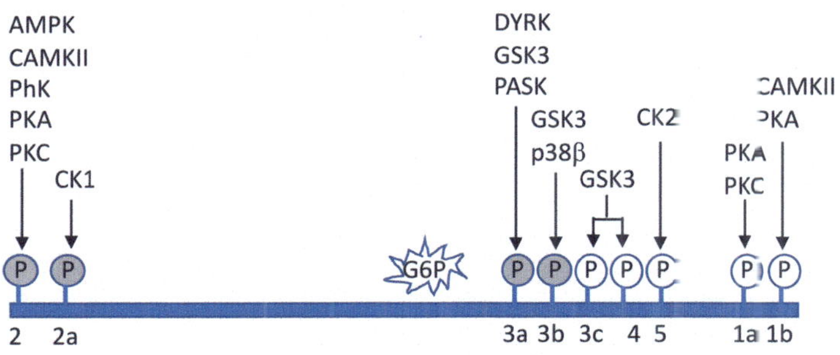

Fig. 3 Glycogen synthase (GYS1) protein structure. Residue 582 in mouse GYS1 is required for glucose-6-P sensitivity. Phosphorylation sites correspond to the following residues in mouse GYS1: site 2, 8; site 2a, 11; site 3a, 641; site 3b, 645; site 3c, 649; site 4, 653; site 5, 657; site 1a, 698; site 1b, 711. *AMPK* AMP-activated protein kinase, *CAMKII* calcium/calmodulin-dependent protein kinase, *CK1* casein kinase 1, *CK2* casein kinase 2, *DYRK* dual-specificity tyrosine-phosphorylated and -regulated kinase, *GSK3* glycogen synthase kinase 3, *p38beta* mitogen-activated protein kinase 11, *PASK* PAS domain-containing serine/threonine kinase *PhK* phosphorylase kinase, *PKA* protein kinase A, *PKC* protein kinase C

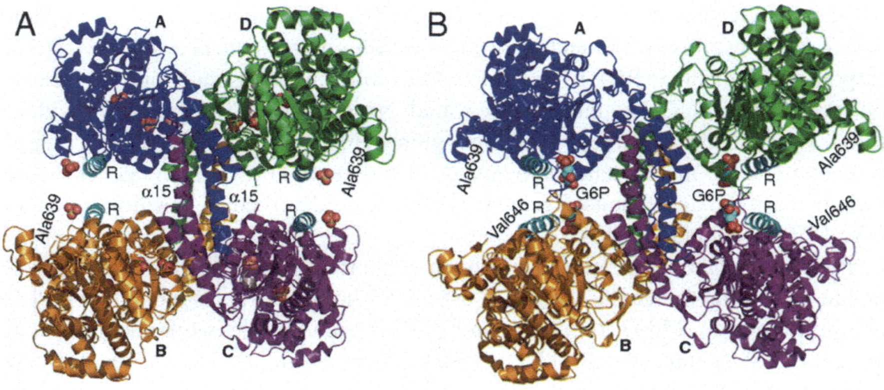

Fig. 4 Glycogen synthase crystal structure. Structures of the basal and activated states of Gsy2p. (**a**) Ribbon diagram of the basal state conformation in which the individual subunits are labeled (A–D) and colored separately. The regulatory helices (α22) are colored cyan and labeled R, while the intersubunit helices are labeled α15 as is the last ordered residue at the **c** terminus of subunits B and D. Ordered sulfate ions in this structure are represented using space filling atoms. (**b**) Ribbon diagram of the activated state of Gsy2p. The color scheme, subunit, and regulatory helix labeling is identical to panel **a**. The bound glucose-6-P molecules at the interface are represented using space filling atoms and labeled G6P. [Produced using Pymol (DeLano 2002) for Windows.] Used with permission from (Baskaran et al. 2010)

enzyme to a low activity state (state I or T state). The low activity state is not as sensitive to glucose-6-P binding, but with saturating levels of this metabolite, glycogen synthase is converted to the high activity state. When the tetrameric structure of the enzyme is taken into account, where hybrid complexes of phosphorylated and unphosphorylated subunits are present, there may be more than three activity states. In addition, while yeast Gsy2p and mammalian GYS1 are both activated by glucose-6-P and inhibited by phosphorylation, the yeast enzyme has 3 phosphorylation sites compared to the 9 found in the mammalian enzyme (Hanashiro and Roach 2002). These additional sites may provide for more complex regulation via phosphorylation/dephosphorylation.

Catalytic mechanism. The catalytic mechanism for glycogen synthase is not well understood. Unlike another glycosyltransferase, glycogenin, glycogen synthase is metal ion- independent, instead using hydrogen bonds to amino acids to stabilize UDP (reviewed in Roach et al. 2012). The particular residues in the yeast glycogen synthase, Gsy2p, implicated in catalysis are Arg320, Lys326, residues 513–521, and Glu509.

In the brain, glycogen synthase was first reported in rat (Leloir et al. 1959) and subsequently partially purified from sheep (Basu and Bachhawat 1961). Kinetic characterization of brain glycogen synthase was conducted with enzyme from rat and pig. In rat, the K_m for UDP-glucose was 0.17 mM and 0.36 mM in the absence and presence of glucose-6-P, respectively (Goldberg and O'Toole 1969). For glycogen synthase partially purified from pig brain, increasing UDP-glucose levels decreased the K_m for glucose-6-P while increased glucose-6-P levels decreased the

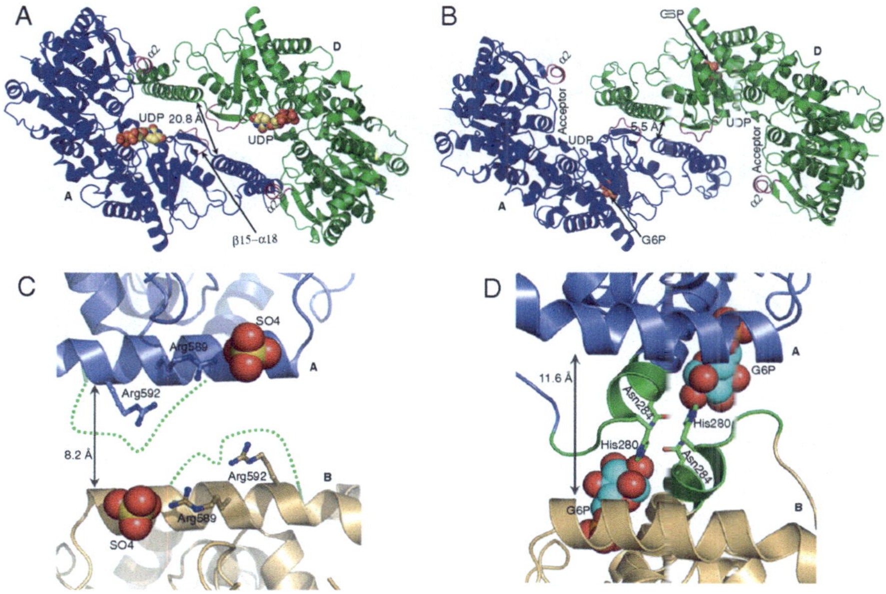

Fig. 5 Comparison of the basal and activated state conformations. (**a**) Ribbon representation of the AD dimer pair in the basal state with bound UDP represented using space filling atoms and ribbon coloring identical to Fig. 4. The loop region β15-α18 and helix α2 are highlighted in purple and labeled. The distance between equivalent positions of the α15 helices is provided. (**b**) Ribbon representation of the AD dimer pair in the activated state using the same coloring scheme as in panel **a**. The glucose-6-phosphate molecules are represented using space filling atoms and labeled, as are the relative positions of UDP and structural elements contributing to glycogen acceptor binding. (**c**) Ribbon representation of the regulatory interface between subunits A and B in the basal state. The individual subunits are indicated with labels and distinct ribbon colors. The bound sulfate molecule near the N terminus of each regulatory helix is displayed using space filling atoms. The disordered region between residues 277–285 is represented as a dashed green line solely to indicate the connectivity. (**d**) Ribbon representation of the regulatory helix interface between subunits A and B in the activated state of the R589A/R592A mutant. The coloring scheme is identical to that in panel **c** and the bound glucose-6-P is displayed using space filling atoms. The residues interacting with glucose-6-P from the 277–285 loop region are represented using green coloring. [Produced using Pymol (DeLano 2002) for Windows.] Used with permission from (Baskaran et al. 2010)

K_m for UDP-glucose for both the dependent and independent forms of the enzyme (Passonneau et al. 1975).

The primary function of glycogen synthase is catalyzing the addition of glucose from UDP-glucose to an elongating glucose chain in a glycogen molecule. However, glycogen synthase also incorporates the beta phosphate of UDP-glucose into glycogen as glucose-6-P (Fig. 7) at a frequency of ~1 phosphate per 10,000 glucose residues (Tagliabracci et al. 2011). This phosphate incorporation into glycogen is one of the mechanisms hypothesized to explain the abnormal glycogen accumulation in Lafora disease (Roach 2015; Tagliabracci et al. 2008). Lafora disease is characterized by myoclonus epilepsy beginning during early adolescence (Lafora and Glueck

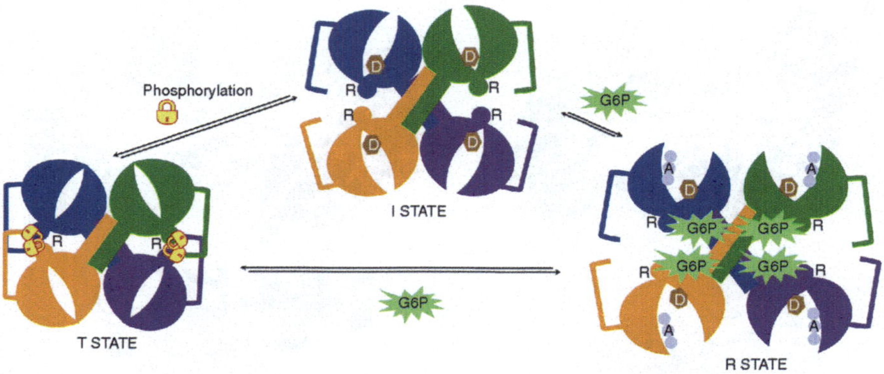

Fig. 6 Glycogen synthase three state model. Schematic representation of the conformational states underlying regulation of Gsy2p activity. The individual subunits are colored according to the coloring scheme in Fig. 4. The regulatory helices containing the arginines are labeled R and the approximate positions of nucleotide-donor sugar (D) and glycogen acceptor (A) are shown for the I- and R-states, respectively. Phosphorylation of Thr668 is shown as "locking" the enzyme in the T-state conformation through intersubunit interactions across the regulatory interface. Glucose-6-P binding frees these constraining interactions to fully activate the enzyme in the R-state conformation. Used with permission from Baskaran et al. (2010)

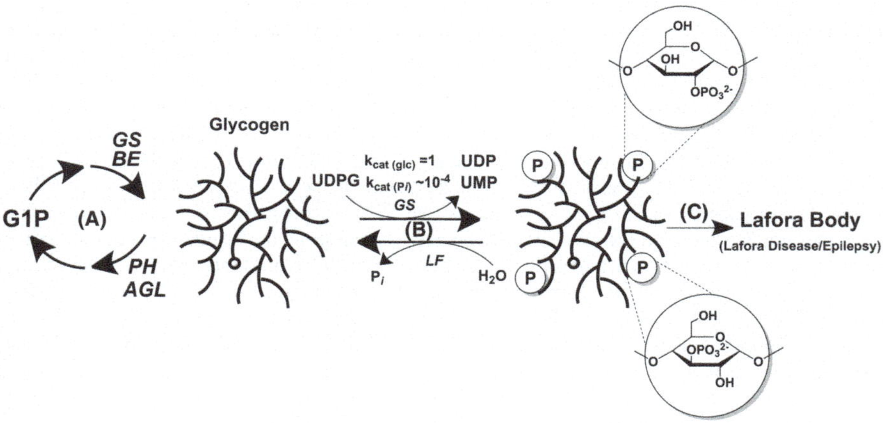

Fig. 7 Model for Glycogen Phosphate Metabolism and Lafora Body Formation. (**a**) Glycogen is synthesized by glycogen synthase (GS) and branching enzyme (BE) and degraded by glycogen phosphorylase (PH) and debranching enzyme (AGL). (**b**) Glycogen synthase infrequently (1 in ~10,000) incorporates phosphate residues into glycogen. The kcat values given simply denote the relative rates of glucose versus glucose phosphate incorporation. Excessive incorporation of phosphate as C2 or C3 phosphomonesters, which disrupts glycogen structure, is normally kept in check by the action of the laforin phosphatase (LF). (**c**) When laforin is defective, excessive phosphorylation results in the formation of Lafora bodies and Lafora disease. Used with permission from Tagliabracci et al. (2011)

1911). Over the course of ~10 years, cognitive function declines, seizure frequency increases, and death typically occurs (Minassian 2001; Serratosa et al. 1999; Turnbull et al. 2016). In these patients, high levels of poorly branched, hyperphosphorylated glycogen, termed polyglucosan or Lafora bodies, accumulate in many tissues including the brain (Minassian 2001; Striano et al. 2008). Neuronal cell bodies and dendrites, but not axons, in all brain regions are affected. Approximately 50% of Lafora disease cases (The Lafora Progressive Myoclonus Epilepsy Mutation and Polymorphism Database; http://projects.tcag.ca/lafora/) are the result of mutations in the *EPM2A* gene which encodes laforin (Minassian et al. 1998; Serratosa et al. 1999). Apart from one patient with a mutation in the *PRMD8* gene (Turnbull et al. 2012), the remainder of Lafora disease cases are due to mutations in the *EPM2B/NHLRC1* gene which encodes malin (Chan et al. 2003), an E3 ubiquitin ligase (Gentry et al. 2005). Laforin, a dual specificity phosphatase, binds to glycogen through a carbohydrate binding domain (Chan et al. 2004; Wang et al. 2002; Wang and Roach 2004) and dephosphorylates glycogen (Tagliabracci et al. 2007) In a mouse model of Lafora disease, created by disruption of Epm2a, muscle glycogen contained ~fivefold higher levels of phosphate as compared to wild type controls (Tagliabracci et al. 2007, 2008). This phosphate is located at the C2-, C3- and C6-positions (DePaoli-Roach et al. 2015). The mechanism for incorporation of phosphate at C2- and C3- can be explained by the action of glycogen synthase, but the mechanism for incorporation at the C6- position is unknown (Chikwana et al. 2013; Contreras et al. 2016; Tagliabracci et al. 2011). Phosphorylation of glycogen decreases branching and solubility, though the mechanism is unclear (Roach 2015). Roach (2011) proposed that the incorporation of phosphate into glycogen may be a catalytic error and that laforin acts as a repair enzyme. An alternative hypothesis for the formation of Lafora bodies is that glycogen synthase activity is increased relative to branching enzyme activity, leading to synthesis of poorly branched glycogen (Vilchez et al. 2007). Both of these hypotheses have recently been challenged. Glycogen synthase activity was normal in both malin and laforin knockout mice (DePaoli-Roach et al. 2010; Tagliabracci et al. 2008) and the overexpression of phosphatase-inactive laforin prevented the formation of Lafora bodies in laforin knockout mice (Gayarre et al. 2014). In this latter model, glycogen remains hyperphosphorylated while the abnormal glycogen chain length pattern characteristic of Lafora bodies is corrected (Nitschke et al. 2017). While the mechanism underlying the formation of Lafora bodies remains an active area of investigation, genetically modified mouse models demonstrate the importance of glycogen synthesis in the pathology of Lafora disease. Disruption of PTG, an activator of glycogen synthase, in a mouse model resulted in a 70% reduction in brain glycogen (Turnbull et al. 2011). When these mice were crossed with a laforin deficient mouse model, formation of Lafora bodies was largely prevented and both myoclonic epilepsy and neurodegeneration were resolved. Similarly, in mice disrupted for both GYS1 and laforin, Lafora bodies and neurodegeneration were absent and seizure susceptibility was normal (Pederson et al. 2013). Also, disrupting GYS1 in malin deficient mice prevented the deleterious effects associate with Lafora disease (Duran et al. 2014). These studies suggest that inhibiting glycogen synthase may be therapeutic for patients with this fatal disease.

4 Glycogen Synthase Localization

Regional localization. Within the mouse brain, glycogen synthase mRNA is expressed widely with the highest levels found in the olfactory bulb, the granular layer of the cerebellum, and the hippocampus (CA1 to CA4 and dentate gyrus) and lower levels in the cerebral cortex and striatum (Pellegri et al. 1996). Likewise, glycogen synthase protein is widely expressed in rat and canine brain with the highest levels observed in hippocampus, cerebral cortex, caudate-putamen, and cerebellar cortex (Inoue et al. 1988). Glycogen synthase enzymatic activity was detected histochemically in both white and gray matter in rat brain (Takeuchi 1965). In humans, glycogen synthase activity in hippocampus was higher than in either white or gray matter, though it should be noted that these measurements were made in tissue from patients with epilepsy (Dalsgaard et al. 2007). Among eight different areas of the rabbit central nervous system, glycogen synthase activity was highest in the molecular layer of the cerebellum (Breckenridge and Crawford 1961). In rat brain, the highest total glycogen synthase activity was measured in cerebellum, and lower levels of activity in hippocampus, medulla oblongata, striatum, midbrain, hypothalamus, and cortex (Knull and Khandelwal 1982). The percent of the glycogen synthase that was in the active form was also highest in the cerebellum. Taken together there is generally an agreement between glycogen synthase expression, glycogen synthase activity, and glycogen content (Oe et al. 2016; Sagar et al. 1987).

Cellular and intracellular localization. Within the brain, glycogen synthase mRNA (Pellegri et al. 1996) and protein (Inoue et al. 1988) were present in both astrocytes and neurons. Higher levels of glycogen synthase mRNA were expressed in astrocytes than neurons (Pellegri et al. 1996). In contrast, higher levels of glycogen synthase protein were expressed in neurons than astrocytes (Inoue et al. 1988). It is not clear what accounts for this apparent discrepancy. Glycogen synthase activity in rat brain was highest in the cytoplasmic fraction, followed by a threefold lower activity in the syntaptoplasmic fraction (Knull and Khandelwal 1982). Much lower activity levels were found in nuclear and mitochondrial fractions. Glycogen synthase mRNA was present in both cytosol and nucleus of cultured astrocytes and neurons from rat hippocampus (Mamczur et al. 2015). Regardless of the activity state of the enzyme, glycogen synthase is associated with glycogen (reviewed in Roach et al. 2012). This occurs despite the lack of a distinct carbohydrate binding motif. The yeast glycogen synthase, Gsy2p, contains a carbohydrate binding site integrated into the catalytic domain (Baskaran et al. 2011). Within neurons, glycogen synthase in mouse embryonic neuronal cultures was present as aggregates in the cytoplasm (Saez et al. 2014). Using fluorescence in situ hybridization (FISH) analysis on cultured cortical neurons, glycogen synthase mRNA was detected in both the soma and processes (Pfeiffer-Guglielmi et al. 2014). In the latter, the mRNA had a granular appearance. The role of glycogen synthase in the neuronal processes is not clear, but has been suggested to perhaps be important in energy demanding processes such as growing, repair, and memory consolidation or potentially in a moonlighting function such as translational control, as proposed by Fuchs et al. (2011).

5 Glycogen Synthase Regulation

5.1 Gene Regulation

Regulation of the glycogen synthase gene has received limited attention. In C2C12 mytotubes, but not HEK293 cells, GYS1 promoter activity was modestly decreased 24 h after insulin treatment (Fredriksson et al. 2004). In cultured cortical astrocytes, but not neurons, glycogen synthase mRNA expression was upregulated by two neurotransmitters, vasoactive intestinal peptide (VIP) and noradrenaline (NA) (Pellegri et al. 1996). This is consistent with the observation that glycogen resynthesis triggered by VIP or NA is dependent on DNA transcription and protein synthesis (Sorg and Magistretti 1992). The neuromodulator, adenosine, accumulates in the extracellular space following increased neuronal activity (Mitchell et al. 1993). Adenosine increased glycogen levels in cultured cortical astrocytes and this effect was blocked when transcription was inhibited (Allaman et al. 2003). Adenosine increased mRNA expression of protein targeting to glycogen (PTG) as well as C/EBP beta and C/EBP delta. Though not examined, the effects of adenosine could work through these mechanisms to potentially affect expression and/or enzymatic activity of glycogen synthase.

5.2 Protein Regulation

The accumulation of glycogen is determined by the balance between synthesis by glycogen synthesis and degradation by glycogen phosphorylase. In rabbit brain, glycogen phosphorylase activity is ~30 fold higher than glycogen synthase (Breckenridge and Crawford 1961), while in mouse brain the concentration of glycogen phosphorylase is 200 fold higher than that of glycogen synthase (Mamczur et al. 2015). These findings highlight the importance of regulation of both of these enzymes. This is especially interesting in light of the unique nature of the brain due to its location within the rigid skull. In contrast, the liver can, and does in certain glycogen storage diseases, expand when it accumulates large amounts of glycogen. This potential for expansion is greatly limited for the brain. The most studied mechanisms for glycogen synthase regulation are activation by glucose-6-P and inhibition by covalent phosphorylation. Glucose-6-P overcomes phosphorylation-induced inactivation and can restore maximal activity. In early studies of the enzyme, the enzyme was classified as either the I form or the D form. The I, independent form, is active without glucose-6-P, while the D, dependent form, is relatively inactive in the absence of glucose-6-P. Taking advantage of this phenomenon, assays to monitor glycogen synthase enzymatic activity are typically conducted in both the presence and absence of added glucose-6-P to provide an index of the phosphorylation state of the enzyme (Roach et al. 2012; Roach and Larner 1977). Alternatively, activity has been measured in the presence of low and high concentrations of glucose-6-P with a low concentration of the substrate UDP-glucose (Guinovart et al. 1979). When these assays are conducted in cell or tissue extract, activity is interpreted as providing an indication of in vivo glycogen synthase phosphorylation. However, the in vivo activity is affected not only by the phosphorylation state, but also by other factors, including but not limited to, the concentration of glycogen, glucose-6-P, and UDP-glucose.

5.2.1 Activation by Glucose-6-phosphate

Glycogen synthase is allosterically activated by glucose-6-P (Leloir et al. 1959). In the rat brain, glucose-6-P activates glycogen synthase six- to ten-fold (Goldberg and O'Toole 1969). Arginine residues in glycogen synthase that confer glucose-6-P sensitivity were identified in *S. cerevisiae* (Pederson et al. 2000) and rabbit skeletal muscle (Hanashiro and Roach 2002). When three of these residues (579,581, and 582) were mutated to alanine in the yeast glycogen synthase (Gsy2p) and expressed in *S. cerevisiae*, glycogen accumulation was reduced as compared to wild type cells (Pederson et al. 2004b). In a knockin mouse expressing constitutively active glycogen synthase kinase 3 (GSK3), insulin was unable to inactivate muscle GSK3 and promote glycogen synthase desphosphorylation and activation. However, these mice still made normal amounts of glycogen (Bouskila et al. 2008). This suggests a major role for glucose-6-P in regulation of glycogen synthesis. A knockin mouse with Arg 582 of GYS1 modified to alanine expressed glycogen synthase that was insensitive to glucose-6-P, had ~80% reduction in muscle glycogen synthesis induced by insulin, and ~50% reduction in skeletal muscle glycogen levels (Bouskila et al. 2010). These studies demonstrate the importance of glucose-6-P in opposition to phosphorylation control, which was intact in this mutant. Unfortunately, glycogen metabolism in the brain was not examined in the either of these mouse models.

5.2.2 Inhibition by Covalent Phosphorylation

Mammalian glycogen synthase was one of the earliest enzymes found to be phosphorylated at multiple sites (Smith et al. 1971). The much studied muscle isoenzyme has 9 phosphorylation sites, with mutagenesis studies in rabbit GYS1 indicating that sites 2, 2a, 3a, and 3b are most important for inhibiting enzymatic activity (Skurat and Roach 1995; Skurat et al. 1994). Phosphorylation of glycogen synthase at Ser640 (site 3a) and Ser7/10 (sites 2 and 2a) is responsible for keeping neuronal glycogen synthase in its normally inactive state (Vilchez et al. 2007). The concept of hierarchal phosphorylation, where the addition of one phosphate enables the addition of a second phosphate, was also derived from studies on glycogen synthase (Roach 1990, 1991). For instance, casein kinase 2 (CK2) must phosphorylate glycogen synthase prior to subsequent phosphorylation by GSK3. The GSK3 beta isoform is itself regulated by phosphorylation and appears to be the most important isoform for phosphorylation of GYS1 in muscle (Patel et al. 2008). Additional kinases that phosphorylate glycogen synthase include AMPK, Ca + 2/calmodulin-dependent protein kinase II (CAMKII), phosphorylase kinase (PhK), PKA, proteins kinase C (PKC), casein kinase I (CK1), Per/Arnt/Som) domain-containing protein kinase (PASK), dual-specificity tyrosine-phosphorylated and regulated kinase (DYRK), stress-activated protein kinase 2b (p38beta), (reviewed in Roach et al. 2012) (Fig. 3). AMPK is perhaps one of the most interesting glycogen synthase kinases due to its role as a cellular energy sensor, the presence of a carbohydrate binding domain which binds glycogen, and its proposed role as a glycogen sensor (McBride and Hardie 2009). AMPK has been proposed to be a site 2 kinase (reviewed in Roach et al. 2012). This kinase phosphorylates glycogen synthase in vitro and a knockout of the alpha 2 subunit of AMPK reduces phosphorylation of glycogen synthase at site 2 plus 2a and increases activation of glycogen synthase (Jorgensen et al. 2004). Consistent with a role AMPK acting as glycogen sensor, in skeletal muscle from mice that lack glycogen due to global disruption of GYS1, AMPK phosphorylation is increased, leading to a more active kinase (Pederson et al. 2005b). In contrast, in the brains of mice lacking GYS1 specifically in the central nervous system, no changes in the activation state of AMPK were observed (Duran et al. 2013). Glycogen synthase kinases in the rat brain were found to be largely (~90%) cAMP-independent (Schlender and Reimann 1977). Glycogen synthase purified from canine brain was phosphorylated by brain calmodulin kinase resulting in inactivation of the enzyme (Inoue et al. 1987). Calmodulin kinase was also purified from rat brain which phosphorylated rabbit muscle glycogen synthase at sites 1b and 2 (Schworer et al. 1985) and caused inactivation of the enzyme (Iwasa et al. 1984). Both CK1 and CK2 activities were detected in the soluble fraction from rat brain (Singh and Huang 1985). p38beta bound to glycogen synthase from brain and phosphorylated the enzyme at 4 sites, including 2 sites that were not reported to be phosphorylated by any other kinases (Bouskila et al. 2008). This phosphorylation did not inactivate glycogen synthase. DYRK1A phosphorylated glycogen synthase at site 3 in COS cells resulting in inactivation of the enzyme (Skurat and Dietrich 2004). DYRK1A is expressed in the developing and adult brain (Hammerle et al. 2003). This kinase is overexpressed in Down's syndrome and mice overexpressing DYRK1A have impaired cognitive function (Altafaj et al. 2001).

Dephosphorylation of glycogen synthase results in activation of the enzyme and thus stimulation of glycogenesis. Dephosphorylation is catalyzed by protein phosphatase 1 (PP1c) bound to one of seven glycogen-targeting subunits (PPP1R3A to PPP1R3G, reviewed in Ceulemans and Bollen 2004; Korrodi-Gregorio et al. 2014). Of the seven subunits, PPP1R3D (R6), PPP1R3F (R3F), PPP1R3G, and PPP1R3C (PTG) are expressed in the brain. PPP1R3D is mainly expressed in the brain, particularly in neurons (Rubio-Villena et al. 2013). PPP1R3D interacts with laforin which targets it for degradation resulting in decreased glycogen synthesis. The importance of the regulation is seen in Lafora disease, as discussed above. PPP1R3F is highly expressed in rodent brain and regulates glycogen synthase in astrocytoma cells (Kelsall et al. 2011). Mutant PPP1R3F is unable to bind PP1c, resulting in hyperphosphorylation of Ser640 (site 3a) and Ser644 (site 3b). PPP1R3G mRNA was detected in human brain (Munro et al. 2005). When disrupted in a mouse model, there was no detectable PPP1R3G protein in the brain, but the effects on brain glycogen metabolism were not described (Zhang et al. 2017). Disruption of PTG resulted in reduced glycogen accumulation in liver and muscle (Zhai et al. 2007) and in the brain (Turnbull et al. 2011, 2014). Overexpression of PTG in cultured mouse astrocytes lead to a large increase in glycogen levels, while down-regulation of PTG via siRNA reduced glycogen levels (Ruchti et al. 2016). PTG mRNA expression is increased in cultured mouse astrocytes by the neurotransmitters noradrenaline and VIP resulting in increased glycogen accumulation (Allaman et al. 2000). This is presumably due to dephosphorylation and activation of glycogen synthase, though this was not measured.

5.2.3 Activation by Insulin

Insulin increased glycogen accumulation in cultured astrocytes (Dringen and Hamprecht 1992) and stimulated glycogen synthase activity (Hamai et al. 1999). These effects were blocked by the phosphoinositide 3-kinase (PI3K) inhibitor, Wortmannin. In another study, the increase in glycogen was shown to occur without an increase in glucose flux across the plasma membrane (Muhic et al. 2015). Studies in skeletal muscle indicated that insulin activates glycogen synthase by promoting dephosphorylation of the protein which reduces the concentration of glucose-6-P required to activate the enzyme (Lawrence 1992). In muscle, phosphorylation at sites 3 (a + b + c) and site 2 and/or 2a are decreased in response to insulin (reviewed in Lawrence and Roach 1997). While early studies suggested that MAPK was important for the insulin stimulated dephosphorylation of glycogen synthase (reviewed in Lawrence and Roach 1997), more recent studies suggest that the process is GSK3-dependent (reviewed in Roach et al. 2012). In vitro, GSK3 phosphorylates sites 3a, 3b, 3c, and 4 (Hemmings et al. 1981). Insulin, acting throughout the protein kinase B (Akt/PKB) pathway promotes phosphorylation of GSK3 at a serine residue that results in inhibition of the kinase (Woodgett 2005). Support for this mechanism in insulin-induced phosphorylation comes from studies in which the beta isoform of GSK3 was either over expressed or disrupted in skeletal muscle. Overexpression led to less active glycogen synthase (Pearce et al. 2004), while disruption resulted in enhanced insulin-stimulated activation of glycogen synthase (Patel et al. 2008). There are two informs of GSK3, both of which are expressed in brain (Yao et al. 2002). In a mouse model where both isoforms of GSK3 had their inactivating phosphorylation site mutated to alanine, GSK3 was constitutively active. In skeletal muscle, this had no effect on the basal activity of glycogen synthase but blunted the insulin-stimulated increase in glycogen synthase activity (McManus et al. 2005). This effect on glycogen synthase activity occurred with the mutation of GSK3 beta but not alpha. The phosphorylation of both GSK3 alpha and beta in mouse cerebral cortex and hippocampus increased with insulin administration (Clodfelder-Miller et al. 2005). Glycogen synthase activity was not assayed in this study. While these studies indicate an important role of GSK3 in insulin-mediated regulation of glycogen synthase, there may be other kinases and PP1c, in conjunction with glycogen targeting subunits, involved (reviewed in Roach et al. 2012).

5.2.4 Other Covalent Modifications

In addition to covalent phosphorylation, modification of glycogen synthase by O-linked attachment of N-acetylglucosamine was reported to inactivate the enzyme (Parker et al. 2003). This glucosylation of glycogen synthase is increased by high glucose levels in 3 T3-L1 adipocytes and by hyperglycemia in mouse adipose tissue. Therefore, this modification may contribute to the reduced activation of glycogen synthase that occurs in insulin resistance. In addition, acetylation of lysine residues in human liver glycogen synthase has been reported (Zhao et al. 2010a). Neither of these modifications has been reported for glycogen synthase in the brain.

5.2.5 Translocation

Another mechanism by which glycogen synthase is regulated, at least in muscle and liver, is by translocation of the enzyme between different sub cellular compartments. There is evidence for the localization of glycogen synthase to be regulated by the concentration of glycogen, glucose, and glucose-6-P and by phosphorylation. Green fluorescent protein (GFP) labeled GYS1 was localized in the nucleus of C2C12 and COS-1 cells in the absence of glucose while in the presence of glucose the enzyme was localized in the cytosol (Ferrer et al. 1997). In contrast, GYS2 in isolated rat hepatocytes was localized in the cytosol without added glucose and translocated to the cell periphery in response to added glucose (Fernandez-Novell et al. 1997). The changes in cellular localization of GYS2 correlated with intracellular levels of glucose-6-P (Fernandez-Novell et al. 1996). Nielsen et al. (2001) found that GYS1 location is influenced by muscle glycogen content. Glycogen synthase translocated from a glycogen enriched membrane fraction to a cytoskeleton fraction with a decrease in glycogen levels. Immunofluorescence of single EDL fibers indicated that larger aggregates of glycogen synthase protein were present when glycogen levels were high. A role of phosphorylation of glycogen synthase in affecting the subcellular distribution of the enzyme is not clear. In resting rabbit muscle, GYS1 was mainly found in the perinuclear region and myofibrillar cross-striations if GYS1 was phosphorylated at site 1a or at sites 3a and 3b (Prats et al. 2005, 2009). However, if GYS1 was phosphorylated at sites 2 and 2a, localization exhibited a diffuse distribution pattern, not associated with the perinuclear region or cross-striations. In contrast, Cid et al. (2005) reported that phosphorylation of GYS1 is not involved in subcellular distribution. Mutation of all 9 serine residues to alanine did not affect localization of glycogen synthase. However, mutation to alanine of the arginine residues involved in conferring glucose-6-P sensitivity to glycogen synthase, resulted in glycogen synthase moving from the nucleus to the cytoplasm more quickly that wild type glycogen synthase after addition of glucose. The nuclear accumulation of GYS1 in cultured cells is in the form of spherical aggregates and GYS1 co-localizes with p80-coilin (Cid et al. 2005). P80-coilin associates with promyelocytic leukemia (PML) bodies that have been implicated in several processes including transcriptional regulation (Ruggero et al. 2000). Cid et al. (2005) suggest that glycogen synthase may act as an energy sensor, translocating to the nucleus to perform a moonlighting function when glycogen stores are depleted. A role for GYS1 in translation has also been proposed. GYS1 phosphorylated on Ser640 was preferentially associated with elongating ribosomes and depletion of GYS1 from HeLa cells affected the translation of a subset of mRNA's (Fuchs et al. 2011). The study's authors suggest that GYS1 provides a feedback loop between the energy status of the cell and translation. In the brain, little has been reported regarding translocation of glycogen synthase. In astrocytes and neurons, there was a decrease in the glycogen synthase nucleus/cytosol ratio when astrocytes and neurons were co-cultured (Mamczur et al. 2015). The significance of this is not known and further studies will be required to determine whether translocation of glycogen synthase within astrocytes and neurons occurs in physiological or pathological conditions.

6 Glycogen Metabolism During Development

Embryonic/fetal development. Claude Bernard first isolated and described the distribution of glycogen in fetal tissues in 1859 (Bernard 1859). Most of the early work was histochemical (reviewed in Cremer 1902; Needham 1931) and it was thought that these glycogen stores were essential for growth and differentiation. This idea waned with reports suggesting that the levels of glycogen in embryonic tissue were not higher than in adults (Needham 1931). However, with the advent of biochemical techniques for quantitative analysis of glycogen, the concentration of glycogen has been found to change during fetal development and/or after birth in many tissues, including the brain (Shelley 1961). In the human fetus, brain glycogen levels decreased from 5 μM/g at 9 weeks to <1 μM/g at 22 weeks (Villee 1954). In embryonic chick brain, glycogen changes with developmental age, increasing tenfold from day 6 until days 18 and 20 (Rinaudo et al. 1969). Another increase in brain glycogen concentration occurs from day 20 of development to hatching at 21 days (Edwards and Rogers 1972; Rinaudo et al. 1969). In the rat, the concentration of glycogen increased ~twofold in 18.5 dpc fetuses as compared to 14.5 fetuses (Gutierrez-Correa et al. 1991). Also in rat brain, brain glycogen increased from 5.5 to 7.5 μmol/g tissue in 17 dpc to 20 dpc fetuses, respectively, before dropping to 6.5 μmol/g tissue at term (22 dpc) (Kohle and Vannucci 1977). Histochemically detectable brain glycogen also increased during rat fetal development. Periodic acid-Schiff (PAS) staining for glycogen in developing rat brain revealed regional differences and age dependent changes (Bruckner and Biesold 1981). Glycogen was observed in epithelial cells of the immature chorioid plexus of ventricles I–IV and radial glial cells in the midbrain and medullary raph6 of the brain stem at 14 dpc

(Bruckner and Biesold 1981). As fetal development progressed, these areas maintained glycogen while radial fibers in the brain stem and the cerebral cortex stored glycogen, though to a lesser degree (Bruckner and Biesold 1981). In mouse choroid plexus, glycogen was detected by electron microscopy beginning at 13.5 dpc and is present throughout the remainder of fetal development where it is first located in what Sturrock (1979) referred to as membrane-bound bags, and later was dispersed in the cytoplasm. In the fetal mouse, brain glycogen levels of ~4 mg/g tissue were measured and were not greatly altered in 14.5 to 18.5 dpc UBC swiss mice (Tye and Burton 1980). In C57BL/6 J fetal mouse brain, glycogen levels were measured over this same gestational time with a trend to increase from the 14.5 to 18.5 days of gestation (author's unpublished observations). In general, large glycogen stores accumulate in tissues late in fetal life and have been postulated to have an important function as fuel sources during asphyxia and after the newborn is separated at birth from glucose provided from the placenta (Dawes 1968).

The key enzymes of glycogen synthesis and degradation have been monitored in developing chick and rat. Glycogen synthase activity in the embryonic chick brain doubled from the sixth to 11th day of development, fell to sixth day levels on day 13 and then slowly returned to peak levels at hatching (Rinaudo et al. 1969). Glycogen phosphorylase activity was low at day 6 and rose sharply from day 11 to 15 and then again from day 20 to hatching (Rinaudo et al. 1969). Levels at hatching were 25% below adult levels (Rinaudo et al. 1969). Glycogen synthesis is more active than phosphorylase for the entirety of chick embryonic development (Rinaudo et al. 1969). In another study in chick brain, glycogen synthase and glycogenin activity exhibited more modest changes during fetal development (Carrizo et al. 1997). In the rat, glycogen phosphorylase activity increased and glycogen synthase tended to be less active in 18.5 dpc as compared to 14.5 fetuses (Gutierrez-Correa et al. 1991).

Postnatal development. In postnatal chicks, Rinaudo and Cussotto (1971) reported that brain glycogen levels remain constant for the first 36 h after hatching (during which time the chick does not eat) as well as after 4 days of consuming a normal diet. In contrast, Edwards and Rogers (Edwards and Rogers 1972) reported that glycogen levels fell by 2 days after hatching and continued to decline up to 21 days of age. The decrease from 1 to 2 days of age occurred in all of the brain regions examined, with the largest drop in the cerebellum. Glycogen synthase activity increased almost twofold from hatching to 8 days of age (Rinaudo 1974). Brain glycogen levels (~3 mg/g tissue) were up to ~threefold higher in fetal rabbit than in 5-day old suckling, 15-day old weaning age, and adult rabbit (Curto et al. 1994). Glycogen levels in rat brain were decreased 10 min following vaginal birth but not cesarean section, suggesting that cerebral hypoxia experienced during delivery is the cause of the decrease in brain glycogen (Kohle and Vannucci 1977). Levels rebounded by 30 min after birth and then fell over the next 7 days where they remained constant for 1 week before rising again at 30 days of age. The change in brain glycogen levels from newborn to adult is region- specific in cats and dogs (Chesler and Himwich 1943). In both species, glycogen levels increased in the cortex and decreased in thalamus, cerebellum, medulla, and spinal cord with aging from newborn to adult.

7 Glycogen Synthase Genetic Modifications

Glycogen storage disease type 0. Two families have been identified with loss of function mutations in GYS1. In one of these families, three siblings had a homozygous stop mutation (R426X). The eldest male child died of cardiac arrest at 10.5 years of age. He experienced an episode of tonic-clonic seizures at age 4. Electroencephalogram (EEG) showed bilateral epileptogenic spikes in the parietal and occipital regions. Subsequent EEG's were normal while his gross motor performance was impaired. A brother exhibited heart abnormalities and muscle fatigue at 11 years of age. He had an IQ of 73 (normal is >70). A 2-year-old sister was asymptomatic. The younger siblings exhibited lack of glycogen, predominance of oxidative fibers, and mitochondrial proliferation (Kollberg et al. 2007). An 8-year-old patient from another family was identified with a homozygous two base pair deletion in exon 2 (c.162-163AG) of GYS1, which is expected to produce a protein 162 amino acids in length (Cameron et al. 2009). He collapsed and died during a bout of exercise. In skeletal muscle, glycogen was depleted and there was a total loss of glycogen synthase protein in patient fibroblasts. There was a proliferation of mitochondria and type 1 fiber predominance in skeletal muscle. There were no neurological issues noted. A 12 year-old girl with intellectual disability that suffered from myoclonic epilepsy since infancy was found to have PAS-negative muscle fibers, consistent with depleted glycogen stores, and increased oxidative enzyme activity (Michelson-Kerman et al. 2003). However, whether this patient had a deficiency in glycogen synthase was not determined. The connection between neurological symptoms observed in some of these patients and the presumed inability to make glycogen in the brain are not clear. In regard to a connection between epilepsy and brain glycogen levels, one patient with the inability to make glycogen had epilepsy while 19 patients with epilepsy had elevated levels of glycogen in the hippocampus, as compared to gray and white matter. This finding lead Dalsgaard et al. (2007) to propose that brain glycogen is a prerequisite for sustained neuronal activity as occurs in epileptic seizures. However, in mice lacking GYS1 in the brain, kainate-induced seizure susceptibility was increased as compared to wild type controls (Lopez-Ramos et al. 2015). Regarding, a connection between brain glycogen and cognitive function, patients that presented with cognitive impairment are consistent with studies in rodents where disruption of glycogen synthesis or inhibition of glycogen utilization resulted in learning impairment (Duran et al. 2013; Suzuki et al. 2011). Additionally, it will be interesting to determine whether the lack of glycogen in the brain leads to metabolic changes in this organ as was reported in skeletal muscle from human patients with GYS1 mutations and in a mouse model disrupted for GYS1 (Pederson et al. 2005b).

Animal models with elevated glycogen storage. Animal models with increased glycogen accumulation demonstrate the interplay between the proteins and metabolites involved in glycogen metabolism. A key question regarding glycogen synthesis was the role of glucose availability versus glycogen synthase activity, a push versus pull concept. This concept has not been explored in the brain but has been studied in skeletal muscle. When a mutant form of GYS1 that cannot be effectively inhibited by phosphorylation was over expressed in skeletal muscle, glycogen accumulation increased up to fivefold along with a decrease in UDP-glucose levels and increase in glucose-6-P levels (Azpiazu et al. 2000; Manchester et al 1996). There was also a threefold increase in total glycogen phosphorylase activity. GLUT4 levels either decreased or were unchanged as was basal and insulin-stimulated uptake of 2-deoxyglucose (Lawrence Jr. et al. 1997). The glycogen was less branched showing the importance of the ratio of branching enzyme to glycogen synthase (Pederson et al. 2003). An activating mutation in GYS1 (R309H) identified in horses with polysaccharide storage myopathy (PSSM), a neuromuscular disease, leads to glycogen accumulation in skeletal muscle (McCue et al. 2008). This mutation increases activity in the absence of glucose-6-P. Muscle from affected horses paradoxically had increased phosphorylation of sites 2 + 2a, but had a lower K_m for UDP-glucose both in the presence and absence of glucose-6-P (Maile et al. 2017). These findings suggest that the mutation prevents normal inactivating effects of phosphorylation. These studies demonstrate that regulation of glycogen synthase activity impacts glycogen levels. Overexpression of GLUT1 in skeletal muscle increased glucose uptake and glycogen accumulation, without affecting glycogen synthase activity (Ren et al. 1993). This study indicated the importance of glucose uptake for glycogen synthesis. On the other hand, GLUT4 over expression in muscle increased insulin-stimulated glucose uptake but did not affect glycogen levels (Brozinick et al. 1996; Hansen et al. 1995). Another model proposed by Shulman et al. (1995) suggests that glucose transport and phosphorylation by hexokinase controls glycogen synthesis and that the role of phosphorylation control of glycogen synthase is to regulate the concentration of glucose-6-P.

Increased levels of brain glycogen were achieved in rats using an inhibitor of glycogen phosphorylase (Suh et al. 2007). The ~twofold increase in brain glycogen was associated with protection against hypoglycemia-induced neuronal degeneration. Overexpression of constitutively active glycogen synthase specifically in neurons in mouse and Drosophila resulted in glycogen accumulation, loss of neurons, impaired locomotion, and shortened lifespan (Duran et al. 2012). In horses with PSSM, there were no histological abnormalities in brains from horses with this disorder, though it is not clear whether brain glycogen was monitored (Valentine et al. 1997).

Animal models with impaired glycogen synthesis. A global knockout of GYS1 in mice (muscle glycogen synthase knockout MGSKO mice) revealed the importance of glycogen synthase in newborn survival (Pederson et al. 2004a). Approximately ninety percent of newborn pups lacking GYS1 died shortly after birth due to cardiopulmonary defects. Surviving MGSKO animals lack glycogen in tissues where GYS1 is expressed, including brain (Pederson et al. 2005b). Overexpressing GYS1 under the control of the muscle creatine kinase promoter in MGSKO mice resulted in mice (MGSKO/GSL30) with restored glycogen synthesis in cardiac and skeletal muscle, but not in brain (Pederson et al. 2004a). There is no perinatal lethality observed in these mice, indicating that brain glycogen is not required for newborn survival. This conclusion is supported by normal survival in a mouse model

(GYS1Nestin-KO) with central nervous system-specific disruption of glycogen synthase (Duran et al. 2013). In this model, brain glycogen synthase protein and activity, as well as glycogen, were undetectable and brain morphology was normal. The protein expression levels of several enzymes involved in glycogen metabolism were also monitored in the brain. The lack of brain GYS1 did not affect the levels of the brain or muscle isoforms of glycogen phosphorylase, total or phosphorylated GSK3, or glycogenin. Glycogen debranching enzyme expression was decreased and AMPK expression was increased. However, the levels of phosphorylated AMPK were normal. This is in contrast to what was observed in skeletal muscle from MGSKO mice where the levels of AMPK and Acetyl-CoA carboxylase phosphorylation were increased suggesting an increased ability to oxidize fatty acids (Pederson et al. 2005b). Because fatty acids are generally not considered to be a significant fuel for the brain, this difference between skeletal muscle and brain may not be surprising. However, it is also interesting to consider AMPK activity in light of the hypothesis that AMPK is a glycogen sensor (McBride and Hardie 2009). Muscle and brain uptake of glucose was decreased in MGSKO mice after a euglycemic-hyperglycemic clamp suggesting insulin resistance. Numerous other changes in gene expression were found in muscle from MGSKO animals (Parker et al. 2006). It will be interesting to examine what adaptations occur in the brain when the ability to synthesize glycogen has been impaired.

Despite lacking glycogen in skeletal muscle, heart, and brain, MGSKO animals had normal exercise capacity (Pederson et al. 2005a). This is of interest in light of the finding that brain glycogen decreases in rats during exercise and is hypothesized to play a role in central fatigue (Matsui et al. 2011). GYS1Nestin-KO mice were impaired in the Skinner box associative learning task and exhibited decreased long-term potentiation. This is consistent with the memory impairment reported in an inhibitory avoidance test for rats treated with an inhibitor of glycogen phosphorylase (Suzuki et al. 2011). In contrast, MGSKO and MGSKO/GSL30 mice are not impaired in the inhibitory avoidance test (author's unpublished observations). In addition, in the author's hands, learning impairment on this test by mice with central nervous-specific disruption of glycogen synthase is dependent on the strength of the aversive stimulus (author's unpublished observations). These findings provide further impetus for examining adaptations in mice lacking brain glycogen synthase. They also highlight a challenge for interpreting studies in mice that are genetically modified during development. They are beneficial models for genetic diseases but may not be as good a model for examining acquired defects. This has been demonstrated for the disruption of glycogen synthase in muscle. A tamoxifen-inducible model was generated to disrupt of GYS1 specifically in skeletal muscle of adult mice (Xirouchaki et al. 2016). This allowed the examination of the effects of reducing GYS1 while mitigating possible adaptations during development without GYS1. The 85% reduction in GYS1 protein resulted in a 70% reduction in muscle glycogen. The mice displayed a decrease in insulin-stimulated glucose uptake in muscle indicative of insulin resistance, as observed in MGSKO mice. However, in contrast to MGSKO mice, exercise performance was impaired in the tamoxifen inducible model.

Studies by Sinadinos et al. (2014) suggest that glycogen synthesis may contribute to neurological decline with aging. Global disruption of GYS1 in mouse prevents the formation of poorly branched polysaccharide-based aggregates, resembling corpora amylacae (CA) that are normally found in aged human brain. When glycogen synthase was reduced specifically in neurons in drosophila, the result was neurological improvement with age and an extended lifespan.

8 Glycogen Synthase in Disease

Deficiencies in the enzymes involved in glycogen synthesis and utilization result in glycogen storage diseases, some of which are described in this and other chapters of this volume. In addition, glycogen synthesis is affected in several disease states including diabetes, cancer, depression, Alzheimer's disease, and Huntington disease.

Diabetes. Brain glycogen synthesis has been examined in diabetes. Newly synthesized glycogen in the occipital lobe of patients with type 1 diabetes was reduced (Oz et al. 2012), perhaps due to reduced glycogen synthase activity. In cortex and retina, glycogen synthase activity in the presence of glucose-6-P was increased 50–100% in a streptozotocin-induced rat model of type 1 diabetes (Sanchez-Chavez et al. 2008). Brain glycogen supercompensation has been examined as a mechanism for hypoglycemia unawareness that occurs in this disease. A rebound, above normal levels, in brain glycogen levels has been reported after a single bout of hypoglycemia in humans (Oz et al. 2009) and in two rodent studies (Canada et al. 2011; Choi et al. 2003) but not in another rodent study (Herzog et al. 2008). In humans, this was associated with higher levels of newly synthesized brain glycogen after hypoglycemia (Oz et al. 2009). In the Otsuka Long–Evans Tokushima fatty rat model of type 2 diabetes, glycogen synthase protein expression was elevated in the hippocampus, while the phosphorylation state was not affected (Shima et al. 2017). Insulin-induced activation of glycogen synthase was impaired in patients with type 2 diabetes and in subjects with increased risk of developing the disease (reviewed in Fredriksson et al. 2004). Associations between GYS1 and type 2 diabetes have been reported. For example, two polymorphic alleles were found in the glycogen synthase gene, which were enriched in patients with type 2 diabetes (Groop et al. 1993). However, the concentration of the glycogen synthase protein in muscle biopsies was normal indicating that these alleles are markers rather than causes of type 2 diabetes. In another study, insulin-induced GYS1 mRNA expression was blunted in muscle biopsies from diabetic patients versus non-diabetic controls, as was the increased in fractional activity (Huang et al. 2000). These effects were interpreted to be secondary to the chronic hyperglycemia experienced by the patients. However, these and other studies have yielded conflicting results, thus it is not clear whether GYS1 expression is influenced by insulin (reviewed in Fredriksson et al. 2004).

Cancer. High levels of glycogen have been reported in several types of cancer, including brain tumors (reviewed in Cloix et al. 2008). Glycogen synthase activity was elevated in human glioma tumors (Beckner et al. 2005; Lowry et al. 1983) as was the level of GYS1 mRNA (Kroes et al. 2007). Decreasing glycogen synthase activity via antisense GS cDNA reduced invasiveness of human glioblastoma multiform cells and the tumor size in nude mice (Ardourel et al. 2007).

Depression. In a mouse model of depression, induced by long term exposure to high glucocorticoid levels, both glycogen concentration and glycogen synthase activity were decreased in the hippocampus (Zhang et al. 2015a). The study's authors suggest that this effect of glucocorticoids on glycogen metabolism induces the depression-like behavior in this model. In contrast, rats that were subjected to prenatal acute stress to induce depression (Detka et al. 2014) exhibited increased glycogen levels in the hippocampus and frontal cortex. The selective serotonin reuptake inhibitor (SSRI) fluoxetine, which is used to create depression, increases glycogen content in astrocytes when administered at low doses, but decreases glycogen content at higher doses (Bai et al. 2017). Thus, it is not clear what role brain glycogen may play in this disorder.

Alzheimer's disease. Several pieces of evidence lead to the hypothesis that glycogen synthesis will be inhibited in Alzheimer's disease (AD). GSK3β, which phosphorylates and inactivates glycogen synthase, is upregulated in AD (Llorens-Martin et al. 2014). In addition, there is up regulation of A2A receptors in AD (Gomes et al. 2011). Binding of adenosine to this receptor promotes glycogen breakdown in cultured astrocytes (Xu et al. 2014). Activation of astrocytic insulin receptors promotes glycogen storage (Heni et al. 2011) while insulin resistance occurs in AD. Amyloid-beta (1–42) impairs glycogen storage in human astrocytes (Zhang et al. 2015b), consistent with its effect of increasing GSK3β activity (King et al. 2014), and impairs learning in day-old chicks (Gibbs 2015). Increasing brain glycogen levels with pyruvate supplementation improved spatial learning, but impaired associative learning, in a mouse model of AD (Koivisto et al. 2016), leaving open the question as to whether boosting brain glycogen levels would be beneficial overall. In addition, CA's appear with aging and accumulates to a greater extent in AD (Inoue et al. 1996; Rai et al. 2018; Rohn 2015). Glucose contained in CA's is thought to be relatively unavailable as a fuel source. Whether CA's contribute to the cause or are a result of AD are unknown, but interestingly, in a mouse model unable to synthesize glycogen, these CA's do not appear with aging (Sinadinos et al. 2014). It is not clear whether manipulating glycogen metabolism in the brain would impact the pathology of AD.

Huntington disease. An intriguing role for glycogen synthesis in neurons in the context of Huntington disease was recently proposed (Rai et al. 2018). Overexpression of toxic huntingtin protein induced glycogen synthesis in neurons by activating glycogen synthase. A similar effect was observed after subjection of neurons to oxidative stress or blockage of the proteasome. In each case this increase in glycogen synthase activity was neuroprotective. The increase in glycogen synthase activity induced by the huntingtin protein increased neuronal autophagic flux which removed the cytoxic protein. This study suggests that activating glycogen synthase could be therapeutic in neurodegenerative disorders.

9 Conclusion

Much has been learned about glycogen synthesis since the identification of glycogen synthase more than 60 years ago. The study of this metabolic pathway has largely focused on its role in muscle and liver, with less investigation in the brain. How well findings in these other tissues translate to the brain is in many cases fertile soil for exploration. Recent discoveries showing important roles for brain glycogen make this an exciting time to study the metabolism of this seductive dynamic glucose polymer.

Acknowledgements The author thanks Dr. Peter J. Roach for kindly providing comments that have improved this chapter. Studies from the author's laboratory have been funded in part by NIH grant DK078370.

References

Akman HO, Sheiko T, Tay SK, Finegold MJ, Dimauro S, Craigen WJ (2011) Generation of a novel mouse model that recapitulates early and adult onset glycogenosis type IV. Hum Mol Genet 20:4430–4439

Allaman I, Lengacher S, Magistretti PJ, Pellerin L (2003) A2B receptor activation promotes glycogen synthesis in astrocytes through modulation of gene expression. Am J Physiol Cell Physiol 284:C696–C704

Allaman I, Pellerin L, Magistretti PJ (2000) Protein targeting to glycogen mRNA expression is stimulated by noradrenaline in mouse cortical astrocytes. Glia 30:382–391

Altafaj X, Dierssen M, Baamonde C, Marti E, Visa J, Guimera J, Oset M, Gonzalez JR, Florez J, Fillat C, Estivill X (2001) Neurodevelopmental delay, motor abnormalities and cognitive deficits in transgenic mice overexpressing Dyrk1A (minibrain), a murine model of Down's syndrome. Hum Mol Genet 10:1915–1923

Ardourel M, Blin M, Moret JL, Dufour T, Duc HT, Trojan J, Cloix JF (2007) A new putative target for antisense gene therapy of glioma: glycogen synthase. Cancer Biol Ther 6:719–723

Azpiazu I, Manchester J, Skurat AV, Roach PJ, Lawrence JC (2000) Control of glycogen synthesis is shared between glucose transport and glycogen synthase in skeletal muscle fibers. Am J Physiol Endocrinol Metab 278:E234–EE43

Bai Q, Song D, Gu L, Verkhratsky A, Peng L (2017) Bi-phasic regulation of glycogen content in astrocytes via Cav-1/PTEN/PI3K/AKT/GSK-3beta pathway by fluoxetine. Psychopharmacology 234:1069–1077

Barbetti F, Rocchi M, Bossolasco M, Cordera R, Sbraccia P, Finelli P, Consalez GG (1996) The human skeletal muscle glycogenin gene: cDNA, tissue expression and chromosomal localization. Biochem Biophys Res Commun 220:72–77

Baskaran S, Chikwana VM, Contreras CJ, Davis KD, Wilson WA, DePaoli-Roach AA, Roach PJ, Hurley TD (2011) Multiple glycogen-binding sites in eukaryotic glycogen synthase are required for high catalytic efficiency toward glycogen. J Biol Chem 286:33999–34006

Baskaran S, Roach PJ, DePaoli-Roach AA, Hurley TD (2010) Structural basis for glucose-6-phosphate activation of glycogen synthase. Proc Natl Acad Sci U S A 107:17563–17568

Basu DK, Bachhawat BK (1961) Purification of uridine diphosphoglucose-glycogen transglucosylase from sheep brain. Biochim Biophys Acta 50:123–128

Beckner ME, Gobbel GT, Abounader R, Burovic F, Agostino NR, Laterra J, Pollack IF (2005) Glycolytic glioma cells with active glycogen synthase are sensitive to PTEN and inhibitors of PI3K and gluconeogenesis. Lab Investig 85:1457–1470

Bernard C (1859) De la matiere glycogene consideree comme condition de developpement de certains tissues chez le foetus avant l'apparition de la fonction glycogenique de foie. C R Hebd Seances Acad Sci 48:673–684

Bjorbaek C, Echwald SM, Hubricht P, Vestergaard H, Hansen T, Zierath J, Pedersen O (1994) Genetic variants in promoters and coding regions of the muscle glycogen synthase and the insulin-responsive GLUT4 genes in NIDDM. Diabetes 43:976–983

Bouskila M, Hirshman MF, Jensen J, Goodyear LJ, Sakamoto K (2008) Insulin promotes glycogen synthesis in the absence of GSK3 phosphorylation in skeletal muscle. Am J Physiol Endocrinol Metab 294:E28–E35

Bouskila M, Hunter RW, Ibrahim AF, Delattre L, Peggie M, van Diepen JA, Voshol PJ, Jensen J, Sakamoto K (2010) Allosteric regulation of glycogen synthase controls glycogen synthesis in muscle. Cell Metab 12:456–466

Breckenridge BM, Crawford EJ (1961) The quantitative histochemistry of the brain. J Neurochem 7:234–240

Browner MF, Nakano K, Bang AG, Fletterick RJ (1989) Human muscle glycogen synthase cDNA sequence: a negatively charged protein with an asymmetric charge distribution. Proc Natl Acad Sci U S A 86:1443–1447

Brozinick JT Jr, Yaspelkis BB, Wilson CM 3rd, Grant KE, Gibbs EM, Cushman SW, Ivy JL (1996) Glucose transport and GLUT4 protein distribution in skeletal muscle of GLUT4 transgenic mice. Biochem J 313(Pt 1):133–140

Bruckner G, Biesold D (1981) Histochemistry of glycogen deposition in perinatal rat brain: importance of radial glial cells. J Neurocytol 10:749–757

Cameron JM, Levandovskiy V, MacKay N, Utgikar R, Ackerley C, Chiasson D, Halliday W, Raiman J, Robinson BH (2009) Identification of a novel mutation in GYS1 (muscle-specific glycogen synthase) resulting in sudden cardiac death, that is diagnosable from skin fibroblasts. Mol Genet Metab 98:378–382

Canada SE, Weaver SA, Sharpe SN, Pederson BA (2011) Brain glycogen supercompensation in the mouse after recovery from insulin-induced hypoglycemia. J Neurosci Res 89:585–591

Carrizo ME, Romero JM, Miozzo MC, Brocco M, Panzetta P, Curtino JA (1997) Biosynthesis of proteoglycogen: modulation of glycogenin expression in the developing chicken. Biochem Biophys Res Commun 240:142–145

Ceulemans H, Bollen M (2004) Functional diversity of protein phosphatase-1, a cellular economizer and reset button. Physiol Rev 84:1–39

Chan EM, Ackerley CA, Lohi H, Ianzano L, Cortez MA, Shannon P, Scherer SW, Minassian BA (2004) Laforin preferentially binds the neurotoxic starch-like polyglucosans, which form in its absence in progressive myoclonus epilepsy. Hum Mol Genet 13:1117–1129

Chan EM, Young EJ, Ianzano L, Munteanu I, Zhao X, Christopoulos CC, Avanzini G, Elia M, Ackerley CA, Jovic NJ, Bohlega S, Andermann E, Rouleau GA, Delgado-Escueta AV, Minassian BA, Scherer SW (2003) Mutations in NHLRC1 cause progressive myoclonus epilepsy. Nat Genet 35:125–127

Chesler A, Himwich HE (1943) The glycogen content of various parts of the central nervous system of dogs and cats at different ages. Arch Biochem 2:175

Chikwana VM, Khanna M, Baskaran S, Tagliabracci VS, Contreras CJ, Depaoli-Roach A, Roach PJ, Hurley TD (2013) Structural basis for 2′-phosphate incorporation into glycogen by glycogen synthase. Proc Natl Acad Sci U S A 110:20976–20981

Choeiri C, Staines W, Messier C (2002) Immunohistochemical localization and quantification of glucose transporters in the mouse brain. Neuroscience 111:19–34

Choi IY, Seaquist ER, Gruetter R (2003) Effect of hypoglycemia on brain glycogen metabolism in vivo. J Neurosci Res 72:25–32

Chown EE (2018) Identification of therapeutic targets in a glycogen storage disease type IV mouse model. University of Toronto, Toronto, ON
Cid E, Cifuentes D, Baque S, Ferrer JC, Guinovart JJ (2005) Determinants of the nucleocytoplasmic shuttling of muscle glycogen synthase. FEBS J 272:3197–3213
Clodfelder-Miller B, De Sarno P, Zmijewska AA, Song L, Jope RS (2005) Physiological and pathological changes in glucose regulate brain Akt and glycogen synthase kinase-3. J Biol Chem 280:39723–39731
Cloix JF, Ardourel M, Hevor T (2008) Glycogen metabolism and brain pathologies. Cent Nerv Syst Agents Med Chem 8:187–197
Contreras CJ, Segvich DM, Mahalingan K, Chikwana VM, Kirley TL, Hurley TD, DePaoli-Roach AA, Roach PJ (2016) Incorporation of phosphate into glycogen by glycogen synthase. Arch Biochem Biophys 597:21–29
Cremer M (1902) Physiologie des Glykogens. Ergebnisse der Physiol 1:803
Cruz NF, Dienel GA (2002) High glycogen levels in brains of rats with minimal environmental stimuli: implications for metabolic contributions of working astrocytes. J Cereb Blood Flow Metab 22:1476–1489
Curto M, Piccinini M, Mioletti S, Mostert M, Bruno R, Ricciardi MP, Rinaudo MT (1994) Levels of carnitine and glycogen in rabbit tissues during development. Int J Biochem 26:163–169
Dalsgaard MK, Madsen FF, Secher NH, Laursen H, Quistorff B (2007) High glycogen levels in the hippocampus of patients with epilepsy. J Cereb Blood Flow Metab 27:1137–1141
Daran JM, Dallies N, Thines-Sempoux D, Paquet V, Francois J (1995) Genetic and biochemical characterization of the UGP1 gene encoding the UDP-glucose pyrophosphorylase from Saccharomyces cerevisiae. Eur J Biochem 233:520–530
Dawes GS (1968) Energy metabolism in foetus and after birth. In: Foetal and neonatal physiology. Year Book Medical Publishers Inc., Chicago
DeLano WL (2002) The PyMOL molecular graphics system. DeLano Scientific, Palo Alto, CA
DePaoli-Roach AA, Contreras CJ, Segvich DM, Heiss C, Ishihara M, Azadi P, Roach PJ (2015) Glycogen phosphomonoester distribution in mouse models of the progressive myoclonic epilepsy, Lafora disease. J Biol Chem 290:841–850
DePaoli-Roach AA, Tagliabracci VS, Segvich DM, Meyer CM, Irimia JM, Roach PJ (2010) Genetic depletion of the malin E3 ubiquitin ligase in mice leads to lafora bodies and the accumulation of insoluble laforin. J Biol Chem 285:25372–25381
Detka J, Kurek A, Basta-Kaim A, Kubera M, Lason W, Budziszewska B (2014) Elevated brain glucose and glycogen concentrations in an animal model of depression. Neuroendocrinology 100:178–190
Dickinson ME, Flenniken AM, Ji X, Teboul L, Wong MD, White JK, Meehan TF, Weninger WJ, Westerberg H, Adissu H, Baker CN, Bower L, Brown JM, Caddle LB, Chiani F, Clary D, Cleak J, Daly MJ, Denegre JM, Doe B, Dolan ME, Edie SM, Fuchs H, Gailus-Durner V, Galli A, Gambadoro A, Gallegos J, Guo S, Horner NR, Hsu CW, Johnson SJ, Kalaga S, Keith LC, Lanoue L, Lawson TN, Lek M, Mark M, Marschall S, Mason J, McElwee ML, Newbigging S, Nutter LM, Peterson KA, Ramirez-Solis R, Rowland DJ, Ryder E, Samocha KE, Seavitt JR, Selloum M, Szoke-Kovacs Z, Tamura M, Trainor AG, Tudose I, Wakana S, Warren J, Wendling O, West DB, Wong L, Yoshiki A, Consortium International Mouse Phenotyping, Laboratory Jackson, Institut Clinique de la Souris Infrastructure Nationale Phenomin, Laboratories Charles River, M. R. C. Harwell, Toronto Centre for Phenogenomics, Institute Wellcome Trust Sanger, Riken BioResource Center, MacArthur DG, Tocchini-Valentini GP, Gao X, Flicek P, Bradley A, Skarnes WC, Justice MJ, Parkinson HE, Moore M, Wells S, Braun RE, Svenson KL, de Angelis MH, Herault Y, Mohun T, Mallon AM, Henkelman RM, Brown SD, Adams DJ, Lloyd KC, McKerlie C, Beaudet AL, Bucan M, Murray SA (2016) High-throughput discovery of novel developmental phenotypes. Nature 537:508–514
DiNuzzo M, Giove F, Maraviglia B, Mangia S (2015) Monoaminergic control of cellular glucose utilization by glycogenolysis in neocortex and hippocampus. Neurochem Res 40:2493–2504

Dringen R, Hamprecht B (1992) Glucose, insulin, and insulin-like growth factor I regulate the glycogen content of astroglia-rich primary cultures. J Neurochem 58:511–517
Duran J, Gruart A, Garcia-Rocha M, Delgado-Garcia JM, Guinovart JJ (2014) Glycogen accumulation underlies neurodegeneration and autophagy impairment in Lafora disease. Hum Mol Genet 23:3147–3156
Duran J, Saez I, Gruart A, Guinovart JJ, Delgado-Garcia JM (2013) Impairment in long-term memory formation and learning-dependent synaptic plasticity in mice lacking glycogen synthase in the brain. J Cereb Blood Flow Metab 33:550–556
Duran J, Tevy MF, Garcia-Rocha M, Calbo J, Milan M, Guinovart JJ (2012) Deleterious effects of neuronal accumulation of glycogen in flies and mice. EMBO Mol Med 4:719–729
Edwards C, Rogers KJ (1972) Some factors influencing brain glycogen in the neonate chick. J Neurochem 19:2759–2766
Farrell CL, Pardridge WM (1991) Blood-brain barrier glucose transporter is asymmetrically distributed on brain capillary endothelial lumenal and ablumenal membranes: an electron microscopic immunogold study. Proc Natl Acad Sci U S A 88:5779–5783
Fernandez-Novell JM, Bellido D, Vilaro S, Guinovart JJ (1997) Glucose induces the translocation of glycogen synthase to the cell cortex in rat hepatocytes. Biochem J 321(Pt 1):227–231
Fernandez-Novell JM, Roca A, Bellido D, Vilaro S, Guinovart JJ (1996) Translocation and aggregation of hepatic glycogen synthase during the fasted-to-refed transition in rats. Eur J Biochem 238:570–575
Ferrer JC, Baque S, Guinovart JJ (1997) Muscle glycogen synthase translocates from the cell nucleus to the cystosol in response to glucose. FEBS Lett 415:249–252
Franken P, Gip P, Hagiwara G, Ruby NF, Heller HC (2003) Changes in brain glycogen after sleep deprivation vary with genotype. Am J Physiol Regul Integr Comp Physiol 285:R413–R419
Fredriksson J, Ridderstrale M, Groop L, Orho-Melander M (2004) Characterization of the human skeletal muscle glycogen synthase gene (GYS1) promoter. Eur J Clin Investig 34:113–121
Fuchs G, Diges C, Kohlstaedt LA, Wehner KA, Sarnow P (2011) Proteomic analysis of ribosomes: translational control of mRNA populations by glycogen synthase GYS1. J Mol Biol 410:118–130
Gayarre J, Duran-Trio L, Criado Garcia O, Aguado C, Juana-Lopez L, Crespo I, Knecht E, Bovolenta P, Rodriguez de Cordoba S (2014) The phosphatase activity of laforin is dispensable to rescue Epm2a−/− mice from Lafora disease. Brain 137:806–818
Gentry MS, Worby CA, Dixon JE (2005) Insights into Lafora disease: malin is an E3 ubiquitin ligase that ubiquitinates and promotes the degradation of laforin. Proc Natl Acad Sci U S A 102:8501–8506
Ghajar JB, Plum F, Duffy TE (1982) Cerebral oxidative metabolism and blood flow during acute hypoglycemia and recovery in unanesthetized rats. J Neurochem 38:397–409
Gibbons BJ, Roach PJ, Hurley TD (2002) Crystal structure of the autocatalytic initiator of glycogen biosynthesis, glycogenin. J Mol Biol 319:463–477
Gibbs M (2015) Reflections on glycogen and beta-amyloid: why does glycogenolytic beta2-adrenoceptor stimulation not rescue memory after beta-amyloid? Metab Brain Dis 30:345–352
Goldberg ND, O'Toole AG (1969) The properties of glycogen synthetase and regulation of glycogen biosynthesis in rat brain. J Biol Chem 244:3053–3061
Gomes CV, Kaster MP, Tome AR, Agostinho PM, Cunha RA (2011) Adenosine receptors and brain diseases: neuroprotection and neurodegeneration. Biochim Biophys Acta 1808:1380–1399
Groop LC, Kankuri M, Schalin-Jantti C, Ekstrand A, Nikula-Ijas P, Widen E, Kuismanen E, Eriksson J, Franssila-Kallunki A, Saloranta C et al (1993) Association between polymorphism of the glycogen synthase gene and non-insulin-dependent diabetes mellitus. N Engl J Med 328:10–14
Guinovart JJ, Salavert A, Massague J, Ciudad CJ, Salsas E, Itarte E (1979) Glycogen synthase: a new activity ratio assay expressing a high sensitivity to the phosphorylation state. FEBS Lett 106:284–288

Gutierrez-Correa J, Hod M, Passoneau JV, Freinkel N (1991) Glycogen and enzymes of glycogen metabolism in rat embryos and fetal organs. Biol Neonate 59:294–302
Hamai M, Minokoshi Y, Shimazu T (1999) L-Glutamate and insulin enhance glycogen synthesis in cultured astrocytes from the rat brain through different intracellular mechanisms. J Neurochem 73:400–407
Hammerle B, Elizalde C, Galceran J, Becker W, Tejedor FJ (2003) The MNB/DYRK1A protein kinase: neurobiological functions and Down syndrome implications. J Neural Transm Suppl:129–137
Hanashiro I, Roach PJ (2002) Mutations of muscle glycogen synthase that disable activation by glucose 6-phosphate. Arch Biochem Biophys 397:286–292
Hansen PA, Gulve EA, Marshall BA, Gao J, Pessin JE, Holloszy JO, Mueckler M (1995) Skeletal muscle glucose transport and metabolism are enhanced in transgenic mice overexpressing the Glut4 glucose transporter. J Biol Chem 270:1679–1684
Hansen PA, Marshall BA, Chen M, Holloszy JO, Mueckler M (2000) Transgenic overexpression of hexokinase II in skeletal muscle does not increase glucose disposal in wild-type or Glut1-overexpressing mice. J Biol Chem 275:22381–22386
Hemmings BA, Yellowlees D, Kernohan JC, Cohen P (1981) Purification of glycogen synthase kinase 3 from rabbit skeletal muscle. Copurification with the activating factor (FA) of the (Mg-ATP) dependent protein phosphatase. Eur J Biochem 119:443–451
Heni M, Hennige AM, Peter A, Siegel-Axel D, Ordelheide AM, Krebs N, Machicao F, Fritsche A, Haring HU, Staiger H (2011) Insulin promotes glycogen storage and cell proliferation in primary human astrocytes. PLoS One 6:e21594
Herzog RI, Chan O, Yu S, Dziura J, McNay EC, Sherwin RS (2008) Effect of acute and recurrent hypoglycemia on changes in brain glycogen concentration. Endocrinology 149:1499–1504
Heyen CA, Tagliabracci VS, Zhai L, Roach PJ (2009) Characterization of mouse UDP-glucose pyrophosphatase, a Nudix hydrolase encoded by the Nudt14 gene. Biochem Biophys Res Commun 390:1414–1418
Huang X, Vaag A, Hansson M, Weng J, Laurila E, Groop L (2000) Impaired insulin-stimulated expression of the glycogen synthase gene in skeletal muscle of type 2 diabetic patients is acquired rather than inherited. J Clin Endocrinol Metab 85:1584–1590
Hutchins DA, Rogers KJ (1970) Physiological and drug-induced changes in the glycogen content of mouse brain. Br J Pharmacol 39:9–25
Imagawa E, Osaka H, Yamashita A, Shiina M, Takahashi E, Sugie H, Nakashima M, Tsurusaki Y, Saitsu H, Ogata K, Matsumoto N, Miyake N (2014) A hemizygous GYG2 mutation and Leigh syndrome: a possible link? Hum Genet 133:225–234
Inoue M, Yagishita S, Itoh Y, Amano N, Matsushita M (1996) Coexistence of paired helical filaments and polyglucosan bodies in the same neuron in an autopsy case of Alzheimer's disease. Acta Neuropathol 92:511–514
Inoue N, Iwasa T, Fukunaga K, Matsukado Y, Miyamoto E (1987) Phosphorylation and inactivation of brain glycogen synthase by a multifunctional calmodulin-dependent protein kinase. J Neurochem 48:981–988
Inoue N, Matsukado Y, Goto S, Miyamoto E (1988) Localization of glycogen synthase in brain. J Neurochem 50:400–405
Iwasa T, Fukunaga K, Yamamoto H, Tanaka E, Miyamoto E (1984) Ca2+, calmodulin-dependent phosphorylation, and inactivation of glycogen synthase by a brain protein kinase. Arch Biochem Biophys 235:212–217
James AP, Barnes PD, Palmer TN, Fournier PA (2008) Proglycogen and macroglycogen: artifacts of glycogen extraction? Metabolism 57:535–543
Jensen VF, Bogh IB, Lykkesfeldt J (2014) Effect of insulin-induced hypoglycaemia on the central nervous system: evidence from experimental studies. J Neuroendocrinol 26:123–150
Jorgensen SB, Nielsen JN, Birk JB, Olsen GS, Viollet B, Andreelli F, Schjerling P, Vaulont S, Hardie DG, Hansen BF, Richter EA, Wojtaszewski JF (2004) The alpha2-5'AMP-activated protein kinase is a site 2 glycogen synthase kinase in skeletal muscle and is responsive to glucose loading. Diabetes 53:3074–3081

Kaslow HR, Lesikar DD (1984) Isozymes of glycogen synthase. FEBS Lett 172:294–298
Kaslow HR, Lesikar DD, Antwi D, Tan AW (1985) L-type glycogen synthase. Tissue distribution and electrophoretic mobility. J Biol Chem 260:9953–9956
Kelsall IR, Voss M, Munro S, Cuthbertson DJ, Cohen PT (2011) R3F, a novel membrane-associated glycogen targeting subunit of protein phosphatase 1 regulates glycogen synthase in astrocytoma cells in response to glucose and extracellular signals. J Neurochem 118:596–610
King MK, Pardo M, Cheng Y, Downey K, Jope RS, Beurel E (2014) Glycogen synthase kinase-3 inhibitors: rescuers of cognitive impairments. Pharmacol Ther 141:1–12
Knull HR, Khandelwal RL (1982) Glycogen metabolizing enzymes in brain. Neurochem Res 7:1307–1317
Kohle SJ, Vannucci RC (1977) Glycogen metabolism in fetal and postnatal rat brain: influence of birth. J Neurochem 28:441–443
Koivisto H, Leinonen H, Puurula M, Hafez HS, Barrera GA, Stridh MH, Waagepetersen HS, Tiainen M, Soininen P, Zilberter Y, Tanila H (2016) Chronic pyruvate supplementation increases exploratory activity and brain energy reserves in young and middle-aged mice. Front Aging Neurosci 8:41
Kollberg G, Tulinius M, Gilljam T, Ostman-Smith I, Forsander G, Jotorp P, Oldfors A, Holme E (2007) Cardiomyopathy and exercise intolerance in muscle glycogen storage disease 0. N Engl J Med 357:1507–1514
Kong J, Shepel PN, Holden CP, Mackiewicz M, Pack AI, Geiger JD (2002) Brain glycogen decreases with increased periods of wakefulness: implications for homeostatic drive to sleep. J Neurosci 22:5581–5587
Korrodi-Gregorio L, Esteves SL, Fardilha M (2014) Protein phosphatase 1 catalytic isoforms: specificity toward interacting proteins. Transl Res 164:366–391
Kroes RA, Dawson G, Moskal JR (2007) Focused microarray analysis of glyco-gene expression in human glioblastomas. J Neurochem 103(Suppl 1):14–24
Lafora GR, Glueck B (1911) Beitrag zur histopathologie der myoklonischen epilepsie. Z Gesamte Neurol Psychiatr 6:1–14
Lawrence JC Jr (1992) Signal transduction and protein phosphorylation in the regulation of cellular metabolism by insulin. Annu Rev Physiol 54:177–193
Lawrence JC Jr, Roach PJ (1997) New insights into the role and mechanism of glycogen synthase activation by insulin. Diabetes 46:541–547
Lawrence JC Jr, Skurat AV, Roach PJ, Azpiazu I, Manchester J (1997) Glycogen synthase: activation by insulin and effect of transgenic overexpression in skeletal muscle. Biochem Soc Trans 25:14–19
Lee YC, Chang CJ, Bali D, Chen YT, Yan YT (2011) Glycogen-branching enzyme deficiency leads to abnormal cardiac development: novel insights into glycogen storage disease IV. Hum Mol Genet 20:455–465
Lehto M, Stoffel M, Groop L, Espinosa R 3rd, Le Beau MM, Bell GI (1993) Assignment of the gene encoding glycogen synthase (GYS) to human chromosome 19, band q13.3. Genomics 15:460–461
Leino RL, Gerhart DZ, van Bueren AM, McCall AL, Drewes LR (1997) Ultrastructural localization of GLUT 1 and GLUT 3 glucose transporters in rat brain. J Neurosci Res 49:617–626
Leloir LF, Olavarria JM, Goldemberg SH, Carminatti H (1959) Biosynthesis of glycogen from uridine diphosphate glucose. Arch Biochem Biophys 81:508–520
Llorens-Martin M, Jurado J, Hernandez F, Avila J (2014) GSK-3beta, a pivotal kinase in Alzheimer disease. Front Mol Neurosci 7:46
Lomako J, Lomako WM, Whelan WJ, Dombro RS, Neary JT, Norenberg MD (1993) Glycogen synthesis in the astrocyte: from glycogenin to proglycogen to glycogen. FASEB J 7:1386–1393
Lopez-Ramos JC, Duran J, Gruart A, Guinovart JJ, Delgado-Garcia JM (2015) Role of brain glycogen in the response to hypoxia and in susceptibility to epilepsy. Front Cell Neurosci 9:431

Lowry OH, Berger SJ, Carter JG, Chi MM, Manchester JK, Knor J, Pusateri ME (1983) Diversity of metabolic patterns in human brain tumors: enzymes of energy metabolism and related metabolites and cofactors. J Neurochem 41:994–1010
Lowry OH, Passonneau JV, Hasselberger FX, Schulz DW (1964) Effect of Ischemia on known substrates and cofactors of the glycolytic pathway in brain. J Biol Chem 239 18–30
Lust WD, Passonneau JV (1976) Cyclic nucleotides in murine brain: effect of hypothermia on adenosine 3′,5′ monophosphate, glycogen phosphorylase, glycogen synthase and metabolites following maximal electroshock or decapitation. J Neurochem 26:11–16
Lust WD, Passonneau JV, Veech RL (1973) Cyclic adenosine monphosphate, metabolites, and phosphorylase in neural tissue: a comparison a methods of fixation. Science 181:280–282
Magistretti PJ, Sorg O, Martin JL (1993) Regulation of glycogen metabolism in astrocytes:physiological, pharmacological, and pathological aspects. San Diego, Academic Press, Inc.
Mahalingan KK, Baskaran S, DePaoli-Roach AA, Roach PJ, Hurley TD (2017) Redox switch for the inhibited state of yeast glycogen synthase mimics regulation by phosphorylation. Biochemistry 56:179–188
Maile CA, Hingst JR, Mahalingan KK, O'Reilly AO, Cleasby ME, Mickelson JR, McCue ME, Anderson SM, Hurley TD, Wojtaszewski JFP, Piercy RJ (2017) A highly prevalent equine glycogen storage disease is explained by constitutive activation of a mutant glycogen synthase. Biochim Biophys Acta 1861:3388–3398
Malfatti E, Nilsson J, Hedberg-Oldfors C, Hernandez-Lain A, Michel F, Dominguez-Gonzalez C, Viennet G, Akman HO, Kornblum C, Van den Bergh P, Romero NB Engel AG, DiMauro S, Oldfors A (2014) A new muscle glycogen storage disease associated with glycogenin-1 deficiency. Ann Neurol 76:891–898
Mamczur P, Borsuk B, Paszko J, Sas Z, Mozrzymas J, Wisniewski JR, Gizak A, Rakus D (2015) Astrocyte-neuron crosstalk regulates the expression and subcellular localization of carbohydrate metabolism enzymes. Glia 63:328–340
Manchester J, Skurat AV, Roach P, Hauschka SD, Lawrence JC Jr (1996) Increased glycogen accumulation in transgenic mice overexpressing glycogen synthase in skeletal muscle. Proc Natl Acad Sci U S A 93:10707–10711
Mathiisen TM, Lehre KP, Danbolt NC, Ottersen OP (2010) The perivascular astroglial sheath provides a complete covering of the brain microvessels: an electron microscopic 3D reconstruction. Glia 58:1094–1103
Matsui T, Soya S, Okamoto M, Ichitani Y, Kawanaka K, Soya H (2011) Brain glycogen decreases during prolonged exercise. J Physiol 589:3383–3393
McBride A, Hardie DG (2009) AMP-activated protein kinase—a sensor of glycogen as well as AMP and ATP? Acta Physiol (Oxf) 196:99–113
McCue ME, Valberg SJ, Miller MB, Wade C, DiMauro S, Akman HO, Mickelson JR (2008) Glycogen synthase (GYS1) mutation causes a novel skeletal muscle glycogenosis. Genomics 91:458–466
McEwen BS, Reagan LP (2004) Glucose transporter expression in the central nervous system: relationship to synaptic function. Eur J Pharmacol 490:13–24
McManus EJ, Sakamoto K, Armit LJ, Ronaldson L, Shpiro N, Marquez R, Alessi DR (2005) Role that phosphorylation of GSK3 plays in insulin and Wnt signalling defined by knockin analysis. EMBO J 24:1571–1583
Melendez-Hevia E, Waddell TG, Shelton ED (1993) Optimization of molecular design in the evolution of metabolism: the glycogen molecule. Biochem J 295(Pt 2):477–483
Michelson-Kerman M, Watemberg N, Nissenkorn A, Gilad E, Sadeh M, Lerman-Sagie T (2003) Muscle glycogen depletion and increased oxidative phosphorylation following status epilepticus. J Child Neurol 18:876–878
Minassian BA (2001) Lafora's disease: towards a clinical, pathologic, and molecular synthesis. Pediatr Neurol 25:21–29

Minassian BA, Lee JR, Herbrick JA, Huizenga J, Soder S, Mungall AJ, Dunham I, Gardner R, Fong CY, Carpenter S, Jardim L, Satishchandra P, Andermann E, Snead OC 3rd, Lopes-Cendes I, Tsui LC, Delgado-Escueta AV, Rouleau GA, Scherer SW (1998) Mutations in a gene encoding a novel protein tyrosine phosphatase cause progressive myoclonus epilepsy. Nat Genet 20:171–174

Mitchell JB, Lupica CR, Dunwiddie TV (1993) Activity-dependent release of endogenous adenosine modulates synaptic responses in the rat hippocampus. J Neurosci 13:3439–3447

Morgenthaler FD, Koski DM, Kraftsik R, Henry PG, Gruetter R (2006) Biochemical quantification of total brain glycogen concentration in rats under different glycemic states. Neurochem Int 48:616–622

Moslemi AR, Lindberg C, Nilsson J, Tajsharghi H, Andersson B, Oldfors A (2010) Glycogenin-1 deficiency and inactivated priming of glycogen synthesis. N Engl J Med 362:1203–1210

Mu J, Skurat AV, Roach PJ (1997) Glycogenin-2, a novel self-glucosylating protein involved in liver glycogen biosynthesis. J Biol Chem 272:27589–27597

Muhic M, Vardjan N, Chowdhury HH, Zorec R, Kreft M (2015) Insulin and insulin-like growth factor 1 (IGF-1) modulate cytoplasmic glucose and glycogen levels but not glucose transport across the membrane in astrocytes. J Biol Chem 290:11167–11176

Munro S, Ceulemans H, Bollen M, Diplexcito J, Cohen PT (2005) A novel glycogen-targeting subunit of protein phosphatase 1 that is regulated by insulin and shows differential tissue distribution in humans and rodents. FEBS J 272:1478–1489

Needham J (1931) Chemical embryology. University Press, Cambridge

Nelson SR, Schulz DW, Passonneau JV, Lowry OH (1968) Control of glycogen levels in brain. J Neurochem 15:1271–1279

Nielsen JN, Derave W, Kristiansen S, Ralston E, Ploug T, Richter EA (2001) Glycogen synthase localization and activity in rat skeletal muscle is strongly dependent on glycogen content. J Physiol 531:757–769

Nilsson J, Halim A, Moslemi AR, Pedersen A, Nilsson J, Larson G, Oldfors A (2012) Molecular pathogenesis of a new glycogenosis caused by a glycogenin-1 mutation. Biochim Biophys Acta 1822:493–499

Nitschke F, Sullivan MA, Wang P, Zhao X, Chown EE, Perri AM, Israelian L, Juana-Lopez L, Bovolenta P, Rodriguez de Cordoba S, Steup M, Minassian BA (2017) Abnormal glycogen chain length pattern, not hyperphosphorylation, is critical in Lafora disease. EMBO Mol Med 9:906–917

Nordstrom CH, Siesjo BK (1978) Effects of phenobarbital in cerebral ischemia. Part I: cerebral energy metabolism during pronounced incomplete ischemia. Stroke 9:327–335

Oe Y, Baba O, Ashida H, Nakamura KC, Hirase H (2016) Glycogen distribution in the microwave-fixed mouse brain reveals heterogeneous astrocytic patterns. Glia 64:1532–1545

Ondruskova N, Honzik T, Vondrackova A, Tesarova M, Zeman J, Hansikova H (2014) Glycogen storage disease-like phenotype with central nervous system involvement in a PGM1-CDG patient. Neuro Endocrinol Lett 35:137–141

Orhan Akman H, Emmanuele V, Kurt YG, Kurt B, Sheiko T, DiMauro S, Craigen WJ (2015) A novel mouse model that recapitulates adult-onset glycogenosis type 4. Hum Mol Genet 24:6801–6810

Orho M, Nikula-Ijas P, Schalin-Jantti C, Permutt MA, Groop LC (1995) Isolation and characterization of the human muscle glycogen synthase gene. Diabetes 44:1099–1105

Oz G, Kumar A, Rao JP, Kodl CT, Chow L, Eberly LE, Seaquist ER (2009) Human brain glycogen metabolism during and following hypoglycemia. Diabetes 58:1978–1985

Oz G, Tesfaye N, Kumar A, Deelchand DK, Eberly LE, Seaquist ER (2012) Brain glycogen content and metabolism in subjects with type 1 diabetes and hypoglycemia unawareness. J Cereb Blood Flow Metab 32:256–263

Parker GE, Pederson BA, Obayashi M, Schroeder JM, Harris RA, Roach PJ (2006) Gene expression profiling of mice with genetically modified muscle glycogen content. Biochem J 395:137–145

Parker GJ, Lund KC, Taylor RP, McClain DA (2003) Insulin resistance of glycogen synthase mediated by o-linked N-acetylglucosamine. J Biol Chem 278:10022–10027
Passonneau JV, Schwartz JP, Rottenberg DA (1975) The partial purification and properties of pig brain glycogen synthase. J Biol Chem 250:2287–2292
Patel S, Doble BW, MacAulay K, Sinclair EM, Drucker DJ, Woodgett JR (2008) Tissue-specific role of glycogen synthase kinase 3beta in glucose homeostasis and insulin action. Mol Cell Biol 28:6314–6328
Pearce NJ, Arch JR, Clapham JC, Coghlan MP, Corcoran SL, Lister CA, Llano A, Moore GB, Murphy GJ, Smith SA, Taylor CM, Yates JW, Morrison AD, Harper AJ, Roxbee-Cox L, Abuin A, Wargent E, Holder JC (2004) Development of glucose intolerance in male transgenic mice overexpressing human glycogen synthase kinase-3beta on a muscle-specific promoter. Metabolism 53:1322–1330
Pederson BA, Chen H, Schroeder JM, Shou W, DePaoli-Roach AA, Roach PJ (2004a) Abnormal cardiac development in the absence of heart glycogen. Mol Cell Biol 24:7179–7187
Pederson BA, Cheng C, Wilson WA, Roach PJ (2000) Regulation of glycogen synthase. Identification of residues involved in regulation by the allosteric ligand glucose-6-P and by phosphorylation. J Biol Chem 275:27753–27761
Pederson BA, Cope CR, Schroeder JM, Smith MW, Irimia JM, Thurberg BL, Depaoli-Roach AA, Roach PJ (2005a) Exercise capacity of mice genetically lacking muscle glycogen synthase: in mice, muscle glycogen is not essential for exercise. J Biol Chem 280:17260–17265
Pederson BA, Csitkovits AG, Simon R, Schroeder JM, Wang W, Skurat AV, Roach PJ (2003) Overexpression of glycogen synthase in mouse muscle results in less branched glycogen. Biochem Biophys Res Commun 305:826–830
Pederson BA, Schroeder JM, Parker GE, Smith MW, Depaoli-Roach AA, Roach PJ (2005b) Glucose metabolism in mice lacking muscle glycogen synthase. Diabetes 54:3466–3473
Pederson BA, Turnbull J, Epp JR, Weaver SA, Zhao X, Pencea N, Roach PJ, Frankland P, Ackerley CA, Minassian BA (2013) Inhibiting glycogen synthesis prevents lafora disease in a mouse model. Ann Neurol 74:297–300
Pederson BA, Wilson WA, Roach PJ (2004b) Glycogen synthase sensitivity to glucose-6-P is important for controlling glycogen accumulation in Saccharomyces cerevisiae. J Biol Chem 279:13764–13768
Pellegri G, Rossier C, Magistretti PJ, Martin JL (1996) Cloning, localization and induction of mouse brain glycogen synthase. Brain Res Mol Brain Res 38:191–199
Pescador N, Villar D, Cifuentes D, Garcia-Rocha M, Ortiz-Barahona A, Vazquez S, Ordonez A, Cuevas Y, Saez-Morales D, Garcia-Bermejo ML, Landazuri MO, Guinovart J, del Peso L (2010) Hypoxia promotes glycogen accumulation through hypoxia inducible factor (HIF)-mediated induction of glycogen synthase 1. PLoS One 5:e9644
Petit JM, Tobler I, Kopp C, Morgenthaler F, Borbely AA, Magistretti PJ (2010) Metabolic response of the cerebral cortex following gentle sleep deprivation and modafinil administration. Sleep 33:901–908
Pfeiffer-Guglielmi B, Dombert B, Jablonka S, Hausherr V, van Thriel C, Schobel N, Jansen RP (2014) Axonal and dendritic localization of mRNAs for glycogen-metabolizing enzymes in cultured rodent neurons. BMC Neurosci 15:70
Poitry-Yamate C, Lei H, Gruetter R (2009) The rate-limiting step for glucose transport into the hypothalamus is across the blood-hypothalamus interface. J Neurochem 109(Suppl 1):38–45
Prats C, Cadefau JA, Cusso R, Qvortrup K, Nielsen JN, Wojtaszewski JF, Hardie DG, Stewart G, Hansen BF, Ploug T (2005) Phosphorylation-dependent translocation of glycogen synthase to a novel structure during glycogen resynthesis. J Biol Chem 280:23165–23172
Prats C, Helge JW, Nordby P, Qvortrup K, Ploug T, Dela F, Wojtaszewski JF (2009) Dual regulation of muscle glycogen synthase during exercise by activation and compartmentalization. J Biol Chem 284:15692–15700

Rai A, Singh PK, Singh V, Kumar V, Mishra R, Thakur AK, Mahadevan A, Shankar SK, Jana NR, Ganesh S (2018) Glycogen synthase protects neurons from cytotoxicity of mutant huntingtin by enhancing the autophagy flux. Cell Death Dis 9:201
Ren JM, Marshall BA, Gulve EA, Gao J, Johnson DW, Holloszy JO, Mueckler M (1993) Evidence from transgenic mice that glucose transport is rate-limiting for glycogen deposition and glycolysis in skeletal muscle. J Biol Chem 268:16113–16115
Reynolds TH 4th, Pak Y, Harris TE, Manchester J, Barrett EJ, Lawrence JC Jr (2005) Effects of insulin and transgenic overexpression of UDP-glucose pyrophosphorylase on UDP-glucose and glycogen accumulation in skeletal muscle fibers. J Biol Chem 280:5510–5515
Rinaudo MT (1974) Proceedings: regulation of the metabolism of carbohydrates in the chick embryo. Ital J Biochem 23:266–267
Rinaudo MT, Giunta C, Bozzi ML, Bruno R (1969) Enzymes of glycogen synthesis and of glycogenolysis in skeletal muscle, heart and brain of chicken embryo. Enzymologia 36:321–331
Rinaudo MT, Cussotto L (1971) Livelli di glicogeno in tessuti di embrioni di polio di 19 giorni e di pulcini alla schiusa, di 36 e di 132 ore. Boll Soc Ital Biol Sper 48:5–7
Roach PJ (1990) Control of glycogen synthase by hierarchal protein phosphorylation. FASEB J 4:2961–2968
Roach PJ (1991) Multisite and hierarchal protein phosphorylation. J Biol Chem 266:14139–14142
Roach PJ (2011) Are there errors in glycogen biosynthesis and is laforin a repair enzyme? FEBS Lett 585:3216–3218
Roach PJ (2015) Glycogen phosphorylation and Lafora disease. Mol Asp Med 46:78–84
Roach PJ, Depaoli-Roach AA, Hurley TD, Tagliabracci VS (2012) Glycogen and its metabolism: some new developments and old themes. Biochem J 441:763–787
Roach PJ, Skurat AV (1997) Self-glucosylating initiator proteins and their role in glycogen biosynthesis. Prog Nucleic Acid Res Mol Biol 57:289–316
Roach RJ, Larner J (1977) Covalent phosphorylation in the regulation glycogen synthase activity. Mol Cell Biochem 15:179–200
Rohn TT (2015) Corpora amylacea in neurodegenerative diseases: cause or effect? Int J Neurol Neurother 2
Rubio-Villena C, Garcia-Gimeno MA, Sanz P (2013) Glycogenic activity of R6, a protein phosphatase 1 regulatory subunit, is modulated by the laforin-malin complex. Int J Biochem Cell Biol 45:1479–1488
Ruchti E, Roach PJ, Depaoli-Roach AA, Magistretti PJ, Allaman I (2016) Protein targeting to glycogen is a master regulator of glycogen synthesis in astrocytes. IBRO Rep 1:46–63
Ruggero D, Wang ZG, Pandolfi PP (2000) The puzzling multiple lives of PML and its role in the genesis of cancer. BioEssays 22:827–835
Saez I, Duran J, Sinadinos C, Beltran A, Yanes O, Tevy MF, Martinez-Pons C, Milan M, Guinovart JJ (2014) Neurons have an active glycogen metabolism that contributes to tolerance to hypoxia. J Cereb Blood Flow Metab 34:945–955
Sagar SM, Sharp FR, Swanson RA (1987) The regional distribution of glycogen in rat brain fixed by microwave irradiation. Brain Res 417:172–174
Sanchez-Chavez G, Hernandez-Berrones J, Luna-Ulloa LB, Coffe V, Salceda R (2008) Effect of diabetes on glycogen metabolism in rat retina. Neurochem Res 33:1301–1308
Schlender KK, Reimann EM (1977) Glycogen synthase kinases. Distribution in mammalian tissues of forms that are independent of cyclic AMP. J Biol Chem 252:2384–2389
Schworer CM, McClure RW, Soderling TR (1985) Calmodulin-dependent protein kinases purified from rat brain and rabbit liver. Arch Biochem Biophys 242:137–145
Serratosa JM, Gomez-Garre P, Gallardo ME, Anta B, de Bernabe DB, Lindhout D, Augustijn PB, Tassinari CA, Malafosse RM, Topcu M, Grid D, Dravet C, Berkovic SF, de Cordoba SR (1999) A novel protein tyrosine phosphatase gene is mutated in progressive myoclonus epilepsy of the Lafora type (EPM2). Hum Mol Genet 8:345–352
Shelley HJ (1961) Glycogen reserves and their changes at birth and in anoxia. Br Med Bull 17:137–143

Shima T, Matsui T, Jesmin S, Okamoto M, Soya M, Inoue K, Liu YF, Torres-Aleman I, McEwen BS, Soya H (2017) Moderate exercise ameliorates dysregulated hippocampal glycometabolism and memory function in a rat model of type 2 diabetes. Diabetologia 60:597–606

Shulman RG, Bloch G, Rothman DL (1995) In vivo regulation of muscle glycogen synthase and the control of glycogen synthesis. Proc Natl Acad Sci U S A 92:8535–8542

Shulman RG, Rothman DL (2001) The "glycogen shunt" in exercising muscle: a role for glycogen in muscle energetics and fatigue. Proc Natl Acad Sci U S A 98:457–461

Sinadinos C, Valles-Ortega J, Boulan L, Solsona E, Tevy MF, Marquez M, Duran J, Lopez-Iglesias C, Calbo J, Blasco E, Pumarola M, Milan M, Guinovart JJ (2014) Neuronal glycogen synthesis contributes to physiological aging. Aging Cell 13:935–945

Singh TJ, Huang KP (1985) Glycogen synthase (casein) kinase-1: tissue distribution and subcellular localization. FEBS Lett 190:84–88

Skurat AV, Dietrich AD (2004) Phosphorylation of Ser640 in muscle glycogen synthase by DYRK family protein kinases. J Biol Chem 279:2490–2498

Skurat AV, Dietrich AD, Roach PJ (2006) Interaction between glycogenin and glycogen synthase. Arch Biochem Biophys 456:93–97

Skurat AV, Roach PJ (1995) Phosphorylation of sites 3a and 3b (Ser640 and Ser644) in the control of rabbit muscle glycogen synthase. J Biol Chem 270:12491–12497

Skurat AV, Wang Y, Roach PJ (1994) Rabbit skeletal muscle glycogen synthase expressed in COS cells. Identification of regulatory phosphorylation sites. J Biol Chem 269:25534–25542

Smith CH, Brown NE, Larner J (1971) Molecular characteristics of the totally dependent and independent forms of glycogen synthase of rabbit skeletal muscle. II. Some chemical characteristics of the enzyme protein and of its change on interconversion. Biochim Biophys Acta 242:81–88

Smythe C, Cohen P (1991) The discovery of glycogenin and the priming mechanism for glycogen biogenesis. Eur J Biochem 200:625–631

Sorg O, Magistretti PJ (1992) Vasoactive intestinal peptide and noradrenaline exert long-term control on glycogen levels in astrocytes: blockade by protein synthesis inhibition. J Neurosci 12:4923–4931

Stojkovic T, Vissing J, Petit F, Piraud M, Orngreen MC, Andersen G, Claeys KG, Wary C, Hogrel JY, Laforet P (2009) Muscle glycogenosis due to phosphoglucomutase 1 deficiency. N Engl J Med 361:425–427

Striano P, Zara F, Turnbull J, Girard JM, Ackerley CA, Cervasio M, De Rosa G, Del Basso-De Caro ML, Striano S, Minassian BA (2008) Typical progression of myoclonic epilepsy of the Lafora type: a case report. Nat Clin Pract Neurol 4:106–111

Sturrock RR (1979) A morphological study of the development of the mouse choroid plexus. J Anat 129:777–793

Suh SW, Bergher JP, Anderson CM, Treadway JL, Fosgerau K, Swanson RA (2007) Astrocyte glycogen sustains neuronal activity during hypoglycemia: studies with the glycogen phosphorylase inhibitor CP-316,819 ([R-R*,S*]-5-Chloro-N-[2-hydroxy-3-(methoxymethylamino)-3-oxo-1-(phenylmet hyl)propyl]-1H-indole-2-carboxamide). J Pharmacol Exp Therap 321:45–50

Suzuki A, Stern SA, Bozdagi O, Huntley GW, Walker RH, Magistretti PJ, Alberini CM (2011) Astrocyte-neuron lactate transport is required for long-term memory formation. Cell 144:810–823

Swaab DF (1971) Pitfalls in the use of rapid freezing for stopping brain and spinal cord metabolism in rat and mouse. J Neurochem 18:2085–2092

Tagliabracci VS, Girard JM, Segvich D, Meyer C, Turnbull J, Zhao X, Minassian BA, Depaoli-Roach AA, Roach PJ (2008) Abnormal metabolism of glycogen phosphate as a cause for Lafora disease. J Biol Chem 283:33816–33825

Tagliabracci VS, Heiss C, Karthik C, Contreras CJ, Glushka J, Ishihara M, Azadi P, Hurley TD, DePaoli-Roach AA, Roach PJ (2011) Phosphate incorporation during glycogen synthesis and Lafora disease. Cell Metab 13:274–282

Tagliabracci VS, Turnbull J, Wang W, Girard JM, Zhao X, Skurat AV, Delgado-Escueta AV, Minassian BA, Depaoli-Roach AA, Roach PJ (2007) Laforin is a glycogen phosphatase, deficiency of which leads to elevated phosphorylation of glycogen in vivo. Proc Natl Acad Sci U S A 104:19262–19266

Takeuchi T (1965) '[Histochemical observations on glycogen metabolism in nerve tissue]', Shinkei Kenkyu No Shimpo. Adv Neurologic Sci 9:695–699

Tegtmeyer LC, Rust S, van Scherpenzeel M, Ng BG, Losfeld ME, Timal S, Raymond K, He P, Ichikawa M, Veltman J, Huijben K, Shin YS, Sharma V, Adamowicz M, Lammens M, Reunert J, Witten A, Schrapers E, Matthijs G, Jaeken J, Rymen D, Stojkovic T, Laforet P, Petit F, Aumaitre O, Czarnowska E, Piraud M, Podskarbi T, Stanley CA, Matalon R, Burda P, Seyyedi S, Debus V, Socha P, Sykut-Cegielska J, van Spronsen F, de Meirleir L, Vajro P, DeClue T, Ficicioglu C, Wada Y, Wevers RA, Vanderschaeghe D, Callewaert N, Fingerhut R, van Schaftingen E, Freeze HH, Morava E, Lefeber DJ, Marquardt T (2014) Multiple phenotypes in phosphoglucomutase 1 deficiency. N Engl J Med 370:533–542

Teschemacher AG, Gourine AV, Kasparov S (2015) A role for astrocytes in sensing the brain microenvironment and neuro-metabolic integration. Neurochem Res 40:2386–2393

Testoni G, Duran J, Garcia-Rocha M, Vilaplana F, Serrano AL, Sebastian D, Lopez-Soldado I, Sullivan MA, Slebe F, Vilaseca M, Munoz-Canoves P, Guinovart JJ (2017) Lack of glycogenin causes glycogen accumulation and muscle function impairment. Cell Metab 26(256–66):e4

Tolmasky DS, Labriola C, Krisman CR (1998) Glycogen brain branching enzyme. Cell Mol Biol (Noisy-le-Grand) 44:455–460

Tolmasky DS, Mendonca MH, Salmoral EM, Cura JA, Krisman C (1991) A new enzymatic activity participating in the initiation of glycogen biosynthesis in rat brain. Cell Mol Biol 37:433–444

Turnbull J, DePaoli-Roach AA, Zhao X, Cortez MA, Pencea N, Tiberia E, Piliguian M, Roach PJ, Wang P, Ackerley CA, Minassian BA (2011) PTG depletion removes Lafora bodies and rescues the fatal epilepsy of Lafora disease. PLoS Genet 7:e1002037

Turnbull J, Epp JR, Goldsmith D, Zhao X, Pencea N, Wang P, Frankland PW, Ackerley CA, Minassian BA (2014) PTG protein depletion rescues malin-deficient Lafora disease in mouse. Ann Neurol 75:442–446

Turnbull J, Girard JM, Lohi H, Chan EM, Wang P, Tiberia E, Omer S, Ahmed M, Bennett C, Chakrabarty A, Tyagi A, Liu Y, Pencea N, Zhao X, Scherer SW, Ackerley CA, Minassian BA (2012) Early-onset Lafora body disease. Brain 135:2684–2698

Turnbull J, Tiberia E, Striano P, Genton P, Carpenter S, Ackerley CA, Minassian BA (2016) Lafora disease. Epileptic Disord 18:38–62

Tye LM, Burton AF (1980) Glycogen deposition in fetal mouse tissues and the effect of dexamethasone. Biol Neonate 38:265–269

Valentine BA, Credille KM, Lavoie JP, Fatone S, Guard C, Cummings JF, Cooper BJ (1997) Severe polysaccharide storage myopathy in Belgian and Percheron draught horses. Equine Vet J 29:220–225

Vannucci SJ, Maher F, Simpson IA (1997) Glucose transporter proteins in brain: delivery of glucose to neurons and glia. Glia 21:2–21

Veech RL, Harris RL, Veloso D, Veech EH (1973) Freeze-blowing: a new technique for the study of brain in vivo. J Neurochem 20:183–188

Vilchez D, Ros S, Cifuentes D, Pujadas L, Valles J, Garcia-Fojeda B, Criado-Garcia O, Fernandez-Sanchez E, Medrano-Fernandez I, Dominguez J, Garcia-Rocha M, Soriano E, Rodriguez de Cordoba S, Guinovart JJ (2007) Mechanism suppressing glycogen synthesis in neurons and its demise in progressive myoclonus epilepsy. Nat Neurosci 10:1407–1413

Villar-Palasi C, Larner J (1958) A uridine coenzyme-linked pathway of glycogen synthesis in muscle. Biochim Biophys Acta 30:449

Villee CA (1954) The intermediary metabolism of human fetal tissues. Cold Spring Harb Symp Quant Biol 19:186–199

Walls AB, Heimburger CM, Bouman SD, Schousboe A, Waagepetersen HS (2009) Robust glycogen shunt activity in astrocytes: effects of glutamatergic and adrenergic agents. Neuroscience 158:284–292

Wang J, Stuckey JA, Wishart MJ, Dixon JE (2002) A unique carbohydrate binding domain targets the lafora disease phosphatase to glycogen. J Biol Chem 277:2377–2380
Wang W, Roach PJ (2004) Glycogen and related polysaccharides inhibit the laforin dual-specificity protein phosphatase. Biochem Biophys Res Commun 325:726–730
Wilson JE (2003) Isozymes of mammalian hexokinase: structure, subcellular localization and metabolic function. J Exp Biol 206:2049–2057
Woodgett JR (2005) Recent advances in the protein kinase B signaling pathway. Curr Opin Cell Biol 17:150–157
Xirouchaki CE, Mangiafico SP, Bate K, Ruan Z, Huang AM, Tedjosiswoyo BW, Lamont B, Pong W, Favaloro J, Blair AR, Zajac JD, Proietto J, Andrikopoulos S (2016) Impaired glucose metabolism and exercise capacity with muscle-specific glycogen synthase 1 (gys1) deletion in adult mice. Mol Metab 5:221–232
Xu J, Song D, Bai Q, Zhou L, Cai L, Hertz L, Peng L (2014) Role of glycogenolysis in stimulation of ATP release from cultured mouse astrocytes by transmitters and high K+ concentrations. ASN Neuro 6:e00132
Yao HB, Shaw PC, Wong CC, Wan DC (2002) Expression of glycogen synthase kinase-3 isoforms in mouse tissues and their transcription in the brain. J Chem Neuroanat 23:291–297
Zeqiraj E, Sicheri F (2015) Getting a handle on glycogen synthase - its interaction with glycogenin. Mol Asp Med 46:63–69
Zeqiraj E, Tang X, Hunter RW, Garcia-Rocha M, Judd A, Deak M, von Wilamowitz-Moellendorff A, Kurinov I, Guinovart JJ, Tyers M, Sakamoto K, Sicheri F (2014) Structural basis for the recruitment of glycogen synthase by glycogenin. Proc Natl Acad Sci U S A 111:E2831–E2840
Zhai L, Choi CS, Irimia-Dominguez J, McGuire AC, Kim S, Bock CB, Roach PJ, Shulman GI, DePaoli-Roach AA (2007) Enhanced insulin sensitivity and energy expenditure in PPP1R3C (PTG) deleted mice. Diabetes 56:A62
Zhai L, Mu J, Zong H, DePaoli-Roach AA, Roach PJ (2000) Structure and chromosomal localization of the human glycogenin-2 gene GYG2. Gene 242:229–235
Zhang H, Zhao Y, Wang Z, Huang Y (2015a) Chronic corticosterone exposure reduces hippocampal glycogen level and induces depression-like behavior in mice. Biomed Biotechnol 16:62–69
Zhang Q, Guo S, Zhang X, Tang S, Wang L, Han X, Shao W, Cong L, Du Y (2015b) Amyloid beta oligomer-induced ERK1/2-dependent serine 636/639 phosphorylation of insulin receptor substrate-1 impairs insulin signaling and glycogen storage in human astrocytes. Gene 561:76–81
Zhang WM, Browner MF, Fletterick RJ, DePaoli-Roach AA, Roach PJ (1989) Primary structure of rabbit skeletal muscle glycogen synthase deduced from cDNA clones. FASEB J 3:2532–2536
Zhang Y, Gu J, Wang L, Zhao Z, Pan Y, Chen Y (2017) Ablation of PPP1R3G reduces glycogen deposition and mitigates high-fat diet induced obesity. Mol Cell Endocrinol 439:133–140
Zhao S, Xu W, Jiang W, Yu W, Lin Y, Zhang T, Yao J, Zhou L, Zeng Y, Li H, Li Y, Shi J, An W, Hancock SM, He F, Qin L, Chin J, Yang P, Chen X, Lei Q, Xiong Y, Guan KL (2010a) Regulation of cellular metabolism by protein lysine acetylation. Science 327:1000–1004
Zhao Y, Fung C, Shin D, Shin BC, Thamotharan S, Sankar R, Ehninger D, Silva A, Devaskar SU (2010b) Neuronal glucose transporter isoform 3 deficient mice demonstrate features of autism spectrum disorders. Mol Psychiatry 15:286–299

The Structure and the Regulation of Glycogen Phosphorylases in Brain

Cécile Mathieu, Jean-Marie Dupret, and Fernando Rodrigues-Lima

Abstract Glycogen constitutes the main store of glucose in animal cells. Being present at much lower concentrations in the brain than in liver and muscles, brain glycogen has long been considered as an emergency source of glucose, mobilized under stress conditions (including hypoglyceamia). Nevertheless, over the past decade, multiple studies have shed a new light on the roles of brain glycogen, being notably an energy supply critical for high-cognitive processes such as learning and memory consolidation. Glycogen phosphorylase (GP) is the key enzyme regulating the mobilization of glycogen in cells. It is found in humans as three isozymes: muscle (mGP), liver (lGP) and brain GP (bGP). In the brain, astrocytes express both mGP and bGP while neurons only express the brain isoform. Although GP isozymes are very similar, their distinct regulatory features confer them distinct metabolic functions that are strongly related to the roles of glycogen in different tissues. Here, we provide an overview of the functions, the regulations and the structures of GPs in the brain and their relation to the specific roles of glycogen in astrocytes and neurons. We also discuss novel findings concerning the specific regulations of bGP by oxidative stress, and the potential of these enzymes as therapeutic targets in the brain.

Keywords Glycogen phosphorylase · Energy metabolism · Crystal structure
Allosteric regulation · Redox regulation · Isozyme

C. Mathieu (✉)
Department of Cell & Molecular Biology, St. Jude Children's Research Hospital, Memphis, TN, USA
e-mail: Cecile.Mathieu@StJude.org

J.-M. Dupret · F. Rodrigues-Lima (✉)
Unité BFA, CNRS UMR 8251, Université de Paris, Paris, France

UFR Sciences du Vivant, Université de Paris, Paris, France
e-mail: fernando.rodrigues-lima@univ-paris-diderot.fr

M. DiNuzzo, A. Schousboe (eds.), *Brain Glycogen Metabolism*, Advances in Neurobiology 23, https://doi.org/10.1007/978-3-030-27480-1_4

1 Glycogen and its Metabolism

Glycogen is a globular polymer of glucose which constitutes the only form of glucose storage in animal cells. Discovered by Claude Bernard in 1857, glycogen constitutes the second energy store of human body, after fat tissues (Young 1957; Berg et al. 2002; Adeva-Andany et al. 2016), In human, glycogen is found in three main tissues: the liver, the muscles (including the heart), and to a lesser extent, in the brain (Berg et al. 2002; Adeva-Andany et al. 2016; Roach 2002). In the brain, glycogen is mostly located in the grey matter, in areas rich in synapses, and more particularly in the hippocampus and the *locus coeruleus*, two regions involved in learning and memory (Sagar et al. 1987; Dalsgaard et al. 2007; Kong et al. 2002). Human glycogen is associated to various functions depending on its location. In the liver, it is metabolized to glucose which regulates glycaemia. In muscles, glycogen provides the energy needed for muscle contraction. In the brain, glycogen has long been considered as an emergency store of glucose, mobilized to protect neurons from stress conditions including hypoglycemia or hypoxic stress (Berg et al. 2002; Adeva-Andany et al. 2016; Roach 2002; Obel et al. 2012; Öz et al. 2015).

Yet, it has become clear that cerebral glycogen supports neuronal activity and normal brain functions as it is notably involved in the regulation of awake-sleep cycles as well as in high cognitive processes such as learning and memory consolidation (Kong et al. 2002; Bellesi et al. 2018; Gibbs 2015; Gibbs and Hutchinson 2012; Gibbs et al. 2007) (DiNuzzo et al., this issue). Consequently, the absence or the lack of mobilization of glycogen in brain alter the various processes in which it is involved (Gibbs et al. 2006). On the other hand, glycogen accumulation is also deleterious for cells, and particularly for neurons (Wolfsdorf and Weinstein 2003). These cells are extremely sensitive to the accumulation of glycogen in their cytoplasm, leading readily to their degeneration (Hedberg-Oldfors and Oldfors 2015; Duran and Guinovart 2015). Moreover, disease-related mutations in proteins involved in glycogen metabolism and resulting in the accumulation of glycogen (due to over-synthesis or insufficient degradation), is associated to a broad spectrum of diseases, termed glycogen storage diseases (GSD). These diseases are usually caused by mutations in a single gene and can affect one or several tissues (Wolfsdorf and Weinstein 2003). Lafora's disease and the adult polyglucosan body disease (APBD) are two GSD affecting the brain. They are both characterized by the accumulation of poorly branched glycogen granules (Duran and Guinovart 2015; Roach 2015; Klein 1993; Moses and Parvari 2002). The limited number of ramifications significantly reduces their solubility and favors a fibrillary structure, allowing them to aggregate, but also to be resistant to degradation (Hedberg-Oldfors and Oldfors 2015; Cavanagh 1999). Therefore, a tight regulation of glycogen metabolism is required to ensure the maintenance of a functional but not deleterious pool of glycogen in the cells.

The metabolism of glycogen is the same in all tissues and is ensured by the concerted action of multiple proteins and enzymes. On one hand, glycogen is mainly synthesized from blood-derived glucose. It is successively metabolized into glucose-6-phosphate, glucose-1-phosphate and UDP-glucose, by the hexokinase,

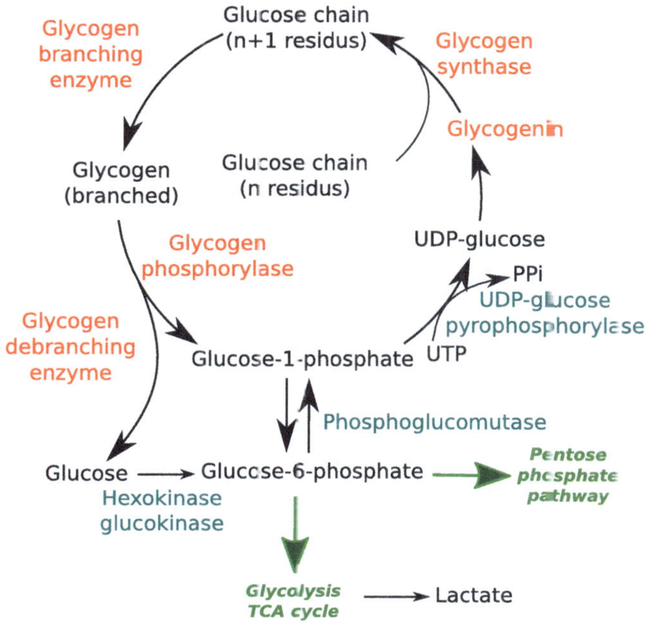

Fig. 1 Metabolism of glycogen. Schematic representation of the metabolism of glycogen and its connection with the different energy metabolism pathways

phosphoglucomutase and UDP-glucose pyrophosphorylase, respectively. In addition, glycogen may also be synthesized from glucose-6-phosphate generated from the gluconeogenesis route, an anabolic pathway that synthesizes glucose from non-carbohydrate precursors including lactate or amico-acid (Pelley 2007). UDP-glucose is then incorporated in the glycogen molecule by the glycogenin and glycogen synthase (GS) in a process is referred as glycogenesis (Fig. 1).

On the other hand, the mobilization of glycogen, or glycogenolysis, into glucose-1-phosphate is under the control of the glycogen phosphorylase (GP) and the glycogen debranching enzyme. The resulting glucose-1-phosphate can then be used in the different energy metabolism pathway, in particular the glycolysis and the TCA cycle, as well as the pentose phosphate pathway (Berg et al. 2002; Adeva-Andany et al. 2016; Roach 2002; Yip et al. 2017) (Fig. 1).

2 The Glycogen Phosphorylases

2.1 The Reaction of Phosphorolysis

At the molecular level, glycogen molecules are composed of branches of glucose subunits linked together by α,1–4 glucosidic bonds. Ramifications are formed through the establishment of α,1–6 glucosidic bonds, resulting in the formation of glycogen granules (Hedberg-Oldfors and Oldfors 2015; Meléndez-Hevia et al. 1993). GP (EC 2.4.1.1) catalyzes the reaction termed phosphorolysis, which corresponds to the cleavage of the α,1–4 bond linking two subunits of glucose and results in the formation of one subunit of glucose-1-phosphate (Fig. 2). To this end, the enzyme associates with the glycogen molecule through the non-reducing end of one branch (Berg et al. 2002; Newgard et al. 1989). However, when only 4 subunits of glucose separate the extremity of a glycogen branch from a ramification, the presence of an α,1–6 glucosidic bond blocks the degradation of the glycogen molecule by GP. At this stage, the glycogen debranching enzyme ensures the transfer of 3 of the 4 subunits of glucose on an adjacent branch and releases a subunit of glucose through the cleavage of the α,1–6 glucosidic link (Taylor et al. 1975).

Because GP is the only protein capable of degrading the glycogen molecules in cells, this enzyme appears to be a key player in glycogen metabolism. GP has raised a lot of attention in the twentieth century, as the study of this enzyme led to discovery of two important regulatory mechanisms and to the awarding of two Nobel prizes: one to Carl and Gerty Cori, in 1947, for the identification of GP, and the other one to Edmond H. Fischer and Edwin G. Krebs, in 1992, for their discovery of the reversible regulation of GP by phosphorylation. The recent findings of a functional pool of glycogen in the brain raised a new interest for GP, but also for its specific regulation in this tissue.

The phosphorolysis of glycogen also involves a cofactor: the pyridoxal phosphate (PLP). This cofactor is covalently linked to lysine 680 of the active site of GP through a Schiff base. The PLP acts as a proton donor during catalysis. Indeed, the phosphate group of PLP protonates inorganic phosphate, which then protonates the hemiacetal oxygen of the α,1–4 glucosidic bond, leading to the formation of a glucosyl carbocation. The nucleophilic attack of the inorganic phosphate on the carbocation then lead to the formation of glucose-1-phosphate (Berg et al. 2002; Newgard et al. 1989) (Fig. 2). Interestingly, GP is also capable of catalyzing the opposite reaction of synthesis of glycogen from glucose-1-phosphate. Nevertheless, this reaction does not occur *in vivo*, due to the high intracellular ratio of inorganic

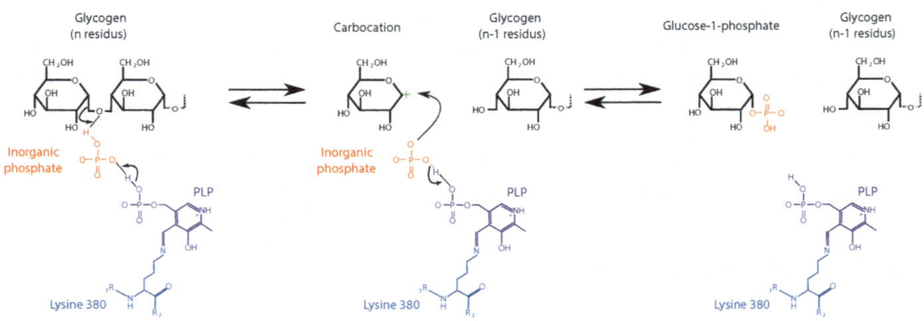

Fig. 2 Phosphorolysis of glycogen. Schematic representation of the reaction catalyzed by GP, involving the PLP as a cofactor

phosphate/glucose-1-phosphate which favors the glycogenolysis (Berg et al. 2002; Newgard et al. 1989).

2.2 *The Regulation of Glycogen Phosphorylase Isozymes*

In human, GP exists as three isoforms, encoded by three different genes and named after the tissue where they predominate: muscle (GPMM or mGP), liver (GPLL or lGP) and brain GP (GPBB or bGP). These three isoenzymes are highly conserved, sharing between 80 and 85% of sequence identity, as well as close 3D structures (Berg et al. 2002; Adeva-Andany et al. 2016; Newgard et al. 1987, 1988, 1989; Lebo et al. 1984). Nevertheless, despite their high sequence identity, the three isoforms of GP can be distinguished by their structural, regulatory and functional properties. As we will discuss later, both the muscle and the brain isoforms of GP are expressed in the brain, contrary to the liver and muscles that mainly express one isoform (Newgard et al. 1989; Saez et al. 2014; Pfeiffer-Guglielmi et al. 2003). The expression of two isoforms in the brain suggest very complex functions of glycogen in this tissue that are strongly related to the specific regulations of these enzymes.

All three isoforms of GP are allosterically-regulated enzymes, implying that they are found in at least an active state (also named R-state) and an inactive state (T-state) (Roach 2002; Newgard et al. 1989). Two modes of regulation allow the transition between the active and inactive states of the enzyme and can occur simultaneously:

- Phosphorylation of Serine 14
- Binding of allosteric effectors.

2.2.1 The Regulation of GPs by Phosphorylation

In the absence of any allosteric effector, the dephosphorylated form of the GP is inactive (Berg et al. 2002; Newgard et al. 1989; Barford et al. 1991; Barford and Johnson 1989). After phosphorylation of serine 14 by its specific kinase, the phosphorylase kinase (EC 2.7.11.19), the phosphorylated GP switches to its active state. The dephosphorylation and subsequent inactivation of GP, however, is permitted by the protein phosphatase 1 (PP1) associated to specific regulatory subunits. Indeed, PP1 is involved in the regulation of a large diversity of cellular processes, including cell division, metabolism or the regulation of ion pumps, transporter and ion channels (Heilmeyer 1987, 1991; Ceulemans and Bollen 2004). The specific regulation of glycogen metabolism by PP1 is mediated by a glycogen-targeting subunit of the PPP1R3 family. Seven different subunits with various patterns of expression, have been identified in mammals. These subunits allow the differential regulation of glycogen synthase, glycogen phosphorylase and phosphorylase kinase by PP1. Notably, PPP1R3 (also known as PPP1R3A) subunit is mainly expressed in muscles while

PPP1R4 (also known as PPP1R3B) is predominantly found in liver. On the other hand, PPP1R5/PTG as well as PPP1R6 harbor a wide distribution of expression (Doherty et al. 1995, 1996; Printen 1997; Newgard et al. 2000). Interestingly, in the brain, PPP1R6 (also known as PPP1R3D) as well as PPP1R5/PTG (also known as PPP1R3C) subunit have been shown to influence glycogen accumulation (Rubio-Villena et al. 2013; Vilchez et al. 2007).

The phosphorylation of GP is controlled by the glycogenolytic cascade, a phosphorylation cascade activated by the binding of extracellular ligands, such as glucagon or noradrenaline, to their specific receptors. The activation of the glycogenolytic cascade occurs first through the production of cAMP by the adenylate cyclase, and the subsequent activation of PKA. This kinase then initiates the phosphorylation cascade and the activation of the phosphorylase kinase. In addition, extracellular signals triggering calcium release from the endoplasmic reticulum also lead to the activation of the glycogenolytic cascade through the direct interaction of calcium ions with adenylate cyclase and phosphorylase kinase and their subsequent activation (Heilmeyer 1991) (Fig. 3).

Together, these mechanisms allow a regulation of GP according to global energy needs. In the liver, for instance, the regulation of GP by insulin and glucagon controls the release of glucose from liver glycogen and participates in the control of glycemia (van de Werve et al. 1977; Agius 2015). In muscles, the depolarization of the membrane as well as the release of noradrenaline stimulate the glycogenolytic cascade, providing energy to muscle cells during a prolonged exercise (Heilmeyer 1987; Jensen and Richter 2012). Finally, in the brain, extracellular signals including neurotransmitters and neuromodulators such as noradrenaline, serotonin, dopamine, histamine, but also VIP or extracellular variations of potassium allow the mobilization of the astrocytic store of glycogen, in response to brain activity (Allaman et al. 2003; Hutchins and Rogers 1970; Magistretti et al. 1981; Sorg and Magistretti 1991, 1992; Xu et al. 2013, 2014; Coggan et al. 2018).

2.2.2 The Regulation by Allosteric Effectors

GP activity is also regulated through the binding of allosteric effectors including AMP, ATP, glucose or glucose-6-phosphate, which reflect the local energy state of the cell. GP has multiple allosteric sites that can simultaneously bind allosteric effectors with agonistic or antagonistic effects, leading to multiple intermediate activation states, and allowing a fine control of the enzyme activity according to cellular energy requirements (Newgard et al. 1989; Barford et al. 1991; Sprang et al. 1991; Oikonomakos et al. 1995; Lukacs et al. 2006).

As a consequence, a high energy demand with a high AMP/ATP ratio results in the preferential binding of AMP in its allosteric site (the AMP binding site) and the transition/stabilization of GP in its active R-state (Barford et al. 1991; Sprang et al. 1991). The enzyme will then be able to degrade glycogen into glucose-1-phosphate

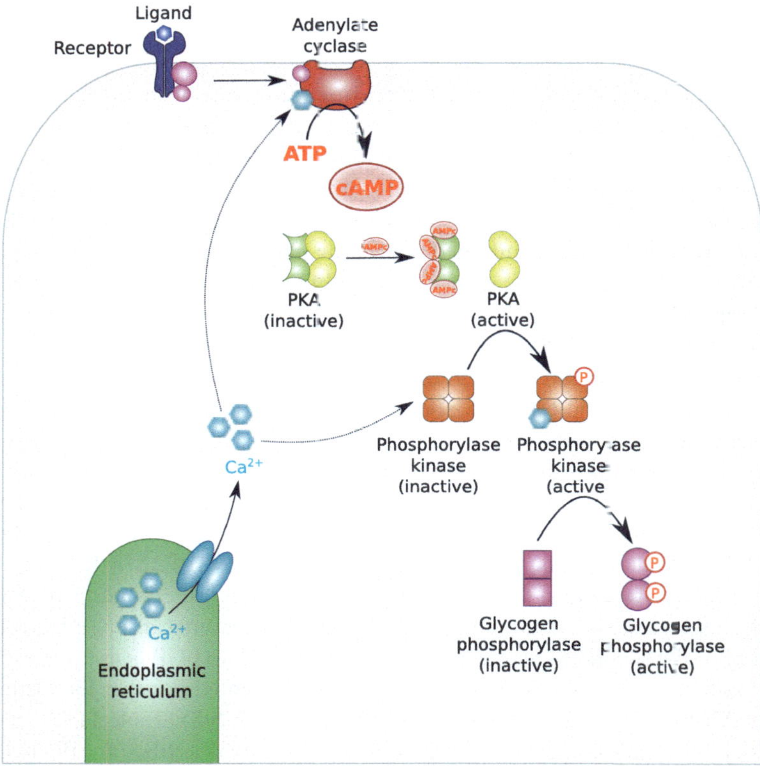

Fig. 3 Glycogenolytic cascade. The activation of GP by phosphorylation is under the control of a phosphorylation cascade, triggered by the binding of extracellular factors, including glucagon and noradrenaline, on their receptor and ultimately resulting in the activation of the enzyme

which is metabolized by the glycolytic pathway and the TCA cycle to produce ATP. On the contrary, ATP or glucose-6-phosphate, which reflect a favorable cellular energy state, can bind the same allosteric site and trigger the inactivation of the enzyme and subsequent replenishment of glycogen stores (Newgard et al. 1989).

Similarly, the binding of glucose in the catalytic site of GP results in the transition/stabilization of the enzyme to its T-state. In addition to its allosteric effect, glucose, but also glucose-1-phosphate, inhibit the activity of GP by competing with glycogen for the binding in the catalytic site (Newgard et al. 1989).

Additionally, as we mentioned above, GP can also catalyze the reaction of glycogen synthesis from glucose-1-phosphate *in vitro*. Although it might be counterintuitive, this reaction is accompanied by the transition of the GP to its R-state, triggered by the binding of glucose-1-phosphate, which acts as the reaction substrate, but also as an allosteric activator. Conversion of GP to its R-state by glucose-1-phosphate allows the activation of the enzyme and the glycogen synthesis (Newgard et al. 1989).

2.3 The Differential Regulation of the GPs

Although all three isoforms of GP are very similar in terms of sequences and structures, they differ in their sensitivity to the different modes of regulation, which appears to be related to the functions of glycogen in the concerned tissues. Indeed, in liver for instance, lGP is poorly sensitive to its regulation by allosteric effectors and responds almost exclusively to phosphorylation (Newgard et al. 1989; Lowry et al. 1964a) (Table 1). As mentioned above, the phosphorylation of GP is in direct link with both insulin and glucagon, and underlines the role of liver glycogen as an energy supply that can be mobilized to provide glucose to the whole body (van de Werve et al. 1977; Agius 2015).

The two isoforms found in the brain, mGP and bGP, are both very sensitive to the binding of the allosteric effectors and are largely regulated by these small molecules, supporting the local function of brain glycogen. Nevertheless, mGP is also strongly activated by phosphorylation, resulting in the over-activation of the enzyme in presence of AMP, while phosphorylation moderately activates bGP (Newgard et al. 1989; Crerar et al. 1995; Müller et al. 2015; Helmreich et al. 1967; Mathieu et al. 2016a). In addition, the behavior of bGP and mGP toward AMP differs. Indeed, as we will describe below, GP is a homodimer. Although the binding of AMP to mGP is cooperative (i.e. the binding of AMP on one monomer of mGP leads to the activation of the second monomer), the binding of AMP to bGP is not (Crerar et al. 1995; Mathieu et al. 2016a). Moreover, it must be pointed out that the affinity of bGP for AMP is higher than that of mGP (Crerar et al. 1995) (Table 1).

Finally, an additional layer of regulation was recently described for the bGP and appears unique to this isoform. Unlike mGP and lGP, this enzyme is highly sensitive to reactive oxygen species (ROS) such as H_2O_2, but also to environmental oxidative compounds, resulting in its reversible inhibition and the redox regulation of the glycogenolysis (Mathieu et al. 2016b, 2017a). This catalytic regulation involves the formation of an intramolecular disulfide bond that prevents the binding of AMP, the allosteric activator, and avoids the subsequent activation of the bGP, without affecting its regulation by phosphorylation. This disulfide bond involves two cysteines residues, Cys318 and Cys326, strongly conserved among mammalian bGPs, the latter being unique to the brain isoform (Mathieu et al. 2016b).

Table 1 Relative activation of the different GP isoenzymes by phosphorylation of Ser14 and AMP (Newgard et al. 1989; Lowry et al. 1964a; Crerar et al. 1995; Helmreich et al. 1967; Mathieu et al. 2016a)

Condition	Liver GP % activity	Muscle GP % activity	Brain GP % activity
Phosphorylation of Ser14	100%	100%	60%
Binding of AMP	10%	80%	100%
Phosphorylation of Ser14 + AMP binding	100%	110%	100%

3 The Structure of the Glycogen Phosphorylase

The first tridimensional structure of a GP isozyme has first been published in 1974, and corresponded to rabbit mGP (Johnson et al. 1974). Over the next two decades, multiples crystal structures of the phosphorylated or non-phosphorylated mGP, as well as crystal structures of the enzyme in complex with different allosteric effectors, led to better understanding of the structural mechanisms underlying the regulation of this GP isozyme (Barford and Johnson 1989; Sprang et al. 1991). The determination of the structure of lGP in 2000, and of bGP in 2016, further allowed a deeper understanding of the specificities of regulation of each isoforms at the structural level (Mathieu et al. 2016a; Rath et al. 2000).

3.1 General Features

The overall structure of GP is very conserved between the different isozymes. The three proteins are homodimeric (2 × 98 kDa), each dimer comprising two faces: the catalytic face and the regulatory face (Newgard et al. 1987; Mathieu et al. 2016a; Johnson et al. 1974; Rath et al. 2000). The catalytic face of GP consists of the glycogen storage site which orientates the substrate to the catalytic site. On the other side, the regulatory face comprises regions involved in the regulation of the activity, including the phosphorylation site, the AMP binding site as well as amino-acids involved in the dimeric interface and responsible for the transmission of the allosteric signal (Barford and Johnson 1989; Sprang et al. 1991; Mathieu et al. 2017a; Rath et al. 2000) (Fig. 4).

In their active state, bGP and mGP dimers may self-associate, leading to the tetramerization of the protein (Barford et al. 1991; Mathieu et al. 2016a, 2017b; Barford and Johnson 1992). This tetramer is inactive because the catalytic site is no longer accessible. Nevertheless, this interaction is very unlikely to occur *in vivo*. Indeed, the binding of glycogen results in the dissociation of the tetramer leading to two active dimers.

3.2 The Structural Basis of mGP and bGP Activation and Inactivation

During activation, GP enzymes undergo structural changes that ultimately lead to the opening of the catalytic site by the gate loop (or 280 loop), which adopts either a close or open conformation, thus regulating the access to the catalytic site (Barford and Johnson 1989; Rath et al. 2000; Mathieu et al. 2017b).

mGP and bGP are believed to share similar structural changes during activation and inactivation. These changes mostly affect the dimeric interface which undergo

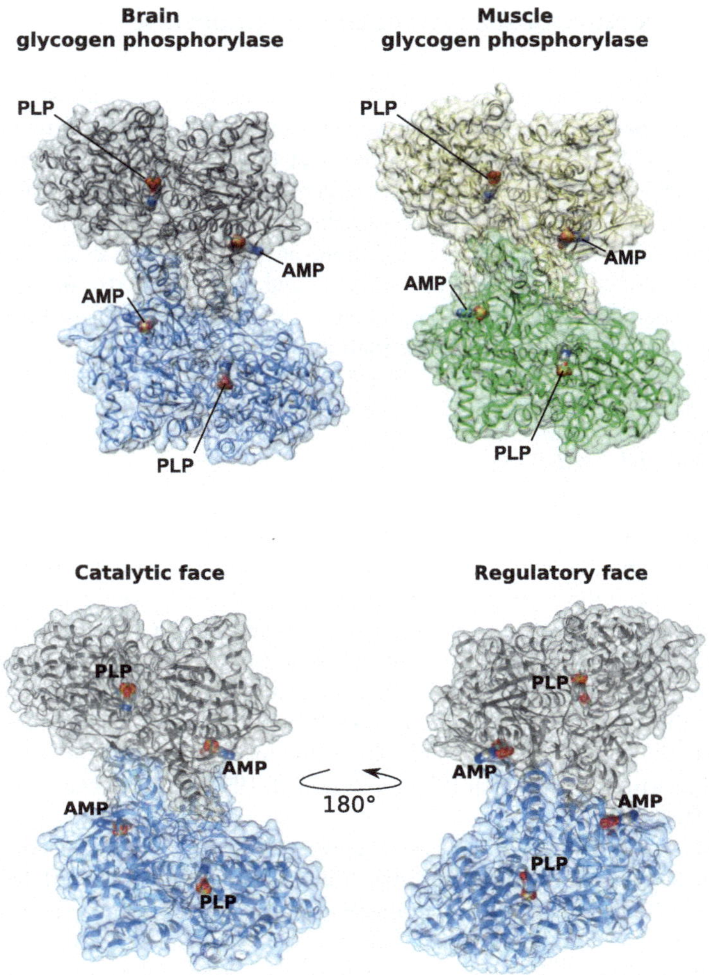

Fig. 4 Overall structures of bGP and mGP. Surface and ribbon representation of the brain (blue and grey) and muscle (green and yellow) GP dimers, showing the strongly conserved overall structures of the two isoforms. The PLP and AMP are shown in their binding site, as surface representation (upper panel). Surface and ribbon representation of the catalytic and regulatory faces of bGP. The catalytic face comprises the catalytic site as well as a glycogen binding site which orientate the substrate to the catalytic site. Regions including the phosphorylation site as well as the AMP-binding site are located on the regulatory face (lower panel)

strong rearrangements, which are particularly evident regarding the relative orientation of the Tower helices. The Tower helices corresponds to the helix 7 of each monomer and are located at the dimer interface, on the catalytic face. They harbor an anti-parallel association and are directly connected to the gate loop of each monomer, thus governing the dimerization of the protein as well as its activation

(Barford and Johnson 1989; Sprang et al. 1991; Mathieu et al. 2016a). In mGP, these two helices undergo a 50° rotation upon activation, resulting in a crossover angle of 75°. The rotation of the Tower helices leads to the destabilization of the gate loop, initially stabilized in a closed conformation, and results in the opening of the catalytic site (Barford and Johnson 1989; Sprang et al. 1991) (Fig. 5). Although the structure of the inactive bGP form is still unknown, the Tower helices of the active bGP have a crossover angle of 85° very similar to mGP and suggest similar activation mechanisms (Mathieu et al. 2016a). Indeed, lGP does not present such rearrangement because of the more rigid body of the dimer. Instead, lGP activation relies on the compaction of the core of the enzyme, comprising notably the Tower helices and the gate loop, but also other secondary structures such as Helix 8, as well as the 250 and 380 loop. The activation of lGP is accompanied with the order to disorder transition of multiple residues in these structures, which results in both a larger dimer interface and the stabilization of the gate loop in open conformation (Rath et al. 2000).

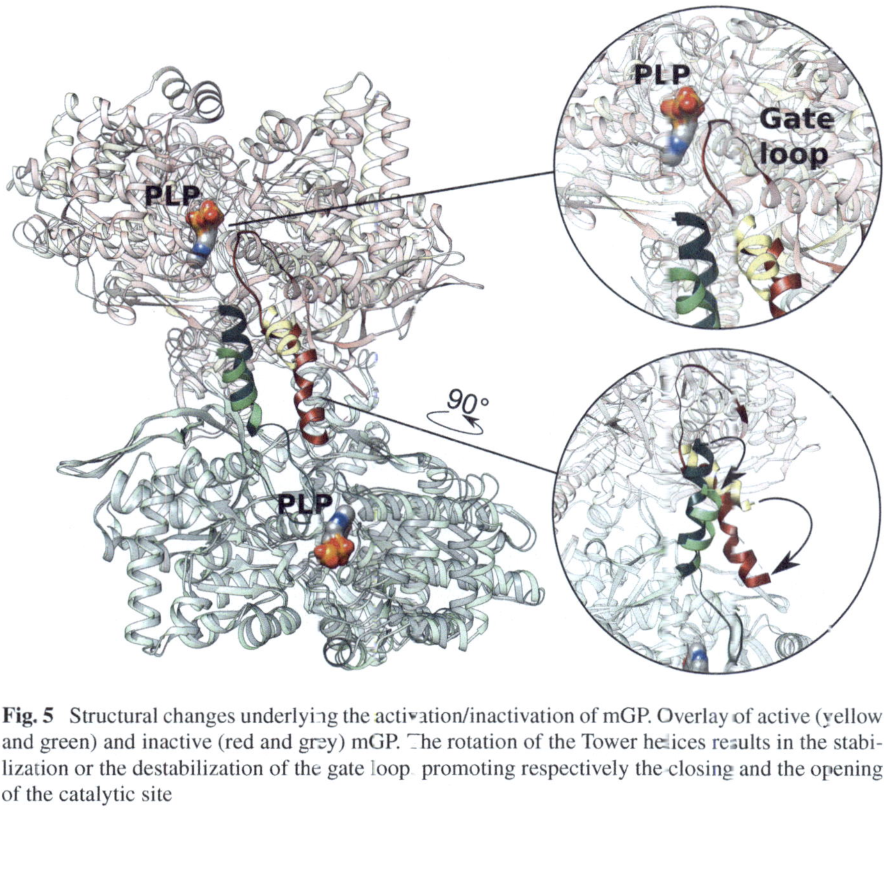

Fig. 5 Structural changes underlying the activation/inactivation of mGP. Overlay of active (yellow and green) and inactive (red and grey) mGP. The rotation of the Tower helices results in the stabilization or the destabilization of the gate loop, promoting respectively the closing and the opening of the catalytic site

3.3 *How Structures Give Insight into the Differential Regulation of GP Isozyme*

The phosphorylation and AMP binding sites, crucial in GP regulation, are proximal and close to the dimeric interface allowing these effectors to trigger the structural rearrangements responsible for the activation of the enzyme.

3.3.1 The Phosphorylation Peptide

Serine 14 is located on the phosphorylation peptide, a stretch of 21 residues (Young 1957; Berg et al. 2002; Adeva-Andany et al. 2016; Roach 2002, 2015; Sagar et al. 1987; Dalsgaard et al. 2007; Kong et al. 2002; Obel et al. 2012; Öz et al. 2015; Bellesi et al. 2018; Gibbs 2015; Gibbs and Hutchinson 2012; Gibbs et al. 2006, 2007; Wolfsdorf and Weinstein 2003; Hedberg-Oldfors and Oldfors 2015; Duran and Guinovart 2015; Klein 1993; Moses and Parvari 2002; Cavanagh 1999) that is disordered in the inactive state, but becomes stabilized or partially stabilized in mGP upon activation (by either phosphorylation of ser 14 or binding of AMP), through its interaction with the dimeric interface (Barford and Johnson 1989; Sprang et al. 1991). These interactions further stabilize the enzyme in its R-state and remains active (Fig. 6). In bGP, this peptide is not stabilized although the enzyme is in the active state. This may be due to multiple substitution at the dimer interface that may reduce the stabilization of the phosphorylation peptide and could explain, at least in part, the reduced sensitivity of the brain form to phosphorylation (Mathieu et al. 2016a).

3.3.2 The AMP Binding Site

bGP is more sensitive to the activation by AMP and has a higher affinity for this effector, compared to mGP (Crerar et al. 1995). The AMP binding site can be divided into three regions that bind the phosphate, the ribose and the adenine moieties of AMP, respectively (Barford et al. 1991; Barford and Johnson 1989; Mathieu et al. 2016a; Rath et al. 2000). In bGP, the binding of AMP is governed by the phosphate group that establishes most of the contacts with the protein. These contacts include notably a hydrogen bond specific from the brain form of GP, provided by the substitution of Phe196 into a Tyr residue. This brain isozyme-specific substitution allows a direct interaction of AMP with the dimeric interface, which can participate to the high sensitivity and high affinity of bGP for AMP (Mathieu et al. 2016a) (Fig. 6).

The AMP binding site comprises secondary structures of each monomer: the helices 2 and 8, as well as the β4/β5 loop and the adenine loop of one subunit, and the Cap' loop of the other subunit. Each secondary structure performs one or several functions. Thus, the helix 8 connects the AMP binding site to the catalytic site,

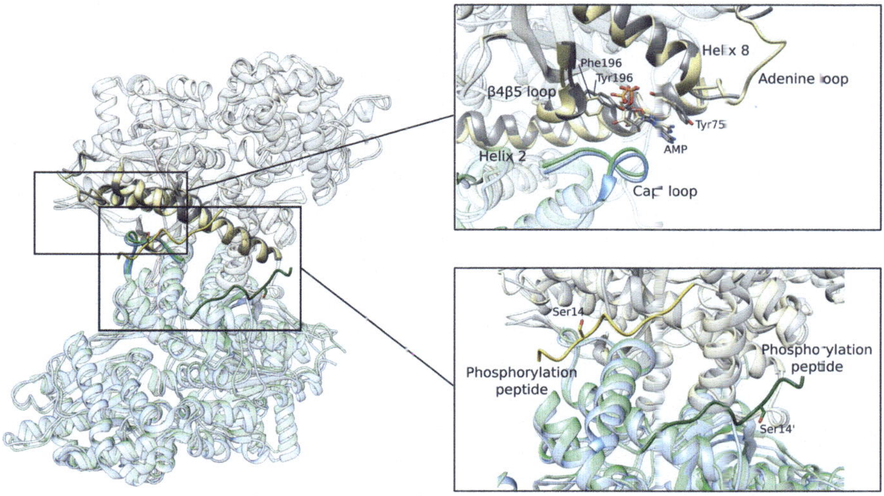

Fig. 6 The AMP-binding site and phosphorylation peptide. The AMP-binding site (upper panel) and the phosphorylation peptides (lower panel) are the two regions of GP involved in the activation of the protein. The AMP-binding site is located at the dimer interface and comprises multiple secondary structures responsible for the binding of the allosteric effector, as well as the transmission of the allosteric signal in the whole dimer. The phosphorylation peptide presents the phosphorylated serine residue. Upon AMP binding, the peptide is stabilized in mGP while it is not in bGP

allowing the transmission of the allosteric signal to the catalytic site. The helix 2 extends across the width of the protein. It ensures the connection between the two AMP binding sites of the dimer and the transmission of the cooperative signal. Moreover, in combination with the adenine loop, the helix 2 allows the transmission of the allosteric signal to the whole dimer. Finally, the concerted action of the helix 2, the Cap' loop and the dimer interface allow the structural rearrangements required for the activation/inactivation process (Barford and Johnson 1989; Sprang et al. 1991; Mathieu et al. 2016a; Rath et al. 2000) (Fig. 6).

In mGP, AMP interacts strongly with the helix 2, particularly through the establishment of a hydrogen bond with the Tyr75, as well as with multiple residues from the adenine loop (Sprang et al. 1991). These interactions are lost in bGP, because of the particular geometry of AMP in its binding site. Indeed, AMP undergoes a 24° rotation of the nucleotide moiety compared to AMP-bound mGP, that results in the different conformation of Tyr75 side chain in the brain enzyme, the loss of hydrogen bonds between AMP and the helix 2, but also precludes the stabilization of the adenine loop along AMP (Mathieu et al. 2016a). The loss of strong interactions between AMP and these secondary structures in bGP might thus explain the non-cooperative binding of AMP to this isoform (Fig. 6).

The adenine loop of bGP is also the central secondary structure responsible for the redox regulation of the enzyme. At the structural level, the formation of a disulfide bridge between cysteines 318 and 326 strongly reduces the mobility of the loop and impairs the binding of AMP, resulting in the insensitivity of the enzyme to its allosteric activator. However, the high flexibility of the loop allows the enzyme to remain highly flexible. The formation of the intramolecular disulfide bond thus does not interfere with the phosphorylation dependent activation of bGP (Mathieu et al. 2016b).

Interestingly, although AMP-binding site residues are mostly conserved between all three isoforms, many of the interactions with AMP, seen in mGP and bGP, are absent in lGP, primarily because of the more rigid body of this isoform. Moreover, similarly to bGP, the conformation of Tyr75 combined to substitutions in the composition of the adenine loop, preclude its interaction with AMP, probably resulting in the loss of activation of this isoform by AMP and the non-cooperative binding of this effector (Rath et al. 2000; Mathieu et al. 2017b).

4 Two Isoforms for Multiple Functions

It is intriguing to notice that both skeletal muscles and liver mainly express one specific isoform of GP, whereas both muscle and brain isozymes are found in the heart and in the brain (Newgard et al. 1987; Saez et al. 2014; Pfeiffer-Guglielmi et al. 2003, 2014; Schmid et al. 2009). The expression of two isozymes regulated differently in these organs suggests that each isoform fulfills common but also specific functions that can be put in parallel to the functions of glycogen in these tissues. In the brain, glycogen is primarily found in glial cells, especially in astrocytes (Saez et al. 2014; Cataldo and Broadwell 1986; Richter et al. 1996). This almost exclusive localization of glycogen in astrocytic cells combined to its very low amount in brain compared to other tissues has long supported the function of brain glycogen only as an emergency store of glucose. Indeed, early studies revealed that brain glycogen was notably mobilized during ischemia, protecting neurons against irreversible damage due to stress conditions (Sagar et al. 1987; Lowry et al. 1964b). Moreover, other stress conditions including hypoglycemia, physical exercise or hypoxia also results in the mobilization of glycogen store in the brain (Saez et al. 2014; Hossain et al. 2014; López-Ramos et al. 2015; Suh et al. 2007; Wender et al. 2000; Brown et al. 2005; Matsui et al. 2012).

However, more recent studies have put in evidence the presence of an active metabolism of glycogen in neurons, and a critical role of glycogen in normal conditions, including awake-sleep cycles, sensorial stimulation, but also in sustaining high cognitive processes such as learning and memory consolidation (Kong et al. 2002; Obel et al. 2012; Bȩllesi et al. 2018; Gibbs and Hutchinson 2012; Gibbs et al. 2006; Swanson et al. 1992; Duran et al. 2013) (DiNuzzo et al., this issue). At the cellular level, glycogen is involved in all the steps underlying neurotransmission, being for instance an energy store for neurons, notably through the astrocyte-neuron lactate shuttle, where astrocytes provide energy to neurons in the form of lactate, derived from astrocytic glycogen. In parallel, astrocytic glycogen also provides the energy required for the uptake of glutamate, used as neurotransmitter (Pellerin and Magistretti 2012; Dringen et al. 1993; Suzuki et al. 2011; Newman et al. 2011) (Fig. 7). In addition, astrocytic glycogen also participates in the maintenance of ions homeostasis, including potassium uptake, and provides the metabolites involved in the synthesis of two neurotransmitters *i.e.* glutamate and GABA (Gibbs et al. 2007; Xu et al. 2013; DiNuzzo et al. 2013; Müller et al. 2014; Sickmann et al. 2012).

However, the function of glycogen in neurons remains poorly understood. Indeed, several publications initially mentioned a limited glycolytic activity in neurons, and the primary use of astrocyte-derived lactate as a source of energy, indicating that glycogen might not primarily serve as a source of energy in these cells (Pellerin and Magistretti 2012; Bouzier-Sore et al. 2003; Bélanger et al. 2011). Yet, recent reports have shown evidence of a glycolytic activity in neurons during stimulation, raising the possibility that glycogen serves also as an energy store in neuronal cells (Díaz-García et al. 2017; Dienel 2017). In addition, neuronal glycogen appears to protect neurons, providing rapidly glucose-6-phosphate and feeding the pentose-phosphate pathway, responsible for the regeneration of reducing power and the detoxification of ROS (Saez et al. 2014; López-Ramos et al. 2015).

These observations clearly highlight the differential roles of astrocytic *versus* neuronal glycogen. Interestingly, astrocytes and neurons do not have the same expression pattern for GP. Astrocytes express both the muscle and the brain forms whereas neurons only express bGP (Saez et al. 2014; Pfeiffer-Guglielmi et al. 2003, 2014). Because of its main regulation by allosteric effectors, bGP appears to predominantly have a local function, allowing the use of glycogen store for the needs of the cell itself, and is consistent with the protective role of neuronal glycogen. On the contrary, mGP, which is also regulated by phosphorylation, also allows the mobilization of glycogen stores in response to neurotransmitters and brain activity

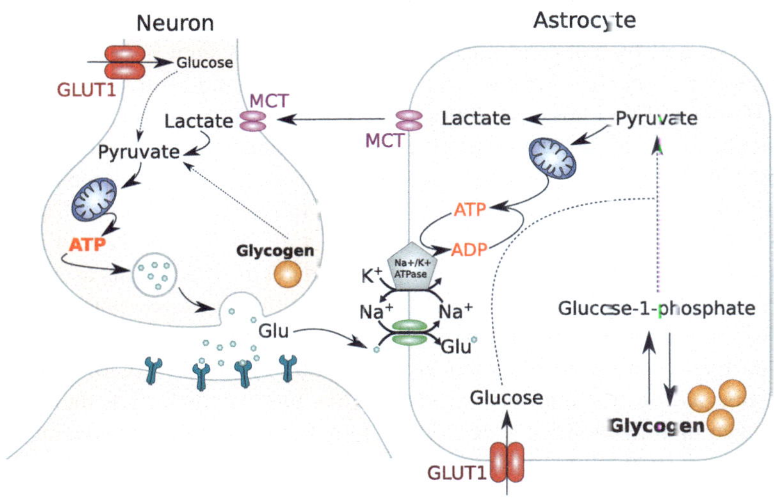

Fig. 7 The role of glycogen in the Astrocyte-Neuron Lactate Shuttle (ANSL). Glycogen stores in astrocytes are metabolized into lactate, which is transferred to neurons and used by neurons as an energy source

(Crerar et al. 1995; Müller et al. 2015; Nadeau et al. 2018). These specificities emphasize the multiple functions of the astrocytic pool of glycogen, being an energy supply for both astrocytes and neurons, but also responding to cerebral activity through the synthesis and the uptake of neurotransmitters as well as the maintenance of ion homeostasis.

5 Concluding Remarks

GPs constitute an extensively studied family of enzymes that play a key role in glycogen metabolism. Because of their differential regulations and particular structural features, both mGP and bGP appear to perform common and but also specific functions in brain glycogen metabolism (Crerar et al. 1995; Müller et al. 2015; Nadeau et al. 2018). Nevertheless, although the allosteric regulation of GPs by allosteric effectors and phosphorylation have been known for decades, the discovery of an additional regulation of bGP by ROS has complexified the understanding of glycogen metabolism in the brain. Indeed, the isoform-specific regulation of bGP by ROS allows the cells to specifically mobilize glycogen in response to extracellular signals without being affected by the cellular energy state (Mathieu et al. 2016b; Nadeau et al. 2018). The exact functions of the redox-regulation of bGP in cells, in normal and pathological conditions, still remain to be elucidated. Previous studies have already pointed an effect of oxidative stress on brain glycogen metabolism, although the exact effects of such stress remains under debate, leading to the activation of glycogen mobilization on one hand, and the accumulation of glycogen on the other hand (Saez et al. 2014; Hossain et al. 2014; Rahman et al. 2000).

Several recent studies shed a new light on glycogen functions in the brain, under both normal and pathological conditions. Cerebral glycogen is crucial for brain functions and cognitive processes such as learning and long-term memory consolidation (Gibbs 2015; Gibbs et al. 2006; Suzuki et al. 2011; Newman et al. 2011). On the other hand, the accumulation of glycogen is particularly deleterious for neurons (Duran et al. 2012, 2014). Glycogen accumulation is observed in a large number of brain diseases such as Alzheimer's disease, Lafora's disease and amyotrophic lateral sclerosis (ALS), as well as in aging (Saez et al. 2014; Sickmann et al. 2012; Sato and Morishita 2015; Dodge et al. 2013; Gertz et al. 1985). This accumulation is suspected of contributing in the loss of cognitive functions associated with these pathologies and the death of neurons. Glycogen metabolism thus appears to be a potential target for the development of therapeutic approaches (Cloix and Hévor 2011). Because of their allosteric regulation, low turnover as well as their key role in glycogen mobilization, GPs are primary targets for the pharmacological control of glycogenolysis (Gaboriaud-Kolar and Skaltsounis 2013). Currently, many studies aim to develop GP effectors in the treatment of diabetes and cancer. To date, most studies have led to the development of GP inhibitors (Gaboriaud-Kolar and Skaltsounis 2013; Zois and Harris 2016). However, restoring the activity of GPs also offers promises for the treatment of GSD, but also in other pathologies associated with glycogen accumulation.

Determining the structures of mGP and lGP has been crucial in this process as they allowed the identification of binding sites for physiological and pharmacological ligands. The determination of the bGP structure now offers a global and comprehensive view of the structure and regulation of the different isoforms, and provides a molecular basis for the development of isoform-specific therapeutic molecules, thus limiting the unwanted side effects.

References

Adeva-Andany MM, González-Lucán M, Donapetry-García C, Fernández-Fernández C, Ameneiros-Rodríguez E (2016) Glycogen metabolism in humans. BBA Clin 5:85–100

Agius L (2015) Role of glycogen phosphorylase in liver glycogen metabolism. Mol Asp Med 46:34–45

Allaman I, Lengacher S, Magistretti PJ, Pellerin L (2003) A2B receptor activation promotes glycogen synthesis in astrocytes through modulation of gene expression. Am J Physiol Cell Physiol 284(3):C696–C704

Barford D, Johnson LN (1989) The allosteric transition of glycogen phosphorylase. Nature 340(6235):609–616

Barford D, Johnson LN (1992) The molecular mechanism for the tetrameric association of glycogen phosphorylase promoted by protein phosphorylation. Protein Sci Publ Protein Soc 1(4):472–493

Barford D, Hu SH, Johnson LN (1991) Structural mechanism for glycogen phosphorylase control by phosphorylation and AMP. J Mol Biol 218(1):233–260

Bélanger M, Allaman I, Magistretti PJ (2011) Brain energy metabolism: focus on astrocyte-neuron metabolic cooperation. Cell Metab 14(6):724–738

Bellesi M, de Vivo L, Koebe S, Tononi G, Cirelli C (2018) Sleep and wake affect glycogen content and turnover at perisynaptic astrocytic processes. Front Cell Neurosci 12:308. https //www.frontiersin.org/article/10.3389/fncel.2018.00308/full

Berg JM, Tymoczko JL, Stryer L, Clarke ND (2002) Biochemistry, 5th edn.. 4 Print. New York, NY, W. H. Freeman, p 1000

Bouzier-Sore A-K, Voisin P, Canioni P, Magistretti PJ, Pellerin L (2003) Lactate is a preferential oxidative energy substrate over glucose for neurons in culture. J Cereb Blood Flow Metab 23(11):1298–1306

Brown AM, Sickmann HM, Fosgerau K, Lund TM, Schousboe A, Waagepetersen HS et al (2005) Astrocyte glycogen metabolism is required for neural activity during aglycemia or intense stimulation in mouse white matter. J Neurosci Res 79(1–2):74–80

Cataldo AM, Broadwell RD (1986) Cytochemical identification of cerebral glycogen and glucose-6-phosphatase activity under normal and experimental conditions. II. Choroid plexus and ependymal epithelia, endothelia and pericytes. J Neurocytol 15(4):511–524

Cavanagh JB (1999) Corpora-amylacea and the family of polyglucosan diseases. Brain Res Brain Res Rev 29(2–3):265–295

Ceulemans H, Bollen M (2004) Functional diversity of protein phosphatase-1, a cellular economizer and reset button. Physiol Rev 84(1):1–39

Cloix J-F, Hévor T (2011) Glycogen as a putative target for diagnosis and therapy in brain pathologies. ISRN Pathol 2011:1–17

Coggan JS, Keller D, Calì C, Lehväslaiho H, Markram H, Schürmann F et al (2018) Norepinephrine stimulates glycogenolysis in astrocytes to fuel neurons with lactate Bouzier-Sore A-K, editor. PLoS Comput Biol 14(8):e1006392

Crerar MM, Karlsson O, Fletterick RJ, Hwang PK (1995) Chimeric muscle and brain glycogen phosphorylases define protein domains governing isozyme-specific responses to allosteric activation. J Biol Chem 270(23):13748–13756

Dalsgaard MK, Madsen FF, Secher NH, Laursen H, Quistorff B (2007) High glycogen levels in the hippocampus of patients with epilepsy. J Cereb Blood Flow Metab 27(6):1137–1141

Díaz-García CM, Mongeon R, Lahmann C, Koveal D, Zucker H, Yellen G (2017) Neuronal stimulation triggers neuronal glycolysis and not lactate uptake. Cell Metab 26(2):361–374.e4

Dienel GA (2017) Lack of appropriate stoichiometry: strong evidence against an energetically important astrocyte-neuron lactate shuttle in brain. J Neurosci Res 95(11):2103–2125

DiNuzzo M, Mangia S, Maraviglia B, Giove F (2013) Regulatory mechanisms for glycogenolysis and K+ uptake in brain astrocytes. Neurochem Int 63(5):458–464

Dodge JC, Treleaven CM, Fidler JA, Tamsett TJ, Bao C, Searles M et al (2013) Metabolic signatures of amyotrophic lateral sclerosis reveal insights into disease pathogenesis. Proc Natl Acad Sci U S A 110(26):10812–10817

Doherty MJ, Moorhead G, Morrice N, Cohen P, Cohen PT (1995) Amino acid sequence and expression of the hepatic glycogen-binding (GL)-subunit of protein phosphatase-1. FEBS Lett 375(3):294–298

Doherty MJ, Young PR, Cohen PT (1996) Amino acid sequence of a novel protein phosphatase 1 binding protein (R5) which is related to the liver- and muscle-specific glycogen binding subunits of protein phosphatase 1. FEBS Lett 399(3):339–343

Dringen R, Gebhardt R, Hamprecht B (1993) Glycogen in astrocytes: possible function as lactate supply for neighboring cells. Brain Res 623(2):208–214

Duran J, Guinovart JJ (2015) Brain glycogen in health and disease. Mol Asp Med 46:70–77

Duran J, Tevy MF, Garcia-Rocha M, Calbó J, Milán M, Guinovart JJ (2012) Deleterious effects of neuronal accumulation of glycogen in flies and mice. EMBO Mol Med 4(8):719–729

Duran J, Saez I, Gruart A, Guinovart JJ, Delgado-García JM (2013) Impairment in long-term memory formation and learning-dependent synaptic plasticity in mice lacking glycogen synthase in the brain. J Cereb Blood Flow Metab 33(4):550–556

Duran J, Gruart A, García-Rocha M, Delgado-García JM, Guinovart JJ (2014) Glycogen accumulation underlies neurodegeneration and autophagy impairment in Lafora disease. Hum Mol Genet 23(12):3147–3156

Gaboriaud-Kolar N, Skaltsounis A-L (2013) Glycogen phosphorylase inhibitors: a patent review (2008–2012). Expert Opin Ther Pat 23(8):1017–1032

Gertz HJ, Cervos-Navarro J, Frydl V, Schultz F (1985) Glycogen accumulation of the aging human brain. Mech Ageing Dev 31(1):25–35

Gibbs ME (2015) Role of Glycogenolysis in memory and learning: regulation by noradrenaline, Serotonin and ATP. Front Integr Neurosci 9:70

Gibbs ME, Hutchinson DS (2012) Rapid turnover of glycogen in memory formation. Neurochem Res 37(11):2456–2463

Gibbs ME, Anderson DG, Hertz L (2006) Inhibition of glycogenolysis in astrocytes interrupts memory consolidation in young chickens. Glia 54(3):214–222

Gibbs ME, Lloyd HGE, Santa T, Hertz L (2007) Glycogen is a preferred glutamate precursor during learning in 1-day-old chick: biochemical and behavioral evidence. J Neurosci Res 85(15):3326–3333

Hedberg-Oldfors C, Oldfors A (2015) Polyglucosan storage myopathies. Mol Asp Med 46:85–100

Heilmeyer LMG (1987. [cited 2016 Apr 27]) Signal Transduction and Protein Phosphorylation [Internet]. Springer US, Boston, MA. https://doi.org/10.1007/978-1-4757-0166-1

Heilmeyer LM (1991) Molecular basis of signal integration in phosphorylase kinase. Biochim Biophys Acta 1094(2):168–174

Helmreich E, Michaelides MC, Cori CF (1967) Effects of substrates and a substrate analog on the binding of 5′-adenylic acid to muscle phosphorylase a. Biochemistry 6(12):3695–3710

Hossain MI, Roulston CL, Stapleton DI (2014) Molecular basis of impaired glycogen metabolism during ischemic stroke and hypoxia. PLoS One 9(5):e97570
Hutchins DA, Rogers KJ (1970) Physiological and drug-induced changes in the glycogen content of mouse brain. Br J Pharmacol 39(1):9–25
Jensen TE, Richter EA (2012) Regulation of glucose and glycogen metabolism during and after exercise. J Physiol 590(5):1069–1076
Johnson LN, Madsen NB, Mosley J, Wilson KS (1974) The crystal structure of phosphorylase b at 6 Å resolution. J Mol Biol 90(4):703–717
Klein CJ (1993) Adult polyglucosan body disease. In: Pagon RA, Adam MP, Ardinger HH, Wallace SE, Amemiya A, Bean LJ et al (eds) GeneReviews(®) [Internet]. University of Washington, Seattle, WA. [cited 2016 Apr 13]. http://www.ncbi.nlm.nih.gov/books/NBK5300/
Kong J, Shepel PN, Holden CP, Mackiewicz M, Pack AI, Geiger JD (2002) Brain glycogen decreases with increased periods of wakefulness: implications for homeostatic drive to sleep. J Neurosci 22(13):5581–5587
Lebo RV, Gorin F, Fletterick RJ, Kao FT, Cheung MC, Bruce BD et al (1984) High-resolution chromosome sorting and DNA spot-blot analysis assign McArdle's syndrome to chromosome 11. Science 225(4657):57–59
López-Ramos JC, Duran J, Gruart A, Guinovart JJ, Delgado-García JM (2015) Role of brain glycogen in the response to hypoxia and in susceptibility to epilepsy. Front Cell Neurosci 9:431
Lowry OH, Schulz DW, Passonneau JV (1964a) Effects of adenylic acid on the kinetics of muscle phosphorylase a. J Biol Chem 239:1947–1953
Lowry OH, Passonneau JV, Hasselberger FX, Schulz DW (1964b) Effect of ischemia on known substrates and cofactors of the glycolytic pathway in brain. J Biol Chem 239:18–30
Lukacs CM, Oikonomakos NG, Crowther RL, Hong L-N, Kammlott RU, Levin W et al (2006) The crystal structure of human muscle glycogen phosphorylase a with bound glucose and AMP: an intermediate conformation with T-state and R-state features. Proteins 63(4):1123–1126
Magistretti PJ, Morrison JH, Shoemaker WJ, Sapin V, Bloom FE (1981) Vasoactive intestinal polypeptide induces glycogenolysis in mouse cortical slices: a possible regulatory mechanism for the local control of energy metabolism. Proc Natl Acad Sci U S A 78(10):6535–6539
Mathieu C, Li de la sierra-Gallay I, Duval R, Xu X, Cocaign A, Léger T et al (2016a) Insights into brain glycogen Metabolism: the structure of human brain glycogen phosphorylase. J Biol Chem 291(35):18072–18083
Mathieu C, Duval R, Cocaign A, Petit E, Bui L-C, Haddad I et al (2016b) An isozyme-specific redox switch in human brain glycogen phosphorylase modulates its allosteric activation by AMP. J Biol Chem 291(46):23842–23853
Mathieu C, Bui L-C, Petit E, Haddad I, Agbulut O, Vinh J et al (2017a) Molecular mechanisms of allosteric inhibition of brain glycogen phosphorylase by neurotoxic Dithiocarbamate chemicals. J Biol Chem 292(5):1603–1612
Mathieu C, Dupret J-M, Rodrigues Lima F (2017b) The structure of brain glycogen phosphorylase-from allosteric regulation mechanisms to clinical perspectives. FEBS J 284(4):546–554
Matsui T, Ishikawa T, Ito H, Okamoto M, Inoue K, Lee M-C et al (2012) Brain glycogen supercompensation following exhaustive exercise. J Physiol 590(Pt 3):607–616
Meléndez-Hevia E, Waddell TG, Shelton ED (1993) Optimization of molecular design in the evolution of metabolism: the glycogen molecule. Biochem J 295(Pt 2):477–483
Moses SW, Parvari R (2002) The variable presentations of glycogen storage disease type IV: a review of clinical, enzymatic and molecular studies. Curr Mol Med 2(2):177–188
Müller MS, Fox R, Schousboe A, Waagepetersen HS, Bak LK (2014) Astrocyte glycogenolysis is triggered by store-operated calcium entry and provides metabolic energy for cellular calcium homeostasis. Glia 62(4):526–534
Müller MS, Pedersen SE, Walls AB, Waagepetersen HS, Bak LK (2015) Isoform-selective regulation of glycogen phosphorylase by energy deprivation and phosphorylation in astrocytes. Glia 63(1):154–162
Nadeau OW, Fontes JD, Carlson GM (2018) The regulation of glycogenolysis. J Biol Chem 293:7099–7107

Newgard CB, Fletterick RJ, Anderson LA, Lebo RV (1987) The polymorphic locus for glycogen storage disease VI (liver glycogen phosphorylase) maps to chromosome 14. Am J Hum Genet 40(4):351–364

Newgard CB, Littman DR, van Genderen C, Smith M, Fletterick RJ (1988) Human brain glycogen phosphorylase. Cloning, sequence analysis, chromosomal mapping, tissue expression, and comparison with the human liver and muscle isozymes. J Biol Chem 263(8):3850–3857

Newgard CB, Hwang PK, Fletterick RJ (1989) The family of glycogen phosphorylases: structure and function. Crit Rev Biochem Mol Biol 24(1):69–99

Newgard CB, Brady MJ, O'Doherty RM, Saltiel AR (2000) Organizing glucose disposal: emerging roles of the glycogen targeting subunits of protein phosphatase-1. Diabetes 49(12):1967–1977

Newman LA, Korol DL, Gold PE (2011) Lactate produced by glycogenolysis in astrocytes regulates memory processing. PLoS One 6(12):e28427

Obel LF, Müller MS, Walls AB, Sickmann HM, Bak LK, Waagepetersen HS et al (2012) Brain glycogen-new perspectives on its metabolic function and regulation at the subcellular level. Front Neuroenerg 4:3

Oikonomakos NG, Zographos SE, Johnson LN, Papageorgiou AC, Acharya KR (1995) The binding of 2-deoxy-D-glucose 6-phosphate to glycogen phosphorylase b: kinetic and crystallographic studies. J Mol Biol 254(5):900–917

Öz G, DiNuzzo M, Kumar A, Moheet A, Seaquist ER (2015) Revisiting glycogen content in the human brain. Neurochem Res 40(12):2473–2481

Pellerin L, Magistretti PJ (2012) Sweet sixteen for ANLS. J Cereb Blood Flow Metab 32(7):1152–1166

Pelley JW (2007) Gluconeogenesis and glycogen metabolism. In: Elsevier's Integrated Biochemistry [Internet]. Elsevier, pp 65–71. [cited 2019 Mar 22]. https://linkinghub.elsevier.com/retrieve/pii/B9780323034104500146

Pfeiffer-Guglielmi B, Fleckenstein B, Jung G, Hamprecht B (2003) Immunocytochemical localization of glycogen phosphorylase isozymes in rat nervous tissues by using isozyme-specific antibodies. J Neurochem 85(1):73–81

Pfeiffer-Guglielmi B, Dombert B, Jablonka S, Hausherr V, van Thriel C, Schöbel N et al (2014) Axonal and dendritic localization of mRNAs for glycogen-metabolizing enzymes in cultured rodent neurons. BMC Neurosci 15:70

Printen JA (1997) PTG, a protein phosphatase 1-binding protein with a role in glycogen Metabolism. Science 275(5305):1475–1478

Rahman B, Kussmaul L, Hamprecht B, Dringen R (2000) Glycogen is mobilized during the disposal of peroxides by cultured astroglial cells from rat brain. Neurosci Lett 290(3):169–172

Rath VL, Ammirati M, LeMotte PK, Fennell KF, Mansour MN, Danley DE et al (2000) Activation of human liver glycogen phosphorylase by alteration of the secondary structure and packing of the catalytic core. Mol Cell 6(1):139–148

Richter K, Hamprecht B, Scheich H (1996) Ultrastructural localization of glycogen phosphorylase predominantly in astrocytes of the gerbil brain. Glia 17(4):263–273

Roach PJ (2002) Glycogen and its metabolism. Curr Mol Med 2(2):101–120

Roach PJ (2015) Glycogen phosphorylation and Lafora disease. Mol Asp Med 46:78–84

Rubio-Villena C, Garcia-Gimeno MA, Sanz P (2013) Glycogenic activity of R6, a protein phosphatase 1 regulatory subunit, is modulated by the laforin-Malin complex. Int J Biochem Cell Biol 45(7):1479–1488

Saez I, Duran J, Sinadinos C, Beltran A, Yanes O, Tevy MF et al (2014) Neurons have an active glycogen metabolism that contributes to tolerance to hypoxia. J Cereb Blood Flow Metab 34(6):945–955

Sagar SM, Sharp FR, Swanson RA (1987) The regional distribution of glycogen in rat brain fixed by microwave irradiation. Brain Res 417(1):172–174

Sato N, Morishita R (2015) The roles of lipid and glucose metabolism in modulation of β-amyloid, tau, and neurodegeneration in the pathogenesis of Alzheimer disease. Front Aging Neurosci 7:199

Schmid H, Pfeiffer-Guglielmi B, Dolderer B, Thiess U, Verleysdonk S, Hamprecht B (2009) Expression of the brain and muscle isoforms of glycogen phosphorylase in rat heart. Neurochem Res 34(3):581–586

Sickmann HM, Waagepetersen HS, Schousboe A, Benie AJ, Bouman SD (2012) Brain glycogen and its role in supporting glutamate and GABA homeostasis in a type 2 diabetes rat model. Neurochem Int 60(3):267–275

Sorg O, Magistretti PJ (1991) Characterization of the glycogenolysis elicited by vasoactive intestinal peptide, noradrenaline and adenosine in primary cultures of mouse cerebral cortical astrocytes. Brain Res 563(1–2):227–233

Sorg O, Magistretti PJ (1992) Vasoactive intestinal peptide and noradrenaline exert long-term control on glycogen levels in astrocytes: blockade by protein synthesis inhibition. J Neurosci 12(12):4923–4931

Sprang SR, Withers SG, Goldsmith EJ, Fletterick RJ, Madsen NB (1991) Structural basis for the activation of glycogen phosphorylase b by adenosine monophosphate. Science 254(5036):1367–1371

Suh SW, Bergher JP, Anderson CM, Treadway JL, Fosgerau K, Swanson RA (2007) Astrocyte glycogen sustains neuronal activity during hypoglycemia: studies with the glycogen phosphorylase inhibitor CP-316,819 ([R-R*,S*]-5-chloro-N-[2-hydroxy-3-(methoxymethylamino)-3-oxo-1-(phenylmethyl)propyl]-1H-indole-2-carboxamide). J Pharmacol Exp Ther 321(1):45–50

Suzuki A, Stern SA, Bozdagi O, Huntley GW, Walker RH, Magistretti PJ et al (2011) Astrocyte-neuron lactate transport is required for long-term memory formation. Cell 144(5):810–823

Swanson RA, Morton MM, Sagar SM, Sharp FR (1992) Sensory stimulation induces local cerebral glycogenolysis: demonstration by autoradiography. Neuroscience 51(2):451–461

Taylor C, Cox AJ, Kernohan JC, Cohen P (1975) Debranching enzyme from rabbit skeletal muscle. Purification, properties and physiological role. Eur J Biochem FEBS 51(1):105–115

Vilchez D, Ros S, Cifuentes D, Pujadas L, Vallès J, García-Fojeda B et al (2007) Mechanism suppressing glycogen synthesis in neurons and its demise in progressive myoclonus epilepsy. Nat Neurosci 10(11):1407–1413

Wender R, Brown AM, Fern R, Swanson RA, Farrell K, Ransom BR (2000) Astrocytic glycogen influences axon function and survival during glucose deprivation in central white matter. J Neurosci 20(18):6804–6810

van de Werve G, Hue L, Hers HG (1977) Hormonal and ionic control of the glycogenolytic cascade in rat liver. Biochem J 162(1):135–142

Wolfsdorf JI, Weinstein DA (2003) Glycogen storage diseases. Rev Endocr Metab Disord 4(1):95–102

Xu J, Song D, Xue Z, Gu L, Hertz L, Peng L (2013) Requirement of glycogenolysis for uptake of increased extracellular K+ in astrocytes: potential implications for K+ homeostasis and glycogen usage in brain. Neurochem Res 38(3):472–485

Xu J, Song D, Bai Q, Cai L, Hertz L, Peng L (2014) Basic mechanism leading to stimulation of glycogenolysis by isoproterenol, EGF, elevated extracellular K+ concentrations, or GABA. Neurochem Res 39(4):661–667

Yip J, Geng X, Shen J, Ding Y (2017) Cerebral Gluconeogenesis and diseases. Front Pharmacol 7:521. [cited 2019 Mar 22];http://journal.frontiersin.org/article/10.3389/fphar.2016.00521/full

Young FG (1957) Claude Bernard and the discovery of glycogen; a century of retrospect. Br Med J 1(5033):1431–1437

Zois CE, Harris AL (2016) Glycogen metabolism has a key role in the cancer microenvironment and provides new targets for cancer therapy. J Mol Med (Berl) Ger 94(2):137–154

Regional Distribution of Glycogen in the Mouse Brain Visualized by Immunohistochemistry

Yuki Oe, Sonam Akther, and Hajime Hirase

Abstract Considering that the brain constantly consumes a substantial amount of energy, the nature of its energy reserve is an important issue. Although the brain is rich in lipid content encompassing membranes, myelin sheath, and astrocytic lipid droplets, it is devoid of adipose tissue which serves as an energy reserve. Notably, glycogen represents the major energy store in the brain. While glycogen has been observed mainly in astrocytes for decades by electron microscopy, glycogen distribution in the brain has only been partially documented. The involvement of glycogen metabolism in memory consolidation, demonstrated by several research groups, has reiterated the functional significance of this macromolecule and the need for description of its comprehensive distribution in the brain. The combination of focused microwave-assisted brain fixation and glycogen immunohistochemistry permits assessment of glycogen distribution in the rodent brain. In this article, we describe glycogen distribution in the mouse brain using glycogen immunohistochemistry. We find heterogeneous glycogen storage patterns at multiple spatial scales. The heterogeneous glycogen distribution patterns may underlie local energy metabolism or synaptic activity, and its mechanistic understanding should extend our knowledge on brain metabolism in health and disease.

Keywords Glycogen immunohistochemistry · Forebrain · Olfactory bulb · Hippocampus · Cortex · Striatum

Y. Oe (✉)
RIKEN Center for Brain Science, Wako, Saitama, Japan
e-mail: yuki.oe@riken.jp

S. Akther · H. Hirase (✉)
RIKEN Center for Brain Science, Wako, Saitama, Japan

Saitama University Brain Science Institute, Saitama, Japan
e-mail: hajime.hirase@riken.jp

M. DiNuzzo, A. Schousboe (eds.), *Brain Glycogen Metabolism*, Advances in Neurobiology 23, https://doi.org/10.1007/978-3-030-27480-1_5

1 Introduction

The energy consumption of the brain is exceptionally high compared to other organs (Raichle and Gusnard 2002). The fundamental energy substrate—adenosine triphosphate (ATP)—is primarily derived from glucose that is supplied by blood circulation in the brain. According to the astrocyte-to-neuron lactate shuttle hypothesis, glucose from blood circulation is transported to astrocytes and broken down into lactate which, in turn, is shuttled to neighboring neuronal compartments. Inside neurons, lactate is converted into pyruvate and further metabolized by the tricarboxylic acid (TCA) cycle to produce ATP molecules. As in other peripheral organs such as the liver or muscles, a portion of glucose is stored primarily inside astrocytes in the form of glycogen. While described in detail in other chapters of the present volume, brain glycogen metabolism supports: (1) on-demand supply of increased energy and (2) promotion of long-term synaptic plasticity. The degrees to which these functionalities are recruited have not been explicitly described and they likely depend on brain region and state. However, the examination of glycogen distribution will enable us to assess the capacity of astrocytes to execute such functions in distinct brain regions.

There are several ways to determine brain glycogen distribution. First, brain regions of interest can be dissected and biochemical assays can be performed on them. Quantitative biochemical glycogen assays include phenol sulfuric acid and OxiRed fluorescent probe methods which react with the amount of (oxidized) glucose generated by enzymatic breakdown of glycogen. The signal strength reflects the amount of glycosyl units that constitute glycogen molecules in the sample. Although these biochemical methods are quantitative, they require a considerable amount of sample and hence are not suitable for assessing the spatial distribution of glycogen finer than broad brain regions. For instance, the BioVision glycogen assay kit K646–100 recommends 10 mg of tissue. Considering the low glycogen content in the brain (an order of magnitude smaller than the liver or skeletal muscle content of a few 100 μM/g wet tissue), a larger sample amount warrants accurate measurements. Since a mouse brain weighs only about half a gram, dissection and biochemical assay should be performed with great care. In the past, glycogen quantification has been performed with the rat brain which typically weighs over 2 g.

Second, glycogen granules are observed as electron-dense puncta in lead- or osmium-intensified sections by electron microscopy. Therefore, electron microscopy provides a direct assessment of subcellular distribution of glycogen. However, electron microscopy is limited by the size of observable area, which is typically in the range of several micrometers in horizontal dimensions of a thickness of tens of nanometers. Even with the development of serial block-face scanning electron microscopy or automated serial microscopy with ultrathin sections mounted on tapes, the volume is practically limited to a single cortical column. The small volume of observation makes approaches with electron microscopy unrealistic to assess brain-wide glycogen distribution.

Third, nuclear magnetic resonance spectroscopy (MRS) is a method that detects unique resonance frequencies glycogen molecules in a magnetic field. The advantage of MRS is that this method is quantitative and can be readily performed on live subjects. For example, liver and skeletal muscle glycogen levels have been quantified by MRS in human subjects (Cline et al. 1999; Magnusson et al. 1992; Öz et al. 2003). However, MRS is limited by its spatial resolution. Since the brain stores an order of magnitude lower amount of glycogen than the liver or skeletal muscles, a typical voxel used to assess glycogen content is larger than a centimeter in each dimension and requires the summation of scans for extended periods of time to obtain reliable signals (Öz et al. 2007). While this method can possibly address glycogen distribution of a human brain with the coarse resolution, it is not suitable for characterizing the glycogen distribution of individual brain regions or brains of smaller animals.

Fourth, compared with the above methods, histological examination by light microscopy is a suitable approach to characterize the regional cellular, and subcellular distribution of glycogen. A critical point to consider for the histological approach is how specific glycogen is labeled in brain sections. Glycogen has traditionally been stained by the periodic acid-Schiff (PAS) histochemical staining procedure, particularly for clinical research purposes. PAS staining is used to indicate glycogen in liver samples where hepatocytes store high amounts of glycogen. However, PAS staining also labels glycoproteins and proteoglycans which are abundant in the brain. Therefore, PAS staining is not ideal for the evaluation of brain glycogen distribution in terms of specificity. Immunohistochemistry is a method which achieves relatively high specificity, given a specific primary antibody. Additionally, immunohistochemistry permits labeling of secondary or tertiary molecules which are useful in determining cell types that store glycogen. In this chapter, we describe brain glycogen distribution using two monoclonal antibodies against glycogen

After describing the procedure for glycogen immunohistochemistry, we describe brain glycogen distribution at three levels: (1) macroscopic distribution by distinct brain region, (2) microscopic (subcellular) distribution within astrocytes, the principal glycogen storage cell type, and (3) mesoscopic distribution in brain areas of high glycogen content.

2 Glycogen Immunohistochemistry

Glycogen is a storage of glucose and its size varies dynamically according to usage or storage of glucosyl units. For instance, when the tissue oxygen level is low, cell energy metabolism shifts to anaerobic metabolism whereby glycogen is broken down into glucose-1-phosphate molecules. Likewise, when excess glucose is available, glycogenesis occurs in cells specialized to store glycogen including hepatocytes, myocytes, and in the central nervous system, astrocytes. Brain glycogen has also been reported to be consumed during prolonged aerobic exercise (Matsui et al. 2011).

Brain immunohistochemistry is typically performed on perfusion-fixed tissue. Perfusion fixation is generally performed in two steps: transcardial perfusion with physiological solution (e.g., saline, phosphate buffered saline, Ringer's solution, etc.), followed by transcardial perfusion of fixative (e.g., 4% paraformaldehyde in 0.1 M phosphate buffer). While the purpose of the first step is to remove blood from the brain, this step typically lasts for a few minutes or longer, which inevitably puts the brain under an anoxic state. As a result, glycogen is vastly degraded and little glycogen remains in the perfusion-fixed brain. A possible way to prepare a fixed brain sample without perfusion is to fix an acutely prepared alive brain slice. Acute brain slice preparation is a common procedure practiced in *in vitro* electrophysiology laboratories and slices are usually recovered in oxygenated artificial cerebral spinal fluid for an hour or so before experiments. Indeed, it has been shown that glycogen in hippocampal acute slices is emptied immediately after slice preparation already (Fiala et al. 2003; Takano et al. 2014). Brain dissection and slice cutting surely induce trauma and resultant extracellular potassium elevation. High extracellular potassium has previously demonstrated to trigger glycogenolysis via activation of soluble adenylyl cyclase (Choi et al. 2012).

Because glycogen is very sensitive to brain insult, a method that instantly preserves the *in vivo* state of metabolism is essential for proper histological visualization of glycogen. Rapid freezing of the brain in liquid nitrogen is a common procedure for biochemistry and in situ hybridization. Focused microwave irradiation is another procedure practiced in biochemical investigation of brain metabolism or protein phosphorylation. The former approach aims to preserve the metabolic state of the brain by halting enzymatic activity by lowering the temperature, whereas the latter approach denatures and inactivates metabolic enzymes by instantaneously heating the sample to 80–90 °C.

Previous biochemical examinations of brain glycogen showed that microwaved preparations preserve significantly higher amounts of glycogen (> ~twice) than rapid freezing (Sagar et al. 1987). More recently, Sugiura et al. (2014) compared the abundance of metabolic substrates in hippocampal samples prepared by post-euthanasia whole-brain freezing, in situ freezing, or focused microwave irradiation. Accordingly, they found that focused microwave irradiated samples contained higher amounts of ATP, adenosine diphosphate (ADP) and pyruvate compared with the other methods, which had more degraded purine and lactate levels. This result shows that focused microwave irradiation preserves the aerobic respiratory state of the brain (which is the default *in vivo* respiration mode) and is adequate for glycogen immunohistochemistry.

While brain samples prepared by focused microwave irradiation have been used for biochemical evaluation of glycogen content (Sagar et al. 1987) and PAS staining (Kong et al. 2002), glycogen immunohistochemistry has not been performed until recently. Here we summarize the glycogen immunohistochemistry procedures used in our laboratory. Currently, two mouse monoclonal IgM antibodies have been verified to be suitable for glycogen immunohistochemistry (Fig. 1). The ESG1A9 antibody (ESG) is raised against enzymatically synthetized glycogen (hence the name ESG), and it tends to have higher affinity to large glycogen molecules (Nakamura-Tsuruta et al. 2012). The IV58B6 antibody (IV) has affinity for glycogen

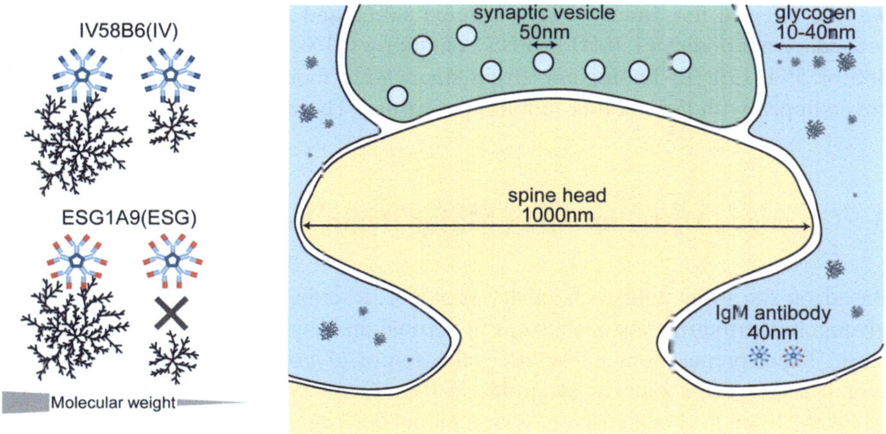

Fig. 1 Consideration of spatial scales in glycogen immunohistochemistry. Left panel: The IV58B6 antibody is considered to have an affinity for glycogen of all sizes, whereas the ESG1A9 antibody has an affinity for large glycogen macromolecules. Right panel: Sketch of a synapse and perisynaptic astrocytic processes incorporating realistic spatial scales

of all sizes (Baba 1993). The exact manner in which these two antibodies express the differential affinity patterns has not been identified. It is supposed that enzymatically synthesized glycogen has a different branching pattern from that of natural glycogen, and the epitopes are associated with the branching pattern.

Adult mice are euthanized by focused microwave irradiation using a MMW-05 microwave fixation system (Muromachi Kikai, Tokyo, Japan). Considering circadian modulation of astrocytic and metabolic functions, it is a reasonable idea to sacrifice mice in a defined time of the day. For instance, all mice studied in our recent study were sacrificed between 6:00 and 8:00 p.m., i.e., within 2 h before the end of diurnal period (Oe et al. 2016). Briefly, a mouse was confined in a specialized tubular animal holder with a hollow wall filled with water (WJM-24 or WJM-28, Muromachi Kikai). The holder with the animal is then placed inside MMW-05. High-energy microwave (5 kW) was focused at the head of the mouse for 0.94–1.05 s. Immediately after focused microwave euthanasia, the brain was removed from the skull and immersed in fixative (4% paraformaldehyde in 0.1 M phosphate buffer) for overnight at 4 °C.

Brain sections of 60 μm thickness were prepared in PB using a microslicer (Pro-7 Linear Slicer, DSK, Japan). After washing in phosphate buffered saline (PBS), the sections were incubated in PBS containing 0.1% Triton X-100 and primary antibodies for 24 hours at 4 °C while gently shaking. The concentrations of ESG and IV antibodies are 1:50 (15 μg/ml) and 1:300 (30 μg/ml), respectively. As mentioned previously, both ESG and IV antibodies are IgM which are several times larger than IgG in molecular size. Therefore, penetration of the primary antibodies should be performed extensively. The authors are not aware of an IgG antibody for glycogen that can be used for immunohistochemistry at this time. The sections were then washed three to five times in Tris-buffered saline and incubated with fluorescent secondary antibodies (1:1000 in PBS containing 0.1% Triton X-100, Alexa Fluor 488 or 594, Life Technologies). Brain slices were mounted on slide glasses and coverslipped with PermaFluor mounting medium (Thermo Scientific).

3 Glycogen Distribution in Major Brain Regions

Based on the immunohistochemistry protocol described in the previous section, glycogen distribution was visualized in coronal and sagittal sections of the mouse brain. These preparations allow the evaluation of glycogen distribution in macroscopic scales. As shown in examples in Fig. 2, the IV (pan-glycogen) antibody labels the brain fairly extensively with regional heterogeneity. For example, the hippocampus generally has high-intensity signals while the principal cell body layers of the hippocampus are low in signal intensities. Moreover, the white matter has been found to be generally low in signal intensity. Accordingly, the corpus callosum, hippocampal fimbria-fornix, medial forebrain bundle, and arbor vitae of the cerebellum are distinguished as low-intensity structures. Moreover, stripe patterns

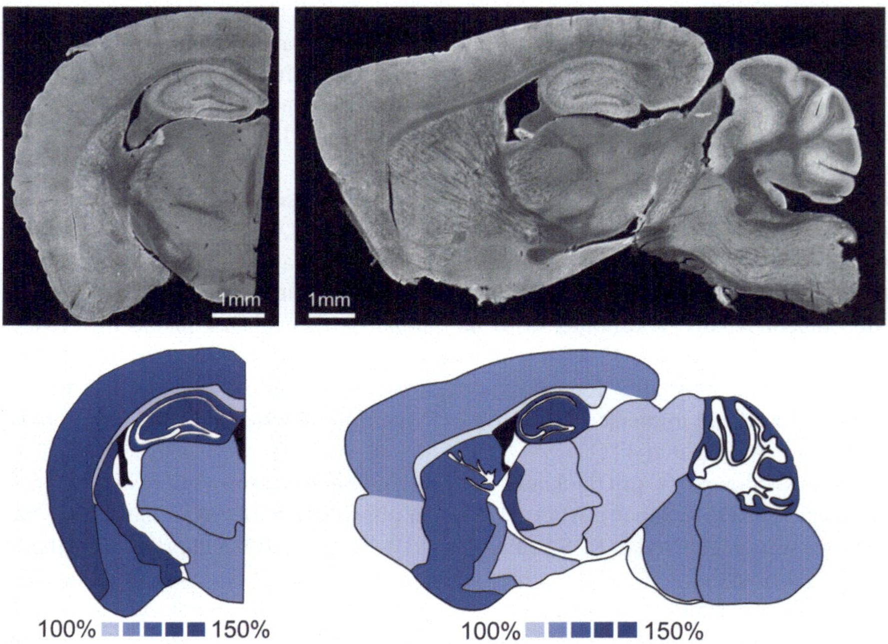

Fig. 2 Glycogen immunohistochemistry on microwave-fixed mouse brain at microscopic scale using IV58B6. Upper panel: Examples of glycogen immunohistochemistry of coronal (left) and sagittal (right) sections. Lower panel: Relative glycogen immunohistochemical signal averaged over 9 mice. Reproduced from Oe et al. (2016)

of axon bundles labeled in low intensity are apparent in sagittal sections of the striatum. Amongst major gray matter structures, the cerebellar cortex, hippocampus, thalamic reticular nucleus, striatum, globus pallidus, substantia nigra (pars reticulata), and pons are labelled well with IV. On the other hand, subcortical structures including the dorsal thalamus and hypothalamus generally were found to be low in IV signals.

The ESG antibody labels the brain rather modestly and signals are barely discernible from intrinsic fluorescence in many brain areas where green-fluorescent secondary antibody is used for visualization. However, where signals were found high with the IV antibody, ESG signals are also detected with high intensity. For instance, ESG signals are easily detected in the hippocampus, striatum and enriched in some parts of allocortex (e.g., entorhinal, perirhinal, and olfactory cortices). In the neocortex and cerebellar cortex, sparse labeling patterns on superficial layers are observed. We note that substantia nigra pars reticulata and ventral tegmental area, the two main dopaminergic brain areas, are labeled in high intensities.

It is imperative to verify the signal strength of microwave fixation-assisted glycogen immunohistochemistry with the actual glycogen content. To this end, we have measured the glycogen contents in the microwave-treated cortex, striatum, hippocampus and cerebellum using a biochemical assay kit (K646–100, BioVision) As a result, we found a reasonable correspondence between glycogen content and relative immunohistochemical signal, except for the cerebellum (Oe et al. 2016). The cerebellar glycogen content was measured to be lower than relative immunohistochemical signals, most likely because the biochemical cerebellar samples included a large proportion of the white matter (arbor vitae) compared with other forebrain structures. Multiple studies have addressed glycogen content by brain region using biochemical assays on microwave fixed samples. We have rearranged published results from rats (Kong et al. 2002; Matsui et al. 2011; Sagar et al. 1987) and mice (Oe et al. 2016) in Fig. 3. These results agree well and the mouse glycogen immunohistochemistry data show that the hippocampus, cerebellum, and the cortex store relatively high levels of glycogen. Interestingly, striatal glycogen is found to be relatively low in both studies, which differs from our mouse study. It is possible that species differences exist in the storage and usage of glycogen. Additionally, regional glycogen storage may depend on metabolic states of the brain, including sleep cycle and circadian rhythm.

4 Astrocytes Store Glycogen in the Forebrain

The forebrain is the most recently developed division of the vertebrate brain and is responsible for higher order functions, such as perception, memory, and most likely, consciousness. The forebrain includes cortical structures and the basal forebrain, basal ganglia, thalamus, and hypothalamus. While there are regional variabilities in glycogen storage within the forebrain, glycogen signals are predominantly found in astrocytes in all areas of the forebrain. As such, IV glycogen immunohistochemistry results in astrocytic patterns. Neurons and microglia are predominantly devoid of IV

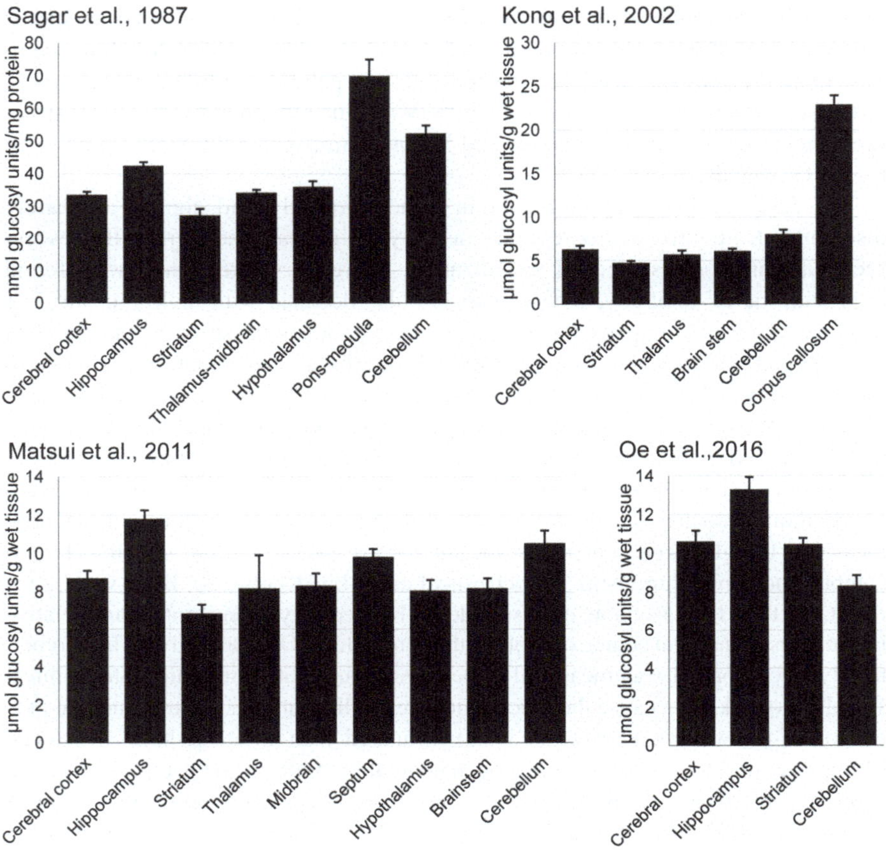

Fig. 3 Biochemical assessment of glycogen content by major brain region. Four published studies are presented. These measurements are made using microwave-fixed brains. Measurements are from rats for Sagar et al. (1987), Kong et al. (2002), and Matsui et al. (2011) while measurements by Oe et al. are from mice. Bar graph are replotted from respective original publications with bar orders modified for comparison purposes

signals. Astrocytic IV signals are characterized by high-intensity and dense punctate patterns (Fig. 4). Less intense and relatively smooth signals are observable in astrocytic cytosol in between IV puncta. By contrast, ESG signals are exclusively punctate and there is hardly any discernible signals outside puncta (Oe et al. 2016). Whether the punctate patterns labeled by IV and ESG overlap has not been addressed; however, these punctate patterns are emphasized in astrocytic microprocesses (Fig. 4a) and endfeet (Fig. 4b), rather than somata or primary branches. The distinct punctate distribution pattern supports the supposed role of glycogen as intermediate energy storage and the source of lactate for long-term synaptic plasticity. Moreover, enriched accumulation of glycogen in perisynaptic microprocesses and endfeet has been confirmed by a recent study using serial block-face scanning electron microscopy (Calì et al. 2016).

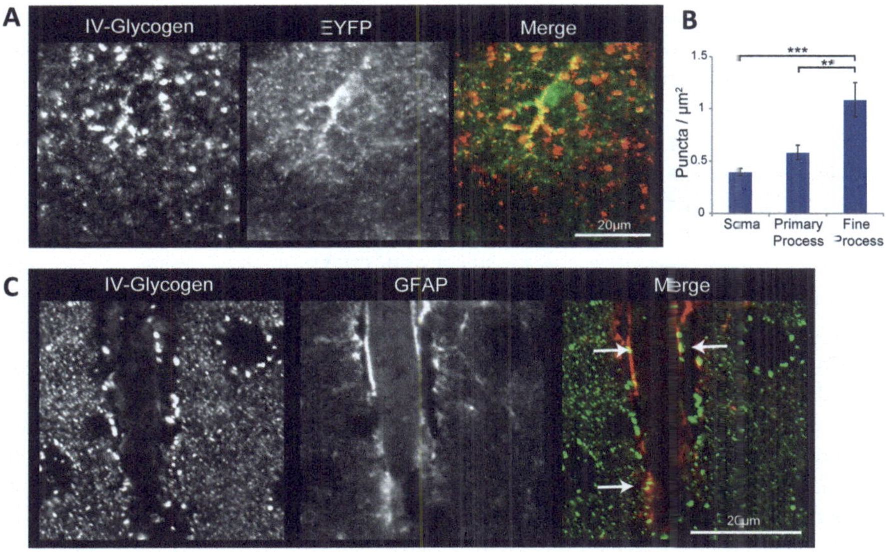

Fig. 4 Glycogen distribution in the cerebral cortex gray matter of the mouse. (**a**) Double immunohistochemistry showing IV glycogen antibody-labeling and YFP-labeled single astrocyte morphology. IV glycogen immunohistochemistry shows that glycogen is preferentially localized in fine processes of astrocytes. (**b**) Quantification of immunolabeled puncta shows preferential localization of glycogen in fine process of astrocytes. (**c**) Perivascular glycogen distribution. Large and intense glycogen puncta are observable along the vasculature. GFAP immunohistochemistry labels astrocytic endfeet which delineate the vasculature. Arrows indicate large and intense glycogen puncta that overlap with GFAP signal. Reproduced from Oe et al. (2016). GFAP: glial fibrillary acidic protein

Glycogen macromolecules are known to be organized in multiple levels of complexity; beta particles which are essentially single-core glycogen molecules, and alpha particles which are an aggregation of multiple glycogen molecules. Each beta particle has a dimer of glycogenin as the core where glycogen chains are originated. Beta particles have a maximum diameter of ~50 nm when glucose is fully stored, whereas alpha particles can be as large as a few hundred nanometers and observable in the liver (for concise reviews, Obel et al. 2012; Prats et al. 2018). Interestingly, a recent report showed that glycogenin is not a requirement of glycogen particle formation and paradoxically, glycogenin deficiency results in higher glycogen accumulation with larger particles (Testoni et al. 2017). It remains unknown if all brain glycogen particles are strictly associated with glycogenin.

An important point to consider is that both ESG and IV antibodies are IgM type and hence have a diameter of 40 nm, a size similar to a medium-size beta glycogen particle (Fig. 2). Puncta signals obtained by glycogen immunohistochemistry has aggregated appearances hinting at the existence of beta particle aggregates. Indeed classical papers by Phelps and Koizumi exemplify astrocytic glycogen particle aggregates in their electronmicrographs (Koizumi 1974; Phelps 1972). Whether such aggregates qualify to be alpha particles is an open question. It is further tempting to speculate that the epitope for the ESG antibody is somewhat associated with a conformation of beta particle aggregates. Further structural studies are required to reveal the true nature of antibody affinity.

Classical electron microscopy studies have additionally identified electron-dense glycogen granules in pericytes, which are scaffold cells of cerebral capillaries and venules (Cataldo and Broadwell 1986). IV glycogen immunohistochemistry shows strong condensation of particles around blood vessels (Fig. 4b). A majority of perivascular glycogen signals are co-localized with GFAP-positive processes, indicating astrocytic endfeet. Co-registration of electron microscopy and ^{13}C-labeled glucose by ultra-high spatial resolution ion microprobe imaging (NanoSIMS) showed incorporation of glucose to glycogen in endfeet and other parts of astrocytes (Takado et al. 2015), though the amount of de novo glucosyl unit incorporation is 25 times smaller than that of hepatocytes in fasted mice. Additionally, there are scattered perivascular signals, which may be of pericyte origin. Pericytic glycogen distribution has not been characterized in detail, possibly because there is not an antibody that uniquely labels the morphology of pericytes. Fortunately, the proteoglycan NG2 promoter-driven transgenic mouse expresses DsRed in both oligodendrocyte precursor cells (so-called NG2-positive glia) and pericytes (Zhu et al. 2007). Future studies should be able to address pericytic glycogen distribution with this mouse strain.

5 Glycogen Distribution in Distinct Mouse Brain Regions

5.1 *Cerebral Cortex*

As stated previously, the total amount of cerebral cortical glycogen is relatively high compared with diencephalic structures such as the thalamus and hypothalamus. Moreover, large glycogen puncta are observable by ESG glycogen immunohistochemistry. Generally, the temporal lobe accumulates higher amounts of glycogen than other lobes. Within the temporal lobe, the piriform cortex stands out in glycogen content. Outside the temporal lobe, glycogen levels are modestly rich and regional variability across different areas of the cortex is modest.

A common characteristic of cortical glycogen distribution is layer dependency. Generally, superficial layers accumulate more glycogen than deep layers. In particular, layer 1 shows the most intense glycogen immunohistochemistry signal with large glycogen puncta which are also labeled by the ESG antibody (Fig. 5a). This may be related to the fact that synapse density is the highest in layer 1 and decreases with depth in the barrel cortex of developed mice (Chandrasekaran et al. 2015), in particular the synapse density of asymmetrical (excitatory) synapses (DeFelipe et al. 2002). Similarly, glycogen signal decreases with cortical depth and is the lowest in layer 6 (Fig. 5b), where the lowest synapse density within the cortical

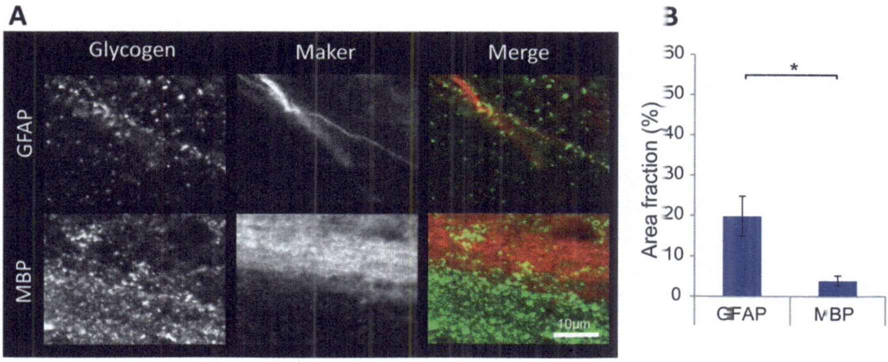

Fig. 5 Glycogen distribution in the cerebral cortex white matter. (**a**) In the upper panel, the corpus callosum is immunolabeled for glycogen (IV) and GFAP. Glycogen puncta are observed along the GFAP positive structure, suggesting astrocytic localization of glycogen. The lower panel represents the corpus callosum/alveus and hippocampal gray matter border. Strong myelin basic protein (MBP, oligodendrocyte marker) signals are seen in the cortical white matter, where glycogen signals are generally low. (**b**) Quantification of glycogen puncta by confocal microscopy indicates a preferential localization in astrocytes to oligodendrocytes (reproduced from Oe et al. (2016))

parenchyma has been reported (Chandrasekaran et al. 2015; DeFelipe et al. 2002; Schuz and Palm 1989). The positive correlation between glycogen storage and synapse density supports the idea that neuronal synapses are the consumers of glycogen metabolites (e.g., lactate). It is imperative to note however, that the synaptic organization of the mouse neocortex is considerably different from that of the rat or the human (Anton-Sanchez et al. 2014; DeFelipe et al. 2002; Palomero-Gallagher and Zilles 2017). Likewise, the periallocortex (e.g. entorhinal cortex) may have a different profile due to its distinct synaptic organization.

A natural question that arises is if the layer-dependent glycogen storage reflects the metabolic rate of a given layer. In an attempt to seek the relationship between constitutive metabolic activity and cortical layer, aerobic metabolism capability was visualized by cytochrome-c oxidase staining. Cytochrome-c oxidase signals are classically known to be obtained where energy demand is supposedly high, such as in visual areas projected from the intact eye after monocular deprivation (Wong-Riley 1979) and in layer 4 barrels of the whisker somatosensory cortex (Wong-Riley and Welt 1980). Comparison of cytochrome-c oxidase and glycogen signals in consecutive cortex sections, however, did not appear to be closely related. Rather, glycogen signals appeared relatively independent of cytochrome-c oxidase signals (Oe et al. 2016). For instance, the barrel structures were hardly discernible from IV glycogen immunohistochemistry. Therefore, it is reasonable to postulate that *constitutive* metabolic activity in the cortex might not rely heavily on glycogenolysis and is supported by glucose supply that comes from the dense network of blood circulation. Nevertheless, it is noted that *evoked* intense neuronal activity by sensory stimulation leads to local glycogenolysis in the corresponding somatosensory area, as shown by ^{14}C(3,4)glucose autoradiography (Swanson et al. 1992). One model that has gained experimental support is that glycogenolysis, and the resultant metabolite shuttle to nearby synapses, is a key mediator of synaptic plasticity and learning (Gibbs et al. 2006; Newman et al. 2011; Suzuki et al. 2011; Wang et al. 2017). According to this view, glycogen is consumed in an on-demand manner and can be regarded as a reserve metabolite storage rather than constant supply of energy substrate.

Relatively large immunolabeled puncta (>0.5 μm diameter) are observable with the ESG antibody, especially in superficial layers. Large ESG puncta are distributed in a pattern with relatively busy clusters and sparse areas. This patchy trend is more obvious in the hippocampus and striatum as described in the next sections. Meanwhile, it is noteworthy that superficial layers, especially layer 1, show patchy (though less pronounced) patterns. Of note, experience-dependent synaptic plasticity has been observed in the adult brain in layer 1 (Trachtenberg et al. 2002).

The cortical white matter—the corpus callosum—shows glycogen immunoreactivity, albeit at lower levels than the gray matter. We present both IV and ESG immunoreactivity in the corpus callosum and hippocampal alveus in Fig. 6. As analyzed

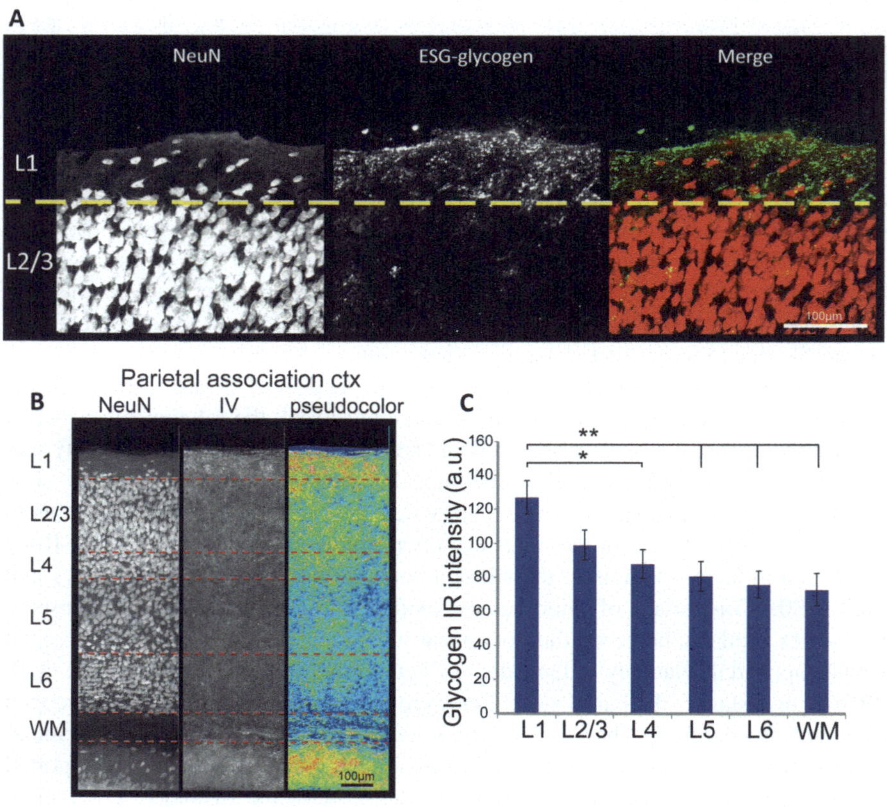

Fig. 6 Layer-dependent glycogen distribution in the mouse cortex. (**a**) ESG glycogen immunohistochemistry shows an enriched condensation of glycogen in layer 1. (**b** and **c**) IV glycogen immunohistochemistry shows a layer-dependent glycogen distribution pattern where glycogen signals are higher in superficial layers (reproduced from Oe et al. (2016))

by Oe et al. (2016), the majority of glycogen signals are detected in astrocytes. Oligodendrocytic localization of glycogen particles has been evaluated in myelin basic protein-positive structures, showing an order of magnitude smaller signal strengths, which may be signal leak from white matter astrocytes due to the resolution limit of confocal microscopy. Indeed, published electron microscopic observations have reported lack of glycogen in cortical and callosal oligodendrocytes (Caley and Maxwell 1968; Mori and Leblond 1969) while callosal astrocytes store abundant amounts of glycogen (Mori and Leblond 1969). However, in peripheral nerves, myelin-forming Schwann cells have been reported to store glycogen (Brown et al. 2012). The functional significance of white matter glycogen has been established as energy source mobilized during energy-scarce conditions (e.g., anoxia and hypoglycemia) in the optic nerve (Brown et al. 2003; Wender et al. 2000).

5.2 *Hippocampus*

The hippocampus is an allocortical structure known for its critical role in episodic memory formation. Hippocampal glycogenolysis has been shown to be crucial for long-term plasticity and consolidation of memory in the recent decade by pharmacological inhibition of glycogen phosphorylase using 1,4-dideoxy-1,4-imino-d-arabinitol (Newman et al. 2011; Suzuki et al. 2011). Glycogenolysis and the downstream metabolism have been suggested to be promoted by the beta-2 noradrenergic receptor (Gao et al. 2016). Hippocampal glycogen levels decrease by prolonged exhaustive exercises, during which multiple major monoamine levels (dopamine, noradrenaline, serotonin) are elevated (Matsui et al. 2011). These monoamines are released by volume transmission and activate astrocytic metabotropic pathways via respective G protein-coupled receptors (for a review, Hirase et al. 2014). Of note, recent papers suggest co-release of noradrenaline and dopamine from classical noradrenergic fibers originating in the locus coeruleus (Kempadoo et al. 2016; Takeuchi et al. 2016). Given the involvement of glycogen in synaptic plasticity and memory consolidation, its distribution in the hippocampus is an important topic and gains further insights into how glycogen metabolism is organized in behaving animals.

The hippocampus is a glycogen-rich structure and the IV antibody labels both the hippocampus proper and dentate gyrus intensely. Glycogen amount has a layer dependence. Principal cell body layers are low in glycogen content most likely due to the highly condensed neuronal somatic presence and hence relatively scarce astrocytic somata and processes (Ogata and Kosaka 2002). By contrast, layers populated by principal cell dendrites have high glycogen levels. These layers include strata oriens, radiatum, lucidum (CA3 only), and lacunosum-moleculare of the hippocampus proper; stratum moleculare and hilus of the dentate gyrus. Glycogen distribution appears seamless across CA1, CA2, and CA3. There is an impression that the CA3 stratum lucidum has subtly lower glycogen signals than other neuropil layers.

Remarkably, the ESG antibody labels the hilus more intensely than all other layers. The functional significance of high glycogen signals in the hilus remains elusive, yet there is an intriguing coincidence that the hilus receives the most intense noradrenergic innervation amongst all hippocampal layers (Loy et al. 1980), hinting at a noradrenergic drive of glycogen metabolism. The hilus is also known as the polymorphic layer and contains unmyelinated axons from dentate granule cells and various types of GABAergic interneurons. Hilar interneurons have been reported to discharge at high rates during dentate spikes—spontaneous, massive, synchronized neuronal activities originating in the dentate gyrus (Bragin et al. 1995). Moreover, neurogenesis continues to occur in the subgranular zone in adulthood, making the hilus subject to axonal proliferation which accompanies ATP-dependent actin polymerization. These features are suggestive of high energy demands in the hilus, and glycogen storage may guarantee sustained energy supply.

Similar to the cortex, IV antibody shows high-intensity punctate patterns with smooth low-intensity signals in the cytosol, the former presumably labeling beta glycogen aggregates. ESG glycogen immunohistochemistry shows a striking pattern in young adult mice in that large puncta are distributed in a "patchy" pattern consisting of puncta-busy and puncta-sparse areas (Fig. 7). The patchy patterns are present in the neuropil areas of the hippocampus proper and dentate gyrus, while patchy patterns are more obvious in the hippocampus proper. The patches are delineated by the morphology of individual astrocytes: there are astrocytes with dense ESG-labeled glycogen puncta and another population of astrocytes devoid of ESG labeling (Oe et al. 2016). Whether these glycogen-puncta-rich and -poor astrocytes comprise two molecularly distinct astrocyte populations or the glycogen storage difference is induced by local environmental factors is not known. Interestingly, the patchy glycogen pattern generally disappears in aged mice. IV glycogen immunohistochemistry shows more uniform distribution in the hippocampus of aged mice, whereas ESG glycogen immunohistochemical signals become low with sporadic high-signal spots reminiscent of polyglucosan bodies. Such alterations in glycogen IHC signals in the aged mouse brain could reflect the known aging-induced increase in *corpora amylacea*.

5.3 Striatum

The striatum, a major subcortical structure in rodents, is known for its function in motor executive functions, reward cognition, reinforcement learning, and motivational salience. Two major afferents to the striatum are the glutamatergic and dopaminergic projections from the cortex and substantia nigra, respectively. The striatum is distinct from other forebrain structures in that the principal cells are GABAergic. In fact, GABAergic medial spiny neurons represent >95% of the neuronal population. Medial spiny neurons are classified into D1 and D2 dopamine receptor-dominant types, which form the direct and indirect afferent pathways, respectively. There does not seem to be a strict spatial organization of D1 and D2 medial spiny neurons and they

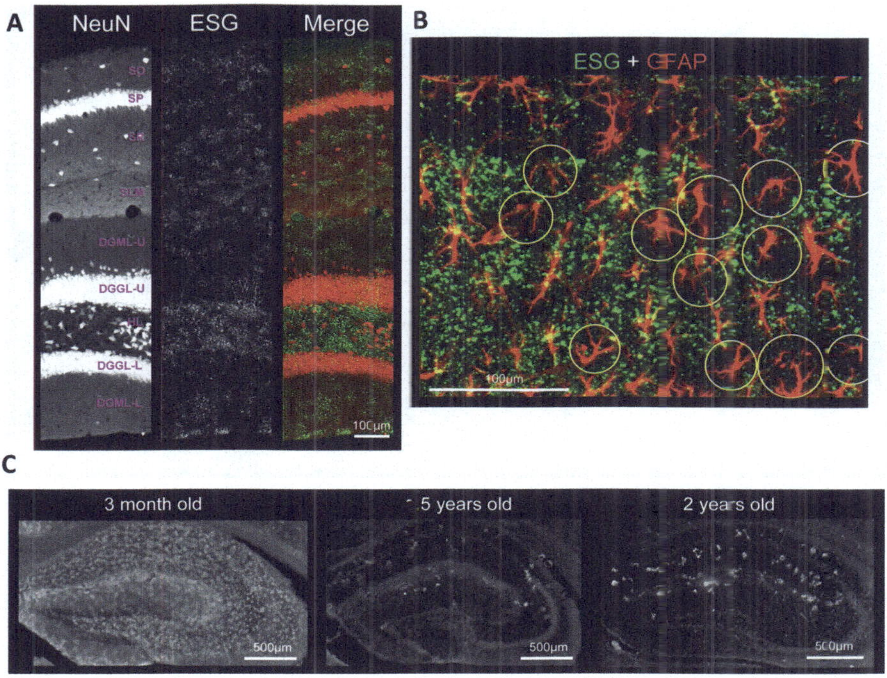

Fig. 7 Patchy glycogen distribution in the mouse hippocampus. (**a**) ESG glycogen immunohistochemistry across hippocampal layers. ESG signals are clustered in patches in dendritic layers (SO, SP, SR, SLM) of dorsal CA1. SO: stratum oriens, SP: stratum pyramidale, SR: stratum radiatum, SLM: stratum lacunosum-moleculare, DGML-U/L: dentate gyrus molecular layer upper/lower blade, DGGL-U/L: dentate gyrus granular layer upper/lower blade, HL hilus. (**b**) Double immunohistochemistry of ESG-labeled glycogen (green) and GFAP (red) shows that patchy glycogen distribution is organized by individual astrocytes. Yellow circles indicate astrocytes that comprise off-patch areas. The sample is from CA1 SP (upper ~15%) and SR (lower ~85%). (**c**) Patchy distribution disappear in aged mice and sporadic high-signal-intensity spots are evident. Reproduced from Oe et al. (2016) with modifications

appear to be homogeneously intermingled. Striatal cytoarchitecture is characterized by the frequent presence of myelinated fiber bundles of cortical and thalamic origins that pierce through the neuropil in stripes. In addition to the white/gray matter segregation, the striatum has histologically identified blobs (e.g., by μ-opioid receptor expression) called the striosomes (also known as the "patches," but here we avoid this term—not to be confused with the patchy glycogen distribution), while the remaining part is referred to as the matrix. In mice coronal sections, the striosomes appear as irregularly shaped areas of diagonal size 100–300 μm and are 300–600 μm apart.

Despite the vastly different cytoarchitecture, mouse striatal glycogen distribution share similar patterns to that of hippocampus proper neuropil. The neuropil is labeled well by IV and ESG antibodies, whereas the fiber bundle white matter yields relatively low glycogen signals. Remarkably, patchy glycogen puncta distributions

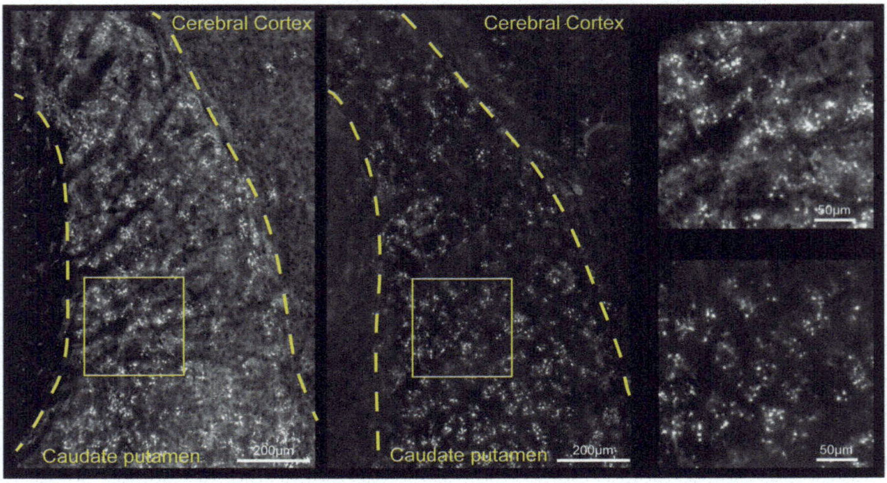

Fig. 8 Glycogen distribution in the striatum. IV (left) and ESG (center) glycogen immunohistochemistry in the mouse striatum (caudate putamen). Dashed lines indicate the borders with the internal capsule (to the left, basal ganglia white matter) and cerebral cortex (to the right). IV glycogen immunohistochemistry shows that striatal fiber bundles (white matter) are devoid of glycogen signals. Patchy patterns are observable with ESG signals. Areas marked by yellow squares are magnified in the right panels

are observed throughout the extent of striatal neuropil. As in the hippocampus, the patchy distribution is organized by intermingled distribution of ESG-immunoreactive and relatively ESG-free astrocytes (Fig. 8). Striatal patchy patterns are organized with a patch size of 50–150 μm, which is smaller than striosomes. Whether the glycogen patches are related to the distribution of D1 and D2 medial spiny neurons remains to be addressed. However, since medial spiny neurons are fairly densely distributed (< 50 μm apart) and their dendrites do not distinguish territories, heterogeneous astrocytic glycogen amounts are not deemed to be defined by proximity to D1 or D2 medial spiny neurons. A recent study suggested that striatal astrocytes are functionally classified into subtypes depending on their preference to signal to either medial spiny neuron subtype (Martin et al. 2015). It remains possible that this astrocytic diversity is related to the patchy glycogen pattern.

5.4 *Olfactory Bulb*

The olfactory bulb is a specialized allocortical structure that processes odor information. It is stratified with seven cytoarchitecturally distinct layers, namely: olfactory nerve layer; glomerular layer, external plexiform layer; mitral cell layer; internal plexiform layer; granule cell layer; and subependymal layer. It receives projections from the olfactory epithelium that terminate in the glomerular layer via the olfactory nerve layer.

The principal cells of the olfactory bulb are mitral and tufted cells whose dendrites comprise glomeruli, the characteristic neuropil-rich structures that receive olfactory nerve projections. Each glomerulus represents an elementary feature of a smell, and the combination of glomerular output is considered to encode complex olfactory information that is passed onto the olfactory cortex.

Overall, the olfactory bulb glycogen signal level is comparable to that in the cerebral cortex according to the strength of IV signal. IV glycogen immunohistochemistry delineates cytoarchitectural characteristics of the olfactory bulb, which is an indication that the storage of glycogen depends on cell type. Evidently, the mitral cell layer is virtually devoid of glycogen signals and glomeruli are outlined with a strip of low signal intensity area that coincide with the territory of juxtaglomerular (aka periglomerular) neurons. The granule cell layer is generally low in glycogen levels except for a mesh-like structure presumably representing astrocytic syncytium. Unlike the hippocampus, patchy patterns are not observed in the neuropil of external and internal plexiform layers or within a glomerulus. Glomeruli are easily distinguishable because of their relatively high glycogen content compared to external plexiform layer neuropil (Fig. 9). These relative glycogen levels are analogous to the expression of glycogen phosphorylase activity observed in the young rat (Coopersmith and Leon 1987). The olfactory bulb receives heavy neuromodulatory projections including noradrenergic, serotonergic, and cholinergic fibers. In addition, some 10% of juxtaglomerular neurons are dopaminergic. Noradrenergic activity-induced glycogenolysis in the rat olfactory bulb has been demonstrated using a biochemical glycogen assay (Coopersmith and Leon 1995). According to this work, beta adrenergic receptor activation is the main trigger of glycogenolysis while alpha adrenergic receptor (presumably alpha-1) activation becomes gradually dominant as the rat matures.

5.5 *Non-forebrain Areas*

Outside the forebrain, the cerebellum and brain stem are high in glycogen storage. Neuronal localization of glycogen in non-forebrain areas has been reported previously. For example, in the facial nucleus of the rat brain stem, glycogen immunohistochemical puncta are mainly found in motor neurons as identified by the expression of NR3B (NMDA receptor subunit 3B) cells (Takezawa et al. 2015). The same study also reports relative scarceness of astrocytic glycogen, although we shall note that glycogen immunohistochemistry was performed on perfusion-fixed brains. Also, we find that brain stem neurons with large somata (>30 μm diameter) tend to accumulate visible amounts of glycogen puncta.

Most literature agrees that the cerebellum is a rich, if not the richest, site of glycogen storage. The cerebellar cortex consists of three layers: the molecular, Purkinje, and granular layer. The Purkinje layer is the cell body layer of Purkinje cells (the GABAergic principal output cells) and Bergmann glia that enwrap the synapses of Purkinje cells in full (Grosche et al. 1999). Purkinje cells' dendrites are formed in

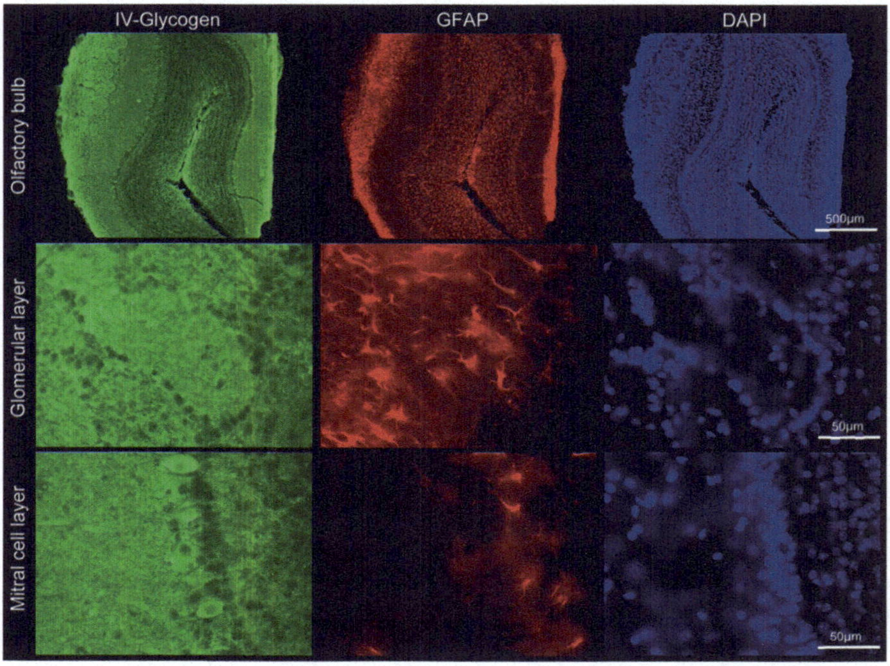

Fig. 9 Glycogen distribution in the olfactory bulb. IV glycogen immunohistochemistry (left) was performed together with GFAP immunohistochemistry (center) and DAPI staining (right). Glomerular and mitral cell layers are magnified in middle and bottom panels, respectively (the olfactory nerve layer is to the left). Glycogen distributions show a characteristic layer-dependent pattern. Glomeruli show high glycogen signals. Whereas periglomerular somata are low in glycogen signal. Mitral cell somata are low in glycogen, however, a population of neuron-like somata show visible green fluorescence

the molecular layer which is superficial to the Purkinje layer. A Purkinje cell has up to 200,000 spines that form synapses with parallel fibers (Napper and Harvey 1988), projecting axons from granule cells that run perpendicularly to the Purkinje cell dendrite plane. In addition, each Purkinje cell is innervated by a climbing fiber which provides powerful excitatory input. Finally, the granular layers are where glutamatergic granule cell bodies are located. Instead of Bergmann glia, the granular layer is populated with so-called velate astrocytes, a kind of protoplasmic astrocytes that ensheath granule cells. Glycogen immunohistochemistry reveals that the molecular layer of the cerebellar cortex has the most intensive glycogen immunohistochemical signals in the mouse brain. Moreover, the granular layer shows scattered distribution of punctate glycogen signals, showing signs of velate astrocytic localization. Bergmann glia accumulate glycogen as evidenced by IV and ESG signals from relatively small somata located in between Purkinje cells. Further characterization of glycogen distribution in the cerebellar cortex is desired and we are currently conducting a more detailed examination of cerebellar glycogen distribution in the mouse.

6 Concluding Remarks

In this chapter, we have described murine brain glycogen distribution according to glycogen immunohistochemistry. Glycogen immunohistochemistry allowed the assessment of regional variability. As a result, we were able to find layer differences in cortical structures and patchy patterns in the hippocampus, striatum, and to a smaller extent, superficial layers of the cortex.

An important issue to remember is that glycogen is a dynamic molecule whose size depends on the metabolic state, which is most likely governed by brain states. Brain glycogen immunohistochemistry data presented here and in Oe et al. (2016) are from mice sacrificed at the end of the light phase of 12/12 h light/dark housing condition. Glycogen distribution may differ in animals prepared at different hours, behavioral conditions, and levels of maturity. In fact, we found that the patchy ESG-labeled glycogen patterns disappear in aged mice. Instead, the ESG antibody labels much larger clots, presumably polyglucosan bodies, that appear frequently (5–30 per hippocampal slice depending on age) in the hippocampus. A previous report suggested that such polyglucosan accumulation is observed in neurons (Sinadinos et al. 2014). More recently, proteomic analyses in the hippocampus revealed a shift of glycogen metabolism enzymes' expression from astrocytes to neurons as mice age (Drulis-Fajdasz et al. 2018). Such a change in metabolism may underlie the disappearance of patchy pattern in aged mice, while functional significances of patchy glycogen distribution remain elusive. Another situation where glycogen storage is altered in astrocytes is inflammation (Shimizu and Hamuro 1958). An electron microscopy study reported that astrocytes accumulate higher amounts of glycogen after radioactive irradiation (Maxwell and Kruger 1965). Further investigations are needed to identify the dynamic nature of brain glycogen distribution in health and disease.

Acknowledgements This work was supported by the RIKEN Brain Science Institute, KAKENHI grants (26117520, 16H01888, 18H05150), and HFSP (RGP0036/2014). We thank members of the laboratory for comments on earlier versions of the manuscript. The authors declare no competing financial interests. SA is supported by the RIKEN IPA program.

References

Anton-Sanchez L, Bielza C, Merchán-Pérez A, Rodríguez J-R, DeFelipe J, Larrañaga P (2014) Three-dimensional distribution of cortical synapses: a replicated point pattern-based analysis. Front Neuroanat 8:85

Baba O (1993) Production of monoclonal antibody that recognizes glycogen and its application for immunohistochemistry. Kokubyo Gakkai Zasshi 60:264–287

Bragin A, Jando G, Nadasdy Z, van Landeghem M, Buzsaki G (1995) Dentate EEG spikes and associated interneuronal population bursts in the hippocampal Hilar region of the rat. J Neurophysiol 73:1691–1705

Brown AM, Tekkök SB, Ransom BR (2003) Glycogen regulation and functional role in mouse white matter. J Physiol 549:501–512

Brown AM, Evans RD, Black J, Ransom BR (2012) Schwann cell glycogen selectively supports myelinated axon function. Ann Neurol 72:406–418

Caley DW, Maxwell DS (1968) An electron microscopic study of the neuroglia during postnatal development of the rat cerebrum. J Comp Neurol 133:45–69

Calì C, Baghabra J, Boges DJ, Holst GR, Kreshuk A, Hamprecht FA, Srinivasan M, Lehväslaiho H, Magistretti PJ (2016) Three-dimensional immersive virtual reality for studying cellular compartments in 3D models from EM preparations of neural tissues. J Comp Neurol 524:23–38

Cataldo AM, Broadwell RD (1986) Cytochemical identification of cerebral glycogen and glucose-6-phosphatase activity under normal and experimental conditions. II. Choroid plexus and ependymal epithelia, endothelia and pericytes. J Neurocytol 15:511–524

Chandrasekaran S, Navlakha S, Audette NJ, McCreary DD, Suhan J, Bar-Joseph Z, Barth AL (2015) Unbiased, high-throughput electron microscopy analysis of experience-dependent synaptic changes in the neocortex. J Neurosci 35:16450–16462

Choi HB, Gordon GRJ, Zhou N, Tai C, Rungta RL, Martinez J, Milner TA, Ryu JK, McLarnon JG, Tresguerres M, Levin LR, Buck J, MacVicar BA (2012) Metabolic communication between astrocytes and neurons via bicarbonate-responsive soluble adenylyl cyclase. Neuron 75:1094–1104

Cline GW, Petersen KF, Krssak M, Shen J, Hundal RS, Trajanoski Z, Inzucchi S, Dresner A, Rothman DL, Shulman GI (1999) Impaired glucose transport as a cause of decreased insulin-stimulated muscle glycogen synthesis in type 2 diabetes. N Engl J Med 341:240–246

Coopersmith R, Leon M (1987) Glycogen phosphorylase activity in the olfactory bulb of the young rat. J Comp Neurol 261:148–154

Coopersmith R, Leon M (1995) Olfactory bulb glycogen metabolism: noradrenergic modulation in the young rat. Brain Res 674:230–237

DeFelipe J, Alonso-Nanclares L, Arellano JI (2002) Microstructure of the neocortex: comparative aspects. J Neurocytol 31:299–316

Drulis-Fajdasz D, Gizak A, Wójtowicz T, Wiśniewski JR, Rakus D (2018) Aging-associated changes in hippocampal glycogen metabolism in mice. Evidence for and against astrocyte-to-neuron lactate shuttle. Glia 66:1481–1495

Fiala JC, Kirov SA, Feinberg MD, Petrak LJ, George P, Goddard CA, Harris KM (2003) Timing of neuronal and glial ultrastructure disruption during brain slice preparation and recovery in vitro. J Comp Neurol 465:90–103

Gao V, Suzuki A, Magistretti PJ, Lengacher S, Pollonini G, Steinman MQ, Alberini CM (2016) Astrocytic β2-adrenergic receptors mediate hippocampal long-term memory consolidation. Proc Natl Acad Sci 113:8526–8531

Gibbs ME, Anderson DG, Hertz L (2006) Inhibition of glycogenolysis in astrocytes interrupts memory consolidation in young chickens. Glia 54:214–222

Grosche J, Matyash V, Moller T, Verkhratsky A, Reichenbach A, Kettenmann H (1999) Microdomains for neuron-glia interaction: parallel fiber signaling to Bergmann glial cells. Nat Neurosci 2:139–143

Hirase H, Iwai Y, Takata N, Shinohara Y, Mishima T (2014) Volume transmission signalling via astrocytes. Philos Trans R Soc B 369:20130604

Kempadoo KA, Mosharov EV, Choi SJ, Sulzer D, Kandel ER (2016) Dopamine release from the locus coeruleus to the dorsal hippocampus promotes spatial learning and memory. Proc Natl Acad Sci 113:14835–14840

Koizumi J (1974) Glycogen in the central nervous system. Prog Histochem Cytochem 6:1–37

Kong J, Shepel PN, Holden CP, Mackiewicz M, Pack AI, Geiger JD (2002) Brain glycogen decreases with increased periods of wakefulness: implications for homeostatic drive to sleep. J Neurosci 22:5581–5587

Loy R, Koziell DA, Lindsey JD, Moore RY (1980) Noradrenergic innervation of the adult rat hippocampal formation. J Comp Neurol 189:699–710

Magnusson I, Rothman DL, Katz LD, Shulman RG, Shulman GI (1992) Increased rate of gluconeogenesis in type II diabetes mellitus. A 13C nuclear magnetic resonance study. J Clin Invest 90:1323–1327

Martin R, Bajo-Graneras R, Moratalla R, Perea G, Araque A (2015) Circuit-specific signaling in astrocyte-neuron networks in basal ganglia pathways. Science 349:730–734

Matsui T, Soya S, Okamoto M, Ichitani Y, Kawanaka K, Soya H (2011) Brain glycogen decreases during prolonged exercise. J Physiol 589:3383–3393

Maxwell DS, Kruger L (1965) The fine structure of astrocytes in the cerebral cortex and their response to focal injury produced by heavy ionizing particles. J Cell Biol 25:141–157

Mori S, Leblond CP (1969) Electron microscopic features and proliferation of astrocytes in the corpus callosum of the rat. J Comp Neurol 137:197–225

Nakamura-Tsuruta S, Yasuda M, Nakamura T, Shinoda E, Furuyashiki T, Kakutani R, Takata H, Kato Y, Ashida H (2012) Comparative analysis of carbohydrate-binding specificities of two anti-glycogen monoclonal antibodies using ELISA and surface plasmon resonance. Carbohydr Res 350:49–54

Napper RMA, Harvey RJ (1988) Number of parallel fiber synapses on an individual Purkinje cell in the cerebellum of the rat. J Comp Neurol 274:168–177

Newman LA, Korol DL, Gold PE (2011) Lactate produced by glycogenolysis in astrocytes regulates memory processing. PLoS One 6:e28427

Obel LF, Müller MS, Walls AB, Sickmann HM, Bak LK, Waagepetersen HS, Schousboe A (2012) Brain glycogen-new perspectives on its metabolic function and regulation at the subcellular level. Front Neuroenerg 4:3

Oe Y, Baba O, Ashida H, Nakamura KC, Hirase H (2016) Glycogen distribution in the microwave-fixed mouse brain reveals heterogeneous astrocytic patterns. Glia 64:1532–1545

Ogata K, Kosaka T (2002) Structural and quantitative analysis of astrocytes in the mouse hippocampus. Neuroscience 113:221–233

Öz G, Henry PG, Seaquist ER, Gruetter R (2003) Direct, noninvasive measurement of brain glycogen metabolism in humans. Neurochem Int 43:323–329

Öz G, Seaquist ER, Kumar A, Criego AB, Benedict LE, Rao JP, Henry P-G, Van De Moortele P-F, Gruetter R (2007) Human brain glycogen content and metabolism: implications on its role in brain energy metabolism. Am J Physiol Metab 292:E946–E951

Palomero-Gallagher N, Zilles K (2017) Cortical layers: cyto-, myelo-, receptor- and synaptic architecture in human cortical areas. NeuroImage 197:716–741

Phelps CH (1972) Barbiturate-induced glycogen accumulation in brain. An electron microscopic study. Brain Res 39:225–234

Prats C, Graham TE, Shearer J (2018) The dynamic life of the glycogen granule. J Biol Chem 293(19):7089–7098

Raichle ME, Gusnard DA (2002) Appraising the brain's energy budget. Proc Natl Acad Sci U S A 99:10237–10239

Sagar SM, Sharp FR, Swanson RA (1987) The regional distribution of glycogen in rat brain fixed by microwave irradiation. Brain Res 417:172–174

Schuz A, Palm G (1989) Density of neurons and synapses in the cerebral cortex of the mouse. J Comp Neurol 286:442–455

Shimizu N, Hamuro Y (1958) Deposition of glycogen and changes in some enzymes in brain wounds. Nature 181:781–782

Sinadinos C, Valles-Ortega J, Boulan L, Solsona E, Tevy MF, Marquez M, Duran J, Lopez-Iglesias C, Calbó J, Blasco E, Pumarola M, Milán M, Guinovart JJ (2014) Neuronal glycogen synthesis contributes to physiological aging. Aging Cell 13:935–945

Sugiura Y, Honda K, Kajimura M, Suematsu M (2014) Visualization and quantification of cerebral metabolic fluxes of glucose in awake mice. Proteomics 14:829–838

Suzuki A, Stern SA, Bozdagi O, Huntley GW, Walker RH, Magistretti PJ, Alberini CM (2011) Astrocyte-neuron lactate transport is required for long-term memory formation. Cell 144:810–823

Swanson RA, Morton MM, Sagar SM, Sharp FR (1992) Sensory stimulation induces local cerebral glycogenolysis: demonstration by autoradiography. Neuroscience 51:451–461

Takado Y, Knott G, Humbel BM, Escrig S, Masoodi M, Meibom A, Comment A (2015) Imaging liver and brain glycogen metabolism at the nanometer scale. Nanomedicine 11:239–245

Takano T, He W, Han X, Wang F, Xu Q, Wang X, Oberheim Bush NA, Cruz N, Dienel GA, Nedergaard M (2014) Rapid manifestation of reactive astrogliosis in acute hippocampal brain slices. Glia 62:78–95

Takeuchi T, Duszkiewicz AJ, Sonneborn A, Spooner PA, Yamasaki M, Watanabe M, Smith CC, Fernández G, Deisseroth K, Greene RW, Morris RGM (2016) Locus coeruleus and dopaminergic consolidation of everyday memory. Nature 537:357–362

Takezawa Y, Baba O, Kohsaka S, Nakajima K (2015) Accumulation of glycogen in axotomized adult rat facial motoneurons. J Neurosci Res 93:913–921

Testoni G, Duran J, García-Rocha M, Vilaplana F, Serrano AL, Sebastián D, López-Soldado I, Sullivan MA, Slebe F, Vilaseca M, Muñoz-Cánoves P, Guinovart JJ (2017) Lack of Glycogenin causes glycogen accumulation and muscle function impairment. Cell Metab 26:256–266.e4

Trachtenberg JT, Chen BE, Knott GW, Feng G, Sanes JR, Welker E, Svoboda K (2002) Long-term in vivo imaging of experience-dependent synaptic plasticity in adult cortex. Nature 420:788–794

Wang J, Tu J, Cao B, Mu L, Yang X, Cong M, Ramkrishnan AS, Chan RHM, Wang L, Li Y (2017) Astrocytic l -lactate signaling facilitates amygdala-anterior cingulate cortex synchrony and decision making in rats. Cell Rep 21:2407–2418

Wender R, Brown AM, Fern R, Swanson RA, Farrell K, Ransom BR (2000) Astrocytic glycogen influences axon function and survival during glucose deprivation in central white matter. J Neurosci 20:6804–6810

Wong-Riley M (1979) Changes in the visual system of monocularly sutured or enucleated cats demonstrable with cytochrome oxidase histochemistry. Brain Res 171:11–28

Wong-Riley MT, Welt C (1980) Histochemical changes in cytochrome oxidase of cortical barrels after vibrissal removal in neonatal and adult mice. Proc Natl Acad Sci U S A 77:2333–2337

Zhu X, Bergles DE, Nishiyama A (2007) NG2 cells generate both oligodendrocytes and gray matter astrocytes. Development 135:145–157

Technical and Comparative Aspects of Brain Glycogen Metabolism

Long Wu, Nicholas J. M. Butler, and Raymond A. Swanson

Abstract It has been known for over 50 years that brain has significant glycogen stores, but the physiological function of this energy reserve remains uncertain. This uncertainty stems in part from several technical challenges inherent in the study of brain glycogen metabolism, and may also stem from some conceptual limitations. Factors presenting technical challenges include low glycogen content in brain, non-homogenous labeling of glycogen by radiotracers, rapid glycogenolysis during postmortem tissue handling, and effects of the stress response on brain glycogen turnover. Here, we briefly review aspects of glycogen structure and metabolism that bear on these technical challenges, and discuss ways these can be overcome. We also highlight physiological aspects of glycogen metabolism that limit the conditions under which glycogen metabolism can be useful or advantageous over glucose metabolism. Comparisons with glycogen metabolism in skeletal muscle provide an additional perspective on potential functions of glycogen in brain.

Keywords 2-deoxyglucose · Brain · Carbon-13 · Glycogen · Immunohistochemistry · Microwave fixation · Radioisotope

1 Introduction

It has been known for over 50 years that brain has significant glycogen stores, but the physiological function of this energy reserve remains uncertain. This uncertainty stems in part from several technical difficulties inherent in the study of brain glycogen metabolism, and may also stem from some conceptual limitations. This chapter will briefly outline basic structural and bioenergetic aspects of glycogen metabolism, discuss how these lead to certain technical challenges for the study of brain glycogen, and compare what is known about glycogen metabolism in brain with skeletal muscle.

L. Wu · N. J. M. Butler · R. A. Swanson (✉)
Department of Neurology, University of California, San Francisco,
and San Francisco Veterans Affairs Health Care System, San Francisco, CA, USA
e-mail: Raymond.swanson@ucsf.edu

M. DiNuzzo, A. Schousboe (eds.), *Brain Glycogen Metabolism*, Advances in Neurobiology 23, https://doi.org/10.1007/978-3-030-27480-1_6

2 Glycogen Structure and Distribution

Glycogen is a large, branched polymer of glucose. The polymer form eliminates the high osmolarity that would result from an equimolar concentration of free glucose. It also protects the otherwise free "reducing" ends of glucose from auto-oxidation (Fig. 1), and provides a means of intracellular localization of the energy store. Glycogen polymers are roughly spherical and organized into concentric tiers, with the inner tiers composed by chains that normally contain two branches, and the

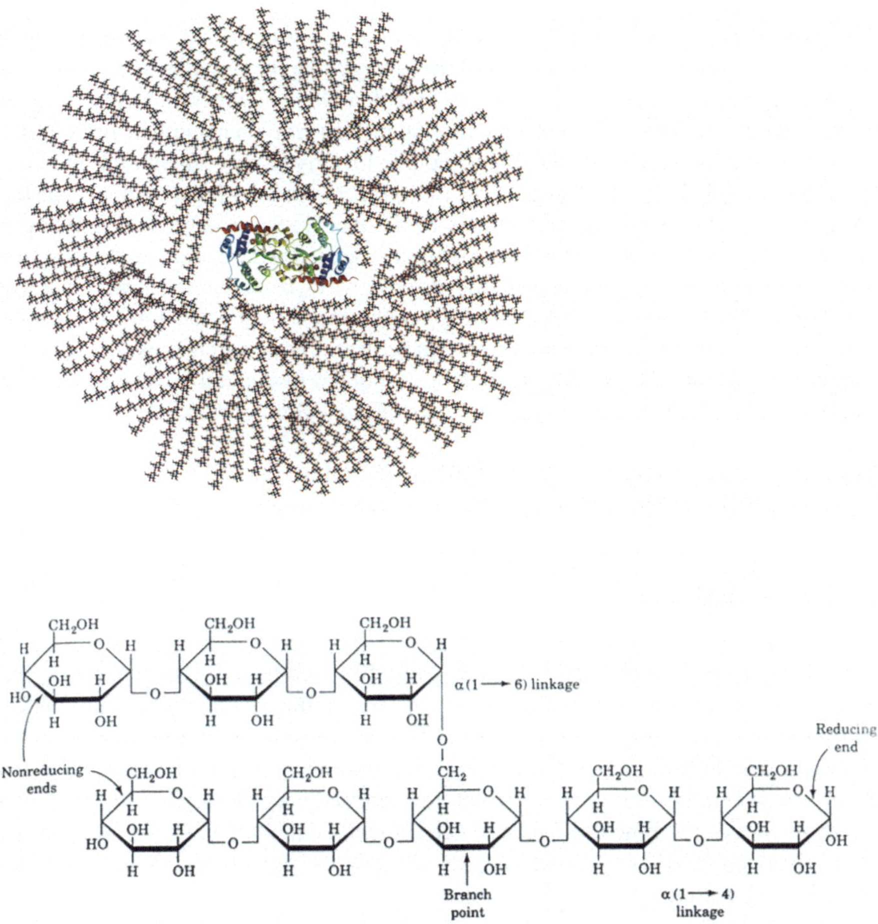

Fig. 1 Schematic two-dimensional cross-sectional view of glycogen. A core protein of glycogenin is surrounded by branches of glucose units. The entire globular granule may contain around 30,000 glucose units. The individual glucose moieties of glycogen are linked by α-1, 4-glycosidic bonds, with branch points at approximately every 12 glucose residues linked by α-1, 6-glycosidic bonds. The exposed ends of all glycogen chains are non-reducing. Image from Haggstrom (2014)

outer tiers composed of unbranched chains (Gunja-Smith et al. 1970). The linear chains of glucose residues are joined by α-1,4-glycosidic bonds, with branch points at approximately every 10–14 glucose residues linked by α-1,6-glycosidic bonds (Calder 1991). The exposed ends of all glycogen chains are non-reducing. Mathematical modeling suggests that the size of a glycogen molecule is limited to 12 tiers, which means a total of approximately 53,000 glucose residues and a radius of 21 nm. In accordance with this modeling, the glycogen particles found in the mouse brain and resting human skeletal muscle ranged from 10 to 44 nm in diameter (Cataldo and Broadwell 1986; Wender et al. 2000; Marchand et al. 2002) The average diameter of these glycogen particles is 20–30 nm, which is estimated to be 7–8 tiers (Goldsmith et al. 1982; Melendez-Hevia et al. 1993).

Glycogen polymers ("granules") are present in the cytosol, endoplasmic reticulum (ER)/ sarcoplasmic reticulum (SR) and lysosomes (Cardell Jr. 1977; Stapleton et al. 2010; Geddes et al. 1992). In liver cells, approximately 10% of all glycogen particles are found in lysosomes (Jiang et al. 2010), where they undergo slow degradation by acid maltase. Almost all tissues contain some detectable glycogen, but the relative amounts of glycogen vary enormously. Glycogen levels in the adult rat under physiological conditions are as follows: liver > > skeletal muscle > cardiac muscle > brain > kidney (Table 1). Glycogen in mammalian brain is localized primarily to astrocytes (Cali et al. 2016; Cataldo and Broadwell 1986; Wender et al. 2000; Koizumi 1974), but much smaller amounts are also found in meningeal cells,

Table 1 Glycogen content in different tissue types

Tissue	Glycogen (mg/g tissue)	Citation
Liver	30.23 ± 2.5	Vissing et al. (1989)
Liver	43.12 ± 5.5 *	Kusunoki et al. (2002)
Liver	32.3 ± 2.0	Khandelwal et al. (1979)
Heart	4.51 ± 0.33	Vissing et al. (1989)
Heart	4.16 ± 0.23	Conlee et al. (1989)
Kidney	0.06 ± 0.01	Khandelwal et al. (1979)
Kidney	0.03 ± 0.01 *	Nannipieri et al. (2001)
Muscle (GPS) [c]	5.94 ± 0.25 [a]	Baker et al. (2005)
Muscle (white gastrocnemius)	5.79 ± 0.72	Garetto et al. (1984)
Muscle (white gastrocnemius)	7.01 ± 0.17	Vissing et al. (1989)
Brain (cortex)	0.68 ± 0.03 [b]	Sagar et al. (1987)
Brain (cortex)	0.60 ± 0.03	Kong et al. (2002)
Brain (cortex)	2.09 ± 0.27 *	Cruz and Dienel (2002)
Brain (whole brain)	2.06 ± 0.21	Oe et al. (2016)

Glycogen content in adult rat or mouse tissues under physiological conditions were shown in the table. Data are reported as means ± SD (*) or means ± SEM

[a]Reported as mmol/kg dry weight; converted here using values of 160.3 mg /mmol as molecular weight of glycogen and 76% as the water content of muscle (Pivarnik and Palmer 1994)

[b]Reported as mmol/mg protein; converted here using values of 122 mg protein/gram wet weight in brain (Banay-Schwartz et al. 1992)

[c]*GPS* gastrocnemius-plantaris-soleus muscle complex

endothelial cells, and other cell types. Neurons contain appreciable amounts of glycogen during development, but this falls to very low levels in the mature brain except in certain brainstem neurons (Koizumi 1974; Ibrahim 1975; Cavalcante et al. 1996; Saez et al. 2014; Cataldo and Broadwell 1986; Borke and Nau 1984; Oe et al. 2016).

3 Glycogen Metabolism

The synthesis of a *de novo* glycogen granule is thought to be initiated by glycognenin, although recent studies with glycogenin deficient mice indicate other mechanisms of glycogen initiation are likely possible (Testoni et al. 2017). Glycogenin glycosylates itself at a tyrosine residue and catalyzes the extension of glucan chains (Smythe and Cohen 1991; Cao et al. 1993). The glucan chains serve as primer for glycogen synthetase, which catalyzes the formation of α-1, 4-glycosidic linkages of glycogen. Glycogen synthesis requires uridine diphosphate glucose (UDP-glucose) as a substrate, which is formed from uridine triphosphate and glucose 1-phosphate by UDP-glucose pyrophosphorylase (Fig. 2). α-1, 6-glycosidic branch points are subsequently produced by glycogen branching enzyme (1,4-alpha-glucan-branching enzyme) at approximately every 12 glucose residues.

Glycogenolysis is mediated by glycogen phosphorylase, which hydrolyzes glucose residues at α-1, 4 linkage points to generate glucose 1-phosphate. Glycogen debranching enzyme linearizes glycogen chains near the α-1, 6 branch points to provide linear substrate for glycogen phosphorylase (Nakayama et al. 2001). Glycogen phosphorylase is thought to be the rate limiting enzyme in glycogen breakdown. The activity of glycogen phosphorylase is regulated by changes in energy state through allosteric actions of AMP, which accelerates activity; and by ATP and glucose-6-phosphate, which slow enzymatic activity. Glycogen phosphorylase is also regulated by its phosphorylation state, through the action of glycogen phosphorylase kinase. Glycogen phosphorylase kinase is in turn regulated by a variety of signaling pathways through phosphorylation and allosteric interactions. These regulatory actions provide a mechanism for "anticipatory" glycogen mobilization to prevent any actual decline of cellular energy state. For example, glycogen phosphorylase kinase is activated by epinephrine through epinephrine-induced elevations in cAMP and activation of protein kinase A. Glycogen phosphorylase kinase can also be partly activated by elevated levels of Ca^{2+}, via binding to its calmodulin subunit. This mechanism is particularly important in skeletal muscle where muscle contraction triggers the release of Ca^{2+} from the sarcoplasmic reticulum.

Glycogen phosphorylase liberates glucose residues from glycogen as glucose-1-phosphate, which is freely converted to glucose-6 phosphate (Fig. 2). The shuttling of each glucose moiety on and off glycogen requires the net consumption of one ATP equivalent. This bioenergetic expense has implications for the potential physiological functions of brain glycogen, as discussed later in this chapter. It may also be significant that formation of glucose-6 phosphate from glycogen does not consume ATP, unlike the initial hexokinase step of glycolysis.

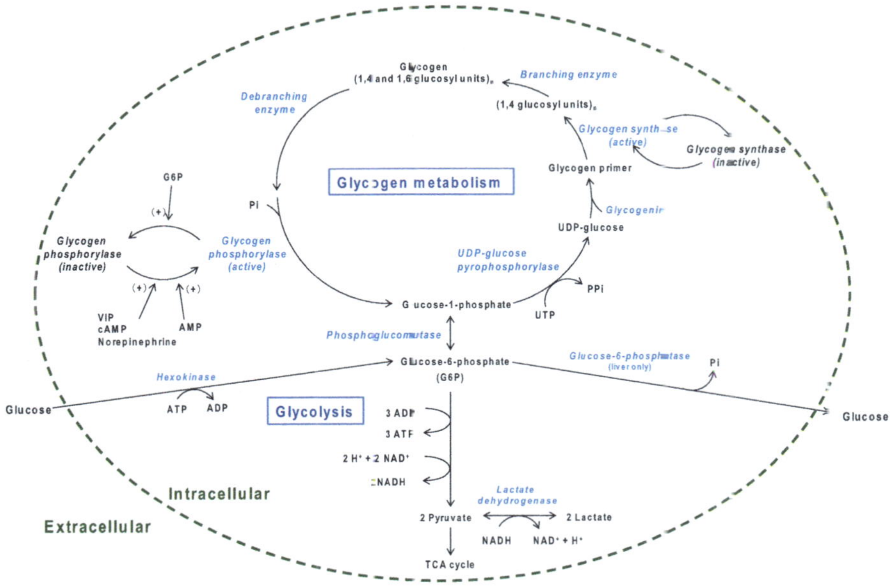

Fig. 2 Bioenergetics and regulation of glycogen metabolism. Glycogen synthase extends an existing glucosan chain of α-1, 4-glycosidic linkages using UDP glucose as substrate. Glycogen branching enzyme subsequently forms α-1, 6-glycosidic bonds to create branch points. Glycogen degradation is mediated by debranching enzyme and glycogen phosphorylase. Glycogen phosphorylase is regulated allosterically in response to hormones, e.g. norepinephrine and vasoactive intestinal peptide (VIP); by changes in energy state (AMP, glucose-6-phosphate (G6P), and others), and by second messengers such as cAMP. The immediate product of glycogen phosphorylase is glucose 1-phosphate, which is freely converted to glucose-6-phosphate. Hepatocytes (but not other cell types) can rapidly dephosphorylate glucose-6-phosphate to generate free glucose for export. There is a net cost of one ATP per molecule of glucose-6-phosphate that is cycled onto and off of the glycogen polymer, as 2ATP equivalents are consumed in forming UDP glucose from glucose-1-phosphate and 1 ATP equivalent is gained back at the formation of glucose-1-phosphate from a glucose residue and inorganic phosphate (Pi) at the glycogen phosphorylase step

4 Brain-Specific Aspects of Glycogen Metabolism

Although neurons are thought to be the primary energy consuming cells in brain, astrocytes contain the vast majority of brain glycogen. Electron microscopy identifies glycogen granules throughout astrocyte cell bodies and processes, particularly near axonal boutons and dendritic spines (Cali et al. 2016). Glutamate uptake is an energy-intensive astrocyte function, and interestingly glycogen phosphorylase has been found to be associated with the astrocyte glutamate transporter, GLT-1 (Genda et al. 2011). Astrocyte glycogen is quickly degraded under conditions of energy failure (Swanson et al. 1989; Suh et al. 2007a; Lowry et al. 1964), as would be expected. Glycogen metabolism in astrocytes is also induced by a several neurotransmitters and other signaling molecules, including vasoactive intestinal peptide (VIP), noradrenaline, arachidonic acid, glutamate, cAMP, and K^+ (Magistretti

1988; Sorg et al. 1995; Cummins et al. 1983; Cambray-Deakin et al. 1988a, b; Subbarao et al. 1995; Subbarao and Hertz 1990; Walls et al. 2009). These signaling molecules serve to couple astrocyte glycogen metabolism to neuronal activity. For VIP and noradrenaline in particular, the anatomical organization of these inputs provides a framework for coordinated signaling. The narrow radial pattern of arborization of intracortical VIP neuron and the tangential intracortical trajectory of the noradrenergic fibers suggests that these two systems may function in a complementary fashion: VIP regulating energy metabolism locally, within individual columnar modules, and norepinephrine exerting a more global effect (Magistretti et al. 1981).

Direct autoradiographic and biochemical measures of glycogen turnover show it to be increased by sensory neuron stimulation in the awake rat (Swanson et al. 1992; Dienel et al. 2007). Conversely, conditions causing focally or globally reduced neuronal activity lead to a corresponding local or global increases in glycogen content, suggesting reduced glycogen utilization (Swanson 1992b). These conditions include focal brain injury, anesthetics, slow wave sleep, and hibernation (Pudenz et al. 1975; Nelson et al. 1968; Phelps 1972; Watanabe and Passonneau 1973; Swanson 1992b).

5 Experimental and Technical Considerations in Study of Brain Glycogen

Low glycogen content in brain. Several aspects of glycogen structure and regulation in brain pose unique challenges for experimental observation. First among these is the relatively low concentration of glycogen in brain. Many of the classical histochemical methods for detecting glycogen were developed using liver or muscle tissues, in which glycogen content is far higher than in brain. For example, the periodic acid schiff (PAS) method of staining polysaccharides works well in liver and muscle, but it in brain the low glycogen content and relatively higher content of glycoproteins and glycolipids, which also react with PAS, makes this approach less useful, even when coupled with dimedone blocking of aldehydes (Cammermeyer and Fenton 1981). Immunohistochemical methods using glycogen antibodies are far more sensitive (Oe et al. 2016). Nonspecific glycogen labeling by either of these methods can be identified as label that persists in tissues after digestion with amylase or rendered ischemic for a few minutes prior to fixation. Low glycogen levels similarly pose a challenge for magnetic resonance studies of glycogen, as recently reviewed (Soares et al. 2017).

Use of radiotracers to assess glycogen turnover. A challenge common to many studies of metabolism is that there is no fixed relationship between the metabolite flux and measured metabolite concentrations. Glycogen turnover in particular—i.e. the breakdown and resynthesis of individual glycogen molecules—may accelerate, decelerate, or stop altogether with little or no change in net glycogen content (Fig. 3). Radiolabeled or isotope enriched substrates are therefore widely used for assessments of metabolic rates. However, accurate quantification by these

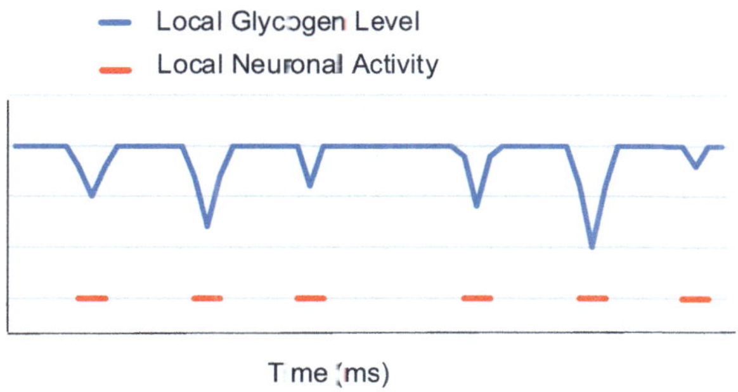

Fig. 3 Schematic relationship between neuronal activity and astrocyte glycogen metabolism. Short bursts of neuronal activity induce mobilization of astrocyte glycogen. The glycogen is resynthesized between bursts, resulting in glycogen turnover with no net change in glycogen content over time. It has been proposed that glycogenolysis is triggered by neuronal activity to provide substrate for increased glycolytic demand over intervals too short to be met by changes in local blood flow

approaches requires that the labeled molecules be homogenously distributed among the unlabeled molecules. This is difficult to achieve in glycogen because individual glucose moieties are sequentially added to and removed from the outer glycogen tiers, and have widely variable dwell times in the glycogen polymer (Youn and Bergman 1987; Elsner et al. 2002). The relationships between rates of glycogenolysis and label release therefore depend upon the patterns of glycogen synthesis and breakdown at the level of individual glycogen granules (as diagrammed in Fig. 4). These limitations do not render assessments of glycogen turnover impossible, but they do limit the precision of these measures and often require certain assumptions. For example, very prolonged administration of tracers can be used to improve homogeneity of glycogen labeling, and this can be further refined by modeling patterns of glycogen turnover (DiNuzzo 2013; Oz et al. 2015; Soares et al. 2017).

The use of radiolabeled glucose to monitor glycogen metabolism is also complicated by the fact that normal glucose metabolism rapidly distributes the label to amino acids, lipids and many other cell constituents other than glycogen. This problem can be surmounted in part by physical isolation of glycogen prior to isotope analysis, or by the use of 2-deoxyglucose, which is incorporated into and released from glycogen but is much more slowly metabolized to other molecules (Nelson et al. 1984; Kai Kai and Pentreath 1981). Glucose labeled at the 3 or 4 carbon positions also mitigates this problem (Swanson et al. 1992), because glucose entering the tricarboxylic acid cycle (from which most other metabolites are formed) loses the 3 and 4 carbons in the pyruvate dehydrogenase reaction. In vivo magnetic resonance spectroscopy studies can in some cases use native ^{13}C abundance to assess brain glycogen concentrations, but the low abundance of ^{13}C coupled with low concentrations of glycogen in brain currently limit the temporal and spatial resolution of this approach (Soares et al. 2017).

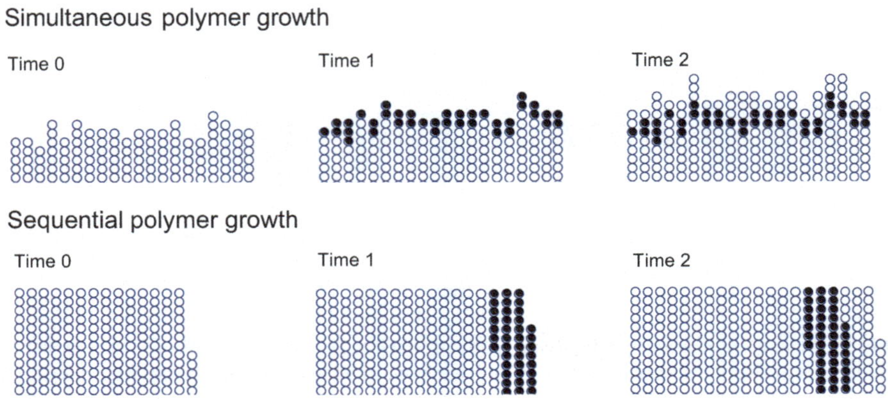

Fig. 4 Results of radiolabeling tracer experiments are influenced by the mode of glycogen polymer growth and breakdown. Schematic diagram showing distribution in glycogen of labeled glucose (filled circles) injected early during glycogen synthesis. Each column of circles represents an individual glycogen polymer. In one scenario, all polymers simultaneously add glucose moieties. In the opposite scenario, each polymer is synthesized to its maximum size before a second one begins to expand. Many other intermediates or more complex patterns are also possible, as are differing patterns of glycogen polymer breakdown. These differing patterns produce a different relationship between rates of glycogen turnover and rates of label release. Redrawn from Youn and Bergman (1987)

Rapid post-mortem glycogenolysis. Glycogenolysis is triggered by very small elevations in AMP, befitting its role as an emergency energy store. This process begins almost instantly with cessation of blood flow, and continues until the brain is frozen, acidified, or otherwise treated to inactivate glycogen phosphorylase. (For frozen brains, the process resumes upon thawing unless other measures are also taken.) This presents a major challenge to histochemical detection of glycogen, particularly with the low concentration of glycogen in brain. Post-mortem glycogenolysis can be prevented by "funnel freezing" with direct application of liquid nitrogen to the animal skull (Ponten et al. 1973; Dienel et al. 2002), or more conveniently by high-energy microwave fixation of brain *in situ*, in which brain temperature is elevated to levels that denature glycogen metabolizing enzymes within 1 s or less (Medina et al. 1975; Sagar et al. 1987). The microwave approach has limitations, however. The high heat induced by and required for microwave fixation can reduce the water content of the brain, thus artificially elevating the glycogen content expressed per gram wet weight. This can be avoided by normalizing to mg protein rather than to wet weight, but special care must be taken to ensure that the fixed (denatured) protein is fully solubilized for protein determination. Failure to correct for brain dehydration or fully solubilize proteins leads to erroneously elevated glycogen measurements in microwave fixed brain (Swanson and Sagar, unpublished observations). A second limitation is that the heating and water vapor formation often distort normal cell architecture and destroy immunoreactivity of many antigens; however, these problems can be overcome by careful titration of microwave power (Oe et al. 2016; Oe et al., this volume).

It is often useful to express glycogen measurements in terms of glucose molecule equivalents. The molecular weight of glucose is 180.16 g/mol however, the molecular weight of each glucose moiety in glycogen is somewhat less, because the glycosidic linkages between the glucose moieties in glycogen each subtract the equivalent of one H_2O from their molecular weight. There is one α-1,4 glycosidic bond between each glucose moiety in the linear chains, and there is in addition one α-1,6 glycosidic bond at each branch point, which occurs approximately every tenth glucose residue. The calculated molecular weight for glycogen can thus be estimated to be 160.3 g/mol glucose equivalent. The true value will vary slightly depending upon the degree of polymer branching.

Hormonal and neurotransmitter effects on glycogenolysis. This aspect of glycogen metabolism may be uniquely problematic for studies of glycogen in brain because it involves sensory experience of the subject. There is a generally inverse relationship between neuronal activity and glycogen levels, as evidenced by elevated brain glycogen levels during anesthesia, hibernation, and slow wave sleep. Conversely, as outlined above, glycogen phosphorylase is activated by a number of signaling molecules to initiate glycogenolysis in anticipation of actual energy demand. For example, epinephrine and norepinephrine released during the stress response are potent glycogenolytic signaling molecules. It follows that animals that are stressed during the interval before brain harvest may have regionally lower brain glycogen levels than non-stressed animals. Experimental evidence supports this concern (Cruz and Dienel 2002). This observation suggests that, like glucose utilization, glycogen levels and turnover rates are likely influenced in a regionally discrete manner by the subject experience and response near the time of the observation.

6 Physiological Functions of Glycogen in Brain

Function under energy failure conditions. There is no question that glycogen can serve as an energy reserve in brain. Electrophysiology experiments using optic nerve and corpus callosum have shown that astrocyte glycogen can sustain axon function during glucose deprivation or during high frequency stimulation (Wender et al. 2000; Brown et al. 2005, 2012). These and other studies also suggest that lactate or pyruvate derived from astrocyte glycogen can be used by neurons for oxidative metabolism (Dringen et al. 1993; Poitry-Yamate et al. 1995; Pellerin et al. 1998; Tekkok et al. 2005). Other studies have demonstrated that increased glycogen in astrocytes improves survival of neurons under conditions of oxygen deprivation or glucose deprivation, both in culture and *in vivo*.

The total glycogen store in brain is approximately 3 mM (as expressed in glucose equivalents). Given estimates of basal brain glucose utilization rate (Sokoloff et al.

1977), this amount of glycogen would be expected to fuel brain metabolism for less than 15 min if it were the only energy supply available. However, there is a compensatory reduction in the rate of glucose utilization and synaptic activity during energy compromise (Suda et al. 1990). In addition, it is rare to have a complete cessation of glucose delivery to brain. During severe insulin-induced hypoglycemia, for example, the flux of glucose from blood to brain is reduced but not zero, and brain glycogen is consumed very gradually (Ratcheson et al. 1981; Choi et al. 2003). A quantitative analysis suggests that normal levels of astrocyte glycogen should be able to support brain metabolism for about 100 min during hypoglycemia (Gruetter 2003). In agreement with this assessment, an experimental study demonstrated that an 80% increase in rat brain glycogen at the onset of severe hypoglycemia extended the period of brain electrical activity by 90 min (Suh et al. 2007b). The rats with elevated initial brain glycogen concentrations also had improved neuronal survival.

Glycogen is far less useful as an energy store in brain under ischemic conditions, in which only anaerobic (glycolytic) energy metabolism is possible. Under these conditions each glucose molecule stored in glycogen can generate only 3 ATP, as opposed to the 34–36 ATP generated under aerobic conditions. Accordingly, glycogen stores are very rapidly depleted (within 3 min) during complete ischemia (Lowry et al. 1964). Moreover, there is no possibility of transferring usable energy substrates from astrocytes to neurons under anoxic conditions because substrates that can exit from astrocytes, such as lactate and pyruvate, can only be metabolized aerobically. However in focal ischemia (as results from occlusion of a cerebral artery), there is often a border zone area with partially preserved blood supply in which energy supply is barely adequate to maintain cell viability. Spreading depression from the ischemic core can strikingly increase energy demands in this "penumbra" area and exacerbate ischemic injury (von Bornstadt et al. 2015). Astrocyte glycogen stores are rapidly consumed during spreading depression (Feuerstein et al. 2016), and increased glycogen stores delay its advancing wave front (Seidel and Shuttleworth 2011). A study of focal brain ischemia showed reduced infarct size in rats that had elevated brain glycogen content (Swanson et al. 1990).

While the aggregate evidence that glycogen can serve as an emergency energy store in brain is convincing, it nevertheless does not establish this as the physiological, evolutionarily directed function of brain glycogen. (As an analogy, the catabolism of skeletal muscle to support nutrition during extreme starvation does not demonstrate this as the primary function of muscle.) In fact, several aspects of brain glycogen are poorly explained by this interpretation. These include the continuous turnover of brain glycogen under normal conditions; the complex neuromodulatory influences that promote astrocyte glycogenolysis in the absence of energy failure; and perhaps most fundamentally, the location of glycogen in astrocytes rather than in neurons.

Given the energetic cost of shuttling glucose into and out of glycogen polymers, it is useful to delineate the conditions under which this might carry advantages over direct glucose utilization. One obvious condition would be to provide buffering for transient, local insufficiencies in glucose supply, as might happen for example in the brief interval between a burst of neuronal activity and subsequent increase in local blood flow (Fig. 3). This idea was proposed by Swanson (Swanson 1992a) and more formally developed by Shulman and colleagues as the "glycogen shunt" hypothesis (Shulman et al. 2001; Shulman and Rothman 2001). A related idea proposes that astrocyte glycogenolysis occurs preemptively in activated tissues as a glucose sparing mechanism, to prevent reduction in available glucose that would otherwise occur (Dienel and Cruz 2015; DiNuzzo et al. 2010). Although these concepts have gained wide support, there is uncertainty as to when or if such a transient glucose insufficiency occurs under physiological conditions. Glucose concentration will not become rate-limiting for energy metabolism unless it falls near the glucose Km of hexokinase, which is less than 0.1 mM (Thompson and Bachelard 1977). Glucose concentration in normal rodent brain is approximately 2–2.5 mM, which can also be taken as the approximate concentration of glucose in the intracellular space. (Values for humans are about 30% lower). Can neuronal activity outstrip supply to the extent that local glucose concentrations fall near the hexokinase Km value? The classic study by Silver and Ericinska (Silver and Erecinska 1994) showed that spreading depression, which causes nearly simultaneous bursting of neurons at its wave front, reduced extracellular glucose concentrations from 2.4 mM to 2 mM, a value well above the hexokinase Km. Similarly, studies of brain energy metabolites during pentylenetetrazole-induced status epilepticus showed reductions in brain glucose concentration to no lower than 0.9 mM (McCandless et al. 1987; Folbergrova et al. 1985; Ingvar et al. 1984), excluding settings in which cerebral blood flow was impaired. Though not definitive, these results obtained using spreading depression and seizure activity, both of which massively increase brain energy demands, cast some doubt on the concept that changes in brain glucose utilization resulting from normal activity could require buffering by local mobilization of glycogen stores.

7 Comparisons to Liver and Skeletal Muscle

Assessment of glycogen functions in other tissues may suggest alternative possible roles for glycogen utilization in brain. Liver has a specialized role as a glycogen storage depot for systemic use. Hepatocytes express high concentrations of glucose-6-phosphatase for liberating free glucose into the blood circulation for use by other organs. However, astrocytes do not have high levels of glucose-6-phosphatase, and it follows that astrocytes do not have a role analogous to hepatocytes as a store of free glucose for use by other cells. Skeletal muscle contains glycogen levels intermediate between liver and brain, and like brain the stored glycogen in muscle undergoes continuous turnover and is not released to the systemic circulation. Skeletal muscle fiber types can be classified as Type 1 "red" or Type 2 'white", with the red fibers containing far more mitochondria and white fibers containing far more glycogen (Nielsen et al. 2011). The Type 2 white fibers contract more quickly but fatigue more rapidly than Type 1 red fibers. Muscle glycogen turnover increases during exercise (Nielsen et al. 2011; Shulman and Rothman 2001), similar to the glycogen turnover increase induced by neuronal activity in brain. However, contracting muscle has repetitive intervals of ischemia corresponding to muscle contractions, during which glycolytic metabolism is essential for ATP production, and this does not have a correlate in brain. Moreover, glycogen in muscle is localized to the contacting myocytes themselves, and there is no evidence that glycogen is metabolized in one cell to fuel activity in another cell as has been proposed to occur in brain.

Muscle fatigue correlates with glycogen depletion (Allen et al. 2008; Matsui et al. this volume), but, surprisingly, the mechanistic link between these events has not been established (Allen et al. 2008; Ortenblad et al. 2013). This point is exemplified by observations in McCardle's disease, an inherited condition caused by a genetic deficiency in muscle glycogen phosphorylase. Individuals with MCardle's disease are generally well, but develop early fatigue with exercise. This occurs even though myocyte glucose levels *increase* with exercise (MacLean et al. 1999; Sahlin 1990). Moreover, muscle performance in McCardle's disease is not significantly improved by hyperglycemia, as should occur if glycogen serves only to buffer intervals of reduced glucose availability. The biochemical defect in these muscles appears not to be a simple lack of glycolytic substrate, but rather a reduced capacity to sequester inorganic phosphate along with other changes that reduce the free energy (ΔG') provided by ATP hydrolysis (Malucelli et al. 2011).

8 Summary

Studies of brain glycogen metabolism are complicated by factors that stem directly from specific structural and regulatory features of glycogen. These factors must be addressed and considered in interpreting study results. Bioenergetic aspects of glycogen metabolism place certain constraints on the settings under which glycogen utilization may be advantageous to brain. These aspects, in conjunction with some aspects of muscle glycogen metabolism, suggest that current concepts of brain glycogen physiology may need to be expanded.

References

Allen DG, Lamb GD, Westerblad H (2008) Skeletal muscle fatigue: cellular mechanisms. Physiol Rev 88(1):287–332. https://doi.org/10.1152/physrev.00015.2007

Baker DJ, Timmons JA, Greenhaff PL (2005) Glycogen phosphorylase inhibition in type 2 diabetes therapy: a systematic evaluation of metabolic and functional effects in rat skeletal muscle. Diabetes 54(8):2453–2459

Banay-Schwartz M, Kenessey A, DeGuzman T, Lajtha A, Palkovits MJA (1992) Protein content of various regions of rat brain and adult and aging human brain. 15(2):51–54. https://doi.org/10.1007/bf02435024

Borke RC, Nau ME (1984) Glycogen, its transient occurrence in neurons of the rat CNS during normal postnatal development. Brain Res 318(2):277–284 von Bornstadt D, Houben T, Seidel JL, Zheng Y, Dilekoz E, Qin T, Sandow N, Kura S, Eikermann-Haerter K, Endres M, Boas DA, Moskowitz MA, Lo EH, Dreier JP, Woitzik J, Sakadzic S, Ayata C (2015) Supply-demand mismatch transients in susceptible peri-infarct hot zones explain the origins of spreading injury depolarizations. Neuron 85(5):1117–1131. https://doi.org/10.1016/j.neuron.2015.02.007

Brown AM, Sickmann HM, Fosgerau K, Lund TM, Schousboe A, Waagepetersen HS, Ransom BR (2005) Astrocyte glycogen metabolism is required for neural activity during aglycemia or intense stimulation in mouse white matter. J Neurosci Res 79(1–2):74–80

Brown AM, Evans RD, Black J, Ransom BR (2012) Schwann cell glycogen selectively supports myelinated axon function. Ann Neurol 72(3):406–418. https://doi.org/10.1002/ana.23607

Calder PC (1991) Glycogen structure and biogenesis. Int J Biochem 23(12):1335–1352

Cali C, Baghabra J, Boges DJ, Holst GR, Kreshuk A, Hamprecht FA, Srinivasan M, Lehvaslaiho H, Magistretti PJ (2016) Three-dimensional immersive virtual reality for studying cellular compartments in 3D models from EM preparations of neural tissues. J Comp Neurol 524(1):23–38. https://doi.org/10.1002/cne.23852

Cambray-Deakin M, Pearce B, Morrow C, Murphy S (1988a) Effects of extracellular potassium on glycogen stores of astrocytes in vitro. J Neurochem 51(6):1846–1851

Cambray-Deakin M, Pearce B, Morrow C, Murphy S (1988b) Effects of neurotransmitters on astrocyte glycogen stores in vitro. J Neurochem 51(6):1852–1857

Cammermeyer J, Fenton IM (1981) Improved preservation of neuronal glycogen by fixation with iodoacetic acid-containing solutions. Exp Neurol 72(2):429–445

Cao Y, Mahrenholz AM, DePaoli-Roach AA, Roach PJ (1993) Characterization of rabbit skeletal muscle glycogenin. Tyrosine 194 is essential for function. J Biol Chem 268(20):14687–14693

Cardell RR Jr (1977) Smooth endoplasmic reticulum in rat hepatocytes during glycogen deposition and depletion. Int Rev Cytol 48:221–279

Cataldo AM, Broadwell RD (1986) Cytochemical identification of cerebral glycogen and glucose-6-phosphatase activity under normal and experimental conditions. II. Choroid plexus and ependymal epithelia, endothelia and pericytes. J Neurocytol 15(4):511–524

Cavalcante LA, Barradas PC, Vieira AM (1996) The regional distribution of neuronal glycogen in the opossum brain, with special reference to hypothalamic systems. J Neurocytol 25(8):455–463

Choi IY, Seaquist ER, Gruetter R (2003) Effect of hypoglycemia on brain glycogen metabolism in vivo. J Neurosci Res 72(1):25–32

Conlee RK, Berg TL, Han DH, Kelly KP, Barnett DW (1989) Cocaine does not alter cardiac glycogen content at rest or during exercise. Metab Clin Exp 38(11):1039–1041

Cruz NF, Dienel GA (2002) High glycogen levels in brains of rats with minimal environmental stimuli: implications for metabolic contributions of working astrocytes. J Cereb Blood Flow Metab 22(12):1476–1489. https://doi.org/10.1097/01.WCB.0000034362.37277.C0

Cummins CJ, Lust WD, Passonneau JV (1983) Regulation of glycogen metabolism in primary and transformed astrocytes in vitro. J Neurochem 40(1):128–136

Dienel GA, Cruz NF (2015) Contributions of glycogen to astrocytic energetics during brain activation. Metab Brain Dis 30(1):281–298. https://doi.org/10.1007/s11011-014-9493-8

Dienel GA, Wang RY, Cruz NF (2002) Generalized sensory stimulation of conscious rats increases labeling of oxidative pathways of glucose metabolism when the brain glucose-oxygen uptake ratio rises. J Cereb Blood Flow Metab 22(12):1490–1502

Dienel GA, Ball KK, Cruz NF (2007) A glycogen phosphorylase inhibitor selectively enhances local rates of glucose utilization in brain during sensory stimulation of conscious rats: implications for glycogen turnover. J Neurochem 102(2):466–478. https://doi.org/10.1111/j.1471-4159.2007.04595.x

DiNuzzo M (2013) Kinetic analysis of glycogen turnover: relevance to human brain 13C-NMR spectroscopy. J Cereb Blood Flow Metab 33(10):1540–1548. https://doi.org/10.1038/jcbfm.2013.98

DiNuzzo M, Mangia S, Maraviglia B, Giove F (2010) Glycogenolysis in astrocytes supports blood-borne glucose channeling not glycogen-derived lactate shuttling to neurons: evidence

from mathematical modeling. J Cereb Blood Flow Metab 30(12):1895–1904. https://doi.org/10.1038/jcbfm.2010.151
Dringen R, Gebhardt R, Hamprecht B (1993) Glycogen in astrocytes: possible function as lactate supply for neighboring cells. Brain Res 623(2):208–214
Elsner P, Quistorff B, Hansen GH, Grunnet N (2002) Partly ordered synthesis and degradation of glycogen in cultured rat myotubes. J Biol Chem 277(7):4831–4838. https://doi.org/10.1074/jbc.M108226200
Feuerstein D, Backes H, Gramer M, Takagaki M, Gabel P, Kumagai T, Graf R (2016) Regulation of cerebral metabolism during cortical spreading depression. J Cereb Blood Flow Metab 36(11):1965–1977. https://doi.org/10.1177/0271678x15612779
Folbergrova J, Ingvar M, Nevander G, Siesjo BK (1985) Cerebral metabolic changes during and following fluorothyl-induced seizures in ventilated rats. J Neurochem 44(5):1419–1426
Garetto LP, Richter EA, Goodman MN, Ruderman NB (1984) Enhanced muscle glucose metabolism after exercise in the rat: the two phases. Am J Phys 246(6 Pt 1):E471–E475. https://doi.org/10.1152/ajpendo.1984.246.6.E471
Geddes R, Jeyarathan P, Taylor JA (1992) Molecular and metabolic aspects of lysosomal glycogen. Carbohydr Res 227:339–349
Genda EN, Jackson JG, Sheldon AL, Locke SF, Greco TM, O'Donnell JC, Spruce LA, Xiao R, Guo W, Putt M, Seeholzer S, Ischiropoulos H, Robinson MB (2011) Co-compartmentalization of the astroglial glutamate transporter, GLT-1, with glycolytic enzymes and mitochondria. J Neurosci 31(50):18275–18288. https://doi.org/10.1523/JNEUROSCI.3305-11.2011
Goldsmith E, Sprang S, Fletterick R (1982) Structure of maltoheptaose by difference Fourier methods and a model for glycogen. J Mol Biol 156(2):411–427
Gruetter R (2003) Glycogen: the forgotten cerebral energy store. J Neurosci Res 74(2):179–183
Gunja-Smith Z, Marshall JJ, Mercier C, Smith EE, Whelan WJ (1970) A revision of the Meyer-Bernfeld model of glycogen and amylopectin. FEBS Lett 12(2):101–104
Haggstrom M (2014) Glycogen structure. Wiki J Med 1(2):8. https://doi.org/10.15347/wjm/2014.008
Ibrahim MZ (1975) Glycogen and its related enzymes of metabolism in the central nervous system. Adv Anat Embryol Cell Biol 52(1):3–89
Ingvar M, Soderfeldt B, Folbergrova J, Kalimo H, Olsson Y, Siesjo BK (1984) Metabolic, circulatory, and structural alterations in the rat brain induced by sustained pentylenetetrazole seizures. Epilepsia 25(2):191–204
Jiang S, Heller B, Tagliabracci VS, Zhai L, Irimia JM, DePaoli-Roach AA, Wells CD, Skurat AV, Roach PJ (2010) Starch binding domain-containing protein 1/genethonin 1 is a novel participant in glycogen metabolism. J Biol Chem 285(45):34960–34971. https://doi.org/10.1074/jbc.M110.150839
Kai Kai MA, Pentreath VW (1981) High resolution analysis of [3H]2-deoxyglucose incorporation into neurons and glial cells in invertebrate ganglia: histological processing of nervous tissue for selective marking of glycogen. J Neurocytol 10(4):693–708
Khandelwal RL, Zinman SM, Knull HR (1979) The effect of streptozotocin-induced diabetes on glycogen metabolism in rat kidney and its relationship to the liver system. Arch Biochem Biophys 197(1):310–316
Koizumi J (1974) Glycogen in the central nervous system. Prog Histochem Cytochem 6(4):1–37
Kong J, Shepel PN, Holden CP, Mackiewicz M, Pack AI, Geiger JD (2002) Brain glycogen decreases with increased periods of wakefulness: implications for homeostatic drive to sleep. J Neurosci 22(13):5581–5587
Kusunoki M, Tsutsumi K, Hara T, Ogawa H, Nakamura T, Miyata T, Sakakibara F, Fukuzawa Y, Suga T, Kakumu S, Nakaya Y (2002) Correlation between lipid and glycogen contents in liver and insulin resistance in high-fat-fed rats treated with the lipoprotein lipase activator NO-1886. Metab Clin Exp 51(6):792–795
Lowry OH, Passonneau JV, Hasselberger FX, Schulz DW (1964) Effect of ischemia on known substrates and cofactors of the glycolytic pathway in brain. J Biol Chem 239:18–30

MacLean DA, Bangsbo J, Saltin B (1999) Muscle interstitial glucose and lactate levels during dynamic exercise in humans determined by microdialysis. J Appl Physiol (1985) 87(4):1483–1490. https://doi.org/10.1152/jappl.1999.87.4.1483

Magistretti PJ (1988) Regulation of glycogenolysis by neurotransmitters in the central nervous system. Diabetes Metab 14(3):237–246

Magistretti PJ, Morrison JH, Shoemaker WJ, Sapin V, Bloom FE (1981) Vasoactive intestinal polypeptide induces glycogenolysis in mouse cortical slices: a possible regulatory mechanism for the local control of energy metabolism. Proc Natl Acad Sci U S A 78(10):6535–6539

Malucelli E, Iotti S, Manners DN, Testa C, Martinuzzi A, Barbiroli B, Lodi R (2011) The role of pH on the thermodynamics and kinetics of muscle biochemistry: an in vivo study by (31) P-MRS in patients with myo-phosphorylase deficiency. Biochim Biophys Acta 1807(9):1244–1249. https://doi.org/10.1016/j.bbabio.2011.06.013

Marchand I, Chorneyko K, Tarnopolsky M, Hamilton S, Shearer J, Potvin J, Graham TE (2002) Quantification of subcellular glycogen in resting human muscle: granule size, number, and location. J Appl Physiol (1985) 93(5):1598–1607. https://doi.org/10.1152/japplphysiol.00585.2001

McCandless DW, Dworsky S, Modak AT, Stavinoha WB (1987) Pentylenetetrazole-induced changes in cerebral energy metabolism in Tupaia glis. Epilepsia 28(2) 184–189

Medina MA, Jones DJ, Stavinoha WB, Ross DH (1975) The levels of labile intermediary metabolites in mouse brain following rapid tissue fixation with microwave irradiation. J Neurochem 24(2):223–227

Melendez-Hevia E, Waddell TG, Shelton ED (1993) Optimization of molecular design in the evolution of metabolism: the glycogen molecule. Biochem J 295(Pt 2):477–483

Nakayama A, Yamamoto K, Tabata S (2001) Identification of the catalytic residues of bifunctional glycogen debranching enzyme. J Biol Chem 276(31):28824–28828. https://doi.org/10.1074/jbc.M102192200

Nannipieri M, Lanfranchi A, Santerini D, Catalano C, Van de Werve G, Ferrannini E (2001) Influence of long-term diabetes on renal glycogen metabolism in the rat. Nephron 87(1):50–57. https://doi.org/10.1159/000045884

Nelson SR, Schulz DW, Passonneau JV, Lowry OH (1968) Control of glycogen levels in brain. J Neurochem 15(11):1271–1279

Nelson T, Kaufman EE, Sokoloff L (1984) 2-Deoxyglucose incorporation into rat brain glycogen during measurement of local cerebral glucose utilization by the 2-deoxyglucose method. J Neurochem 43(4):949–956

Nielsen J, Holmberg HC, Schroder HD, Saltin B, Ortenblad N (2011) Human skeletal muscle glycogen utilization in exhaustive exercise: role of subcellular localization and fibre type. J Physiol 589(Pt 11):2871–2885. https://doi.org/10.1113/jphysiol.2010.204487

Oe Y, Baba O, Ashida H, Nakamura KC, Hirase H (2016) Glycogen distribution in the microwave-fixed mouse brain reveals heterogeneous astrocytic patterns. Glia 64(9):1532–1545. https://doi.org/10.1002/glia.23020

Ortenblad N, Westerblad H, Nielsen J (2013) Muscle glycogen stores and fatigue. J Physiol 591(18):4405–4413. https://doi.org/10.1113/jphysiol.2013.251629

Oz G, DiNuzzo M, Kumar A, Moheet A, Seaquist ER (2015) Revisiting glycogen content in the human brain. Neurochem Res 40(12):2473–2481. https://doi.org/10.1007/s11064-015-1664-4

Pellerin L, Pellegri G, Bittar PG, Charnay Y, Bouras C, Martin JL, Stella N, Magistretti PJ (1998) Evidence supporting the existence of an activity-dependent astrocyte-neuron lactate shuttle. Dev Neurosci 20(4–5):291–299

Phelps CH (1972) Barbiturate-induced glycogen accumulation in brain. An electron microscopic study. Brain Res 39(1):225–234

Pivarnik JM, Palmer RA (1994) In: Wolinsky I, Hickson JF (eds) Water and electrolyte balance during rest and exercise. Nutrition in exercise and sport. CRC Press, Boca Raton, FL

Poitry-Yamate CL, Poitry S, Tsacopoulos M (1995) Lactate released by Muller glial cells is metabolized by photoreceptors from mammalian retina. J Neurosci 15(7 Pt 2):5179–5191

Ponten U, Ratcheson RA, Siesjo BK (1973) Metabolic changes in the brains of mice frozen in liquid nitrogen. J Neurochem 21(5):1211–1216

Pudenz RH, Bullara LA, Jacques S, Hambrecht FT (1975) Electrical stimulation of the brain. III The neural damage model. Surg Neurol 4(4):389–400

Ratcheson RA, Blank AC, Ferrendelli JA (1981) Regionally selective metabolic effects of hypoglycemia in brain. J Neurochem 36(6):1952–1958

Saez I, Duran J, Sinadinos C, Beltran A, Yanes O, Tevy MF, Martinez-Pons C, Milan M, Guinovart JJ (2014) Neurons have an active glycogen metabolism that contributes to tolerance to hypoxia. J Cereb Blood Flow Metab 34(6):945–955. https://doi.org/10.1038/jcbfm.2014.33

Sagar SM, Sharp FR, Swanson RA (1987) The regional distribution of glycogen in rat brain fixed by microwave irradiation. Brain Res 417(1):172–174

Sahlin K (1990) Muscle glucose metabolism during exercise. Ann Med 22(3):85–89

Seidel JL, Shuttleworth CW (2011) Contribution of astrocyte glycogen stores to progression of spreading depression and related events in hippocampal slices. Neuroscience 192:295–303. https://doi.org/10.1016/j.neuroscience.2011.05.006

Shulman RG, Rothman DL (2001) The "glycogen shunt" in exercising muscle: a role for glycogen in muscle energetics and fatigue. Proc Natl Acad Sci U S A 98(2):457–461. https://doi.org/10.1073/pnas.98.2.457

Shulman RG, Hyder F, Rothman DL (2001) Cerebral energetics and the glycogen shunt: neurochemical basis of functional imaging. Proc Natl Acad Sci U S A 98(11):6417–6422

Silver IA, Erecinska M (1994) Extracellular glucose concentration in mammalian brain: continuous monitoring of changes during increased neuronal activity and upon limitation in oxygen supply in normo-, hypo-, and hyperglycemic animals. J Neurosci 14(8):5068–5076

Smythe C, Cohen P (1991) The discovery of glycogenin and the priming mechanism for glycogen biogenesis. Eur J Biochem 200(3):625–631

Soares AF, Gruetter R, Lei H (2017) Technical and experimental features of magnetic resonance spectroscopy of brain glycogen metabolism. Anal Biochem 529:117–126. https://doi.org/10.1016/j.ab.2016.12.023

Sokoloff L, Reivich M, Kennedy C, Des Rosiers MH, Patlak CS, Pettigrew KD, Sakurada O, Shinohara M (1977) The [14C]deoxyglucose method for the measurement of local cerebral glucose utilization: theory, procedure, and normal values in the conscious and anesthetized albino rat. J Neurochem 28(5):897–916

Sorg O, Pellerin L, Stolz M, Beggah S, Magistretti PJ (1995) Adenosine triphosphate and arachidonic acid stimulate glycogenolysis in primary cultures of mouse cerebral cortical astrocytes. Neurosci Lett 188(2):109–112

Stapleton D, Nelson C, Parsawar K, McClain D, Gilbert-Wilson R, Barker E, Rudd B, Brown K, Hendrix W, O'Donnell P, Parker G (2010) Analysis of hepatic glycogen-associated proteins. Proteomics 10(12):2320–2329. https://doi.org/10.1002/pmic.200900628

Subbarao KV, Hertz L (1990) Effect of adrenergic agonists on glycogenolysis in primary cultures of astrocytes. Brain Res 536(1–2):220–226

Subbarao KV, Stolzenburg JU, Hertz L (1995) Pharmacological characteristics of potassium-induced, glycogenolysis in astrocytes. Neurosci Lett 196(1–2):45–48

Suda S, Shinohara M, Miyaoka M, Lucignani G, Kennedy C, Sokoloff L (1990) The lumped constant of the deoxyglucose method in hypoglycemia: effects of moderate hypoglycemia on local cerebral glucose utilization in the rat. J Cereb Blood Flow Metab 10(4):499–509

Suh SW, Bergher JP, Anderson CM, Treadway JL, Fosgerau K, Swanson RA (2007a) Astrocyte glycogen sustains neuronal activity during hypoglycemia: studies with the glycogen phosphorylase inhibitor CP-316,819 ([R-R∗,S∗]-5-chloro-N-[2-hydroxy-3-(methoxymethylamino)-3-oxo-1-(phenylmet hyl)propyl]-1H-indole-2-carboxamide). J Pharmacol Exp Ther 321(1):45–50. https://doi.org/10.1124/jpet.106.115550

Suh SW, Bergher JP, Anderson CM, Treadway JL, Fosgerau K, Swanson RA (2007b) Astrocyte glycogen sustains neuronal activity during hypoglycemia: studies with the glycogen phosphorylase inhibitor CP-316,819 ([R-R∗,S∗]-5-chloro-N-[2-hydroxy-3-(methoxymethylamino)-3-oxo-

1-(phenylmethyl)pro pyl]-1H-indole-2-carboxamide). J Pharmacol Exp Ther 321(1):45–50. https://doi.org/10.1124/jpet.106.115550
Swanson RA (1992a) Physiologic coupling of glial glycogen metabolism to neuronal activity in brain. Can J Physiol Pharmacol 70(Suppl):S138–S144
Swanson RA (1992b) Physiologic coupling of glial glycogen metabolism to neuronal activity in brain. Can J Physiol Pharmacol 70(Suppl):S138–S144
Swanson RA, Sagar SM, Sharp FR (1989) Regional brain glycogen stores and metabolism during complete global ischaemia. Neurol Res 11(1):24–28
Swanson RA, Shiraishi K, Morton MT, Sharp FR (1990) Methionine sulfoximine reduces cortical infarct size in rats after middle cerebral artery occlusion. Stroke 21(2):322–327
Swanson RA, Morton MM, Sagar SM, Sharp FR (1992) Sensory stimulation induces local cerebral glycogenolysis: demonstration by autoradiography. Neuroscience 51(2):451–461
Tekkok SB, Brown AM, Westenbroek R, Pellerin L, Ransom BR (2005) Transfer of glycogen-derived lactate from astrocytes to axons via specific monocarboxylate transporters supports mouse optic nerve activity. J Neurosci Res 81(5):644–652
Testoni G, Duran J, Garcia-Rocha M, Vilaplana F, Serrano AL, Sebastian D, Lopez-Soldado I, Sullivan MA, Slebe F, Vilaseca M, Munoz-Canoves P, Guinovart JJ (2017) Lack of glycogenin causes glycogen accumulation and muscle function impairment. Cell Metab 26(1):256–266. e254. https://doi.org/10.1016/j.cmet.2017.06.008
Thompson MF, Bachelard HS (1977) Differences in catalytic properties between cerebral cytoplasmic and mitochondrial hexokinases. Biochem J 161(3):593–598
Vissing J, Wallace JL, Galbo H (1989) Effect of liver glycogen content on glucose production in running rats. J Appl Physiol (1985) 66(1):318–322. https://doi.org/10.1152/jappl.1989.66.1.318
Walls AB, Heimburger CM, Bouman SD, Schousboe A, Waagepetersen HS (2009) Robust glycogen shunt activity in astrocytes: effects of glutamatergic and adrenergic agents. Neuroscience 158(1):284–292. https://doi.org/10.1016/j.neuroscience.2008.09.058
Watanabe H, Passonneau JV (1973) Factors affecting the turnover of cerebral glycogen and limit dextrin in vivo. J Neurochem 20(6):1543–1554
Wender R, Brown AM, Fern R, Swanson RA, Farrell K, Ransom BR (2000) Astrocytic glycogen influences axon function and survival during glucose deprivation in central white matter. J Neurosci 20(18):6804–6810
Youn JH, Bergman RN (1987) Patterns of glycogen turnover in liver characterized by computer modeling. Am J Phys 253(4 Pt 1):E360–E369

Metabolism of Glycogen in Brain White Matter

Angus M. Brown, Laura R. Rich, and Bruce R. Ransom

Abstract Brain glycogen is a specialized energy buffer, rather than a conventional reserve. In the rodent optic nerve, a central white matter tract, it is located in astrocytes, where it is converted to lactate, which is then shuttled intercellularly from the astrocyte to the axon. This basic pathway was elucidated from non-physiological experiments in which the nerve was deprived of exogenous glucose. However, this shuttling also occurs under physiological conditions, when tissue energy demand is increased above baseline levels in the presence of normoglycemic concentrations of glucose. The signaling mechanism by which axons alert astrocytes to their increased energy requirement is likely to be elevated interstitial K^+, the inevitable consequence of increased neuronal activity.

Keywords Lactate · Glycogen · Glucose · Optic nerve · Compound action potential

1 General Energy Requirements of the Brain

All living cells share the requirement for energy. For our single celled ancestors who emerged from the sea they evolved in, this was a relatively straightforward process, as they were bathed in a soup of easily accessible energy substrates. However, as multi cellular organisms evolved, more complex methods of delivering energy to cells had to adapt to this increased morphological complexity, since not all cells were in direct contact with the sea (Lane 2015). When life moved from the sea, which could be considered a soup of metabolic substrates, onto land, thus, limiting the availability of easily accessible energy sources, mechanisms had to evolve to

A. M. Brown (✉) · B. R. Ransom
School of Life Sciences, University of Nottingham, Nottingham, UK

Department of Neurology, University on Washington, Seattle, WA, USA
e-mail: ambrown@nottingham.ac.uk

L. R. Rich
School of Life Sciences, University of Nottingham, Nottingham, UK

M. DiNuzzo, A. Schousboe (eds.), *Brain Glycogen Metabolism*, Advances in Neurobiology 23, https://doi.org/10.1007/978-3-030-27480-1_7

deliver energy to every cell in the body. The energy burden placed on complex multi-cellular organisms facilitated the development of mitochondria (Davies 1998), which increased the energy yield by adapting to oxidative phosphorylation, driven by the increase in atmospheric oxygen resulting from the emergence of blue green algae about 2 billion years ago (Gale 2009). The emergence of multi-cellular land dwelling organisms required an efficient means by which to deliver oxygen and energy substrate to all cells. The development of the cardiovascular system in which the lungs, heart and blood vessels worked in concert to transfer atmospheric oxygen to the blood exploited the avid reaction between oxygen and iron, contained in haemoglobin in red blood cells (Boron and Boulpaep 2009). The heart pumps blood to all tissues, thus, the system may be simplistically viewed as delivery of glucose as the prime substrate to tissues, with oxygen employed to ensure that the glucose efficiently metabolised (Boron and Boulpaep 2009). In humans the increase in brain volume requires an exponentially greater blood flow since maintaining the transmembrane gradients that underlie electrical conduction (Attwell and Laughlin 2001), coupled with the additional computing power (Wang et al. 2008) came at a high metabolic cost (Karbowski 2007). The dogmatic view of brain energy metabolism is that the systemic blood supplies the brain with oxygen and glucose, the cerebrospinal fluid glucose concentration being about two thirds that of the systemic concentration (Boron and Boulpaep 2009). The cerebrospinal fluid contacts the interstitial fluid via the Virchow Robin space (Boron and Boulpaep 2009), with interstitial glucose concentrations determined by glucose delivery and localised energy requirement. This complex balance can be regarded as a flux between delivery, which is governed by, but not limited to, systemic glucose concentration, density of glucose transporters, interstitial volume, rate of cerebrospinal fluid flow, and glucose uptake. The interstitial glucose is taken up by all cells in the brain to fuel their individual energy requirements, which superficially at least appears logical, since there is no dispute that interstitial glucose can be in the millimolar range and bathes every cell in the brain (Hu and Wilson 1997a).

2 Consequences of Shortfall in Glucose Delivery to the Brain

From the whole organ point of view the brain must be considered unique as it receives a fixed proportion of cardiac output (20%) irrespective of its immediate requirements (McKenna et al. 2006). Thus, unlike other organs, such as muscle and gut, which can temporarily increase their blood flow tenfold (Pocock and Richards 2006), the brain cannot increase the volume of blood it receives, although it should be borne in mind that the brain receives ten times more blood than it contributes to body weight, and at any instant the adult human brain contains about 2 pints of blood. Given this fixed delivery of blood flow it is vital that the brain receives an excess of substrate relative to its needs, i.e. supply of substrate must always exceeds demand. The consequences of failing to meet this demand can be appreciated by considering two scenarios. For example, standing up suddenly from a sitting

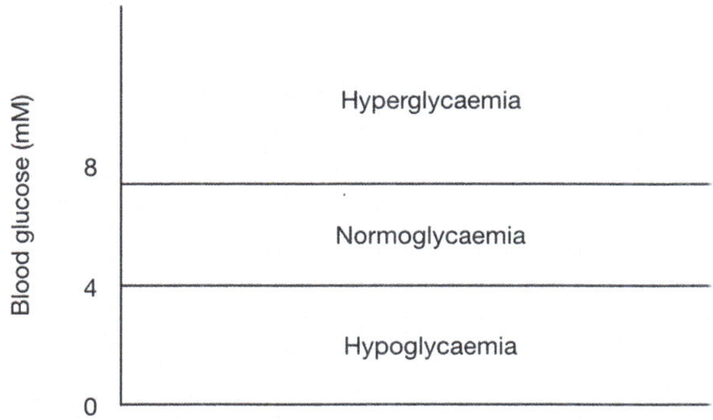

Fig. 1 Blood glucose concentrations associated with hypoglycaemia. The normoglycaemic range is 4.0–7.2 mM. Above this lies the hyperglycaemic range associated with diabetes. Below 4.0 mM lies the hypoglycaemic range, with decreasing blood glucose concentrations evoking the following responses. 3.8 mM; counter regulatory hormone release of glucagon and adrenaline. 3.8–3.2 mM; onset of autonomic and neuroglycopenic symptoms. 3.0–2.4 mM; neurophysiological dysfunction. 3.0 mM; widespread EEG changes. 2.8 mM; cognitive dysfucniton. <1.5 mM; reduced consciousness, convulsions and coma. Adapted from Fig. 6.1 (Frier et al. 2014)

position, may result in a temporary feeling of 'light-headedness', and an impending feeling of faintness. This is due to low blood pressure known as orthostatic hypotension, resulting in a case of insufficient blood flow to the brain leading to temporary mismatch between demand and supply (Silverthorn 2007). A more extreme example occurs during insulin-induced hypoglycaemia (IIH) experienced by type 1 diabetic patients who mismatch prevailing glucose levels with injected insulin to the extent that blood glucose levels are driven to hypoglycaemic levels resulting in inadequate supply of glucose to the brain, the consequences being altered brain function involving loss of consciousness, seizure, coma and even death (Frier and Fisher 2007). The case of IIH is informative as it indicates there are no substrates equivalent to glucose present in the systemic circulation that can substitute for glucose in the event of systemic hypoglycaemia. However, events within the brain at a cellular level are more controversial (Fig. 1).

3 Metabolic Glia-to-Neurone Lactate Shuttling

It was originally thought, and universally accepted, that each cell within the brain, be it neurone or glial cell, directly took up glucose from the interstitial space. This was supported by the distribution of glucose transporters, whereby all cells studied expressed at least one subtype of glucose transporter, as did the blood brain barrier (BBB) to facilitate uptake of glucose from the systemic circulation into the brain parenchyma (Vannucci et al. 1997). A general conclusion reached was of neuronal GLUT3 and astrocytic GLUT1 expression (Vannucci et al. 1997). GLUT1 and GLUT3 are insulin independent and have a high affinity for glucose, and are thus responsible for continuous uptake of glucose (Vannucci et al. 1997). However, this simplistic arrangement has been questioned in recent decades (Thorens and Mueckler 2010). Initial studies relating glucose uptake with oxygen consumption were inconsistent with complete glucose oxidation, with the ratio found to be in the region of 5.5 at rest, rather than 6 expected from compete oxidation (Dienel et al. 2009). The implied inefficiency with regard to glucose oxidation was at odds with the high demand for energy which is a key feature of the brain. However, this phenomenon is supported by the presence of significant interstitial lactate (Mangia et al. 2009a). Lactate has long been viewed as a toxic waste compound resulting from incomplete glucose oxidation, its accumulation, particularly in the brain, an indicator of pathology (Tymoczko et al. 2015). However, in muscle, during intense exercise, the consumption of glucose is not matched by the uptake of oxygen required to fully metabolise all of the lactate produced, leading to an accumulation of lactate and a build up in systemic circulation of up to ten times that of resting levels (Tymoczko et al. 2015). A more productive role for lactate was proposed in the mid 1990s by Pellerin and Magistretti, who linked increases in synaptic activity with increased astrocytic metabolism, in a scheme where increased synaptic activity leads to increased interstitial glutamate, which is buffered by Na^+-coupled glutamate uptake in astrocytes. This leads to increased glucose uptake into the astrocytes, which is glycolytically converted to lactate. The two molecules of ATP thus produced fuel the Na^+ pump required to restore the trans-membrane Na^+ gradient, and convert glutamate to glutamine. The glutamine and lactate are separately shuttled to the neurones where the glutamine is converted to glutamate and the lactate is used as a substrate for oxidative metabolism (Pellerin and Magistretti 1994). Such a scheme is stoichiometrically attractive but has met with fierce criticism: a representative critique of both sides of the argument are summarised in the following reviews (Pellerin and Magistretti 2012; Dienel 2012; DiNuzzo et al. 2010; Mangia et al. 2009b; Korf 2006). The scheme proposed by Pellerin and Magisteretti makes several important assumptions, the first of which is that neurones can use lactate as a substrate, which presumes that neurones can take up, metabolise and efficiently metabolise lactate. It has been presumed that the BBB offers an efficient barrier to lactate uptake into the brain (Simpson et al. 2007). However, whole body studies have suggested that significant lactate can cross the BBB and be used by the brain (Dalsgaard 2006; Dalsgaard et al. 2004).

4 The Rodent Optic Nerve Model of Central White Matter

For the scheme i.e. lactate transfer between glia and neurones, to be accepted it must be viewed on a more general level i.e. what are the advantages of lactate transfer between the cell types? Numerous papers have been published that demonstrate lactate release from astrocytes and lactate uptake by neurones (Dringen et al. 1993, 1995). The controversial issues arise in inconsistencies of not what can happen, but what does happen, under physiological conditions. These considerations have led to a polarisation of the field with opinions divided as to whether the lactate shuttle operates under physiological conditions. These problems arose mainly from the experimental models used to illustrate lactate fluxes. Culture models are accompanied by the legitimate argument that they exist under non *in vivo* conditions (Dringen et al. 1993, 1995), of particular concern the infinitely large interstitial space that separates cells, which under *in vivo* conditions would be closely apposed, separated by a narrow interstitial gap (Ransom 2009). Even certain *in vitro* models are less than ideal as stimuli may induce non-physiological behaviour. Additional complications derive from consideration of the source of lactate, which may be glucose, glycogen or fructose (Dringen et al. 1993; Meakin et al. 2007; Magistretti and Allaman 2018; Pellerin et al. 2002). For the purposes of this review we will focus on the role of glycogen in the CNS white matter. The presence of glycogen in the brain has been known since the invention and utilisation of electron microscopy (Koizumi 1974; Koizumi and Shiraishi 1970a, b; Phelps 1972, 1975) and its measurement by biochemical assay championed by Oliver Lowry (Passonneau et al. 1967). However, these initial exploratory studies were not followed up by any functional studies. The reasons for this are likely related to the paucity of glycogen present in the brain being considered insignificant as an energy store in the manner it is in liver and skeletal muscle (Nelson et al. 1968). It is perhaps useful at this point to describe the established roles of glycogen present in liver and skeletal muscle. In skeletal muscle glycogen acts as a localised energy reserve for immediate use upon increased work by muscles. Its breakdown, particularly in response to increased energy demand, can lead to increased lactate production due to incomplete oxidative metabolism of the glycogen (Stryer 1995). Liver glycogen acts as an energy reserve to sustain normoglycaemic blood glucose. Falls in systemic glucose are compensated for by glycogen breakdown and release of glycogen-derived glucose into the systemic circulation (Stryer 1995). Thus, although it may be viewed that the role for liver glycogen is to supply all tissue with glucose, since the brain is more sensitive to shortfalls in glucose that other organs, liver glycogen's principle role is to supply the brain with glucose—it is thus prescient to ask 'what then is the role of brain glycogen?'

Glycogen in the adult mammalian brain is located almost exclusively in astrocytes (Cataldo and Broadwell 1986). Although glycogen related enzymes are found in neurones (Vilchez et al. 2007) and axons (Pfeiffer-Guglielmi et al. 2006, 2007), glycogen itself is expressed in neurones only under pathological conditions such as Lafora's Disease (Vilchez et al. 2007), thus, we can safely assume that in the adult mammalian brain glycogen is exclusively present in astrocytes in both grey and white matter. Such a location immediately prompts the key question, why is the glycogen not present in the neurones since they consume more energy than astrocytes? In addition, the molecular apparatus required to transport the lactate between cells must be present and functioning. The most plausible reason for the astrocytic location of glycogen is to relocate some energy burden away from neurones and onto astrocytes. This has a precedent with regard to glutamate uptake. Whereas for neurotransmitters such as serotonin, the reuptake mechanism occurs in the presynaptic terminal once the neurotransmitter has been released from the post-synaptic receptor (Siegel et al. 2006), and is the target of selective serotonin reuptake inhibitors that are at the core of the treatment of depression (Ferguson 2001), glutamate is taken up into the astrocyte via Na^+-coupled transporters (Sonnewald et al. 1997). It is subsequently processed in the astrocyte to glutamine prior to release as an inactive compound into the interstitial fluid for uptake and regeneration into glutamate in the neurones (Schousboe et al. 1997). This moves the metabolic burden from oxidative neurones to glycolytic astrocytes (Tymoczko et al. 2015) as the Na^+ influx that accompanies the glutamate movement must be re-equilibrated via the ATP dependent Na^+ pump, and the enzyme glutamine synthase, which converts glutamine to glutamate, requires ATP.

It is known that the majority of glucose uptake in the brain occurs in astrocytes, yet the majority of energy produced is via neurones (Nehlig et al. 2004; Chuquet et al. 2010). These apparently dichotomous findings can be allied if we accept that lactate transfer between astrocytes and neurones occurs.

5 Establishing the Presence and Elementary Role of Glycogen in Rodent Optic Nerve

We first started investigating glycogen in white matter over 20 years using the rodent optic nerve as our model system. The *in vitro* rodent optic nerve lacks the complications of glutamatergic synapses and cell bodies and as such is a relatively simple model of central white matter consisting of axons, astrocytes and oligodendrocytes (Ransom and Orkand 1996). Recordings of the stimulus evoked compound action potential (CAP) can be used to assess the conduction of the nerve (Stys et al. 1991), since the CAP consists of the summed contribution of all conducting axons (Cummins et al. 1979), thus, a decrease in CAP area is indicative of a loss of contributing axons (Stys et al. 1991). The investigations were prompted by the following experimental evidence. In the *in vitro* rodent model of central white matter, in which the tissue is maintained in a superfusion chamber perfused with oxygenated aCSF at 37 °C (Stys et al. 1991), replacement of oxygen with nitrogen or adding cyanide (Stys et al. 1992) causes a rapid failure of the CAP within a few minutes, indicating the inability for CAP conduction to persist in the absence of oxidative phosphorylation. However, when glucose was withdrawn, but with oxygen still present, the CAP took up to 30 min before it began to fail after which time it was rapidly abolished (Ransom and Fern 1997). There were several possible explanations for this observation: (1) the tissue survives on the available glucose-derived metabolites present in the tissue, (2) the tissue survives on glucose present in the intra- and extracellular milieu, (3) there is an endogenous energy substrate present that supports the tissue temporarily. The third scenario appears most likely based on the following evidence. Survival of cultured neurones was improved if they were co-cultured with astrocytes, although at the time of the study, the nature of this support afforded by the astrocytes was not investigated (Whatley et al. 1981). Subsequent studies demonstrated similar findings but revealed that the key factor that determined neuronal survival was the presence of glycogen in astrocytes, since depleting astrocytes of glycogen attenuated neuronal survival (Swanson and Choi 1993). Subsequent tissue culture studies demonstrated that astrocytes release lactate when supplied with glucose, a property not demonstrated by neurones (Dringen et al. 1995). With this backdrop we proceeded in our studies on rodent optic nerve to understand the role of glycogen in central white matter tissue. Due to historical precedent our first paper was carried out in rat optic nerve (RON) (Wender et al. 2000), but all subsequent studies were carried out on mouse optic nerve (MON). The presence of glycogen in the RON was confirmed using both EM and biochemical assay (Wender et al. 2000), demonstrating a low concentration of glycogen in comparison to those in the liver or skeletal muscle. The ability of glycogen to support axon conduction was monitored as its ability to support the stimulus evoked CAP in the absence of exogenously applied glucose, with any loss of the CAP area being considered as a loss of conducting axons (Stys et al. 1991). The glycogen content was measured at about 10 pmol ug protein^{-1}. Removal of glucose caused a reduction in glycogen content to a nadir of about 2 pmol ug protein^{-1}. This temporally correlated with the failure of the CAP after about 30 min. This encouraging information prompted us to progress with a more detailed assessment of glycogen. The lability of glycogen content was investigated by incubating the tissue with 10 mM glucose, 25 mM fructose or 1 mM noradrenaline plus 1 mM glucose. These were chosen for the following reasons, since we would anticipate that 25 mM glucose should increase glycogen, but equivalent fructose should have no effect on glycogen since it does not contribute to glycogen content (Champe and Harvey 2008). Glucose surplus to immediate requirements of astrocytes and axons is converted to glycogen in the astrocytes to form a readily available pool of lactate. Noradrenaline previously has been shown to deplete glycogen (Sorg and Magistretti 1992). The results were as expected, demonstrating that glycogen can be up or down regulated rapidly. As well as the glycogen content being measured under such circumstances the latency to CAP failure was measured, as this would give an indication as to whether there were any functional consequences to manipulating glycogen levels.

6 Glycogen-Derived Lactate Supports Axon Conduction in Rodent Optic Nerve

The latency to CAP failure increased from 30 to about 40 min after pre-incubation for 2 h in 25 mM glucose, but with 25 mM fructose there was no effect. This was confirmation that loading the interstitial space with utilisable substrates did not augment latency to CAP failure. Noradrenaline led to a significant decrease in the latency to CAP failure, thus, these results were indicative of a role for glycogen in supporting CAP conduction in the absence of exogenously applied glucose (Wender et al. 2000). Since glycogen is too large a molecule to be transported between astrocytes and axons it must first be converted to a transportable conduit in astrocytes prior to shuttling to axons. The most likely candidate for this is lactate, as it is considered the end product of glycolysis (Schurr and Payne 2007) and it is the most visible substrate to be released by glycogen-rich astrocytes (Dringen et al. 1993). If the conduit is lactate, then exogenously applied lactate in the absence of glucose should support axon conduction, and this is what we found, 20 mM lactate supporting the CAP for over 2 h. As a logical consequence of this there must exist transport mechanisms for lactate efflux from astrocytes and uptake into axons, and this is most likely to be via monocarboxylate transporters (MCTs), which are inhibited by various compounds, including quercitin, CIN and pCMBS. Quercetin, related to saffron, accelerated the CAP failure during aglycemia but did not affect the ability of CAP conduction in the presence of 20 mM lactate suggesting it blocks lactate efflux from astrocytes, but does not affect lactate influx into axons. The addition of CIN or PCMBs accelerated CAP failure during aglycemia and partially inhibited the CAP supported by 20 mM lactate, suggesting they block lactate uptake into axons. This introductory study established important benchmarks with regards to CNS white matter glycogen, which can be summarised as follows; (1) Glycogen is present in CNS white matter astrocytes, (2) Glycogen content can be up and down regulated, thereby providing a labile pool from which lactate can be released, (3) Glycogen content at the onset of aglycaemia determines latency to CAP failure, (4) Lactate can support CAP conduction, which is blocked if lactate transport is interrupted (Wender et al. 2000).

7 Establishing Role of Glycogen during Hypoglycaemia and Increased Energy Demand in Rodent Optic Nerve

These intriguing results encouraged us to carry out a more detailed study on the role of CNS white matter glycogen, but we chose the mouse as our model for the following reasons; investigations into the effects of anoxia on rat optic nerve (RON) led us to conclude that the diffusion distance to the centre of the RON under our experimental conditions, where the nerve is placed in a superfusion chamber, was too great for adequate perfusion of glucose, and the centre of the tissue is almost certainly ischaemic (Baltan Tekkök et al. 2003). The smaller MON, and its smaller diffusion distance bypasses this problem, as well as offering the advantage of being compatible with the development of transgenic models, which are almost exclusively murine in origin. The MON CAP displayed a qualitatively similar trend to that of the RON in that on withdrawal of glucose it failed, but failure was accelerated compared to the rat at 20 versus 30 min, with the glycogen content correspondingly lower at 8 pmol ug protein^{-1} (Brown et al. 2003). A period of aglycemia depleted glycogen, which could be replenished when the nerve was subsequently exposed to glucose-containing aCSF, but not lactate-containing, as shown by a vastly accelerated CAP failure following a period of aglycaemia after which 20 mM lactate aCSF was added to rescue the CAP. The glycogen content was shown to be labile as it was affected by bathing the tissue for 2 h in various concentrations of glucose from 2 to 30 mM, with the higher the ambient glucose, the higher the glycogen content. Subsequent exposure of MONs treated with increasing glucose concentrations led to increased latency to CAP failure following introduction of aglycaemia, with the higher the ambient glucose, and thus, the higher glycogen levels, the longer the latency to CAP failure. Plotting glycogen content prior to aglycaemia versus latency to CAP failure resulted in a linear relationship: the importance of this relationship is that knowing the glycogen content immediately prior to aglycemia allows one to predict with a high degree of accuracy the latency to CAP failure. Translating this to the clinical environment, where type 1 diabetic patients who mismatch insulin administration relative to ambient glucose can lead to hypoglycaemic episodes (Frier and Fisher 2007), implies that increasing the glycogen content in these patients would extend the period of time before brain disruption and could potentially stave off brain damage and allow the patient and clinical staff additional time to intervene (Fig. 2).

Thus, far the role of glycogen that has been deduced from these experiments is to support axon conduction in the absence of glucose, which is admittedly an artificially experimental environment. However, this information is useful in establishing key roles for glycogen and the associated mechanisms involved (Brown et al. 2003). However, we next sought to determine if glycogen played a role under more physiological conditions. Thus, we investigated the contribution of glycogen to supporting CAP conduction in hypoglycaemic concentrations of glucose. Superfusion of CSF containing 2 mM glucose supported the CAP for 2 h, which was an unexpected result given that 2 mM glucose in the systemic circulation is considered hypoglycaemic (Frier and Fisher 2007) and incapable of supporting brain function. However, in our MON model, depleting glycogen with a brief period of aglycemia revealed

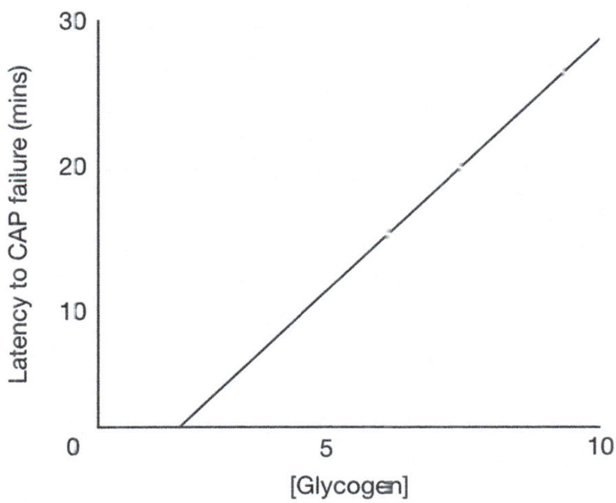

Fig. 2 Relationship between latency to CAP failure after onset of aglycaemia and concentration of glycogen (pmol μg protein^{-1}) present in the MON. Adapted from Fig. 2D (Brown et al. 2003)

that 2 mM glucose could no longer support the CAP, a result reinforced by the ability of CIN to reduce the CAP area when superfusing 2 mM glucose. The conclusion from these experiments was that 2 mM glucose alone is insufficient to support the CAP and is supplemented by glycogen, however, depleting glycogen or interfering with lactate transport results in loss of the CAP, confirming a supporting role for glycogen in the presence of hypoglycaemic concentrations of glucose, where glycogen is metabolised to provide supplemental lactate to ensure the energy demands of the tissue are met by the supply of glucose and lactate (Frier and Fisher 2007). These results suggested that glycogen can play a role in brain function under more physiological conditions, and this was further explored by imposing high frequency stimulus on the MON in order to increase the energy demand of the tissue to determine how this was met with regards to substrate supply. Imposing 100 Hz stimulus for 4 min caused an elongation of the CAP profile, no doubt due to the resulting perturbation of trans-membrane ion gradients affecting Na^+ influx (Stys and Lopachin 1996), but the CAP was maintained above baseline levels. However, the addition of CIN caused a rapid fall in the CAP, as did quercetin, supportive of glycogen metabolism being used to provide supplemental lactate to the axons in order to meet this increased energy demand. Addition of CIN with 10 mM glucose under control conditions of 1 stimulus every 30 s had no effect on the CAP. This role for glycogen in supporting increased energy demand in the presence of normoglycaemic concentrations of glucose has profound implications in proposing a physiological role(s) for glycogen. Imposing 100 Hz stimulus for periods of time ranging from 1 to 4 min prior to exposing the tissue to aglycemia under baseline stimulus protocol resulted in accelerated CAP failure, with the degree of attenuation directly correlated to the duration of high frequency stimulus (Fig. 3).

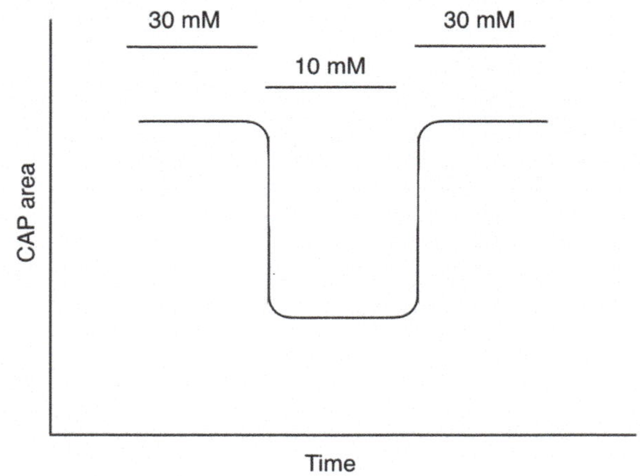

Fig. 3 CAP area is affected by substrate concentration during high frequency stimulus. The CAP area is maintained in 30 mM glucose but falls when exposed to 10 mM glucose in an MON exposed to 100 Hz. Adapted from Fig. 4C (Brown et al. 2003)

8 Inhibiting Glycogen Metabolism and Blocking Lactate Uptake

Whereas glycogen in the liver and skeletal muscle act as a reservoir of energy substrate that can be used for long term sustenance e.g. liver glycogen is reported to be exhausted after 24 h of starvation (Stryer 1995), and muscle glycogen can support extended duration physical exercise such as marathon running (Tymoczko et al. 2015), the brains glycogen must be viewed as fulfilling a fundamentally different role. We have already discussed that the role of liver glycogen is to maintain normal glycaemic concentrations of glucose in order to supply the brain with adequate glucose. The feeding of the other organs/tissues is a consequence of a systemic blood supply. Thus, the role of brain glycogen is not as one would expect to supply the brain with glucose but to act as a short-term energy buffer to deliver energy substrate in the form of lactate during periods of need such as a/hypoglycaemia and increased energy demand under normal glycaemic conditions (Fig. 4).

The development of compounds that inhibit glycogen phosphorylase, the key enzyme that initiates glycogen metabolism, allowed further investigation into the role of brain glycogen. Although predominately of a confirmatory nature such data provided a strong foundation from which established roles of glycogen emerged. Addition of DAB (Andersen et al. 1999; Fosgerau et al. 2000; Walls et al. 2008) or isofagomine (Andersen et al. 2001; Waagepetersen et al. 2000), compounds that inhibit glycogen phosphorylase, resulted in an accelerated latency to CAP failure in MONs exposed to aglycemia (Brown et al. 2005). The effect of these compounds was independent of the glycogen content, and indeed in the presence of DAB the glycogen content increased significantly but could not be utilised. This provided an indirect indication as to the relative turnover rate of glycogen, since application of DAB for 2 h increased glycogen content by over a third (Walls et al. 2008). This suggests an active turnover of glycogen in the MON (see later). The compound CIN is known to inhibit pyruvate uptake into mitochondria, which could cloud results, so

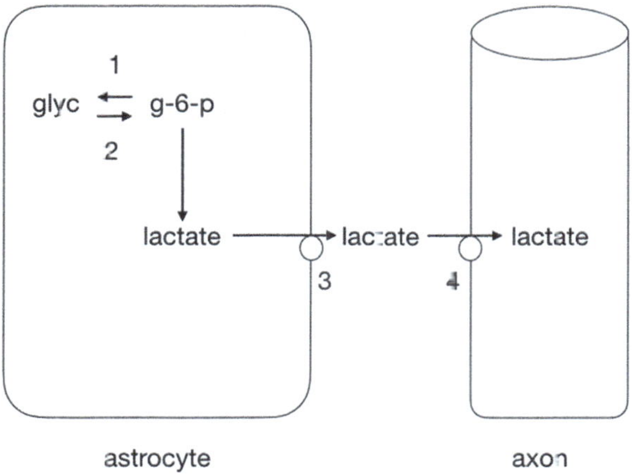

Fig. 4 Glycogen-derived lactate fuels axon conduction during aglycaemia. In the absence of exogenously applied glucose astrocytic glycogen is metabolised to lactate prior to shuttling to the axon via monocarboxylate transporters (3 and 4). Glycogen synthase controls glycogen synthesis (1) and glycogen phosphorylase its metabolism (2)

we used D-lactate, the inactive stereoisomer of L-lactate, which acts as a competitive compound at the MCT but is not metabolised intracellularly, to block lactate uptake into axons. D-lactate at 20 mM had no effect on the CAP area when administered with 10 mM glucose, but accelerated latency to CAP failure at the onset of aglycaemia supporting a role for lactate transfer (Tekkok et al. 2005). Such transfer must require the appropriate transporters to shuttle lactate between cells, plus there must be the capability to transport lactate out of astrocytes and into axons. Immunohistochemical studies using antibodies specific for MCT subtypes showed that the MCT1 was present on axons, since these were co-labeled with the axon specific antibody neurofilament, whereas the MCT 2 was co-localised with astrocyte specific stain GFAP (Tekkok et al. 2005). The MCT co-transports lactate with H^+ thus, lactate movement is associated with alterations in pH. We investigated these phenomena in a study in which we determined the injury mechanisms incurred by the MON on exposure to aglycemia (Yang et al. 2014). Lactate biosensors recorded the lactate concentration on the periphery of the pial-glia boundary, and thus did not measure interstitial lactate, and pH sensitive microelectrodes were used to construct a scheme following exposure of the MON to aglycemia. At the onset of aglycemia there was a rapid and complete fall in lactate, supporting a flux of lactate from astrocytes to axons, which is abolished at the onset of aglycemia, since the axons are without glucose and take up all available lactate. The pH results were more difficult to interpret. The baseline pH under superfusion with 10 mM glucose is 7.25, considerably more acidic then the pH of the perfusing aCSF, at 7.45. This is likely due to the hydrogen ions accompanying a tonic release of lactate by the astrocytes. Upon introduction of aglycaemia there was a two-step process with an initial rapid pH jump to about 7.4 while the CAP was maintained. However, coinciding with the subsequent CAP failure, which signifies the exhaustion of glycogen, the pH jumped to 7.45 i.e. equivalent to bath pH. We interpret this as signifying the glycogen is exhausted and thus no more lactate is produced or transported, and hence there is no outward hydrogen flux to acidify the interstitial fluid (Yang et al. 2014) (Fig. 5).

9 Lactate Biosensors Reveal Tonic Lactate Efflux from Rodent Optic Nerve

The lack of effect of CIN on the CAP in the presence of 10 mM glucose and limiting uptake of lactate into axons has no significant effect on axon conduction, which can be supported without the need of lactate transfer from astrocytes, whether the source is glucose or glycogen (Brown et al. 2003). However, the introduction of lactate sensors into our technical armoury used to study the roles of glycogen revealed a finding that is consistent with previous reports, namely that there exists in the interstitial space significant concentrations of lactate (Yang et al. 2014). Since the lactate sensor is placed outside of the nerve at the pia glial border, a stable signal actually indicates a steady efflux from the tissue. The reason why there is a tonic lactate efflux from the tissue in the presence of normoglycaemia is consistent with a

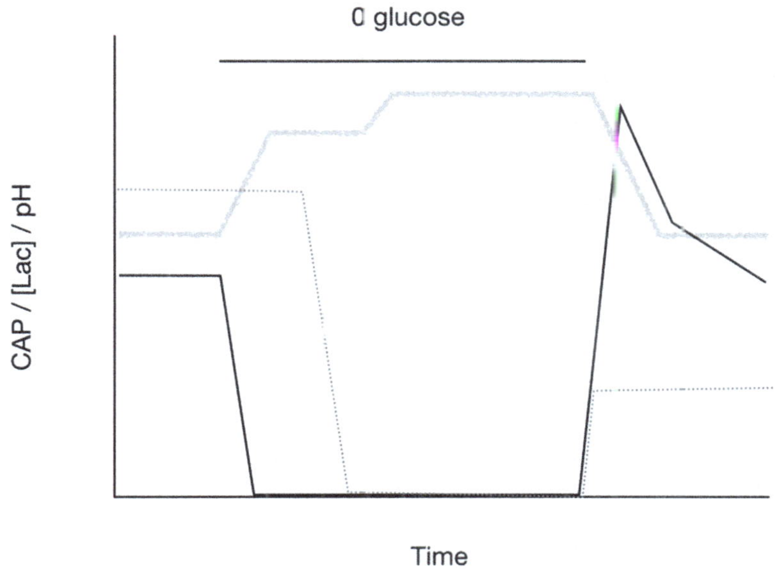

Fig. 5 Relationship between pH, $[Lac]_o$ and CAP during aglycaemia. Resting pH (grey line) was about 7.2 under baseline conditions. At the onset of aglycaemia there was a two-stage alkalization of pH, the first stage coincided with a decrease in [Lac] (black line) measured at the pial glial border, and the second phase coincided with the loss of the CAP (dotted line). Adapted from Fig. 4C (Yang et al. 2014)

glucose utilisation/oxygen uptake ratio of less than 6, but the fundamental consequences are unknown. Is it the case that there is a degree of versatility in the substrates used by axons under baseline conditions, and that blocking the route of lactate uptake results in a corresponding increase in glucose uptake? In addition, the fact that we record lactate at the outer boarders of the nerve is key evidence that it is not utilised and may be considered surplus to immediate requirements. This is supported by data recorded during aglycaemia when the lactate levels are abolished rapidly (Yang et al. 2014), even although electrophysiological evidence points to the fact that lactate is being transferred from astrocytes to axons within the nerve, but in the absence of exogenously applied glucose all lactate present in the interstitial fluid is avidly taken up by axons leaving none to move from the interstitial compartment to the outer border of the nerve.

10 Lactate Release from Astrocytes

Using state of the art FRET sensors the release of lactate from astrocytes has been studied in both in vitro and in vivo conditions. Lactate pools are present in astrocytes and are released through lactate channels as well as via the conventional MCT

(Sotelo-Hitschfeld et al. 2015). The lactate pool is estimated to be about 1.5 mM, thus, a substantial reserve of lactate is present for immediate release. The channel is not dependent upon the balance of trans-membrane lactate and H^+ concentrations, thus, lactate efflux occurs even when it is electrochemically unfavourable. The release of lactate is dependent upon astrocyte membrane potential, depolarisation favouring release. Thus, computational studies that show a near instantaneous increase in interstitial lactate in response to stimulus support the presence of a large, rapidly available pool of intracellular lactate (Aubert et al. 2007, 2005).

11 Astrocyte Heterogeneity

In studies of this kind astrocytes are considered a homogeneous cell population sharing identical functions. However, it should be borne in mind that there are specialised subtypes of astrocytes e.g. Muller cells and Bergmann glia (Farmer and Murai 2017). In addition, the heterogeneous nature of astrocytes is related to the brain region in which they reside, their phenotype determined, in part, by developmental cues (Farmer and Murai 2017). Although in its infancy recent studies are attempting to relate astrocyte phenotype with function suggesting at least three main astrocyte subpopulations (Chia-Ching et al. 2017). Also of relevance is the fact that virtually all brain tumours are of glial origin, the malignant phenotype contributing to the pathology (Sontheimer 2008).

Given the reliance on the astrocyte membrane potential sensitivity to K^+, which acts as a sensor of interstitial K^+ and hence axonal activity (Ransom and Goldring 1973), one may expect that astrocytes would express similar resting membrane potentials, but this was not found to be the case in both grey and white matter. In hippocampus recorded using patch pipettes there was a wide range of resting membrane potentials, which may have its origins in the method of recordings. High resistance sharp microelectrodes tended to record more hyperpolarised potentials, whereas patch pipettes recorded more depolarised potentials (McKhann 2nd et al. 1997). However, in MON a bipolar distribution of membrane potential was recorded using sharp microelectrodes, with the morphology of the recorded cells falling into two categories (Bolton et al. 2006). Thus, there does not appear to be a uniform astrocyte resting membrane potential close to E_K, but rather there is a range of RMPs, which may signify different functions of astrocytes. Although currently not known one would suspect that astrocytes which do not show a Nernstian response to K^+ i.e. astrocytes that also show a permeability to Na^+ and/or Cl^- (Kandel et al. 2013) would not be as sensitive in their response to neuronal activity, thus, the role of responding to physiological elevations in K^+ and subsequently supplying lactate to neurones may be the responsibility of astrocytes that are exclusively permeable to K^+.

12 Implications from Other Models on the Functions of Central White Matter Glycogen

In the last decade or so there have been many studies investigating the role of glycogen in the nervous system, although most were carried out in grey matter. Nonetheless some of the data reported, and conclusions reached, can be extrapolated to white matter to reveal potential roles of glycogen and lactate shuttling that await experimental confirmation.

Computational studies, which are based on experimentally derived data, are important in yielding information that may be experimentally very difficult to obtain. A multi compartment model assembled using the MATLAB programming environment has been used to investigate how compartmentally segregated lactate concentrations are affected under physiologically relevant conditions. The initial model showed a significant resting lactate concentration in the interstitial fluid. On introduction of stimulus there was transient decrease in this lactate concentration before it recovered and steadily increased until cessation of the stimuli (Aubert et al. 2005). This was a critical finding in that it reproduced experimental data obtained under in vivo conditions using lactate sensors to record interstitial lactate (Hu and Wilson 1997b). The significance of the transient decrease is that it indicates that at the onset on stimulus the energy requirement of the axons is met by uptake of interstitial lactate, and crucially that astrocytic production of the lactate lags behind the axonal requirement. The transient decrease, in the order of between 10 and 15 s (Fig. 1, Hu and Wilson 1997b), rapidly reverses, an indication that whatever signalling mechanisms that exists between axons and astrocytes has commenced, and suitable delivery of lactate in excess of immediate axonal requirements is being delivered to the interstitial space. A clue to the signalling mechanism was revealed in a further experiment in which repetitive stimuli were imposed, and the lactate concentration increased in an attenuating manner (Aubert et al. 2005), similar to the effect of repetitive stimuli on the *Necturus* glial cell membrane potential in Kuffler's classic experiments (Orkand et al. 1966). Efflux of K^+ from axons during stimulus was hypothesised (Hodgkin and Huxley 1952) and subsequently measured (Hodgkin and Huxley 1953) in experiments on squid axons, and is universally accepted as a consequence of axonal action potential firing (Hille 2001), thus, there is convincing evidence that the signalling pathway whereby neurones/axons signal astrocytes during periods of increased activity is via increased interstitial K^+ (Choi et al. 2012), to which astrocytes are exquisitely sensitive. The higher the degree of neuronal activity the greater amount of lactate is shuttled from astrocytes to neurones/axons (Fig. 6).

13 Effects of Pathology on Glycogen Content

A variety of pathological conditions are known to affect glycogen content in grey matter, a consequence that may have implications to white matter function. The maturation of astrocytes is temporally correlated with expression of glycogen such that undifferentiated neural stem cells do not express glycogen (Brunet et al. 2010),

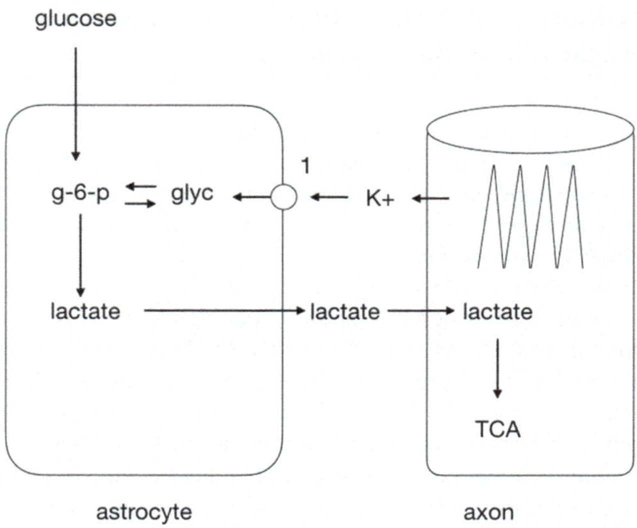

Fig. 6 K^+ stimulated glycogenolysis. Axonal action potentials cause release of K^+ to the interstitium, which act upon the Na^+ coupled bicarbonate transporter (1) leading to intracellular alkalization of the astrocyte. pH sensitive adenyl cyclase activates glycogen phosphorylase promoting glycogen metabolism, ultimately leading to lactate transfer to axons. Adapted from Fig. 7 (Choi et al. 2012)

whereas mature astrocytes do, thus, the expression of glycogen can be used as an indicator of astrocyte differentiation. In addition, stress in known to be a factor that regulates glycogen expression in hippocampus. Stress causes a reshaping of astrocyte morphology and a subsequent decrease in glycogen expression, where glycogen degradation outweighs glycogen synthesis despite reports that stress increases expression of both glycogen synthase and phosphorylase (Zhao et al. 2017).

There is a connection between glycogen expression and the migraine associated with cortical spreading depression (CSD). Glycogen metabolism is required by astrocytes to maintain low interstitial K^+, a vital buffering function that maintains the trans-membrane ion gradients optimal for neuronal excitability and action potential firing. It has been reported that sleep deprivation causes glycogen synthesis rather than degradation, which would interfere with K^+ buffering. An increase in K^+ can result in activation of Pannexin-1 channels, whose activation is thought to underlie the migraine aura that accompanies CSD. Inhibition of glycogen metabolism causes activation of Pannexin-1 channels with sleep deprivation causing Pannexin-1 channel opening and reduced CSD threshold. The reduced threshold could be reversed by application of glucose or lactate implying glycogen metabolism is vital to maintaining K^+ homeostasis, with disruptions of this vital function leading to defined pathological consequences (Kilic et al. 2018).

14 Conclusions

In the MON glycogen is present in astrocytes where its ability to support axons during periods of aglycaemia led to defining its physiological role as providing supplemental lactate to axons during periods of increased energy demand under normoglycaemic concentrations of glucose. However, since glycogen reserves are limited the tissue benefits only for a short period of time. It appears that astrocytic glycogen may play a vital role in the metabolic glia to axon signalling that occurs during periods of increased neuronal activity, when K^+ released from active axons acts as an indicator of increased activity, to which astrocytes are exquisitely sensitive in their response, supplying glycogen derived lactate to the hungry axons. Unresolved issues remain - under what physiological conditions is lactate required by axons and how is this signalled and translated into action by the astrocytes. Such issues may be addressed with higher resolution measurements of real time interstitial lactate. From the pathological viewpoint intercellular lactate shuttling may underlie diabetic neuropathy, memory impairment, migraine and age related neurovascular disorders.

References

Andersen B, Rassov A, Westergaard N, Lundgren K (1999) Inhibition of glycogenolysis in primary rat hepatocytes by 1, 4-dideoxy-1,4-imino-D-arabinitol. Biochem J 342:545–550

Andersen B, Mortensen E, Winther GM, Westergaard N (2001) Characterization of isofagomine's inhibitory effect on glycogen phosphorylase and glycogenolysis in primary rat hepatocytes. Diabetologia 44(Suppl 1):Poster 0618

Attwell D, Laughlin SB (2001) An energy budget for signaling in the grey matter of the brain. J Cereb Blood Flow Metab 21:1133–1145

Aubert A, Costalat R, Magistretti PJ, Pellerin L (2005) Brain lactate kinetics: modeling evidence for neuronal lactate uptake upon activation. Proc Natl Acad Sci U S A 102:16448–16453

Aubert A, Pellerin L, Magistretti PJ, Costalat R (2007) A coherent neurobiological framework for functional neuroimaging provided by a model integrating compartmentalized energy metabolism. Proc Natl Acad Sci U S A 104:4188–4193

Baltan Tekkök S, Brown AM, Ransom BR (2003) Axon function persists during anoxia in mammalian white matter. J Cereb Blood Flow Metab 23:1340–1348

Bolton S, Greenwood K, Hamilton N, Butt AM (2006) Regulation of the astrocyte resting membrane potential by cyclic AMP and protein kinase a. Glia 54:316–328

Boron WF, Boulpaep EL (2009) Medical physiology. Elsevier

Brown AM, Tekkok SB, Ransom BR (2003) Glycogen regulation and functional role in mouse white matter. J Physiol 549:501–512

Brown AM et al (2005) Astrocyte glycogen metabolism is required for neural activity during aglycemia or intense stimulation in mouse white matter. J Neurosci Res 79:74–80

Brunet JF, Allaman I, Magistretti PJ, Pellerin L (2010) Glycogen metabolism as a marker of astrocyte differentiation. J Cereb Blood Flow Metab 30:51–55

Cataldo AM, Broadwell RD (1986) Cytochemical identification of cerebral glycogen and glucose-6-phosphatase activity under normal and experimental conditions. I. Neurons and glia. J Elec Micro Tech 3:413–437

Champe PC, Harvey RA (2008) Biochemistry, 4th edn. Lippincott Williams & Wilkins, Baltimore, MD
Chia-Ching JL et al (2017) Identification of diverse astrocyte populations and their malignant analogs. Nat Neurosci 20:396–405
Choi HB et al (2012) Metabolic communication between astrocytes and neurons via bicarbonate-responsive soluble adenylyl cyclase. Neuron 75:1094–1104
Chuquet J, Quilichini P, Nimchinsky EA, Buzsaki G (2010) Predominant enhancement of glucose uptake in astrocytes versus neurons during activation of the somatosensory cortex. J Neurosci 30:15298–15303
Cummins KL, Perkel DH, Dorfman LJ (1979) Nerve fiber conduction-velocity distributions. I. Estimation based on the single-fiber and compound action potentials. Electroencephalogr Clin Neurophysiol 46:634–646
Dalsgaard MK (2006) Fuelling cerebral activity in exercising man. J Cereb Blood Flow Metab 26:731–750
Dalsgaard MK et al (2004) A reduced cerebral metabolic ratio in exercise reflects metabolism and not accumulation of lactate within the human brain. J Physiol 554:571–578
Davies P (1998) The fifth miracle. Penguin Press, London
Dienel GA (2012) Brain lactate metabolism: the discoveries and the controversies. J Cereb Blood Flow Metab 32:1107–1138
Dienel GA, Byrne JH, Roberts JL (eds) (2009) From Molecules to Networks: An Introduction to cellular and molecular neuroscience. Chapter 3. Academic Press, pp 49–110
DiNuzzo M, Mangia S, Maraviglia B, Giove F (2010) Glycogenolysis in astrocytes supports blood-borne glucose channeling not glycogen-derived lactate shuttling to neurons: evidence from mathematical modeling. J Cereb Blood Flow Metab 30:1895–1904
Dringen R, Gebhardt R, Hamprecht B (1993) Glycogen in astrocytes: possible function as lactate supply for neighboring cells. Brain Res 623:208–214
Dringen R, Peters H, Wiesinger H, Hamprecht B (1995) Lactate transport in cultured glial cells. Dev Neurosci 17:63–69
Farmer WT, Murai K (2017) Resolving astrocyte heterogeneity in the CNS. Front Cell Neurosci 11:300
Ferguson JM (2001) SSRI antidepressant medications: adverse effects and tolerability. Prim Care Companion J Clin Psychiatry 3:22–27
Fosgerau K et al (2000) Kinetic and functional characterization of 1,4-dideoxy-1, 4-imino-d- arabinitol: a potent inhibitor of glycogen phosphorylase with anti- hyperglyceamic effect in ob/ob mice. Arch Biochem Biophys 380:274–284
Frier BM, Fisher BM (2007) Hypoglycaemia in clinical diabetes, 2nd edn. John Wiley and Sons, Ltd., New York, pp 1–346
Frier BM, Heller SR, McCrimmon RJ (2014) Hypoglycaemia in clinical diabetes, 3rd edn. Wiley and Sons, Chichester
Gale J (2009) Astrobiology of Earth. Oxford University Press, Oxford
Hille B (2001) Ion channels of excitable membranes. Sinauer Associates Inc., Sunderland, MA
Hodgkin AL, Huxley AF (1952) A quantitative description of membrane current and its application to conduction and excitation in nerve. J Physiol 117:500–544
Hodgkin AL, Huxley AF (1953) Movement of radioactive potassium and membrane current in a giant axon. J Physiol 121:403–414
Hu Y, Wilson GS (1997a) Rapid changes in local extracellular rat brain glucose observed with an in vivo glucose sensor. J Neurochem 68:1745–1752
Hu Y, Wilson GS (1997b) A temporary local energy pool coupled to neuronal activity: fluctuations of extracellular lactate levels in rat brain monitored with rapid-response enzyme-based sensor. J Neurochem 69:1484–1490
E. R. Kandel, J. H. Schwartz, T. M. Jessell, S. A. Siegelbaum, A. J. Hudspeth, in Principles of neural science, E. R. Kandel, J. H. Schwartz, T. M. Jessell, S. A. Siegelbaum, A. J. Hudspeth, Eds. (McGraw Hill, New York, 2013), pp. 126–147

Karbowski J (2007) Global and regional brain metabolic scaling and its functional consequences. BMC Biol 5:1–11
Kilic K et al (2018) Inadequate brain glycogen or sleep increases spreading depression susceptibility. Ann Neurol 83:61–73
Koizumi J (1974) Glycogen in the central nervous system. Prog Histochem Cytochem 6:1–37
Koizumi J, Shiraishi H (1970a) Ultrastructural appearance of glycogen in the hypothalamus of the rabbit following chlorpromazine administration. Exp Brain Res 10:276–282
Koizumi J, Shiraishi H (1970b) Glycogen accumulation in dendrites of the rabbit pallidum following trifluoperazine administration. Exp Brain Res 11:387–391
Korf J (2006) Is brain lactate metabolized immediately after neuronal activity through the oxidative pathway? J Cereb Blood Flow Metab 26:1584–1586
Lane N (2015) The vital question. Profile Publishing, London
Magistretti PJ, Allaman I (2018) Lactate in the brain: from metabolic end-product to signalling molecule. Nat Rev Neurosci 19:235–249
Mangia S et al (2009a) Metabolic and hemodynamic events after changes in neuronal activity: current hypotheses, theoretical predictions and in vivo NMR experimental findings. J Cereb Blood Flow Metab 29:441–463
Mangia S, Simpson IA, Vannucci SJ, Carruthers A (2009b) The in vivo neuron-to-astrocyte lactate shuttle in human brain: evidence from modeling of measured lactate levels during visual stimulation. J Neurochem 109(Suppl 1):55–62
McKenna MC, Gruetter R, Sonnewald U, Waagepetersen HS, Schousboe A (2006) In: Siegel GJ, Albers RW, Brady ST, Price DL (eds) Basic neurochemistry. Academic, San Diego, CA, pp 531–558
McKhann GM 2nd, D'Ambrosio R, Janigro D (1997) Heterogeneity of astrocyte resting membrane potentials and intercellular coupling revealed by whole-cell and gramicidin-perforated patch recordings from cultured neocortical and hippocampal slice astrocytes. J Neurosci 17:6850–6863
Meakin PJ et al (2007) Fructose metabolism in the adult mouse optic nerve, a central white matter tract. J Cereb Blood Flow Metab 27:86–99
Nehlig A, Wittendorp-Rechenmann E, Lam CD (2004) Selective uptake of [^{14}C]2-deoxyglucose by neurons and astrocytes: high-resolution microautoradiographic imaging by cellular ^{14}C-trajectography combined with immunohistochemistry. J Cereb Blood Flow Metab 24:1004–1014
Nelson SR, Schulz DW, Passonneau JV, Lowry OH (1968) Control of glycogen levels in brain. J Neurochem 15:1271–1279
Orkand RK, Nicholls JG, Kuffler SW (1966) Effect of nerve impulses on the membrane potential of glial cells in the central nervous system of amphibia. J Neurophysiol 29:788–806
Passonneau JV, Gatfield PD, Schulz DW, Lowry OH (1967) An enzymic method for measurement of glycogen. Anal Biochem 19:315–326
Pellerin L, Magistretti PJ (1994) Glutamate uptake into astrocytes stimulates aerobic glycolysis: a mechanism coupling neuronal activity to glucose utilization. Proc Natl Acad Sci U S A 91:10625–10629
Pellerin L, Magistretti PJ (2012) Sweet sixteen for ANLS. J Cereb Blood Flow Metab 32:1152–1166
Pellerin L, Bonvento G, Chatton JY, Pierre K, Magistretti PJ (2002) Role of neuron-glia interaction in the regulation of brain glucose utilization. Diabetes Nutr Metab 15 268–273; discussion 273
Pfeiffer-Guglielmi B et al (2006) Immunocytochemical analysis of rat vagus nerve by antibodies against glycogen phosphorylase isozymes. Brain Res 1110:23–29
Pfeiffer-Guglielmi B, Francke M, Reichenbach A, Hamprecht B (2007) Glycogen phosphorylase isozymes and energy metabolism in the rat peripheral nervous system—an immunocytochemical study. Brain Res 1136:20–27
Phelps CH (1972) Barbiturate-induced glycogen accumulation in brain. An electron microscopic study. Brain Res 39:225–234

Phelps CH (1975) An ultrastructural study of methionine sulphoximine-induced glycogen accumulation in astrocytes of the mouse cerebral cortex. J Neurocytol 4:479–490
Pocock G, Richards CD (2006) Human physiology: the basis of medicine, 3rd edn. OUP, Oxford
Ransom B (2009) In: Boron WF, Boulpaep EL (eds) Medical physiology. Saunders Elsevier, Philadelphia, PA, pp 289–309
Ransom BR, Fern R (1997) Does astrocytic glycogen benefit axon function and survival in CNS white matter during glucose deprivation? Glia 21:134–141
Ransom BR, Goldring S (1973) Slow depolarization in cells presumed to be glia in cerebral cortex of cat. J Neurophysiol 36:869–878
Ransom BR, Orkand RK (1996) Glial-neuronal interactions in non-synaptic areas of the brain: studies in the optic nerve. TINS 19:352–358
Schousboe A et al (1997) Trafficking between glia and neurons of TCA cycle intermediates and related metabolites. Glia 21:99–105
Schurr A, Payne RS (2007) Lactate, not pyruvate, is neuronal aerobic glycolysis end product: an in vitro electrophysiological study. Neuroscience 147:613–619
Siegel GJ, Albers RW, Brady ST, Price DL (2006) Brain neurochemistry, 7th edn. Elsevier, London
Silverthorn DU (2007) Human physiology. Chapter 15. Pearson, San Francisco, CA, pp 500–533
Simpson IA, Carruthers A, Vannucci SJ (2007) Supply and demand in cerebral energy metabolism: the role of nutrient transporters. J Cereb Blood Flow Metab 27:1766–1791
Sonnewald U, Westergaard N, Schousboe A (1997) Glutamate transport and metabolism in astrocytes. Glia 21:56–63
Sontheimer H (2008) A role for glutamate in growth and invasion of primary brain tumors. J Neurochem 105:287–295
Sorg O, Magistretti PJ (1992) Vasoactive intestinal peptide and noradrenaline exert long-term control on glycogen levels in astrocytes: blockade by protein synthesis inhibition. J Neurosci 12:4923–4931
Sotelo-Hitschfeld T et al (2015) Channel-mediated lactate release by K^+-stimulated astrocytes. J Neurosci 35:4168–4178
Stryer L (1995) Biochemistry, 4th edn. W.H. Freeman & Co., New York
Stys PK, Lopachin RM Jr (1996) Elemental composition and water content of rat optic nerve myelinated axons during in vitro post-anoxia reoxygenation. Neuroscience 73:1081–1890
Stys PK, Ransom BR, Waxman SG (1991) Compound action potential of nerve recorded by suction electrode: a theoretical and experimental analysis. Brain Res 546:18–32
Stys PK, Waxman SG, Ransom BR (1992) Ionic mechanisms of anoxic injury in mammalian CNS white matter: role of Na^+ channels and Na^+-Ca^{2+} exchanger. J Neurosci 12:430–439
Swanson RA, Choi DW (1993) Glial glycogen stores affect neuronal survival during glucose deprivation *in vitro*. J Cereb Blood Flow Met 13:162–169
Tekkok SB, Brown AM, Westenbroek R, Pellerin L, Ransom BR (2005) Transfer of glycogen-derived lactate from astrocytes to axons via specific monocarboxylate transporters supports mouse optic nerve activity. J Neurosci Res 81:644–652
Thorens B, Mueckler MM (2010) Glucose transporters in the 21st century. Am J Physiol Endocrinol Metab 298:E141–E145
Tymoczko JL, Berg JM, Stryer L (2015) Biochemistry: a short course, 3rd edn. W.H. Freeman and Co., New York
Vannucci SJ, Maher F, Simpson IA (1997) Glucose transporter proteins in brain: delivery of glucose to neurons and glia. Glia 21:2–21
Vilchez D et al (2007) Mechanism suppressing glycogen synthesis in neurons and its demise in progressive myoclonus epilepsy. Nat Neurosci 10:1407–1413
Waagepetersen HS, Westergaard N, Schousboe A (2000) The effects of isofagomine, a potent glycogen phosphorylase inhibitor, on glycogen metabolism in cultured mouse cortical astrocytes. Neurochem Int 36:435–440
Walls AB et al (2008) Characterization of 1,4-dideoxy-1,4-imino-d-arabinitol (DAB) as an inhibitor of brain glycogen shunt activity. J Neurochem 105:1462–1470

Wang SS et al (2008) Shaping of white matter composition by biophysical scaling constraints. J Neurosci 28:4047–4056
Wender R et al (2000) Astrocytic glycogen influences axon function and survival during glucose deprivation in central white matter. J Neurosci 20:6804–6810
Whatley SA, Hall C, Lim L (1981) Hypothalamic neurons in dissociated cell culture: the mechanism of increased survival times in the presence of non-neuronal cells. J Neurochem 36:2052–2056
Yang X et al (2014) Novel hypoglycemic injury mechanism: N-methyl-D-aspartate receptor-mediated white matter damage. Ann Neurol 75:492–507
Zhao Y et al (2017) Decreased glycogen content might contribute to chronic stress-induced atrophy of hippocampal astrocyte volume and depression-like behavior in rats. Sci Rep 7:43192

Glycogenolysis in Cerebral Cortex During Sensory Stimulation, Acute Hypoglycemia, and Exercise: Impact on Astrocytic Energetics, Aerobic Glycolysis, and Astrocyte-Neuron Interactions

Gerald A. Dienel and Douglas L. Rothman

Abstract Most glycogen in cerebral cortex is located in astrocytes, and the importance of glycogenolysis for critical functions, including neurotransmission and memory consolidation, is strongly supported by many studies. However, specific mechanisms through which glycogen sustains essential functions remain to be established by rigorous, quantitative studies. Cerebral cortical glycogen concentrations are in the range of 10–12 μmol/g in carefully-handled animals, and the calculated rate of glycogenolysis ($CMR_{glycogen}$) during sensory stimulation is approximately 60% that of glucose utilization (CMR_{glc}) by all cells, with lower rates during acute hypoglycemia and exercise to exhaustion. $CMR_{glycogen}$ is at least fourfold higher when the volume fraction of astrocytes is taken into account. Inclusion of glycogen consumed during sensory stimulation in calculation of the oxygen-glucose index ($OGI = CMR_{O2}/CMR_{glc}$, which has a theoretical maximum of 6 when no other substrates are metabolized) reduces OGI from 5.0 to 2.8. Thus at least 53% of the carbohydrate is not oxidized, suggesting that glycogen mobilization supports astrocytic glycolysis, not neuronal oxidation of glycogen-derived lactate that would cause OGI to exceed 6. Failure of glycogenolysis to dilute the specific activity of lactate formed from blood-borne [6-^{14}C]glucose indicates compartmentation of glycolytic metabolism of glucose and glycogen and the rapid release from cerebral cortex of glycogen-derived lactate. Together, these findings invalidate the conclusion by others that glycogen-derived lactate is a major fuel for neurons during

G. A. Dienel (✉)
Department of Neurology, University of Arkansas for Medical Sciences, Little Rock, AR, USA

Department of Cell Biology and Physiology, University of New Mexico, Albuquerque, NM, USA
e-mail: gadienel@uams.edu

D. L. Rothman
Magnetic Resonance Research Center and Department of Diagnostic Radiology, Yale University, New Haven, CT, USA
e-mail: douglas.rothman@yale.edu

M. DiNuzzo, A. Schousboe (eds.), *Brain Glycogen Metabolism*, Advances in Neurobiology 23, https://doi.org/10.1007/978-3-030-27480-1_8

neurotransmission, memory consolidation, and exercise to exhaustion. Alternative mechanisms, including glucose sparing for neurons, are presented as testable explanations for data interpreted as lactate shuttling.

Keywords Astrocyte · Cerebral cortex · Brain activation · Exercise · Glucose utilization · Glucose sparing · Glycogen · Glycogen shunt · Glycogen turnover · Hypoglycemia · Lactate · Lactate shuttling · Oxygen consumption · Neuron · Neurotransmission · Memory consolidation · Sensory stimulation

Abbreviations

$(A\text{-}V)_{substrate}$	Arteriovenous difference across the brain for the identified substrate
Asp	Aspartate
BAY U6751	4-(2-Chlorophenyl)-L-ethyl-1,4-dihydro-6-methyl-2,3,5-pyridinetricarboxylic acid 5-isopropyl ester disodium salt hydrate
CBF	Cerebral blood flow rate
CMR	Cerebral metabolic rate for substrate of interest = $CBF(A\text{-}V)_{substrate}$
CMR_{glc}	Cerebral metabolic rate for glucose = $CBF(A\text{-}V)_{glc}$
$CMR_{glycogen}$	Cerebral metabolic rate for glycogen = Δ[glycogen]/time
CMR_{O2}	Cerebral metabolic rate for oxygen $CBF(A\text{-}V)_{O2}$
CP-316,819	[R-R∗,S∗]-5-chloro-N-[2-hydroxy-3-(methoxymethylamino)-3-oxo-L-(phenylmethyl)propyl]-1H-indole-2-carboxamide
DAB	1,4-Dideoxy-1,4-imino-D-arabinitol
DG	2-deoxy-D-glucose
DMSO	Dimethyl sulfoxide
FDG	2-fluoro-2-deoxy-D-glucose
Glc	Glucose
Glc-6-P	Glucose-6-phosphate
Gln	Glutamine
Glu	Glutamate
GLUT	Glucose transporter; GLUT1 in vascular endothelium and astrocytes, and GLUT3 and GLUT4 in neurons
GPR81	G-protein-coupled lactate receptor81, also known as HCAR1
HCAR1	Hydroxycarboxylic acid receptor1 also known as GPR81
KO	Knockout
Lac	Lactate
LTP	Long-term potentiation
MCT	Monocarboxylic acid transporter; MCT1 and MCT4 isoforms are mainly astrocytic, whereas MCT2 is predominantly neuronal
NMDA	*N*-Methyl-D-aspartate

OCI	Oxygen carbohydrate index = $CMR_{O2}/[CMR_{glc} + 0.5CMR_{lac} + CMR_{glycogen}] = (A\text{-}V)_{O2}/((A\text{-}V)_{glc} + 0.5(A\text{-}V)_{lac} + \Delta[glycogen])$, where lactate and [glycogen] are expressed in glucosyl units (2Lac = 1Glc)
OGI	Oxygen-glucose index = $CMR_{O2}/CMR_{glc} = (A\text{-}V)_{O2}/(A\text{-}V)_{glc}$ (CBF cancels out). This calculation assumes no other substrates are oxidized
PAPs	Peripheral astrocytic processes
RSA	Relative specific activity (SA) = ratio of the SA of a compound of interest to the SA of a reference compound, e.g., SA lactate/SA glucose
SA	Specific activity (dpm/μmol)
TCA	Tricarboxylic acid

1 Introduction

A range of studies have shown the importance of glycogenolysis for critical brain functions, including neurotransmission and memory consolidation (reviewed by (Duran and Guinovart 2015; Bak et al. 2018)). However, the understanding of how glycogenolysis supports specific brain functional processes is complicated by its compartmentation primarily in astrocytes (Cataldo and Broadwell 1986), with small amounts in neurons (Saez et al. 2014; Pfeiffer-Guglielmi et al. 2003). In other organs with high transient increases in energy demand, such as skeletal muscle and heart, glucose-6-phosphate (Glc-6-P) from glycogenolysis directly provides energy via glycolytic ATP to all of the cells. In contrast, ATP from glycolytic metabolism of astrocytic glycogen cannot directly support neuronal energy demands. In the cerebral cortex the fraction of total energy demand from neurons is approximately 2–3 times greater than astrocytes (Yu et al. 2018; Lanz et al. 2013), yet the glucose is stored as glycogen mainly in astrocytes and its consumption is triggered by specific neurotransmitters and other compounds related to neuronal signaling activity (Obel et al. 2012). Among its functions, glycogen can preferentially fuel ATPases for pumping of K^+ (Hertz et al. 2015a, b) and Ca^{++} (Müller et al. 2014) that probably contribute to its roles in higher brain functions.

1.1 *Energetic Advantage of Glycogenolysis*

Glc-6-P is a 'branch-point' metabolite that can either continue down the glycolytic pathway, enter the pentose phosphate shunt pathway, or be stored by its incorporation into glycogen, predominantly in astrocytes (Fig. 1). Glycogen synthesis involves several enzymatic steps: the phosphoglucomutase reaction transforms Glc-6-P to Glc-1-P which is then converted to UDP-glc by UDP-glc pyrophosphorylase prior to incorporation of the glucosyl group into glycogen by glycogen synthase.

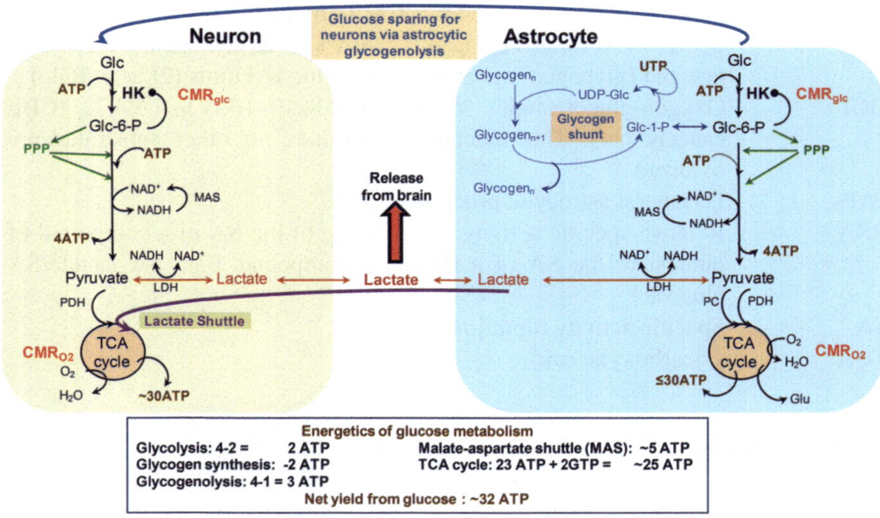

Fig. 1 Models for the glycogen shunt and fate of glycogen carbon. Glucose (Glc) is metabolized to form pyruvate via the glycolytic pathway with a net yield of 2 ATP, and another 30 ATP are produced by the oxidative pathway and the malate-aspartate shuttle (MAS) that transfers NADH reducing equivalents from the cytoplasm to mitochondria. MAS activity is required to produce pyruvate as an oxidative and anaplerotic substrate because regeneration of NAD^+ by the lactate dehydrogenase (LDH) reaction causes loss of pyruvate as lactate, which must be released from the cell for glycolysis to continue. The hexokinase (HK) reaction is the first irreversible step in the glycolytic pathway, and it is feedback inhibited by glucose-6-phosphate (Glc-6-P) (denoted by the 'ball'). The flux through the hexokinase step represents the total rate of glucose utilization (CMR_{glc}) that can be compared with the rate of oxygen consumption (CMR_{O2}). The stoichiometry of oxygen-to-glucose utilization can be quantified by the oxygen-glucose index (OGI = CMR_{O2}/CMR_{glc}, according to the reaction $1\ Glc + 6O_2 \rightarrow 6\ CO_2 + 6H_2O$). As long as no other substrates are metabolized the theoretical maximal OGI is 6. Some Glc-6-P can enter the pentose phosphate shunt pathway (PPP), and in normal adult brain, most of the carbon re-enters the glycolytic pathway at the fructose-6-P and glyceraldehyde-3-P steps. Most glycogen in brain is located in astrocytes, although very small amounts are present in neurons (not shown). Glycogen synthesis requires 1 ATP equivalent as UTP, so incorporation of 1 glucosyl unit into glycogen ($glycogen_n \rightarrow glycogen_{n+1}$) uses 2 ATP. Glycogenolysis has a net yield of 3 ATP, for an overall yield of 1 ATP per turnover of 1 glucosyl unit. The advantage of storage of glucose in glycogen is that synthesis can be carried out when energy demand is lower and degradation can quickly fulfill increased ATP demand. The yield from pyruvate oxidation via the pyruvate dehydrogenase (PDH) reaction in astrocytes is the same as neurons, but if some pyruvate enters the TCA cycle via the pyruvate carboxylase (PC), the ATP yield is reduced because 1 ATP is used for CO_2 fixation, and the resulting oxaloacetate serves as a catalytic or anaplerotic compound. The anaplerotic capability of astrocytes to synthesize glutamate de novo from glucose requires the PC reaction to make oxaloacetate and the PDH reaction to add one acetyl CoA to oxaloacetate to make a 'new' 6-carbon compound, citrate, that is metabolized further to α-ketoglutarate and transaminated to form a new molecule of glutamate (Glu). The lactate shuttle is proposed to transfer glucose- and/or glycogen-derived lactate from astrocytes to neurons where it can be oxidized. This process requires stoichiometric balance of CMR_{O2} with total CMR_{glc} plus $CMR_{glycogen}$, which is not observed experimentally (see Sect. 3.1). An alternative model proposed by DiNuzzo et al. (2010a) is that glycogenolysis reduces astrocytic demand for blood-borne glucose during brain activation, sparing an equivalent amount of glucose for neuronal utilization (see Sect. 4). This model is evaluated in our companion paper (Rothman and Dienel 2019). Modified from Fig. 9 in Dienel (2019a) © American Physiological Society, with permission

Glycogen is degraded by phosphorylase to produce Glc-1-P, followed by its conversion to Glc-6-P, then metabolism by glycolysis or the pentose shunt Glycogen has the energetic advantage that its synthesis can occur during periods of lower ATP demand and its mobilization does not require the ATP to generate Glc-6-P, thereby providing a net yield of 3 glycolytic ATP per glucosyl moiety compared with 2 ATP from glucose. Thus, the cost of 1 ATP plus 1 UTP required to incorporate one glucosyl unit into glycogen is 'pre-paid', and glycogenolysis quickly generates Glc-6-P for metabolism via energy- or NADPH-producing pathways when local demand is accelerated. Cycling of Glc-6-P through glycogen instead of its direct entry into downstream pathways is called the glycogen shunt (Shulman et al. 2001; Shulman and Rothman 2001) (Fig. 1).

1.2 Roles of Glycogen in Astrocytic and Neuronal Glycolysis

Insight into the mechanism by which glycogenolysis may additionally support neuronal energetics comes from consideration of the two pathways of astrocytic Glc-6-P formation, hexokinase acting on glucose and glycogen phosphorylase acting on glycogen. Hexokinase is the first irreversible enzymatic step of the glycolytic pathway (Fig. 1), and Glc-6-P regulates hexokinase activity by noncompetitive feedback inhibition (Crane and Sols 1954). Hexokinase type I is the predominant isoform in brain, with some type II, and little of the major liver isozyme, type III hexokinase or glucokinase (Grossbard and Schimke 1966). The type III isoform has been detected in nuclear periphery of brain cells, including cerebellar Purkinje neurons (Wilson 2003; Sebastian et al. 1999; Preller and Wilson 1992). Hexokinase Type I is 50% inhibited by Glc-6-P in the range of 0.02–0-.07 mM for soluble, particulate, and homogenate preparations (Newsholme et al. 1968; Wilson 2003). Normal brain Glc-6-P concentration is about 0.1–0.2 mM (Veech et al. 1973), and brain hexokinase activity in vivo is only ~3% of its maximal activity assayed in vitro, due mainly, but not exclusively, to inhibition by Glc-6-P (Lowry and Passonneau 1964; Bachelard and Goldfarb 1969).

Inhibition of hexokinase by Glc-6-P may be important for two reasons. First, passage of Glc-6-P through astrocytic gap junctions is highly restricted, in contrast to glyceraldehyde-3-P, lactate, NADH, NADPH, anionic fluorescent dyes, amino acids, and various electrolytes that can readily diffuse through the syncytium ((Gandhi et al. 2009b) and references cited therein). Retention of Glc-6-P in the astrocyte where it was produced from glucose or glycogen can contribute to regulation of the rate utilization of blood-borne glucose (CMR_{glc}) in each astrocyte by governing its hexokinase rate. Second, DiNuzzo and colleagues (2010b, 2011, 2012) proposed the novel hypothesis based on metabolic modeling that Glc-6-P produced from glycogen during brain activation may further inhibit hexokinase, thereby reducing astrocytic utilization of blood-borne glucose and sparing an equivalent amount of this glucose for use by neurons. In this model, glycogenolysis bypasses the hexokinase step and provides the Glc-6-P for glycolytic metabolism in astrocytes, whereas blood-borne glucose fuels the incremental rise in glycolytic flux in activated neurons (see Sect. 4).

1.3 Glc-6-P and Regulation of Glycogen Phosphorylase Isoforms in Brain Cells

Glycogen phosphorylase exists as three major isoforms, liver, muscle, and brain. Most studies of the complex regulation of these enzymes by small ligands and phosphorylation are based on muscle and liver phosphorylase, and much less is known about regulation of brain phosphorylase (Nadeau et al. 2018). Muscle phosphorylase b (AMP-dependent, non-phosphorylated, inactive state) is activated by AMP (as well as Ca^{++} and phosphorylation), and Glc-6-P inhibits this activation by competing with AMP for the same binding site, whereas Glc-6-P has little effect on muscle phosphorylase a (AMP-independent, phosphorylated, active state) (Griffiths et al. 1976; Danchin and Buc 1973; Morgan and Parmeggiani 1964). In hepatocytes, regulation by Glc-6-P is more complex, and it can govern the subcellular distribution of phosphorylase a and act in conjunction with glucose to inhibit phosphorylase a (Aiston et al. 2003, 2004).

The brain and muscle glycogen phosphorylase isoforms are more similar to each other than to the liver isoform (Hudson et al. 1993). Both the brain and muscle isozymes are expressed in astrocytes (Pfeiffer-Guglielmi et al. 2000, 2003), whereas only the brain isoform is in neurons (Pfeiffer-Guglielmi et al. 2003; Saez et al. 2014). The brain and muscle (and liver) isoforms differ in terms of their activation by phosphorylation and AMP and other properties (Crerar et al. 1995; Mayer et al. 1992)}, as well as their responses to norepinephrine vs. AMP (Müller et al. 2015) and to glycolytic supercompensation (Jakobsen et al. 2017). Brain phosphorylase a is inhibited by Glc-6-P with a K_i of ~5 mM, and its kinetics with brain phosphorylase b are complex, showing activation by Glc-6-P up to 0.5 mM and inhibition at concentrations >2 mM (Mayer et al. 1992). These findings suggest that any increases in brain Glc-6-P concentration during glycogenolysis may preferentially further suppress hexokinase activity, as opposed to inhibiting phosphorylase, but many factors govern phosphorylase activity. To our knowledge, changes in Glc-6-P concentration in astrocytes have not been evaluated during glycogen mobilization, and future studies to evaluate the DiNuzzo hypothesis require detailed metabolic analyses in astrocytes after the onset of glycogenolysis, including hexokinase activity and changes in the concentrations of Glc-6-P and Glc.

1.4 *Technical Difficulties in Evaluating Glycogen Level and Turnover in Brain Tissue*

Measurements of glycogen concentration, glycogen turnover, and the magnitude of the glycogen shunt in living brain are extremely difficult for many reasons. First, glycogenolysis is activated by various neurotransmitters, especially norepinephrine, when subjects experience stress, alerting, handling, sensory stimulation, and other situations. Second, phosphorylase is very rapidly activated by increased energy demand during normal brain activation, as well by hypoglycemia, hypoxia, anoxia, and ischemia. Assays of glycogen concentration and labeling, therefore, require rapid enzyme inactivation (e.g., by in situ freezing or microwave fixation) in conjunction with appropriate handling care and extraction and analytical procedures. Third, glycogen synthesis can help maintain overall glycogen concentration in tissue during activation, and glycogen shunt flux can minimize net incorporation of label into glycogen due to greater turnover of the outer tiers compared with limit dextrin (Watanabe and Passonneau 1973). Fourth, complexity of glycogen turnover also arises from the possibility that any single glycogen molecule may turn over independently of other glycogen granules in the same cell, and both synthesis and degradation of glycogen in different regions (e.g., endfeet, cytosol, large processes, and perisynaptic peripheral astrocytic processes [PAPs]) of one astrocyte may take place simultaneously. Also, glycogen turnover in different astrocytes in the same brain region of interest may be governed by conditions in each cell as well as by neurotransmitters. Thus, regulation of glycogen level and turnover can be governed simultaneously at global, local, and cellular levels.

1.5 *Glycogenolysis in Activated Cerebral Cortex*

This review focuses on selected, quantitative studies of glycogen in cerebral cortex of awake rats to emphasize the importance of further studies of the poorly-understood roles of glycogen in astrocytic energetics in living brain. Evidence is presented for high rates of glycogenolysis, on-going turnover during activation, and predominant conversion to lactate that is released from cerebral cortex. Glycogen consumption during acute hypoglycemia serves to compensate for the glucose supply-demand mismatch, rates are lower than during sensory stimulation or exhaustive exercise. The notion of glycogen-derived lactate shuttling to neurons (Fig. 1) to maintain neurotransmission and memory consolidation or to fuel the brain during exhaustive exercise is challenged, and alternative mechanisms are presented to explain reported data. Strong emphasis is placed on the importance of comprehensive, quantitative assays of interrelated metabolic fluxes and testing of predicted stoichiometries of models for roles of glycogen in brain functions. The following sections describe glycogen utilization during activation, evidence for glycolytic metabolism of glycogen, release of glycogen-derived lactate from cerebral cortex, the concept of glucose sparing for neurons by astrocytic glycogenolysis, roles for glycogen in neurotransmission and learning/memory, and evaluation of glycogenolysis rates during acute hypoglycemia and exercise to exhaustion. Because aspects of these discussions require background in metabolism to appreciate the technical complexities of analysis and interpretation of roles of glycogen in brain function, the introductory and summary material is intended to highlight 'take-home' concepts for non-experts.

2 Glycogen Level, Glycogenolytic Rate, and Turnover During and After Sensory Stimulation of Awake Rats

Glycogen is an important storage site for glucose in brain, and its utilization in astrocytes during physiological activation is a major component of overall brain energetics.

2.1 Mobilization of Unlabeled Glycogen

Comparison of the net rate of glycogenolysis ($CMR_{glycogen}$) in cerebral cortex with CMR_{glc} during and after sensory stimulation of awake rats reveals that $CMR_{glycogen}$ is a large fraction of the rate of utilization of blood-borne glucose by all cells. In two independent experiments carried out a year apart, the mean concentration of glycogen in resting cerebral cortex of very carefully-handled awake rats was 12.4 μmol/g (Table 1), similar to levels reported in other studies (Kong et al. 2002; Oe et al. 2016) and several fold higher than reported in earlier studies (see references cited in (Cruz and Dienel 2002)). It was further shown that the lower values in earlier studies could be quantitatively explained by losses that occurred in the extraction procedures used (Cruz and Dienel 2002).

After 5–6 min of generalized sensory stimulation, glycogen concentration fell 23%, and decreased further at 15 min recovery for a net decrease of 4 μmol/g, or 32% (Table 1). CMR_{glc} was determined in the Dienel et al. (2002) study by measuring the rate of incorporation of plasma [6-^{14}C]glucose into total metabolites. CMR_{glc} increased 26% during stimulation and normalized within 15 min after termination of stimulation. However, this value for calculated CMR_{glc} during activation is probably underestimated due to rapid efflux of [^{14}C]lactate from tissue (Cruz et al. 1999, 2007; Adachi et al. 1995) and because it was based on label accumulation mainly in TCA cycle derivatives when only one triose was labeled by [6-^{14}C]glucose. If instead [1,6-^{14}C]glucose were used as the precursor, labeling of derivatives of pyruvate and calculated CMR_{glc} would be about twice as high.

Table 1 CMR_{glc}, glycogen levels, and calculated $CMR_{glycogen}$ in cerebral cortex of awake rats during rest, sensory stimulation, and recovery

Condition	CMR_{glc} (μmol/g/min)	Unlabeled glycogen (μmol/g)	Mean change in [glycogen] during activity interval (μmol/g)	[a]$CMR_{glycogen}$ (μmol/g/min) [%CMR_{glc}]	[b]Unlabeled glycogen if astrocytes 25% of brain mass (μmol/g)	$CMR_{glycogen}$ if astrocytes 25% of brain mass (μmol/g/min) [%CMR_{glc}]
Rest	0.74[c]	12.3[c], 12.5[d] (12.4)	—	–	49.6	–
Activation, 5 min	0.93[c]*	9.6[c], 9.6[d] (9.6)	2.8	0.56 [60%]	38.4	2.24 [241%]
Recovery, 15 min	0.73[c]	8.2[c], 8.5[d]* (8.4)	1.2	0.08 [11%]	33.4	0.32 [44%]

*$p < 0.05$ vs. rest
[a]$CMR_{glycogen}$ was calculated by dividing the mean net change in glycogen concentration during the indicated experimental interval by the duration of the interval. [b]Average tissue volume fraction occupied by astrocytes was stated by Hertz and colleagues (Hertz et al. 2007) to vary regionally and estimated to be *at most* 20–25% (for further discussion and cited references, see the section in Hertz et al. 2007 entitled, "High oxidative rates in astrocytes in the brain in vivo"); also see Sect. 2.1 and (Gundersen et al. 2015). Mean values in awake rats are from [c](Dienel et al. 2002) that used a 5 min sensory stimulation duration and metabolic labeling with [6-^{14}C]glucose, and [d](Cruz and Dienel et al. 2002) that analyzed samples from our previous study (Madsen et al. 1999) that used a 6 min sensory stimulation duration. Means of the two studies are indicated in parentheses and used to calculate changes in [glycogen] during activity intervals

$CMR_{glycogen}$ during activation and recovery was 0.56 and 0.08 μmol/g/min, equivalent to 60% and 11%, respectively, of calculated CMR_{glc} in all cerebral cortical cells using [6-^{14}C]glucose (Table 1). For comparison, glycogenolysis rates determined in earlier literature under various conditions were lower and ranged from 0.003 to 0.1 μmol/g/min, but increased to ~2–3 μmol/g/min during ischemia (see Table 3 in (Dienel 2012)).

Based on an estimated astrocytic volume fraction in cerebral cortex of *at most* 20–25% (Hertz et al. 2007), resting glycogen concentration in astrocytes would be at least fourfold higher, ~50 μmol/g (Table 1), approaching that in resting gastrocnemius muscle, ~70 μmol/g (Price et al. 1994). When an estimated astrocytic mass of 25% is taken into account, $CMR_{glycogen}$ exceeds calculated CMR_{glc} by 2.4-fold during activation, and it rises to 44% of CMR_{glc} during recovery (Table 1). Use of ~25% mass fraction is conservative, and $CMR_{glycogen}$ per unit astrocytic mass may be much higher because Gundersen et al. (2015) stated in their "Section A. Neuroglia" that the glial (mostly astrocytic processes) contribution to gray matter neuropil in CA1 hippocampus ranges from 4% to 6% in rat and human, and rarely exceeds 20%. These remarkably-high rates of glycogenolysis at the cellular level during activation and recovery indicate that glycogen fulfills important, unidentified functions. Furthermore, they suggest that glycolytic ATP from glycogen derived Glc-6-P may be the major energy source for astrocytes during intense transient increases in metabolic demand associated with neuronal activation.

2.2 Estimated Glycolytic ATP Production Rates from Glycogenolysis

The rate of glycolytic ATP generation from $CMR_{glycogen}$ during sensory stimulation (Table 1) is 0.56 μmol/g/min × 3 ATP/glucosyl moiety = 1.68 μmol ATP/g/min in cerebral cortex, and it is at least 3 × 2.24 = 6.72 μmol ATP/g/min in astrocytes based on their estimated mass fraction. For comparison to overall ATP production from glucose during stimulation, the OGI fell from about 5.5 to 5 (see Sect. 3.1.1 and Table 4), indicating that 5/6 or 83% was oxidized and 17% metabolized glycolytically. CMR_{glc} = 0.93 μmol/g/min and oxidative ATP production rate is 0.93 × 0.83 × 30 μmol ATP/glc = 23.2 μmol/g/min, whereas glycolytic ATP rate is 0.93 × 0.17 × 2 ATP/glc = 0.32 μmol/g/min, for a total of 23.5 μmol ATP/g/min. Thus, estimated astrocytic glycolytic ATP production rate from glycogen is ~5–21 times higher than glucose-derived glycolytic ATP formation in all cells (i.e., 1.68/0.32 or 6.72/0.32), and it corresponds to ~7–29% to total ATP production from blood-borne glucose during activation. These numbers are regarded as tentative because the moment-to-moment metabolic rates and the true $CMR_{glycogen}$ in astrocytes are not known, but they provide an initial data set to evaluate the importance of glycogenolysis to astrocytic energetics.

2.3 Metabolic Labeling Reveals that the Glycogen Shunt is Active During Sensory Stimulation of Awake Rats

The glycogen shunt (Shulman et al. 2001; Shulman and Rothman 2001) was proposed to explain why astrocytes use glycogen during activation even though adequate glucose supplies are available. In analogy to muscle, glycogenolysis is used when there are brief needs for rapid ATP production during high levels of neuronal activity (Fig. 1). In between these periods glucose is shunted from Glc-6-P to resynthesize lost glycogen. Evidence for shunting of Glc-6-P through glycogen during activation can be obtained by comparison of the fractional release of label from pre-labeled glycogen with the magnitude of change in total glycogen concentration. When glycogen was pre-labeled with [1-^{14}C]glucose followed by multi-modal sensory stimulation involving gentle brushing of the whiskers and body along with acoustic and visual stimulation of awake rats, the percent decrease in labeled glycogen exceeded that of the net decrease in unlabeled glycogen concentration in three brain regions (Table 2). Label release was detectable in cerebral cortex, inferior colliculus, and superior colliculus, whereas glycogen concentration fell only in

Table 2 Release of label from glycogen exceeds fall in concentration of unlabeled glycogen during activation

Brain region	Unlabeled glycogen (μmol/g)			Labeled glycogen (nCi/g)		
	Rest	Activation	% change	Rest	Activation	% change
Cerebral cortex	4.6	4.0*	−13	2.3	1.75*	−24
Inferior colliculus	4.1	4.0	−2	2.4	2.0*	−17
Superior colliculus	3.7	3.7	0	2.0	1.8*	−10

*$p < 0.05$ vs. rest. From Fig. 1 of (Dienel et al. 2007a). The amount of ^{14}C-label recovered in purified glycogen from each brain was normalized to its respective integrated specific activity of arterial plasma glucose to take into account any differences in precursor availability during the labeling period

cerebral cortex. Thus, glycogen synthesis was on-going even as it was degraded, and synthesis matched degradation in the two colliculi. Net incorporation of [6-^{14}C] glucose into cerebral cortical glycogen during rest, activation, and recovery, i.e., 0.003, 0.002, and 0.005 μmol/g/min, respectively, confirmed the continuous but low net rates of incorporation of blood-borne glucose into glycogen in cerebral cortex during all stages of the activity cycle (Dienel et al. 2002).

Label release exceeded concentration change by 10–17%, and using mean glycogen level of ~4 μmol/g (Table 2) the outer tier degradation rate during the 10 min stimulation interval is estimated to be 0.04–0.07 μmol/g/min (i.e., (4 μmol/g) × (0.1–0.17)/10 min), a rate that can be compared with CMR_{glc} in these three regions of ~0.9–2 μmol/g/min during activation (Dienel et al. 2002, 2007a). As a very rough estimate, glycogen shunt flux in this experimental paradigm, based on outer tier turnover, is ~20–40 times higher than net incorporation of label from blood-borne glucose into glycogen during activation, but it is a small fraction (<5–10%) of glucose utilization by all cells. However as pointed out by Öz, DiNuzzo, and colleagues (Öz et al. 2015, 2017; DiNuzzo 2013) the rate of glycogen synthesis may be substantially greater than estimated from total label incorporation due to the complexity of the glycogen turnover process which does not homogeneously incorporate label across the glycogen particles present.

2.4 *Compensatory Increases in CMR_{glc} When Glycogenolysis is Inhibited*

The significance of glycogenolysis to energetics of sensory stimulation is underscored by the regionally-heterogeneous increases in utilization of blood-borne glucose in brain of awake rats when glycogenolysis was inhibited with CP-316,819 (Dienel et al. 2007a). This study used the protocol of Suh et al. (2007) who showed an 88% increase in glycogen level in vivo after treatment with CP-316,819 treatment. This protocol also prolonged EEG activity and reduced neuronal death during severe hypoglycemia, consistent with release of phosphorylase inhibition by CP-316,819 that requires glucose concentrations in the normoglycemic range (see discussion and references cited by Suh et al.). In this experiment it was hypothesized that if glycogenolysis were activated by sensory stimulation, blocking glycogen phosphorylase would result in its breakdown being replaced by an equivalent amount of glucose consumption above the level measured in control animals without the inhibitor. As illustrated in Table 3, sensory stimulation caused especially-large responses in layer 4 plus adjacent tissue of parietal and sensory cortex, with net rises in CMR_{glc} above resting rates of 0.9 and 1.37 μmol/g/min, respectively. These net increases approximate or exceed their resting CMR_{glc}, and are 2.9- and 1.8-fold higher than the respective responses in vehicle-treated rats. Furthermore, the net compensatory increases in CMR_{glc} in layer 4 of sensory and parietal cortex over and above those in vehicle-treated rats during stimulation (~0.60 μmol/g/min, Table 3) are similar to $CMR_{glycogen}$, 0.56 μmol/g/min, for the entire dorsal cerebral cortex and correspond to 50–55% of resting CMR_{glc} by all cells (Table 3). If the regional differences are set aside, these findings suggest that most of the compensatory increases in CMR_{glc} may occur mainly in astrocytes to replace the energy that in control animals was supplied by glycogenolysis. Furthermore, as discussed in Sect. 4, the large increase in glucose consumption when glycogen phosphorylase is inhibited supports the hypothesis that astrocytic glycogenolysis spares glucose during intense activity for the neuron.

2.5 *Summary: Contribution of Glycogen in Energetics*

Glycogen is present in cerebral cortex at high levels, its utilization rate during sensory stimulation is a large fraction of total carbohydrate consumption (blood-borne glucose utilization plus glycogenolysis) by all cells. Glycolytic ATP production rate by glycogenolysis exceeds the amount of glycolytic ATP generated from blood-borne glucose by all cells, glycogen synthesis and degradation take place continuously but at rates that vary with physiological state. Consistent with its proposed key energetic role inhibition of glycogenolysis causes a large increase in utilization of blood-borne glucose.

3 Fate of Glycogen Carbon Skeleton

Glycolytic metabolism of glycogen produces pyruvate in astrocytes that can have various alternative metabolic fates: oxidation via the TCA cycle, transamination to alanine, carboxylation to oxaloacetate by pyruvate carboxylase, and reduction to lactate that can be released from the activated tissue to blood or perivascular fluid or shuttled to neurons for oxidation as supplemental fuel. Detailed analyses of metabolite concentration changes, metabolic pathway fluxes, and metabolic

Table 3 Selected compensatory increases in CMR_{glc} when glycogenolysis in awake rats is inhibited with CP-316,819

Brain region	CMR_{glc} in vehicle-treated rats (μmol/g/min)			CMR_{glc} in CP-316,819-treated rats (μmol/g/min)			Net increase CP-316,819 / vehicle	Net increase (CP—vehicle)	Net increase % of resting CMR_{glc}
	Rest	Activation	Net increase	Rest	Activation	Net increase			
Sensory cortex	1.06	1.66	0.60	1.01	2.02	1.01	1.68	0.41	38.7
Layer 4 + surround	1.20	1.97	0.77	1.01	2.38	1.37	1.77	0.60	50.0
Parietal cortex	0.99	1.34	0.35	0.96	1.56	0.60	1.71	0.25	25.3
Layer 4 + surround	1.06	1.37	0.31	0.99	1.89	0.90	2.90	0.59	55.7

Data are from (Dienel et al. 2007a) where CMR_{glc} in many brain regions are reported. Rats were pre-treated with an inhibitor of glycogenolysis (CP-316,819) or vehicle, then CMR_{glc} was assayed during a 30 min interval in both groups with [^{14}C]deoxyglucose during rest and generalized sensory stimulation. Cortical layer 4 is the location of the thalamo-cortical input that is highly activated in the whisker-to-barrel cortex when the whiskers are brushed labeling patterns described in the following sections provide strong evidence that most glycogen is not oxidized in astrocytes or neurons and is instead released from activated tissue.

3.1 Glycogen-Derived Pyruvate/Lactate is Not Completely Oxidized: Oxygen Consumption is Too Low

3.1.1 Oxygen-Glucose Index (OGI) Falls During Activation

Evidence for oxidation of glycogen-derived pyruvate in living brain is very difficult to obtain and distinguish from glucose oxidation. However, an indirect approach is to evaluate the oxygen-to-carbohydrate stoichiometry during activation and recovery. We previously determined the oxygen-glucose index (OGI), the ratio of the rate of oxygen utilization (CMR_{O2}) to CMR_{glc} ($OGI = CMR_{O2}/CMR_{glc} = CBF(A\text{-}V)_{O2}/CBF(A\text{-}V)_{glc} = (A\text{-}V)_{O2}/(A\text{-}V)_{glc}$, where CBF = cerebral blood flow rate) by measuring the arteriovenous differences (A-V) for oxygen, glucose, and lactate (lac) in the same blood samples during rest, activation, and recovery (Madsen et al. 1999). OGI fell from 5.5 to 5.0 during activation and rose to 7.2 during recovery (Table 3). When no other substrates are metabolized, the theoretical maximum for OGI is 6.0 because 6 O_2 are required for complete oxidation of one glucose molecule ($6\ O_2 + 1\ Glc \rightarrow 6\ CO_2 + 6\ H_2O$).

3.1.2 Inclusion of Glycogen Reduces Oxygen-Carbohydrate Index (OCI) Below OGI

To assess whether CMR_{O2} can fully support glycogen oxidation, the amount of glycogen expressed in glucosyl units consumed during a 1-minute interval (corresponding to the duration of the (A-V) sampling interval) and lactate expressed in glucose equivalents was included in the OGI calculation. The resulting oxygen-carbohydrate index ($OCI = CMR_{O2}/(CMR_{glc} + (CMR_{lac}/2) + CMR_{glycogen}) = (A\text{-}V)_{O2}/((A\text{-}V)_{glc} + [(A\text{-}V)_{lac}/2] + \Delta[glycogen]$ in 1 min) fell from 4.95 to 2.80 during stimulation and from 7.2 to 6.1 during recovery (Table 3). Note that inclusion of three major carbohydrate pools in the OCI calculation (Glc, Lac, glycogen) brings the ratio closer to 6.0 during rest and recovery. However, the glycogen level is still subnormal after 15 min recovery and replenishment will consume glucose without oxygen to depress OGI (see Sect. 3.2). In addition to these carbohydrates, about 20% of the glutamate taken up into astrocytes during excitatory neurotransmission is oxidized (Hertz and Rothman 2017). Based on the rate of the glutamate-glutamine cycle (neurotransmission rate, V_{NT}) in awake rats of 0.57 μmol/g/min (Lanz et al. 2013), astrocytic glutamate oxidation would be 0.57∗0.2 = 0.11 μmol/g/min, and

Table 4 Influence of glycogenolysis on magnitude of fall in oxygen-glucose index (OGI) during sensory stimulation and recovery and duration of reduced glycogen concentration during recovery

Stage of activity cycle	[a]Glycogen concentration (μmol/g)	Δ[glycogen] per min (μmol/g)	[b]$(A\text{-}V)_{O2}$ (μmol/g)	[b]$(A\text{-}V)_{glc}$ (μmol/g)	[b]$[(A\text{-}V)_{lac}]/2$ (μmol/g)	[b]OGI	[c]OCI	Time to restore initial glycogen level (hours)
Rest	12.3	–	3.75	0.68	−0.04	5.51	5.86	
Activation	9.6	0.45	2.97	0.60	0.01	4.95	2.80	
Recovery	8.5	0.07	3.82	0.53	−0.025	7.21	6.11	
Net consumed	3.8							
Estimated glycogen synthesis rate (μmol/g/min)								
0.005[d]								12.5
0.037[e]								1.7
0.073[e]								0.9

[a] From (Cruz and Dienel et al. 2002) that used a 6 min sensory stimulation duration, and glycogen was measured in ethanol extracts of the same brains of animals in which [b] mean arteriovenous differences (A-V) for oxygen (O_2), glucose (glc), and lactate (lac) were determined by (Madsen et al. 1999). These mean values were used to calculate the oxygen-glucose index as $OGI = (A\text{-}V)_{O2}/(A\text{-}V)_{glc}$ (Note: these values differ somewhat from the means of the individual OGIs determined in each animal) and the [c]oxygen-carbohydrate index, $OCI = (A\text{-}V)_{O2}/[(A\text{-}V)_{glc} + ((A\text{-}V)_{lac})/2 + \Delta[glycogen/min])$, where the lactate was converted to glucose equivalents by dividing by 2 and glycogen consumed during the 1 min interval of the arteriovenous sample was calculated as 1 min × (Δ[glycogen]/assay interval duration); the interval was 6 min for activation and 15 min for recovery. [d]Measured rate of net incorporation of blood-borne glucose into glycogen during recovery from a 5 min stimulation, i.e., 0.005 μmol/g/min (Dienel et al. 2002). [e]Estimated as either 5 or 10% of measured CMR_{glc} (i.e., 0.73 μmol/g/min (Dienel et al. 2002)) during recovery (i.e., 0.037 or 0.073 μmol/g/min)

inclusion glutamate oxidized (as glucose equivalents: 0.11∗[5/6]) in 1 min would further reduce OCI to 2.6 during activation (ignoring the complexities of truncated vs. complete oxidation of glutamate and conversion to glucose equivalents).

If all of the glycogen consumed during 1 min of activation (0.45 μmol/g) were oxidized, 2.70 μmol O_2/g would be required (6 × 0.45). However, only a small excess of this amount, 2.97 μmol O_2/g, was actually taken up into during activation and cannot support complete oxidation of the glucose taken up (Table 4), indicating that lactate is probably generated from both glucose and glycogen. Based on an OCI of 2.8, about 47% of the total carbohydrate was oxidized (2.8/6). The partitioning between oxidation of glucose and glycogen cannot be determined in this study, but if glycogenolysis is assumed to replace all of the astrocyte's oxidative ATP from glucose (basal plus activated) the following calculation can be made. Based on the calculations in Sect. 3.3.2 (second paragraph), if glycogen fully replaces oxidation of glucose, then the rate of lactate production from glycogen in astrocytes is 0.4 μmol/g/min (i.e., 2∗[0.54–0.34]). From Table 5, the net rise in cerebral cortical lactate level during activation is 1.1 μmol/g divided by 5 min equals a rate of 0.22 μmol lactate/g/min. Because glycogen-derived lactate does not dilute blood-borne glucose-derived lactate (see below, Sect. 3.3.2 and Table 5), most of the lactate from glycogen must be quickly released at a rate about twice that (0.4/0.22) of lactate accumulation in cerebral cortex that presumably occurs mainly in neurons.

3.1.3 Glycogen Carbon Utilization to Synthesize Amino Acids and Other Compounds

Alanine was not included in the above calculations because its cellular distribution is not known, the increase in its concentration during activation was small (0.2 μmol/g), and, although its labeling doubled, it was a minor fraction of the total metabolite pool; serine level and labeling were unchanged from rest to activation (Dienel et al. 2002). Pyruvate carboxylase activity was not measured, but is typically small compared to glucose oxidation in neurons and astrocytes (Yu et al. 2018; Lanz et al. 2013). Mass balance calculations based on changes in concentrations of measured metabolites and the number of carbons per compound indicates that the fall in glycogen carbon greatly exceeded the sum of the incremental changes in the other metabolites retained in brain (Dienel et al. 2002), consistent with metabolite release from cerebral cortex.

In day-old chick brain, strong circumstantial evidence supports the conclusion that glycogen is a precursor for glutamate synthesis during specific stages of memory consolidation (Gibbs and Hertz 2005; Gibbs et al. 2007; Hertz and Gibbs 2009; Gibbs 2016), indicating that some glycogen-derived pyruvate enters the anaplerotic astrocytic oxidative pathway via pyruvate carboxylase and pyruvate dehydrogenase (Fig. 1). However, the actual quantity of glycogen carbon used for glutamate synthesis in chick brain needs to be determined. In view of the much higher glutamate levels in neurons compared with astrocytes (Ottersen et al. 1992; Storm-Mathisen et al. 1983), the relatively small perisynaptic compartments that are associated with glutamate-glutamine fluxes, and the oxidation of an estimated 20% of the glutamate upon its uptake into astrocytes (McKenna et al. 1996; Hertz and Rothman 2017), the replenishment of transmitter glutamate associated with memory stages with carbon from glucose and/or glycogen probably involves small pools and needs to be examined in more detail.

Table 5 Influence of glycogenolysis on lactate specific activity in cerebral cortex during sensory stimulation

Compound	Concentration (μmol/g)		Specific activity (SA) (dpm/μmol)		Lactate SA relative to Glucose SA (RSA)		Lactate SA if glycogen → Lac	Lactate RSA if glycogen → Lac
	Rest	Activation	Rest	Activation	Rest	Activation	Activation	Activation
Glucose	1.9	2.0	88	78	1.0	1.0		
Lactate	0.6	1.7	46	46	0.55	0.59		
Glycogen	12.3	9.6	0.29	0.27	0.003	0.003		
Δ[Glycogen] → Lac								
Total		2.7 → 5.4 Lac					11	0.14
Amount in 1 min		0.54 → 1.08 Lac					28	0.36

Mean values are from Tables 3 and 5 of (Dienel et al. 2002) that used a 5 min stimulation duration in which brain metabolites were labeled by intravenous injection of [6-^{14}C]glucose for 5 min. Because [6-^{14}C]glucose produces one labeled and one unlabeled lactate, the specific activity (SA) of lactate would be half that of glucose if there were no dilution or loss because 1 mole of glucose generates 2 moles of lactate with the same dpm as glucose; a relative specific activity (RSA) of 0.5 is the theoretical maximum. The calculated lactate SAs and RSAs that include the lactate equivalents of glycogen consumed are the values that would be expected if all of the glycogen-derived lactate were retained in the brain or if the amount of lactate produced in one min were retained in the brain. Calculated dilutions would occur if the lactate remained in astrocytes throughout the experimental interval, then diluted the blood-borne glucose-derived [^{14}C]lactate upon homogenization of tissue, or if glycogen-derived lactate continuously mixed with and diluted the blood-borne [^{14}C]glucose-derived [^{14}C] lactate

3.2 Glycogen Resynthesis During Recovery Contributes to Low CMR_{O2}/CMR_{glc}

Aerobic glycolysis is the preferential upregulation of carbohydrate utilization compared with oxygen consumption even though oxygen level and delivery to brain is normal (Dienel and Cruz 2016), and aerobic glycolysis can be quantified by the extent of the fall in OGI or OCI. Not only does glycogenolysis substantially increase the magnitude of aerobic glycolysis during activation, glycogen level is still below normal after the 15 min recovery interval (Table 4). All of the stimulus-evoked changes in concentrations of ^{14}C-labeled and unlabeled cerebral cortical metabolites except glycogen normalized within 15 min of recovery from sensory stimulation (Dienel et al. 2002). Re-synthesis of 4 μmol/g glycogen during the ensuing recovery period consumes glucose without oxygen, causing aerobic glycolysis to become manifest and persist. The time required to replace glycogen was first estimated using the measured net rate of incorporation of glucose into glycogen during recovery, and it would take nearly 13 h (Table 4). If glycogen synthesis subsequently increased to 5 or 10% of glucose utilization, restoration of glycogen content would still take about 2 or 1 h, respectively. Thus, if subjects are stressed or activated prior to or during metabolic assays, glycogen will be consumed and replaced, enhancing the magnitude of and prolonging the duration of aerobic glycolysis.

3.3 Lactate Production from Glycogen and Its Release from Cerebral Cortex

Section 3.1 presented evidence that the oxygen consumed during activation cannot support oxidation of all of the glucose taken up, let alone the glycogen metabolized. Glycogen-derived lactate shuttling to neurons involves oxidation and is therefore ruled out as being a major pathway. Based on mass balance of major metabolites, net conversion of total pyruvate carbon (from glucose and glycogen) to other compounds is small compared to utilization of glycogen. This leaves lactate as the most likely end product of glycogenolysis, with its release to blood and perivascular fluid. The following sections present detailed analyses of metabolite concentrations and metabolic labeling that support the conclusion that lactate production with release is a major fate of glycogen.

3.3.1 Glycogen-Derived Lactate Release from Cultured Astrocytes

Dringen et al. (1993) showed that stimulation of glycogenolysis in cultured astrocytes is accompanied by release of lactate to the culture medium in amounts nearly stoichiometric with glycogen concentration. In addition, their demonstration that release of glucose to the medium could not be detected was very important because it revealed a major difference in the fate of glycogen carbon in astrocytes compared with hepatocytes that supply glucose to blood. However, the amount of lactate released from cultured cells is probably inflated due to the equilibrative lactate dehydrogenase and lactate transport reactions. The released lactate is continuously diluted by large culture medium volume compared with intracellular volume, thereby 'pulling' lactate from cells into the medium and enhancing glycolytic flux (Hertz and Dienel 2005).

3.3.2 Glycogen-Derived Unlabeled Lactate Does Not Dilute Lactate Labeled by Blood-Borne Glucose or Accumulate in Cerebral Cortex

Analysis of labeling of rat cerebral cortical glucose, lactate, and glycogen by an intravenous injection of [6-^{14}C]glucose during rest and activation provides evidence for compartmentation of glycolytic metabolism of glucose and glycogen and for rapid release of glycogen-derived lactate. The concentration of glucose did not change from rest to activation, indicating supply-demand coupling, whereas lactate level rose almost threefold, and glycogen content fell 22% (Table 5). The specific activity (SA = dpm/μmol) of cerebral cortical lactate was about half that of glucose, i.e., equivalent to the theoretically-maximal relative SA (RSA = $SA_{lactate}/SA_{glucose}$). Metabolism of [6-^{14}C]glucose to lactate produces two molecules, only one of which is labeled, causing lactate SA to fall by 50%. Glycogen labeling was very low and its SA was more than 100-fold lower than that of lactate (Table 5). The high RSA of lactate indicates that essentially all of the lactate retained in the cerebral cortex was derived from blood-borne glucose and that dilution by glycogenolysis was negligible.

If all of the glycogen consumed during activation were converted to lactate (12.3–9.6 = 2.7 glucosyl units × 2 = lactate equivalents), 5.4 μmol/g lactate would be produced, an amount 4.9 times the measured rise in cerebral cortical lactate level of 1.1 μmol/g (Table 5). Thus, even if all of the lactate retained in the cerebral cortex were considered to be from glycogen (which can't be the case, based on lactate RSA), more than 80% of the glycogen-derived lactate would have to have been oxidized and/or released from cerebral cortex. The calculations in Sect. 3.3.3 indicate that efflux from cerebral cortex must be the predominant fate of the glycogen-derived lactate, not oxidation in neurons.

3.3.3 Compartmentation of Glycogen and Glucose Glycolytic Fluxes

If all of the glycogen-derived lactate rapidly and completely mixed with the labeled lactate formed from labeled glucose, the resulting dilution of the lactate SA and RSA can be calculated by adding the 5.4 μmol/g lactate equivalents of glycogen consumed to the measured cerebral cortical lactate concentration during activation. The recalculated lactate SA would fall from 46 to 11 (75% dilution of the SA) and the RSA from 0.59 to 0.14 (Table 5). The time course of glycogenolysis is not known, but if assumed to be constant, the amount of lactate equivalents formed during the last minute of activation just prior to tissue harvest (1.08 μmol/g) would still reduce the lactate SA and RSA to 28 (39% dilution of SA) and 0.36, respectively. Thus, using either the total lactate or the lactate equivalents per minute, the lactate SA would be diluted well below the observed values if glycogen-derived lactate were retained in tissue and mixed with that from glucose.

Another approach to evaluate dilution of lactate by glycogenolysis is to consider cerebral cortical lactate as a single pool that arises from rapid lactate exchange across all cellular membranes so that total astrocytic and neuronal glycolytic fluxes continuously contribute to it. Based on Fig. 8 of Lanz et al. (Lanz et al. 2013) the total astrocytic oxidative consumption of glucose in resting awake brain can be calculated as Vpc + Vg/2, where Vpc is the rate of the pyruvate carboxylase reaction (with units of μmol glucose/g/min) and Vg is the rate of the astrocytic TCA cycle (with units of μmol triose/g/min; dividing Vg by two converts the rate to units of glucose oxidation). $CMR_{glc\text{-}oxidation\text{-}astrocyte}$ = 0.14 + (0.4/2) = 0.34 μmol/g/min. $CMR_{glycogen}$ = 2.7 μmol/g/5 min (Table 3) = 0.54 μmol/g/min. If glycogen fully replaces glucose oxidation in astrocytes during activation, 63% (0.34/0.54) of the glycogen would be oxidized (with no net change in oxygen consumption, only substituting the source of pyruvate). An increase in astrocytic oxidation rate during activation is neglected in this calculation, but it may be about 20% based on the rise in acetate oxidation during photic stimulation of awake rats (Dienel et al. 2007b). The rate of astrocytic lactate production from glycogen can then be calculated by difference: 2∗(0.54–0.34) = 0.4 μmol/g/min. In the 5 min stimulation interval, this would produce 2 μmol/g leaving 3.5 μmol/g (5.4–2) or 65% of the lactate unaccounted for in the absence of its efflux from tissue.

Lactate dilution by fluxes of glucose and glycogen into the total lactate pool is calculated as follows. From the total rate of glucose utilization of 0.93 μmol/g/min (Table 1), total pyruvate/lactate production from blood-borne glucose would be 2∗0.93 = 1.86 μmol/g/min. For one lactate pool fed by glucose and glycogen, the dilution of lactate would be 1-(1.86/[1.86 + 0.4]) = 16% dilution, which could have been detected experimentally. On the other hand, if none of the glycogen were oxidized, the rate of lactate production from glycogen would be 2∗0.54 = 1.08 μmol/g/min. Then the dilution would be 1-(1.86/[1.86 + 1.08]) = 37%, a value equivalent to that calculated above by adding the glycogen-derived lactate to the lactate recovered in cerebral cortex and recalculating the specific activity (1-[28/46]) = 39%. One of the uncertainties in the above analysis is that the fraction of glycogen oxidized or converted to lactate has not been measured, and this is an important issue for future work.

Failure of glycogenolysis to dilute lactate SA and RSA during a 5 min stimulation indicates (1) lactate recovered in cerebral cortical extracts was derived from blood glucose, and (2) products of glycolytic metabolism of blood-borne glucose and glycogen are segregated. Lactate compartmentation is not due to retention of all or the per minute amount of glycogen-derived lactate in astrocytes during activation because SA dilution would have occurred when the tissue was homogenized and extracted. Most of the astrocytic glycogen-derived lactate must, therefore, have been either oxidized, most likely replacing glucose as an oxidative source for the astrocytes due to hexokinase inhibition, and/or released from cerebral cortex, presumably via perivascular endfeet, without mixing with [^{14}C]glucose-derived lactate. Oxidation was ruled out by assay of oxygen consumption, and efflux needs to be examined directly.

The basis for compartmentation remains to be established, but one possibility is that most or all of the [^{14}C]glucose-derived lactate is in neurons and unlabeled glycogen-derived lactate in astrocytes. This means that shuttling of glycogen-derived lactate to neurons where it would mix with and dilute the neuronal pyruvate/lactate pools (Fig. 1) must be a relatively small flux compared with efflux of lactate away from the activated region and from cerebral cortex. Conceivably, the smaller fractional volume of astrocytes could result in a higher intracellular lactate concentration in astrocytes during rapid glycogenolysis that would favor its release from endfeet to blood as an infinite sink. In contrast, at the cerebral cortical lactate level of 1.7 μmol/g (Table 5) lactate uptake into neurons would be 71% of Vmax (assuming Michaelis-Menten kinetics, v = Vmax S/[Km + S], where S = substrate concentration, v = transport rate, Vmax = maximal transport rate) because the Km of neuronal monocarboxylic acid transporter, MCT2, for lactate is about 0.7 mmol/L, compared with about 3.5 and 28 mmol/L for the astrocytic MCT1 and MCT4, respectively (Manning Fox et al. 2000). Also, *net* neuronal uptake would be governed by its oxidation rate that is much slower than transport (Hertz and Dienel 2005; Dienel and Hertz 2001) whereas release to blood would be favored due to its dilution (arterial plasma lactate level was about 0.7 mmol/L in the same animals (Dienel et al. 2002) and high blood flow rate that increased from 1.1 mL/g/min at rest to 1.8 mL/g/min after 6 min of sensory stimulation (Madsen et al. 1998).

Previous work from our laboratory has demonstrated rapid release of glucose-derived labeled lactate to cerebral venous blood during spreading depression (Cruz et al. 1999) and lactate release to perivascular fluid and meninges during acoustic stimulation (Ball et al. 2010). However, efflux of unlabeled glycogen-derived lactate to these two drainage systems has not, to our knowledge, been directly demonstrated. This is even more technically difficult than assays of labeled lactate release because it would be necessary to measure the lactate specific activity in the drainage fluid and compare it to lactate SA in the source tissue and show a dilution by unlabeled glycogen-derived lactate that could be prevented with a phosphorylase inhibitor.

3.4 The Majority of Lactate Elevation During Brain Activation is Most Likely Due to Neuronal Nonoxidative Glycolysis and Not Astrocytic Glycolysis/Glycogenolysis

The increase in brain lactate level during activation is often *assumed* (based on tissue culture assays) to be produced by astrocytes from glucose and/or glycogen, but this notion has never been directly evaluated and proven. Indeed, at least some, and possibly the majority, of the lactate is probably of neuronal origin. The findings that [^{14}C]lactate increases in direct proportion to unlabeled lactate and the relative specific activity (RSA: SA = dpm/µmol, and RSA = $SA_{lactate}/SA_{[6\text{-}14C]glucose}$) of cerebral cortical lactate is ~0.5 (because one lactate is labeled and one is unlabeled giving the same dpm as glucose but 2 moles of lactate/mole glucose) during rest, activation, and recovery in normal cerebral cortex and during spreading cortical depression (Dienel and Cruz 2009) indicate that virtually all lactate recovered in extracts is derived from blood-borne [^{14}C]glucose. These data are most simply explained if lactate is generated in stimulated neurons where it is not diluted by unlabeled glycogen-derived lactate.

Neuronal metabolism of most of the blood-borne glucose during activation is predicted by the DiNuzzo hypothesis (see Sect. 4) that glycogenolysis spares blood glucose for neurons during activation (Fig. 1), and supported by a recent study demonstrating increased neuronal glycolysis with lactate production and probable export during activation (Yellen 2018; Diaz-Garcia et al. 2017). Further support for increased neuronal glycolysis during activation comes from rapid mobilization of the glucose transporter GLUT4 from intracellular membrane stores to the hippocampal synaptic plasma membrane (within seconds after triggering action potentials) to support synaptic vesicle cycling (Ashrafi et al. 2017; Ashrafi and Ryan 2017) and hippocampal memory formation (Pearson-Leary et al. 2018; Pearson-Leary and McNay 2016). Also, glycolytic enzymes are mobilized at presynaptic sites of *C. elegans* during hypoxia-induced energy stress, and disruption of assembly of the glycolytic metabolon blocks synaptic vesicle cycling and synaptic recovery, and disrupts locomotion (Jang et al. 2016). Collectively, these independent metabolic studies experimentally support an increase in neuronal glycolysis during activation and are consistent with glucose sparing for neurons by glycogenolysis. Clearly, more work is required to directly measure the cellular contributions to total glucose utilization during activation.

3.5 Summary: Major Metabolic Products of Glycogen

The fate of glycogen carbon is especially difficult to evaluate in living brain. However, indirect evidence from parallel, quantitative metabolic labeling studies and analysis of metabolite concentration changes indicates that most of the glycogen cannot be oxidized. Incorporation of glycogen carbon into amino acids and oxaloacetate is also small. Lactate production must, therefore, be the predominant fate of glycogen, with rapid removal from the region of generation.

Two independent lines of evidence, no dilution of lactate SA and further reduction of OCI by glycogenolysis during activation, support the following inferences: (1) glycogen-derived lactate is rapidly released from activated cerebral cortex, (2) the proportions of glycogen oxidized and converted to lactate are estimated to be 60/40, (3) based on the limiting supply of oxygen during activation, glycogen oxidation would require that an equivalent amount of blood-borne glucose be converted to lactate and released instead of being oxidized, (4) little, if any, glycogen-derived lactate is shuttled to neurons and oxidized, and (5) a major function of glycogenolysis is rapid production of oxidative and glycolytic ATP for use by astrocytes. Obviously, more work is required to evaluate these conclusions, but they underscore the importance of glycogen as an active participant in energetics of astrocytic activation. An important issue is the degree to which glycogen totally or partially replaces glycolytic and oxidative metabolism of glucose in astrocytes, and in doing so how much blood-borne glucose is spared for neurons.

4 Two Aspects of Glycogenolysis: Fueling Astrocytes and Sparing Glucose for Neurons

Glycogen is located mainly in astrocytes where its utilization can provide 'fast' glycolytic ATP for ion pumping and at the same time replace astrocytic glycolytic metabolism of blood-borne glucose. The shift from glucose to glycogen could spare an equivalent amount of blood-borne glucose for neurons.

4.1 Glucose Sparing by Glycogen

Based on metabolic modeling, DiNuzzo and colleagues proposed that Glc-6-P generated glycogenolysis increases feedback inhibition of astrocytic hexokinase, thereby reducing the amount of blood-borne glucose consumed by astrocytes and sparing an equivalent amount of glucose for neurons. In this model, Glc-6-P derived from glycogen in astrocytes would support some or all of fuel the incremental increase in flux through astrocytic glycolytic pathway when glycogenolysis is activated, not blood glucose. On the other hand, blood glucose would be the source for the increase in glucose utilization during activation when glycogen is consumed. This means that the increase in the metabolism of labeled glucose during the interval of intense glycogenolysis would be predominantly in neurons, consistent with increased labeling of the large glutamate pool located in neurons when glycogen is consumed (Dienel et al. 2002), and suggesting that most or all of the labeled lactate recovered in cerebral cortex and released from cerebral cortex would be of neuronal origin.

A shift in utilization of blood-borne glucose from being the almost exclusive energy source in astrocytes and neurons under resting fed conditions (Nehlig et al. 2004) to potentially predominantly neuronal utilization (based on the glucose-sparing hypothesis) has important implications for interpretation of metabolic brain images assayed with [^{14}C]deoxyglucose (DG) and [^{18}F]fluorodeoxyglucose-positron emission tomography (FDG-PET) (Reivich et al. 1979; Sokoloff et al. 1977). The increase in CMR_{glc} reflected by brain images during activation would reflect neuronal, not astrocytic, glucose utilization, with the inference that small increases in CMR_{glc} in diseased compared with normal brain may predominantly reflect neuronal deficits. This is a very important issue because the cellular basis of brain images is not known, and most previous reports have ascribed the images to astrocytic glucose uptake that fuel the neurons through the astrocyte-neuron lactate shuttle.

The astrocyte-neuron lactate shuttle model is based on the premise that glutamate uptake into astrocytes stimulates astrocytic glycolysis, with lactate shuttling to neurons for oxidation (Pellerin and Magistretti 1994). Compartmentation of glycolysis in astrocytes was suggested to be the basis for DG and FDG-PET images (Magistretti and Pellerin 1996), and a subsequent study claimed that the FDG-PET images are driven by glutamate uptake (Zimmer et al. 2017). However, the stoichiometry of metabolism related to glutamate uptake in astrocytes does not support the lactate shuttle model (Dienel 2017a), there are many other serious flaws in the model, including failure of the rise in CMR_{O2} to match $CMR_{glc+glycogen}$, that invalidate its translation from tissue culture to in vivo situations (reviewed in (Dienel 2019a)), and the FDG-PET images assigned to glutamate uptake are not valid because the method does not have cellular resolution and contributions of neuronal activation were not taken into account (Dienel et al. 2018). Careful evaluation of the hypothesis by DiNuzzo and colleagues (see our companion paper, Rothman and Dienel 2019) is important for understanding of cell-cell interactions in activated brain and for interpretation of metabolic brain images in human brain.

4.2 *Tentative Experimental Support for the Glycogen Glucose-Sparing Model*

To our knowledge, there is no direct evidence supporting the DiNuzzo model. However, this hypothesis is supported by data obtained by Prebil et al. (2011) who reported that adrenergic treatment of cultured astrocytes evoked an increase in cytosolic glucose concentration. The rise in glucose level was blocked by inhibition of glycogenolysis, consistent with suppression of glucose utilization secondary to glycogen breakdown and an increased concentration of Glc-6-P. Reduced hexokinase activity would cause intracellular glucose level to rise as it re-equilibrated with medium glucose because the intracellular glucose level is the net result of the rate of influx, efflux, and metabolism, and it will rise when metabolism falls. This experimental paradigm could be used in future studies with [^{14}C]deoxyglucose assays to measure medium glucose utilization at the hexokinase step as glycogenolysis is triggered and to evaluate the net change in Glc-6-P and intracellular Glc concentrations after glycogen is mobilized.

Another line of evidence for lactate in brain tissue being derived from non-glycogen sources comes from the study of Fray et al. (1996). Extracellular glucose and lactate were monitored by microdialysis in freely-moving rats when glycogenolysis was elicited by a β-adrenergic agonist or a second messenger, a cAMP analog, both of which increased microdialysate glucose level with no change in lactate level. Treatment with a β-adrenergic antagonist blocked the agonist-evoked rise in glucose level. Also, a tail pinch stimulated increases in both glucose and lactate levels, but only the glucose increase was blocked by a β-adrenergic antagonist, indicating differential regulation of glucose and lactate levels. An inference is that glucose level rose due to glucose sparing by glycogenolysis, whereas the lactate was derived from blood-borne glucose. These possibilities need to be tested and can be distinguished by metabolic labeling assays.

4.3 Summary: Astrocytic Glycogenolysis Can Free Up Glucose for Neurons

The model for glucose sparing for neurons by glycogenolysis in astrocytes does not involve astrocyte-neuron shuttling of lactate or other metabolites. Instead, hexokinase activity is further suppressed by glycogen-derived Glc-6-P, allowing unmetabolized blood-borne glucose to diffuse within the interstitial fluid after its uptake into the basal lamina (Mangia et al. 2009, 2011; Simpson et al. 2007) and to diffuse throughout the astrocytic syncytium formed by gap junction-coupled astrocytes to neurons (Gandhi et al. 2009a).

5 Summary of Major Attributes and Functional Implications of Glycogenolysis in Cerebral Cortex During Physiological Stimulation of Awake Rats

Rodent cerebral cortical glycogen concentrations are much higher (10–12 μmol/g) than historically recognized (2–6 μmol/g) when the animals are sequestered or very carefully handled to avoid alerting, stress, and sensory-motor stimulation that occurs when more than one animal is in a cage, cages are transported, and animals are handled prior to tissue harvest. This two- to six-fold difference in concentration probably represents the very labile fraction of the outer tiers of glycogen, since >20% of the glycogen was consumed within 5–6 min of sensory stimulation at a rate 60% of CMR_{glc} by all cells. The outer tiers are estimated to turn over via the glycogen shunt at a rate ~20–40 times the rate of net incorporation of [^{14}C]glucose into glycogen. Remarkably, when the approximate volume fraction of astrocytes is taken into account, cerebral cortical glycogen level approaches that in resting muscle, and $CMR_{glycogen}$ is at least twice CMR_{glc}. The ~1.8–2.9-fold compensatory increases in utilization of blood-borne glucose in the activated sensory and parietal cortex support the conclusion that glycogenolysis makes a major contribution to astrocytic energetics even in the presence of adequate glucose levels and glucose delivery to the brain.

Lack of dilution of cerebral cortical lactate by glycogenolysis reveals compartmentation of glycolytic metabolism of glucose and glycogen, and indicates that glycogen-derived lactate must be quickly released from cerebral cortex without shuttling to and oxidation in neurons or mixing with lactate labeled by blood-borne [^{14}C]glucose. Inclusion of glycogen carbon in the oxygen-carbohydrate index substantially reduces its value, and glycogen re-synthesis substantially prolongs the duration of aerobic glycolysis. Thus, most of the glycogen-derived pyruvate that is not oxidized in astrocytes must be converted to lactate and rapidly discharged from cerebral cortex.

There may also be increased entry of some glycogen-derived Glc-6-P into the pentose shunt pathway (PPP) to produce NADPH during activation. PPP flux from glycogen increases in cultured astrocytes to detoxify peroxides even in the presence of glucose in the medium (Rahman et al. 2000). The glutathione peroxidase/glutathione reductase system consumes NADPH to detoxify hydrogen peroxide produced by monoamine oxidase or superoxide dismutase during brain activation (Dringen et al. 2005, 2007), and during acoustic stimulation, PPP flux in vivo increases 3.5-fold in the major tonotopic band in the inferior colliculus (Dienel 2012). Carbon one of Glc-6-P is decarboxylated in the PPP, and, in normal adult brain, most of the carbon probably re-enters the glycolytic pathway at the fructose-6-P or glyceraldehyde-3-P steps, with generation of glycolytic ATP as the major outcome of glycogenolysis, along with some NADPH production.

Two major points strongly challenge the notion that shuttling of glycogen-derived lactate to neurons as fuel is required to support neuronal energetics of neurotransmission, memory consolidation, and exercise to exhaustion: (1) the quantity of oxygen consumed during activation cannot support complete oxidation of all of the glycogen and glucose consumed, indicating that increased quantities of lactate must be produced and released from cerebral cortex, and (2) lack of lactate specific activity dilution. More work is required to test directly whether these characteristics of sensory stimulation apply to other experimental paradigms, including neurotransmission, learning/memory and exercise.

6 Roles of Brain Glycogenolysis in Neurotransmission and Memory Consolidation and Brain Slices

Knockout (KO) of brain glycogen synthase impairs memory consolidation and hippocampal long-term potentiation (LTP), clearly proving a major role for glycogen in cognitive activities (Duran et al. 2013). Because neurons have very low levels of glycogen (Saez et al. 2014) the effects of glycogen synthase KO may be ascribed mainly to astrocytic glycogen. However, a fascinating recent discovery presented by Jordi Duran) at the International Conference in Brain Energy Metabolism (Valdivia, Chile, March, 2018) was that KO of *neuronal* glycogen synthase also disrupts memory and hippocampal LTP, indicating that the effects of glycogen on cognitive functions are quite complex and involve actions in both astrocytes and neurons (Duran et al. 2019).

Shuttling of glycogen-derived lactate from astrocytes to neurons as fuel has been proposed to be necessary to maintain excitatory, glutamatergic neurotransmission in vitro and memory consolidation in vivo. However, based on the above sensory stimulation studies, it is highly unlikely that much, if any, glycogen-derived lactate was transported to neurons and oxidized. Furthermore, none of the studies that claimed that lactate shuttling fuels neurons actually measured glycogen, astrocyte-neuron lactate shuttling, neuronal lactate metabolism, lactate transport or its inhibition, or demonstrated that the all of the rise in extracellular lactate was actually derived from glycogen. Instead, indirect assays were used, including inhibition of glycogenolysis, extremely high-dose lactate injections to rescue memory, and impairment of lactate transport with inhibitors or monocarboxylic acid transporter (MCT) knockdown. These issues are discussed in detail by Dienel (2019a, b) and in recent reviews (Dienel 2017a, b; Dienel and Cruz 2016). A brief synopsis of key issues with testable, alternative explanations for observed findings is provided below.

6.1 *Disruption of Glycogenolysis Impairs Neurotransmission in Cultured Cells*

6.1.1 Electrophysiological and Developmental Considerations

A number of studies have shown that inhibition of glycogen phosphorylase with different compounds interferes with glutamatergic neurotransmission. For example, when miniature excitatory postsynaptic currents (mEPSC) were measured in neuronal monocultures and astrocyte-neuron co-cultures, the amplitudes of the currents were twice as high in the co-cultures and reduced by 40% by treatment with the glycogen phosphorylase inhibitor 4-(2-Chlorophenyl)-1-ethyl-1,4-dihydro-6-methyl-2,3,5-pyridinetricarboxylic acid 5-isopropyl ester disodium salt hydrate (BAY U6751) in the co-cultures but not neuronal cultures (Mozrzymas et al. 2011). In hippocampal slices from 1-month-old rats, BAY U6751 disrupted the late phase of long-term potentiation (LTP) whereas in slices from 20- to 22-month-old rats there was a tendency for enhancement of LTP (Drulis-Fajdasz et al. 2015). These effects were suggested as possibly due to age-dependent lactate shuttling and neuronal fueling by other pathways (Drulis-Fajdasz et al. 2018). However, lactate shuttling was not measured and other explanations are likely.

6.1.2 Vesicular Glutamate Release and Glutamate Re-uptake in Co-cultures

Neurotransmission was evaluated in cerebellar astrocyte-neuron co-cultures by pre-loading the neurons with tracer amounts of D-[^{3}H]aspartate (D-Asp) as non-metabolizable analog to track vesicular glutamate release and re-uptake during superfusion with a glucose-containing medium and brief pulses of *N*-methyl-D-aspartate (NMDA), glycine, and K^+ (Sickmann et al. 2009). Inhibition of glycogenolysis with [R-R∗,S∗]-5-chloro-N-[2-hydroxy-3-(methoxymethylamino)-3-oxo-1-(phenylmethyl)propyl]-1H-indole-2-carboxamide (CP-316,819) impaired D-[^{3}H]Asp release and reuptake, whereas glycogenolysis blockade with 1,4-dideoxy-1,4-imino-D-arabinitol (DAB) *did not* alter release and reuptake. Inclusion of 1 mmol/L D-lactate along with CP-316,819 to inhibit neuronal uptake of lactate via MCT2 enhanced the observed effects, leading to the authors' conclusion that lactate transfer to neurons was necessary for energetics of excitatory neurotransmission. However, it was not established why only CP-316,819 has these effects and not DAB, weakening this conclusion. Furthermore, previous work, most from the same lab, demonstrated that cerebellar neurons in monoculture are highly glycolytic, releasing about half of the glucose metabolized as lactate under resting and depolarizing conditions (Waagepetersen et al. 2000; Jekabsons et al. 2017; Gebril et al. 2016). In addition, pulses of NMDA stimulate cerebellar neuronal monocultures to release D-Asp, increase deoxyglucose phosphorylation (i.e., CMR_{glc}), enhance lactate release to the medium, and preferentially oxidize glucose, not lactate (Bak et al. 2006, 2009, 2012). It is not clear why there should be neuronal dependence on lactate shuttling and oxidation to sustain neurotransmission in co-culture when lactate oxidization is not upregulated during NMDA pulses in monoculture. Alternative mechanisms and explanations for the observed findings include the following:

Disruption of Ca^{++} Homeostasis by DAB: Gliotransmitter Release

When DAB is used to inhibit glycogenolysis in cultured astrocytes, regulation of intracellular Ca^{2+} homeostasis is disrupted (Fig. 2a). Ca^{2+} dysregulation is due to glycogen being the preferred substrate (in the presence of glucose in the medium) for the ATPase that pumps Ca^{2+} into the endoplasmic reticulum (Müller et al. 2014). The resulting elevated [Ca^{2+}] can cause release of the gliotransmitter glutamate from astrocytes to alter neuronal activity (Araque et al. 1998a, b; Zorec et al. 2012; Parpura et al. 1994), thereby decreasing glutamate and D-[^{3}H]Asp release. It is not known if CP-316,819 blockade of glycogenolysis has the same effect on the calcium pump as DAB, but if it does, reduced D-[^{3}H]Asp release may be the consequence of gliotransmitter action on neurons, not absence of lactate shuttling.

Disruption of K^+ Uptake and Astrocyte Signaling by DAB

The extent of glycogenolysis is directly proportional to the concentration of K^+ in the medium of mouse cortical slices (Hof et al. 1988), and K^+ uptake is inhibited by DAB, thereby interfering with clearance of K^+ from extracellular fluid during neurotransmission (Xu et al. 2014). In addition, vesicular release of ATP from astrocytes is stimulated by glutamate, adenosine, or elevated [K^+], and this release is abolished by DAB, demonstrating a role for glycogenolysis in astrocytic signaling (Xu et al. 2014). The consequences of disruption of K^+ clearance to prolong its extracellular concentration during neuronal signaling (Fig. 2a) and ATP signaling on excitatory neurotransmission remain to be evaluated.

Secondary Effects of Blockade of Neuronal Lactate Transport: Inhibition of Mitochondrial Pyruvate Transport and HCAR1 Activation

D-Lactate may not only impair lactate transport through MCT2, but also inhibit pyruvate uptake into mitochondria (Fig. 2b), lowering oxidative ATP production in both neurons and astrocytes, and altering energy-dependent processes related to neurotransmission, including synaptic vesicle cycling to interfere with glutamate and D-[^{3}H]Asp release. If oxidative metabolism is depressed by 1 mmol/L D-lactate, a compensatory stimulation of glycolysis is anticipated, with increased lactate release from astrocytes to suppress neuronal firing via the G-protein-coupled lactate receptor (HCAR1 or GPR81; Fig. 2b). Also, release of increased amounts of L-lactate generated within neurons when pyruvate oxidation is depressed may be impaired by D-lactate blockade of neuronal MCT2, perhaps inhibiting glycolysis, leading to energy deficiency in neurons (Fig. 2b). In contrast, this concentration of D-lactate would not inhibit astrocytic MCT1 and MCT4 that have higher Km values, and increased astrocytic glycolysis in astrocytes and enhanced lactate release may further disrupt neuronal energetics and glutamate and D-[^{3}H]Asp release via HCAR1.

Disruption of Movements of Peripheral Astrocytic Processes

Glutamate exposure of cultured astrocytes elicits [Ca^{2+}] elevations and stimulates extension of peripheral astrocytic processes (PAPs) mediated by actin cycling (Cornell-Bell et al. 1990a, b). The fuel for activity-driven PAP movements is not known, but they are packed with glycogen (Oe et al. 2016), enriched with metabotropic glutamate receptors, glutamate transporters and glutamine synthetase (Derouiche and Frotscher 1991; Lavialle et al. 2011; Derouiche and Rauen 1995), and contain small mitochondria (Derouiche et al. 2015). PAPs are dynamic structures that advance and retract to and from synapses (Reichenbach et al. 2010), and

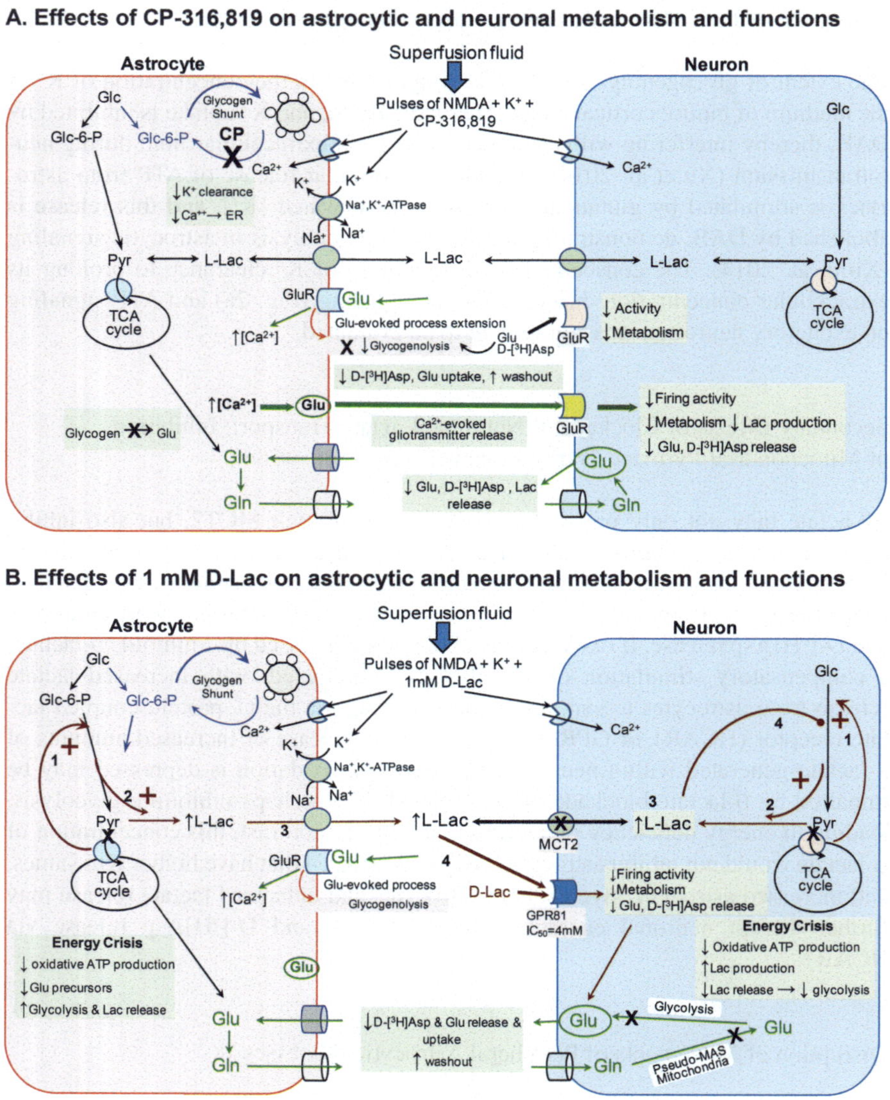

Fig. 2 Alternative mechanisms that may help explain consequences of inhibition of glycogenolysis without astrocyte-neuron lactate shuttling. There are testable alternative mechanisms that may contribute to or explain observed experimental findings without involving lactate shuttling if effects of impairing astrocytic processes that are preferentially fueled by glycogen or to side effects of lactate uptake inhibitors are taken into account. (**a**) In the study by Sickmann et al. (2009), cerebellar astrocyte-neuron co-cultures were used to evaluate glutamatergic neurotransmission (see Sect. 6.1.2) by using D-[^{3}H]Asp, a glutamate analog that is used to track neuronal vesicular Glu release and (mainly astrocytic) Glu reuptake. To depolarize cells and cause glutamate and D-[^{3}H] Asp release, the co-cultures were superfused with pulses of NMDA + glycine + KCl with or without CP-316,819 (CP) to inhibit glycogenolysis and affect different processes (denoted by X). Glycogenolysis preferentially fuels Na^+,K^+-ATPase and Ca^{2+} ATPase that pumps Ca^{2+} into the

their movements are regulated (reviewed by (Jackson and Robinson 2018)). Closer proximity of PAPs to active synapses may facilitate K^+ and glutamate uptake into astrocytes from extracellular fluid. A testable hypothesis is that inhibition of glycogenolysis impairs fueling of astrocytic PAP advances toward synapses releasing glutamate, thereby reducing uptake of glutamate and D-[^{3}H]Asp and increasing D-[^{3}H]Asp washout in the superfusate (Fig. 2a). Prolonged elevation of extracellular glutamate may also contribute to suppression of presynaptic glutamate (and D-[^{3}H]Asp) release via metabotropic glutamate receptors (Moussawi et al. 2011; Scanziani et al. 1997).

Fig. 2 (continued) ER. K^+ uptake into astrocytes after release from neurons would be impaired, prolonging the extracellular rise in [K^+] during neurotransmission. Intracellular Ca^{2+} concentration elevations cause gliotransmitter glutamate release, leading to its interactions with neuronal glutamate receptors, thereby altering neuronal firing activity, reducing Glu and D-[^{3}H]Asp release, and secondarily altering neuronal metabolism and neuronal lactate production and release. If glutamate-evoked movements of peripheral astrocytic processes toward active synapses are fueled by glycogenolysis that is inhibited, astrocytic uptake from extracellular fluid of glutamate, D-[^{3}H]Asp, and K^+ may be reduced, causing increased washout of D-[^{3}H]Asp in the superfusate. In addition, reduced glutamate uptake from the synaptic cleft may allow its interaction with glutamate receptors to downregulate neuronal activity and metabolism, also reducing neuronal lactate release. These secondary consequences of glycogenolysis inhibition may explain reduced D-[^{3}H]Asp release and less reuptake with greater washout. Blockade of glycogenolysis may also deprive the astrocytes of precursor for anaplerotic synthesis of glutamate (see Sect. 3.1.3), interfering with neurotransmission. Major consequences of inhibition of glycogenolysis are indicated by text in green highlights and may also contribute to abnormal electrophysiological responses observed in co-cultures and brain slices (see Sect. 6.1.1). (**b**) The co-cultures were superfused in the presence or absence of 1 mM D-lactate to block neuronal uptake of putative glycogen-derived lactate via MCT2 without impairing astrocytic lactate fluxes through MCT1 and MCT4. However, D-lactate probably also inhibits mitochondrial pyruvate uptake in all cells, suppressing ATP production by the oxidative pathways, and causing an energy crisis that triggers a compensatory rise in glycolysis in all cells (response #1, brown arrow). Astrocytes will produce (arrow #2) and release (arrow #3) lactate that interacts with neuronal GPR81 (arrow #4) to reduce neuronal firing, glutamate and D-[^{3}H]Asp release, and neuronal metabolism. The metabolic crisis may also reduce reuptake of D-[^{3}H]Asp into astrocytes and provision of glutamate precursors. In neurons, the compensatory rise in glycolysis (#1, brown arrow) will increase production of L-lactate (arrow #2) that now accumulates (#3) due to D-lactate-mediated inhibition of efflux to the medium via MCT2. Lactate accumulation would eventual inhibit neuronal glycolysis (line #4, with bal at end), exaggerating neuronal energy failure, causing functional decline. Interference with oxidative metabolism and glycolysis in neurons may also disrupt conversion of glutamine to glutamate via the pseudo-malate-aspartate shuttle (MAS) (Hertz and Chen 2017) and with vesicular packaging of glutamate (Ueda 2016) (see Sect. 6.1.2.3). Major potential consequences of D-lactate treatment are indicated by text in green highlights. Experiments using CP-316,819 plus 1 mM D-lactate would exhibit the combined effects illustrated in panels **a** and **b**. Reproduced from Fig. 2 of (Dienel 2019b) [©2019 Wiley Periodicals, Inc., with permission] that was modified from Fig. 9B of (Dienel 2019a) © American Physiological Society, with permission. **Abbreviations**: *Glc* glucose, *Glc-6-P* glucose-6-phosphate, *Pyr* pyruvate, *Lac* lactate, *TCA* tricarboxylic acid, *Glu* glutamate, *GluR* glutamate receptor, *Gln* glutamine, *NMDA* *N*-methyl-D-aspartate, *Asp* aspartate, *ER* endoplasmic reticulum, *CP* CP-316,819, *DAB* 1,4-dideoxy-1,4-imino-D-arabinitol, *GPR81* G-protein-coupled lactate receptor81 also known as *HCAR1*, hydroxycarboxylic acid receptor1, IC_{50} concentration at which the response is reduced by half, *MAS* malate-aspartate shuttle, *MCT* monocarboxylic acid transporter

6.1.3 Summary: Roles of Glycogen in Neurotransmission

Disruption of Ca^{2+} and K^+ homeostasis, gliotransmitter release, filopodial movements, and oxidative metabolism secondary to inhibition of glycogenolysis and MCT2 may explain or contribute to the observed interference with glutamate release and reuptake without glycogen-derived lactate shuttling to neurons. These consequences may also contribute to electrophysiological abnormalities in astrocyte-neuron co-cultures and brain slices treated with BAY U6751, but secondary effects of BAY U6751 on astrocytes have not, to our knowledge, been evaluated to the same extent as DAB and CP-16,819. Furthermore, the differential effects of DAB and CP-16,819 on co-cultures underscore the importance of using more than two glycogen phosphorylase inhibitors that may have different, unrecognized side effects. Sickmann et al. (2009) did take into account the poor cell permeability of DAB that was previously evaluated in detail by Walls et al. (2008), but effects of CP-16,819 on K^+ and Ca^{2+} homeostasis need to be confirmed. A unique aspect of glycogen phosphorylase inhibition by CP-16,819 is that it requires normoglycemic glucose levels, and inhibition is released under hypoglycemic conditions, an important attribute for its potential use in diabetic patients (see (Suh et al. 2007) and cited references). Thus, if this inhibitor is used when glucose supply is reduced or demand markedly increased, glycogenolysis must be assessed.

6.2 Disruption of Glycogenolysis Impairs Learning/Memory-Related Activities in Rodents In Vivo

Memory-evoking situations transiently increase extracellular lactate concentrations in microdialysates by ~20% to ~2-fold. Inhibition of glycogenolysis with DAB prevented the rise in lactate level and impaired memory consolidation and LTP, leading to the *assumption* that the lactate was astrocytic glycogen-derived and lactate transfer to neurons was required for memory and LTP (Newman et al. 2011; Zhang et al. 2016; Alberini et al. 2018; Gao et al. 2016; Steinman et al. 2016; Suzuki et al. 2011). However, examination the experimental findings indicates that the data supporting the conclusion that lactate fuels neurons are incomplete and inconclusive, and alternative mechanisms may contribute to or explain the findings.

As discussed in Sect. 6.1, disruption of astrocytic intracellular Ca^{2+} homeostasis when glycogenolysis is inhibited by DAB is anticipated to cause release of gliotransmitter glutamate and alter neuronal activity (and secondarily neuronal metabolism and lactate production), potentially impairing the early stages of memory consolidation (Fig. 3a). Also, movements of PAPs to and from active synapses, glutamate and K^+ uptake into astrocytes, and glutamate-glutamine cycling during neuronal signaling (Fig. 3a), as well as astrocytic ATP release, are probably disrupted by DAB treatment, potentially altering the functional sequelae initiated by a memory-triggering event.

6.2.1 Rescue by Massive Doses of L-Lactate

There are important differences in the experimental design of the in vivo studies cited above, but memory could be 'rescued' (i.e., its consolidation was preserved as opposed to memory that was lost and regained) by microinjection of 0.5 μL of 50 but not 30 mmol/L L-lactate (Newman et al. 2011), and memory and LTP rescued by 1 μL of 100 but not 10 mmol/L L-lactate (Suzuki et al. 2011; Zhang et al. 2016). In the Suzuki et al. (2011) study, the volume of distribution of L-lactate was estimated by the hippocampal volume labeled by diffusion of Chicago Sky blue at 1 h after a 1 μL microinjection, and the average tissue concentrations for 10 and 100 mmol/L L-lactate were estimated to be 0.5 and 5 mmol/L. If this 20-fold decrease also applies to D-lactate, the 20 mmol/L injection may fall to a mean level of 1 mmol/L. However, the anionic blue dye is probably restricted to extracellular space whereas D- and D-lactate can enter and leave cells and be discharged to blood via MCTs, so the dye diffusion volumes are unlikely to represent those of lactate. The actual concentration gradients and time courses of lactate clearance from tissue are unknown. This is important because for several reasons:

Unidentified Thresholds and Role(s) for Lactate in Memory Consolidation and Its Rescue

Normal resting brain lactate concentration is in the range of 0.5–1 mmol/L, and if the above tentative average tissue concentration based on dye labeling is accepted, injection of the 10 and 30 mmol/L L-lactate doses that were *ineffective* in memory rescue cause increases in brain lactate level by 0.5 and 1.5 mmol/L, respectively, i.e., 50–300% increases, depending on the initial baseline value. These percent increases are comparable to the rise in extracellular lactate in microdialysate observed during memory-evoking events, suggesting that (a) the rise in lactate concentration during the memory-evoking event is not sufficient to contribute to memory formation, and (b) the massive, pathological dose of 100 mmol/L lactate has effects unrelated to the true extracellular lactate concentration during and shortly after a memory-evoking event.

The extremely high doses of lactate could be expected to cause intracellular acidification and increase the NADH/NAD$^+$ ratio, both of which should impair glycolysis in neurons and astrocytes (Fig. 3b). Because neuronal vesicular loading of glutamate preferentially uses glycolysis (Ueda 2016), excitatory neurotransmission could be altered. Similarly, astrocytic energetics that depend on glycolysis may be disrupted. Consequences of massive lactate doses are much more complex than simply providing oxidative fuel for neurons.

Receptor-Mediated Inhibition of Neuronal Firing by L- and D-Lactate

Treatment of cultured excitatory and inhibitory neurons with L- or D-lactate inhibits firing via HCAR1 with an IC_{50} of 4.2 mmol/L (Bozzo et al. 2013), indicating that the estimated average hippocampal lactate level of ~5 mmol/L after injection of

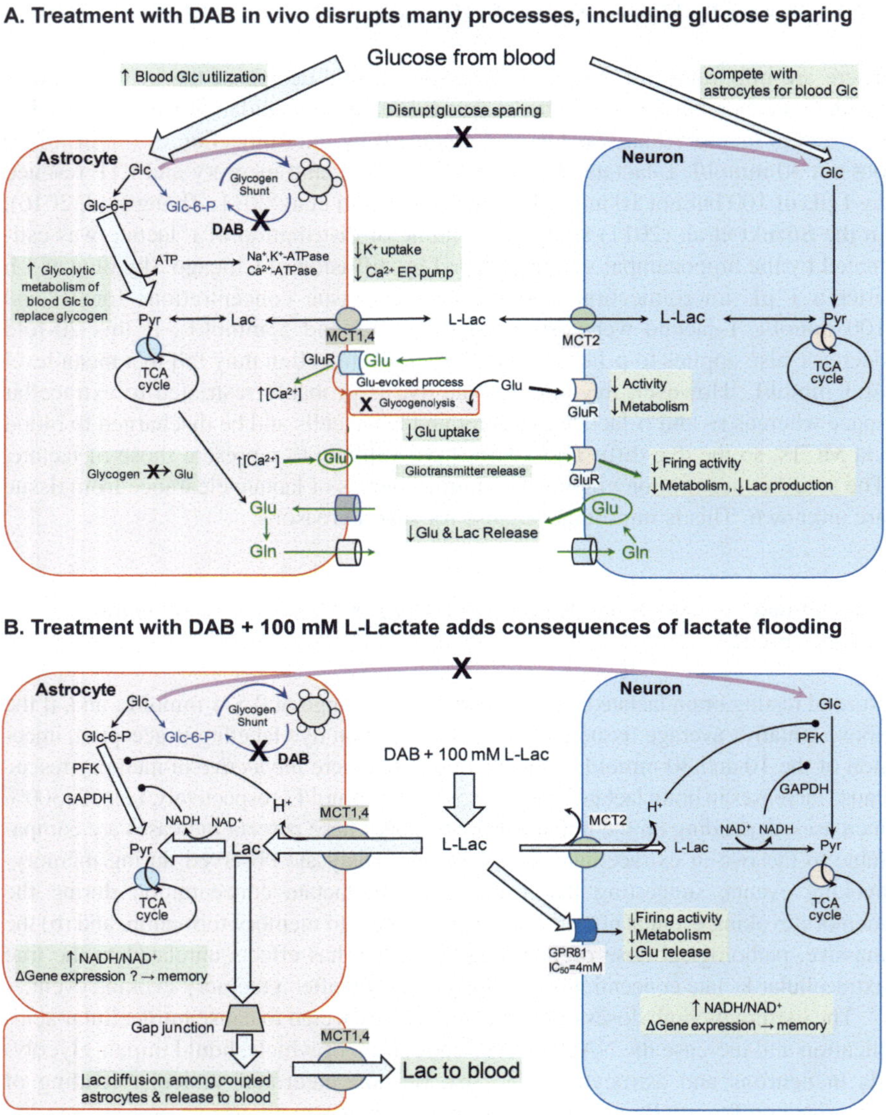

Fig. 3 Alternative mechanisms to explain impairment of memory consolidation in rats by inhibition of glycogenolysis or lactate transport and memory rescue by massive lactate doses. Three interventions were used to evaluate memory formation, impairment, and rescue in rat brain (see Sect. 6.2): (1) DAB was injected into the brain to inhibit astrocytic glycogen phosphorylase (denoted by X); (2) DAB plus an extremely large amount of sodium lactate was injected into brain with the goal of replacing glycogen-derived lactate and increase its oxidative metabolism in neurons; and (3) 20 mM D-Lactate or 60 μM α-cyano-4-hydroxycinnamate (4-CIN) was injected to block neuronal L-lactate uptake via MCT2. Notably, these procedures have additional consequences that confound data interpretation (figure text in green highlights). Consequences of blockade of MCT2-mediated lactate transport with impairment of mitochondrial pyruvate uptake are

100 mmol/L (Suzuki et al. 2011) can cause 50% inhibition of hippocampal firing, with more suppression in neurons closer to the injection site (Fig. 3b) Furthermore, in vivo administration of L-lactate via microdialysis probes to give an extracellular level of 12.5 mmol/L caused a reversible 88% inhibition of hippocampal pyramidal neuron firing (Gilbert et al. 2006). HCAR1 is enriched in pre- and postsynaptic membranes of excitatory synapses in the hippocampus and cerebellum (Lauritzen et al. 2014) and is in a position to influence energetics at the synapse. It is, therefore, important to establish the magnitude and duration of the lactate concentration gradient from the injection site and to evaluate receptor-mediated inhibition of neuronal function with high-dose lactate injections. This information would help evaluate whether LTP can be evoked in the presence of high lactate concentrations or whether

Fig. 3 (continued) similar to those described for 1 mM D-lactate in astrocyte-neuron co-cultures (see Fig. 2b), but probably greater in magnitude due to higher concentration of D-lactate and the very low K_i of 4-CIN for mitochondrial pyruvate transport (6.3 μM (Halestrap 1975)). (**a**). Inhibition of glycogenolysis is also expected to disrupt glucose sparing and cause a compensatory rise in astrocytic glycolytic metabolism of blood-borne glucose. Activated astrocytes and neurons now compete for the glucose entering brain. Because endfeet surround the vasculature, astrocytes have greater access to the incoming glucose, potentially depriving neurons of fuel. In addition, inhibition of glycogenolysis with DAB in vivo would impair the Na^+,K^+-ATPase and Ca^{2+} ATPase, to cause the same effects described in Fig. 2a for CP-316,819 and highlighted in green text boxes. Impairment of glutamate-evoked movements of peripheral astrocytic processes would reduce astrocytic uptake of glutamate and K^+ during neurotransmission. Elevated glutamate levels in the synaptic cleft may activate glutamate receptors to reduce neuronal activity that would secondarily reduce metabolism and neuronal lactate release. (**b**). DAB and a massive dose of sodium L-lactate was injected into brain with the intention of replacing putative glycogen-derived lactate as neuronal fuel (see Sect. 6.2.1). Increased extracellular lactate levels would reduce neuronal activity via GPR81, thereby suppressing metabolism and potentially reducing glutamate and lactate release from neurons. Lactate influx into all cells is concentration-driven and is accompanied by a proton to cause intracellular acidification, inhibit phosphofructokinase (PFK) and reduce glycolytic flux (inhibition denoted by black lines with ball at end). Massive lactate influx would also cause a large increase in the NADH/NAD^+ ratio and reduce NAD^+ availability for glyceraldehyde-3-phosphate dehydrogenase (GAPDH), also impairing glycolytic flux. The increase in NADH/NAD^+ ratio may also induce expression of memory-related genes. Because astrocytic MCT1 and MCT4 have higher Km values for lactate than neuronal MCT2 (3.5–7, 28–35, and 0.7 mM, respectively (Manning Fox et al. 2000; Bröer et al. 1999)) more lactate will enter astrocytes, diffuse throughout gap junction-coupled cells, followed by discharge from endfeet to blood. Near-saturation of MCT2 by high lactate concentrations (denoted by black arrow returning to interstitial fluid) should maintain lactate influx into neurons at near-maximal velocity. High lactate through-flux via astrocytes to blood means that astrocytes handle most of the removal of the injected lactate. Reproduced from Fig. 4 of (Dienel 2019b) [©2019 Wiley Periodicals, Inc., with permission] that was modified from Fig. 9B of (Dienel 2019a) © American Physiological Society, with permission. **Abbreviations**: *Glc* glucose, *Glc-6-P* glucose-6-phosphate, *Pyr* pyruvate, *Lac* lactate, *MCT* monocarboxylic acid transporter, *DAB* 1,4-dideoxy-1,4-imino-D-arabinitol, *4-CIN* α-cyano-4-hydroxycinnamate, *TCA* tricarboxylic acid, *Glu* glutamate, *GluR* glutamate receptor, *Gln* glutamine, *ER* endoplasmic reticulum, *GPR81* G-protein-coupled lactate receptor81 also known as *HCAR1*, hydroxycarboxylic acid receptor1, IC_{50} concentration at which the response is reduced by half *MCT* monocarboxylic acid transporter, *PFK* phosphofructokinase, *GAPDH* glyceraldehyde-3-phosphate dehydrogenase lactate clearance from tissue had already occurred within the 30 min interval between 100 mmol/L lactate injection and high-frequency stimulation to evoke LTP (Suzuki et al. 2011). The role(s) of transient HCAR1-mediated inhibition of neuronal activity on brain function and memory consolidation remain to be established.

Redox Shifts Caused by Extremely-High Lactate Levels

Another consequence of pathologically-high injections of L-lactate is a large shift in the intracellular NADH/NAD$^+$ ratio (Fig. 3b) due to mass action-driven entry of lactate into all brain cells to push the lactate dehydrogenase reaction towards pyruvate plus NADH. Yang et al. (2014) showed that immature cultured neurons grown in high-glucose media and exposed to extremely-high lactate levels express memory-related genes, indicating that responses evoked by large redox changes can be unrelated to true memory-triggering events yet give the appearance of 'rescue'. Notably, the values for the control NADH/NAD$^+$ ratio of 55.97 (highly reduced) and the 250% rise upon addition of 20 mmol/L L-lactate reported by Yang et al. (2014) are much too high and cannot be correct (see the Methods in Yang et al.). These abnormally-high redox ratios in cell extracts contrast sharply with NADH/NAD$^+$ of about 0.3 (oxidized) in extracts of cultured astrocytes (see (Wilhelm and Hirrlinger 2011) and discussion and cited references therein).

Of related interest, exercise causes modest increases in mouse hippocampal lactate concentration (but it is not clear whether brain enzymes were inactivated by in situ freezing or microwave fixation) that leads to induction of brain-derived neurotrophic factor (BDNF) by a sirtuin 1 (SIRT 1) deacetylase-dependent pathway and improves memory/learning (El Hayek et al. 2019). Intraperitoneal lactate injections had similar effects, suggesting that a rise in hippocampal lactate level is sufficient to improve memory and expression of memory-related genes. The implication is that aerobic glycolysis and intracerebral lactate production during brain activation in association with a memory evoking test (e.g., footshock or maze) or lactate injections can initiate this or related memory consolidation processes independent of lactate serving as fuel.

Lactate as Fuel and Anaplerotic Precursor

If lactate-derived pyruvate is used as oxidative fuel in neurons and astrocytes after lactate flooding injections, some may also be used to synthesize glutamate during memory processing in rats, as in day-old chick brain (Gibbs and Hertz 2005; Gibbs et al. 2007; Hertz and Gibbs 2009; Gibbs 2016). Conceivably, very high doses of lactate may substitute for glycogen-derived pyruvate and contribute to 'rescue' of memory consolidation (see Sect. 3.1.3).

6.2.2 Putative Inhibition of Glycogen-Derived Lactate Shuttling to Neurons

D-Lactate (20 mmol/L), α-cyano-4-hydroxycinnamate (4-CIN, 6, 20, or 60 μmol/L), or MCT knockdown experiments were used to try to interfere with transfer of glycogen-derived lactate from astrocytes to neurons. As mentioned above (Sect. 6.1.2.3), 1 mmol/L D-lactate could be expected to inhibit MCT2 and pyruvate oxidation, with the same outcomes as described for in vitro experiments (Fig. 2b). At the doses used, 4-CIN is likely to have similar effects as D-lactate. A caveat is that 4-CIN was dissolved in 1% dimethyl sulfoxide (DMSO) (Newman et al. 2011), and propranolol and co-injected drugs were dissolved in 10% DMSO (Gao et al. 2016). DMSO is a solvent that has significant effects on metabolism and glutamatergic neurotransmission over a very wide range of concentrations, 0.00025% to 0.25% (see (Nasrallah et al. 2008) and cited references), and data obtained with DMSO must be interpreted with caution.

Injection of antisense oligodeoxynucleotides to knockdown specific MCT isoforms caused about a 50% decrease in total transporter protein level at 12 h after the injection (given 1 h before training), but lactate transport was not measured and proven to be inhibited. If the fall in MCT protein amount is linear with time, the rate would be ~4%/h, for a decrease of 8% by1h after the memory-evoking event. This decrement is unlikely to disrupt lactate transport because transport capacity exceeds its metabolism by >5-fold in cultured neurons and by >50-fold in cultured astrocytes (Dienel and Hertz 2001).

6.3 Summary: Many Potential Roles for Glycogen in Learning/Memory

There is no question that glycogen has complex roles in neurotransmission, LTP, and cognition, and that inhibition of glycogenolysis and deletion of glycogen synthase has profound effects on brain function. However, the simple conclusion that glycogen-derived lactate fuels neuronal energy demands is quite unlikely to be correct or quantitatively important. The bottom line is that regardless of the approaches used, lactate transport, inhibition, and oxidation were never measured and quantified. In fact, the origin of extracellular lactate has not been proven to be from astrocytic glycogen, although this notion has been asserted as fact. The mechanisms remain to be established and alternatives that do not involve lactate shuttling need to be tested experimentally. If glycogen is not used as a major oxidative fuel, what are its contributions to astrocytic energetics? These issues are discussed in our companion paper (Rothman and Dienel 2019).

7 Brain Glycogenolysis During Acute Hypoglycemia and Exercise to Exhaustion

To this point, this chapter has emphasized glycogenolysis in cerebral cortex in normal, awake, normoglycemic rats that had been sequestered in a shelter for about 3 h before being abruptly exposed to the laboratory environment and given sensory stimulation. Alerting and attentiveness to a novel situation probably activated widespread noradrenergic signaling throughout the brain via the locus coeruleus to trigger glycogenolysis, followed by cellular metabolic responses to increased signaling activities in response to sensory stimulation. Three important points need to be taken into account when interpreting the functions and fate of glycogen in the presence of adequate supply and delivery of glucose to the brain. First, about 25% of the basal resting glycogen concentration was rapidly consumed (within 5–6 min) during sensory stimulation at rates that are about 60% of CMR_{glc} (Table 1). Second, inclusion of glycogen in the oxygen-carbohydrate index causes the ratio to fall further, indicating that most of the glycogen (plus glucose) is not oxidized (Table 4). Third, glycogenolysis does not dilute the specific activity of brain lactate that is derived almost exclusively from blood-borne labeled glucose (Table 5). These properties indicate that glycogenolysis probably mainly satisfies the energetic demands of astrocytes, and glycogen-derived lactate is rapidly released from brain (because it does not accumulate in brain) without mixing with lactate formed from blood-borne glucose. This section evaluates rates of glycogenolysis during acute hypoglycemia and during exhaustive exercise that causes hypoglycemia and increases release of neurotransmitters that stimulate glycogenolysis.

7.1 Acute Hypoglycemia Triggers Glycogenolysis

When the arterial plasma glucose concentration of α-chloralose-anesthetized rats was clamped at <2 mM during an intravenous infusion of [^{13}C]glucose, the brain [^{13}C]glucose concentration gradually fell from an initial level of 3.2 μmol/g because metabolic demand for glucose in brain exceeded its supply (Choi et al. 2003). The brain [^{13}C]glucose level approached the limit of detection by MRS (~0.25–0.5 μmol/g) within about an hour, indicating that most of the glucose entering brain was phosphorylated immediately after its uptake. Brain [^{13}C]glycogenolysis commenced when brain [^{13}C]glucose was 0.1 μmol/g, and during the following hour when brain [^{13}C]glucose was 'zero' (i.e., not detectable) [^{13}C]glycogen was consumed at a rate of 0.04–0.05 μmol/g/min (Table 6), corresponding to about 8–11% of cerebral cortical CMR_{glc} in non-stimulated, similarly-anesthetized but normoglycemic rats (Nakao et al. 2001) (Table 6). Choi et al. (2003) suggested that glycogenolysis accounted for most or all of the deficit in glucose supply to brain during hypoglycemia.

Table 6 Effects of acute hypoglycemia and treadmill running exercise on brain glycogen and glucose utilization rates

Physiological state	Ref.	Glycogen concentration (μmol/g)	$CMR_{glycogen}$[a] (μmol/g/min)	Plasma or blood glucose level (μmol/ml)	Brain glucose level (μmol/g)	Brain lactate level (μmol/g)	Ref.	Plasma glucose level (μmol/ml)	CMR_{glc} (μmol/g/min)	
		Initial → final		Initial → final	Initial → final	Initial → final			Rest	Run
α-chloralose anesthesia, acute hypoglycemia	Choi et al. (2003)	5 → 1.8	Cx 0.04–0.05[b]	Plasma 10 → <2	3.2 → ~0 → limit of detection =0.25–0.5		Nakao et al. (2001)[c]	6.5	Whisker barrel Cx 0.45	
Awake, acute hypoglycemia	Herzog et al. (2008)	5.4 → 2	Parietal Cx 0.18[d]	Plasma 10 → 1.9	2 → 0.4		Suda et al. (1990)[e]	2.4	Parietal Cx 0.91	
Awake, running 28 m/min ~85% V_{O2Max}							Vissing et al. (1996)[f]	9–10	Cx 1.21 Hip 0.87	1.51 1.05
Awake, running 20 m/min ~50–70% V_{O2Max} Untreated	Matsui et al. (2011) (Figure 2, Table 1)[g]	60 → 120 min Cx 9.5 → 5 Hip14.5 → 8	60 → 120 min 0.08 0.11	Blood 60 → 20 min Glc 6 → 3.8 Lac 2 → 5.5	60 → 120 min Cx 3.2 → 1.3 Hip 3.5 → 1.6	60 → 120 min Cx 1.9 → 2.5 Hip 2.0 → 2.7				
Awake, running 20 m/min ~50–70% V_{O2Max} Untreated	Matsui et al. (2017) (Figures 1 and 3)[h]	0 → 84 min Cx 8.2 → 2 Hip 14 → 4	60 → 84 min 0.26 0.42	Blood 0 → 84 min Glc 5.5 → 1.8 Lac 0.5 → 13	0 → 84 min Cx 3.1 → 0.5 Hip 3.2 → 0.4	0 → 84 min Cx 2.1 → 3 Hip 2.6 → 3.1				

(continued)

Table 6 (continued)

Physiological state	Ref.	Glycogen concentration (μmol/g)	$CMR_{glycogen}$[a] (μmol/g/min)	Plasma or blood glucose level (μmol/ml)	Brain glucose level (μmol/g)	Brain lactate level (μmol/g)	Ref.	Plasma glucose level (μmol/ml)	CMR_{glc} (μmol/g/min)	
		Initial → final		Initial → final	Initial → final	Initial → final			Rest	Run
Awake, running 20 m/min ~50–70% V_{O2Max} Vehicle control	Matsui et al. (2017) (Figures 5, S2, and S3)[i]	60 → 90 min Cx 11 → 4 Hip 15 → 7	60 → 90 min 0.23 0.27	Blood 60 → 90 min Glc 7 → 1.8 Lac 4 → 13		60 → 90 min Cx 1.2 → 2.2 Hip 1.5 → 2.4				
Awake, running 20 m/min ~50–70% V_{O2Max} + DAB	Matsui et al. (2017) (Figures 5, S2, and S3)[j]	60 → 73 min Cx 11 → 4 Hip 15 → 11	60 → 73 min 0.54 0.31	Blood 60 → 73 min Glc 5.5 → 2 Lac 4 → 11		60 → 73 min Cx 1.2 → 1.8 Hip 1.5 → 2				

Values were taken from tabular data or estimated from graphical presentations in the cited references; the latter values are considered approximate

[a]$CMR_{glycogen}$ was calculated by dividing the net change in glycogen concentration by the experimental duration

[b][^{13}C]Glycogenolysis started when brain (mainly cerebral cortex [Cx]) [^{13}C]glucose level in α-chloralose anesthetized rats approached zero (i.e., the limit of detection by MRS, which is about 0.25–0.5 μmol/g). $CMR_{glycogen}$ was estimated over about a 1 h interval during which brain [^{13}C]glucose levels were not detectable. There was, however, continuous influx of glucose into brain from blood (See Table 8), and $CMR_{glycogen}$ was about 9–11% of CMR_{glc} determined in normoglycemic rats by Nakao et al. (1999)

[c]Glucose utilization rates (CMR_{glc}) were measured in non-stimulated whisker barrel (somatosensory) cortex

[d]Glycogen concentrations in parietal cortex fell during the 3 h hypoglycemic episode and correspond to about 18% of CMR_{glc} under similar hypoglycemic conditions determined by Suda et al. (1990). Plasma glucose levels were not reported but were calculated to be about 10 mM, based on brain/plasma distribution ratio for glucose of 0.2 (Holden et al. 1991) and brain glucose concentration of 2 μmol/g

[e]Regional CMR_{glc} was measured in awake normo- and moderately-hypoglycemic rats Hypoglycemia caused modest reductions in cortical CMR_{glc} ranging from −3 to −20%, with a non-significant decrease in parietal cortex

[f]Rats were first trained on a treadmill at 28 m/min running speed for 35 min/day for 7 days. CMR_{glc} was assayed during running for 30 min, and the arithmetic average of regional cerebral cortical CMR_{glc} rose 26%, and hippocampal (Hip) CMR_{glc} increased 23% during exercise

[g]Rats were trained on a treadmill at increasing running speeds from 5 to 25 m/min for 30 min/day in five sessions over 6 days. Experiments were carried out 2 days after training with fasting for 3 h and running at 20 m/min for different durations. Because brain glycogen, glucose, and lactate levels were similar at 0, 30, and 60 min running time and their levels changed at 120 min, values for the 60 and 120 min intervals are tabulated, and $CMR_{glycogen}$ calculated using 60 min (60–120) as the experimental duration. It is not clear when or if the rats in this study became exhausted, and the actual duration for change of metabolite levels is uncertain. Exhaustion in normal or vehicle-injected controls occurred at 80–90 min in the Matsui et al. (2017) study, suggesting that the rats at 120 min in the Matsui et al. (2011) study became exhausted, then recovered somewhat at 120 min, consistent with the higher blood glucose and lower blood lactate at 120 min compared to values at exhaustion in the Matsui et al. (2017) study

[h]Rats were trained as above, and exercised at 20 m/min to exhaustion when they could not keep the pace, lay flat on the treadmill, and stayed in place. Brain glycogen, glucose, and lactate levels were similar at 0, 30, and 60 min running time Matsui et al. (2011), and exhaustion occurred at 84 min, so the 24 min interval (60–84) was used as the duration to calculate $CMR_{glycogen}$

[i]Rats were treated as above, given vehicle injections, and exercised to exhaustion that occurred at about 90 min. Blood glucose levels were stable at 30 and 60 min, then fell at 90 min. A 30 min interval (60–90) was used for calculation of $CMR_{glycogen}$

[j]Rats were treated as above and given intracerebroventricular injections of 1,4-dideoxy-1,4-imino-D-arabinitol (DAB) to inhibit glycogen phosphorylase. Rats were exercised to exhaustion that occurred at about 73 min. Blood glucose levels were stable at 30 and 60 min, then fell at 73 min, and a 13 min interval (73–60) was used for calculation of $CMR_{glycogen}$

When awake rats were rendered hypoglycemic by an insulin injection to reduce arterial plasma glucose level from 10 to 1.9 mM, the rate of glycogenolysis in parietal cortex over a 3 h interval was 0.18 μmol/g/min (Herzog et al. 2008), or 20% of parietal CMR_{glc} in awake, acutely hypoglycemic rats (Suda et al. 1990) (Table 6).

Together, these two sets of findings demonstrate that hypoglycemia reduces delivery of glucose from blood to brain and decreases glucose utilization rate due to supply-demand mismatch. As a compensatory measure, consumption of glycogen, the brain's stored glucose, increases. Most of the brain's glycogen is in astrocytes, and $CMR_{glycogen}$ in hypoglycemic astrocytes ranged from ~10–20% of utilization of blood-borne glucose by all brain cells determined in separate studies under similar conditions (Table 6). The glycogenolytic response to hypoglycemia is a much smaller fraction of CMR_{glc} than during sensory stimulation of awake, normoglycemic rats (60%, Table 1).

7.2 Exercise to Exhaustion

7.2.1 Treadmill Running Increases Glucose and Glycogen Utilization in Rat Brain

Vissing et al. (1996) showed that 30 min of treadmill running at a rate corresponding to about 85% of maximal oxygen utilization capacity (V_{O2Max}) increased CMR_{glc} throughout the brain, with a 23–26% average increases in regions of cerebral cortex and hippocampus (Table 6) and 50–100% and 2-fold increases in regions of the thalamus and cerebellum, respectively (not shown).

A series of very interesting and important studies by Matsui and colleagues (Matsui et al. 2011, 2012, 2017) established that glycogen consumption also increased in all brain regions during running to exhaustion. The high baseline glycogen levels obtained in these reports are similar to those obtained in earlier studies in rats (Dienel et al. 2002; Cruz and Dienel 2002; Kong et al. 2002; Oe et al. 2016) and humans (Öz et al. 2015). In these studies, the treadmill speed was 50–70% V_{O2Max}, and $CMR_{glycogen}$ in cerebral cortex and hippocampus ranged between 0.08–0.42 μmol/g/min in untreated or vehicle-treated rats (Table 6). The lowest rates were in the Matsui et al. (2011) study that were calculated based on a 60 min duration (from 60–120 min running). However, these rates have the highest uncertainty because the rats in subsequent studies (Matsui et al. 2012, 2017) became exhausted at 80–90 min, suggesting that the rats became exhausted and recovered somewhat between 90 and 120 min. Partial recovery may account for the higher blood and brain glucose levels at 120 min compared with the values at exhaustion at 80–90 min (see Table 6 and its legend).

The glycogenolysis rate of about 0.25 μmol/g/min is 17% of mean cerebral cortical CMR_{glc} in rats running for 30 min at a higher speed, and hippocampal rates of 0.27 and 0.42 μmol/g/min correspond to about 25–40% of CMR_{glc} in that structure (Table 6). In these experiments, brain glucose, lactate, and glycogen concentrations were stable at 30 and 60 min of running, and after 60 min when plasma glucose fell towards 2 mM (secondary to depletion of glycogen in liver and muscle), glucose supply to the brain became inadequate, cortical and hippocampal glucose levels fell from ~3 to ~0.5 μmol/g, and glycogen was consumed during the interval between 60 min and onset of exhaustion. Blood and brain glucose levels were positively correlated with regional glycogen level and negatively correlated with lactate concentration in most regions (Matsui et al. 2011).

The exercising rats were previously habituated to running on the treadmill over a 6 day period, so they are unlikely to experience alerting and activation of locus coeruleus noradrenergic signaling as were the sheltered rats with abrupt exposure to the environment prior to sensory stimulation. These factors may contribute to the rapid, twofold higher $CMR_{glycogen}$ in stimulated rats (Table 1) compared with the smaller, delayed rise in exercising rats (Table 6). However, $CMR_{glycogen}$ during exercise may be underestimated because the actual time course of glycogen consumption during the 60–90 min interval is not known.

At the point of exhaustion, brain ATP concentrations were normal, but levels of its metabolites increased indicating higher turnover. Also, brain levels of metabolites of noradrenaline and serotonin rose during exercise and were negatively correlated with brain glycogen level in cortex (Matsui et al. 2011), indicating higher release and metabolism of these neurotransmitters that are known to stimulate glycogenolysis. Thus, there are two triggers to enhance glycogen mobilization during exercise to exhaustion, hypoglycemia and neurotransmission. The rates of glycogenolysis in exercising rats are higher than in sedentary hypoglycemic rats that had lower brain glucose levels (Table 6), presumably due to stimulatory effects of monoamine neurotransmitters and other complex metabolic changes in brain and peripheral organs during exercise to exhaustion.

7.2.2 Exercise Increases Blood and Brain Lactate Levels and Estimated $CMR_{lactate}$

Arterial plasma lactate levels are relatively low (0.5–1 mM) prior to sensory stimulation of relatively sedentary rats, and they increase by a small amount (0.5–1.5 mM) when the animals are stimulated and move around (Dienel et al. 2002, 2007a; Madsen et al. 1999). In sharp contrast, exercise to exhaustion in humans causes very large increases in arterial plasma lactate level (from 0.7–1.1 mM up to 15–21 mM) due to lactate release from muscle; these high concentrations drive lactate down its concentration gradient into brain and provide supplemental oxidative fuel (van Hall et al. 2009; Quistorff et al. 2008). During human exercise to exhaustion, the oxygen-carbohydrate index ($OCI = [A\text{-}V]_{O2}/([A\text{-}V]_{glc} + 0.5[A\text{-}V]_{lac})$ falls from close to 6 at rest to as low as 1.7–3 when determined by arteriovenous (A-V) differences for oxygen, glucose and lactate, indicating that 50–72% of the glucose plus lactate taken up across the blood-brain barrier is not oxidized (Quistorff et al. 2008). Consumption of endogenous brain glycogen during exercise to exhaustion is not detected by A-V difference, and if glycogen consumed were included in the calculation the OCI would fall (e.g., from 4.95 to 2.8 during sensory stimulation, Table 4) below 1.7–3 because measured net $[A\text{-}V]_{O2}$ accounts for oxidation of all substrates.

In exercising rats in the Matsui et al. (2017) study, blood lactate levels rose (non-significantly) from about 0.5 mM at rest to 2–4 mM at 30 and 60 min of running, then increased to about 13 mM at the point of exhaustion (Table 6). At the same time, the brain lactate increased by very modest amounts (<1 μmol/g). When converted to hexose units (divide by two), the rise in brain lactate level corresponded to 10–20% of the fall in brain glucose level and 4–7% of the fall in brain glycogen concentration (Table 6).

Because blood lactate concentration greatly exceeded that in brain after 60 min of running (Table 6), calculated influx of lactate into cerebral cortex during the 60–90 or 60–120 min interval increased by an estimated 1.2–2-fold at exhaustion, depending on the values used for V_{max} and K_T for lactate transport (Table 7). Calculated $CMR_{blood\ lactate}$ in cerebral cortex at exhaustion (73–90 min) was about 75–85% of the rate of unidirectional lactate uptake (Table 7), indicating that most of the incoming lactate is oxidized. If the rate of unidirectional uptake in excess of oxidation rate between 60 and 90 min is 0.05 μmol/g/min (Table 7), lactate influx in excess of metabolism during this 30 min interval could contribute 1.5 μmol lactate/g, which exceeds the net rise in brain lactate concentration of 0.5–1 μmol/g lactate at exhaustion (Table 6). Blood lactate influx is, therefore, the source of some or all of the rise in brain lactate level.

Calculated $CMR_{blood\ lactate}$ rose up to about twofold from 60 min to exhaustion (depending on the values of V_{max} and K_T), and after conversion to hexose units, the rate is about 0.1–0.3 μmol/g/min (Table 7), equivalent to about 50–100% of $CMR_{glycogen}$ in the same animals (Table 7), and about 10–20% of CMR_{glc} during shorter, more vigorous exercise (Table 6). For comparison, the cumulated uptake of lactate into human brain during exhaustive exercise was about 20% of cumulated glucose uptake, but at the point of exhaustion $[A\text{-}V]_{lac}$ was a higher fraction of $[A\text{-}V]_{glc}$ compared with earlier times even though most of the glucose plus lactate is not oxidized (Quistorff et al. 2008) (see above, fall in OCI in exercising humans). Thus, as in humans, blood lactate supplements glucose as oxidative fuel in brain of exercising rats, particularly between 60 min and exhaustion, but blood lactate is also consumed at low rates once its level exceeds that in brain.

Neuronal MCT2 has a Km for lactate of 0.7 mM (Bröer et al. 1999), and based on brain lactate levels of 2–3 mM at exhaustion (Table 6), it can be calculated using simple Michaelis-Menten kinetics ($v = V_{max}\, S/(K_T + S)$ instead of reversible kinetics because intracellular and extracellular levels in different cell types are not known), neuronal lactate transport rate is 75–80% of V_{max}. Near saturation of MCT2 may underlie its 50% increase in levels during exhaustive exercise (Matsui et al. 2017) to raise neuronal lactate transport capacity. Because the trans-membrane lactate gradients are not known, the direction of lactate fluxes cannot be determined. Conceivably, neuronal lactate oxidation could spare neuronal glucose for glycolysis (and lactate production and release), since glycolysis is the preferred source of ATP for loading glutamate into synaptic vesicles (Ueda 2016). In resting humans, blood-borne

Table 7 Comparison of calculated lactate influx rate into cerebral cortex and cortical $CMR_{lactate}$ with cortical $CMR_{glycogen}$ during exercise to exhaustion

Experimental condition	Lactate influx (μmol/g/min) (Triose units)	$CMR_{lactate}$ (μmol/g/min) (Triose units)	$CMR_{glycogen}$[a] (μmol/g/min) (Hexose units)
Matsui et al. (2011)	60 → 120 min	60 → 120 min	60 → 120 min
Untreated	0.09 → 0.17[b]	0 → 0.09[b]	0.08
	0.23 → 0.39[c]	0.01 → 0.21[c]	
Matsui et al. (2017)	0 → 84 min	0 → 84 min	60 → 84 min
Untreated	0.03 → 0.25[b]	−0.08 → 0.19[b]	0.26
	0.07 → 0.53[c]	−0.24 → 0.41[c]	
Vehicle-control	60 → 90 min	60 → 90 min	60 → 90 min
	0.16 → 0.26[b]	0.11 → 0.21[b]	0.23
	0.39 → 0.55[c]	0.27 → 0.46[c]	
	0.46 → 0.61[d]	0.37 → 0.55[d]	
DAB-treated	60 → 73 min	60 → 73 min	60 → 73 min
	0.16 → 0.25[b]	0.11 → 0.21[b]	0.54
	0.39 → 0.54[c]	0.27 → 0.45[c]	

Lactate influx rates from blood into cerebral cortex and cortical $CMR_{lactate}$ were calculated from data tabulated in Table 6 using reversible Michaelis-Menten kinetics as described by Boumezbeur et al. (2010) and Benveniste et al. (2018):

$V_{in} = V_{max}$ [Lac]b / (K_T + [Lac]b + [Lac]Cx), and

$CMR_{lactate} = V_{in} - V_{out} = V_{max}$ ([Lac]b − [Lac]Cx)/(K_T + [Lac]b + [Lac]Cx),

where V_{in} is the influx rate, V_{out} is the efflux rate, V_{max} is the maximal transport rate, K_T is the concentration at half-maximal transport, [Lac]b and [Lac]Cx are lactate concentrations in blood and cerebral cortex, respectively. Calculated rates correspond to the time intervals in the exercise studies, where 0 represents the resting state, 60 min is the time after which blood and brain glucose begin to fall and blood lactate increases sharply, and the longest times (84, 90, and 73 min) are the points of exhaustion. Negative values for $CMR_{lactate}$ indicate efflux from brain to blood. Note that lactate influx rate and $CMR_{lactate}$ must be divided by two to convert the triose units to hexose units for comparison to $CMR_{glycogen}$ and CMR_{glc}

[a]Values for $CMR_{glycogen}$ are from Table 6

[b]V_{max} and K_T were taken to be 0.4 μmol/g/min and 5.1 mM, respectively, as reported for adult human brain with reversible kinetics (Boumezbeur et al. 2010)

[c]We are unaware of determination of K_T and V_{max} in awake rats using reversible kinetics in the literature. To evaluate the impact of different values for kinetic constants, rates were calculated using unpublished preliminary data for V_{max} = 0.75 μmol/min/g and K_T = 2.5 mM

[d]Blood and plasma lactate levels are equal in resting humans, but plasma lactate levels rise more than that in whole blood during exercise, and when blood lactate increased to 4 mM, plasma was 6 mM, and when blood lactate was 10 mM, plasma was 16 mM (Hildebrand et al. 2000). These plasma/blood ratios were used to convert the blood levels reported in Fig. S2B of Matsui et al. (2017) to estimated plasma levels, and calculate lactate influx and CMR_{lac} in one example to illustrate the effects of higher lactate levels in plasma. These corrections yielded plasma lactate levels of 6 and 20.8 mM from reported blood levels of 4 and 13 mM (Table 6)

lactate is oxidized in astrocytes and neurons in the same proportion as glucose, but the relative proportions are not known during exercise. Conceivably, glycolytic metabolism of glycogen in astrocytes to generate lactate may raise intracellular lactate level and reduce its uptake from interstitial fluid, which could spare blood-borne lactate for oxidation in neurons, secondarily sparing neuronal glucose for the glycolytic pathway. These are intriguing issues that require further study.

Note that the calculated rates of lactate influx and $CMR_{lactate}$ are not as accurate as desired because the kinetic constants (V_{max} and K_T) in awake adult rat brain determined with more accurate reversible Michaelis-Menten kinetics are not known. As a first approximation, we used values determined in human brain (V_{max} = 0.4 μmol/g/min and K_T = 5.1 mM) (Boumezbeur et al. 2010) and preliminary unpublished estimates of values for rats (V_{max} = 0.75 μmol/g/min and K_T = 2.5 mM; see legend to Table 7). Note that the rates are directly proportional to V_{max} and inversely related to K_T plus the sum of blood and brain lactate levels (Table 7, legend), so the V_{max} ratio of 0.75/0.4 = 1.9 is the main driving force to increase the rates. Another issue is that Matsui et al. (2011, 2017) reported blood lactate levels, and plasma lactate is the source for uptake into brain. At rest, human blood and plasma lactate levels are equal, but during exercise plasma levels rise higher than blood levels (Hildebrand et al. 2000). Conversion of blood to estimated plasma levels increased influx and $CMR_{lactate}$ rates even more (Table 7). In spite of uncertainties in the absolute values for these rates, the trends for each set of kinetic constants are similar, and relative differences among the groups and time intervals are considered reasonably predictive.

The origin of the increase in brain lactate during exercise cannot be determined from the data in Table 6 because the elevated brain lactate is due, at least in part, to uptake from blood (Table 7), and some may also be derived from glycolytic metabolism of glucose and/or glycogen. The negative correlations between brain glycogen and brain lactate concentrations (Matsui et al. 2011) could be due to either or both glycogenolysis and lactate uptake into brain from blood, which increases as blood and brain glucose fall (Tables 6 and 7). Although metabolic pathway fluxes and compartmentation probably differ during exhaustive exercise and sensory stimulation, the failure of glycogen consumed to accumulate in brain as lactate (the mean change in cortical glycogen level in the Matsui et al. (2011, 2017) was 6.2 μmol/g (Table 6), equivalent to 12.4 μmol/g lactate, whereas lactate level rose <1 μmol/g), and failure to be oxidized during sensory stimulation suggests that the exercise-induced rise in brain lactate is predominantly from brain metabolism of blood-borne glucose and from lactate taken up from blood. Future metabolic labeling studies are required to evaluate the sources of brain lactate during exhaustive exercise.

7.2.3 Functions of Glycogenolysis During Exhaustive Exercise

Is Glycogen-Derived Lactate Shuttled to Neurons as Oxidative Fuel?

Neurons do not take up glucose- or glycogen-derived lactate when activated in brain slices incubated in 10 mM glucose or in normoglycemic animals in vivo; instead, they increase glycolytic metabolism of glucose (Yellen 2018; Diaz-Garcia and Yellen 2019; Diaz-Garcia et al. 2017). However, metabolic demands and fuel usage probably differ when blood lactate is substantially elevated and the brain is hypoglycemic to limit neuronal glycolysis.

Detailed analysis of metabolic changes in brain, muscle, and liver by Matsui et al. (2017) included demonstration that levels of the neuronal lactate transporter, MCT2, in cerebral cortex and hippocampus increased by about 50%, and the glucose transporter GLUT1 level increased in astrocytes and endothelium by 25% and 50%, respectively, in cortex but not in hippocampus. Also, the neuronal glucose transporter GLUT3 level increased 60% in cortex with no change in hippocampus. Thus, glucose transport capacity across the blood brain barrier and across plasma membranes of astrocytes and neurons is increased during exhaustive exercise. Based on these findings and literature studies that claimed shuttling of glycogen-derived lactate during memory-evoking events (Suzuki et al. 2011; Zhang et al. 2016; Newman et al. 2011), Matsui et al. (2017) suggested that glycogen-derived lactate may fuel neurons during exhaustive exercise. Instead, however, under conditions with high blood lactate, enhanced lactate influx and $CMR_{lactate}$, and very low brain and blood glucose (Tables 6 and 7), MCT2 probably facilitates entry of plasma-derived lactate into neurons for oxidation to compensate for inadequate glucose supply.

Furthermore, none of the above-cited memory studies that concluded there was shuttling of glycogen-derived lactate to neurons actually measured lactate transport from astrocytes to neurons, none identified the metabolic and cellular origin of the lactate in extracellular fluid, and alternative interpretations of observed data that do not involve astrocyte-neuron lactate shuttling (Dienel 2019b) were not tested and ruled out (see Sect. 6 and Figs. 2 and 3). It is clear from the Matsui et al. series of studies that glycogen is an important fuel for brain during exhaustive exercise, but its cellular roles and metabolic fate remain to be established. Liver and muscle glycogen is also consumed during exercise, but the reported concentrations in these organs (Matsui et al. 2011, 2012, 2017) were probably underestimated due to postmortem ischemia. In contrast, postmortem levels of glycogen and other labile brain metabolites were appropriately preserved by microwave fixation to inactivate brain enzymes.

Roles of Glycogenolysis and MCT2 in Exercise Endurance

The hypothesis that glycogen mobilization contributes to endurance and may have a role in central fatigue was tested by Matsui et al. (2017) by pre-treating rats with an intracerebroventricular (icv) injection of an inhibitor of glycogen phosphorylase, 1,4-dideoxy-1,4-imino-D-arabinitol (DAB) 15 min prior to exercise. Hippocampal (but not cortical) glycogenolysis was impaired and the rise in lactate level was reduced by a small amount (~0.4 μmol/g, Table 6), ATP concentration fell by about 12% at exhaustion, and time to exhaustion was accelerated to about 73 min vs. 90 min with vehicle injection.

To evaluate whether lactate transport was involved, the rats were given an icv injection of an MCT2 inhibitor, α-cyano-4-hydroxycinnamate (4-CIN), which reduced hippocampal ATP level by 30% and the time to exhaustion by 35% from 80 to about 50 min. These findings were interpreted as both glycogenolysis and neuronal lactate transport being key factors in fatigue and in brain energetics during exhaustive exercise.

Interpretive Caveats with DAB and 4-CIN Treatments

The icv injections consisted of 10 μl volumes of 150 mM DAB in 0.9% saline or 36 mM 4-CIN in 40% dimethylsulphoxide (DMSO) in 0.9% saline (Matsui et al. 2017). The authors assumed a 100-fold dilution of these injectant concentrations in brain to give estimated final concentrations of 1.5 mM DAB, 0.36 mM 4-CIN, and 0.4% DMSO. The distribution an icv injectant in brain was visualized with Chicago sky blue, which appeared to label mainly the perivascular space in the major blood vessels (Fig. S4 of Matsui et al. (2017)), a labeling pattern similar to that of Evans blue-bound albumin, amyloid-β, glucose, lactate, and other metabolites (Dienel 2012; Ball et al. 2010), suggesting that much of the injectant in the cerebrospinal fluid may have been cleared from the tissue via the perivascular-lymphatic drainage system without mixing with interstitial fluid.

Inhibition of glycogenolysis in cultured astrocytes by pretreatment with 300 μM DAB for 60 min impaired fueling of the ATPase involved in Ca^{++} uptake into the endoplasmic reticulum by glycogen (Müller et al. 2014) (see Sect. 6.1.2.1 and Fig. 2a). Also, DAB (10 mM for 2 min) interferes with K^+ clearance from extracellular fluid and ATP release by cultured astrocytes (Hertz et al. 2015a; Xu et al. 2013, 2014) (see Sect. 6.1.2.2). Dysregulation of Ca^{++} and K^+ homeostasis can have secondary effects involving gliotransmitter release to alter neuronal activity and metabolism (reviewed in (Dienel 2019a, b)), perhaps contributing to changes in hippocampal lactate and ATP levels and fatigue, independent of any lactate shuttling to neurons as oxidative fuel (see Sect. 6 and Figs. 2 and 3).

4-CIN (estimated 360 μM) not only inhibits MCT2 (IC_{50} = 24 μM (Bröer et al. 1999)), but it is also a potent inhibitor of pyruvate transport into mitochondria (K_i = 6.3 μM (Halestrap 1975)). 4-CIN can enter erythrocytes, it blocks transport of pyruvate into and out of isolated mitochondria, and in perfused rat heart that has MCT1 and other MCTs (Halestrap and Price 1999) 200 μM 4-CIN increased the output of lactate and pyruvate, indicating inhibition of mitochondrial pyruvate transport and oxidation enhances glycolysis (Halestrap and Denton 1975; Halestrap 1975). Thus, 360 μM 4-CIN is likely to block oxidative metabolism of pyruvate *in all brain cells* within its distribution zone after icv injection, causing energy failure and contributing to the observed fall in ATP concentration and accelerated onset of fatigue (see Sects. 6.1.2.3 and 6.2.2 and Fig. 2b). This conclusion is supported by findings of McKenna et al. (2001) who showed that 250 μM 4-CIN reduced oxidation of glucose to CO_2 by 45–60% in cultured astrocytes and neurons without affecting glucose uptake from the medium. Another confound is that 0.4% DMSO modulates brain metabolism of glucose and pyruvate (Nasrallah et al. 2008), and its effects on metabolism during exhaustive exercise, while taken into account by vehicle treatments, are not known.

A major effect of 4-CIN treatment during exhaustive exercise is probably inhibition of oxidation of glycogen- and glucose-derived pyruvate, with additional effects on neuronal plasma membrane lactate fluxes. If neuronal pyruvate oxidation is impaired, neuronal glycolysis should be stimulated to provide more ATP. If the efflux of lactate from these neurons is impaired by 4-CIN intracellular lactate would accumulate, eventually suppressing neuronal glycolysis and enhancing neuronal energy failure and onset of fatigue. On the other hand, enhanced glycolysis in astrocytes would lead to lactate efflux to blood and perivascular fluid as well as release to extracellular fluid via MCT1 and MCT4, where it could interact with the neuronal lactate receptor HCAR1 to suppress neuronal firing (Bozzo et al. 2013; Lauritzen et al. 2014) and secondarily reduce neuronal metabolism. For a more detailed discussion of these complex issues, see Sect. 6 and (Dienel 2019a, b). Inhibition of mitochondrial pyruvate transport and oxidation by 4-CIN, besides MCT2 inhibition, could explain the observed fall in ATP concentration and faster onset of fatigue without any lactate shuttling to fuel neurons.

7.3 Summary

Brain glycogen is consumed during hypoglycemia and prolonged exercise to exhaustion and contributes to cellular energetics and time of onset of fatigue. MCT2-mediated lactate transport may also be involved, but the directions and magnitudes of cellular lactate fluxes remain to be established. As in the studies of sensory stimulation and exhaustive exercise in humans where the oxygen-carbohydrate index falls to a very low value, the likelihood is that most glycogen is metabolized glycolytically to support astrocytic energetics, with lactate release to blood and perivascular-lymphatic drainage systems. There is currently no compelling evidence that glycogen-derived lactate fuels brain during exhaustive exercise. Alternative explanations and mechanisms for roles of glycogen and the sources of brain lactate need to be tested and evaluated.

8 Concluding Comments

Glycogen is clearly a very important active participant in astrocytic energetics, with glycolytic ATP production being important for support of different energy-dependent processes during activation. Na^+,K^+-ATPase activity for K^+ uptake from extracellular space is a major energy demand for astrocytes and neurons (DiNuzzo et al. 2017) that can be preferentially satisfied by glycogenolysis, even in the presence of glucose (Xu et al. 2013). In addition, glycolytic ATP generated by glycogenolysis may be utilized to maintain Ca^{++} homeostasis and fuel its pumping into the endoplasmic reticulum (Müller et al. 2014), and it may help fuel filopodial movements during neurotransmission (Reichenbach et al. 2010). Sparing of blood-borne glucose for use by neurons when glycogen is mobilized in astrocytes (DiNuzzo et al. 2010b) is a novel alternative to the notion of glycogen-derived lactate shuttling. A stoichiometric model for the role of the glycogen shunt for glucose sparing during sensory stimulation and intense activation is evaluated in our companion paper (Rothman and Dienel 2019). This review has emphasized the roles of glycogen in the functions and energetics of cerebral cortex, and there are probably regional differences in functions of glycogen (e.g., in hippocampus and other structures involved in learning/memory), as well as differences between gray and white matter. Future quantitative studies of roles of glycogen in these processes and related metabolic pathway fluxes are required to better understand the functions of glycogen in living brain.

Acknowledgements

Conflict of interest: The authors declare that they have no conflicts of interest.

Sources of funding: DLR: R01MH109159, R01 NS087568, R01 NS100106.

References

Adachi K, Cruz NF, Sokoloff L, Dienel GA (1995) Labeling of metabolic pools by [6-^{14}C]glucose during K(+)-induced stimulation of glucose utilization in rat brain. J Cereb Blood Flow Metab 15(1):97–110. https://doi.org/10.1038/jcbfm.1995.11

Aiston S, Andersen B, Agius L (2003) Glucose 6-phosphate regulates hepatic glycogenolysis through inactivation of phosphorylase. Diabetes 52(6):1333–1339

Aiston S, Green A, Mukhtar M, Agius L (2004) Glucose 6-phosphate causes translocation of phosphorylase in hepatocytes and inactivates the enzyme synergistically with glucose. Biochem J 377(Pt 1):195–204. https://doi.org/10.1042/bj20031191

Alberini CM, Cruz E, Descalzi G, Bessières B, Gao V (2018) Astrocyte glycogen and lactate: New insights into learning and memory mechanisms. Glia 66(6):1244–1262. https://doi.org/10.1002/glia.23250

Araque A, Parpura V, Sanzgiri RP, Haydon PG (1998a) Glutamate-dependent astrocyte modulation of synaptic transmission between cultured hippocampal neurons. Eur J Neurosci 10(6):2129–2142. https://doi.org/10.1046/j.1460-9568.1998.00221.x

Araque A, Sanzgiri RP, Parpura V, Haydon PG (1998b) Calcium elevation in astrocytes causes an NMDA receptor-dependent increase in the frequency of miniature synaptic currents in cultured hippocampal neurons. J Neurosci 18(17):6822–6829

Ashrafi G, Ryan TA (2017) Glucose metabolism in nerve terminals. Curr Opin Neurobiol 45:156–161. https://doi.org/10.1016/j.conb.2017.03.007

Ashrafi G, Wu Z, Farrell RJ, Ryan TA (2017) GLUT4 mobilization supports energetic demands of active synapses. Neuron 93(3):606–615.e603. https://doi.org/10.1016/j.neuron.2016.12.020

Bachelard HS, Goldfarb PS (1969) Adenine nucleotides and magnesium ions in relation to control of mammalian cerebral-cortex hexokinase. Biochem J 112(5):579–586

Bak LK, Schousboe A, Sonnewald U, Waagepetersen HS (2006) Glucose is necessary to maintain neurotransmitter homeostasis during synaptic activity in cultured glutamatergic neurons. J Cereb Blood Flow Metab 26(10):1285–1297. https://doi.org/10.1038/sj.jcbfm.9600281

Bak LK, Walls AB, Schousboe A, Ring A, Sonnewald U, Waagepetersen HS (2009) Neuronal glucose but not lactate utilization is positively correlated with NMDA-induced neurotransmission and fluctuations in cytosolic Ca2+ levels. J Neurochem 109(Suppl 1):87–93. https://doi.org/10.1111/j.1471-4159.2009.05943.x

Bak LK, Obel LF, Walls AB, Schousboe A, Faek SA, Jajo FS, Waagepetersen HS (2012) Novel model of neuronal bioenergetics: post-synaptic utilization of glucose but not lactate correlates positively with Ca2+ signaling in cultured mouse glutamatergic neurons. ASN Neuro 4(3):151–160. https://doi.org/10.1042/an20120004

Bak LK, Walls AB, Schousboe A, Waagepetersen HS (2018) Astrocytic glycogen metabolism in the healthy and diseased brain. J Biol Chem 293(19):7108–7116. https://doi.org/10.1074/jbc.R117.803239

Ball KK, Cruz NF, Mrak RE, Dienel GA (2010) Trafficking of glucose, lactate, and amyloid-beta from the inferior colliculus through perivascular routes. J Cereb Blood Flow Metab 30(1):162–176. https://doi.org/10.1038/jcbfm.2009.206

Benveniste H, Dienel G, Jacob Z, Lee H, Makaryus R, Gjedde A, Hyder F, Rothman DL (2018) Trajectories of brain lactate and re-visited oxygen-glucose index calculations do not support elevated non-oxidative metabolism of glucose across childhood. Front Neurosci 12:631. https://doi.org/10.3389/fnins.2018.00631

Boumezbeur F, Petersen KF, Cline GW, Mason GF, Behar KL, Shulman GI, Rothman DL (2010) The contribution of blood lactate to brain energy metabolism in humans measured by dynamic 13C nuclear magnetic resonance spectroscopy. J Neurosci 30(42):13983–13991. https://doi.org/10.1523/jneurosci.2040-10.2010

Bozzo L, Puyal J, Chatton JY (2013) Lactate modulates the activity of primary cortical neurons through a receptor-mediated pathway. PLoS One 8(8):e71721. https://doi.org/10.1371/journal.pone.0071721

Bröer S, Bröer A, Schneider HP, Stegen C, Halestrap AP, Deitmer JW (1999) Characterization of the high-affinity monocarboxylate transporter MCT2 in Xenopus laevis oocytes. Biochem J 341:529–535

Cataldo AM, Broadwell RD (1986) Cytochemical identification of cerebral glycogen and glucose-6-phosphatase activity under normal and experimental conditions. I. Neurons and glia. J Electron Microsc Tech 3(4):413–437. https://doi.org/10.1002/jemt.1060030406

Choi IY, Seaquist ER, Gruetter R (2003) Effect of hypoglycemia on brain glycogen metabolism in vivo. J Neurosci Res 72(1):25–32. https://doi.org/10.1002/jnr.10574

Cornell-Bell AH, Finkbeiner SM, Cooper MS, Smith SJ (1990a) Glutamate induces calcium waves in cultured astrocytes: long-range glial signaling. Science 247(4941):470–473

Cornell-Bell AH, Thomas PG, Smith SJ (1990b) The excitatory neurotransmitter glutamate causes filopodia formation in cultured hippocampal astrocytes. Glia 3(5):322–334. https://doi.org/10.1002/glia.440030503

Crane RK, Sols A (1954) The non-competitive inhibition of brain hexokinase by glucose-6-phosphate and related compounds. J Biol Chem 210(2):597–606

Crerar MM, Karlsson O, Fletterick RJ, Hwang PK (1995) Chimeric muscle and brain glycogen phosphorylases define protein domains governing isozyme-specific responses to allosteric activation. J Biol Chem 270(23):13748–13756

Cruz NF, Dienel GA (2002) High glycogen levels in brains of rats with minimal environmental stimuli: implications for metabolic contributions of working astrocytes. J Cereb Blood Flow Metab 22(12):1476–1489. https://doi.org/10.1097/00004647-200212000-00008

Cruz NF, Adachi K, Dienel GA (1999) Rapid efflux of lactate from cerebral cortex during K+ – induced spreading cortical depression. J Cereb Blood Flow Metab 19(4):380–392. https://doi.org/10.1097/00004647-199904000-00004

Cruz NF, Ball KK, Dienel GA (2007) Functional imaging of focal brain activation in conscious rats: impact of [(14)C]glucose metabolite spreading and release. J Neurosci Res 85(15):3254–3266. https://doi.org/10.1002/jnr.21193

Danchin A, Buc H (1973) Affinity labeling of the adenosine 5′-monophosphate binding site of rabbit muscle glycogen phosphorylase b with an adenosine 5′-monophosphate-cobalt(3) complex. J Biol Chem 248(9):3241–3247

Derouiche A, Frotscher M (1991) Astroglial processes around identified glutamatergic synapses contain glutamine synthetase: evidence for transmitter degradation. Brain Res 552(2):346–350

Derouiche A, Rauen T (1995) Coincidence of L-glutamate/L-aspartate transporter (GLAST) and glutamine synthetase (GS) immunoreactions in retinal glia: evidence for coupling of GLAST and GS in transmitter clearance. J Neurosci Res 42(1):131–143. https://doi.org/10.1002/jnr.490420115

Derouiche A, Haseleu J, Korf HW (2015) Fine astrocyte processes contain very small mitochondria: glial oxidative capability may fuel transmitter metabolism. Neurochem Res 40(12):2402–2413. https://doi.org/10.1007/s11064-015-1563-8

Diaz-Garcia CM, Yellen G (2019) Neurons rely on glucose rather than astrocytic lactate during stimulation. J Neurosci Res 97(8):883–889. https://doi.org/10.1002/jnr.24374

Diaz-Garcia CM, Mongeon R, Lahmann C, Koveal D, Zucker H, Yellen G (2017) Neuronal stimulation triggers neuronal glycolysis and not lactate uptake. Cell Metab 26(2):361–374. e364. https://doi.org/10.1016/j.cmet.2017.06.021

Dienel GA (2012) Fueling and imaging brain activation. ASN Neuro 4(5):267–321. art:e00093. https://doi.org/10.1042/an20120021

Dienel GA (2017a) Lack of appropriate stoichiometry: strong evidence against an energetically-important astrocyte-neuron lactate shuttle in brain. J Neurosci Res 95(11):2103–2125. https://doi.org/10.1002/jnr.24015

Dienel GA (2017b) The metabolic trinity, glucose–glycogen–lactate, links astrocytes and neurons in brain energetics, signaling, memory, and gene expression. Neurosci Lett 637:18–25. https://doi.org/10.1016/j.neulet.2015.02.052

Dienel GA (2019a) Brain glucose metabolism: Integration of energetics with function. Physiol Rev 99(1):949–1045. https://doi.org/10.1152/physrev.00062.2017

Dienel GA (2019b) Does shuttling of glycogen-derived lactate from astrocytes to neurons take place during neurotransmission and memory consolidation? J Neuro Res 97(8):863–882. https://doi.org/10.1002/jnr.24387

Dienel GA, Cruz NF (2009) Exchange-mediated dilution of brain lactate specific activity: implications for the origin of glutamate dilution and the contributions of glutamine dilution and other pathways. J Neurochem 109(Suppl 1):30–37. https://doi.org/10.1111/j.1471-4159.2009.05859.x

Dienel GA, Cruz NF (2016) Aerobic glycolysis during brain activation: adrenergic regulation and influence of norepinephrine on astrocytic metabolism. J Neurochem 138(1):14–52. https://doi.org/10.1111/jnc.13630

Dienel GA, Hertz L (2001) Glucose and lactate metabolism during brain activation. J Neurosci Res 66(5):824–838

Dienel GA, Wang RY, Cruz NF (2002) Generalized sensory stimulation of conscious rats increases labeling of oxidative pathways of glucose metabolism when the brain glucose-oxygen uptake ratio rises. J Cereb Blood Flow Metab 22(12):1490–1502. https://doi.org/10.1097/00004647-200212000-00009

Dienel GA, Ball KK, Cruz NF (2007a) A glycogen phosphorylase inhibitor selectively enhances local rates of glucose utilization in brain during sensory stimulation of conscious rats: implications for glycogen turnover. J Neurochem 102(2):466–478. https://doi.org/10.1111/j.1471-4159.2007.04595.x

Dienel GA, Schmidt KC, Cruz NF (2007b) Astrocyte activation in vivo during graded photic stimulation. J Neurochem 103(4):1506–1522. https://doi.org/10.1111/j.1471-4159.2007.04859.x

Dienel GA, Behar KL, Rothman DL (2018) Cellular origin of [^{18}F]FDG-PET imaging signals during ceftriaxone-stimulated glutamate uptake: astrocytes and neurons. Neuroscientist 24(4):316–328

DiNuzzo M (2013) Kinetic analysis of glycogen turnover: relevance to human brain ^{13}C-NMR spectroscopy. J Cereb Blood Flow Metab 33(10):1540–1548. https://doi.org/10.1038/jcbfm.2013.98

DiNuzzo M, Mangia S, Maraviglia B, Giove F (2010a) Changes in glucose uptake rather than lactate shuttle take center stage in subserving neuroenergetics: evidence from mathematical modeling. J Cereb Blood Flow Metab 30(3):586–602. https://doi.org/10.1038/jcbfm.2009.232

DiNuzzo M, Mangia S, Maraviglia B, Giove F (2010b) Glycogenolysis in astrocytes supports blood-borne glucose channeling not glycogen-derived lactate shuttling to neurons: evidence from mathematical modeling. J Cereb Blood Flow Metab 30(12):1895–1904. https://doi.org/10.1038/jcbfm.2010.151

DiNuzzo M, Maraviglia B, Giove F (2011) Why does the brain (not) have glycogen? BioEssays 33(5):319–326. https://doi.org/10.1002/bies.201000151

DiNuzzo M, Mangia S, Maraviglia B, Giove F (2012) The role of astrocytic glycogen in supporting the energetics of neuronal activity. Neurochem Res 37:2432–2438. https://doi.org/10.1007/s11064-012-0802-5

DiNuzzo M, Giove F, Maraviglia B, Mangia S (2017) Computational flux balance analysis predicts that stimulation of energy metabolism in astrocytes and their metabolic interactions with neurons depend on uptake of K^+ rather than glutamate. Neurochem Res 42(1):202–216. https://doi.org/10.1007/s11064-016-2048-0

Dringen R, Gebhardt R, Hamprecht B (1993) Glycogen in astrocytes: possible function as lactate supply for neighboring cells. Brain Res 623(2):208–214

Dringen R, Pawlowski PG, Hirrlinger J (2005) Peroxide detoxification by brain cells. J Neurosci Res 79(1–2):157–165. https://doi.org/10.1002/jnr.20280

Dringen R, Hoepken HH, Minich T, Ruedig C (2007) Pentose phosphate pathway and NADPH metabolism. In: Gibson GE, Dienel GA (eds) Brain energetics. Integration of molecular and cellular processes. Handbook of neurochemistry and molecular neurobiology, 3rd edn. Springer, Berlin, pp 41–62

Drulis-Fajdasz D, Wojtowicz T, Wawrzyniak M, Wlodarczyk J, Mozrzymas JW, Rakus D (2015) Involvement of cellular metabolism in age-related LTP modifications in rat hippocampal slices. Oncotarget 6(16):14065–14081. https://doi.org/10.18632/oncotarget.4188

Drulis-Fajdasz D, Gizak A, Wojtowicz T, Wisniewski JR, Rakus D (2018) Aging-associated changes in hippocampal glycogen metabolism in mice. Evidence for and against astrocyte-to-neuron lactate shuttle. Glia 66(7):1481–1495. https://doi.org/10.1002/glia.23319

Duran J, Guinovart JJ (2015) Brain glycogen in health and disease. Mol Asp Med 46:70–77. https://doi.org/10.1016/j.mam.2015.08.007

Duran J, Gruart A, Varea O, López-Soldado I, Delgado-García JM, Guinovart JJ (2019) Lack of Neuronal Glycogen Impairs Memory Formation and Learning-Dependent Synaptic Plasticity in Mice. Frontiers in Cellular Neuroscience 13

Duran J, Saez I, Gruart A, Guinovart JJ, Delgado-Garcia JM (2013) Impairment in long-term memory formation and learning-dependent synaptic plasticity in mice lacking glycogen synthase in the brain. J Cereb Blood Flow Metab 33(4):550–556. https://doi.org/10.1038/jcbfm.2012.200

El Hayek L, Khalifeh M, Zibara V, Abi Assaad R, Emmanuel N, Karnib N, El-Ghandour R, Nasrallah P, Bilen M, Ibrahim P, Younes J, Abou Haidar E, Barmo N, Jabre V, Stephan JS, Sleiman SF (2019) Lactate mediates the effects of exercise on learning and memory through SIRT1-dependent activation of hippocampal brain-derived Neurotrophic factor (BDNF). J Neurosci 39(13):2369–2382. https://doi.org/10.1523/jneurosci.1661-18.2019

Fray AE, Forsyth RJ, Boutelle MG, Fillenz M (1996) The mechanisms controlling physiologically stimulated changes in rat brain glucose and lactate: a microdialysis study. J Physiol 496(1):49–57. https://doi.org/10.1113/jphysiol.1996.sp021664

Gandhi GK, Cruz NF, Ball KK, Dienel GA (2009a) Astrocytes are poised for lactate trafficking and release from activated brain and for supply of glucose to neurons. J Neurochem 111(2):522–536. https://doi.org/10.1111/j.1471-4159.2009.06333.x

Gandhi GK, Cruz NF, Ball KK, Theus SA, Dienel GA (2009b) Selective astrocytic gap junctional trafficking of molecules involved in the glycolytic pathway: impact on cellular brain imaging. J Neurochem 110(3):857–869. https://doi.org/10.1111/j.1471-4159.2009.06173.x

Gao V, Suzuki A, Magistretti PJ, Lengacher S, Pollonini G, Steinman MQ, Alberini CM (2016) Astrocytic beta2-adrenergic receptors mediate hippocampal long-term memory consolidation. Proc Natl Acad Sci U S A 113(30):8526–8531. https://doi.org/10.1073/pnas.1605063113

Gebril HM, Avula B, Wang YH, Khan IA, Jekabsons MB (2016) (13)C metabolic flux analysis in neurons utilizing a model that accounts for hexose phosphate recycling within the pentose phosphate pathway. Neurochem Int 93:26–39. https://doi.org/10.1016/j.neuint.2015.12.008

Gibbs ME (2016) Role of glycogenolysis in memory and learning: regulation by noradrenaline, serotonin and ATP. Front Integr Neurosci 9:70. https://doi.org/10.3389/fnint.2015.00070

Gibbs ME, Hertz L (2005) Importance of glutamate-generating metabolic pathways for memory consolidation in chicks. J Neurosci Res 81(2):293–300. https://doi.org/10.1002/jnr.20548

Gibbs ME, Lloyd HG, Santa T, Hertz L (2007) Glycogen is a preferred glutamate precursor during learning in 1-day-old chick: biochemical and behavioral evidence. J Neurosci Res 85(15):3326–3333. https://doi.org/10.1002/jnr.21307

Gilbert E, Tang JM, Ludvig N, Bergold PJ (2006) Elevated lactate suppresses neuronal firing in vivo and inhibits glucose metabolism in hippocampal slice cultures. Brain Res 1117(1):213–223. https://doi.org/10.1016/j.brainres.2006.07.107

Griffiths JR, Dwek RA, Radda GK (1976) Heterotropic interactions of ligands with phosphorylase b. Eur J Biochem 61(1):243–251

Grossbard L, Schimke RT (1966) Multiple hexokinases of rat tissues. Purification and comparison of soluble forms. J Biol Chem 241(15):3546–3560

Gundersen V, Storm-Mathisen J, Bergersen LH (2015) Neuroglial Transmission. Physiol Rev 95(3):695–726. https://doi.org/10.1152/physrev.00024.2014

Halestrap AP (1975) The mitochondrial pyruvate carrier. Kinetics and specificity for substrates and inhibitors. Biochem J 148(1):85–96

Halestrap AP, Denton RM (1975) The specificity and metabolic implications of the inhibition of pyruvate transport in isolated mitochondria and intact tissue preparations by alpha-Cyano-4-hydroxycinnamate and related compounds. Biochem J 148(1):97–106

Halestrap AP, Price NT (1999) The proton-linked monocarboxylate transporter (MCT) family: structure, function and regulation. Biochem J 343(Pt 2):281–299

van Hall G, Stromstad M, Rasmussen P, Jans O, Zaar M, Gam C, Quistorff B, Secher NH, Nielsen HB (2009) Blood lactate is an important energy source for the human brain. J Cereb Blood Flow Metab 29(6):1121–1129. https://doi.org/10.1038/jcbfm.2009.35

Hertz L, Chen Y (2017) Integration between glycolysis and glutamate-glutamine cycle flux may explain preferential glycolytic increase during brain activation, requiring glutamate. Front Integr Neurosci 11:18. https://doi.org/10.3389/fnint.2017.00018

Hertz L, Dienel GA (2005) Lactate transport and transporters: general principles and functional roles in brain cells. J Neurosci Res 79(1–2):11–18. https://doi.org/10.1002/jnr.20294

Hertz L, Gibbs ME (2009) What learning in day-old chickens can teach a neurochemist: focus on astrocyte metabolism. J Neurochem 109(Suppl 1):10–16. https://doi.org/10.1111/j.1471-4159.2009.05939.x

Hertz L, Rothman DL (2017) Glutamine-glutamate cycle flux is similar in cultured astrocytes and brain and both glutamate production and oxidation are mainly catalyzed by aspartate aminotransferase. Biology (Basel) 6(1):17. https://doi.org/10.3390/biology6010017

Hertz L, Peng L, Dienel GA (2007) Energy metabolism in astrocytes: high rate of oxidative metabolism and spatiotemporal dependence on glycolysis/glycogenolysis. J Cereb Blood Flow Metab 27(2):219–249. https://doi.org/10.1038/sj.jcbfm.9600343

Hertz L, Gerkau NJ, Xu J, Durry S, Song D, Rose CR, Peng L (2015a) Roles of astrocytic Na(+),K(+)-ATPase and glycogenolysis for K(+) homeostasis in mammalian brain. J Neurosci Res 93(7):1019–1030. https://doi.org/10.1002/jnr.23499

Hertz L, Xu J, Song D, Du T, Li B, Yan E, Peng L (2015b) Astrocytic glycogenolysis: mechanisms and functions. Metab Brain Dis 30(1):317–333. https://doi.org/10.1007/s11011-014-9536-1

Herzog RI, Chan O, Yu S, Dziura J, McNay EC, Sherwin RS (2008) Effect of acute and recurrent hypoglycemia on changes in brain glycogen concentration. Endocrinology 149(4):1499–1504. https://doi.org/10.1210/en.2007-1252

Hildebrand A, Lormes W, Emmert J, Liu Y, Lehmann M, Steinacker JM (2000) Lactate concentration in plasma and red blood cells during incremental exercise. Int J Sports Med 21(07):463–468. https://doi.org/10.1055/s-2000-7412

Hof PR, Pascale E, Magistretti PJ (1988) K^+ at concentrations reached in the extracellular space during neuronal activity promotes a Ca^{2+}-dependent glycogen hydrolysis in mouse cerebral cortex. J Neurosci 8(6):1922–1928

Holden JE, Mori K, Dienel GA, Cruz NF, Nelson T, Sokoloff L (1991) Modeling the dependence of hexose distribution volumes in brain on plasma glucose concentration: implications for estimation of the local 2-deoxyglucose lumped constant. J Cereb Blood Flow Metab 11(2):171–182. https://doi.org/10.1038/jcbfm.1991.50

Hudson JW, Golding GB, Crerar MM (1993) Evolution of allosteric control in glycogen phosphorylase. J Mol Biol 234(3):700–721. https://doi.org/10.1006/jmbi.1993.1621

Jackson JG, Robinson MB (2018) Regulation of mitochondrial dynamics in astrocytes: mechanisms, consequences, and unknowns. Glia 66(6):1213–1234. https://doi.org/10.1002/glia.23252

Jakobsen E, Bak LK, Walls AB, Reuschlein AK, Schousboe A, Waagepetersen HS (2017) Glycogen shunt activity and glycolytic supercompensation in astrocytes may be distinctly mediated via the muscle form of glycogen phosphorylase. Neurochem Res 42(9):2490–2494. https://doi.org/10.1007/s11064-017-2267-z

Jang S, Nelson Jessica C, Bend Eric G, Rodríguez-Laureano L, Tueros Felipe G, Cartagenova L, Underwood K, Jorgensen Erik M, Colón-Ramos Daniel A (2016) Glycolytic enzymes localize to synapses under energy stress to support synaptic function. Neuron 90(2):278–291. https://doi.org/10.1016/j.neuron.2016.03.011

Jekabsons MB, Gebril HM, Wang YH, Avula B, Khan IA (2017) Updates to a 13C metabolic flux analysis model for evaluating energy metabolism in cultured cerebellar granule neurons from neonatal rats. Neurochem Int 109:54–67. https://doi.org/10.1016/j.neuint.2017.03.020

Kong J, Shepel PN, Holden CP, Mackiewicz M, Pack AI, Geiger JD (2002) Brain glycogen decreases with increased periods of wakefulness: implications for homeostatic drive to sleep. J Neurosci 22(13):5581–5587

Lanz B, Gruetter R, Duarte JM (2013) Metabolic flux and compartmentation analysis in the brain in vivo. Front Endocrinol (Lausanne) 4:156. https://doi.org/10.3389/fendo.2013.00156

Lauritzen KH, Morland C, Puchades M, Holm-Hansen S, Hagelin EM, Lauritzen F, Attramadal H, Storm-Mathisen J, Gjedde A, Bergersen LH (2014) Lactate receptor sites link neurotransmission, neurovascular coupling, and brain energy metabolism. Cereb Cortex 24(10):2784–2795. https://doi.org/10.1093/cercor/bht136

Lavialle M, Aumann G, Anlauf E, Prols F, Arpin M, Derouiche A (2011) Structural plasticity of perisynaptic astrocyte processes involves ezrin and metabotropic glutamate receptors. Proc Natl Acad Sci U S A 108(31):12915–12919. https://doi.org/10.1073/pnas.1100957108

Lowry OH, Passonneau JV (1964) The relationships between substrates and enzymes of glycolysis in brain. J Biol Chem 239:31–42

Madsen PL, Linde R, Hasselbalch SG, Paulson OB, Lassen NA (1998) Activation-induced resetting of cerebral oxygen and glucose uptake in the rat. J Cereb Blood Flow Metab 18(7):742–748. https://doi.org/10.1097/00004647-199807000-00005

Madsen PL, Cruz NF, Sokoloff L, Dienel GA (1999) Cerebral oxygen/glucose ratio is low during sensory stimulation and rises above normal during recovery: excess glucose consumption during stimulation is not accounted for by lactate efflux from or accumulation in brain tissue. J Cereb Blood Flow Metab 19(4):393–400. https://doi.org/10.1097/00004647-199904000-00005

Magistretti PJ, Pellerin L (1996) Cellular mechanisms of brain energy metabolism. Relevance to functional brain imaging and to neurodegenerative disorders. Ann N Y Acad Sci 777:380–387

Mangia S, Simpson IA, Vannucci SJ, Carruthers A (2009) The in vivo neuron-to-astrocyte lactate shuttle in human brain: evidence from modeling of measured lactate levels during visual stimulation. J Neurochem 109(Suppl 1):55–62. https://doi.org/10.1111/j.1471-4159.2009.06003.x

Mangia S, DiNuzzo M, Giove F, Carruthers A, Simpson IA, Vannucci SJ (2011) Response to 'comment on recent modeling studies of astrocyte-neuron metabolic interactions': much ado about nothing. J Cereb Blood Flow Metab 31(6) 1346–1353. https://doi.org/10.1038/jcbfm.2011.29

Manning Fox JE, Meredith D, Halestrap AP (2000) Characterisation of human monocarboxylate transporter 4 substantiates its role in lactic acid efflux from skeletal muscle. J Physiol 529(Pt 2):285–293

Matsui T, Soya S, Okamoto M, Ichitani Y, Kawanaka K, Soya H (2011) Brain glycogen decreases during prolonged exercise. J Physiol 589(Pt 13):3383–3393. https://doi.org/10.1113/jphysiol.2010.203570

Matsui T, Ishikawa T, Ito H, Okamoto M, Inoue K, Lee MC, Fujikawa T, Ichitani Y, Kawanaka K, Soya H (2012) Brain glycogen supercompensation following exhaustive exercise. J Physiol 590(3):607–616. https://doi.org/10.1113/jphysiol.2011.217919

Matsui T, Omuro H, Liu YF, Soya M, Shima T, McEwen BS, Soya H (2017) Astrocytic glycogen-derived lactate fuels the brain during exhaustive exercise to maintain endurance capacity. Proc Natl Acad Sci U S A 114(24):6358–6363. https://doi.org/10.1073/pnas.1702739114

Mayer D, Seelmann-Eggebert G, Letsch I (1992) Glycogen phosphorylase isoenzymes from hepatoma 3924A and from a non-tumorigenic liver cell line. Comparison with the liver and brain enzymes. Biochem J 282(Pt 3):665–673

McKenna MC, Sonnewald U, Huang X, Stevenson J, Zielke HR (1996) Exogenous glutamate concentration regulates the metabolic fate of glutamate in astrocytes. J Neurochem 66(1):386–393

McKenna MC, Hopkins IB, Carey A (2001) Alpha-cyano-4-hydroxycinnamate decreases both glucose and lactate metabolism in neurons and astrocytes: implications for lactate as an energy substrate for neurons. J Neurosci Res 66(5):747–754

Morgan HE, Parmeggiani A (1964) Regulation of glycogenollysis in muscle. 3. Control of muscle glycogen phosphorylase activity. J Biol Chem 239:2440–2445

Moussawi K, Riegel A, Nair S, Kalivas PW (2011) Extracellular glutamate: functional compartments operate in different concentration ranges. Front Syst Neurosci 5:94. https://doi.org/10.3389/fnsys.2011.00094

Mozrzymas J, Szczęsny T, Rakus D (2011) The effect of glycogen phosphorolysis on basal glutaminergic transmission. Biochem Biophys Res Commun 404(2):652–655. https://doi.org/10.1016/j.bbrc.2010.12.033

Müller MS, Fox R, Schousboe A, Waagepetersen HS, Bak LK (2014) Astrocyte glycogenolysis is triggered by store-operated calcium entry and provides metabolic energy for cellular calcium homeostasis. Glia 62(4):526–534. https://doi.org/10.1002/glia.22623

Müller MS, Pedersen SE, Walls AB, Waagepetersen HS, Bak LK (2015) Isoform-selective regulation of glycogen phosphorylase by energy deprivation and phosphorylation in astrocytes. Glia 63(1):154–162. https://doi.org/10.1002/glia.22741

Nadeau OW, Fontes JD, Carlson GM (2018) The regulation of glycogenolysis in the brain. J Biol Chem 293(19):7087–7088. https://doi.org/10.1074/jbc.R117.803023

Nakao Y, Gotoh J, Kuang TY, Cohen DM, Pettigrew KD, Sokoloff L (1999) Cerebral blood flow responses to somatosensory stimulation are unaffected by scopolamine in unanesthetized rat. J Pharmacol Exp Ther 290(2):929–934

Nakao Y, Itoh Y, Kuang TY, Cook M, Jehle J, Sokoloff L (2001) Effects of anesthesia on functional activation of cerebral blood flow and metabolism. Proc Natl Acad Sci U S A 98(13):7593–7598. https://doi.org/10.1073/pnas.121179898

Nasrallah FA, Garner B, Ball GE, Rae C (2008) Modulation of brain metabolism by very low concentrations of the commonly used drug delivery vehicle dimethyl sulfoxide (DMSO). J Neurosci Res 86(1):208–214. https://doi.org/10.1002/jnr.21477

Nehlig A, Wittendorp-Rechenmann E, Lam CD (2004) Selective uptake of [^{14}C]2-deoxyglucose by neurons and astrocytes: high-resolution microautoradiographic imaging by cellular ^{14}C-trajectography combined with immunohistochemistry. J Cereb Blood Flow Metab 24(9):1004–1014. https://doi.org/10.1097/01.wcb.0000128533.84196.d8

Newman LA, Korol DL, Gold PE (2011) Lactate produced by glycogenolysis in astrocytes regulates memory processing. PLoS One 6(12):e28427. https://doi.org/10.1371/journal.pone.0028427

Newsholme EA, Rolleston FS, Taylor K (1968) Factors affecting the glucose 6-phosphate inhibition of hexokinase from cerebral cortex tissue of the guinea pig. Biochem J 106(1):193–201

Obel LF, Müller MS, Walls AB, Sickmann HM, Bak LK, Waagepetersen HS, Schousboe A (2012) Brain glycogen-new perspectives on its metabolic function and regulation at the subcellular level. Front Neuroenerg 4:3. https://doi.org/10.3389/fnene.2012.00003
Oe Y, Baba O, Ashida H, Nakamura KC, Hirase H (2016) Glycogen distribution in the microwave-fixed mouse brain reveals heterogeneous astrocytic patterns. Glia 64(9):1532–1545. https://doi.org/10.1002/glia.23020
Ottersen OP, Zhang N, Walberg F (1992) Metabolic compartmentation of glutamate and glutamine: morphological evidence obtained by quantitative immunocytochemistry in rat cerebellum. Neuroscience 46(3):519–534
Öz G, DiNuzzo M, Kumar A, Moheet A, Seaquist ER (2015) Revisiting glycogen content in the human brain. Neurochem Res 40(12):2473–2481. https://doi.org/10.1007/s11064-015-1664-4
Öz G, DiNuzzo M, Kumar A, Moheet A, Khowaja A, Kubisiak K, Eberly LE, Seaquist ER (2017) Cerebral glycogen in humans following acute and recurrent hypoglycemia: Implications on a role in hypoglycemia unawareness. J Cereb Blood Flow Metab 37(8):2883–2893. https://doi.org/10.1177/0271678x16678240
Parpura V, Basarsky TA, Liu F, Jeftinija K, Jeftinija S, Haydon PG (1994) Glutamate-mediated astrocyte-neuron signalling. Nature 369(6483):744–747
Pearson-Leary J, McNay EC (2016) Novel roles for the insulin-regulated glucose transporter-4 in hippocampally dependent memory. J Neurosci 36(47):11851–11864. https://doi.org/10.1523/jneurosci.1700-16.2016
Pearson-Leary J, Jahagirdar V, Sage J, McNay EC (2018) Insulin modulates hippocampally-mediated spatial working memory via glucose transporter-4. Behav Brain Res 338:32–39. https://doi.org/10.1016/j.bbr.2017.09.033
Pellerin L, Magistretti PJ (1994) Glutamate uptake into astrocytes stimulates aerobic glycolysis: a mechanism coupling neuronal activity to glucose utilization. Proc Natl Acad Sci U S A 91(22):10625–10629
Pfeiffer-Guglielmi B, Bröer S, Bröer A, Hamprecht B (2000) Isozyme pattern of glycogen phosphorylase in the rat nervous system and rat astroglia-rich primary cultures: electrophoretic and polymerase chain reaction studies. Neurochem Res 25(11):1485–1491
Pfeiffer-Guglielmi B, Fleckenstein B, Jung G, Hamprecht B (2003) Immunocytochemical localization of glycogen phosphorylase isozymes in rat nervous tissues by using isozyme-specific antibodies. J Neurochem 85(1):73–81
Prebil M, Vardjan N, Jensen J, Zorec R, Kreft M (2011) Dynamic monitoring of cytosolic glucose in single astrocytes. Glia 59(6):903–913. https://doi.org/10.1002/glia.21161
Preller A, Wilson JE (1992) Localization of the type III isozyme of hexokinase at the nuclear periphery. Arch Biochem Biophys 294(2):482–492
Price TB, Taylor R, Mason GF, Rothman DL, Shulman GI, Shulman RG (1994) Turnover of human muscle glycogen with low-intensity exercise. Med Sci Sports Exerc 26(8):983–991
Quistorff B, Secher NH, Van Lieshout JJ (2008) Lactate fuels the human brain during exercise. FASEB J 22(10):3443–3449. https://doi.org/10.1096/fj.08-106104
Rahman B, Kussmaul L, Hamprecht B, Dringen R (2000) Glycogen is mobilized during the disposal of peroxides by cultured astroglial cells from rat brain. Neurosci Lett 290(3):169–172
Reichenbach A, Derouiche A, Kirchhoff F (2010) Morphology and dynamics of perisynaptic glia. Brain Res Rev 63(1–2):11–25. https://doi.org/10.1016/j.brainresrev.2010.02.003
Reivich M, Kuhl D, Wolf A, Greenberg J, Phelps M, Ido T, Casella V, Fowler J, Hoffman E, Alavi A, Som P, Sokoloff L (1979) The [18F]fluorodeoxyglucose method for the measurement of local cerebral glucose utilization in man. Circ Res 44(1):127–137
Rothman DL, Dienel GA (2019) Development of a model to test whether glycogenolysis can support astrocytic energy demands of Na+,K+-ATPase and glutamate-glutamine cycling, sparing an equivalent amount of glucose for neurons. in M. DiNuzzo, A. Schousboe (eds), Brain Glycogen Metabolism, Advances in Neurobiology 23. https://doi.org/10.1007/978-3-030-27480-1_1
Saez I, Duran J, Sinadinos C, Beltran A, Yanes O, Tevy MF, Martínez-Pons C, Milán M, Guinovart JJ (2014) Neurons have an active glycogen metabolism that contributes to tolerance to hypoxia. J Cereb Blood Flow Metab 34(6):945–955. https://doi.org/10.1038/jcbfm.2014.33

Scanziani M, Salin PA, Vogt KE, Malenka RC, Nicoll RA (1997) Use-dependent increases in glutamate concentration activate presynaptic metabotropic glutamate receptors. Nature 385(6617):630–634. https://doi.org/10.1038/385630a0

Sebastian S, White JA, Wilson JE (1999) Characterization of the rat type III hexokinase gene promoter. A functional octamer 1 motif is critical for basal promoter activity. J Biol Chem 274(44):31700–31706

Shulman RG, Rothman DL (2001) The "glycogen shunt" in exercising muscle: a role for glycogen in muscle energetics and fatigue. Proc Natl Acad Sci U S A 98(2):457–461. https://doi.org/10.1073/pnas.98.2.457

Shulman RG, Hyder F, Rothman DL (2001) Cerebral energetics and the glycogen shunt: neurochemical basis of functional imaging. Proc Natl Acad Sci U S A 98(11):6417–6422. https://doi.org/10.1073/pnas.101129298

Sickmann HM, Walls AB, Schousboe A, Bouman SD, Waagepetersen HS (2009) Functional significance of brain glycogen in sustaining glutamatergic neurotransmission. J Neurochem 109(Suppl 1):80–86. https://doi.org/10.1111/j.1471-4159.2009.05915.x

Simpson IA, Carruthers A, Vannucci SJ (2007) Supply and demand in cerebral energy metabolism: the role of nutrient transporters. J Cereb Blood Flow Metab 27(11):1766–1791. https://doi.org/10.1038/sj.jcbfm.9600521

Sokoloff L, Reivich M, Kennedy C, Des Rosiers MH, Patlak CS, Pettigrew KD, Sakurada O, Shinohara M (1977) The [^{14}C]deoxyglucose method for the measurement of local cerebral glucose utilization: theory, procedure, and normal values in the conscious and anesthetized albino rat. J Neurochem 28(5):897–916

Steinman MQ, Gao V, Alberini CM (2016) The role of lactate-mediated metabolic coupling between astrocytes and neurons in long-term memory formation. Front Integr Neurosci 10:10. https://doi.org/10.3389/fnint.2016.00010

Storm-Mathisen J, Leknes AK, Bore AT, Vaaland JL, Edminson P, Haug FM, Ottersen OP (1983) First visualization of glutamate and GABA in neurones by immunocytochemistry. Nature 301(5900):517–520

Suda S, Shinohara M, Miyaoka M, Lucignani G, Kennedy C, Sokoloff L (1990) The lumped constant of the deoxyglucose method in hypoglycemia: effects of moderate hypoglycemia on local cerebral glucose utilization in the rat. J Cereb Blood Flow Metab 10(4):499–509. https://doi.org/10.1038/jcbfm.1990.92

Suh SW, Bergher JP, Anderson CM, Treadway JL, Fosgerau K, Swanson RA (2007) Astrocyte glycogen sustains neuronal activity during hypoglycemia: studies with the glycogen phosphorylase inhibitor CP-316,819 ([R-R∗,S∗]-5-chloro-N-[2-hydroxy-3-(methoxymethylamino)-3-oxo-1-(phenylmethyl)pro pyl]-1H-indole-2-carboxamide). J Pharmacol Exp Ther 321(1):45–50. https://doi.org/10.1124/jpet.106.115550

Suzuki A, Stern SA, Bozdagi O, Huntley GW, Walker RH, Magistretti PJ, Alberini CM (2011) Astrocyte-neuron lactate transport is required for long-term memory formation. Cell 144(5):810–823. https://doi.org/10.1016/j.cell.2011.02.018

Ueda T (2016) Vesicular glutamate uptake. Adv Neurobiol 13:173–221. https://doi.org/10.1007/978-3-319-45096-4_7

Veech RL, Harris RL, Veloso D, Veech EH (1973) Freeze-blowing: a new technique for the study of brain in vivo. J Neurochem 20(1):183–188

Vissing J, Andersen M, Diemer NH (1996) Exercise-induced changes in local cerebral glucose utilization in the rat. J Cereb Blood Flow Metab 16(4):729–736. https://doi.org/10.1097/00004647-199607000-00025

Waagepetersen HS, Sonnewald U, Larsson OM, Schousboe A (2000) A possible role of alanine for ammonia transfer between astrocytes and glutamatergic neurons. J Neurochem 75(2):471–479

Walls AB, Sickmann HM, Brown A, Bouman SD, Ransom B, Schousboe A, Waagepetersen HS (2008) Characterization of 1,4-dideoxy-1,4-imino-d-arabinitol (DAB) as an inhibitor of brain glycogen shunt activity. J Neurochem 105(4):1462–1470. https://doi.org/10.1111/j.1471-4159.2008.05250.x

Watanabe H, Passonneau JV (1973) Factors affecting the turnover of cerebral glycogen and limit dextrin in vivo. J Neurochem 20(6):1543–1554

Wilhelm F, Hirrlinger J (2011) The NAD+/NADH redox state in astrocytes: Independent control of the NAD+ and NADH content. J Neurosci Res 89(12):1956–1964. https://doi.org/10.1002/jnr.22638

Wilson JE (2003) Isozymes of mammalian hexokinase: structure, subcellular localization and metabolic function. J Exp Biol 206(Pt 12):2049–2057

Xu J, Song D, Xue Z, Gu L, Hertz L, Peng L (2013) Requirement of glycogenolysis for uptake of increased extracellular K+ in astrocytes: potential implications for K+ homeostasis and glycogen usage in brain. Neurochem Res 38(3):472–485. https://doi.org/10.1007/s11064-012-0938-3

Xu J, Song D, Bai Q, Zhou L, Cai L, Hertz L, Peng L (2014) Role of glycogenolysis in stimulation of ATP release from cultured mouse astrocytes by transmitters and high K+ concentrations. ASN Neuro 6(1):e00132. https://doi.org/10.1042/an20130040

Yang J, Ruchti E, Petit J-M, Jourdain P, Grenningloh G, Allaman I, Magistretti PJ (2014) Lactate promotes plasticity gene expression by potentiating NMDA signaling in neurons. Proc Natl Acad Sci U S A 111(33):12228–12233. https://doi.org/10.1073/pnas.1322912111

Yellen G (2018) Fueling thought: management of glycolysis and oxidative phosphorylation in neuronal metabolism. J Cell Biol 217(7):2235–2246. https://doi.org/10.1083/jcb.201803152

Yu Y, Herman P, Rothman DL, Agarwal D, Hyder F (2018) Evaluating the gray and white matter energy budgets of human brain function. J Cereb Blood Flow Metab 38(8):0339–1353. https://doi.org/10.1177/0271678x17708691

Zhang Y, Xue Y, Meng S, Luo Y, Liang J, Li J, Ai S, Sun C, Shen H, Zhu W, Wu P, Lu L, Shi J (2016) Inhibition of lactate transport erases drug memory and prevents drug relapse. Biol Psychiatry 79(11):928–929. https://doi.org/10.1016/j.biopsych.2015.07.007

Zimmer ER, Parent MJ, Souza DG, Leuzy A, Lecrux C, Kim HI, Gauthier S, Pellerin L, Hamel E, Rosa-Neto P (2017) [^{18}F]FDG PET signal is driven by astroglial glutamate transport. Nat Neurosci 20(3):393–395. https://doi.org/10.1038/nn.4492

Zorec R, Araque A, Carmignoto G, Haydon PG, Verkhratsky A, Parpura V (2012) Astroglial excitability and gliotransmission: an appraisal of Ca2+ as a signalling route. ASN Neuro 4(2):e00080. https://doi.org/10.1042/an20110061

State-Dependent Changes in Brain Glycogen Metabolism

Mauro DiNuzzo, Anne B. Walls, Gülin Öz, Elizabeth R. Seaquist, Helle S. Waagepetersen, Lasse K. Bak, Maiken Nedergaard, and Arne Schousboe

Abstract A fundamental understanding of glycogen structure, concentration, polydispersity and turnover is critical to qualify the role of glycogen in the brain. These molecular and metabolic features are under the control of neuronal activity through the interdependent action of neuromodulatory tone, ionic homeostasis and availability of metabolic substrates, all variables that concur to define the state of the system. In this chapter, we briefly describe how glycogen responds to selected behavioral, nutritional, environmental, hormonal, developmental and pathological conditions. We argue that interpreting glycogen metabolism through the lens of brain state is an effective approach to establish the relevance of energetics in connecting molecular and cellular neurophysiology to behavior.

Keywords Glycogen β-particle · Brain state · Astrocyte-neuron interactions · Ion homeostasis · Metabolic compartmentation

M. DiNuzzo (✉)
Center for Basic and Translational Neuroscience, Faculty of Health and Medical Sciences, University of Copenhagen, Copenhagen, Denmark
e-mail: mauro.dinuzzo@neuroenergetics.org

A. B. Walls · H. S. Waagepetersen · L. K. Bak · A. Schousboe
Department of Drug Design and Pharmacology, Faculty of Health and Medical Sciences, University of Copenhagen, Copenhagen, Denmark

G. Öz
Center for Magnetic Resonance Research, Department of Radiology, University of Minnesota, Minneapolis, MN, USA

E. R. Seaquist
Department of Medicine, University of Minnesota, Minneapolis, MN, USA

M. Nedergaard
Center for Basic and Translational Neuroscience, Faculty of Health and Medical Sciences, University of Copenhagen, Copenhagen, Denmark

Center for Translational Neuromedicine, University of Rochester Medical School, Rochester, NY, USA

M. DiNuzzo, A. Schousboe (eds.), *Brain Glycogen Metabolism*, Advances in Neurobiology 23, https://doi.org/10.1007/978-3-030-27480-1_9

Abbreviations

GABA	γ-Aminobutyric acid
GP	Glycogen phosphorylase
GS	Glycogen synthase
NAD(P)H	Nicotinamide adenine dinucleotide (phosphate)
NMDA	N-methyl-D-aspartate
NMR	Nuclear magnetic resonance
PAP	Peripheral astrocytic process
REM	Rapid eye movement
TCA	Tricarboxylic acid
VIP	Vasoactive intestinal peptide

1 Introduction

The brain incessantly undergoes spontaneous fluctuations in regional cortical activity, neuronal excitability, neuron-glia communications and metabolism that reflect excitatory and inhibitory inputs updating the brain about the internal and external environment as well as plasticity changes coordinating functional connectivity and information processing (Bergmann 2018; Lee and Dan 2012; Quilichini and Bernard 2012; Zagha and McCormick 2014). All these energy-consuming transactions feed a number of recurrent loops between cortical and subcortical structures whose ultimate goal is to make behavioral responses appropriate for adaptation and survival (DiNuzzo and Nedergaard 2017). In the present context, we focus on two main aspects characterizing brain state, namely neuronal excitability (i.e. the capability of responding to incoming stimuli with appropriate behavioral responses) and synaptic plasticity (i.e. the capability of modifying neuronal excitability according to the relevance and information content of incoming stimuli). The brain is an open system and as such it works by exchanging energy with its surroundings. In particular, the metabolism of a living cell, tissue or organism is not separable from its interactions with the environment, i.e. what is commonly referred to as function. Brain state alters tissue energy metabolism in terms of metabolic rates, pathways and substrates (Aalling et al. 2018). Most importantly, not only does cerebral metabolism adapt to specialized cellular energy demands, but metabolic products exert critical signaling functions that in turn can modulate brain activity (DiNuzzo 2016).

The first demonstrations that brain glycogen level parallels a number of altered behavioral states date back to the 1950s (Chance and Yaxley 1950; Kubo and Shimizu 1957; Svorad 1958a, b). Since then, many experimental investigations carried out in humans and other mammals (some of which we review in this chapter) have unequivocally shown that glycogen actively supports brain function in a state-dependent fashion. In the following, we first summarize the metabolic features of brain glycogen that are fundamental to understand its role as glucose reserve. We then enumerate the experimental findings supporting a close relationship between brain state and glycogen metabolism. Our aim is to illustrate the complexity of such relationship, although the literature is not complete under many viewpoints and our discussion is thus necessarily incomplete. Finally, we speculate about the potential relevance of glycogen to the basic neurophysiological mechanisms that concur to shape brain states.

2 Molecular and Metabolic Features of Brain Glycogen

2.1 Structure

Glycogen is a homogeneously branched polysaccharide that has evolved to store large amounts of readily-releaseable glucose residues while circumventing osmolarity issues (Melendez et al. 1997). The structural features of glycogen need to be taken into account in order to understand the biochemical properties underlying its synthesis and degradation patterns (Roach 2002). Briefly, glycogen biosynthesis is thought to start from an initial glucose chain derived from the self-glucosylating protein glycogenin (but see Testoni et al. 2017). An individual glucose chain grows linearly by polymerization through α-1,4-glycosydic linkages and branches out every 4–5 residues through α-1,6-glycosydic linkages. This gives rise to a mature glycogen granule (or β-particle) consisting of 12 concentric tiers of chains accommodating 13 glucose residues on average. Having a branching degree of 2, each successive tier contains twice the amount of glucose of the preceding ones. Cessation of biosynthesis of both the single chain and the whole molecule occurs due to steric factors hindering the catalytic activity of glycogen synthase (GS) and possibly glycogen branching enzyme (Melendez-Hevia et al. 1993; Melendez et al. 1998). Although particle size is thought to be a self-limiting factor, other yet unknown mechanisms might contribute to control the size of individual β-particles. Notably, many unphysiological conditions are associated with excessive elongation of glucosyl chains that might result in a bigger molecular radius of glycogen molecules not necessarily accompanied by higher glucose content. As an illustration, using computational modeling it has been shown that the loss of affinity of GS for inner chains yields glycogen particles with a substantially bigger size compared with "normal" mature particles with the same number of glucose units (DiNuzzo 2013).

Much like synthesis, degradation of glycogen by glycogen phosphorylase (GP) can only occur at the surface of the glycogen granule, i.e. enzymes do not have free access to the bulk of glucose residues. Moreover, GP can exclusively proceed from glucose chains longer than 4 residues, at which point maximally trimmed chains must be transferred to adjacent chains by glycogen debraching enzyme. Therefore, in principle only 34.6% (9/13 of 50%, i.e. last tier content) of the total glucose contained in the glycogen molecule can be mobilized before any debranching occurs. In fact, the effective concentration that GP can "see" at any moment is a complex function depending on the molecular location of the enzyme (DiNuzzo 2013). Glycogen structure is especially relevant to labeling experiments, because the dynamics of labeled glucose cannot be thought as a random process and standard kinetic modeling might provide incorrect results (Oz et al. 2015).

Another fundamental and somewhat overlooked consequence of glycogen structure is the accessibility to degradation. Aberrant hyperphosphorylated glycogen granules called polyglucosans as well as glycogen-based aggregates, such as Lafora bodies and corpora amylacea, are synthesized under both physiological (e.g., aging) and pathological (e.g., epilepsy) conditions (DePaoli-Roach et al. 2015; Roach 2011; Sinadinos et al. 2014; Tagliabracci et al. 2011). Polyglucosans are characterized by excessive chain length and inhomogeneous branching, which makes these compounds water-insoluble and thus unable to undergo enzymatic degradation. There is evidence that insolubility of glycogen is independent of incorporation of C2 and C3 phosphate monoesters by GS and instead largely depends on dysfunctional chain-length regulation (Sullivan et al. 2017, 2019). Overall, the levels of glycogen in the brain are not the sole variable to consider, because the underlying glycogenolytic capacity critically depends on whether the polysaccharide is accessible to degradation. Unfortunately, at present there is considerable lack of experimental tools for the measurements of glycogen structure (e.g., homogeneity and branching). In principle, excessive chain length can be revealed by measuring the glucose released during alternating treatments with glycogen phosphorylase (yielding a limit dextrin) and glycogen debranching enzyme. In addition, comparison of the total glucose released by the latter procedure with that released by lysosomal acid α-glucosidase could possibly reveal the fraction of glycogen resistant to hydrolysis (DiNuzzo et al. 2015b). The presence of phosphate esters can be measured from digested glycogen using inorganic phosphate assay after, for example, recombinant laforin treatment (Tagliabracci et al. 2007). Specific phosphate esters can be identified through matrix-assisted laser desorption ionization time-of-flight mass spectrometry eventually combined with nuclear magnetic resonance (NMR) spectroscopy (Tagliabracci et al. 2011).

2.2 *Concentration*

Glycogen concentration is commonly determined by ex vivo amyloglucosidase-based assay. Glycogen levels from rodent brain are reported to be between 3 and 12 μmol/g (Dienel and Cruz 2015). Such a wide range reflects the exquisite sensitivity of glycogen upon experimental parameters. Animal protocols suffer from several important experimental issues, such as high lability of glycogen and its rapid post-mortem degradation (depending on tissue fixation, e.g., liquid nitrogen versus focused-beam microwave irradiation), stress-induced glycogenolysis due to animal handling before euthanasia, nutritional history (e.g., overnight fasting prior to experiments) and time of the day (predominant sleep or wake periods), among others (Cruz and Dienel 2002; Dienel 2019b). The conditions under which glycogen concentrations are measured in human brain by in vivo ^{13}C magnetic resonance (^{13}C-NMR) spectroscopy are also stressful and require continuous labeled glucose infusion protocols and long acquisition periods (>8 h) (Oz et al. 2007). Values obtained by ^{13}C-NMR spectroscopy may be biased by partial volume effects, for example if the spectroscopic voxel contains several structures known to have different glycogen content. Using ^{13}C-NMR spectroscopy, an average value of nearly 8 μmol/g for glycogen level has been recently reported in the occipital lobe of healthy human subjects (Oz et al. 2015).

Under physiological conditions, glycogen and glycogen-metabolizing enzymes are almost exclusively confined to astrocytes, although neurons also store very low amounts of glycogen (about 10- to 30-fold less than in astrocytes) (Saez et al. 2014). Assuming a volume fraction of astrocytes in cortical gray matter to be in the range 10–15% across species [mouse cerebral cortex <11% (Braitenberg and Schüz 1998); rat hippocampus 8 ± 4% (Mishchenko et al. 2010); cat motor cortex 11.4% (Barron et al. 1988); cat frontal cortex 15.5% (Williams et al 1980); rabbit visual cortex 12.7 ± 2.2% (Schmolke and Schleicher 1989); human parietal cortex 11.4% (Virgintino et al. 1997)] and a basal glycogen level in the gray matter of mammalian brain of 6–10 μmol/g (Cruz and Dienel 2002; Oz et al. 2015), the glycogen concentration in terms of glucose equivalents in astrocytes would be 40–100 mmol/L or at least >30-fold of the free glucose concentration at euglycemia (~1.2 mmol/L). Even though the "effective" glycogen concentration available to enzyme action is actually lower, the amount of glucose that can be released by glycogenolysis is very large, especially considering the rapidity of glycogen breakdown (Obel et al. 2012b). Most of brain glycogen resides in perineuronal peripheral astrocytic processes (PAPs) and in perivascular endfeet, while a minor fraction is found in cell soma and primary astrocytic processes (Oe et al. 2016). In particular, PAPs include a huge number of filopodia and lamellipodia having very small intracellular volume, high cytosolic tortuosity and restricted diffusion (Hertz et al. 2007). Thus, glycogen might be particularly important in cellular regions with high surface-to-volume ratio, a feature that has been suggested to identify a role for glycogenolysis in the control of ion homeostasis and glucose channeling (DiNuzzo 2019).

A poorly understood phenomenon of glycogen metabolism is glycogen supercompensation, a rebound of glycogen synthesis above basal levels occurring whenever glycogen is degraded. Glycogen supercompensation occurs in brain as well as in other tissues (see for example Matsui et al. 2012) in response to both physiological (e.g., physical exercise) or pathological (e.g., hypoglycemia) energy challenges. In the brain, given the high degree of polydispersity (see below) and the relative abundance of immature glycogen particles, it is difficult to predict whether supercompensation takes place in the same particles that underwent partial degradation or also involves recruitment of new granules. Furthermore, it is presently unknown whether glycogen supercompensation is epiphenomenic or functionally meaningful.

2.3 Polydispersity

In the brain, glycogen is almost exclusively found in the form of β-particles. These are quasi-spherical molecules having a theoretical maximal diameter of ~46 nm, corresponding to a molecular weight of ~10^7 Da and a total number of glucose residues of about 53,000 (13 residues per chain times 2^{12} chains, where 2 is the branching degree and 12 the total number of tiers) (DiNuzzo 2013). Glycogen synthesis is thought to occur in discrete steps with associated enzymology, leading first to the formation of proglycogen (i.e. a ~400 kDa polysaccharide, acid-insoluble fraction) and then macroglycogen (i.e. the mature glycogen molecule, acid-soluble fraction) (Lomako et al. 1991, 1993). However, the existence of well-defined glycogen species is presently questioned and there is strong evidence of continuous polydispersity in vivo (Katz 2006; Roach 2002). Glycogen β-particle size is best seen using three-dimensional serial block electron microscopy. Using this method, recent glycogen particle counting from mouse primary motor cortex showed a very broad distribution of molecular size spanning the range 8–58 nm (Bellesi et al. 2018), similar to that reported from human vastus lateralis muscle biopsies (10–44 nm) (Marchand et al. 2002). The distribution of glycogen size observed in the brain was somewhat bimodal, with most of the particles having a diameter of >22 nm and a small but separable peak around 16–19 nm. Notably, this size corresponds to a molecular weight of 200–400 kDa, but whether it indeed represents a distinct glycogen pool cannot be presently established. The study also found an average granule diameter of 27 nm in cortical astrocytes, with few particles exceeding 40 nm (Bellesi et al. 2018), essentially the same value reported in human skeletal muscle fibers (mean diameter 26 nm) (Marchand et al. 2002) or rabbit retina Müller gial cells (mean diameter 23 nm) (Magalhaes and Coimbra 1970). It should be realized that these studies suffer from rapid post-mortem (or post-extraction) glycogenolysis, which most likely shifted the glycogen size distribution towards smaller values. Further ultrastructural electron microscopy experiments employing focused-beam microwave irradiation are required.

In addition to ultrastructural studies employing electron microscopy, estimates of glycogen polydispersity in brain tissue can be obtained through immunohistochemistry, although more knowledge is needed about the affinity of specific antibodies for glycogen particles of different size (Hirase et al. 2019; Oe et al. 2016). Size distribution of purified glycogen molecules in solution can be measured by dynamic light scattering (Zhang et al. 2001), refractive index-based size exclusion chromatography (Deng et al. 2016; Izumi et al. 2013; Sullivan et al. 2014) and time-resolved acid hydrolysis (Deng et al. 2016).

2.4 Turnover

Glycogen turnover is the simultaneous insertion and release of glucose residues in the β-particle, which can lead to steady-state (i.e. no change in glycogen levels) when the underlying reaction rates are identical, and otherwise to net synthesis or net degradation. While in principle glycogen turnover only applies at the level of individual glycogen granule, the situation in vivo is complicated by the fact that different granules can be found in different states of synthesis and degradation. The distribution of glycogen granule size suggests that the fraction of mature β-particles is relatively small. The high degree of polydispersity is consistent with the notions that (1) GP operates very rapidly by targeting the small pool of mature β-particles (i.e. those exhibiting the highest glycogenolytic capacity), and (2) GS is considerably slower, yet it operates on the large pool of incomplete β-particles. The activity of GP or GS can be measured by a spectrophotometric method using the enzyme purified from the brain. This approach employs coupled enzyme systems to measure the optical properties associated with the activity-dependent production of a specific substrate, commonly NAD(P)H. However, this approach only measures enzyme capacity (e.g., sensitive to enzyme phosphorylation state), and it does not take into account other in vivo conditions that critically depend on neuronal activity (e.g., level of allosteric effectors such as AMP).

As mentioned above, glycogen β-particles exhibit highly non-homogeneous turnover. Indeed, synthesis and degradation occur simultaneously at the molecular surface of glycogen. Thus, the "effective" concentration of glycogen changes with the size of the molecule. It is, however, possible to define a molecular turnover that is independent of both the size of individual particles and the total number of particles. By optimizing the fit between measured and predicted time-course of glycogen labeling experiments, it has been estimated that human brain glycogen turns over with a rate of 40 residues per minute under basal conditions (Oz et al. 2015). In order to obtain the turnover rate in the brain, the molecular turnover value needs to be multiplied by the tissue glycogen content, which gives a turnover of 6 nmol/(g min) for a concentration of 8 μmol/g. This number represents the activity of GS and GP averaged over long time periods and over many glycogen granules. Therefore, turnover per se is not informative on the frequency and rate of glycogenolytic bursts.

The concept of glycogen shunt takes into account the temporal separation between glycogen synthesis and degradation (Shulman et al. 2001a, 2001b; Shulman and Rothman 2001, 2015). As an illustration, let us consider a glycogen turnover of 6 nmol/(g min) or 0.1 pmol/(g s). When examined at a temporal resolution of 1 ms, this turnover can be explained by either a continuous process involving 0.1 fmol/(g ms) or a series of discrete bursts, for example lasting 1 ms and occurring at a frequency of 10 Hz, each involving 10 fmol/(g ms), i.e. a rate 100-fold higher than the continuous case. Thus, a rapid burst of net glycogen breakdown for glycolytic ATP is followed by a longer period of net glycogen resynthesis (in the above example having a duration of 99 ms). Overall, such process does 'shunt' glucose through glycogen, which is important as glycogen can meet large and rapid increases in glucose demand that could not be supplied by glucose transport and phosphorylation. These considerations are critical to understand the role of glycogen in the brain, but at present there are no experimental tools to measure glycogen shunt in astrocytes, i.e. glycogenolysis at high spatiotemporal resolution

Glycogen turnover is commonly determined in vivo or in vitro using labeling experiments, e.g. radioassays or NMR spectroscopy employing ^{14}C- or ^{13}C-labeled glucose, respectively. Compared to similar labeling techniques, ^{13}C-NMR spectroscopy has two important advantages. First, it does not require the delicate and time-consuming process of post-mortem tissue extraction/fixation and glycogen isolation. Second, it allows to follow the full temporal dynamics of label incorporation, retention and washout in the same animal/subject in vivo. On the other side, it also has disadvantages: it requires high-tech equipment, whole body labeling with [^{13}C]glucose is expensive, sensitivity may be an issue, and time frames for analysis of changes in glycogen level or labeling differ from other methods. Other promising (though only qualitative) approaches that have been used in other tissues include real-time optical methods, such as 2-NBDG pre-loading (Louzao et al. 2008), although in the cerebral cortex and hippocampus this glucose analogue predominantly labels neurons in situ (unpublished observation).

3 State-Related Factors Affecting Synthesis and Degradation of Brain Glycogen

3.1 Behavioral Factors

3.1.1 Rest or Quiet Waking

Uniquely defining basal activity for the "restless" brain (Raichle 2011) is a difficult, if not impossible, task. Many experimental studies assume that different sub-states of wakefulness can be studied in relative terms, i.e. by examining changes between them. Although this assumption is not inherently incorrect, it is de facto impractical because wakefulness depends on a large number of variables and the actual baseline state, which is only incompletely defined, may significantly alter the functional and metabolic response to a given stimulus (Scholvinck et al. 2010; Shulman et al. 2009, 2007). In the present section, we discuss the relevance of some of these variables in the context of brain glycogen metabolism. In experiments involving humans, rest is defined as a state in which the subjects commonly lie supine, voluntarily abstain from (or substantially reduce) movements and maintain a low vigilance level (i.e. mind-wandering). This state is substantially different from its animal counterpart, where rest does not necessarily involve inactive immobility, and when it does it is normally associated with high levels of restraint-associated stress and/or discomfort. These considerations indicate that the concept of "quiet waking" (reduced locomotion and absence of sensory/cognitive stimulation) is perhaps more appropriate in order to attempt generalizing findings obtained in humans and animals.

Glycogen concentration in resting healthy human subjects is about 6–8 μmol/g (Oz et al. 2015), which is close to the mid-range value of glycogen levels reported in rat brain (4–12 μmol/g) (Cruz and Dienel 2002; Garriga and Cusso 1992; Kong et al. 2002; Swanson et al. 1989a, 1990) and mouse brain (3–14 μmol/g) (Nelson et al. 1968; Oe et al. 2016; Watanabe and Passonneau 1973). The corresponding turnover times are nearly 22 h in humans (Oz et al. 2015, 2007), 7–10 h in rats (Choi and Gruetter 2003) and 4–5 h in mice (Watanabe and Passonneau 1973) (although some reports vary substantially, see for example Prokhorova and Tupikova 1958). These numbers are consistent with turnover rates of 0.35 μmol/(g h) in humans (Oz et al. 2015, 2007), 0.5 in rats (Choi et al. 1999) and 0.6–0.9 in mice (Watanabe and Passonneau 1973), which roughly correspond to 1–2% of cerebral metabolic rates of glucose in these species (Dienel 2019a).

3.1.2 Stimulation or Active Waking

Similar arguments of those mentioned above can be adduced to use the concept of "active waking" instead of "stimulation". Indeed, the latter can involve one or more sensory modalities with different impact on stimulus perception and/or cognitive stimulus processing, which might or might not result in specific behavioral responses (e.g., increased locomotion). Moreover, stimulation consists of different stages (alerting, onset, progression, recovery) that are normally associated with changes in neuromodulatory tone inducing corresponding changes in brain metabolism (Dienel and Cruz 2016). Locomotion per se is not correlated with changes in brain glycogen content (Hutchins and Rogers 1970).

Ex vivo ^{14}C autoradiography demostrated a ~20% label loss upon 10–20 min (but not 2–5 min) of whisker stimulation in rat somatosensory cortex (Swanson et al. 1992) (Fig. 1). Similarly, 5–6 min or 30 s of generalized sensory stimulation in rats decreased the amount of ^{14}C from prelabeled glycogen by 33% or 16%, respectively (Cruz and Dienel 2002). Notably, 5–6 min (but not 30 s) of stimulation were required to observe a decrease (about 22%) in total glycogen concentration, suggesting that the glycogenolytic response to stimulation involves only the metabolically active (i.e. where labeled residues first disappear) surface of the molecule. Moreover, glycogen turnover is thought to increase much more than net degradation during stimulation, which implies concomitant resynthesis (Dienel et al. 2007; Dienel and Cruz 2015). These notions are somewhat contradictory with the finding that no change of ^{13}C-glycogen was found upon 20 min of visual stimulation in the occipital cortex of human subjects when previously-labeled glycogen was allowed to turn over in the presence of unlabeled glucose (DiNuzzo 2013; Oz et al. 2007). It should also be noted that the human studies measure signal from a relatively large volume-of-interest and may have been confounded by partial volume effects. Hence, glycogenolysis may have not been detectable within the sensitivity of the technique due to predominant label loss from the outier tiers of glycogen prior to stimulation in a fraction of the volume. Interestingly, inhibition of glycogenolysis in rats had no effect on resting glucose consumption, but resulted in a regional compensatory rise in glucose consumption during generalized sensory stimulation (about 25–50% above the respective value in stimulated vehicle-treated controls) (Dienel et al. 2007). Overall, glycogen is mobilized upon stimulation with a glycogenolytic rate approaching half of cerebral glucose utilization, or a 25- to 50-fold above the turnover during unstimulated conditions (for an in-depth discussion, see Dienel 2019a). This notion is in agreement with the very large enzymatic capacity of both GS and GP relative to their basal reaction rates. For example, in rat brain maximal activities

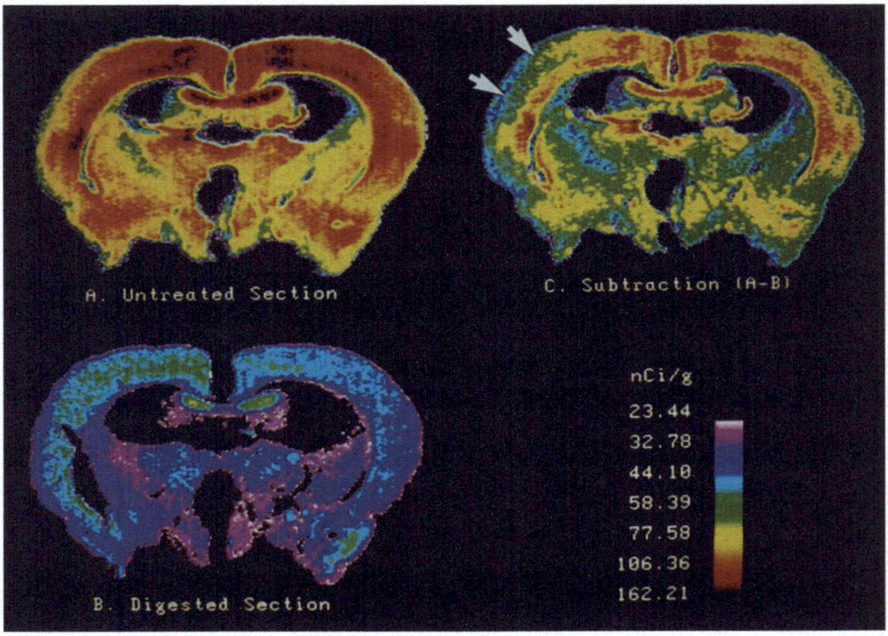

Fig. 1 Glycogen utilization in rat brain during unilateral whisker stimulation. Representative coronal sections including somatosensory cortex showing ^{14}C-labeled glycogen before (**a**) and after (**b**) amyloglucosidase-based glycogen digestion. The difference image (**c**) shows side-to-side ^{14}C activity changes in the barrel field region, indicating stimulation-induced glycogenolysis. Adapted from (Swanson et al. 1992) (with permission)

of GP and GS have been reported to be as high as 300 μmol/(g h) and 26 μmol/(g h), respectively (see Nelson et al. 1968 and references therein). There is some evidence that glycogen supercompensation might occur in the brain after sensory stimulation (Abdul-Ghani and Nahas 1990).

3.1.3 Anesthesia

The effect of anesthesia is generally characterized by a dramatic increase of brain glycogen levels, which is almost independent of body temperature and anesthesia-induced hyperglycemia (Nelson et al. 1968). Changes in brain glycogen metabolism associated with anesthesia are secondary to depression of neuronal activity (i.e. glutamergic and monoaminergic neurotransmission), as anesthetics are without effect when applied directly in astrocytic cultures (Swanson et al. 1989b). Since anesthesia affects energy metabolism much more in gray matter than white matter (McKenna et al. 2012), it is expected that anesthesia-induced changes in glycogen content are correspondingly more elevated in gray matter regions. Accordingly, prolonged (8 h) anesthesia has been found to increase glycogen levels in all cortical layers (200% in layer I and 30–40% in layers II–VI) but not in subjacent white matter of mouse brain (Folbergrova et al. 1970).

Anesthesia-induced rise in brain glycogen content (~20%) has been reported in mice during ether vapor exposure (1 h at 4.5%) (Estler and Heim 1960). Halothane (2 h at 1.5–2.5%) has been observed to increase glycogen concentration by ~40% in rat cerebral cortex and cerebellum with bigger changes in adult compared with young animals (Gip et al. 2002). Similar values have been reported in mouse brain for halothane (50% or 150% increase at 2 or 4 h, respectively) and other volatile anesthetics (up to 200%) (Brunner et al. 1971). Isoflurane (105 min at 2%) has been reported to increase brain glycogen content by 20% in rat brain (Morgenthaler et al. 2006). Pentobarbitone (40 mg/Kg) anesthesia has been found to increase glycogen content in rat cerebral cortex, cerebellum, brainstem and spinal cord (range 50–120%) (Nahas and Abdul-Ghani 1989). Similarly, rats given pentobarbitone (50 mg/Kg) had a 35% increase in whole-brain glycogen content (assayed at 15–25 min) (Dalsgaard et al. 2007). Phenobarbitone (150 mg/Kg) has been reported to increase glycogen concentration in mouse cerebral cortex, hippocampus, cerebellum and brainstem (range 15–50% assayed at 60 min) (Estler 1961; Gatfield et al. 1966). Similarly, mice anesthetized with phenobarbitone (150 mg/Kg) had whole-brain glycogen increasing progressively with time (60–250% during 24 h) and glycogen turnover decreasing to one-half of control animals (Watanabe and Passonneau 1973) (Fig. 2). The effect of phenobarbitone on mouse brain glycogen (assayed at 8 h) has been shown to be dose-dependent (60%, 190% and 220% for 75, 170 and 240 mg/Kg) (Nelson et al. 1968). In another study, phenobarbitone (250 mg/Kg) failed to alter glycogen in rat brain (8% increase was not significant; assayed at 75 min), although it immediately halved glycogen turnover (Strang and Bachelard 1973). An ultrastructural study of the effect of phenobarbitone (150 mg/Kg) on mouse brain glycogen demonstrated substantial accumulation of the polysaccharide in hippocampus and frontal cortex, in large part due to the formation of aggregates of β-particles often included in autophagic vacuoles (Phelps 1972) (Fig. 3). Tranquillizers and other substances depressing noradrenergic signaling have also been associated with increased brain glycogen levels (Hutchins and Rogers 1970; Nelson et al. 1968; Svorad 1959). Overall, the above-mentioned results (often affected by post-mortem glycogen degradation to variable extent) demostrate that anesthesia produces a substantial rise of glycogen content and a decline of glycogen turnover in the brain.

3.1.4 Sleep

Replenishment of brain glycogen stores consumed during waking is the key mechanism underlying the so-called energy hypothesis of sleep (Benington and Heller 1995; Scharf et al. 2008). Initial observations for diurnal changes of brain glycogen content were obtained in rats, whereby lower values were observed during the predominantly-wake compared with the predominantly-sleep period of the day (Hutchins and Rogers 1970; Pronaszko-Kurczynska 1976; Svorad 1958b) (Fig. 4). Thereafter, rapid (i.e. minutes) and substantial (i.e. 70% above basal levels) glycogen accumulation upon natural slow-wave sleep onset and likewise rapid return to

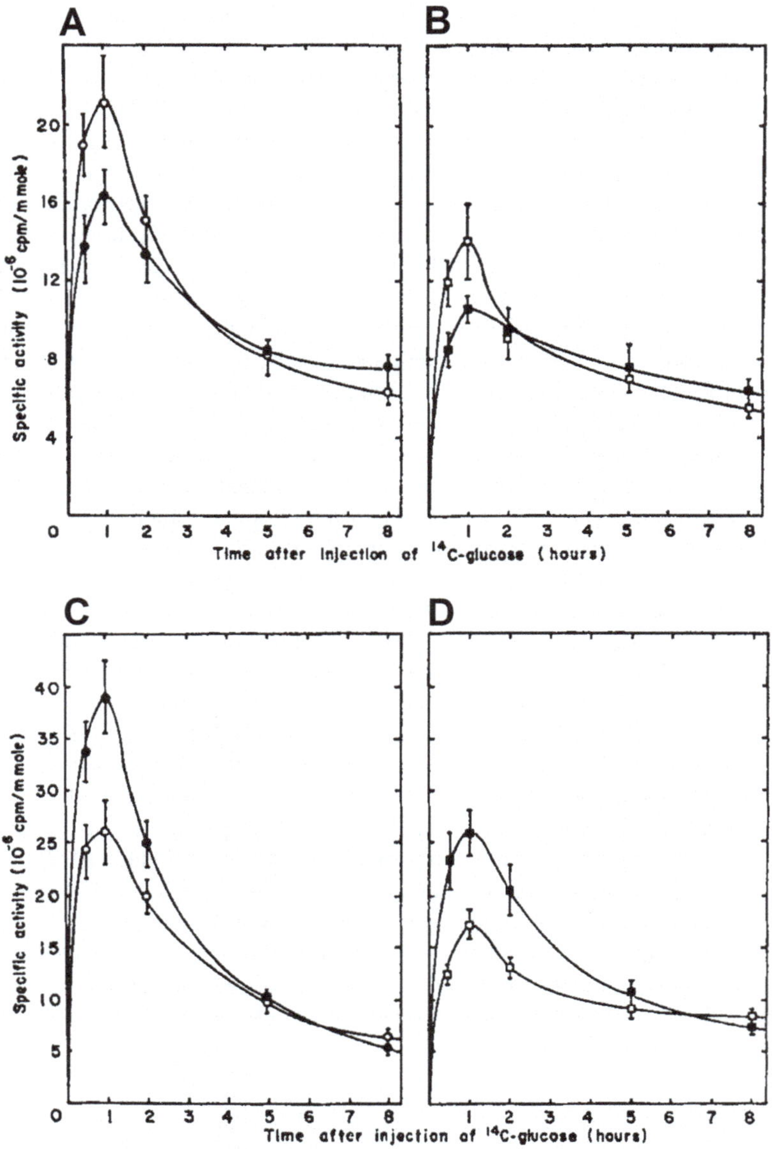

Fig. 2 Incorporation of labeled glucose into mouse brain glycogen. (**a**, **b**) [U-^{14}C]glucose glucose (10 μCi/20 g) was given 5 h after either saline (control, open symbols) or phenobarbital (150 mg/Kg) anesthesia (filled symbols). Label incorporation during phenobarbital anesthesia is showed for glycogen (**a**) and limit dextrin (**b**). (C,D) Labeled glucose was given 5 h after either saline (control, open symbols) or hydrocortisone (25 mg/Kg) anesthesia (filled symbols). Label incorporation during hydrocortisone treatment is showed for glycogen (**c**) and limit dextrin (**d**). The corresponding glycogen turnover decreased in anesthetized animals and increased in hydrocortisone-treated animals, yet both drugs brought about a substantial increase in total glycogen concentration. Adapted from (Watanabe and Passonneau 1973) (with permission)

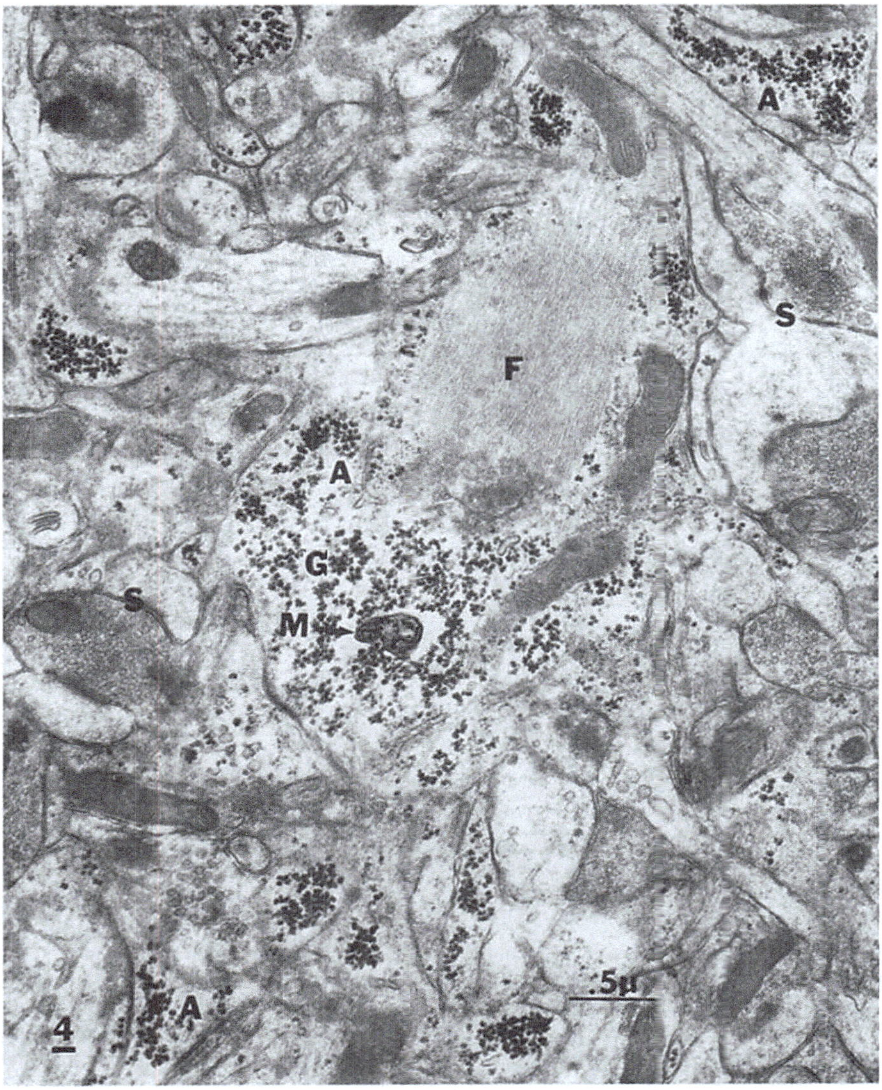

Fig. 3 Glycogen granules in the dentage gyrus of mouse hippocampus after anesthesia. Six hours after phenobarbital (250 mg/Kg) injection glycogen granules (G) were diffusely present in the cytoplasm of peripheral astrocytic processes (A), the latter interspersed among axo-dendritic synapses (S). Several glycogen granules were contained in a membrane-bound organelle (M) Large astrocytic processes containing gliofilaments (F) are also visible in this electromicrograph. High density of glycogen β-particles is a hallmark of the anesthetized brain and is not observed under wakefulness, natural sleep or sleep deprivation (compare with Fig. 5c). Scale bar 0.5 µm. Adapted from (Phelps 1972) (with permission)

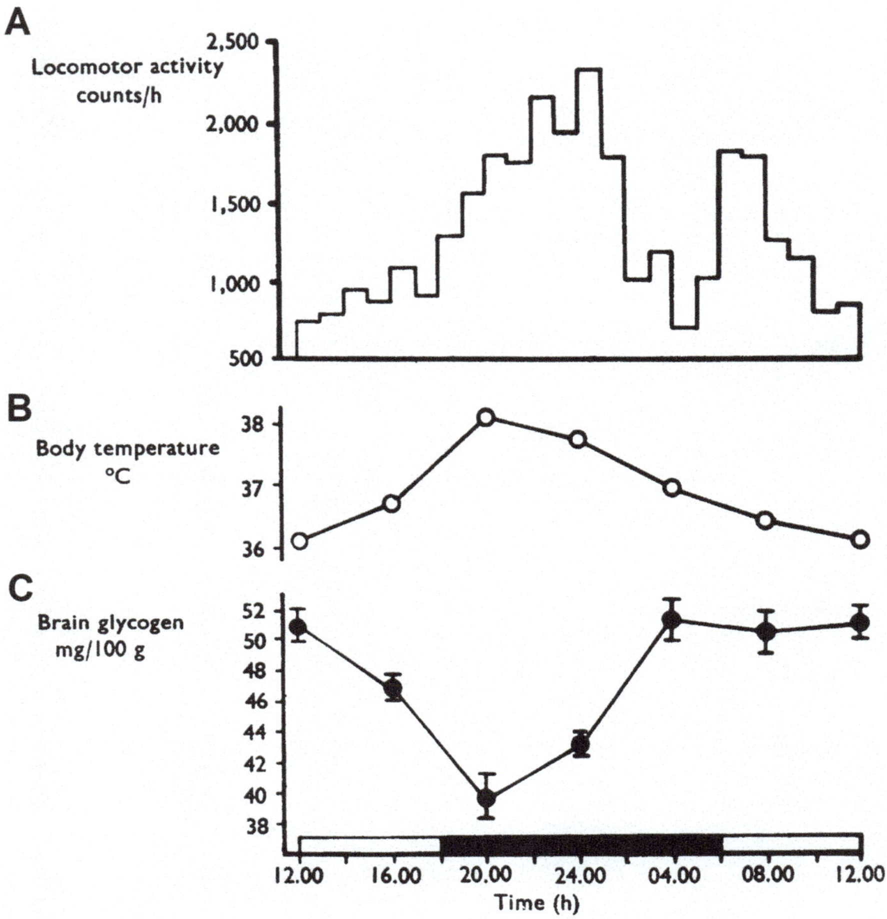

Fig. 4 Typical circadian changes in locomotor activity, body temperature and brain glycogen content in mice. (**a**) Locomotion increased progressively during the dark period (predominantly awake), but animals moved substantially even during the light period (predominantly sleep). Therefore, there was no clear correlation between locomotor activity and either body temperature (**b**) or brain glycogen (**c**). In particular, changes in brain glycogen content (mean ± SEM) appeared to anticipate the light-dark transitions, although the statistical power of the study was limited (3 trials). Adapted from (Hutchins and Rogers 1970) (with permission)

baseline upon awakening were reported in rats (Karnovsky et al. 1983). Notably, these changes in cerebral glycogen content parallel sleep- and wake-related alterations in brain extracellular glucose and lactate concentrations, consistent with reduced glucose utilization and lactate production taking place during sleep (reviewed in DiNuzzo and Nedergaard 2017). Moreover, the rapidity of changes of brain glycogen in following sleep-wake cycle is at odds with recovery of brain glycogen content after sleep deprivation (2 h) or anesthesia (3–9 h) (Choi and Gruetter 2003; Franken et al. 2006; Karnovsky et al. 1983; Watanabe and Passonneau 1973), which evidences the profound differences between these brain states. The role of glycogen resynthesis in sleep is presently debated (Petit et al. 2015). However, a recent electron microscopy study investigating glycogen granules in mouse cerebral cortex during wakefulness and natural sleep strongly supports the notion that brain glycogen is closely linked to the sleep-wake cycle (Bellesi et al. 2018). Specifically, sleep has been associated with reduction in granule number as well as concomitant increase in granule size and distance from the synaptic cleft, resulting in an overall increase of cortical glycogen levels by ~6% during sleep compared with wakefulness (Bellesi et al. 2018) (Fig. 5). As stated above, this study is limited by post-mortem degradation (Lowry et al. 1964) and the sleep-related change in brain glycogen content might be inaccurate. Nonetheless, the sleep-related alteration of glycogen localization serves to illustrate the notion that concentration changes do not exhaust the possible modifications of brain glycogen metabolism taking place under different states. For example, rearrangement of glycogen granules in different subcellular domains of astrocytes might alter the functional relevance of glycogen synthesis and degradation under different brain states. In addition, it has been suggested that sleep might serve to restore glycogen structure (eventually channeling aberrant mocules to lysosomal degradation or lymphatic drainage) rather than to replenish glycogen levels in the brain (DiNuzzo et al. 2015b).

Glycemic levels are unlikely to play a role in the changes of brain glycogen content during the sleep-wake cycle, because the diurnal variations of blood glucose are opposite to those of brain glycogen (Ahlersova et al. 1980). The pattern of brain glycogen oscillations is also reversed compared to what happens in muscle, whereby glycogen is increased during (sedentary) wakefulness and decreased during sleep (although there is also a purely circadian component; see Takahashi et al. 2015 and references therein).

3.1.5 Sleep Deprivation

Sleep deprivation is sometimes assumed to be an extension of waking. However, such assumption is wrong, if anything because external interventions (i.e. stressors) are required to maintain wakefulness (McEwen 2006). A study on Drosophila Melanogaster demonstrated that sleep deprivation is associated with elevated stress levels, which recapitulated the observed whole-body metabolic changes even in the absence of sleep loss (Harbison and Sehgal 2009). The difficulty in separating the effect of stress from sleep loss is evidenced by the finding that glucocorticoids strongly influence brain glycogen levels during sleep deprivation protocols in rats (Gip et al. 2004). Accordingly, gentle sleep deprivation elicits a nearly twofold increase in plasma corticosterone levels (Petit et al. 2010). Although some animal studies have incorporated control stress groups (e.g., see Karadzic and Mrsulja 1969a), it is still questionable whether such controls are appropriately established (Karnovsky et al. 1983).

The complex effects of stressors as well as other variables on brain glycogen metabolism can explain the high degree of discrepancy between different studies (Petit et al. 2015). It has been reported that REM (or paradoxical) sleep depriva-

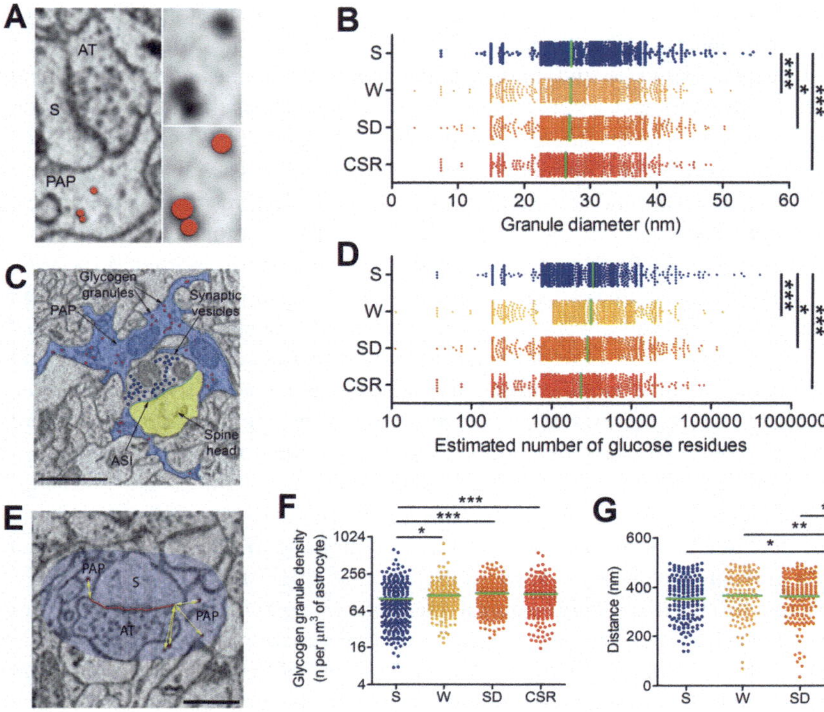

Fig. 5 Brain glycogen granule distribution and subcellular localization in the cerebral cortex of mouse under different states. (**a**, **c**, **e**) Glycogen segmentation (red circles) from a representative electron micrograph of mouse primary motor cortex (layers II/III) containing a peripheral astrocytic process (PAP) surrounding a synapse (dendritic spine, S; axon terminal, AT). Segmentation of cellular elements for exclusion of synaptic vesicles (blue circles) and identification of axon-spine interface (ASI). (**b**) Distribution of granule diameter under sleep (S), wake (W), sleep deprivation (SD, 8 h) or chronic sleep restriction (CSR, 4.5 days). (**d**) Number of glucose units in cortical glycogen estimated from granule diameter and corresponding density of glycogen granules (**f**) and distance from ASI (**g**). Regardless the limitations of the study (see text) it is apparent that brain state affects several features of glycogen, namely granule density and localization, in addition to granule size and turnover. Adapted from (Bellesi et al. 2018). Copyright © 2018 Bellesi, de Vivo, Koebe, Tononi and Cirelli

tion reduces brain glycogen levels in rats and cats (Karadzic and Mrsulja 1969b). Specifically, decreases in glycogen content in hippocampus and subcortical structures (but not in the cerebral cortex) were observed during REM sleep deprivation (Karadzic and Mrsulja 1969a; Mrsulja et al. 1967). Similarly, total sleep deprivation resulted in a decrease of glycogen content in rat hippocampus (9%) and cerebellum (23–24%) but not in cortex at 6 h, while cortical glycogen actually increased after 12 h of sleep deprivation (Gip et al. 2002, 2004). The same outcome was observed in mouse cerebral cortex, with a monotonic glycogen increase during sleep deprivation up to 6 h (Franken et al. 2006). Quite opposite

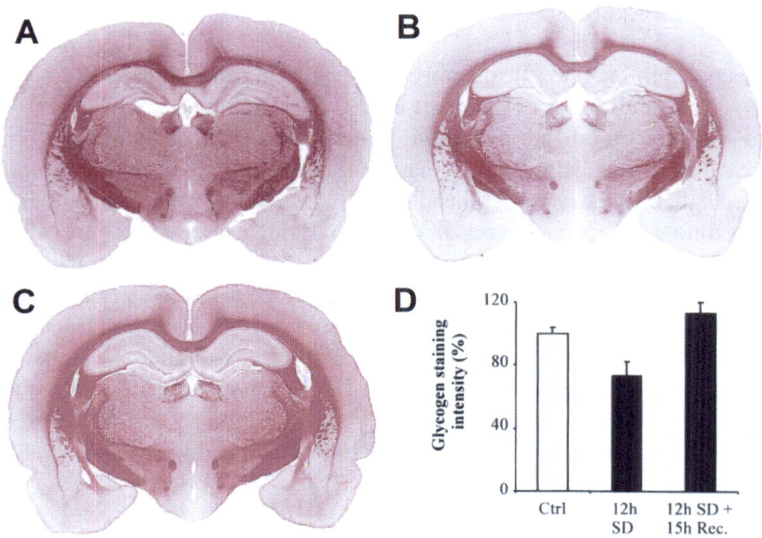

Fig. 6 Effect of sleep deprivation and recovery sleep on rat brain glycogen. Glycogen was histochemically determined in control animals (**a**) as well as in rats sleep deprived for 12 h (**b**) and after 15 h of recovery sleep (**c**). (**d**) Brain glycogen content (mean ± SEM) decreased during sleep deprivation and returned to basal levels after recovery sleep. Adapted from (Kong et al. 2002). Copyright © 2002 Society for Neuroscience

changes were reported in rat brain (excluding cerebellum and brainstem), where glycogen content decreased after 12–24 h but not 6 h of sleep deprivation (Kong et al. 2002) (Fig. 6). No or inconsistent changes in glycogen levels at the end of 6 h of sleep deprivation or sleep fragmentation have been observed in the cerebral cortex of several mouse strains (Baud et al. 2016; Franken et al. 2003; Petit et al. 2010). A recent electron microscopy study revealed a reduction of glycogen in layers II/III of mouse frontal cortex after either 7–8 h of forced wake or 4.5 days of sleep restriction (glycogen decreased by 15% or 29%, respectively) (Bellesi et al. 2018) (Fig. 5), suggesting that changes in cortical glycogen levels might have a regional and/or laminar localization.

Notably, sleep deprivation induces a substantial rise in protein targeting to glycogen, which results in increased activity of both GS and GP (Petit et al. 2002, 2010) and thus affects brain glycogen turnover but not necessarily glycogen levels (Morgenthaler et al. 2009). Therefore, concentration alone could be insufficient to reveal changes in brain glycogen metabolism during sleep deprivation (Gip et al. 2002). The fact that either GP inhibition or sleep loss are associated with inefficient reuptake of extracellular K^+ and increased susceptibility to cortical spreading depression strongly supports the notion that sleep deprivation affects brain glycogen metabolism regardless of any effect on its tissue concentration (Kilic et al. 2018).

3.1.6 Physical Exercise

Glucose consumption and glycogenolysis are both increased during exercise in the brain, liver and muscle (see Matsui et al. 2011 and references therein). Exercise-induced glycogen mobilization in multiple areas of rat brain is mainly due to monoaminergic signaling, while only marginally affected by hypoglycemia after prolonged exhaustive exercise (Matsui et al. 2015, 2011). In particular, noradrenergic and serotonergic tone increases in the brain during exercise and the ensuing rise in glycogenolysis is hypothesized to be implicated in central fatigue (Dalsgaard and Secher 2007; Matsui et al. 2011; Nybo and Secher 2004). Recovery from exercise is associated with supercompensated glycogen levels in both brain and skeletal muscle (6–24 and 24–48 h, respectively), with rates of resynthesis correlated with rates of previous degradation (Matsui et al. 2012). Interestingly, in cultured astrocytes acutely exposed to norepinephrine glycogen is initially degraded and subsequently (observed at 9 h) resynthesized to highly supercompensated levels (Sorg and Magistretti 1991, 1992).

3.2 Nutritional Factors

3.2.1 Glycemia

Under physiological conditions blood glucose concentration always exceeds cerebral demand (Mergenthaler et al. 2013) and mild variations in systemic glucose concentration have normally minor effects on brain glycogen content (Hutchins and Rogers 1970). During severe (e.g., cerebral glucose approaching Km of hexokinase, or 0.05 mmol/L) and prolonged hypoglycemia brain glycogen stores are almost depleted (see Hutchins and Rogers 1970 and references therein). Brain glycogen is also degraded during acute hypoglycemia (Canada et al. 2011; Choi et al. 2003; Herzog et al. 2008; Oz et al. 2009; Suh et al. 2007). This condition leads to initial glycogen degradation followed by its supercompensation (Canada et al. 2011) (but see Herzog et al. 2008). Glycogen supercompensation was proposed as part of a counter-regulatory response to delay energy depletion during subsequent hypoglycemic or hypoxic events. In the context of recurrent hypoglycemia, e.g., in diabetic patients under insulin therapy, such neuroprotective effect would also carry the risk of failure in detecting systemic hypoglycemia, possibly resulting in hypoglycemia unawareness (Choi et al. 2003; Oz et al. 2009). However, ^{13}C-NMR spectroscopy in healthy humans revealed a clear glycogen supercompensatory response following acute but not recurrent hypoglycemic episodes, which suggest that glycogen supercompensation is not involved in hypoglycemia unawareness (Oz et al. 2017). On the other side, prolonged hyperglycemia results in increased brain glycogen content (Nelson et al. 1968; Watanabe and Passonneau 1973), whereas acute (up to 120 min) hyperglycemia has been reported to produce either a fall (Hutchins and Rogers 1970) or a rise (Goldberg and O'Toole 1969) in cerebral glycogen. Glycogen content in cultured astrocytes strongly covaries with ambient glucose (Cummins et al. 1983; Swanson et al. 1989b). Yet, the concentration changes of brain glycogen in vivo are possibly quite small (i.e. in the order of 10-15%) during physiological fluctuations of systemic glucose levels.

3.2.2 Diet

Fasting has been found to reduce brain glycogen (e.g., 25% after 24 h in rats) (Prasannan et al. 1963). The effect of starvation upon brain glycogen is transient (<72 h in rats) and probably dependent on the concomitant decrease of brain glucose, as glycogen turnover appears to be unchanged (Garriga and Cusso 1992; Visweswaran et al. 1969). Normalization of brain glycogen levels during prolonged starvation probably involves gluconeogenesis from amino acids. Preconditioning to gluconeogenesis has been indeed suggested to explain resistance to hypoxia and high rates of glycogenesis occurring in body tissues of rats on high fat diet (Purshottam et al. 1978). Interestingly, ketogenic diet is known to reduce susceptibility to epileptic seizures (Lutas and Yellen 2013) and it has been shown to improve cognitive performances in rats (Murray et al. 2016) and preserve memory in aging mice (Newman et al. 2017) (but see Zhao et al. 2004 for possible detrimental effects of ketogenic diet during development), all features also associated with brain glycogen metabolism. Unfortunately, glycogen levels in the brain during ketogenic diet have not been investigated. However, there are indirect findings supporting a role for β-hydroxybutyrate in stimulating glycogen accumulation, conceivably because of inhibition of glycolysis induced by metabolism of ketones (Matsui et al. 2017; Thurston et al. 1980). Conversely, carbohydrate intake does not affect brain glycogen content, while it significantly correlates with glycogen levels in both muscle and liver (Soya et al. 2018).

3.3 Environmental factors

3.3.1 Temperature

Brain glycogen normally changes in opposite fashion compared to the circadian variation of body temperature (Hutchins and Rogers 1970) (although REM sleep is associated with a small increase in brain temperature; see Wehr 1992) (see Fig. 4). The attenuation of brain activity accompanying hibernation or hypothermia is likewise associated with increases in cerebral glycogen content (Al-Badry and Taha 1983; Koizumi 1974; Lust et al. 1989). This finding could simply reflect the decrease in cerebral blood flow and metabolism during systemic as well as brain-selective hypothermia (Okubo et al. 2001; Walter et al. 2000). Yet, exposure of awake mice to either low (8–10 °C) or high (30–32 °C) ambient temperature significantly decreased brain glycogen content by 15–20% after 4–8 h (Hutchins and Rogers 1970). A caveat is that glycogenolysis may have been stimulated by stress and/or discomfort of the animals under both conditions (Cruz and Dienel 2002; Zhao et al. 2017). Accordingly, it is possible that systemic effects are responsible for ambient temperature-induced changes in brain glycogen concentration and not brain temperature per se. In comparison, muscle glycogen metabolism is largely dependent on environmental temperature. For example, during hypothermia or acute exposure to cold, muscle glycogen is readily depleted (Fuhrman and Fuhrman 1966) and high ambient temperature (32–35 °C) is detrimental to glycogen synthesis, especially during recovery from exercise (Blackwood et al. 2019; Naperalsky et al. 2010). It is noted that fluctuations in brain temperature do normally occur even in constant ambient temperature and are primarily due to intracerebral heat production (Kiyatkin 2010).

3.3.2 Altitude

The brain relies on an uninterrupted supply of oxygen from the bloodstream. Hypobaric and hyperbaric conditions are associated with hypoxemia and hyperoxemia, respectively. Both acute and chronic exposure of rats to hypobaric hypoxia (i.e. high altitude; approximate range 3000–8000 m) leads to a reduction in brain glycogen concentration (Chapler and York 1972; Harik et al. 1995; Purshottam et al. 1977; Woolley and Timiras 1963) followed by supercompensation (Brucklacher et al. 2002; Canada et al. 2011; Cui et al. 2001; Lopez-Ramos et al. 2015), while normobaric or hyperbaric hyperoxia steadily augments brain glycogen deposition (Bronovitskaya and Shapovalova 1957; Dydyk et al. 1978).

3.3.3 Hydrostatic Pressure

For humans and non-acquatic mammals hydrostatic pressure only becomes relevant at depths of more than 100 m or rates >100 m/min, because mechanical compression on nervous system can lead to overstimulated and/or uncoordinated nerve conduction. Specifically, hyperbaric pressure-induced hyperexcitability is thought to depend on NMDA receptor activation, possibly through reduction of Mg^{2+} blockade efficacy (Mor and Grossman 2007, 2010). Thus, any pressure-induced change in brain glycogen is secondary to neuronal hyperexcitation and nonetheless difficult to disentangle from the oxygen toxicity and nitrogen narcosis associated with such extreme conditions.

3.3.4 Ionizing Radiation

Exposure of the head of mice, rats and monkeys to electromagnetic radiation or alpha-particles (helium-4 nuclei) has been repeatedly reported to result in accumulation of glycogen within the whole cerebral cortex or close to the exposed brain area, respectively (Ibrahim et al. 1970; Klatzo et al. 1961; Miquel and Haymaker

1965; Miquel et al. 1963; Wolfe et al. 1962). In rat brain, the X-ray or γ-ray radiation-induced rise in brain glycogen content was found to increase monotonically with the exposure (in the range 0.3–5 C/Kg) and glycogen elevation persisted for up to 96 h after irradiation (Miquel et al. 1966). The effect of ionizing radiation on brain glycogen has been suggested to be a consequence of damage to astrocytic mitochondrial oxidative metabolism (Miquel and Haymaker 1965), increased glucose uptake and phosphorylation (Ibrahim et al. 1970) or selective impairment of glycogen degrading enzymes (Lundgren and Miquel 1970).

3.4 *Hormonal Factors*

3.4.1 Sex

Sex-related differences in brain glycogen metabolism have been poorly investigated. A single report about diurnal changes of glycogen content in the pineal gland of mouse brain showed no difference between males and females during the estrous cycle (Kachi et al. 1973), although this might be region-dependent. It is interesting to notice that in Drosophila Melanogaster sleep patterns are genetically correlated with glycogen metabolism in males and with triglyceride metabolism in females, perhaps reflecting sex-specific metabolic needs (Harbison and Sehgal 2008). Since gonadal steroids affect brain and body bioenergetics (including glucose uptake, glycolysis, tricarboxylic acid cycle, mitochondrial oxidative phosphorylation and oxidative stress; reviewed in (Rettberg et al. 2014), it is likely that brain glycogen metabolism is sexually dimorphic (Tamrakar et al. 2014). Notably, susceptibility to epileptic seizures, which is affected by brain glycogen metabolism (Lopez-Ramos et al. 2015), also appear to differ in males and females (Giorgi et al. 2014; Veliskova and Desantis 2013).

3.4.2 Stress

Exposure to stressors causes the activation of central stress response through rapid (i.e. seconds) release of norepinephrine to multiple brain areas followed by delayed (i.e. minutes) release of glucocorticoids from adrenal cortex to the bloodstream (reviewed in Pearson-Leary et al. 2016). Both acute and chronic stress has been associated to decrements in brain glycogen content (Hashiguchi et al. 1998; Zhao et al. 2017). Yet, the effect of systemic administration of adrenal corticosteroids, which readily cross the blood-brain barrier, was earlier reported to increase the deposition of glycogen in the brain as in other tissues (Abood and Kocsis 1950; Gordan 1956; Gordan et al. 1951; Kosmina 1967; Prasannan et al. 1964; Timiras et al. 1956) (see Fig. 2), possibly via gluconeogenesis from amino acids (Herbet et al. 2019; Woodbury 1972). In vivo and in vitro studies reported increased, decreased or unchanged brain glycogen levels after glucocorticoids administration

(Ibrahim 1975; Tombaugh et al. 1992; Zhang et al. 2015). The exact mechanisms by which chronic stress affects glycogen synthesis and degradation in the brain need to be further investigated.

3.5 *Developmental Factors*

3.5.1 Early Development

Aerobic glycolysis is a crucial metabolic pathway during development, whereby it provides substrate for biosynthetic reactions that are critical for growth and structural maintenance of brain tissue (Bauernfeind and Babbitt 2014; Caravas and Wildman 2014). The involvement of glycogen in neurogenesis/synaptogenesis and early brain development and maturation has not been thoroughly investigated. However, a strong developmental increase of hippocampal and cerebellar glycogen content (peaking at P34 and slowly declining thereafter, i.e. after the switch from cerebral utilization of ketone bodies to carbohydrates) has been reported in rat brain (Gip et al. 2002). A similar decrease of glycogen content in cerebral cortex, cerebellum and diencephalon has been reported in mice after P16 (Schreiber 1981). The glycogenolytic response to norepinephrine is accordingly absent in mouse primary astrocytic cultures at early developmental age (10–14 days) but progressively present thereafter (>18 days) (O'Dowd et al. 1995).

3.5.2 Aging

In both neurons and astrocytes, aging is associated with important reorganizations of major energy metabolic pathways, including glycolytic, glycogenolytic and mitochondrial oxidative metabolism (Drulis-Fajdasz et al. 2018; Ivanisevic et al. 2016). In particular, a global age-related decrease in energy metabolism has been reported in human brain, with glucose uptake decreasing more than oxygen utilization (Goyal et al. 2017) (but see Dienel 2019a for a discussion of the limitations of this study). At least part of the ensuing reduction in aerobic glycolysis can be explained by a decrease in glycogen metabolism due to cerebral accumulation of unmetabolizable corpora amylacea and/or polyglucosans (Duran and Guinovart 2015; Mrak et al. 1997). Interestingly, inhibition of glycogenolysis has been reported to disrupt long term potentiation in hippocampal slices from young mice and enhance it in aged animals (Drulis-Fajdasz et al. 2015), possibly after a shift in the localization of glycogen-metabolizing enzymes from astrocytes to neurons (Drulis-Fajdasz et al. 2018). Appearance of neuronal glycogen seems to be indeed associated with physiological aging (Sinadinos et al. 2014), but the consequences of glycogen synthesis in neurons are controversial (Duran et al. 2012; Saez et al. 2014).

3.6 *Pathological Factors*

3.6.1 Glycogen Storage Diseases

The inherited deficiency or mutation of specific enzymes underlying the metabolic pathways for glycogen synthesis and degradation (or regulation thereof) lead to severe pathological phenotypes involving skeletal muscle, liver, hearth, kidney, brain and spinal cord (Ellingwood and Cheng 2018; Hicks et al. 2011; Ozen 2007). Although the principal clinical outcomes are related to hepatic and muscular function, the relevance of disorders of glycogen metabolism to the central nervous system is apparent in Type II glycogen storage diseases. These diseases are associated with accumulation of glycogen in all body tissues due to the loss-of-function of the lysosomal enzyme α-glucosidase (Pompe disease) or lysosome-associated membrane protein (Danon disease). Depending on the extent of glycogen accumulation in cerebral cortex, Type II glycogen storage diseases can be accompanied by mental retardation, learning difficulties and intellectual disability (Boucek et al. 2011; Kashio et al. 1991; Spiridigliozzi et al. 2017; Teng et al. 2004), although how exactly the impaired lysosomal disposal of glycogen granules impact on brain function is unknown.

3.6.2 Dementia

Recent work in laboratory animals demonstrated that brain glycogenolysis is necessary for memory formation and retention (Duran et al. 2013; Gibbs et al. 2006; Newman et al. 2011; Suzuki et al. 2011; Zhang et al. 2016). Cerebral metabolic changes during sleep and sleep sub-stages are also associated with synaptic plasticity and neuronal firing rate homeostasis (reviewed in DiNuzzo and Nedergaard 2017), possibly subserving memory consolidation (e.g. corticalization of hippocampus-dependent memory) (Almeida-Filho et al. 2018; Puentes-Mestril et al. 2019). Given the involvement of brain glycogen in learning and memory as well as in the effects of sleep and sleep loss, it is plausible that glycogen metabolism plays a role in memory impairment during dementia. Remarkably, Alzheimer's disease-related, age-related as well as vascular dementia are all associated with an increased cerebral accumulation of corpora amylacea, glycoproteinaceous waste aggregates primarily containing glucose polymers (e.g., polyglucosans) (Augé et al. 2018; Hoyaux et al. 2000; Mann et al. 1987; Pisa et al. 2013, 2016; Rohn 2015; Snow et al. 1988). Although the precise role of corpora amylacea in the brain is debated (Augé et al. 2017), their abundance during memory impairment and/or hyperexcitability has been hypothesized to induce dysfunctional glycogen metabolism (not necessarily associated with changes in brain glycogen levels) (DiNuzzo et al. 2015b). Notably, hyperexcitability characterizes the early stage of Alzheimer's disease progression (i.e. mild cognitive impairment) and it has been even suggested to be a primary mechanism contributing to cognitive decline (Palop and Mucke

2009). When one excludes the accumulation of glycogen in the form of corpora amylacea (or similar aggregates), the concentration and/or turnover of brain glycogen in dementia can be hypothesized to be reduced due to the upregulation of glycogen synthase kinase 3β and ensuing inhibition of GS (see Bak et al. 2018; Bass et al. 2015 and references therein).

3.6.3 Lafora Disease and Epilepsy

Lafora disease is the sole pathological condition that unquestionably relates aberrant brain glycogen structure and metabolism with epilepsy and neurodegeneration (although Lafora disease is also associated with alterations in glucose transport, glutamate uptake, autophagy, oxidative stress and inflammatory reactions) (Duran and Guinovart 2015; García-Gimeno et al. 2018; Gentry et al. 2018; Nitschke et al. 2018; Rubio-Villena et al. 2018). Accordingly, glycogen is the key target for treatment, as knockout of GS or depletion of protein targeting to glycogen in animal models of Lafora disease rescued several features of the disease, such as formation of Lafora bodies, neurodegeneration and susceptibility to epileptic seizures (Duran et al. 2014; Pederson et al. 2013; Turnbull et al. 2011, 2014). However, the absence of glycogen synthesis in these animal models is expected to carry important side-effects, including memory impairment and hyperexcitability (Duran et al. 2013). Indeed, glycogen is critical for neurotransmitter and ion homeostasis, thus any disturbance in glycogen synthesis and degradation is expected to affect neuronal excitability and synaptic plasticity (see below). Interestingly, animal models of Lafora disease exhibit episodic memory deficits increasing with age (Criado et al. 2012; Garcia-Cabrero et al. 2012), consistent with the observation that in Lafora disease patients the appearance of epileptic seizures coincides with a progressive decline in intellectual functions (Cukiert et al. 1990; Singh and Ganesh 2012). Aberrant glycogen structure has been suggested to be implicated in epileptogenesis due to impairment of astrocytic glycogenolysis and corresponding deficits in K^+ reuptake and glutamate metabolism (DiNuzzo et al. 2014, 2011).

3.6.4 Traumatic Brain Injury

The effect of traumatic injury on brain glycogen metabolism resembles that of hypoglycemia or hypoxia-ischemia. In particular, in animal models of brain trauma glycogen content decreased and turnover increased within minutes in the involved regions (Watanabe and Passonneau 1974), while a decrease in glycogen turnover and glycogen supercompensation occurred at 24 h after injury (Otori et al. 2004; Watanabe and Passonneau 1974). The trauma-induced supercompensated glycogen levels increase tolerance to ischemia and thus are likely neuroprotective (Otori et al. 2004). Interestingly, sex hormones influence the recovery from injury (Bramlett and Dietrich 2001), although the involvement of sex-specific differences in brain glycogen metabolism during traumatic injury has not been investigated.

4 Relevance of Glycogen to Neuronal Excitability and Synaptic Plasticity

4.1 Ion and Transmitter Homeostasis

Glycogen in astrocytes actively contributes to modulate the electrochemical activity of neurons. A critical function of glial astrocytes is the control of neuroactive compounds in the extracellular space (Verkhratsky and Nedergaard 2018) and glycogen is implicated in regulating the levels of both K^+ and glutamate. Specifically, not only does glycogen-derived energy support K^+ homeostasis (Choi et al. 2012) but active uptake of excess extracellular K^+ after neuronal stimulation is impaired when glycogenolysis is inhibited, even in the presence of glucose (Kilic et al. 2018; Xu et al. 2013). In addition, glycogen-derived pyruvate is a preferred precursor of glutamate through astrocytic pyruvate carboxylation (see below) and glutamate can be an anaplerotic substrate of glycogen through astrocytic gluconeogenesis (Hevor et al. 1986). The roles of glycogen in active K^+ reuptake and glutamate/GABA cycling have direct relevance to neuronal excitability, as confirmed by the increased suceptibility to epileptic seizures (Lopez-Ramos et al. 2015) and to cortical spreading depression (Kilic et al. 2018) upon inhibition of glycogenolysis. Glycogen is also exquisitely sensitive to neuromodulators (reviewed in DiNuzzo et al. 2015a), which is important because the glycogenolytic response to these substances can dissociate glycogen metabolism from underlying neuronal activity (i.e. extracellular K^+ and/or glutamate). For instance, glycogenolysis might represent a novel pathway by which monoaminergic signaling regulates Na^+/K^+-ATPase activity (DiNuzzo 2019) and the ensuing interstitial K^+ concentration in a state-dependent manner (Ding et al. 2016). Rapid glycogen breakdown following monoaminergic signaling has been also suggested to have profound repercussions on the channeling of glucose to neurons, which might be critical for supporting neuronal activity and associated energy demand (Dienel and Cruz 2015; DiNuzzo et al. 2015a).

While studying the role of glycogen in relation to memory formation in a chick learning model (Gibbs et al. 2006, 2007; Hertz et al. 2003) it was found that glycogen may function as an important substrate for de novo synthesis of glutamine and subsequently glutamate serving as a neurotransmitter. This involves degradation of glycogen to lactate/pyruvate via the concerted action of glycogenolysis and glycolysis, as explained in detail elsewhere (Bak et al. 2018), followed by pyruvate carboxylation and synthesis of α-ketoglutarate in the tricarboxylic acid (TCA) cycle. The α-ketoglutarate will act as substrate for aspartate aminotransferase forming glutamate which is subsequently converted to glutamine. Since glycogen is preferentially if not exclusively located in astrocytes (Cataldo and Broadwell 1986; Ibrahim 1975) and the two enzymes pyruvate carboxylase and glutamine synthetase are likewise exclusively expressed in astrocytes (Norenberg and Martinez-Hernandez 1979; Yu et al. 1983), the cellular location of the synthetic machinery required for glycogen acting as a precursor for glutamine is astrocytic. Moreover, glutamine synthesized in astrocytes will act as precursor for neuronal synthesis of the two amino acid neurotransmitters glutamate and GABA since glutamine transporters located in the plasma membranes of astrocytes and neurons will facilitate transfer of glutamine from astrocytes to neurons (see Leke and Schousboe 2016). The finding that an obese rat model exhibiting hampered glycogen metabolism in combination with a lowered cerebral glycogen content had reduced glutamate-glutamine cycle activity (Sickmann et al. 2010) clearly supports the functional importance of astrocytic glycogen for optimal glutamatergic activity. This is in keeping with the notion that filling of the vesicles in glutamatergic neurons is dependent on the cytosolic glutamate concentration which, in turn, depends on an adequate supply of glutamine from surrounding astrocytes. Moreover, it has been demonstrated that inhibition of glycogen-shunt activity results in significant reduction in glutamatergic activity in neuronal/astroglial cultures from mouse brain as well as in normal Sprague-Dawley and diabetic rats (Sickmann et al. 2012, 2009).

Glycogen could also be involved in cortical information processing, although such possibility has hitherto never been directly explored. Specifically, vasoactive intestinal peptide (VIP) is a powerful glycogenolytic agent (Sorg and Magistretti 1991 1992). VIP-positive interneurons mediate the disinhibition of pyramidal neurons in many areas of the cerebral cortex, which is thought to subserve gain modulation in a brain state-dependent manner (Neske and Connors 2016; Pi et al. 2013). Thus, by controlling local energy metabolism via glycogen mobilization, VIP-positive interneurons might transiently alter extracellular levels of K^+, glutamate/GABA, lactate (Magistretti 1990; Magistretti et al. 1998, 1984) and in turn synaptic plasticity rules (DiNuzzo and Nedergaard 2017). The regulation of extracellular K^+ by glycogenolysis might also give astrocytes the control of subthreshold neuronal oscillations and possibly brain state transitions (Bellot-Saez et al. 2018; Ding et al. 2016; Kjaerby et al. 2017).

4.2 Metabolic Compartmentation

Glucose is metabolized through glycolysis to form pyruvate and/or lactate and further to carbon dioxide in the TCA cycle (for details, see Schousboe et al. 2015). This is also the case for glucose or glucose-1-phosphate generated from glycogen by glycogenolysis, but the pools of pyruvate and lactate may not be identical in two scenarios. It has been shown that astrocytes may generate a large amount of lactate which may be derived from both glucose and glycogen (Dringen et al. 1993; Schousboe et al. 1997; Waagepetersen et al. 2000b; Walz and Mukerji 1988). Using cultured astrocytes from mouse cerebral cortex or cerebellum and [U-^{13}C]glucose to label the glucose and glycogen pools it has been demonstrated that lactate originating from glucose and glycogen, respectively, was not present in the same metabolic pool (Sickmann et al. 2005). This underlines the complexity of astrocytic metabolism being compartmentalized not only considering the glycolytic and glycogenolytic processes but also the TCA cycle (Sickmann et al. 2005; Waagepetersen et al. 2001, 2003). This is in keeping with the demonstration of compartmentalized pyruvate metabolism in astrocytes (Obel et al. 2012a; Waagepetersen et al. 2000a). Further in vivo labeling studies of rat brain [6-^{14}C]glucose metabolism during rest, stimulation and recovery showed very high specific activity of recovered lactate, indicating a strong metabolic compartmentation between blood-borne glucose and endogenous glycogen (for details, see Dienel 2019b).

The metabolic basis for pyruvate compartmentation is not known. It has been suggested that pyruvate generated from glycolytic processing of blood-borne extracellular glucose takes place predominantly in neurons, while glycogenolysis-derived pyruvate and its clearance from brain tissue as lactate occurs in astrocytes (Dienel 2019b). This scenario entails that at any given instant an individual astrocyte metabolizes either glucose or glycogen, yielding a corresponding compartmented pool of intracellular glucose-6-phosphate and downstream metabolites, something that has long been discussed in the biochemical literature (see DiNuzzo 2019 and references therein). A complete switch between glucose and glycogen metabolism in astrocytes can be explained by the so-called glucose-sparing mechanism involving glucose phosphates (DiNuzzo et al. 2011). Specifically, glycogenolysis induces a transient elevation in intracelullar glucose-6-phospate (and possibly glucose-1,6-bisphosphate), which inhibits glucose phosphorylation and channels glycogen-derived glucose-1-phosphate into astrocytic metabolism (DiNuzzo et al. 2010, 2011).

As glycogen is essentially restricted to astrocytes, both glucose-sparing and glycogen-derived lactate production endows these cells with the ability to control the extent to which neurons can increase glycolysis or oxidative phosphorylation, respectively. As stated above, metabolism is inextricably linked with function and astrocytic glycogen breakdown may have direct effects on neurotransmission. For example, glycolysis in neurons supports presynaptic vesicular glutamate loading and glutamate exocytotic machinery (Hinckelmann et al. 2016; Ikemoto et al. 2003; Ishida et al. 2009; Zala et al. 2013). In turn, glutamatergic and monoaminergic neurotransmission stimulates filopodial astrocytic movements, possibly fueled by glycogen (Dienel 2019b), that bring fine processes of these cells closer to neuronal synapses (Bellesi et al. 2015). Such a mechanism would configure a metabolic counterpart of astrocytic support to changes in synaptic strength. In particular, the degree of glucose-sparing for neurons by astrocytic glycogenolysis together with participation of glycogen in the glutamate-glutamine cycle would modulate glutamate quantal content as well as release and uptake rates. At the same time, transient NADH changes might be produced in dendritic spines by either increased neuronal glycolysis or neuronal uptake of astrocytic glycogen-derived lactate and subsequent conversion to pyruvate by lactate dehydrogenase. Alternatively, or in addition, lactate taken up by dendritic spines can also be derived from glycolysis in presynaptic terminals (DiNuzzo and Giove 2012). Whatever the mechanism, the ensuing change in redox state is known to be involved in control of gene expression, notably of plasticity-related genes (Yang et al. 2014). Although the functional advantage of compartmentation between glucose and glycogen metabolism is presently unknown, it is intriguing to speculate that the metabolic effects of glycogenolysis in astrocytes might modulate synaptic transactions and plasticity.

5 Concluding Remarks

In summary, glycogen cannot be fully characterized by its tissue concentration, because structural integrity, particle size distribution and turnover also play critical roles for its metabolism. Nonetheless, studies that measured brain glycogen levels clearly indicated that this polysaccharide actively participates in many basic as well as higher brain functions. In particular, neuronal excitability is affected by glycogen metabolism, as glycogen appears to be necessary for ion and transmitter homeostasis and inhibition of glycogenolysis is associated with increased susceptibility to epileptic seizures and cortical spreading depression. Glycogen is required for astrocytic potassium uptake and it is involved in glutamate/GABA synthesis. Moreover, glycogen is implicated in synaptic plasticity, as inhibition of glycogenolysis or glycogenesis impairs learning and memory. Neuronal excitability and synaptic plasticity largely concur to define brain state, as the ability of responding to external stimulations and to make new memories are state-dependent (e.g., consciousness versus sleep/anesthesia), and so is brain glycogen metabolism. Given the multifaceted effects of glycogen synthesis and degradation to brain function in health and disease, it can be predicted that glycogen research will be at the frontiers of neuroscience in the years to come.

Acknowledgments This project has received funding from the European Union's Horizon 2020 research and innovation programme under the Marie Sklodowska-Curie grant agreement No 701635. The content is solely the responsibility of the authors and does not necessarily represent the official views of the European Union.

Competing interests: The authors have declared that no competing interests exist.

References

Aalling NN, Nedergaard M, DiNuzzo M (2018) Cerebral metabolic changes during sleep. Curr Neurol Neurosci Rep 18(9):57

Abdul-Ghani AS, Nahas N (1990) Effect of afferent brachial stimulation and photic stimulation on glycogen concentration in cortical brain tissue. Neurochem Int 16(2):115–118

Abood LG, Kocsis JJ (1950) Effect of ACTH on glycogenesis and glycolysis in hypophysectomized rats. Proc Soc Exp Biol Med 75(1):55–58

Ahlersova E, Ahlers I, Smajda B, Ivancinova O, Paulikova E (1980) Circadian rhythm of blood glucose and tissue glycogen in fed and fasted rats. Physiol Bohemoslov 29(6):515–523

Al-Badry KS, Taha HM (1983) Hibernation-hypothermia and metabolism in hedgehogs. Changes in some organic components. Comp Biochem Physiol A Comp Physiol 74(1):143–148

Almeida-Filho DG, Queiroz CM, Ribeiro S (2018) Memory corticalization triggered by REM sleep: mechanisms of cellular and systems consolidation. Cell Mol Life Sci 75(20):3715–3740

Augé E, Cabezón I, Pelegrí C, Vilaplana J (2017) New perspectives on corpora amylacea in the human brain. Sci Rep 7:41807

Augé E, Duran J, Guinovart JJ, Pelegrí C, Vilaplana J (2018) Exploring the elusive composition of corpora amylacea of human brain. Sci Rep 8(1):13525

Bak LK, Walls AB, Schousboe A, Waagepetersen HS (2018) Astrocytic glycogen metabolism in the healthy and diseased brain. J Biol Chem 293(19):7108–7116

Barron KD, Dentinger MP, Kimelberg HK, Nelson LR, Bourke RS, Keegan S, Mankes R, Cragoe EJ (1988) Ultrastructural features of a brain injury model in cat. Acta Neuropathol 75(3):295–307
Bass B, Upson S, Roy K, Montgomery EL, Jalonen TO, Murray IVJ (2015) Glycogen and amyloid-beta: key players in the shift from neuronal hyperactivity to hypoactivity observed in Alzheimer's disease? Neural Regeneration Res 10(7):1023–1025
Baud MO, Parafita J, Nguyen A, Magistretti PJ, Petit JM (2016) Sleep fragmentation alters brain energy metabolism without modifying hippocampal electrophysiological response to novelty exposure. J Sleep Res 25(5):583–590
Bauernfeind AL, Babbitt CC (2014) The appropriation of glucose through primate neurodevelopment. J Hum Evol 77:132–140
Bellesi M, de Vivo L, Koebe S, Tononi G, Cirelli C (2018) Sleep and wake affect glycogen content and turnover at perisynaptic astrocytic processes. Front Cell Neurosci 12:308
Bellesi M, de Vivo L, Tononi G, Cirelli C (2015) Effects of sleep and wake on astrocytes: clues from molecular and ultrastructural studies. BMC Biol 13:66
Bellot-Saez A, Cohen G, van Schaik A, Ooi L, Morley JW, Buskila Y (2018) Astrocytic modulation of cortical oscillations. Sci Rep 8(1):11565
Benington JH, Heller HC (1995) Restoration of brain energy metabolism as the function of sleep. Prog Neurobiol 45(4):347–360
Bergmann TO (2018) Brain state-dependent brain stimulation. Front Psychol 9:2108
Blackwood SJ, Hanya E, Katz A (2019) Effect of postexercise temperature elevation on postexercise glycogen metabolism of isolated mouse soleus muscle. J Appl Physiol (1985) 126(4):1103–1109
Boucek D, Jirikowic J, Taylor M (2011) Natural history of Danon disease. Genet Med 13(6):563–568
Braitenberg V, Schüz A (1998) Cortex: statistics and geometry of neuronal connectivity, 2nd, thoroughly rev. edition. Springer, Berlin; New York
Bramlett HM, Dietrich WD (2001) Neuropathological protection after traumatic brain injury in intact female rats versus males or ovariectomized females. J Neurotrauma 18(9):891–900
Bronovitskaya ZG, Shapovalova NS (1957) The glucose and glycogen of the brain in animals under raised oxygen pressure. Ukr Biokh Zh 29:20–24
Brucklacher RM, Vannucci RC, Vannucci SJ (2002) Hypoxic preconditioning increases brain glycogen and delays energy depletion from hypoxia-ischemia in the immature rat. Dev Neurosci 24(5):411–417
Brunner EA, Passonneau JV, Molstad C (1971) The effect of volatile anaesthetics on levels of metabolites and on metabolic rate in brain. J Neurochem 18(12):2301–2316
Canada SE, Weaver SA, Sharpe SN, Pederson BA (2011) Brain glycogen supercompensation in the mouse after recovery from insulin-induced hypoglycemia. J Neurosci Res 89(4):585–591
Caravas J, Wildman DE (2014) A genetic perspective on glucose consumption in the cerebral cortex during human development. Diabetes Obes Metab 16(Suppl 1):21–25
Cataldo AM, Broadwell RD (1986) Cytochemical identification of cerebral glycogen and glucose-6-phosphatase activity under normal and experimental conditions. II. Choroid plexus and ependymal epithelia, endothelia and pericytes. J Neurocytol 15(4):511–524
Chance MR, Yaxley DC (1950) Central nervous function and changes in brain metabolite concentration. I. Glycogen and lactate in convulsing mice. J Exp Biol 27(3–4):311–323
Chapler CK, York DH (1972) The effect of hypoxia on glycogen stores in the adult cat brain. Brain Res 45(1):321–324
Choi HB, Gordon GR, Zhou N, Tai C, Rungta RL, Martinez J, Milner TA, Ryu JK, McLarnon JG, Tresguerres M, Levin LR, Buck J, MacVicar BA (2012) Metabolic communication between astrocytes and neurons via bicarbonate-responsive soluble adenylyl cyclase. Neuron 75(6):1094–1104
Choi IY, Gruetter R (2003) In vivo 13C NMR assessment of brain glycogen concentration and turnover in the awake rat. Neurochem Int 43(4–5):317–322

Choi IY, Seaquist ER, Gruetter R (2003) Effect of hypoglycemia on brain glycogen metabolism in vivo. J Neurosci Res 72(1):25–32

Choi IY, Tkac I, Ugurbil K, Gruetter R (1999) Noninvasive measurements of [1-(13)C]glycogen concentrations and metabolism in rat brain in vivo. J Neurochem 73(3):1300–1308

Criado O, Aguado C, Gayarre J, Duran-Trio L, Garcia-Cabrero AM, Vernia S, San Millan B, Heredia M, Roma-Mateo C, Mouron S, Juana-Lopez L, Dominguez M, Navarro C, Serratosa JM, Sanchez M, Sanz P, Bovolenta P, Knecht E, Rodriguez de Cordoba S (2012) Lafora bodies and neurological defects in malin-deficient mice correlate with impaired autophagy. Hum Mol Genet 21(7):1521–1533

Cruz NF, Dienel GA (2002) High glycogen levels in brains of rats with minimal environmental stimuli: implications for metabolic contributions of working astrocytes. J Cereb Blood Flow Metab 22(12):1476–1489

Cui XY, Li L, An YY, Lu GW (2001) Changes in the contents of glycogen and lactate in the brain and blood during hypoxic preconditioning. Sheng Li Xue Bao 53(4):325–328

Cukiert A, Vilela MM, Scapolan HB, Lefevre BH, Marques-Assis L (1990) Mental deterioration in Lafora's disease. Arq Neuropsiquiatr 48(2):236–240

Cummins CJ, Lust WD, Passonneau JV (1983) Regulation of glycogen metabolism in primary and transformed astrocytes in vitro. J Neurochem 40(1):128–136

Dalsgaard MK, Madsen FF, Secher NH, Laursen H, Quistorff B (2007) High glycogen levels in the hippocampus of patients with epilepsy. J Cereb Blood Flow Metab 27(6):1137–1141

Dalsgaard MK, Secher NH (2007) The brain at work: a cerebral metabolic manifestation of central fatigue? J Neurosci Res 85(15):3334–3339

Deng B, Sullivan MA, Chen C, Li J, Powell PO, Hu Z, Gilbert RG (2016) Molecular structure of human-liver glycogen. PLoS One 11(3):e0150540

DePaoli-Roach AA, Contreras CJ, Segvich DM, Heiss C, Ishihara M, Azadi P, Roach PJ (2015) Glycogen phosphomonoester distribution in mouse models of the progressive myoclonic epilepsy, Lafora disease. J Biol Chem 290(2):841–850

Dienel GA (2019a) Brain glucose metabolism: integration of energetics with function. Physiol Rev 99(1):949–1045

Dienel GA (2019b) Does shuttling of glycogen-derived lactate from astrocytes to neurons take place during neurotransmission and memory consolidation? J Neurosci Res 97(8):863–882

Dienel GA, Ball KK, Cruz NF (2007) A glycogen phosphorylase inhibitor selectively enhances local rates of glucose utilization in brain during sensory stimulation of conscious rats: implications for glycogen turnover. J Neurochem 102(2):466–478

Dienel GA, Cruz NF (2015) Contributions of glycogen to astrocytic energetics during brain activation. Metab Brain Dis 30(1):281–298

Dienel GA, Cruz NF (2016) Aerobic glycolysis during brain activation: adrenergic regulation and influence of norepinephrine on astrocytic metabolism. J Neurochem 138(1):14–52

Ding F, O'Donnell J, Xu Q, Kang N, Goldman N, Nedergaard M (2016) Changes in the composition of brain interstitial ions control the sleep-wake cycle. Science 352(6285):550–555

DiNuzzo M (2013) Kinetic analysis of glycogen turnover: relevance to human brain C-NMR spectroscopy. J Cereb Blood Flow Metab 33(10):1540–1548

DiNuzzo M (2016) Astrocyte-neuron interactions during learning may occur by lactate signaling rather than metabolism. Front Integr Neurosci 10:2

DiNuzzo M (2019) How glycogen sustains brain function: a plausible allosteric signaling pathway mediated by glucose phosphates. J Cereb Blood Flow Metab 39(8):1452–1459

DiNuzzo M, Giove F (2012) Activity-dependent energy budget for neocortical signaling: effect of short-term synaptic plasticity on the energy expended by spiking and synaptic activity. J Neurosci Res 90(11):2094–2102

DiNuzzo M, Giove F, Maraviglia B, Mangia S (2015a) Monoaminergic control of cellular glucose utilization by glycogenolysis in neocortex and hippocampus. Neurochem Res 40(12):2493–2504

DiNuzzo M, Mangia S, Maraviglia B, Giove F (2010) Glycogenolysis in astrocytes supports blood-borne glucose channeling not glycogen-derived lactate shuttling to neurons: evidence from mathematical modeling. J Cereb Blood Flow Metab 30(12):1895–1904

DiNuzzo M, Mangia S, Maraviglia B, Giove F (2014) Physiological bases of the K+ and the glutamate/GABA hypotheses of epilepsy. Epilepsy Res 108(6):995–1012

DiNuzzo M, Mangia S, Maraviglia B, Giove F (2015b) Does abnormal glycogen structure contribute to increased susceptibility to seizures in epilepsy? Metab Brain Dis 30(1):307–316

DiNuzzo M, Maraviglia B, Giove F (2011) Why does the brain (not) have glycogen? BioEssays 33(5):319–326

DiNuzzo M, Nedergaard M (2017) Brain energetics during the sleep-wake cycle. Curr Opin Neurobiol 47:65–72

Dringen R, Gebhardt R, Hamprecht B (1993) Glycogen in astrocytes: possible function as lactate supply for neighboring cells. Brain Res 623(2):208–214

Drulis-Fajdasz D, Gizak A, Wojtowicz T, Wisniewski JR, Rakus D (2018) Aging-associated changes in hippocampal glycogen metabolism in mice. Evidence for and against astrocyte-to-neuron lactate shuttle. Glia 66(7):1481–1495

Drulis-Fajdasz D, Wójtowicz T, Wawrzyniak M, Wlodarczyk J, Mozrzymas JW, Rakus D (2015) Involvement of cellular metabolism in age-related LTP modifications in rat hippocampal slices. Oncotarget 6(16):14065–14081

Duran J, Gruart A, Garcia-Rocha M, Delgado-Garcia JM, Guinovart JJ (2014) Glycogen accumulation underlies neurodegeneration and autophagy impairment in Lafora disease. Hum Mol Genet

Duran J, Guinovart JJ (2015) Brain glycogen in health and disease. Mol Asp Med 46:70–77

Duran J, Saez I, Gruart A, Guinovart JJ, Delgado-Garcia JM (2013) Impairment in long-term memory formation and learning-dependent synaptic plasticity in mice lacking glycogen synthase in the brain. J Cereb Blood Flow Metab 33(4):550–556

Duran J, Tevy MF, Garcia-Rocha M, Calbo J, Milan M, Guinovart JJ (2012) Deleterious effects of neuronal accumulation of glycogen in flies and mice. EMBO Mol Med 4(8):719–729

Dydyk L, Piekarczyk A, Szreter T, Prokopczyk J (1978) Effect of normobaric hyperoxia on the glycogen content in rabbit maturing brain. Neuropatol Pol 16(4):497–506

Ellingwood SS, Cheng A (2018) Biochemical and clinical aspects of glycogen storage diseases. J Endocrinol 238(3):R131–r141

Estler CJ (1961) The glycogen content of the brain of white mice under the effect of phenobarbital and its relations to the blood sugar and body temperature. Med Exp Int J Exp Med 4:30–36

Estler CJ, Heim F (1960) The content of adenine nucleotide, creatine phosphate, coenzyme A, glycogen and lactic acid in the brain of white mice in ether excitation and anesthesia. Med Exp Int J Exp Med 3:241–248

Folbergrova J, Lowry OH, Passonneau JV (1970) Changes in metabolites of the energy reserves in individual layers of mouse cerebral cortex and subjacent white matter during ischaemia and anaesthesia. J Neurochem 17(8):1155–1162

Franken P, Gip P, Hagiwara G, Ruby NF, Heller HC (2003) Changes in brain glycogen after sleep deprivation vary with genotype. Am J Physiol Regul Integr Comp Physiol 285(2):R413–R419

Franken P, Gip P, Hagiwara G, Ruby NF, Heller HC (2006) Glycogen content in the cerebral cortex increases with sleep loss in C57BL/6J mice. Neurosci Lett 402(1-2):176–179

Fuhrman FA, Fuhrman GJ (1966) Effect of temperature on synthesis of glycogen by diaphragm muscle in vitro. Am J Phys 210(6):1225–1228

Garcia-Cabrero AM, Marinas A, Guerrero R, de Cordoba SR, Serratosa JM, Sanchez MP (2012) Laforin and malin deletions in mice produce similar neurologic impairments. J Neuropathol Exp Neurol 71(5):413–421

García-Gimeno MA, Knecht E, Sanz P (2018) Lafora disease: a ubiquitination-related pathology. Cell 7(8):87

Garriga J, Cusso R (1992) Effect of starvation on glycogen and glucose metabolism in different areas of the rat brain. Brain Res 591(2):277–282

Gatfield PD, Lowry OH, Schulz DW, Passonneau JV (1966) Regional energy reserves in mouse brain and changes with ischaemia and anaesthesia. J Neurochem 13(3):185–195
Gentry MS, Guinovart JJ, Minassian BA, Roach PJ, Serratosa JM (2018) Lafora disease offers a unique window into neuronal glycogen metabolism. J Biol Chem 293(19):7117–7125
Gibbs ME, Anderson DG, Hertz L (2006) Inhibition of glycogenolysis in astrocytes interrupts memory consolidation in young chickens. Glia 54(3):214–222
Gibbs ME, Lloyd HG, Santa T, Hertz L (2007) Glycogen is a preferred glutamate precursor during learning in 1-day-old chick: biochemical and behavioral evidence. J Neurosci Res 85(15):3326–3333
Giorgi FS, Galanopoulou AS, Moshe SL (2014) Sex dimorphism in seizure-controlling networks. Neurobiol Dis 72(Pt B):144–152
Gip P, Hagiwara G, Ruby NF, Heller HC (2002) Sleep deprivation decreases glycogen in the cerebellum but not in the cortex of young rats. Am J Physiol Regul Integr Comp Physiol 283(1):R54–R59
Gip P, Hagiwara G, Sapolsky RM, Cao VH, Heller HC, Ruby NF (2004) Glucocorticoids influence brain glycogen levels during sleep deprivation. Am J Physiol Regul Integr Comp Physiol 286(6):R1057–R1062
Goldberg ND, O'Toole AG (1969) The properties of glycogen synthetase and regulation of glycogen biosynthesis in rat brain. J Biol Chem 244(11):3053–3061
Gordan GS (1956) Influence of steroids on cerebral metabolism in man. Recent Prog Horm Res 12:153–170. discussion, 170–4
Gordan GS, Bentinck RC, Eisenberg E (1951) The influence of steroids on cerebral metabolism. Ann N Y Acad Sci 54(4):575–607
Goyal MS, Vlassenko AG, Blazey TM, Su Y, Couture LE, Durbin TJ, Bateman RJ, Benzinger TL, Morris JC, Raichle ME (2017) Loss of brain aerobic glycolysis in normal human aging. Cell Metab 26(2):353–360.e3
Harbison ST, Sehgal A (2008) Quantitative genetic analysis of sleep in Drosophila melanogaster. Genetics 178(4):2341–2360
Harbison ST, Sehgal A (2009) Energy stores are not altered by long-term partial sleep deprivation in Drosophila melanogaster. PLoS One 4(7):e6211
Harik SI, Lust WD, Jones SC, Lauro KL, Pundik S, LaManna JC (1995) Brain glucose metabolism in hypobaric hypoxia. J Appl Physiol (1985) 79(1):136–140
Hashiguchi Y, Molina PE, Boxer R, Naukam R, Abumrad NN (1998) Differential responses of brain, liver, and muscle glycogen to opiates and surgical stress. Surg Today 28(4):471–474
Herbet M, Natorska-Chomicka D, Korga A, Ostrowska M, Izdebska M, Gawronska-Grzywacz M, Piatkowska-Chmiel I, Pawlowski K, Slaska B, Poleszak E, Dudka J (2019) Altered expression of genes involved in brain energy metabolism as adaptive responses in rats exposed to chronic variable stress; changes in cortical level of glucogenic and neuroactive amino acids. Mol Med Rep 19(3):2386–2396
Hertz L, O'Dowd BS, Ng KT, Gibbs ME (2003) Reciprocal changes in forebrain contents of glycogen and of glutamate/glutamine during early memory consolidation in the day-old chick. Brain Res 994(2):226–233
Hertz L, Peng L, Dienel GA (2007) Energy metabolism in astrocytes: high rate of oxidative metabolism and spatiotemporal dependence on glycolysis/glycogenolysis. J Cereb Blood Flow Metab 27(2):219–249
Herzog RI, Chan O, Yu S, Dziura J, McNay EC, Sherwin RS (2008) Effect of acute and recurrent hypoglycemia on changes in brain glycogen concentration. Endocrinology 149(4):1499–1504
Hevor TK, Delorme P, Beauvillain JC (1986) Glycogen synthesis and immunocytochemical study of fructose-1,6-biphosphatase in methionine sulfoximine epileptogenic rodent brain. J Cereb Blood Flow Metab 6(3):292–297
Hicks J, Wartchow E, Mierau G (2011) Glycogen storage diseases: a brief review and update on clinical features, genetic abnormalities, pathologic features, and treatment. Ultrastruct Pathol 35(5):183–196

Hinckelmann M-V, Virlogeux A, Niehage C, Poujol C, Choquet D, Hoflack B, Zala D, Saudou F (2016) Self-propelling vesicles define glycolysis as the minimal energy machinery for neuronal transport. Nat Commun 7:13233

Hirase H, Akther S, Wang X, Oe Y (2019) Glycogen distribution in mouse hippocampus. J Neurosci Res 97(8):923–932

Hoyaux D, Decaestecker C, Heizmann CW, Vogl T, Schafer BW, Salmon I, Kiss R, Pochet R (2000) S100 proteins in Corpora amylacea from normal human brain. Brain Res 867(1-2):280–288

Hutchins DA, Rogers KJ (1970) Physiological and drug-induced changes in the glycogen content of mouse brain. Br J Pharmacol 39(1):9–25

Ibrahim MZ (1975) Glycogen and its related enzymes of metabolism in the central nervous system. Adv Anat Embryol Cell Biol 52(1):3–89

Ibrahim MZ, Atlan H, Miquel J, Castellani P (1970) Synthetic and hydrolytic enzymes of glycogen in the normal and the irradiated rat brain. Radiat Res 43(2):341–356

Ikemoto A, Bole DG, Ueda T (2003) Glycolysis and glutamate accumulation into synaptic vesicles. Role of glyceraldehyde phosphate dehydrogenase and 3-phosphoglycerate kinase. J Biol Chem 278(8):5929–5940

Ishida A, Noda Y, Ueda T (2009) Synaptic vesicle-bound pyruvate kinase can support vesicular glutamate uptake. Neurochem Res 34(5):807–818

Ivanisevic J, Stauch KL, Petrascheck M, Benton HP, Epstein AA, Fang M, Gorantla S, Tran M, Hoang L, Kurczy ME, Boska MD, Gendelman HE, Fox HS, Siuzdak G (2016) Metabolic drift in the aging brain. Aging (Albany NY) 8(5):1000–1020

Izumi Y, Aikawa S, Matsuda F, Hasunuma T, Kondo A (2013) Aqueous size-exclusion chromatographic method for the quantification of cyanobacterial native glycogen. J Chromatogr B Analyt Technol Biomed Life Sci 930:90–97

Kachi T, Matsushima S, Ito T (1973) Diurnal variations in pineal glycogen content during the estrous cycle in female mice. Arch Histol Jpn 35(2):153–159

Karadzic V, Mrsulja B (1969a) Deprivation of paradoxical sleep and brain glycogen. J Neurochem 16(1):29–34

Karadzic V, Mrsulja B (1969b) Glycogenic response of CNS to paradoxical sleep deprivation in cats and rats. Electroencephalogr Clin Neurophysiol 27(5):552

Karnovsky ML, Reich P, Anchors JM, Burrows BL (1983) Changes in brain glycogen during slow-wave sleep in the rat. J Neurochem 41(5):1498–1501

Kashio N, Usuki F, Akamine T, Nakagawa S, Higuchi I, Nakahara K, Okada A, Osame M, Murata F (1991) Cardiomyopathy, mental retardation, and autophagic vacuolar myopathy. Abnormal MRI findings in the head. J Neurol Sci 105(1):1–5

Katz A (2006) Glycogenin, proglycogen, and glycogen biogenesis: what's the story? Am J Physiol Endocrinol Metab 290(4):E757–E758. author reply E758-9

Kilic K, Karatas H, Donmez-Demir B, Eren-Kocak E, Gursoy-Ozdemir Y, Can A, Petit JM, Magistretti PJ, Dalkara T (2018) Inadequate brain glycogen or sleep increases spreading depression susceptibility. Ann Neurol 83(1):61–73

Kiyatkin EA (2010) Brain temperature homeostasis: physiological fluctuations and pathological shifts. Front Biosci (Landmark Ed) 15:73–92

Kjaerby C, Rasmussen R, Andersen M, Nedergaard M (2017) Does global astrocytic calcium signaling participate in awake brain state transitions and neuronal circuit function? Neurochem Res

Klatzo I, Miquel J, Tobias C, Haymaker W (1961) Effects of alpha particle radiation on the rat brain, including vascular permeability and glycogen studies. J Neuropathol Exp Neurol 20:459–483

Koizumi J (1974) Glycogen in the central nervous system. Prog Histochem Cytochem 6(4):1–37

Kong J, Shepel PN, Holden CP, Mackiewicz M, Pack AI, Geiger JD (2002) Brain glycogen decreases with increased periods of wakefulness: implications for homeostatic drive to sleep. J Neurosci 22(13):5581–5587

Kosmina MM (1967) Effect of hydrocortisone on certain indexes of energy metabolism in the brain cortex of guinea pigs. Fiziol Zh 13:252–255

Kubo Z, Shimizu N (1957) Histochemical studies on brain glycogen of the guinea pig and its alteration following electric shock. J Neuropathol Exp Neurol 16(1):40–47
Lee SH, Dan Y (2012) Neuromodulation of brain states. Neuron 76(1):209–222
Leke R, Schousboe A (2016) The glutamine transporters and their role in the glutamate/GABA-glutamine cycle. Adv Neurobiol 13:223–257
Lomako J, Lomako WM, Whelan WJ (1991) Proglycogen: a low-molecular-weight form of muscle glycogen. FEBS Lett 279(2):223–228
Lomako J, Lomako WM, Whelan WJ, Dombro RS, Neary JT, Norenberg MD (1993) Glycogen synthesis in the astrocyte: from glycogenin to proglycogen to glycogen. FASEB J 7(14):1386–1393
Lopez-Ramos JC, Duran J, Gruart A, Guinovart JJ, Delgado-Garcia JM (2015) Role of brain glycogen in the response to hypoxia and in susceptibility to epilepsy. Front Cell Neurosci 9:431
Louzao MC, Espina B, Vieytes MR, Vega FV, Rubiolo JA, Baba O, Terashima T, Botana LM (2008) "Fluorescent glycogen" formation with sensibility for in vivo and in vitro detection. Glycoconj J 25(6):503–510
Lowry OH, Passonneau JV, Hasselberger FX, Schulz DW (1964) Effect of ischemia on known substrates and cofactors of the glycolytic pathway in brain. J Biol Chem 239:18–30
Lundgren PR, Miquel J (1970) The incorporation of isotopic carbon 14C into the cerebral glycogen of normal and X-irradiated rats. J Neurochem 17(9):1383–1386
Lust WD, Wheaton AB, Feussner G, Passonneau J (1989) Metabolism in the hamster brain during hibernation and arousal. Brain Res 489(1):12–20
Lutas A, Yellen G (2013) The ketogenic diet: metabolic influences on brain excitability and epilepsy. Trends Neurosci 36(1):32–40
Magalhaes MM, Coimbra A (1970) Electron microscope radioautographic study of glycogen synthesis in the rabbit retina. J Cell Biol 47(1):263–275
Magistretti PJ (1990) VIP neurons in the cerebral cortex. Trends Pharmacol Sci 11(6):250–254
Magistretti PJ, Cardinaux JR, Martin JL (1998) VIP and PACAP in the CNS: regulators of glial energy metabolism and modulators of glutamatergic signaling. Ann N Y Acad Sci 865:213–225
Magistretti PJ, Morrison JH, Shoemaker WJ, Bloom FE (1984) Morphological and functional correlates of VIP neurons in cerebral cortex. Peptides 5(2):213–218
Mann DM, Sumpter PQ, Davies CA, Yates PO (1987) Glycogen accumulations in the cerebral cortex in Alzheimer's disease. Acta Neuropathol 73(2):181–184
Marchand I, Chorneyko K, Tarnopolsky M, Hamilton S, Shearer J, Potvin J, Graham TE (2002) Quantification of subcellular glycogen in resting human muscle: granule size, number, and location. J Appl Physiol (1985) 93(5):1598–1607
Matsui T, Ishikawa T, Ito H, Okamoto M, Inoue K, Lee MC, Fujikawa T, Ichitani Y, Kawanaka K, Soya H (2012) Brain glycogen supercompensation following exhaustive exercise. J Physiol 590(3):607–616
Matsui T, Omuro H, Liu Y-F, Soya M, Shima T, McEwen BS, Soya H (2017) Astrocytic glycogen-derived lactate fuels the brain during exhaustive exercise to maintain endurance capacity. Proc Natl Acad Sci 114(24):6358
Matsui T, Soya S, Kawanaka K, Soya H (2015) Brain glycogen decreases during intense exercise without hypoglycemia: the possible involvement of serotonin. Neurochem Res 40(7):1333–1340
Matsui T, Soya S, Okamoto M, Ichitani Y, Kawanaka K, Soya H (2011) Brain glycogen decreases during prolonged exercise. J Physiol 589(Pt 13):3383–3393
McEwen BS (2006) Sleep deprivation as a neurobiologic and physiologic stressor: allostasis and allostatic load. Metabolism 55(10 Suppl 2):S20–S23
McKenna MC, Dienel GA, Sonnewald U, Waagepetersen HS, Schousboe A (2012) Energy metabolism of the brain. In: Brady ST, Siegel GJ, Albers RW, Price DL (eds) Basic neurochemistry. Principles of molecular, cellular, and medical neurobiology, 8th edn. Academic Press, Elsevier, Waltham, MA, pp 200–231
Melendez-Hevia E, Waddell TG, Shelton ED (1993) Optimization of molecular design in the evolution of metabolism: the glycogen molecule. Biochem J 295(Pt 2):477–483

Melendez R, Melendez-Hevia E, Cascante M (1997) How did glycogen structure evolve to satisfy the requirement for rapid mobilization of glucose? A problem of physical constraints in structure building. J Mol Evol 45(4):446–455

Melendez R, Melendez-Hevia E, Mas F, Mach J, Cascante M (1998) Physical constraints in the synthesis of glycogen that influence its structural homogeneity: a two-dimensional approach. Biophys J 75(1):106–114

Mergenthaler P, Lindauer U, Dienel GA, Meisel A (2013) Sugar for the brain: the role of glucose in physiological and pathological brain function. Trends Neurosci 36(10):587–597

Miquel J, Haymaker W (1965) Astroglial reactions to ionizing radiation: with emphasis on glycogen accumulation*. In: De Robertis EDP, Carrea R (eds) Progress in brain research. Elsevier, Amsterdam, pp 89–114

Miquel J, Klatzo I, Menzel DB, Haymaker W (1963) Glycogen changes in X-irradiated rat brain. Acta Neuropathol 3:482–490

Miquel J, Lundgren PR, Jenkins JO (1966) Effects of roentgen radiation on glycogen metabolism of the rat brain. Acta Radiol Ther Phys Biol 5:123–132

Mishchenko Y, Hu T, Spacek J, Mendenhall J, Harris KM, Chklovskii DB (2010) Ultrastructural analysis of hippocampal neuropil from the connectomics perspective. Neuron 67(6):1009–1020

Mor A, Grossman Y (2007) High pressure modulation of NMDA receptor dependent excitability. Eur J Neurosci 25(7):2045–2052

Mor A, Grossman Y (2010) The efficacy of physiological and pharmacological N-methyl-D-aspartate receptor block is greatly reduced under hyperbaric conditions. Neuroscience 169(1):1–7

Morgenthaler FD, Koski DM, Kraftsik R, Henry PG, Gruetter R (2006) Biochemical quantification of total brain glycogen concentration in rats under different glycemic states. Neurochem Int 48(6-7):616–622

Morgenthaler FD, Lanz BR, Petit JM, Frenkel H, Magistretti PJ, Gruetter R (2009) Alteration of brain glycogen turnover in the conscious rat after 5h of prolonged wakefulness. Neurochem Int 55(1–3):45–51

Mrak RE, Griffin ST, Graham DI (1997) Aging-associated changes in human brain. J Neuropathol Exp Neurol 56(12):1269–1275

Mrsulja BB, Rakic LM, Radulovacki M (1967) The influence of deprivation of paradoxicalx sleep on glycogen content in various brain structures of the cat. Experientia 23(3):200–201

Murray AJ, Knight NS, Cole MA, Cochlin LE, Carter E, Tchabanenko K, Pichulik T, Gulston MK, Atherton HJ, Schroeder MA, Deacon RMJ, Kashiwaya Y, King MT, Pawlosky R, Rawlins JNP, Tyler DJ, Griffin JL, Robertson J, Veech RL, Clarke K (2016) Novel ketone diet enhances physical and cognitive performance. FASEB J 30(12):4021–4032

Nahas N, Abdul-Ghani AS (1989) Species-directed variation and non-uniform distribution of glycogen in mammalian brains during starvation, diabetes and anesthesia. Neurochem Int 14(1):19–24

Naperalsky M, Ruby B, Slivka D (2010) Environmental temperature and glycogen resynthesis. Int J Sports Med 31(8):561–566

Nelson SR, Schulz DW, Passonneau JV, Lowry OH (1968) Control of glycogen levels in brain. J Neurochem 15(11):1271–1279

Neske GT, Connors BW (2016) Distinct roles of SOM and VIP interneurons during cortical up states. Front Neural Circuits 10:52

Newman JC, Covarrubias AJ, Zhao M, Yu X, Gut P, Ng C-P, Huang Y, Haldar S, Verdin E (2017) Ketogenic diet reduces midlife mortality and improves memory in aging mice. Cell Metab 26(3):547–557.e8

Newman LA, Korol DL, Gold PE (2011) Lactate produced by glycogenolysis in astrocytes regulates memory processing. PLoS One 6(12):e28427

Nitschke F, Ahonen SJ, Nitschke S, Mitra S, Minassian BA (2018) Lafora disease—from pathogenesis to treatment strategies. Nat Rev Neurol 14(10):606–617

Norenberg MD, Martinez-Hernandez A (1979) Fine structural localization of glutamine synthetase in astrocytes of rat brain. Brain Res 161(2):303–310
Nybo L, Secher NH (2004) Cerebral perturbations provoked by prolonged exercise. Prog Neurobiol 72(4):223–261
O'Dowd BS, Barrington J, Ng KT, Hertz E, Hertz L (1995) Glycogenolytic response of primary chick and mouse cultures of astrocytes to noradrenaline across development. Brain Res Dev Brain Res 88(2):220–223
Obel LF, Andersen KM, Bak LK, Schousboe A, Waagepetersen HS (2012a) Effects of adrenergic agents on intracellular Ca2+ homeostasis and metabolism of glucose in astrocytes with an emphasis on pyruvate carboxylation, oxidative decarboxylation and recycling: implications for glutamate neurotransmission and excitotoxicity. Neurotox Res 21(4):405–417
Obel LF, Muller MS, Walls AB, Sickmann HM, Bak LK, Waagepetersen HS, Schousboe A (2012b) Brain glycogen-new perspectives on its metabolic function and regulation at the subcellular level. Front Neuroenerg 4:3
Oe Y, Baba O, Ashida H, Nakamura KC, Hirase H (2016) Glycogen distribution in the microwave-fixed mouse brain reveals heterogeneous astrocytic patterns. Glia 64(9):1532–1545
Okubo K, Itoh S, Isobe K, Kusaka T, Nagano K, Kondo M, Onishi S (2001) Cerebral metabolism and regional cerebral blood flow during moderate systemic cooling in newborn piglets. Pediatr Int 43(5):496–501
Otori T, Friedland JC, Sinson G, McIntosh TK, Raghupathi R, Welsh FA (2004) Traumatic brain injury elevates glycogen and induces tolerance to ischemia in rat brain. J Neurotrauma 21(6):707–718
Oz G, DiNuzzo M, Kumar A, Moheet A, Khowaja A, Kubisiak K, Eberly LE, Seaquist ER (2017) Cerebral glycogen in humans following acute and recurrent hypoglycemia: implications on a role in hypoglycemia unawareness. J Cereb Blood Flow Metab 37(8):2883–2893
Oz G, DiNuzzo M, Kumar A, Moheet A, Seaquist ER (2015) Revisiting glycogen content in the human brain. Neurochem Res 40(12):2473–2481
Oz G, Kumar A, Rao JP, Kodl CT, Chow L, Eberly LE, Seaquist ER (2009) Human brain glycogen metabolism during and after hypoglycemia. Diabetes 58(9):1978–1985
Oz G, Seaquist ER, Kumar A, Criego AB, Benedict LE, Rao JP, Henry PG, Van De Moortele PF, Gruetter R (2007) Human brain glycogen content and metabolism: implications on its role in brain energy metabolism. Am J Physiol Endocrinol Metab 292(3):E946–E951
Ozen H (2007) Glycogen storage diseases: new perspectives. World J Gastroenterol 13(18):2541–2553
Palop JJ, Mucke L (2009) Epilepsy and cognitive impairments in Alzheimer disease. Arch Neurol 66(4):435–440
Pearson-Leary J, Osborne DM, McNay EC (2016) Role of Glia in stress-induced enhancement and impairment of memory. Front Integr Neurosci 9(63)
Pederson BA, Turnbull J, Epp JR, Weaver SA, Zhao X, Pencea N, Roach PJ, Frankland PW, Ackerley CA, Minassian BA (2013) Inhibiting glycogen synthesis prevents Lafora disease in a mouse model. Ann Neurol 74(2):297–300
Petit JM, Burlet-Godinot S, Magistretti PJ, Allaman I (2015) Glycogen metabolism and the homeostatic regulation of sleep. Metab Brain Dis 30(1):263–279
Petit JM, Tobler I, Allaman I, Borbely AA, Magistretti PJ (2002) Sleep deprivation modulates brain mRNAs encoding genes of glycogen metabolism. Eur J Neurosci 16(6):1163–1167
Petit JM, Tobler I, Kopp C, Morgenthaler F, Borbely AA, Magistretti PJ (2010) Metabolic response of the cerebral cortex following gentle sleep deprivation and modafinil administration. Sleep 33(7):901–908
Phelps CH (1972) Barbiturate-induced glycogen accumulation in brain. An electron microscopic study. Brain Res 39(1):225–234
Pi HJ, Hangya B, Kvitsiani D, Sanders JI, Huang ZJ, Kepecs A (2013) Cortical interneurons that specialize in disinhibitory control. Nature 503(7477):521–524

Pisa D, Alonso R, Marina AI, Rábano A, Carrasco L (2018) Human and microbial proteins from Corpora Amylacea of Alzheimer's disease. Sci Rep 8(1):9880

Pisa D, Alonso R, Rábano A, Carrasco L (2016) Corpora Amylacea of brain tissue from neurodegenerative diseases are stained with specific antifungal antibodies. Front Neurosci 10:85–86

Prasannan KG, Rajan R, Subrahmanyam (1964) Effect of glutamic acid on the brain and liver glycogen of normal and cortisone-treated rats. Indian J Med Res 52:208–212

Prasannan KG, Rajan R, Subrahmanyam K (1963) Brain glycogen in the fed and fasting state. Indian J Med Res 51:703–707

Prokhorova MI, Tupikova ZN (1958) Peculiarities of the brain glycogen metabolism in various functional states. Prog Nucl Energy 6 Biol Sci 2(3):416–431

Pronaszko-Kurczynska A (1976) Diurnal and seasonal variations of glycogen content in the rat brain. Acta Physiol Pol 27(2):157–162

Puentes-Mestril C, Roach J, Niethard N, Zochowski M, Aton SJ (2019) How rhythms of the sleeping brain tune memory and synaptic plasticity. Sleep 42(7). https://doi.org/10.1093/sleep/zsz095

Purshottam T, Kaveeshwar U, Brahmachari HD (1977) Changes in tissue glycogen stores of rats under acute and chronic hypoxia and their relationship to hypoxia tolerance. Aviat Space Environ Med 48(4):351–355

Purshottam T, Kaveeshwar U, Brahmachari HD (1978) Hypoxia tolerance in rats in relation to tissue glycogen levels. Aviat Space Environ Med 49(9):1062–1064

Quilichini PP, Bernard C (2012) Brain state-dependent neuronal computation. Front Comput Neurosci 6:77

Raichle ME (2011) The restless brain. Brain Connect 1(1):3–12

Rettberg JR, Yao J, Brinton RD (2014) Estrogen: a master regulator of bioenergetic systems in the brain and body. Front Neuroendocrinol 35(1):8–30

Roach PJ (2002) Glycogen and its metabolism. Curr Mol Med 2(2):101–120

Roach PJ (2011) Are there errors in glycogen biosynthesis and is laforin a repair enzyme? FEBS Lett 585(20):3216–3218

Rohn TT (2015) Corpora Amylacea in neurodegenerative diseases: cause or effect? Int J Neurol Neurother 2(3):031

Rubio-Villena C, Viana R, Bonet J, Garcia-Gimeno MA, Casado M, Heredia M, Sanz P (2018) Astrocytes: new players in progressive myoclonus epilepsy of Lafora type. Hum Mol Genet 27(7):1290–1300

Saez I, Duran J, Sinadinos C, Beltran A, Yanes O, Tevy MF, Martinez-Pons C, Milan M, Guinovart JJ (2014) Neurons have an active glycogen metabolism that contributes to tolerance to hypoxia. J Cereb Blood Flow Metab 34(6):945–955

Scharf MT, Naidoo N, Zimmerman JE, Pack AI (2008) The energy hypothesis of sleep revisited. Prog Neurobiol 86(3):264–280

Schmolke C, Schleicher A (1989) Structural inhomogeneity in the neuropil of lamina II/III in rabbit visual cortex. Exp Brain Res 77(1):39–47

Scholvinck ML, Maier A, Ye FQ, Duyn JH, Leopold DA (2010) Neural basis of global resting-state fMRI activity. Proc Natl Acad Sci U S A 107(22):10238–10243

Schousboe A, Walls AB, Bak LK, Waagepetersen HS (2015) Astroglia and brain metabolism: focus on energy and neurotransmitter amino acid homeostasis. Colloquium Series on Neuroglia in Biology and Medicine: From Physiology to Disease 2(4):1–64

Schousboe A, Westergaard N, Waagepetersen HS, Larsson OM, Bakken IJ, Sonnewald U (1997) Trafficking between glia and neurons of TCA cycle intermediates and related metabolites. Glia 21(1):99–105

Schreiber RA (1981) Developmental changes in brain glucose, glycogen, phosphocreatine, and ATP levels in DBA/2J and C57BL/6J mice, and audiogenic seizures. J Neurochem 37(3):655–661

Shulman RG, Hyder F, Rothman DL (2001a) Cerebral energetics and the glycogen shunt: neurochemical basis of functional imaging. Proc Natl Acad Sci U S A 98(11):6417–6422

Shulman RG, Hyder F, Rothman DL (2001b) Lactate efflux and the neuroenergetic basis of brain function. NMR Biomed 14(7-8):389–396
Shulman RG, Hyder F, Rothman DL (2009) Baseline brain energy supports the state of consciousness. Proc Natl Acad Sci U S A 106(27):11096–11101
Shulman RG, Rothman DL (2001) The "glycogen shunt" in exercising muscle: a role for glycogen in muscle energetics and fatigue. Proc Natl Acad Sci U S A 98(2):457–461
Shulman RG, Rothman DL (2015) Homeostasis and the glycogen shunt explains aerobic ethanol production in yeast. Proc Natl Acad Sci U S A 112(35):10902–10907
Shulman RG, Rothman DL, Hyder F (2007) A bold search for baseline. NeuroImage 36(2):277–281
Sickmann HM, Schousboe A, Fosgerau K, Waagepetersen HS (2005) Compartmentation of lactate originating from glycogen and glucose in cultured astrocytes. Neurochem Res 30(10):1295–1304
Sickmann HM, Waagepetersen HS, Schousboe A, Benie AJ, Bouman SD (2010) Obesity and type 2 diabetes in rats are associated with altered brain glycogen and amino-acid homeostasis. J Cereb Blood Flow Metab 30(8):1527–1537
Sickmann HM, Waagepetersen HS, Schousboe A, Benie AJ, Bouman SD (2012) Brain glycogen and its role in supporting glutamate and GABA homeostasis in a type 2 diabetes rat model. Neurochem Int 60(3):267–275
Sickmann HM, Walls AB, Schousboe A, Bouman SD, Waagepetersen HS (2009) Functional significance of brain glycogen in sustaining glutamatergic neurotransmission. J Neurochem 109(Suppl 1):80–86
Sinadinos C, Valles-Ortega J, Boulan L, Solsona E, Tevy MF, Marquez M, Duran J, Lopez-Iglesias C, Calbo J, Blasco E, Pumarola M, Milan M, Guinovart JJ (2014) Neuronal glycogen synthesis contributes to physiological aging. Aging Cell 13(5):935–945
Singh S, Ganesh S (2012) Phenotype variations in Lafora progressive myoclonus epilepsy: possible involvement of genetic modifiers? J Hum Genet 57(5):283–285
Snow AD, Mar H, Nochlin D, Raskind M, Wight TN (1988) Corpora Amylacea in aging and Alzheimer's brain: immunolocalization of chondroitin sulfate and heparan sulfate proteoglycans. In: Isobe T, Araki S, Uchino F, Kito S, Tsubura E (eds) Amyloid and amyloidosis. Boston, MA, Springer, pp 561–566
Sorg O, Magistretti PJ (1991) Characterization of the glycogenolysis elicited by vasoactive intestinal peptide, noradrenaline and adenosine in primary cultures of mouse cerebral cortical astrocytes. Brain Res 563(1–2):227–233
Sorg O, Magistretti PJ (1992) Vasoactive intestinal peptide and noradrenaline exert long-term control on glycogen levels in astrocytes: blockade by protein synthesis inhibition. J Neurosci 12(12):4923–4931
Soya M, Matsui T, Shima T, Jesmin S, Omi N, Soya H (2018) Hyper-hippocampal glycogen induced by glycogen loading with exhaustive exercise. Sci Rep 8(1):1285
Spiridigliozzi GA, Keeling LA, Stefanescu M, Li C, Austin S, Kishnani PS (2017) Cognitive and academic outcomes in long-term survivors of infantile-onset Pompe disease: a longitudinal follow-up. Mol Genet Metab 121(2):127–137
Strang RH, Bachelard HS (1973) Rates of cerebral glucose utilization in rats anaesthetized with phenobarbitone. J Neurochem 20(4):987–996
Suh SW, Bergher JP, Anderson CM, Treadway JL, Fosgerau K, Swanson RA (2007) Astrocyte glycogen sustains neuronal activity during hypoglycemia: studies with the glycogen phosphorylase inhibitor CP-316,819 ([R-R*,S*]-5-chloro-N-[2-hydroxy-3-(methoxymethylamino)-3-oxo-1-(phenylmethyl)propyl]-1H-indole-2-carboxamide). J Pharmacol Exp Ther 321(1):45–50
Sullivan MA, Nitschke S, Skwara EP, Wang P, Zhao X, Pan XS, Chown EE, Wang T, Perri AM, Lee JPY, Vilaplana F, Minassian BA, Nitschke F (2019) Skeletal muscle glycogen chain length correlates with insolubility in mouse models of polyglucosan-associated neurodegenerative diseases. Cell Rep 27(5):1334–1344.e6
Sullivan MA, Nitschke S, Steup M, Minassian BA, Nitschke F (2017) Pathogenesis of Lafora disease: transition of soluble glycogen to insoluble polyglucosan. Int J Mol Sci 18(8)

Sullivan MA, Powell PO, Witt T, Vilaplana F, Roura E, Gilbert RG (2014) Improving size-exclusion chromatography separation for glycogen. J Chromatogr A 1332:21–29
Suzuki A, Stern SA, Bozdagi O, Huntley GW, Walker RH, Magistretti PJ, Alberini CM (2011) Astrocyte-neuron lactate transport is required for long-term memory formation. Cell 144(5):810–823
Svorad D (1958a) Changes in the topographical distribution of glycogen in the brain during animal hypnosis. Nature 181(4611):775–776
Svorad D (1958b) Diurnal changes in the brain glycogen. Experientia 14(12):452–452
Svorad D (1959) The relation of tranquilizers to some cerebral inhibitory states in topical distribution of brain glycogen. Arch Int Pharmacodyn Ther 121:71–77
Swanson RA, Morton MM, Sagar SM, Sharp FR (1992) Sensory stimulation induces local cerebral glycogenolysis: demonstration by autoradiography. Neuroscience 51(2):451–461
Swanson RA, Sagar SM, Sharp FR (1989a) Regional brain glycogen stores and metabolism during complete global ischaemia. Neurol Res 11(1):24–28
Swanson RA, Shiraishi K, Morton MT, Sharp FR (1990) Methionine sulfoximine reduces cortical infarct size in rats after middle cerebral artery occlusion. Stroke 21(2):322–327
Swanson RA, Yu AC, Sharp FR, Chan PH (1989b) Regulation of glycogen content in primary astrocyte culture: effects of glucose analogues, phenobarbital, and methionine sulfoximine. J Neurochem 52(5):1359–1365
Tagliabracci VS, Heiss C, Karthik C, Contreras CJ, Glushka J, Ishihara M, Azadi P, Hurley TD, DePaoli-Roach AA, Roach PJ (2011) Phosphate incorporation during glycogen synthesis and Lafora disease. Cell Metab 13(3):274–282
Tagliabracci VS, Turnbull J, Wang W, Girard JM, Zhao X, Skurat AV, Delgado-Escueta AV, Minassian BA, Depaoli-Roach AA, Roach PJ (2007) Laforin is a glycogen phosphatase, deficiency of which leads to elevated phosphorylation of glycogen in vivo. Proc Natl Acad Sci U S A 104(49):19262–19266
Takahashi H, Kamei A, Osawa T, Kawahara T, Takizawa O, Maruyama K (2015) (1)(3)C MRS reveals a small diurnal variation in the glycogen content of human thigh muscle. NMR Biomed 28(6):650–655
Tamrakar P, Shrestha P, Briski KP (2014) Sex-specific basal and hypoglycemic patterns of in vivo caudal dorsal vagal complex astrocyte glycogen metabolic enzyme protein expression. Brain Res 1586:90–98
Teng YT, Su WJ, Hou JW, Huang SF (2004) Infantile-onset glycogen storage disease type II (Pompe disease): report of a case with genetic diagnosis and pathological findings. Chang Gung Med J 27(5):379–384
Testoni G, Duran J, Garcia-Rocha M, Vilaplana F, Serrano AL, Sebastian D, Lopez-Soldado I, Sullivan MA, Slebe F, Vilaseca M, Munoz-Canoves P, Guinovart JJ (2017) Lack of glycogenin causes glycogen accumulation and muscle function impairment. Cell Metab 26(1):256–266.e4
Thurston JH, Hauhart RE, Dirgo JA, Jones EM (1980) Mechanisms of increased brain glucose and glycogen after hydrocortisone: possible clinical significance. Ann Neurol 7(6):515–523
Timiras PS, Woodbury DM, Baker DH (1956) Effect of hydrocortisone acetate, desoxycorticosterone acetate, insulin, glucagon and dextrose, alone or in combination, on experimental convulsions and carbohydrate metabolism. Arch Int Pharmacodyn Ther 105(3-4):450–467
Tombaugh GC, Yang SH, Swanson RA, Sapolsky RM (1992) Glucocorticoids exacerbate hypoxic and hypoglycemic hippocampal injury in vitro: biochemical correlates and a role for astrocytes. J Neurochem 59(1):137–146
Turnbull J, DePaoli-Roach AA, Zhao X, Cortez MA, Pencea N, Tiberia E, Piliguian M, Roach PJ, Wang P, Ackerley CA, Minassian BA (2011) PTG depletion removes Lafora bodies and rescues the fatal epilepsy of Lafora disease. PLoS Genet 7(4):e1002037
Turnbull J, Epp JR, Goldsmith D, Zhao X, Pencea N, Wang P, Frankland PW, Ackerley CA, Minassian BA (2014) PTG protein depletion rescues malin-deficient Lafora disease in mouse. Ann Neurol 75(3):442–446

Veliskova J, Desantis KA (2013) Sex and hormonal influences on seizures and epilepsy. Horm Behav 63(2):267–277
Verkhratsky A, Nedergaard M (2018) Physiology of astroglia. Physiol Rev 98(1):239–389
Virgintino D, Monaghan P, Robertson D, Errede M, Bertossi M, Ambrosi G, Roncali L (1997) An immunohistochemical and morphometric study on astrocytes and microvasculature in the human cerebral cortex. Histochem J 29(9):655–660
Visweswaran P, Prasannan KG, Subrahmanyam K (1969) Effect of insulin on the incorporation of 14C of radioactive glucose into glycogen and carbon dioxide in cerebral cortical slices. J Neurochem 16(9):1389–1392
Waagepetersen HS, Sonnewald U, Larsson OM, Schousboe A (2000a) A possible role of alanine for ammonia transfer between astrocytes and glutamatergic neurons. J Neurochem 75(2):471–479
Waagepetersen HS, Sonnewald U, Larsson OM, Schousboe A (2001) Multiple compartments with different metabolic characteristics are involved in biosynthesis of intracellular and released glutamine and citrate in astrocytes. Glia 35(3):246–252
Waagepetersen HS, Sonnewald U, Schousboe A (2003) Compartmentation of glutamine, glutamate, and GABA metabolism in neurons and astrocytes: functional implications. Neuroscientist 9(5):398–403
Waagepetersen HS, Westergaard N, Schousboe A (2000b) The effects of isofagomine, a potent glycogen phosphorylase inhibitor, on glycogen metabolism in cultured mouse cortical astrocytes. Neurochem Int 36(4–5):435–440
Walter B, Bauer R, Kuhnen G, Fritz H, Zwiener U (2000) Coupling of cerebral blood flow and oxygen metabolism in infant pigs during selective brain hypothermia. J Cereb Blood Flow Metab 20(8):1215–1224
Walz W, Mukerji S (1988) Lactate release from cultured astrocytes and neurons: a comparison. Glia 1(6):366–370
Watanabe H, Passonneau JV (1973) Factors affecting the turnover of cerebral glycogen and limit dextrin in vivo. J Neurochem 20(6):1543–1554
Watanabe H, Passonneau JV (1974) The effect of trauma on cerebral glycogen and related metabolites and enzymes. Brain Res 66(1):147–159
Wehr TA (1992) A brain-warming function for REM sleep. Neurosci Biobehav Rev 16(3):379–397
Williams V, Grossman RG, Edmunds SM (1980) Volume and surface area estimates of astrocytes in the sensorimotor cortex of the cat. Neuroscience 5(7):1151–1159
Wolfe LS, Klatzo I, Miquel J, Tobias C, Haymaker W (1962) Effect of alpha-particle irradiation on brain glycogen in the rat. J Neurochem 9:213–218
Woodbury DM (1972) Biochemical effects of adrenocortical steroids on the central nervous system. In: Lajtha A (ed) Handbook of neurochemistry: Volume VII Pathological chemistry of the nervous system. Boston, MA, Springer, pp 255–287
Woolley DE, Timiras PS (1963) Changes in brain glycogen concentration in rats during high altitude (12,470 FT) exposure. Proc Soc Exp Biol Med 114:571–574
Xu J, Song D, Xue Z, Gu L, Hertz L, Peng L (2013) Requirement of glycogenolysis for uptake of increased extracellular K(+) in astrocytes: potential implications for K (+) homeostasis and glycogen usage in brain. Neurochem Res 38(3):472–485
Yang J, Ruchti E, Petit JM, Jourdain P, Grenningloh G, Allaman I, Magistretti PJ (2014) Lactate promotes plasticity gene expression by potentiating NMDA signaling in neurons. Proc Natl Acad Sci U S A 111(33):12228–12233
Yu AC, Drejer J, Hertz L, Schousboe A (1983) Pyruvate carboxylase activity in primary cultures of astrocytes and neurons. J Neurochem 41(5):1484–1487
Zagha E, McCormick DA (2014) Neural control of brain state. Curr Opin Neurobiol 29:178–186
Zala D, Hinckelmann MV, Yu H, Lyra da Cunha MM, Liot G, Cordelieres FP, Marco S, Saudou F (2013) Vesicular glycolysis provides on-board energy for fast axonal transport. Cell 152(3):479–491

Zhang H-Y, Zhao Y-N, Wang Z-I, Huang Y-F (2015) Chronic corticosterone exposure reduces hippocampal glycogen level and induces depression-like behavior in mice. J Zhejiang Univ Sci B 16(1):62–69

Zhang SZ, Zhao FL, Li KA, Tong SY (2001) Determination of glycogen by Rayleigh light scattering. Anal Chim Acta 431(1):133–139

Zhang Y, Xue Y, Meng S, Luo Y, Liang J, Li J, Ai S, Sun C, Shen H, Zhu W, Wu P, Lu L, Shi J (2016) Inhibition of lactate transport erases drug memory and prevents drug relapse. Biol Psychiatry 79(11):928–939

Zhao Q, Stafstrom CE, Fu DD, Hu Y, Holmes GL (2004) Detrimental effects of the ketogenic diet on cognitive function in rats. Pediatr Res 55(3):498–506

Zhao Y, Zhang Q, Shao X, Ouyang L, Wang X, Zhu K, Chen L (2017) Decreased glycogen content might contribute to chronic stress-induced atrophy of hippocampal astrocyte volume and depression-like behavior in rats. Sci Rep 7:43192

Glycogen in Astrocytes and Neurons: Physiological and Pathological Aspects

Jordi Duran, Agnès Gruart, Juan Carlos López-Ramos, José M. Delgado-García, and Joan J. Guinovart

Abstract Brain glycogen is stored mainly in astrocytes, although neurons also have an active glycogen metabolism. Glycogen has gained relevance as a key player in brain function. In this regard, genetically modified animals have allowed researchers to unravel new roles of this polysaccharide in the brain. Remarkably, mice in which glycogen synthase is abolished in the brain, and thus devoid of brain glycogen, are viable, thereby indicating that the polysaccharide in this organ is not a requirement for survival. While there was growing evidence supporting a role of glycogen in learning and memory, these animals have now confirmed that glycogen participates in these two processes.

The association of epilepsy with brain glycogen has also attracted attention. Analysis of genetically modified mice indicates that the relation between brain glycogen and epilepsy is complex. While the formation of glycogen aggregates clearly underlies epilepsy, as in Lafora Disease (LD), the absence of glycogen also favors the occurrence of seizures.

LD is a rare genetic condition that affects children. It is characterized by epileptic seizures and neurodegeneration, and it develops rapidly until finally causing death. Research into this disease has unveiled new aspects of glycogen metabolism.

J. Duran
Institute for Research in Biomedicine (IRB Barcelona), Barcelona Institute of Science and Technology, Barcelona, Spain

Centro de Investigación Biomédica en Red de Diabetes y Enfermedades Metabólicas Asociadas (CIBERDEM), Madrid, Spain

A. Gruart · J. C. López-Ramos · J. M. Delgado-García
Division of Neurosciences, Pablo de Olavide University, Seville, Spain

J. J. Guinovart (✉)
Institute for Research in Biomedicine (IRB Barcelona), Barcelona Institute of Science and Technology, Barcelona, Spain

Centro de Investigación Biomédica en Red de Diabetes y Enfermedades Metabólicas Asociadas (CIBERDEM), Madrid, Spain

Department of Biochemistry and Molecular Biomedicine, University of Barcelona, Barcelona, Spain
e-mail: guinovart@irbbarcelona.org

M. DiNuzzo, A. Schousboe (eds.), *Brain Glycogen Metabolism*, Advances in Neurobiology 23, https://doi.org/10.1007/978-3-030-27480-1_10

Animal models of LD accumulate polyglucosan bodies formed by aberrant glycogen aggregates, called Lafora bodies (LBs). The abolition of glycogen synthase (GS) prevents the formation of LBs and the development of LD, thereby indicating that glycogen accumulation underlies this disease and the associated symptoms, and thus establishing a clear relation between the accumulation of glycogen aggregates and the incidence of seizures.

Although it was initially accepted that LBs were essentially neuronal, it is now evident that astrocytes also accumulate polyglucosan aggregates in LD. However, the appearance and composition of these deposits differs from that observed in neurons. Of note, the astrocytic aggregates in LD models show remarkable similarities with corpora amylacea (CA), a type of polyglucosan aggregate observed in the brains of aged mice and humans. The abolition of GS in mice also impedes the formation of CA with age and at the same time prevents the formation of a number of protein aggregates associated with aging. Therefore CA may play a role in age-related neurological decline.

Keywords Learning · Memory · Long-term potentiation · Epilepsy · Lafora disease · Corpora amylacea · Hypoxia

Abbreviations

ANLS	Astrocyte-neuron lactate shuttle
APBD	Adult polyglucosan body disease
CAL	Corpora amylacea-like
DAB	1,4-dideoxy-1,4-imino-D-arabinitol
EPM2	Progressive myoclonic epilepsy 2
GBE	Glycogen branching enzyme
GP	Glycogen phosphorylase
GS	Glycogen synthase
HFS	High-frequency stimulus
LBs	Lafora bodies
LD	Lafora disease
LFP	Local field potential
LGS	Liver glycogen synthase
LTP	Long-term potentiation
MGS	Muscle glycogen synthase
PGBs	Polyglucosan bodies
PP1	Protein phosphatase 1
PTG	Protein targeting to glycogen
SAMP8	Senescence accelerated mouse prone 8

1 Brain Glycogen in Learning and Epilepsy

Glycogen is produced by GS—the only enzyme able to synthesize glucose polymers in mammals—and degraded by glycogen phosphorylase (GP) in the cytosol and by alpha-glucosidase in the lysosome. Mammals express two isoforms of GS encoded by *GYS1* and *GYS2*. The latter encodes the hepatic isoform (LGS), whose expression is restricted to the liver, while the former encodes the muscle isoform (MGS), which is widely expressed in all organs except the liver (Kaslow et al. 1985). MGS is regulated by phosphorylation at multiple serine residues located in the amino- and carboxy-terminal domains of the enzyme. Phosphorylation by several kinases, including GSK3, induces the inactivation of the enzyme (Roach et al. 1998). Dephosphorylation is facilitated by scaffolding proteins, such as Protein Targeting to Glycogen (PTG), which bring the catalytic subunit of protein phosphatase 1 (PP1) into contact with GS on the glycogen particle, thus causing its dephosphorylation and consequent activation (Vilchez et al. 2007). GS is also allosterically activated by glucose-6-phosphate (G6P) in the brain (Goldberg and O'Toole 1969) and in other tissues (Bouskila et al. 2010; von Wilamowitz-Moellendorff et al. 2013). Glycogen is a branched molecule, which is a crucial property that confers solubility in water. Therefore, the synthesis and degradation of this polysaccharide requires two additional enzymes, namely glycogen branching enzyme, which generates the branching points during the synthesis, and glycogen debranching enzyme (GBE), which removes them during degradation.

Glycogen concentration in the brain is much lower than in the liver or muscle, thus explaining why its function in brain has been largely overlooked. However, a growing body of evidence has accumulated in the last decade indicating that brain glycogen plays an important role in memory formation and learning, in susceptibility to epilepsy, and in many other brain functions. These roles have been elucidated mainly by experiments in which glycogen usage has been impeded by means of specific metabolic inhibitors of GP and more recently confirmed with the use of genetically modified animal models.

The role of glycogen in memory formation was first demonstrated in young chicks with bead discrimination experiments, in which chicks are trained to avoid beads of a specific color (Gibbs et al. 2006). Intracranial injection of 1,4-dideoxy-1,4-imino-D-arabinitol (DAB), a GP inhibitor, or 2-deoxyglucose, which blocks glycolysis, impaired learning of this task. The authors of that study thus concluded that glycogenolysis and aerobic glycolysis are necessary for memory formation (reviewed in (Gibbs 2015)). Other studies performed in rats using DAB injection or knocking down lactate transporters in astrocytes or neurons are consistent with the idea that the mobilization of astrocytic glycogen to generate lactate to feed neurons is critical for memory formation (ANLS hypothesis, see below) (reviewed in (Alberini et al. 2018)).

Epilepsy results from a sudden synchronization of the activity of a group of neurons. During epileptic seizures, neurons and astrocytes dramatically increase their energy consumption. Therefore, the availability of glycogen in this context could be of great importance. In this regard, high glycogen content has been found in biopsies of the brains of epileptic patients (Dalsgaard et al. 2007). Moreover, in the methionine sulfoximine epileptic model, an increase in brain glycogen content has been reported (Phelps 1975; Hevor et al. 1985), and susceptibility of two inbred mouse strains to methionine sulfoximine inversely correlates with their capacity to accumulate glycogen in the brain (Bernard-Helary et al. 2000). Furthermore, the amount of glycogen in epileptic foci is reduced after seizures induced by kainate (Walls et al. 2014) and other epileptogenic agents or situations (Lopez-Ramos et al. 2015). These series of studies have shown a correlation between glycogen content and epilepsy. However, they do not categorically prove that glycogen is directly involved in the disorder.

The use of genetically modified animals in the study of the physiological and pathological aspects of brain glycogen presents clear advantages over methods based on drug treatments, which are hampered by non-specificity and uncontrolled side effects. Furthermore, deletion of specific genes only in the brain of these animals ensures that the results reflect the importance of glycogen exclusively in this organ.. Furthermore and in contrast to pharmacological approaches, genetic manipulation allows researchers to increase the amount of glycogen in specific cell types.

In this regard, the generation of a mouse model that lacks GS specifically in the nervous system (GYS1$^{\text{Nestin-KO}}$), and thus depleted of glycogen in the brain, has shed light on the long-standing question regarding the role of glycogen in this organ (Duran et al. 2013; Lopez-Ramos et al. 2015). The first conclusions drawn from this model is that the animals are viable and that they have a normal lifespan. The same is true for mice lacking GYS1, which are depleted of glycogen in the whole body except the liver (Raben et al. 2001). These results support the notion that brain glycogen is not an absolute requirement for life and that the brain can survive on free glucose. However, GYS1$^{\text{Nestin-KO}}$ animals are not normal, as they present deficiencies in learning and memory and an increased susceptibility to epilepsy (Duran et al. 2013; Lopez-Ramos et al. 2015).

GYS1$^{\text{Nestin-KO}}$ mice show a significant deficit in capacity to learn an instrumental conditioning task (Fig. 1a). They also show deficits in concomitant activity-dependent changes in synaptic strength, the most striking of which is the almost complete absence of Long-Term Potentiation (LTP), the long-lasting increase in the strength of excitatory synapses after a high-frequency stimulation protocol (Duran et al. 2013) (Fig. 1b). LTP is believed to be the cellular process that underlies information storage within neural systems (Martinez and Derrick 1996; Gruart et al. 2006). These observations clearly demonstrate the key role of brain glycogen in the proper and timed acquisition of relatively difficult associative learning tasks. The characterization of GYS1$^{\text{Nestin-KO}}$ mice has also contributed to shedding light on the relation between brain glycogen and epilepsy. Animals lacking brain glycogen are more susceptible to hippocampal seizures after the administration of kainate, a drug widely used to induce epilepsy (Lopez-Ramos et al. 2015) (Fig. 2). This observation thus indicates that glycogen availability contributes to the maintenance of the proper equilibrium between excitatory and inhibitory neurotransmission.

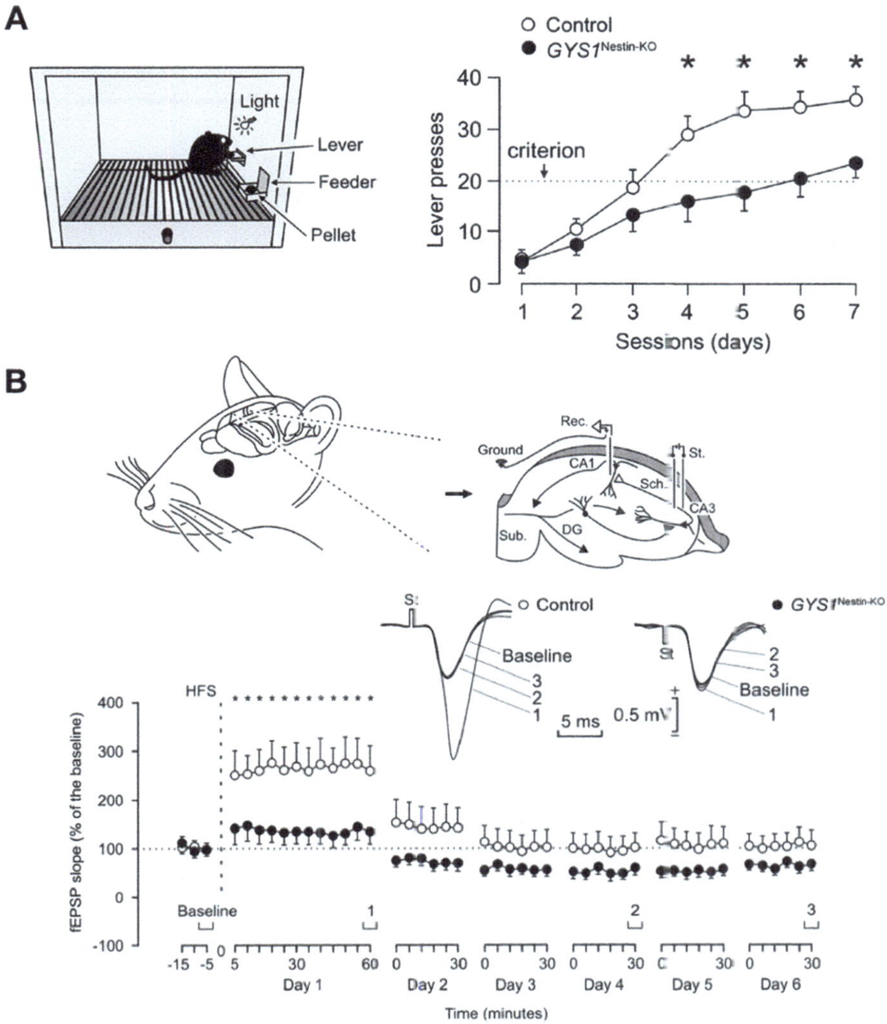

Fig. 1 Adapted from (Duran et al. 2013). (**a**) Impaired performance of GYS1$^{Nestin\text{-}KO}$ mice in an operant conditioning task. Mice were trained in a Skinner box to press a lever to obtain a food pellet (left). Lever presses in the first 7 days of training of task (right). Dotted line corresponds to criterion. (**b**) Animals were chronically implanted with stimulating electrodes in the hippocampal Schaffer collaterals and with a recording electrode in the ipsilateral pyramidal CA1 area. An extra wire was attached to the bone as ground (Top) (DG, dentate gyrus; Sub., subiculum). Time course of long-term potentiation (LTP) evoked in the CA3-CA1 synapse after a high-frequency stimulation (HFS) session (bottom) (mean ± s.e.m. fEPSP slopes given as a percentage of values collected during baseline recordings (100%)). Representative examples of fEPSPs collected at the indicated times are plotted at the top. ∗Statistically significant ($P < 0.05$) differences between the two groups

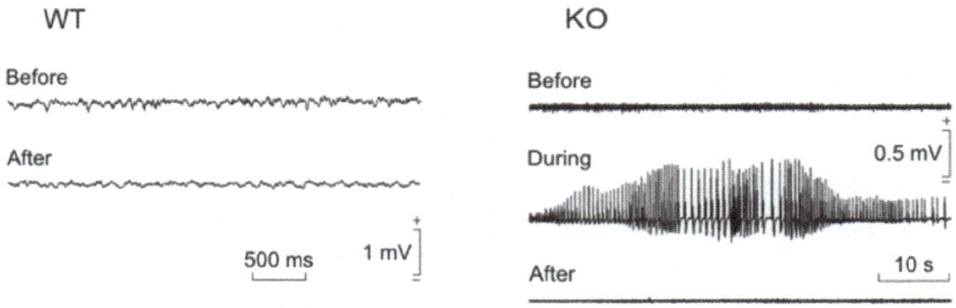

Fig. 2 Adapted from (Lopez-Ramos et al. 2015). Effects of kainate injection on spontaneous electric potential generated by neurons (Local Field Potentials, LFPs) recorded in the hippocampus of behaving mice. Representative examples of LFPs recorded from a wild-type (WT) and a GYS1$^{Nestin\text{-}KO}$ (KO) mouse before and 30 min after a kainate injection (8 mg/kg,i.p.)

2 Astrocytes Vs. Neurons

A further complication in the study of brain glycogen arises from the fact that the polysaccharide is present mainly in astrocytes. Therefore, in brain function, all the physiological roles of glycogen as an energy reserve or contributing to the consolidation of memory have been essentially attributed to astrocytic glycogen. It has been hypothesized that astrocytes accumulate glycogen for the benefit of neurons. However, since astrocytes do not have glucose-6-phosphatase they cannot release free glucose to be taken up by neurons. Therefore, the question as to how neurons can benefit from glycogen stored in astrocytes is puzzling. A simple explanation is that astrocytes use their own glycogen in times of high energy demand, sparing interstitial glucose for neurons. A more sophisticated mechanism implies that astrocytes degrade glycogen to lactate, which is released, and taken up and consumed by neurons—a hypothesis known as the astrocyte-neuron lactate shuttle (ANLS). This hypothesis continues to be debated (reviewed in (Waitt et al. 2017)).

Another difficulty encountered when studying brain glycogen arises from its rapid degradation in post-mortem conditions. This explains why glycogen is difficult to detect in cell types with a low glycogen content, such as neurons, and why its role in these cells has been overlooked (Hertz and Chen 2018). Localization studies performed in microwave-fixed brains (a procedure that stops enzymatic activity immediately and thus preserves the metabolic state) indicate that neurons also contain glycogen, although at much lower concentrations than astrocytes (Oe et al. 2016). Furthermore, the presence of glycogen deposition in neurons in certain conditions such as LD (see below) demonstrates that these cells do synthesize glycogen. In this regard, it has been shown that primary cultured neurons accumulate glycogen, although in small amounts (Fig. 3a), and express GS and GP, thus indicating that these cells have the capacity to synthesize and degrade this polysaccharide (Vilchez et al. 2007; Saez et al. 2014). The availability of glycogen in these neurons is relevant for their tolerance to hypoxia (Fig. 3b). Therefore, a pressing question was to dissect the relative roles of neuronal and astrocyte glycogen in

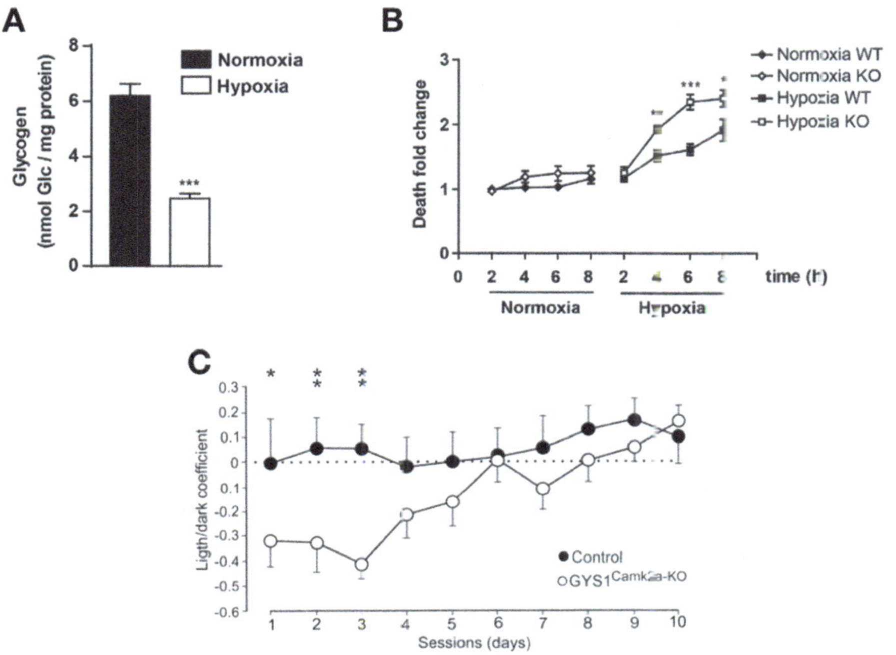

Fig. 3 (**a**) Adapted from (Saez et al. 2014). Neurons contain glycogen, which is mobilized in hypoxia. Glycogen content was determined in control (Normoxia, exposed to environmental 21% O2) and treated neurons (Hypoxia, exposed to 1% O2) and represents the mean ± s.e.m. (**n.**7). ***$P < 0.001$ versus Normoxia. (**b**) Adapted from (Saez et al. 2014). Glycogen synthase (GS) has a protective role in neurons under hypoxia. Death fold change after increasing exposure to hypoxia in GS wild-type (WT) (black) and knockout (KO) (white) neurons. (**c**) Adapted from (Duran et al. 2019). Performance of control and $GYS1^{Camk2a\text{-}KO}$ mice was studied in an operant conditioning task. Mice were placed in a Skinner box and trained to press a lever to obtain a pellet only when a light bulb was switched on. Control mice outperformed $GYS1^{Camk2a\text{-}KO}$ mice [$F_{(9,225)} = 2.82$; $P = 0.01$]

learning and epilepsy. The characterization of a mouse model depleted of glycogen specifically in neurons of the forebrain ($GYS1^{Camk2a\text{-}KO}$) involved in memory and learning has shed light on this long-standing issue (Duran et al. 2019). As in the case of $GYS1^{Nestin\text{-}KO}$ animals, in which GS is depleted from both neurons and astrocytes, $GYS1^{Camk2a\text{-}KO}$ mice showed decreased LTP evoked in the hippocampal CA3-CA1 synapse and a significant deficiency in the acquisition of an instrumental learning task (Fig. 3c). In contrast, they did not present the greater susceptibility to hippocampal seizures and myoclonus observed in the $GYS1^{Nestin\text{-}KO}$ model. These results unequivocally demonstrate the presence of an active glycogen metabolism in neurons in vivo and its fundamental role in the proper acquisition of new motor and cognitive abilities and in the changes in synaptic strength underlying such acquisition.

A conclusion drawn from these studies is that neuronal glycogen is responsible for some of the roles previously attributed exclusively to astrocytic glycogen (Hertz and Chen 2018). Therefore the relevance of neuronal glycogen for brain function should be reconsidered.

3 Lafora Disease

A rare genetic disease, Progressive Myoclonic Epilepsy 2 (EPM2), also known as Lafora disease (LD), has greatly contributed to enhancing our knowledge of glycogen metabolism in the brain and its relation to epilepsy. This disease is an invariably fatal epilepsy that affects both genders equally. Its onset occurs during adolescence, in apparently healthy children, causing absence seizures and/or visual auras. Patients then typically experience generalized tonic-clonic seizures and insidious decline in cognitive function. LD at onset is difficult to distinguish from idiopathic generalized epilepsies. Myoclonic seizures, staring spells, and generalized convulsions follow and escalate over time. LD patients also develop epileptic and non-epileptic visual hallucinations (Gentry et al. 2018; Nitschke et al. 2018).

LD is caused by mutations in either *NHLRC1*, which encodes malin, an E3-ubiquitin ligase, or *EPM2A*, which encodes laforin, a serine-threonine phosphatase. Patients carrying loss-of-function mutations in either of these two genes are indistinguishable. Genetically modified animal models, i.e. the malin knockout (malin^{KO}) (DePaoli-Roach et al. 2010; Valles-Ortega et al. 2011) and the laforin knockout mice (Ganesh et al. 2002), are ideal systems in which to study the disease.

The hallmark of LD is the presence of large inclusions of glycogen aggregates known as polyglucosan bodies (PGBs) or, more specifically, Lafora bodies (LBs), in the brain (Fig. 4) and in other tissues such as muscle and heart (Cavanagh 1999). LBs were traditionally considered to be neuronal inclusions that occur concomitantly with neurodegeneration and epilepsy and that are related to the inexorable worsening of the condition until death in early adulthood (Machado-Salas et al. 2012). However, studies with LD mouse models have demonstrated the presence of PGB also in astrocytes (Valles-Ortega et al. 2011; Auge et al. 2018; Rubio-Villena et al. 2018). Thus, malin^{KO} mice present two types of PGB in the brain, one affecting neurons and the other astrocytes. These two types differ in size, distribution and presence of specific neo-epitopes that are recognized by natural antibodies present in the sera of mammals (Auge et al. 2018). Interestingly, astrocytic, but not neuronal, PGBs also appear in aged control animals, in animal models of accelerated aging and in animals with enhanced glycogen synthesis in the brain (Auge et al. 2018). This observation indicates that the absence of malin triggers the formation of LBs in neurons and enhances the formation of PGBs in astrocytes. Therefore, only the PGBs present in the neurons are specific to LD, while those in astrocytes can be found in other conditions. Given their similarity to corpora amylacea (see below), these astrocytic PGBs could be referred to as corpora amylacea-like granules (CAL) (Fig. 4a).

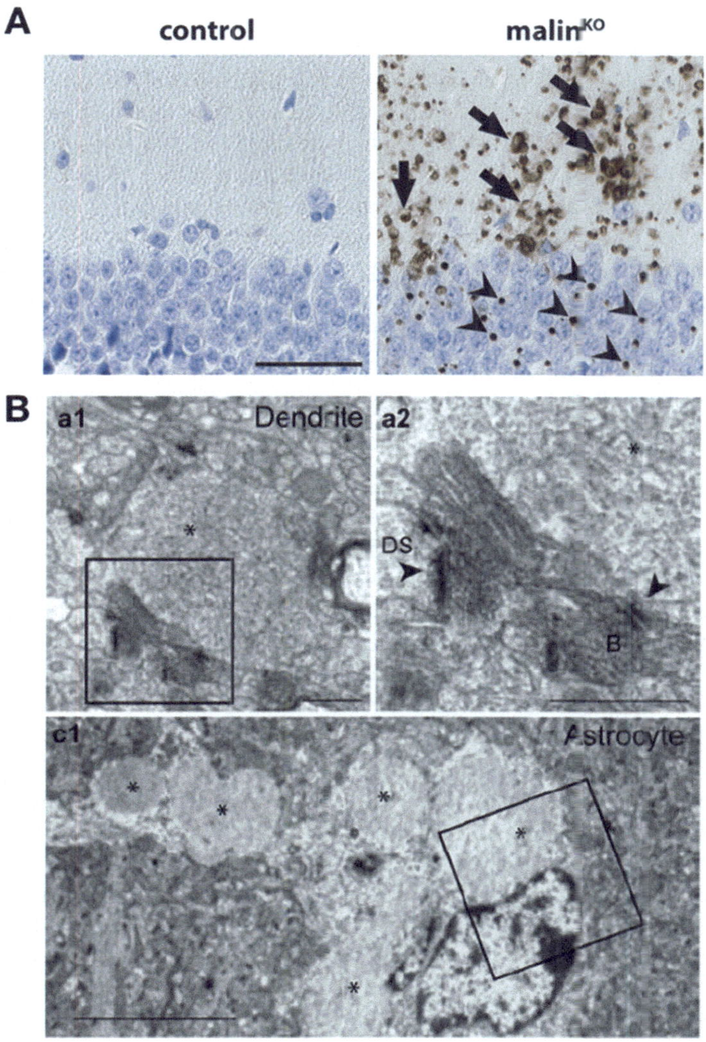

Fig. 4 Lafora bodies in malin^{KO} brain. (**a**) GS immunostaining of control and malin^{KO} brains. Arrows indicate clusters of CAL inclusions, arrowheads indicate nLBs. (**b**) Adapted from (Valles-Ortega et al. 2011). Electron microscopy images of CA1 region. Micrographs depict the presence of LBs and glycogen granules in dendrites (a1, a2) and in astrocytes (c1). *, Lafora Body; black arrowhead: postsynaptic density; B: synaptic bouton; DS: dendritic spine. Scale bars are 5 μm in c1 and 0.5 μm in a1 and a2

The mechanism by which these aggregates form remains unclear. However, given that laforin is able to remove phosphate from glycogen, it has been proposed that the accumulation of phosphorylated glycogen is the main underlying factor involved. This increased phosphate would underlie a change in glycogen structure, decreasing its degree of branching and making it less soluble (Gentry et al. 2009). However, this does not explain the similar accumulation of LBs in the absence of malin, a situation marked by an increase in laforin (Duran et al. 2014). Furthermore, the phosphatase activity of laforin does not appear to be required to prevent LD (Gayarre et al. 2014), and glycogen hyperphosphorylation does not cause the formation of LBs (Nitschke et al. 2017). As mentioned above, while CAL granules are also formed in aging and other conditions in which glycogen synthesis is increased, neuronal LBs appear only in models lacking malin or laforin. However, the presence of high numbers of CAL granules in the malin^{KO} mouse model suggests that the lack of malin alters glycogen metabolism not only in neurons but also in astrocytes. Nevertheless, there is an important difference. While the absence of malin (or laforin) is a requirement to trigger the formation of neuronal LBs, it is not a prerequisite to generate astrocytic CAL granules. However, the absence of malin clearly enhances the formation of these aggregates.

Since the malin–laforin complex had been described to have functions other than that of regulating glycogen synthesis, such as the control of autophagy, the role of glycogen accumulation in the etiopathology of LD was a matter of debate. In this regard, autophagy impairment was hypothesized to underlie neurodegeneration in LD (Criado et al. 2012; Knecht et al. 2012). A first indication that glycogen accumulation is the direct cause of the neurodegeneration and functional impairments seen in LD was that the deletion of PTG prevents the manifestation of LD in the malin^{KO} model (Turnbull et al. 2011). These animals showed reduced glycogen accumulation and the resolution of neurodegeneration and myoclonic epilepsy. However, since PTG is a regulatory protein of PP1, which has many targets, it could not be ruled out that the observed effects were due to changes in the phosphorylation of other targets.

A direct demonstration that glycogen accumulation induces neurodegeneration came from experiments in which a mutant form of MGS in which 9 regulatory serine residues were mutated to alanine, rendering MGS that is not unactivatable by phosphorylation (MGS-9A), was specifically expressed in Purkinje neurons of mice (Duran et al. 2012). Glycogen accumulation in these neurons resulted in a time-dependent loss of these cells by apoptosis and the associated motor impairment (Fig. 5a). Similarly, overexpression of MGS-9A in fly (*Drosophila*) neurons leads to the accumulation of glycogen, reduced lifespan and locomotion defects (Fig. 5b).

However, final evidence of the causal role of glycogen accumulation in the etiopathology of LD was provided by experiments that showed that not only the double laforin-MGS KO (Pederson et al. 2013) and double malin-MGS KO animals (Duran et al. 2014) do not show the accumulation of brain glycogen (Fig. 6a, b) but also that the neurological alterations inherent to LD are rescued. In this regard, a malin-deficient mouse that cannot synthesize glycogen in the brain (malin^{KO} + MGS^{KO}) does not show the increase in markers of neurodegeneration seen in the malin^{KO} model (Fig. 6c). Furthermore, the partial reduction of MGS expression by deleting only one of the GYS1 alleles (malin^{KO} + MGS^{het}) is sufficient to decrease the levels of glycogen (Fig. 6b) and the neurodegeneration markers (Fig. 6c) almost to the levels of control animals (Duran et al. 2014). The increase in LTP of hippocampal synapses and the susceptibility to kainate-induced epilepsy seen in the malin^{KO} model

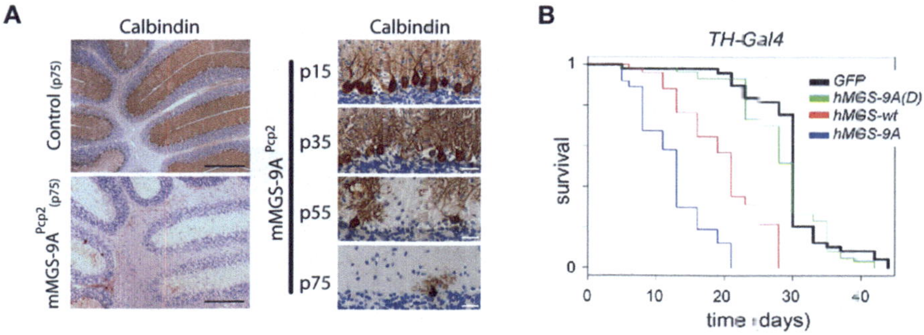

Fig. 5 Adapted from (Duran et al. 2012). (**a**) Progressive loss of Purkinje neurons in animals overexpressing MGS-9A in these cells (mMGS-9A^{Pcp2}). Cerebellar sections of p15-p75 mMGS-9A^{Pcp2} animals labelled to visualize Purkinje cells by Calbindin protein expression (brown). Sections were also labelled with haematoxylin (blue). Scale bars, 500 μm (left) and 30 μm (right). (**b**) Survival assay of control (GFAP) or flies overexpressing wild type MGS (hMGS-wt), MGS-9A (hMGS-9A) or a catalytically inactive form of MGS-9A (hMGS-9A(D)) specifically in neurons

are also reversed in these malin^{KO} + MGS^{het} animals (Fig. 6c). These observations thus indicate that the malin^{KO} mice are rescued from their neurological dysfunctions when GS is diminished. This finding gains relevance in the context of the treatment of LD, as it implies that partial inhibition of GS activity may be sufficient to prevent the progression of the disease. This notion has fueled interest in identifying inhibitors of GS or oligonucleotides able to block its expression in the CNS, as potential treatments for the disease.

At this point, it is interesting to draw attention to a paradox, namely that both the absence and the accumulation of brain glycogen (MGS$^{Nestin\text{-}KO}$ and malin^{KO}, respectively) induce increased susceptibility to epilepsy (Duran et al. 2014; Lopez-Ramos et al. 2015). This paradox could be attributed to non-degradable nature of the glycogen accumulated in LD, thus bringing about a similar situation to that found when there is a lack of glycogen. In this regard, it has been proposed that abnormal glycogen structure contributes to susceptibility to epilepsy (DiNuzzo et al. 2015). In contrast, while the accumulation of brain glycogen in malin^{KO} brains generates an increase in LTP in the CA3-CA1 synapse, the absence of brain glycogen in the MGS$^{Nestin\text{-}KO}$ model results in the loss of LTP in the same synapse (Duran et al. 2013, 2014). Thus, in this case, the increase in LTP in malin^{KO} mice is not a consequence of an incapacity to mobilize glycogen.

As mentioned before, autophagy impairment had been hypothesized to underlie neurodegeneration in LD. Interestingly, autophagy impairment is rescued in malin^{KO} + MGS^{KO} animals, and partially rescued in the malin^{KO} + MGS^{het} animals (Fig. 7a). These observations thus demonstrate that the accumulation of glycogen is not a consequence of autophagy impairment, but rather the cause of it, and show that glycogen accumulation precedes autophagy and not vice versa. To reinforce this idea, other models of glycogen accumulation in the CNS in which malin levels have not been modified (namely animals overexpressing 9A-MGS or PTG) also show impaired autophagy (Duran et al. 2014) (Fig. 7b).

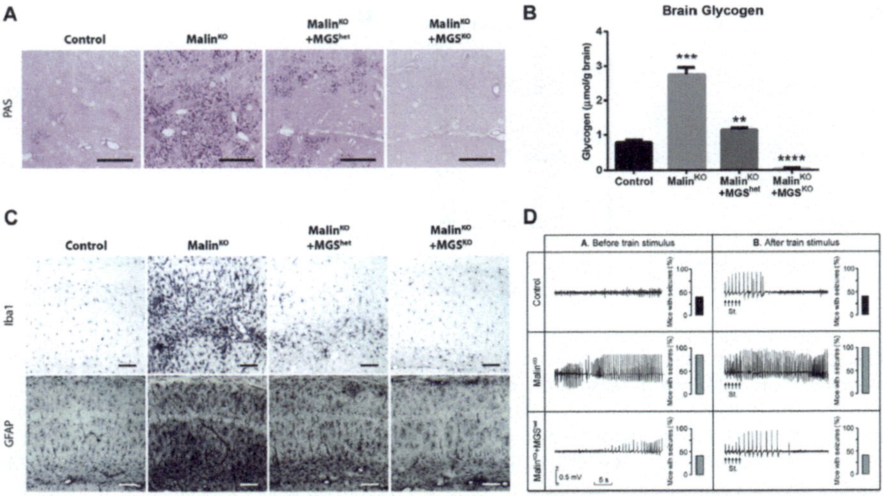

Fig. 6 Adapted from (Duran et al. 2014). (**a**) MGS is indispensable for the formation of LBs. PAS staining shows that LBs accumulated in malin^{KO} brains are absent in malin^{KO} + MGS^{KO} and greatly reduced in malin^{KO} + MGS^{het} brains. (**b**) Graph represents biochemical determination of brain glycogen concentration. Malin^{KO} + MGS^{KO} brains are devoid of glycogen, and the concentration of the polysaccharide is clearly reduced in malin^{KO} + MGS^{het} brains. (**c**) Neurodegeneration is rescued in malin^{KO} mice that cannot synthesize glycogen in the brain. GFAP and Iba1 stainings, as markers of neurodegeneration, show increased staining in the malin^{KO} mice with respect to the controls, thus indicating neurodegeneration. This staining is normalized in the malin^{KO} + MGS^{KO} animals and partially normalized in the malin^{KO} + MGS^{het} animals. (**d**) Increased susceptibility to kainate-induced epilepsy is also dependent on MGS. Representative hippocampal local field potential (LFP) recordings were carried out 30 min after kainic acid injection before and after train stimulation. The percentage (%) of mice presenting seizures within each group is shown to the right of the LFP recordings. Malin^{KO} animals show enhanced susceptibility, which is normalized in malin^{KO} + MGS^{het} animals

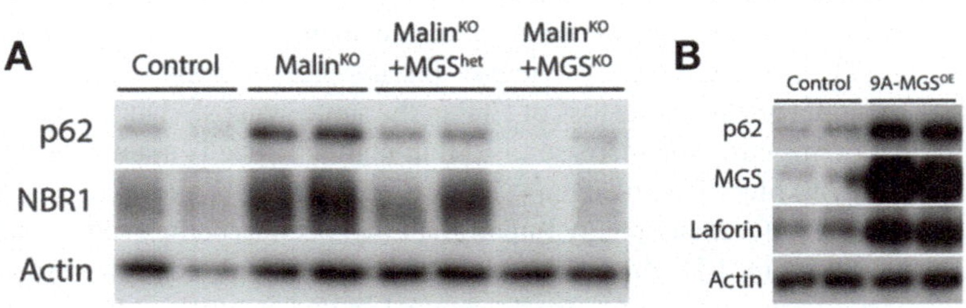

Fig. 7 Adapted from (Duran et al. 2014). (**a**) Analysis of autophagy markers. Accumulation of p62 and NBR1 in malin^{KO} brains was dependent on the expression of MGS. Brain extracts from 11-month-old mice were analyzed by western blot with antibodies against p62 and NBR1. Actin was used as loading control. (**b**) Brain extracts from 3-week-old 9A-MGSOE and littermate controls were analyzed as in **a**

The extensive neuronal loss and severe neuropathological phenotypes observed in patients and animal models of LD suggest that neurons are particularly vulnerable to excess glycogen accumulation (Delgado-Escueta 2007; Valles-Ortega et al. 2011; Duran et al. 2014; Lopez-Gonzalez et al. 2017). This evidences the need for tight control of glycogen synthesis in neurons since an excess of the polysaccharide results in the death of this cell population by apoptosis (Vilchez et al. 2007; Duran et al. 2012). However, the aforementioned accumulation of glycogen in astrocytes may also contribute to the physiopathology of LD.

Nevertheless, LD is not the only condition in which glycogen is deposited in an abnormally manner in nervous tissue. In other diseases like adult polyglucosan body disease and diabetic neuropathy, but also in normal aging, glycogen accumulates in the form of PGBs in neural tissue.

4 Adult Polyglucosan Body Disease

Adult Polyglucosan Body Disease (APBD) is a rare progressive neurodegenerative disorder caused by mutations in (GBE). The lack of GBE produces abnormal glycogen with low solubility due to the lack of branching, thereby leading to the accumulation of PGBs. Loss-of-function mutations of the enzyme cause glycogen storage disease type IV, also known as Andersen disease, a condition that affects neonates and children. In this condition, glycogen accumulates most severely in liver, cardiac and muscle cells, resulting in cirrhosis and death within 5 years. APBD, the adult-onset form of the disease, results from the partial loss of activity of the enzyme (Bruno et al. 1993). It is characterized by the deposition of PGBs in nervous tissue, which induces severe leukodystrophy and atrophy of the spine and medulla (Lossos et al. 1998). The accumulation of glycogen in astrocytes is sufficient to cause the disease (Dainese et al. 2013).

5 Diabetic Neuropathy and Diabetic Retinopathy

Diabetic neuropathy and diabetic retinopathy are two common complications of diabetes mellitus, affecting approximately 60% and 30% of diabetic patients respectively (Vincent and Feldman 2004; Simo et al. 2014). Peripheral nerves are surrounded by the perineurium, which, although acting as a diffusion barrier, is not as efficient as the blood–brain barrier. Consequently, peripheral nerves are in a less well-regulated microenvironment compared to the CNS. In normal conditions, nerve glycogen is present mainly in Schwann cells. However, glycogen accumulation has been described in the axons of peripheral nerves in diabetic patients (Yagihashi and Matsunaga 1979; Mancardi et al. 1985) and in several animal models of diabetes (Moore et al. 1981; Zotova et al. 2008). In fact, intra-axonal glycogen deposition is one of the parameters used to assess diabetic neuropathy (Orloff and Greenleaf 1990).

This glycogen is accumulated in large spherical deposits similar to LBs (Powell et al. 1977). In the retina, glycogen is present in Müller cells and in various types of neuron, especially those of the inner retina (Rungger-Brandle et al. 1996). Although the retina has one of the highest metabolic demands of any tissue, retinal oxygen tension is relatively low, especially in the inner layers. This conditioning might explain why retinal neurons accumulate more glycogen than brain neurons in normal conditions. Interestingly, several reports describe an increased deposition of glycogen in the retina of animal models of diabetic retinopathy (Sosula et al. 1974; Sanchez-Chavez et al. 2008; Osorio-Paz et al. 2012). In the light of the discovery of the toxic role of glycogen accumulation in LD, it is reasonable to hypothesize that glycogen accumulation in the nerves and retinas of diabetic patients could contribute to the physiopathology of diabetic neuropathy and diabetic retinopathy.

6 Aging

The accumulation of PGBs, known as corpora amylacea (CA), has also been observed in the aged human brain (Cavanagh 1999). CA share multiple histological and biochemical characteristics with LBs, including a composition of insoluble, poorly branched polysaccharide, resistance to digestion by amylase, and minor protein content.

Remarkably, the significance of PGB accumulation with regard to aging has been largely overlooked, and only a few authors have proposed that CA have a relevant role in neurodegeneration (Singhrao et al. 1993). Although human brain CA are formed mainly by polyglucosan, the presence of waste elements is a recurrent feature of these structures. This observation suggests that that they are involved in trapping and sequestering potentially hazardous products (Cavanagh 1999; Pirici et al. 2014; Rohn 2015). CA contain a number of neo-epitopes—specific epitopes that are not present in healthy brain structures but appear in situations of cellular stress and tissue damage (Binder 2010; Auge et al. 2017). The neo-epitopes in CA are recognized by natural IgM antibodies, thus revealing the potential role of the natural immune system in the removal of these aggregates (Auge et al. 2017).

In the same way in which CA accumulate with age in the human brain, CAL granules progressively appear in the aging mouse brain. These granules are present in a wide range of mouse strains but are particularly abundant in the senescence-accelerated mouse prone 8 (SAMP8) model (Manich et al. 2016). This model is a non-genetically modified strain of mice with an accelerated aging process and it shares characteristics with aged humans, such as a reduced lifespan, lordosis, hair loss, and reduced physical activity (Takeda 2009). In these animals, CAL granules appear in various regions of the brain as early as 3 months of age, and their number increases faster than in other strains (Jucker et al. 1994a, b; Del Valle et al. 2010). Remarkably, glycogen also accumulates in the brain of aged flies (*Drosophila*). This has relevant consequences for neurological function in vivo. If endogenous glycogen synthase (dGS) is knocked down by RNAi using neuron-specific drivers, glycogen

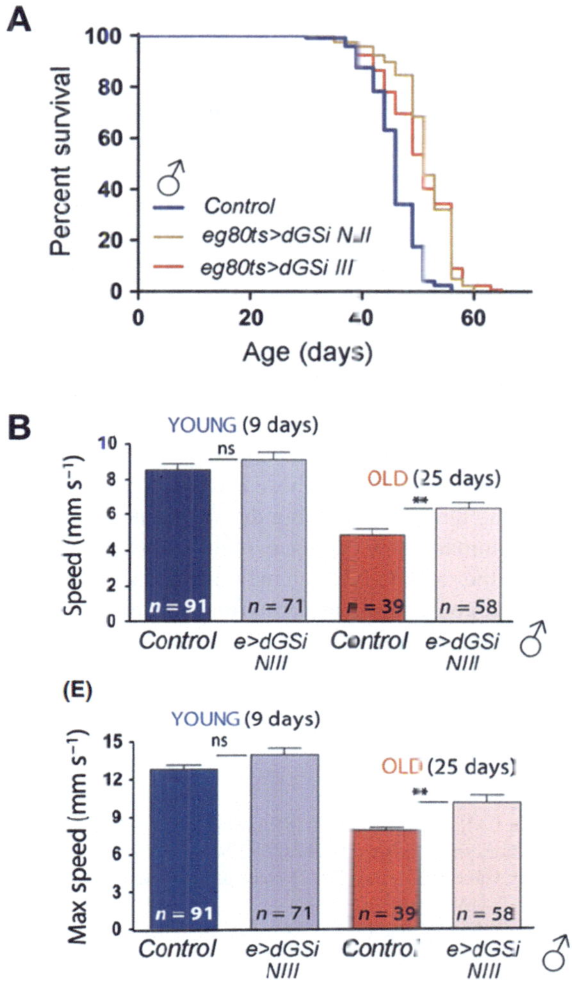

Fig. 8 Adapted from (Sinadinos et al. 2014). Functional consequences of reduced GS in the nervous system of aging Drosophila. (**a**) Survival curves of control and flies with reduced expression of GS (eg80ts > dGSi NIII and eg80ts > dGSi III). (**b**) Average (top graph) maximum (bottom graph) climbing speed of young (9 d, blue) and old (25 d, red) male control flies and flies with reduced expression of GS (e > dGSi NIII)

deposition in the brain is reduced. Adult males expressing dGS-RNAi have a significantly longer median (log-rank test) and maximum lifespan than driver-only-expressing controls (Fig. 8a). Young males expressing dGS-RNAi show a normal climbing response, with no significant change in climbing speed relative to control flies. In contrast, aged males showed a higher average and maximum climbing speed than controls (Fig. 8b). Interestingly, these changes are not observed in female flies.

As indicated earlier, neo-epitopes are present on CAL granules in mouse brains and on CA in human brains (Manich et al. 2016; Auge et al. 2017). Although the composition of CA must be re-examined due to the possible false positive staining in immunohistochemical studies (Auge et al. 2017), there is wide consensus that CA contain waste products (Cavanagh 1999; Pirici et al. 2014) and ubiquitin (Cisse and Schipper 1995; Wilhelmus et al. 2011; Pirici et al. 2014). It has been proposed that the presence of neo-epitopes on CA is related to the removal of these bodies via the natural immune system after their extrusion (Auge et al. 2017). In this regard, CAL granules contain p62 protein, which has a ubiquitin-binding domain and is involved in the sequestration of ubiquitinated proteins and organelles (Liu et al. 2016). As expected, CAL granules are not present in any brain region of MGS^{KO} mice, thereby confirming that glycogen is an essential component of the aggregates. Interestingly, aggregates of alpha-synuclein, hsp70, and ubiquitin, which normally can all be found in the brains of aged controls, are also absent in MGS^{KO} mouse brains (Sinadinos et al. 2014). This observation indicates that glycogen is involved in the formation of protein-based aggregates, such as age-dependent accumulations of aggregation-prone or stress-response proteins. The increased accumulation of these markers in the brains of young $malin^{KO}$ mice is consistent with this proposal.

Collectively, these findings suggest that the progressive accumulation of glycogen aggregates in aged brains is a widespread phenomenon that contributes to neurological decline and that mutations in malin and laforin in LD dramatically increase the rate of this process. These observations thus point to glycogen synthesis as a promising target for reducing the age-related deterioration of the nervous system.

In summary, new evidence, mainly from genetically modified animals, has shed light on the role of glycogen in both astrocytes and neurons. It is now acknowledged that glycogen is required for normal functioning of the brain, but that its overaccumulation induces neurodegeneration, as demonstrated in Lafora disease.

References

Alberini CM, Cruz E et al (2018) Astrocyte glycogen and lactate: new insights into learning and memory mechanisms. Glia 66(6):1244–1262

Auge E, Cabezon I et al (2017) New perspectives on corpora amylacea in the human brain. Sci Rep 7:41807

Auge E, Pelegri C et al (2018) Astrocytes and neurons produce distinct types of polyglucosan bodies in Lafora disease. Glia 66:2094–2107

Bernard-Helary K, Lapouble E et al (2000) Correlation between brain glycogen and convulsive state in mice submitted to methionine sulfoximine. Life Sci 67(14):1773–1781

Binder CJ (2010) Natural IgM antibodies against oxidation-specific epitopes. J Clin Immunol 30(Suppl 1):S56–S60

Bouskila M, Hunter RW et al (2010) Allosteric regulation of glycogen synthase controls glycogen synthesis in muscle. Cell Metab 12(5):456–466

Bruno C, Servidei S et al (1993) Glycogen branching enzyme deficiency in adult polyglucosan body disease. Ann Neurol 33(1):88–93

Cavanagh JB (1999) Corpora-amylacea and the family of polyglucosan diseases. Brain Res Brain Res Rev 29(2–3):265–295

Cisse S, Schipper HM (1995) Experimental induction of corpora amylacea-like inclusions in rat astroglia. Neuropathol Appl Neurobiol 21(5):423–431

Criado O, Aguado C et al (2012) Lafora bodies and neurological defects in Malin-deficient mice correlate with impaired autophagy. Hum Mol Genet 21(7):1521–1533

Dainese L, Monin ML et al (2013) Abnormal glycogen in astrocytes is sufficient to cause adult polyglucosan body disease. Gene 515(2):376–379

Dalsgaard MK, Madsen FF et al (2007) High glycogen levels in the hippocampus of patients with epilepsy. J Cereb Blood Flow Metab 27(6):1137–1141
Del Valle J, Duran-Vilaregut J et al (2010) Early amyloid accumulation in the hippocampus of SAMP8 mice. J Alzheimers Dis 19(4):1303–1315
Delgado-Escueta AV (2007) Advances in lafora progressive myoclonus epilepsy. Curr Neurol Neurosci Rep 7(5):428–433
DePaoli-Roach AA, Tagliabracci VS et al (2010) Genetic depletion of the malin E3 ubiquitin ligase in mice leads to lafora bodies and the accumulation of insoluble laforin. J Biol Chem 285(33):25372–25381
DiNuzzo M, Mangia S et al (2015) Does abnormal glycogen structure contribute to increased susceptibility to seizures in epilepsy? Metab Brain Dis 30(1):307–316
Duran J, Tevy MF et al (2012) Deleterious effects of neuronal accumulation of glycogen in flies and mice. EMBO Mol Med 4(8):719–729
Duran J, Saez I et al (2013) Impairment in long-term memory formation and learning-dependent synaptic plasticity in mice lacking glycogen synthase in the brain. J Cereb Blood Flow Metab 33(4):550–556
Duran J, Gruart A et al (2014) Glycogen accumulation underlies neurodegeneration and autophagy impairment in Lafora disease. Hum Mol Genet 23(12):3147–3156
Duran J, Gruart A et al (2019) Lack of Neuronal Glycogen Impairs Memory Formation and Learning-Dependent Synaptic Plasticity in Mice. Frontiers in Cellular Neuroscience 13
Ganesh S, Delgado-Escueta AV et al (2002) Targeted disruption of the Epm2a gene causes formation of Lafora inclusion bodies, neurodegeneration, ataxia, myoclonus epilepsy and impaired behavioral response in mice. Hum Mol Genet 11(11):1251–1262
Gayarre J, Duran-Trio L et al (2014) The phosphatase activity of laforin is dispensable to rescue Epm2a−/− mice from Lafora disease. Brain 137(Pt 3):806–818
Gentry MS, Dixon JE et al (2009) Lafora disease: insights into neurodegeneration from plant metabolism. Trends Biochem Sci 34(12):628–639
Gentry MS, Guinovart JJ et al (2018) Lafora disease offers a unique window into neuronal glycogen metabolism. J Biol Chem 293(19):7117–7125
Gibbs ME (2015) Role of Glycogenolysis in memory and learning: regulation by noradrenaline, Serotonin and ATP. Front Integr Neurosci 9:70
Gibbs ME, Anderson DG et al (2006) Inhibition of glycogenolysis in astrocytes interrupts memory consolidation in young chickens. Glia 54(3):214–222
Goldberg ND, O'Toole AG (1969) The properties of glycogen synthetase and regulation of glycogen biosynthesis in rat brain. J Biol Chem 244(11):3053–3061
Gruart A, Munoz MD et al (2006) Involvement of the CA3-CA1 synapse in the acquisition of associative learning in behaving mice. J Neurosci 26(4):1077–1087
Hertz L, Chen Y (2018) Glycogenolysis, an astrocyte-specific reaction, is essential for both astrocytic and neuronal activities involved in learning. Neuroscience 370:27–36
Hevor TK, Delorme P et al (1985) Glycogen content and fructose-1, 6-bi phosphatase activity in methionine sulfoximine epileptogenic mouse brain and liver after protein synthesis inhibition. Neuropathol Appl Neurobiol 11(2):129–139
Jucker M, Walker LC et al (1994a) Age-related fibrillar deposits in brains of C57BL/6 mice. A review of localization, staining characteristics, and strain specificity. Mol Neurobiol 9(1–3):125–133
Jucker M, Walker LC et al (1994b) Age-related deposition of glia-associated fibrillar material in brains of C57BL/6 mice. Neuroscience 60(4):875–889
Kaslow HR, Lesikar DD et al (1985) L-type glycogen synthase. Tissue distribution and electrophoretic mobility. J Biol Chem 260(18):9953–9956
Knecht E, Criado-Garcia O et al (2012) Malin knockout mice support a primary role of autophagy in the pathogenesis of Lafora disease. Autophagy 8(4):701–703
Liu WJ, Ye L et al (2016) p62 links the autophagy pathway and the ubiqutin-proteasome system upon ubiquitinated protein degradation. Cell Mol Biol Lett 21:29

Lopez-Gonzalez I, Viana R et al (2017) Inflammation in Lafora disease: evolution with disease progression in Laforin and Malin Knock-out mouse models. Mol Neurobiol 54(5):3119–3130
Lopez-Ramos JC, Duran J et al (2015) Role of brain glycogen in the response to hypoxia and in susceptibility to epilepsy. Front Cell Neurosci 9:431
Lossos A, Meiner Z et al (1998) Adult polyglucosan body disease in Ashkenazi Jewish patients carrying the Tyr329Ser mutation in the glycogen-branching enzyme gene. Ann Neurol 44(6):867–872
Machado-Salas J, Avila-Costa MR et al (2012) Ontogeny of Lafora bodies and neurocytoskeleton changes in Laforin-deficient mice. Exp Neurol 236(1):131–140
Mancardi GL, Schenone A et al (1985) Polyglucosan bodies in the sural nerve of a diabetic patient with polyneuropathy. Acta Neuropathol 66(1):83–86
Manich G, Cabezon I et al (2016) Periodic acid-Schiff granules in the brain of aged mice: from amyloid aggregates to degenerative structures containing neo-epitopes. Ageing Res Rev 27:42–55
Martinez JL Jr, Derrick BE (1996) Long-term potentiation and learning. Annu Rev Psychol 47:173–203
Moore SA, Peterson RG et al (1981) Glycogen accumulation in tibial nerves of experimentally diabetic and aging control rats. J Neurol Sci 52(2–3):289–303
Nitschke F, Sullivan MA et al (2017) Abnormal glycogen chain length pattern, not hyperphosphorylation, is critical in Lafora disease. EMBO Mol Med 9(7):906–917
Nitschke F, Ahonen SJ et al (2018) Lafora disease - from pathogenesis to treatment strategies. Nat Rev Neurol 14(10):606–617
Oe Y, Baba O et al (2016) Glycogen distribution in the microwave-fixed mouse brain reveals heterogeneous astrocytic patterns. Glia 64(9):1532–1545
Orloff MJ, Greenleaf G et al (1990) Reversal of diabetic somatic neuropathy by whole-pancreas transplantation. Surgery 108(2):179–189; discussion 189–190
Osorio-Paz I, Sanchez-Chavez G et al (2012) Control of glycogen content in retina: allosteric regulation of glycogen synthase. PLoS One 7(2):e30822
Pederson BA, Turnbull J et al (2013) Inhibiting glycogen synthesis prevents lafora disease in a mouse model. In: Ann Neurol, vol 74, pp 297–300
Phelps CH (1975) An ultrastructural study of methionine sulphoximine-induced glycogen accumulation in astrocytes of the mouse cerebral cortex. J Neurocytol 4(4):479–490
Pirici I, Margaritescu C et al (2014) Corpora amylacea in the brain form highly branched three-dimensional lattices. Romanian J Morphol Embryol 55(3 Suppl):1071–1077
Powell H, Knox D et al (1977) Alloxan diabetic neuropathy: electron microscopic studies. Neurology 27(1):60–66
Raben N, Danon M et al (2001) Surprises of genetic engineering: a possible model of polyglucosan body disease. Neurology 56(12):1739–1745
Roach PJ, Cheng C et al (1998) Novel aspects of the regulation of glycogen storage. J Basic Clin Physiol Pharmacol 9(2–4):139–151
Rohn TT (2015) Corpora amylacea in neurodegenerative diseases: cause or effect? Int J Neurol Neurother 2(3):031
Rubio-Villena C, Viana R et al (2018) Astrocytes: new players in progressive myoclonus epilepsy of Lafora type. Hum Mol Genet 27(7):1290–1300
Rungger-Brandle E, Kolb H et al (1996) Histochemical demonstration of glycogen in neurons of the cat retina. Invest Ophthalmol Vis Sci 37(5):702–715
Saez I, Duran J et al (2014) Neurons have an active glycogen metabolism that contributes to tolerance to hypoxia. J Cereb Blood Flow Metab 34:945–955
Sanchez-Chavez G, Hernandez-Berrones J et al (2008) Effect of diabetes on glycogen metabolism in rat retina. Neurochem Res 33(7):1301–1308
Simo R, Hernandez C et al (2014) Neurodegeneration in the diabetic eye: new insights and therapeutic perspectives. Trends Endocrinol Metab 25(1):23–33

Sinadinos C, Valles-Ortega J et al (2014) Neuronal glycogen synthesis contributes to physiological aging. Aging Cell 13(5):935–945
Singhrao SK, Neal JW et al (1993) Corpora amylacea could be an indicator of neurodegeneration. Neuropathol Appl Neurobiol 19(3):269–276
Sosula L, Beaumont P et al (1974) Glycogen accumulation in retinal neurons and glial cells of streptozotocin-diabetic rats. Quantitative electron microscopy. Diabetes 23(3):221–231
Takeda T (2009) Senescence-accelerated mouse (SAM) with special references to neurodegeneration models, SAMP8 and SAMP10 mice. Neurochem Res 34(4):639–659
Turnbull J, DePaoli-Roach AA et al (2011) PTG depletion removes Lafora bodies and rescues the fatal epilepsy of Lafora disease. PLoS Genet 7(4):e1002037
Valles-Ortega J, Duran J et al (2011) Neurodegeneration and functional impairments associated with glycogen synthase accumulation in a mouse model of Lafora disease. EMBO Mol Med 3(11):667–681
Vilchez D, Ros S et al (2007) Mechanism suppressing glycogen synthesis in neurons and its demise in progressive myoclonus epilepsy. Nat Neurosci 10(11):1407–1413
Vincent AM, Feldman EL (2004) New insights into the mechanisms of diabetic neuropathy. Rev Endocr Metab Disord 5(3):227–236
Waitt AE, Reed L et al (2017) Emerging roles for glycogen in the CNS. Front Mol Neurosci 10:73
Walls AB, Eyjolfsson EM et al (2014) A subconvulsive dose of kainate selectively compromises astrocytic metabolism in the mouse brain in vivo. J Cereb Blood Flow Metab 34(8):1340–1346
von Wilamowitz-Moellendorff A, Hunter RW et al (2013) Glucose-6-phosphate-mediated activation of liver glycogen synthase plays a key role in hepatic glycogen synthesis. Diabetes 62(12):4070–4082
Wilhelmus MM, Verhaar R et al (2011) Novel role of transglutaminase 1 in corpora amylacea formation? Neurobiol Aging 32(5):845–856
Yagihashi S, Matsunaga M (1979) Ultrastructural pathology of peripheral nerves in patients with diabetic neuropathy. Tohoku J Exp Med 129(4):357–366
Zotova EG, Schaumburg HH et al (2008) Effects of hyperglycemia on rat cavernous nerve axons: a functional and ultrastructural study. Exp Neurol 213(2):439–447

Endurance and Brain Glycogen: A Clue Toward Understanding Central Fatigue

Takashi Matsui, Mariko Soya, and Hideaki Soya

Abstract Brain glycogen stored in astrocytes produces lactate as a neuronal energy source transported by monocarboxylate transporters (MCTs) to maintain neuronal functions, such as hippocampus-regulated memory formation. Although exercise activates brain neurons, the role of astrocytic glycogen in the brain during exercise remains unknown. Since muscle glycogen fuels active muscles during exercise, we hypothesized that astrocytic glycogen plays an energetic role in the brain during exercise to maintain endurance capacity through lactate transport. To explore this hypothesis, we have used a rat model of prolonged exercise, microwave irradiation for the accurate detection of brain glycogen, capillary electrophoresis-mass spectrometry-based metabolomics, and inhibitors of glycogenolysis (1,4-dideoxy-1,4-imino-D-arabinitol; DAB) and lactate transport (α-cyano-4-hydroxycinnamate; 4-CIN). During prolonged exhaustive exercise, muscle glycogen was depleted and brain glycogen decreased when associated with decreased blood glucose levels and increased serotonergic activity known as central fatigue factors, suggesting brain glycogen decrease as an integrative factor for central fatigue. Prolonged exhaustive exercise also increased MCT2 protein in the brain, which takes up lactate in neurons, just as muscle MCTs are increased. Metabolomics revealed that brain but not muscle adenosine triphosphate (ATP) was maintained with lactate and other glycogenolytic and glycolytic sources. Intracerebroventricular (icv) injection of DAB suppressed brain lactate production and decreased hippocampal ATP levels at exhaustion. An icv injection of 4-CIN also decreased hippocampal ATP, resulting in lower endurance capacity. Our findings provide direct evidence that astrocytic glycogen-derived lactate fuels the brain to maintain endurance capacity during exhaustive exercise. Brain ATP levels maintained by glycogen might serve as a possible defense mechanism for neurons in the exhausted state.

Keywords Endurance exercise · Central fatigue · Brain glycogen · Lactate · Metabolomics

---

T. Matsui (✉) · M. Soya · H. Soya (✉)
Sport Neuroscience Division, Faculty of Health and Sport Sciences, Advanced Research Initiative for Human High Performance (ARIHHP), University of Tsukuba, Ibaraki, Japan
e-mail: matsui.takashi.ga@u.tsukuba.ac.jp; soya.hideaki.gt@u.tsukuba.ac.jp

M. DiNuzzo, A. Schousboe (eds.), *Brain Glycogen Metabolism*, Advances in Neurobiology 23, https://doi.org/10.1007/978-3-030-27480-1_11

Abbreviations

4-CIN	α-Cyano-4-hydroxycinnamate
5-HIAA	5-Hydroxyindoleacetic acid
5-HT	5-Hydroxytryptamine (serotonin)
AAA	Aromatic amino acid
AMP	Adenosine monophosphate
ATP	Adenosine triphosphate
BCAA	Branched-chain amino acid
DAB	1,4-Dideoxy-1,4-imino-D-arabinitol
F1-6P	Fructose-1, 6-bisphosphate
fMRI	Functional magnetic resonance imaging
GLUT	Glucose transporter
icv	Intracerebroventricular
IMP	Inosine monophosphate
MCT	Monocarboxylate transporter
MHPG	Methoxyhydroxyphenylglycol
MI	Microwave irradiation
NA	Noradrenaline
PCr	Phosphocreatine
TCA	Tricarboxylic acid

1 Introduction

Glucose derived from the blood is the primary energy source for generating adenosine triphosphate (ATP) in the brain (Sokoloff et al. 1977), but an important brain energy reserve is glycogen synthesized from glucose in astrocytes (Belanger et al. 2011). Astrocytic glycogen is broken down into lactate as a neuronal energy substrate transported by monocarboxylate transporters (MCTs) (Machler et al. 2016). Indeed, brain glycogen decreases during memory tasks (O'Dowd et al. 1994) and in some physiologically exhaustive conditions, such as sleep deprivation (Kong et al. 2002) and hypoglycemia (Herzog et al. 2008). The genetic and/or pharmacologic inhibitions of glycogenolysis and/or lactate transport impair neuronal survival under severe hypoglycemia, axonal transmission, and hippocampus-related memory formation (Swanson and Choi 1993; Newman et al. 2011; Suzuki et al. 2011). Therefore, astrocytic glycogen-derived lactate is a critical energy source for meeting brain energy demands for neuronal functions and/or survival.

Physical exercise activates brain neurons and increases brain energy demand, although to a lesser degree than what occurs in muscles during exercise (Secher et al. 2008). Blood glucose and lactate contribute to brain energetics during moderate or intense exercise (Vissing et al. 1996; Ide et al. 2000; Larsen et al. 2008). Muscle glycogen is an important energy source for maintaining muscle contraction during endurance exercise (Gollnick et al. 1974); however, the role of brain glycogen during exercise remains uncertain. In this article, we describe our recent studies testing the hypothesis that astrocytic glycogen plays an energetic role in the brain during exercise to maintain endurance capacity through lactate transport.

2 Glycogen: The Sole Energy Storage in the Brain

The astrocyte-neuron lactate-shuttle hypothesis proposes that lactate released from astrocytes into the extracellular space is metabolized by neurons (Pellerin and Magistretti 1994; Matsui and Soya 2013) (Fig. 1). To date, it has been thought that the only energy source for the brain is blood-borne glucose. However, previous studies have suggested that lactate released from astrocytes is associated with neuronal activation (Tsacopoulos and Magistretti 1996), and that this lactate is transported to be metabolized predominantly in active neurons via MCTs (Chuquet et al. 2010). Furthermore, culture experiments have revealed that glycogen localized in astrocytes is degraded into lactate by excitatory neurotransmitters, such as noradrenaline (NA) and serotonin (5-hydroxytryptamine; 5-HT) (Magistretti 1988).

Not so long ago, it was believed that brain glycogen levels were so low that they could not play a significant role in cerebral metabolism during several physiological stimulations, including exercise. This is because there were technical difficulties in

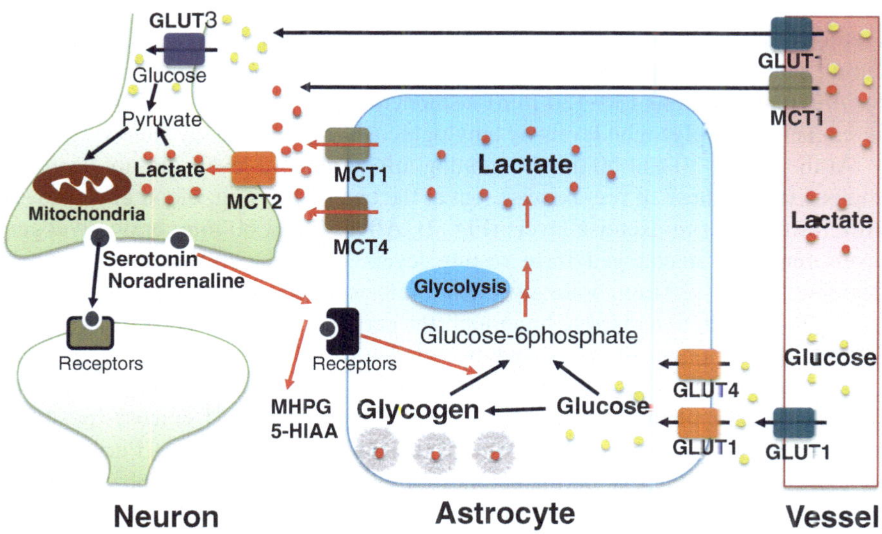

Fig. 1 Glycogen metabolism in the brain: contribution to the astrocyte–neuron lactate shuttle. G-6-P, glucose-6-phosphate; GLUT, Glucose transporter; MCT, monocarboxylic acid transporter. Energy sources for neurons include not only blood glucose but also lactate. Astrocytic glycogen is synthesized from blood glucose and degraded into lactate by excitatory neurotransmitters, such as NA and 5-HT. Lactate is taken up by neurons and changed to pyruvate, which is used for ATP synthesis in the mitochondria. Reproduced from Matsui and Soya (2013) with modifications

determining the post-mortem brain glycogen levels, which was due to the rapid breakdown of brain glycogen following death. A current technique resolves this issue by using high-power (10 kW) microwave irradiation (MI) to inactivate enzymes mediating glycogen metabolism (Kong et al. 2002; Matsui et al. 2011). Kong et al. (2002) indicated that using high-power MI for about 1 s inhibits glycogen metabolism by elevating the brain temperature to approximately 90 °C, which allows the determination of accurate brain glycogen levels.

Animal studies using MI have shown that astrocytic glycogen is degraded into lactate to provide fuel for neurons during hypoglycemia and sleep deprivation (Kong et al. 2002; Herzog et al. 2008). Interestingly, the inhibition of hippocampal glycogen degradation in rats with 1,4-dideoxy-1,4-imino-D-arabinitol (DAB, a glycogen phosphorylase inhibitor) prevented long-term memory formation (Suzuki et al. 2011). Furthermore, hyper brain glycogen contributes to maintaining neuronal activation during hypoglycemia (Suh et al. 2007a). These studies suggest that astrocytic glycogen is an important energy storage for producing lactate for neurons when the glucose provision from blood is insufficient and when there are sudden increases in energy demands during neuronal activation.

3 Brain Glycogen Decreases During Prolonged Exercise

Exercise increases brain energy demand through neuronal activation, and prolonged exercise induces hypoglycemia, leading us to postulate that brain glycogen decreases during exercise. To address this issue, we exercised male Wistar rats on a treadmill for different durations (30–120 min) at moderate intensity (20 m/min) and measured their brain glycogen levels by amyloglucosidase-based assay following MI.

At the end of 30 and 60 min of running, blood glucose levels did not decrease compared with those of pre-exercise, but at the end of 120 min, blood glucose was 46% lower than pre-exercise levels (Fig. 2). After 30 and 60 min, brain glycogen levels remained unchanged from resting levels, but liver and muscle glycogen decreased. After 120 min, brain glycogen levels decreased significantly by an average of 34–60% in five discrete brain loci (the cerebellum 60%, cortex 48%, hippocampus 43%, brainstem 37%, and hypothalamus 34%) compared to those of pre-exercise levels (Fig. 2). The brain glycogen levels after running in all five regions were significantly correlated with the respective blood glucose (positive) and with brain lactate (negative) levels after running (data not shown). Furthermore, in the cortex, the levels of metabolites of NA (methoxyhydroxyphenylglycol; MHPG) and 5-HT (5-hydroxyindoleacetic acid; 5-HIAA), which are potentially involved in the degradation of brain glycogen, increased during prolonged exercise and were negatively correlated with glycogen levels (Fig. 3). These findings indicate that brain glycogen decreases when associated with decreased blood glucose levels and increased serotonergic activity, both of which are known as central fatigue factors (Matsui et al. 2011).

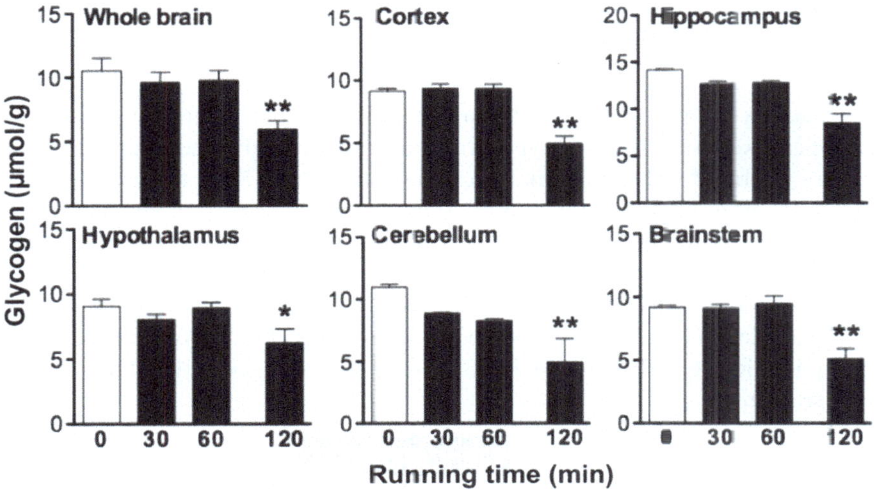

Fig. 2 Brain glycogen decreases during prolonged exercise. The data represent the mean ± SEM (n = 5–6 rats). *P < 0.05; **P < 0.01 compared to pre-exercised rats (Dunnett's *post hoc* test). Reproduced from Matsui et al. (2011) with modifications

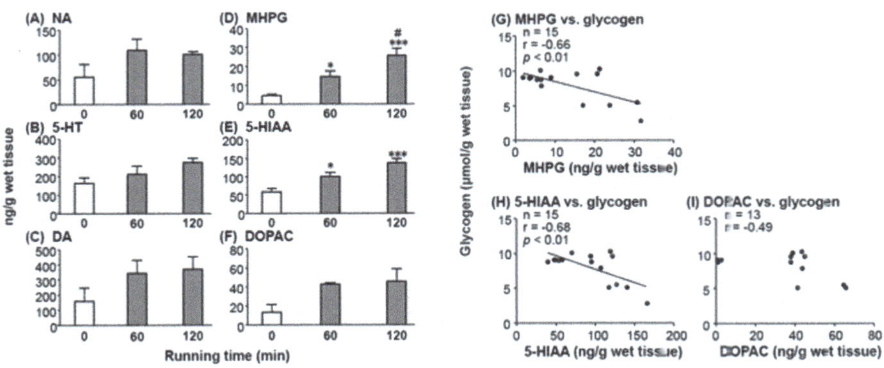

Fig. 3 Monoaminergic activity in the cortex during prolonged exercise. Data represent the mean ± SEM (n = 4–6 rats). (**a**) NA; (**b**) 5-HT; (**c**) dopamine (DA); (**d**) methoxyhydroxyphenylglycol (MHPG); (**e**) 5-hydroxyindoleacetic acid (5-HIAA); and (**f**) 3,4-dihydroxyphenylacetic acid (DOPAC), *P < 0.05; ***P < 0.0001 compared to pre-exercise, †P < 0.05 compared to 60 min (Tucky's post hoc test). (**g**) correlation between MHPG and glycogen levels; (**h**) correlation between 5-HIAA and glycogen levels; and (**i**) correlation between DOPAC and glycogen levels (Pearson's product–moment correlation test). Reproduced from Matsui et al. (2011) with modifications

4 Brain Glycogen Decrease: A Possible Integrative Factor of Central Fatigue

A decrease in brain glycogen may be an integrative factor of central fatigue during prolonged exercise. To date, hypoglycemia with liver glycogen depletion, an increase in brain 5-HT (serotonin hypothesis), and a high brain temperature (hot brain) have been recognized as central fatigue factors during prolonged exercise (Newsholme et al. 1992; Nybo and Secher 2004). These factors are interesting because they are induced from peripheral tissues and have an influence on central fatigue. In addition, hypoglycemia and 5-HT are not only central fatigue factors but also activators for glycogenolysis in astrocytes. Thus, brain glycogen decreases are a possible integrative factor for the development of central fatigue during exercise (Fig. 4).

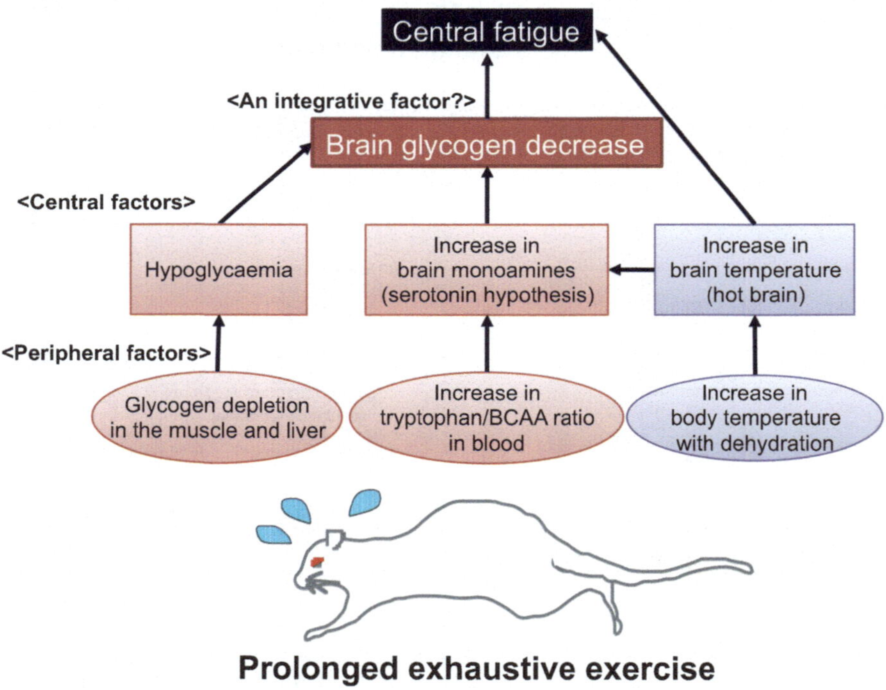

Fig. 4 Hypothetical diagram showing brain glycogen decrease as an integrative factor of central fatigue during prolonged exercise. Prolonged exercise induces glycogen depletion in the muscles and liver, as well as hypoglycemia, which causes peripheral fatigue. Hypoglycemia elicits energy shortages in the brain and likely induces central fatigue. Increase in brain 5-HT due to a rise in the ration of tryptophan to branched-chain amino acid (BCAA) in the blood also induces central fatigue by eliciting lassitude (serotonin hypothesis). Furthermore, increases in body and brain temperature attributed to dehydration induce central fatigue directly and/or indirectly through increases in brain NA and 5-HT. Hypoglycemia and 5-HT are not only inducing factors of central fatigue but also enhancing factors of astrocytic glycogen degradation. Indeed, we observed that brain glycogen levels after running were correlated with the respective blood glucose and increased 5-HT metabolism (Matsui et al. 2011). A decrease in exercise-induced brain glycogen could be an integrative factor of central fatigue. Reproduced from Matsui and Soya (2013)

5 Prolonged Exercise Not Only Decreases Glycogen But Also Increases MCT2 Protein in the Brain

We next examined the effect of exercise on the MCT proteins in the brain (Matsui et al. 2017). Rats were exercised on the treadmill until exhaustion (20 m/min; time to exhaustion 84.4 ± 2.9 min). Blood lactate was significantly increased and glucose levels were significantly decreased compared with the sedentary group ($P < 0.01$). Blood ketone body (β-hydroxybutyrate) levels increased ($P < 0.01$). Exhaustive exercise also caused a depletion (decrease by 97.3%) of muscle glycogen levels ($P < 0.01$). Brain glycogen levels in the cortex and hippocampus were decreased by 75.1 and 66.3%, respectively ($P < 0.01$), but the depletion in the brain did not occur to the same extent as that seen in the skeletal muscles. Concomitantly, MCT2 protein levels in the cortex and hippocampus increased with exhaustive exercise ($P < 0.05$), similar to the increases in MCT proteins observed in the skeletal muscles (Figs. 5a, c, e). Glucose transporter (GLUT) 1 and 3 protein levels increased only in the cortex, and GLUT4 increased in muscles (Figs. 5b, d, f, g). These results imply that MCT2 and GLUT1 and 3 in the brain are very rapidly upregulated with acute prolonged exercise, which is similar to the MCTs in the skeletal muscles (Coles et al. 2004).

6 Metabolomics of the Brain with Exhaustive Exercise

In the next step, we used capillary-electrophoresis-mass-spectrometry-based metabolomics to observe comprehensive energetics of the brain (cortex and hippocampus) and muscle (plantaris) (Matsui et al. 2017). Metabolomics measured 159, 183, and 182 metabolites and revealed that 76, 79, and 72 metabolites were changed significantly in the plantaris muscle, cortex, and hippocampus, respectively, with exhaustive exercise. Principal component analysis and hierarchical cluster analysis clearly indicated the difference in metabolic profiles between sedentariness and exhaustion in all tissues.

The glycolysis map of the plantaris muscle after exhaustive exercise showed depletion of glycogen and glucose and almost total depletion of glycolytic sources and lactate ($P < 0.05$) (Fig. 6a). However, tricarboxylic acid (TCA)-cycle sources increased ($P < 0.05$) (Fig. 6b), suggesting the contribution of β-oxidation of lipids lacking in the brain. Maps of the cortex and hippocampus revealed a decrease in glycogen, glucose, and upstream glycolytic metabolites, including fructose-1, 6-bisphosphate (F1-6P) ($P < 0.05$), but none of these metabolites were depleted. Downstream metabolites, such as F1-6P, together with TCA-cycle sources, were sustained, and the lactate level was increased ($P < 0.01$) (Fig. 6c–f).

Metabolomics-produced purine and pyrimidine maps of the muscle and brain showed that ATP and phosphocreatine (PCr) decreased significantly in the exercise-exhausted group compared with the sedentary group ($P < 0.05$) (Fig. 7a) but that brain

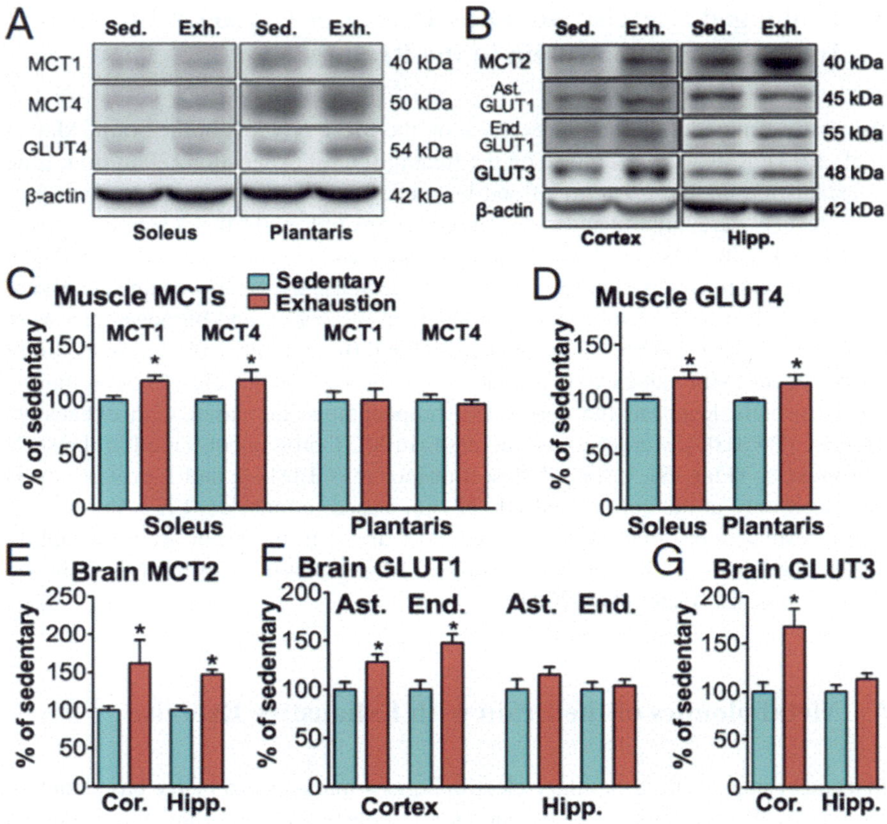

Fig. 5 Exhaustive exercise increases MCT and GLUT protein in muscles and the brain. (**a**) Typical photos of Western blotting bands for MCT1, MCT4, GLUT4, and β-actin in muscles. Exh., exhaustion; Sed., sedentary. (**b**) Typical photos of Western blotting bands for MCT2, astrocytic (Ast.) GLUT1, endothelial (End.) GLUT1, GLUT3, and β-actin in the brain. (**c**) Muscle MCT protein. (**d**) Muscle GLUT4 protein. (**e**) Brain MCT2 protein in the cortex and hippocampus. (**f**) Brain GLUT1 protein. (**g**) Brain GLUT3 protein. Data are expressed as mean ± SE (n = 5 per group). *P < 0.05. Reproduced from Matsui et al. (2017)

ATP and PCr levels were unchanged after exhaustive exercise (Fig. 7b, c). Several downstream sources of purine metabolism, such as adenosine monophosphate (AMP), inosine, inosine monophosphate (IMP), hypoxanthine, and uric acid, increased in both the muscle and brain of exercise-exhausted animals (P < 0.05) (Fig. 7). These data are direct evidence that ATP consumption is increased in both the brain and muscles but that only brain ATP levels are maintained during exhaustive exercise.

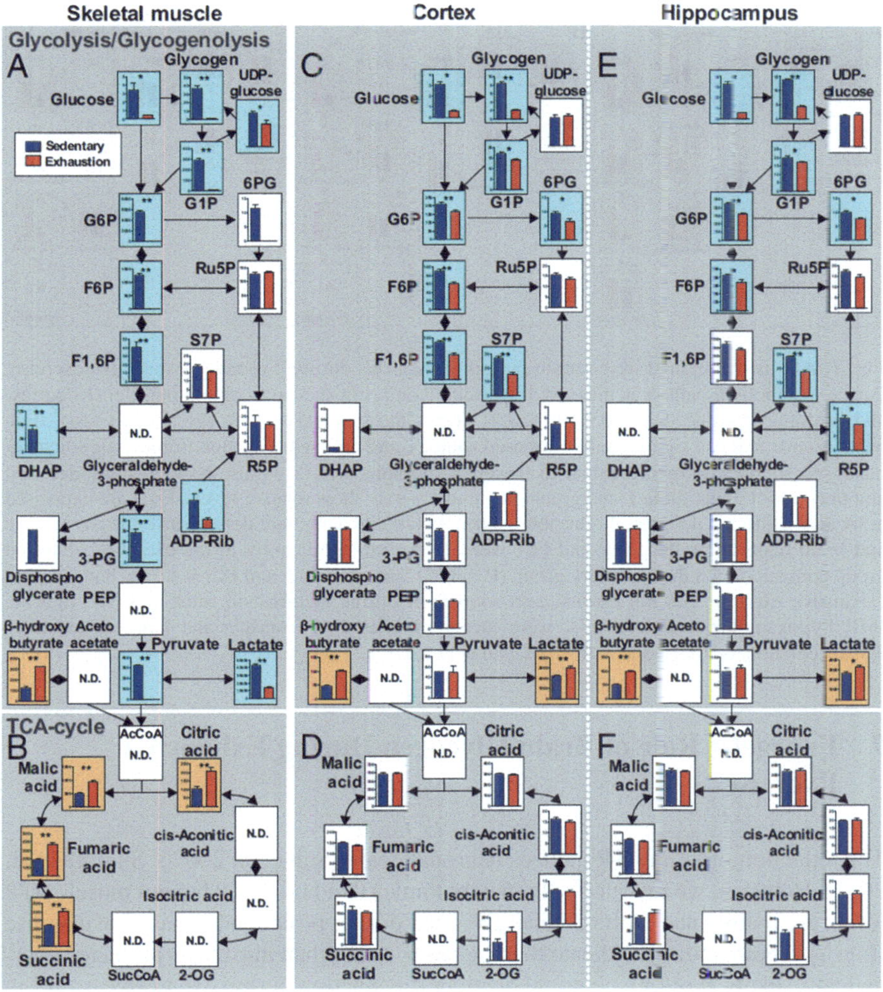

Fig. 6 Lactate increases in the brain but not in muscles during prolonged exhaustive exercise. Glycolytic pathways measured by metabolomics in the (**a**) plantaris muscle, (**c**) cortex, and (**e**) hippocampus. TCA-cycle pathways in the (**b**) plantaris muscle, (**d**) cortex, and (**f**) hippocampus. Glucose and glycogen results are inserted from results of glycogen assays. Data are expressed as mean ± SE (n = 5 per group). *$P < 0.05$, **$P < 0.01$ versus sedentary group (Student's t test); N.D., not determined. Blue backgrounds indicate significantly decreased sources, and orange backgrounds imply significantly increased sources with exhaustive exercise. Graphs with a y-axis show absolute detected amounts (nmol/g wet tissue), and graphs without a y axis show relative levels. The map of the plantaris muscle after exhaustive exercise shows a depletion of glycogen and glucose and almost total depletions of glycolytic sources, including lactate. Maps of the cortex and hippocampus reveal a decrease in glycogen and glucose and upstream glycolytic metabolites, including F1-6P, but none of these metabolites were depleted. Downstream metabolites, such as 3-PG and pyruvate, together with TCA-cycle sources, were sustained and lactate was increased. Reproduced from Matsui et al. (2017)

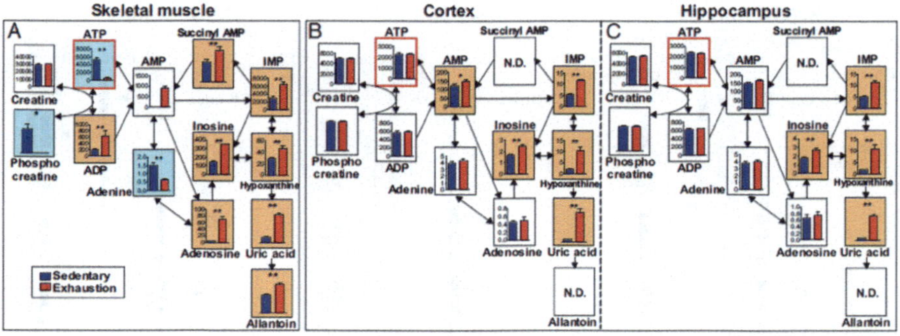

Fig. 7 ATP is maintained in the brain but not in muscles during prolonged exhaustive exercise. Purine metabolism pathways measured by metabolomics in the (**a**) plantaris muscle, (**b**) cortex, and (**c**) hippocampus. $^{*}P < 0.05$, $^{**}P < 0.01$ versus the sedentary group (Student's t test). Blue backgrounds indicate significantly decreased sources, and orange backgrounds imply significantly increased sources following exhaustive exercise. Graphs with a y-axis show absolute detected amounts (nmol/g wet tissue), and graphs without a y-axis show relative levels. Data are expressed as mean ± SE (n = 5 per group). The abbreviated metabolite names are defined in Table S1. Muscle and brain maps show that ATP and PCr were decreased significantly in the exercise-exhausted group compared with the sedentary group ($P < 0.05$), but brain ATP and PCr were unchanged after exhaustive exercise. Several downstream sources of purine metabolism, such as AMP, inosine, IMP, hypoxanthine, and uric acid, were increased in both the muscle and brain of exercise-exhausted animals. Reproduced from Matsui et al. (2017)

7 Energetic Role of Brain Glycogen During Exhaustive Exercise

The reduction of brain ATP induces neuronal death (Suh et al. 2007b; Sugiura et al. 2011). However, we revealed with metabolomics that brain ATP but not muscle ATP levels are maintained with residual glycogen during prolonged exhaustive exercise (Fig. 7). These findings indicate that the brain, rather than muscle, is protected energetically, likely avoiding neuronal death or dysfunction during exhaustive exercise, supporting the "selfish brain" theory regarding energy competition among organs (Peters et al. 2004).

A localized blockade of brain glycogen breakdown (glycogenolysis) using DAB inhibited the increases in hippocampal lactate and decreased hippocampal ATP during exhaustive exercise (Fig. 8a–d). A localized disruption of the MCT2 protein using (α-cyano-4-hydroxycinnamate; 4-CIN) also disrupted the maintenance of hippocampal ATP (Fig. 8f, g). These data indicate that glycogen-derived lactate transported by increased MCT2 is required for brain ATP maintenance during exhaustive exercise, at least in the hippocampus, which is direct evidence for the energetic importance of brain glycogen during endurance exercise. This evidence provides new insight into the strategies that protect and promote brain functions related to performance in animals (e.g., endurance capacity and/or cognitive functions).

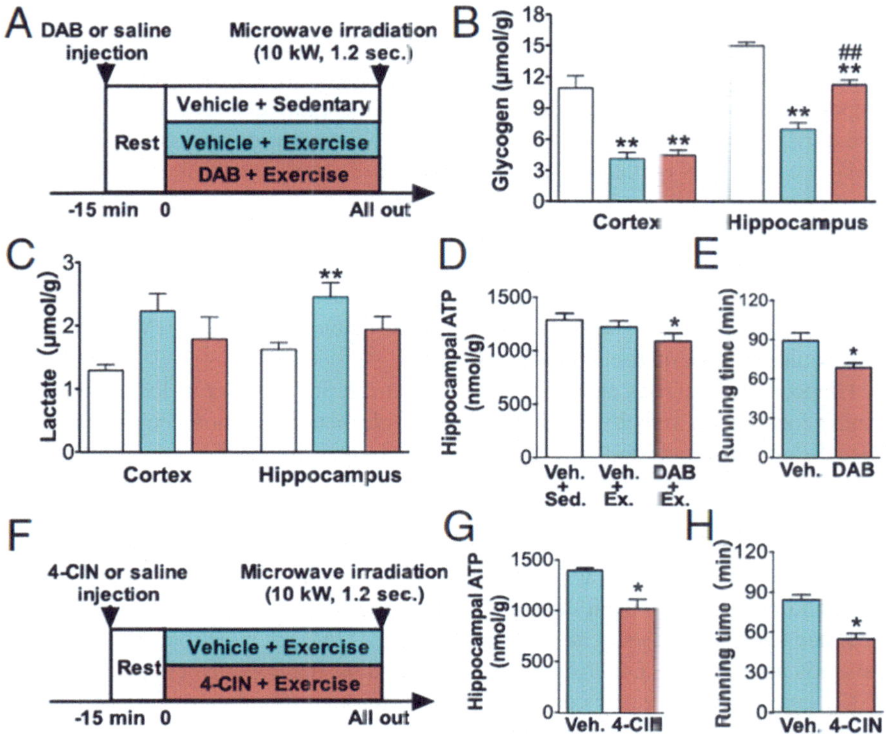

Fig. 8 Blockade of brain glycogenolysis and MCT2 decreases endurance capacity associated with brain ATP. Data are expressed as mean ± SE (n = 5–7 per group). (**a**) Experimental design for exhaustive prolonged exercise with DAB icv injection in rats. (**b**, **c**) Glycogen (**b**) and lactate (**c**) in the cortex and hippocampus. $^{**}P < 0.01$ versus vehicle (Veh.) + sedentary group; $^{\#\#}P < 0.01$ versus Veh. + exercise (by one-way ANOVA with Tukey's post hoc tests). (**d**) ATP in the hippocampus. (**e**) Running time to exhaustion with DAB icv injection. (**f**) Experimental design for exhaustive prolonged exercise with 4-CIN icv injection in rats. (**g**) ATP in the hippocampus. (**h**) Running time to exhaustion with 4-CIN icv injection. $^{*}P < 0.05$ versus vehicle group (by Student's t test). Reproduced from Matsui et al. (2017)

8 Brain Glycogen and Endurance Capacity

The blockade of brain glycogenolysis and lactate transport via MCT2 accelerated exhaustion during prolonged exercise (Fig. 8). These data support the hypothesis that brain lactate derived from astrocytic glycogen plays a role in endurance capacity because of its energetic contribution (Nybo and Secher 2004; Larsen et al. 2008; Matsui et al. 2011). Although DAB affected hippocampal glycogen content, it did not affect cortical glycogen with exhaustive exercise (Fig. 8b). However, intracerebroventricular (icv) inhibition of glycogen phosphorylase blocked glycogen depletion in the hypothalamus, brainstem, and cerebellum, although it only significantly reduced lactate levels in the hippocampus, raising the question of which brain sites signal exhaustion.

DAB inhibited the decrease of glycogen in the hippocampus caused by exhaustive exercise (Fig. 8b). Because hippocampal glycogen-derived lactate acts in memory function (Suzuki et al. 2011), the decreased glycogen utilization in the exhausted hippocampus might be a cause of exercise-induced cognitive fatigue. However, there is methodological difficulty in detecting cognitive functions of exercise-fatigued animals because, due to fatigue, they cannot move when given memory tasks. In humans, exhaustive exercise decreases cognitive functions, including working memory (Perciavalle et al. 2015), but it is unknown whether or not this reflects hippocampus-based cognitive decline. A new system for determining the effects of moderate exercise on pattern separation, a dentate gyrus-specific ability (Suwabe et al. 2017), can be applied together with functional magnetic resonance imaging (fMRI) analysis for fatigue research on this important topic.

Hippocampal neurons also play an important role in the onset of locomotion and exhibit locomotion velocity-dependent firing with theta oscillation (Fuhrmann et al. 2015). Although untested, glycogen-derived lactate might be a contributor to locomotion-dependent hippocampal firing. This postulation could provide novel insight into the significance of the hippocampus not only for memory but also for exercise capacity, implicating the underlying positive relationship between aerobic fitness and cognitive function (Kramer et al. 1999; Erickson et al. 2009; Hyodo et al. 2016).

In addition, the glycogen decrease in the hypothalamus, cerebellum, and brainstem due to exhaustive exercise was inhibited with DAB icv injection. Hypothalamic lactate is an important factor in counterregulation during hypoglycemia (Chan et al. 2013), and brainstem lactate controls of arousal through stimulation of noradrenergic neurons (Tang et al. 2014). Therefore, the blockade of brain glycogenolysis and lactate transport would result in a lower endurance capacity through suppression of brain region-specific functions (e.g., hippocampus: locomotion; hypothalamus: regulation of energy metabolism; cerebellum: motor control; brainstem: arousal control; etc.).

9 Biochemical Insight into the Development of Exhaustion and Central Fatigue during Prolonged Exercise

Fatigue induced by prolonged exercise is separated into muscle and central (brain) factors (Nybo and Secher 2004). Our metabolomics provide insight into the biochemistry behind fatigue during prolonged exercise (Fig. 9). In the muscles of exercise-exhausted rats, ATP, PCr, and glycogen were significantly decreased, whereas hypoxanthine levels were increased due to purine metabolism (Figs. 3 and 4). These findings are consistent with studies on exhausted skeletal muscles (Sahlin et al. 1999). Although undetected in the present study, purine metabolism generates ammonia, which is a known muscle fatigue factor (Broberg and Sahlin 1989). Therefore, the depletion of energy sources and accumulation of inhibitors of muscle contraction (e.g., ammonia) are factors in muscle fatigue (failure of muscle contraction).

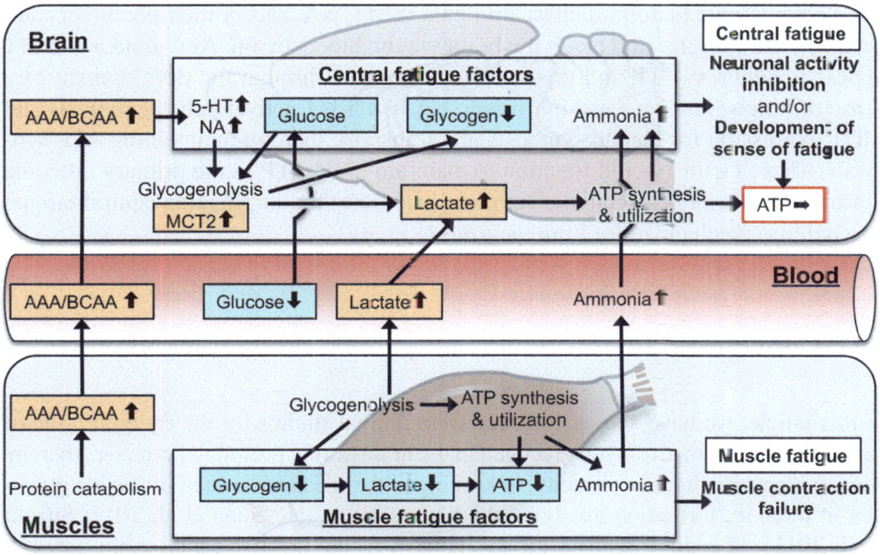

Fig. 9 Schematic figure for muscle–brain coupling in the development of physiological fatigue during exhaustive exercise. In the exhausted skeletal muscle, ATP decrease is associated with glycogen depletion, whereas ammonia production via purine metabolism is increased. These are muscle fatigue factors for the failing muscle contraction. In the exhausted brain, increased uptake of muscle-derived tryptophan as a 5-HT precursor induces a sense of fatigue (tryptophan–serotonin hypothesis). Hypoglycemia, induced together with muscle glycogen depletion, causes a decrease in brain glycogen that is probably compounded by increased NA and 5-HT. In addition, although undetected in the present study, endogenous production of ammonia via purine metabolism in the muscles and brain is probably increased and contributes to inhibition of neuronal activity. Overall, physiological fatigue in the exhausted state is regulated via complex muscle–brain coupling and serves primarily to maintain brain ATP as a biological outcome for neuronal protection. Reproduced from Matsui et al. (2017)

In the brain, ammonia, which is essentially detoxified by astrocytes through the glutamate–glutamine cycle, increases derived from exercising muscles and/or the brain itself and inhibits neuronal activity (Nybo and Secher 2004). The uptake of tryptophan is also increased due to the elevated ratio of branched-chain amino acids (BCAAs) and aromatic amino acids (AAAs) in the blood, which are precursors of NA and 5-HT. Increased brain tryptophan induces elevated 5-HT levels, promoting a "sense of fatigue" and inhibiting neuronal activity (the tryptophan–serotonin hypothesis) (Cotel et al. 2013). Further, hypoglycemia induced by depletion of liver and muscle glycogen creates a lack of energy that further inhibits neuronal activity. NA, 5-HT, and hypoglycemia are also strong activators of astrocytic glycogenolysis. Indeed, increased AAAs, which are likely converted to NA and 5-HT, are correlated with decreased brain glycogen. These factors are also induced not only during fatigue but also by sleep deprivation (Lopez-Rodriguez et al. 2003) and hypoglycemia (Heyes et al. 1990; Herzog et al. 2008), conditions that decrease brain glycogen. Thus, the decrease in brain glycogen is a possible common mechanism for central fatigue.

Central fatigue factors, such as ammonia, 5-HT, NA, and/or their precursors, are derived from muscles and reach the brain via the bloodstream. Ammonia and 5-HT appear to suppress ATP and glycogen consumption through the development of a sense of fatigue and neuronal inhibition. NA and 5-HT activate glycogenolysis and MCT expression for lactate synthesis and transport, thereby maintaining ATP synthesis. These factors would function to maintain brain ATP as the primary outcome through muscle–brain metabolic coupling in exhaustion, implicating central fatigue as a defense mechanism for brain neurons (Fig. 9).

10 Concluding Remarks

In this article, we have shown evidence from animal studies for the energetic role of lactate derived from astrocytic glycogen in the brain during prolonged exercise, thereby contributing to endurance capacity, in keeping with the known role of astrocytic glycogen in memory formation involving the hippocampus (Newman et al. 2011; Suzuki et al. 2011). Shedding light on the mechanism of the positive relationship between endurance and memory, our metabolomics analysis also revealed that the decrease in brain glycogen is a possible factor for exercise-induced central fatigue, which involves muscle–brain metabolic crosstalk. Importantly, the ATP maintenance contributed by brain glycogen likely serves as a neuroprotective mechanism at exhaustion.

Recent human studies using nuclear magnetic resonance (NMR) have also detected brain glycogen metabolism in healthy people and type I diabetes patients (Oz et al. 2009, 2012, 2017). If NMR can be applied to exercise and sports, it may soon be possible to propose new concepts targeting brain glycogen with exercise and/or nutrition to mitigate central fatigue during exercise.

Acknowledgements This research was supported in part by special funds for Education and Research from the Ministry of Education, Culture, Sports, Science and Technology (MEXT) granted to the Human High Performance (HHP) Research Project; the Team "Nippon" Multi-Support project; grants from the Japan Society for the Promotion of Science (JSPS) to the Global Initiative for Sports Neuroscience (GISN): For Development of Exercise Prescription Enhancing Cognitive Functions; and JSPS Grants-in-Aid for Scientific Research A, Challenging Exploratory Research, JSPS Fellow (Superlative Post-Doc), and Young Scientist A.

References

Belanger M, Allaman I, Magistretti PJ (2011) Brain energy metabolism: focus on astrocyte-neuron metabolic cooperation. Cell Metab 14:724–738

Broberg S, Sahlin K (1989) Adenine nucleotide degradation in human skeletal muscle during prolonged exercise. J Appl Physiol (1985) 67:116–122

Chan O, Paranjape SA, Horblitt A, Zhu W, Sherwin RS (2013) Lactate-induced release of GABA in the ventromedial hypothalamus contributes to counterregulatory failure in recurrent hypoglycemia and diabetes. Diabetes 62:4239–4246

Chuquet J, Quilichini P, Nimchinsky EA, Buzsaki G (2010) Predominant enhancement of glucose uptake in astrocytes versus neurons during activation of the somatosensory cortex. J Neurosci 30:15298–15303

Coles L, Litt J, Hatta H, Bonen A (2004) Exercise rapidly increases expression of the monocarboxylate transporters MCT1 and MCT4 in rat muscle. J Physiol 561:253–261

Cotel F, Exley R, Cragg SJ, Perrier JF (2013) Serotonin spillover onto the axon initial segment of motoneurons induces central fatigue by inhibiting action potential initiation. Proc Natl Acad Sci U S A 110:4774–4779

Erickson KI, Prakash RS, Voss MW, Chaddock L, Hu L, Morris KS, White SM, Wojcicki TR, McAuley E, Kramer AF (2009) Aerobic fitness is associated with hippocampal volume in elderly humans. Hippocampus 19:1030–1039

Fuhrmann F, Justus D, Sosulina L, Kaneko H, Beutel T, Friedrichs D, Schoch S, Schwarz MK, Fuhrmann M, Remy S (2015) Locomotion, Theta oscillations, and the speed-correlated firing of hippocampal neurons are controlled by a medial septal glutamatergic circuit. Neuron 86:1253–1264

Gollnick PD, Piehl K, Saltin B (1974) Selective glycogen depletion pattern in human muscle fibres after exercise of varying intensity and at varying pedalling rates. J Physiol 241:45–57

Herzog RI, Chan O, Yu S, Dziura J, McNay EC, Sherwin RS (2008) Effect of acute and recurrent hypoglycemia on changes in brain glycogen concentration. Endocrinology 149:1499–1504

Heyes MP, Papagapiou M, Leonard C, Markey SP, Auer RN (1990) Brain and plasma quinolinic acid in profound insulin-induced hypoglycemia. J Neurochem 54:1027–1033

Hyodo K, Dan I, Kyutoku Y, Suwabe K, Byun K, Ochi G, Kato M, Soya H (2015) The association between aerobic fitness and cognitive function in older men mediated by frontal lateralization. NeuroImage 125:291–300

Ide K, Schmalbruch IK, Quistorff B, Horn A, Secher NH (2000) Lactate, glucose and O_2 uptake in human brain during recovery from maximal exercise. J Physiol 522:159–164

Kong J, Shepel PN, Holden CP, Mackiewicz M, Pack AI, Geiger JD (2002) Brain glycogen decreases with increased periods of wakefulness: implications for homeostatic drive to sleep. J Neurosci 22:5581–5587

Kramer AF, Hahn S, Cohen NJ, Banich MT, McAuley E, Harrison CR, Chason J, Vakil E, Bardell L, Boileau RA, Colcombe A (1999) Ageing, fitness and neurocognitive function. Nature 400:418–419

Larsen TS, Rasmussen P, Overgaard M, Secher NH, Nielsen HB (2008) Non-selective beta-adrenergic blockade prevents reduction of the cerebral metabolic ratio during exhaustive exercise in humans. J Physiol 586:2807–2815

Lopez-Rodriguez F, Wilson CL, Maidment NT, Poland RE, Engel J (2003) Total sleep deprivation increases extracellular serotonin in the rat hippocampus. Neuroscience 121:523–530

Machler P, Wyss MT, Elsayed M, Stobart J, Gutierrez R, von Faber-Castell A, Kaelin V, Zuend M, San Martin A, Romero-Gomez I, Baeza-Lehnert F, Lengacher S, Schneider BL, Aebischer P, Magistretti PJ, Barros LF, Weber B (2016) In vivo evidence for a lactate gradient from astrocytes to neurons. Cell Metab 23:94–102

Magistretti PJ (1988) Regulation of glycogenolysis by neurotransmitters in the central nervous system. Diabete Metab 14:237–246

Matsui T, Soya H (2013) Brain glycogen decrease and supercompensation with exhaustive exercise. In: Hall PA (ed) Social neuroscience and public health. Springer, New York, pp 253–264

Matsui T, Soya S, Okamoto M, Ichitani Y, Kawanaka K, Soya H (2011) Brain glycogen decreases during prolonged exercise. J Physiol 589:3383–3393

Matsui T, Omuro H, Liu YF, Soya M, Shima T, McEwen BS, Soya H (2017) Astrocytic glycogen-derived lactate fuels the brain during exhaustive exercise to maintain endurance capacity. Proc Natl Acad Sci U S A 114:6358–6363

Newman LA, Korol DL, Gold PE (2011) Lactate produced by glycogenolysis in astrocytes regulates memory processing. PLoS One 6:e28427

Newsholme EA, Blomstrand E, Ekblom B (1992) Physical and mental fatigue: metabolic mechanisms and importance of plasma amino acids. Br Med Bull 48:477–495

Nybo L, Secher NH (2004) Cerebral perturbations provoked by prolonged exercise. Prog Neurobiol 72:223–261

O'Dowd BS, Gibbs ME, Ng KT, Hertz E, Hertz L (1994) Astrocytic glycogenolysis energizes memory processes in neonate chicks. Brain Res Dev Brain Res 78:137–141

Oz G, Kumar A, Rao JP, Kodl CT, Chow L, Eberly LE, Seaquist ER (2009) Human brain glycogen metabolism during and after hypoglycemia. Diabetes 58:1978–1985

Oz G, Tesfaye N, Kumar A, Deelchand DK, Eberly LE, Seaquist ER (2012) Brain glycogen content and metabolism in subjects with type 1 diabetes and hypoglycemia unawareness. J Cereb Blood Flow Metab 32:256–263

Oz G, DiNuzzo M, Kumar A, Moheet A, Khowaja A, Kubisiak K, Eberly LE, Seaquist ER (2017) Cerebral glycogen in humans following acute and recurrent hypoglycemia: implications on a role in hypoglycemia unawareness. J Cereb Blood Flow Metab 37:2883–2893

Pellerin L, Magistretti PJ (1994) Glutamate uptake into astrocytes stimulates aerobic glycolysis: a mechanism coupling neuronal activity to glucose utilization. Proc Natl Acad Sci U S A 91:10625–10629

Perciavalle V, Maci T, Perciavalle V, Massimino S, Coco M (2015) Working memory and blood lactate levels. Neurol Sci 36:2129–2136

Peters A, Schweiger U, Pellerin L, Hubold C, Oltmanns KM, Conrad M, Schultes B, Born J, Fehm HL (2004) The selfish brain: competition for energy resources. Neurosci Biobehav Rev 28:143–180

Sahlin K, Tonkonogi M, Soderlund K (1999) Plasma hypoxanthine and ammonia in humans during prolonged exercise. Eur J Appl Physiol Occup Physiol 80:417–422

Secher NH, Seifert T, Van Lieshout JJ (2008) Cerebral blood flow and metabolism during exercise: implications for fatigue. J Appl Physiol 104:306–314

Sokoloff L, Reivich M, Kennedy C, Des Rosiers MH, Patlak CS, Pettigrew KD, Sakurada O, Shinohara M (1977) The [14C]deoxyglucose method for the measurement of local cerebral glucose utilization: theory, procedure, and normal values in the conscious and anesthetized albino rat. J Neurochem 28:897–916

Sugiura Y, Taguchi R, Setou M (2011) Visualization of spatiotemporal energy dynamics of hippocampal neurons by mass spectrometry during a kainate-induced seizure. PLoS One 6:e17952

Suh SW, Bergher JP, Anderson CM, Treadway JL, Fosgerau K, Swanson RA (2007a) Astrocyte glycogen sustains neuronal activity during hypoglycemia: studies with the glycogen phosphorylase inhibitor CP-316,819 ([R-R*,S*]-5-chloro-N-[2-hydroxy-3-(methoxymethylamino)-3-oxo-1-(phenylmethyl)propyl]-1H-indole-2-carboxamide). J Pharmacol Exp Ther 321:45–50

Suh SW, Hamby AM, Swanson RA (2007b) Hypoglycemia, brain energetics, and hypoglycemic neuronal death. Glia 55:1280–1286

Suwabe K, Hyodo K, Byun K, Ochi G, Yassa MA, Soya H (2017) Acute moderate exercise improves mnemonic discrimination in young adults. Hippocampus 27:229–234

Suzuki A, Stern SA, Bozdagi O, Huntley GW, Walker RH, Magistretti PJ, Alberini CM (2011) Astrocyte-neuron lactate transport is required for long-term memory formation. Cell 144:810–823

Swanson RA, Choi DW (1993) Glial glycogen stores affect neuronal survival during glucose deprivation in vitro. J Cereb Blood Flow Metab 13:162–169

Tang F, Lane S, Korsak A, Paton JF, Gourine AV, Kasparov S, Teschemacher AG (2014) Lactate-mediated glia-neuronal signalling in the mammalian brain. Nat Commun 5:3284

Tsacopoulos M, Magistretti PJ (1996) Metabolic coupling between glia and neurons. J Neurosci 16:877–885

Vissing J, Andersen M, Diemer NH (1996) Exercise-induced changes in local cerebral glucose utilization in the rat. J Cereb Blood Flow Metab 16:729–736

Role of Brain Glycogen During Ischemia, Aging and Cell-to-Cell Interactions

Chinthasagar Bastian, John Quinn, Christine Doherty, Caroline Franke, Anna Faris, Sylvain Brunet, and Selva Baltan

Abstract The astrocyte-neuron lactate transfer shuttle (ANLS) is one of the important metabolic systems that provides a physiological infrastructure for glia-neuronal interactions where specialized architectural organization supports the function. Perivascular astrocyte end-feet take up glucose via glucose transporter 1 to actively regulate glycogen stores, such that high ambient glucose upregulates glycogen and low levels of glucose deplete glycogen stores. A rapid breakdown of glycogen into lactate during increased neuronal activity or low glucose conditions becomes essential for maintaining axon function. However, it fails to benefit axon function during an ischemic episode in white matter (WM). Aging causes a remarkable change in astrocyte architecture characterized by thicker, larger processes oriented parallel to axons, as opposed to vertically-transposing processes. Subsequently, aging axons become more vulnerable to depleted glycogen, although aging axons can use lactate as efficiently as young axons. Lactate equally supports function during aglycemia in corpus callosum (CC), which consists of a mixture of myelinated and unmyelinated axons. Moreover, axon function in CC shows greater resilience to a lack of glucose compared to optic nerve, although both WM tracts show identical recovery after aglycemic injury. Interestingly, emerging evidence implies that a lactate transport system is not exclusive to astrocytes, as oligodendrocytes support the axons they myelinate, suggesting another metabolic coupling pathway in WM. Future studies are expected to unravel the details of oligodendrocyte-axon lactate metabolic coupling to establish that all WM components metabolically cooperate and that lactate may be the universal metabolite to sustain central nervous system function.

Keywords D-lactate · White matter · Axon · Ischemia · Ageing · Corpus callosum · End-feet

---

C. Bastian · J. Quinn · C. Doherty · C. Franke · A. Faris · S. Brunet · S. Baltan (✉)
Department of Neurosciences, Lerner Research Institute, Cleveland Clinic Foundation, Cleveland, OH, USA
e-mail: baltans@ccf.org

M. DiNuzzo, A. Schousboe (eds.), *Brain Glycogen Metabolism*, Advances in Neurobiology 23, https://doi.org/10.1007/978-3-030-27480-1_12

1 Introduction

Energy homeostasis in the brain is maintained primarily by oxidative metabolism of glucose to fulfill the energy demands associated with ionic gradients in neurons and astrocytes. The metabolic rate of white matter (WM) is about 50% of the rate of gray matter on the basis of oxygen consumption (Nishizaki et al. 1988), and axons rather than glial cells are the main energy-consuming components of WM (Attwell and Laughlin 2001). It is important to recognize that axon energy metabolism is independent of the neuronal cell body. Because axons extend for great distances from their neuronal cell bodies, they are dependent upon local energy production to maintain ionic homeostasis in order to maintain electrical excitability. Consequently, axons may suffer from energy deprivation independent of the energy status of neuronal cell bodies. The cellular and anatomic organization of astrocytes determine their role in emerging as the main provider to maintain axon function during energy deprivation. Glycogen and glycogen phosphorylase activity are confined to astrocytes (Phelps 1972; Pfeiffer-Guglielmi et al. 2003), and astrocytes are strategically positioned to uptake glucose from the blood circulation via glucose transporter 1 (GLUT 1) (Vannucci et al. 1997; Maher et al. 1994; Simpson et al. 1994) (Fig. 1). Moreover, astrocytic end-feet take up glucose and store it as glycogen

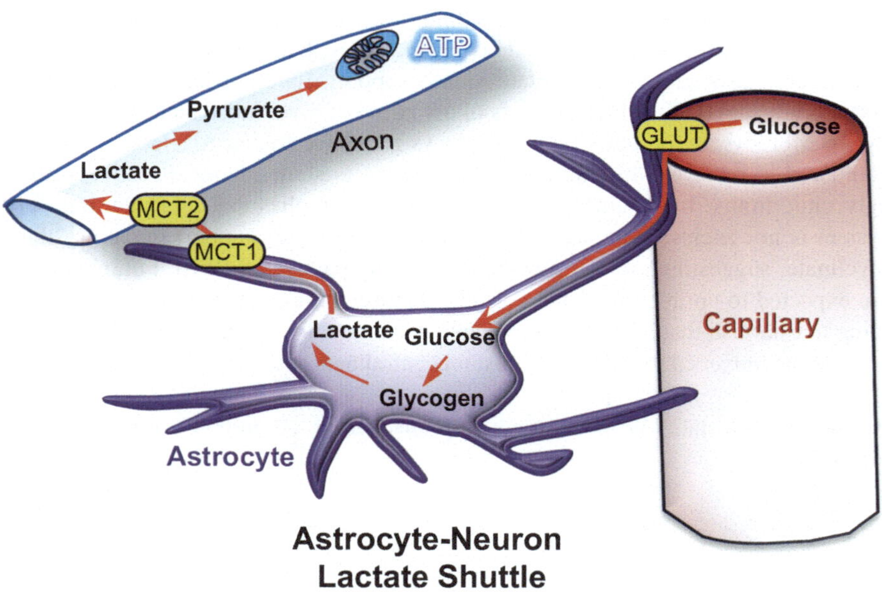

Fig. 1 Schematic diagram of structural and functional components of the astrocyte-neuron lactate shuttle transport (ANLST) in WM. Astrocytes (purple) enwrap capillaries (red) with their end-feet to take up glucose via glucose transporter 1 (GLUT) to store it as glycogen and convert it to lactate via monocarboxylate transporter 1 (MCT1) when ambient glucose levels are low. Axons (blue) take up lactate via MCT2 to derive energy using their mitochondria

(Chih et al. 2001) (Fig. 1). Indeed, astrocytes are the only cells in the mammalian brain that contain significant amounts of glycogen (Cataldo and Broadwell 1986). Based on these properties, astrocytes are the principal cells that are uniquely equipped to maintain energy homeostasis (Magistretti and Pellerin 1996), even for axons extending long distances from their neuronal soma (Brown et al. 2003; Baltan et al. 2008). During increased activity or glucose depletion, glycogen in astrocytes is converted to lactate and extruded via astrocyte monocarboxylate transporter 1 (MCT1) to axons. Axons take up the lactate via monocarboxylate transporter 2 (MCT2) (Fig. 1). Thus, the astrocyte-neuron lactate transfer shuttle (ANLS) efficiently supports axon function. In this paradigm, glycogen serves as a glucose reserve and substrate buffer for local energy demand. The regulation of glycogen in astrocytes is under the control of enzymatic machinery that is exquisitely sensitive to cell energy status. Furthermore, at the cellular level, the sodium-potassium pump (Na^+/K^+ pump) plays a key role in neuron-astrocyte metabolic interactions that form the basis for coupling between axonal activity and energy demands (Munzer et al. 1994; Hertz et al. 2007). The astrocytic Na^+/K^+ pump is intimately sensitive to increases in extracellular K^+ (Henn et al. 1972; Hajek et al. 1996; Honegger and Pardo 1999), and as a result, astrocytes act as sensors during axonal activity when extracellular K^+ is elevated particularly when other ATP-dependent pumps are activated and consume energy during increased axonal activity.

The first established physiological role of brain glycogen proved to be challenging and conflicting. Although Kuffler (Kuffler and Nicholis 1964; Kuffler and Potter 1964; Kuffler and Nicholls 1966; Kuffler et al. 1966) proposed the idea that glial cells participate in maintaining neuronal energy metabolism, it was Tsacopoulus (Tsacopoulos et al. 1994) who established substrate transfer from glial cells to neuronal cells while working with a honeybee retina model system. However, since then, accumulating evidence has shown that there is a functional ANLS in multiple regions of the mammalian brain (Magistretti et al. 1994; Pellerin et al. 2002; Wender et al. 2000). The human brain evolved such that there are equal volumes of glia and neurons and phylogeny indicates that glial density increases with brain complexity and evolution. Therefore, it is no surprise that astrocyte glycogen, which is the sole energy store in brain, can be shared with axons (Brown et al. 2003; Baltan et al. 2008; Tekkok et al. 2005). Faithful conduction of WM axons is crucial for maintaining or regaining brain function after acute or chronic injury. WM tracts comprise axons of various sizes and lengths with varying degrees of myelination. Due to the challenge of meeting metabolic needs of transport and signal conduction in these axons, lactate emerges as a critical energy substrate in WM tracts. Chapter "Metabolism of Glycogen in Brain White Matter" from this book by Ransom and Brown provides the proof-of-concept studies that established the principles of ANLS during glucose deprivation or high-intensity activity in WM. Of particular interest is whether this support system is impaired during aging, which could contribute to age-dependent neurodegenerative diseases. These questions currently remain under investigation. This chapter focuses on (a) whether glia-derived lactate transfer to axons in the optic nerve, a pure WM tract, remains operational to benefit axon function recovery [Figure 2a, Compound Action Potential (CAP), optic nerve]

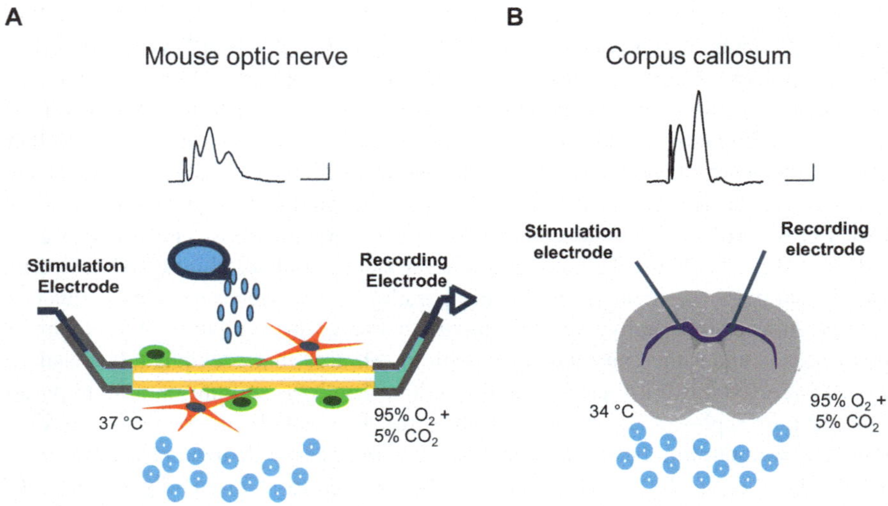

Fig. 2 Electrophysiological recording configuration of axon function. (**a**) Evoked compound action potentials (CAPs) in mouse optic nerve using suction electrodes display a typical three-peaked CAPs at 37 °C. Peaks denote most myelinated and faster-conducting (P1) to lightly myelinated, slow-conducting axons in order (P3). (**b**) Evoked CAPs in corpus callosum show two peaks using extracellular recording configuration at 34 °C. The first peak (P1) represents myelinated, while the second peak (P2) represent the unmyelinated axon response. Recording chambers were oxygenated with 95% O_2 + 5% CO_2

after other metabolic insults such as ischemia induced by combined glucose-oxygen deprivation; (b) whether lactate supports axon function in different WM tracts that consist of myelinated and unmyelinated axons (Fig. 2b, CAP, corpus callosum); and (c) whether aging alters astrocytes and vulnerability of aging axons to glucose deprivation. This chapter will also interrogate whether the glia-axon lactate support mechanism is unique to astrocytes by reviewing evidence supporting oligodendrocyte glycolysis as a potential source of lactate to support axon function in CNS WM. This chapter provides selected published data, while also providing our new and unpublished evidence to focus on the role of lactate as a universal energy buffer for myelinated and unmyelinated axons in young and aging WM.

2 Role of Astrocyte Glycogen During Ischemia in WM

Brain energy metabolism is interrupted in many neurological disorders. Energy failure is the central cause of loss of cerebral function during a hypoglycemic episode or during ischemia. Extensive axonal energy demand and physiological maintenance of glial cells place WM tracts under metabolic challenge, even under physiological conditions (Ransom and Orkand 1996; Harris and Attwell 2012). Moreover, due to their unique vascular network that is less dense and more sparse compared to gray matter, WM components and glial cells have relatively restricted access to energy substrates (Moody et al. 1990). The morphology of WM structure and components further confound energy metabolism because oligodendrocytes wrap myelin around axons to facilitate axon transport, but this inevitably forms a barrier around axons (see Sect. 5 for oligodendrocyte-axon lactate support system) that restricts extracellular metabolites from gaining access to axons (Nave 2010a). Expediently, morphological evidence shows that astrocytes contact axons at nodes of Ranvier (Black and Waxman 1988), presumably to deliver energy substrate to axons via MCT1 (Pellerin et al. 1998). This shuttle system in principle proposes that capillary glucose taken up by astrocyte end-feet is stored as glycogen in astrocytes, to be delivered to axons as lactate on demand (Dringen et al. 1993). Naturally, the first experiments assessed the efficiency of ANLS in WM during glucose depletion. Studies using mouse optic nerve (MON) established a direct correlation between astrocyte glycogen content and axon function (Brown et al. 2003) and further suggested a link between astrocyte-derived lactate and axonal metabolic coupling (Tekkok et al. 2005) during aglycemia. Whether the principles of this shuttle apply to WM during different metabolic insults such as ischemia was investigated. Evidence from neuronal studies showed that glutamate-mediated cortical injury was attenuated by lactate and that neurons preferred lactate as an energy source during excitotoxicity (Alessandri et al. 1996; Ros et al. 2001). However, glycogen-derived lactate under ischemic conditions failed to confer protection to axon function or to improve recovery (Fig. 3a, b, green). After incubating optic nerves in high (30 mM) glucose to upregulate glycogen levels, oxygen and glucose deprivation (OGD) indeed caused a more rapid reduction in axon function following high glucose incubation (Fig. 3b, c, purple), followed by a comparable level of recovery observed

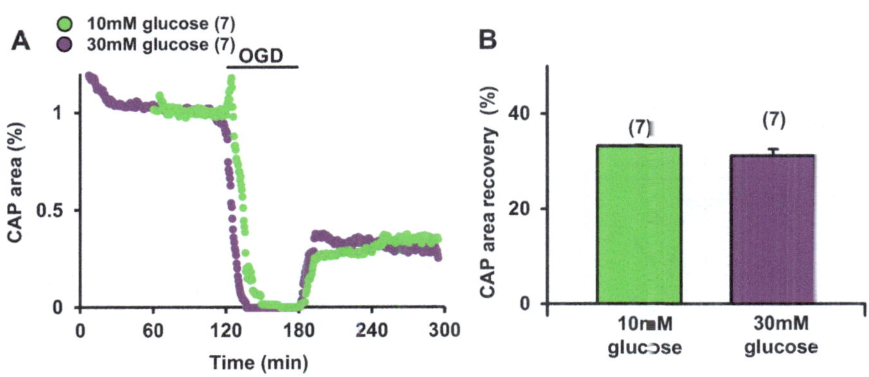

Fig. 3 Lactate cannot be used as a substrate during ischemia. (**a**) Time course of compound action potential (CAPs) area shows that following oxygen glucose deprivation (OGD), axon function recorded in MONs incubated in high (30 mM, purple) glucose declines faster than in MONs incubated in normal (10 mM, green) glucose, but both groups recover to similar levels following 60 min of OGD. (**b**) Histograms show that CAP area recovery is similar between the two groups and is independent of lactate availibility during OGD (60 min). *Figure reproduced from Baltan 2015, Met. Brain Dis*

under normoglycemic conditions (10 mM glucose). In *in vitro* preparations, removal of oxygen and glucose is an accepted form of ischemia and mimics the interruption of blood supply to parts of the brain upon formation of a thrombus. Unlike aglycemia, oxidative phosphorylation is inhibited during ischemia. WM and axon function show an unusual resilience to even prolonged durations of anoxia (Tekkok et al. 2003), such that a subgroup of axons sustain their function solely on glucose. However, during ischemia, glycogen-derived lactate cannot be used as a substrate to support axon function. In contrast, glycogen indeed becomes the preferred metabolite for neuronal elements in the peri-infarct area and neurons benefit from lactate to survive in this region following a stroke (Alessandri et al. 1996, Ros et al. 2001). Studies in the penumbra region after a WM stroke have not yet provided an explanation for this metabolic preference.

3 Lactate Supports Axon Function in Different WM Tracts

The optic nerve consists of fully myelinated axons; therefore, initial experiments inevitably considered ANLS function only in myelinated fibers, raising the subsequent question as to whether this cell-cell interaction is equally important for providing metabolic support to unmyelinated axons. Studies on mammalian peripheral nerves have suggested that Schwann cells operate ANLS to support large myelinated fibers akin to astrocytes in the CNS (A fibers), while glycogen does not benefit the slow-conducting, small diameter unmyelinated axons (C fibers) (Brown et al. 2012). However, lactate is as effective as glucose in supporting sciatic nerve function. We reassessed this phenomenon in corpus callosum (CC), which is a WM tract that consists of a mixture of myelinated and unmyelinated nerves (Fig. 4). Axon function in CC is more sensitive to any form of injury compared to optic nerve. For

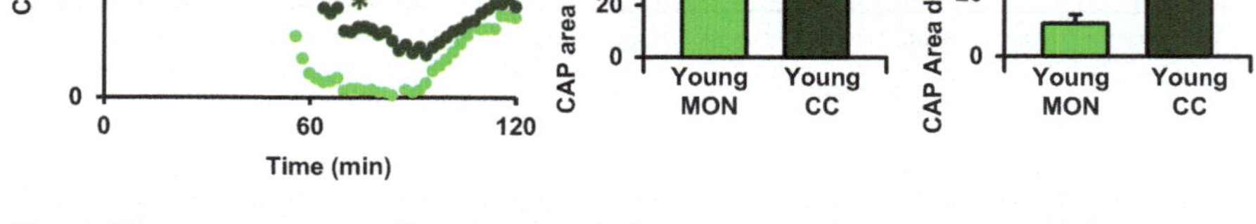

Fig. 4 CC axons are more resilient to removal of glucose than MONs. (**a**) Exposing MONs (light green) and CC (dark green) to aglycemia (0 mM glucose for 60 min) shows a more rapid (green arrows) loss of axon function in MONs compared to CC (**b**), although axon function recovers to similar extent after aglycemia. (**a**, **c**) Interestingly, greater number of axons remain functional during aglycemia in CC (dark green histogram in **c** and asteriks in (**a**)). $^{**}p < 0.01$

instance, exposure to transient OGD for 30 min results in irreversible loss of axon function in CC, while this duration of OGD in optic nerve is followed by ~75% recovery (Tekkok et al. 2007). To our surprise, axon function in CC displayed an unexpected level of resistance to removal of glucose (Fig. 4a, dark green line) compared to optic nerve axons (Fig. 4a, light green line). In addition to a notable delay in axon function loss during aglycemia (Fig. 4a, dark green arrows), a substantial portion of axons sustained their function during glucose deprivation (Fig. 4a, dark green asterisk and c). Moreover, axon function in optic nerve and CC showed identical recovery following 60 min of aglycemia (Fig. 4b). It is interesting to contemplate whether myelinated or unmyelinated axons exhibit this resilience to removal of glucose and whether the subsequent recovery is primarily due to rescue of function of myelinated or unmyelinated axons. Typical evoked compound action potentials (CAPs) in CC exhibit two peaks, one for myelinated axons with a shorter conduction time and a second one for unmyelinated axons with a slower conduction time (Fig. 4b). During aglycemia, the first peak showed an immediate sensitivity to glucose deprivation and was lost first, while the remaining CAP area was mainly due to the second peak. Though the second peak also declined in amplitude over time, a prominent portion was sustained, even at the end of 60 min of aglycemia. Likewise, the recovery was mainly attributable to the second peak. These observations can be interpreted as unmyelinated nerves being more resistant to aglycemia; however, it is also plausible that myelin is the immediate target, so therefore axons become demyelinated and contribute to the unmyelinated group. Further histological and ultrastructural studies are underway to conclusively answer these questions. It is important to note that there is a greater resistance of CC axons to anoxia compared to optic nerve, suggesting that unmyelinated axons and/or the smallest axons with the thinnest myelin sheaths are resistant to anoxia (Baltan 2006). Collectively, these results reveal that CNS WM is remarkably tolerant to glucose deprivation when lactate is available, although there is regional variability in their ability to function and survive this metabolic switch. To achieve optimal protection of the CNS in various neurological diseases, it is critical to understand the properties of regional energy metabolisms and injury mechanisms for successful therapeutic approaches.

4 ANLS Supports Axon Function in Aging WM

What happens to astrocyte structure and function with age? Recent studies performing RNA sequencing of astrocytes from different regions of the brain across the lifespan of the mouse identified age-related transcriptional changes that could contribute to cognitive decline (Clarke et al. 2018). Astrocytes have distinct region-dependent transcriptomal identities and they age in a region dependent manner. For instance, astrocytes in hippocampus and striatum undergo more dramatic transcriptional changes compared to cortical astrocytes. As a result, aged astrocytes in hippocampus or striatum express more reactive neuroinflammatory and ischemia-specific genes together with an activated morphology. On the other hand, the top-down regulated genes in aging astrocytes are involved in mitochondrial function, energy production, and anti-oxidant defence, suggesting that aging astrocytes have impaired energy capacity and low anti-oxidant ability. Since astrocytes are the most abundant class of glial cells, their RNA sequencing profiles may provide clues to a better understanding of the aging brain and age-related changes specific for the blood-brain barrier, for metabolic regulation, and for cognitive function. Because neuron-specific gene expression patterns change relatively little in comparison to astrocytes (Soreq et al. 2017), major changes to astrocytes with age suggest these glial cells are involved in various age-related diseases such as dementia, Alzhemier's disease, Parkinson's disease, as well as various neurodegenerative diseases such as amyotrophic lateral sclerosis and multiple sclerosis. Drastic changes in aging astrocytes also raise the question as to whether these neurodegenerative diseases are neuronal or glial in origin and onset.

Astrocytes are arranged in a specific manner among myelinated axons. While maintaining 3-dimensional cell-to-cell interactions with other glia, astrocytes extend their processes perpendicular to axons and traverse them, interweaving in a bundle of myelinated structures in young WM (Fig. 5a, top row, green for GFAP (+) astrocytes and blue for Sytox (+) nuclei; "magnified" images show cell bodies chosen from "merged" images as indicated by white dotted squares). This architectural organization between astrocytes and axons is such that the tips of astrocyte processes reach in close proximity to nodes of Ranvier, presumably to monitor levels of function, energy demand, and maintenance of ionic homeostasis. This anatomical precision supports the concept of ANLS function. Interestingly, aging causes visible anatomical changes in WM astrocytes in addition to the above transcriptional changes. Aging astrocytes show a prominent increase in soma size and a remarkable increase in the diameter of their processes (Fig. 5a, bottom row, green for GFAP (+) astrocytes and blue for Sytox (+) nuclei; "magnified" images show cell bodies chosen from "merged" images as indicated by white dotted squares). Moreover, these thick processes change their vertical orientation to parallel and follow aging axons along their length. This orientation change of aging astrocyte processes seem to correlate with alterations in Aquaporin 4 (AQP-4) expression. AQP-4 is expressed in the end-feet of young astrocytes, but becomes dispersed throughout the cytoplasm of aging astrocytes (Baltan, data not shown). Loss of perivascular AQP-4 expression is accepted as a factor that renders the aging brain vulnerable to neurodegenerative conditions (Zeppenfeld et al. 2017). Therefore, it is plausible that these structural changes in size and alignment may also imbalance the nutritive function of astrocytes in WM. Indeed, when aging axons (obtained from 12-month-old mice) were deprived of ambient glucose (10 mM), they showed less recovery of CAP area compared to young axons (obtained from 8-week-old mice) following a period of aglycemia (60 min), suggesting that aging axons undergo a metabolic switch to become more dependent upon aerobic metabolism. Indeed, axons undergo complex and profound changes with aging, characterized by an increased diameter enwrapped in thicker myelin and populated with longer and thicker mitochondria that produce less ATP but increased reactive oxidative stress markers (Stahon et al. 2016).

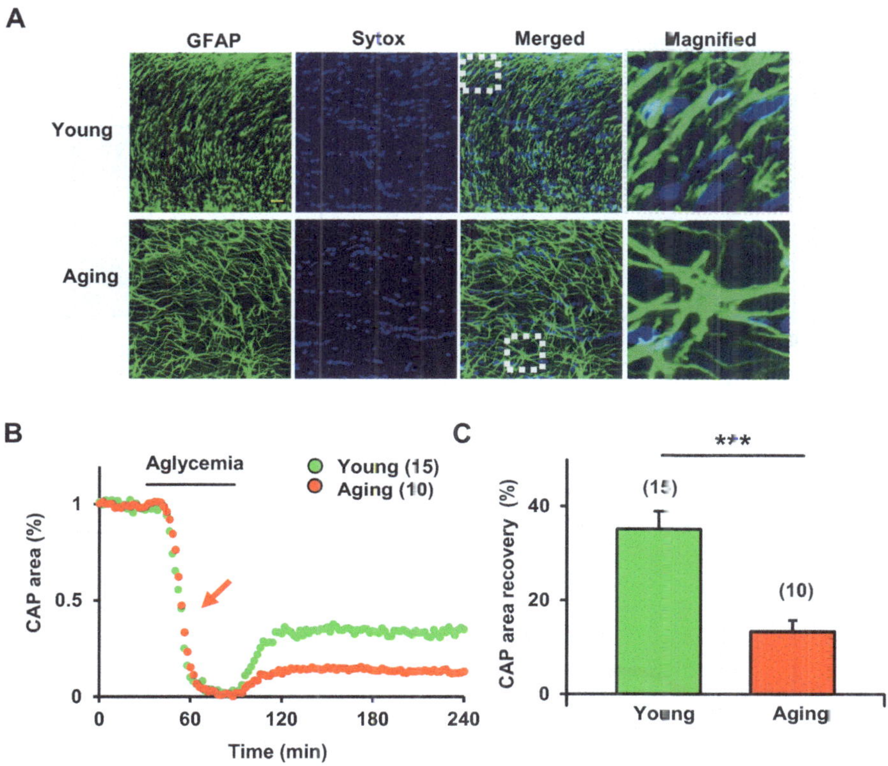

Fig. 5 Aging causes increase in astrocyte size and increased vulnerability to aglycemia in axon function. (**a**) GFAP (+) astrocytes (green) increase in size and display a horizontal orientation with age, while their nuclei remain the same size (Sytox (+) blue). Note the magnified images for compasion between young and aging astrocytes. (**b**, **c**) Aging axons (red line and red histogram) become more vulnerble to removal of glucose compared to young axons (green line and green histograms). ***$p < 0.001$

Mitochondrial dysfunction in aging axons may be the reason why aging axons become more dependent upon glucose (Stahon et al. 2016; Baltan 2014). This metabolic switch may contribute to the increased susceptibility of aging WM to aglycemia. It also suggests that lactate may not be the preferred energy substrate for aging axons, or that lactate delivery to axons may be compromised. Either a decline in MCT2 expression levels by aging axons or reduced availability of lactate provided by aging astrocytes are likely to be the main contributors to dysfunctional ANLS in aging WM. To directly test whether aging axons efficiently take up and use lactate as an alternative energy source, lactate was added (20 mM – carbon equal of glucose) during aglycemia (0 mM). This approach completely disguises the effects of aglycemia in young WM (Fig. 6a, b, green trace and histogram) so that lactate equally maintains axon function (Fig. 6a, b, blue trace and histogram). Surprisingly, lactate equally sustained aging axon function when glucose was

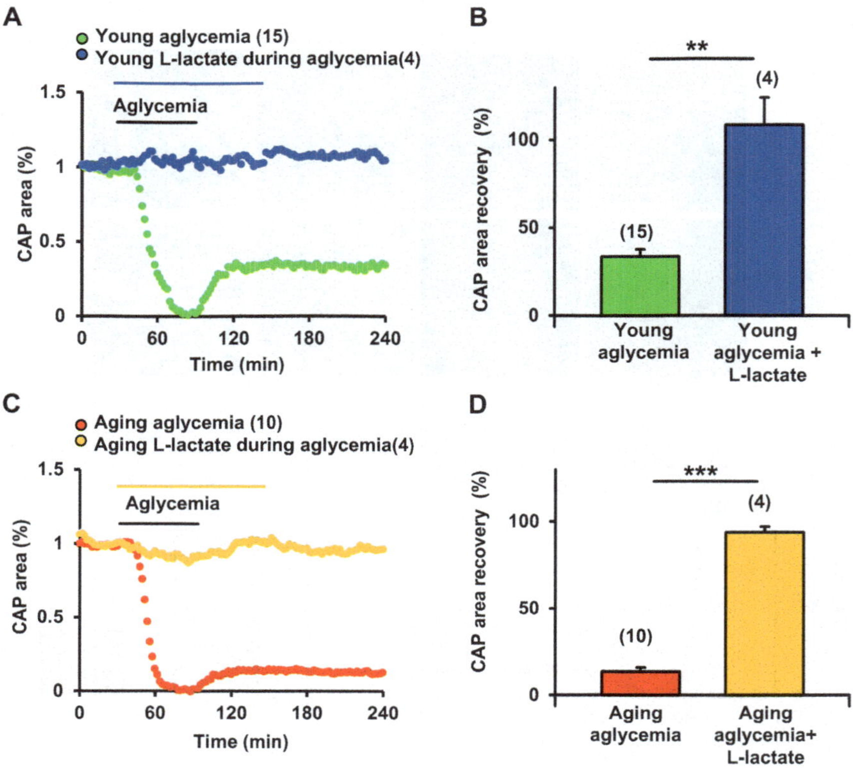

Fig. 6 Aging axons use lactate as an alternative energy substrate as efficiently as young axons. (**a, b**) Replacing glucose (10 mM) with lactate (20 mM, carbon equivalent) fully sustains young axon function during aglycemia (blue line and blue histogram) compared to aglycemia (green line, green histogram). (**c, d**) Aging axons use lactate (yellow line, yellow histogram) when glucose is not available to fully maintain their function as opposed to aglucose (red line and red histogram). $^{**}p < 0.01$, $^{***}p < 0.001$

removed (Fig. 6c, d, yellow trace and histogram compared to red aglycemia trace and histogram), confirming that aging axons can use lactate interchangeably with glucose to maintain full functionality. This ability of aging axons indicate that MCT2 expression and function are maintained in aging axons. Furthermore, when glucose is removed, the rates of loss of young and aging axon function remain identical (Fig. 5b, green trace for young and red trace for aging axon function; red arrow indicates the identical rate of loss of axon function), further verifying that aging axons maintain their molecular machinery relevant to ANLS and can use lactate despite major changes in their mitochondrial structure and function.

Notably in this approach, lactate is provided externally, replacing glucose to verify axonal uptake of lactate is functional. Whether aging astrocytes take up glucose efficiently, store it as glycogen, and convert it to lactate to sustain aging axons is another question. A practical way of testing this ability of astrocytes is to incubate

MONs in 30 mM glucose to upregulate astrocyte glycogen levels and then subject MONs to aglycemia. In young MONs, this experimental design delays loss of axon function during aglycemia and prevents loss of axon function at the end of aglycemia (60 min) due to increased availability of lactate, followed by an improved recovery after aglycemia (Brown et al. 2003; Baltan et al. 2008; Tekkok et al. 2005). A similar approach using aging MONs showed an identical delay in the loss of axon function during aglycemia, preservation of axon function at the end of aglycemia, and improved recovery of axon function following aglycemia. Indeed, aging axons incubated in high glucose before aglycemia recovered better compared to young axons, suggesting that aging astrocytes are able to take up glucose, regulate their glycogen stores, and deliver lactate to axons as efficiently as young astrocytes. In this aspect, aging does not impair the beneficial role of astrocytes or ANLS despite prominent anatomical and transcriptomic alterations. On the other hand, it is plausible that astrocytes alter their shape, size, and orientation to continue to efficiently support ANLS in aging WM. However, a recent study indicates a shift of glycogen metabolism enzymes in their localization from astrocytes to neurons in hippocampus, suggesting that neurons become independent of astrocytes and ANLS during aging (Drulis-Fajdasz et al. 2018). Therefore, aging breaks down the linkage between glycogen and the neural plasticity involved in learning and memory, and aging is associated with cognitive disturbances that are partly attributable to brain metabolic deficiency (Korol and Gold 1998; Abdelhafiz et al. 2015). Likewise, aging astrocytes *in vitro* demonstrate decreased glucose uptake and glutamine synthase activity (Souza et al. 2015). It is interesting that this crosstalk between astrocytes and axons remain undisrupted in aging WM. These findings emphasize the differences in cell-cell interactions between gray matter and WM in young and aging brain, and increases our understanding of how mechanisms of interaction between glia and neurons vary between different brain regions.

5 Is ANLS Unique to Astrocytes?

Oligodendrocytes display metabolic diversity similar to neurons and can utilize both glucose and lactate to extract energy and to synthesize lipids to deposit as myelin along axons (Sanchez-Abarca et al. 2001). Oligodendrocytes contact multiple axons and oligodendroglial cytoplasm extends into multiple myelinating shafts that connects the periaxonal space, establishing a network of access to axons. During the process of myelination, each oligodendrocyte produces large amounts of myelin, making them metabolically demanding during development. Once myelination is complete, preserving the intricate composition of myelin remains as a metabolic challenge due to the slow and challenging process of myelin delivery. This morphological organization is now recognized as the basis of a metabolic collaboration between oligodendrocytes and axons that is more intricate than our current understanding (Funfschilling et al. 2012). Oligodendrocytes are charged with regulating the efficiency of signal propagation through specific myelination of axons as the activity and metabolic needs of the CNS change. It is proposed that glucose supply at the nodes of Ranvier may be adequate to maintain mitochondrial ATP production under steady-state conditions without any glial cell support. On the other hand, with increased activity in longer and larger fibers, both oligodendrocytes and astrocytes supply axons with pyruvate/lactate as the need arises (Lee et al. 2012; Funfschilling et al. 2012; Nave 2010a, b). This supports the suggestion that the lactate shuttle is not restricted to astrocytes and axons, but rather that another specialized lactate shuttle exists between oligodendrocytes and their myelinated axons (Amaral et al. 2013). Recent genetic studies propose that oligodendrocytes sustain axonal energy metabolism by performing aerobic glycolysis and delivering lactate to the myelinated compartments via MCTs (Funfschilling et al. 2012). This metabolic support depends upon non-compacted regions of myelin that connect oligodendrocyte soma to the periaxonal cytosolic space under the myelin sheath. These tube-like cytosolic compartments called myelinic channels run parallel to axons, facing the axonal internode, and assist metabolic exchange via MCTs. Alterations in metabolic support by oligodendrocytes is now accepted to underlie various diseases, ranging from amyotrophic lateral sclerosis to psychiatric diseases (Nave and Ehrenreich 2014).

6 Concluding Remarks

Astrocytes in WM store glycogen that is converted to lactate when glucose concentration is low or when there is increased energy demand under normoglycemic conditions such as increased activity, confirming that CNS WM glycogen plays a crucial role as an energy reserve and buffer store (Brown et al. 2003). Therefore, astrocyte glycogen content governs neuronal and axonal function (Swanson and Choi 1993) and when glycogen content is depleted during prevailing hypoglycemia, brain function fails and suffers irreversible injury (Auer 1986; Frier and Fisher 1999). Glycogen content present in the tissue at the onset of aglycemia determines the duration of axon function (Brown et al. 2003), a relationship that may have profound implications for patients that suffer from type-1 diabetes mellitus and experience frequent silent WM infarcts leading to declining cognitive function. Similarly, axon function at the presynaptic input dictates synaptic transmission and synaptic plasticity underlying learning and memory-related events that decline with age and in neurodegenerative diseases. Therefore, all evidence points toward astrocyte-derived lactate as a crucial substrate and this metabolic coupling is validated in both gray matter and WM portions of the brain. Astrocytes transporting glucose from the vascular side via end feet to neuronal and glial cells also highlights an important mission of the neurovascular unit (NVU). Oligodendrocytes, microglia, oligodendrocytes and vascular cells consisting of endothelial cells, smooth muscle cells, pericytes and extracellular matrix (ECM), astrocytes form a complex intricate arrangement. The dynamic interactions among components of the NVU affect the clinical outcome of numerous central nervous system (CNS) diseases. For instance, in addition to its importance in the field of stroke, NVU dysfunction is now accepted as the cause or the result of other CNS diseases including Alzheimer's disease, vascular dementia, Parkinson's disease, amyotrophic lateral sclerosis, and multiple sclerosis (Arai et al. 2011). Furthermore, the diverse cells comprising the NVU are heterogeneous in developmental origin, physiological properties and function while they respond to injury, maintaining their interaction with each other. Therefore ANLS is an important part of the cell-cell signaling mechanism that contributes to multi-complex functions of NVU under physiological and pathological conditions.

Furthermore, a recent concept that oligodendrocytes support axon function by delivering lactate proposes another metabolic coupling in WM, suggesting that astrocytes are not alone in this nurturing role. The circumstances that trigger this oligodendrocyte-axon lactate shuttle system and its signaling mechanisms are of great interest, particularly because oligodendrocytes aim to cooperate with NVU to repair WM after injury such that injured axons either undergo degeneration or become resistant (Hamanaka et al. 2018). Therefore, the impaired oligodendrocyte-axon lactate shuttle is proposed to contribute to chronic demyelinating and psychiatric diseases (Nave and Ehrenreich 2014). Future studies are expected to unravel the details of oligodendrocyte-axon lactate metabolic coupling to establish that all WM components metabolically cooperate and that lactate may be the universal metabolite to sustain CNS function.

Acknowledgement This work was supported by grants from National Institute of Aging (NIA) to SB and NINDS to SB and SB, as well as a gift from Rose Mary Kubik. Selva Baltan has previously published as Selva Tekkök. The authors thank Dr. Chris Nelson, medical writer, for his help editing this chapter.

References

Abdelhafiz AH, Rodriguez-manas L, Morley JE, Sinclair AJ (2015) Hypoglycemia in older people—a less well recognized risk factor for frailty. Aging Dis 6:156–167

Alessandri B, Landolt H, Langemann H, Gregorin J, Hall J, Gratzl O (1996) Application of glutamate in the cortex of rats: a microdialysis study. Acta Neurochir Suppl 67:6–12

Amaral AI, Meisingset TW, Kotter MR, Sonnewald U (2013) Metabolic aspects of neuron-oligodendrocyte-astrocyte interactions. Front Endocrinol 4:54

Arai K, Lok J, Guo S, Hayakawa K, Xing C, Lo EH (2011) Cellular mechanisms of neurovascular damage and repair after stroke. J Child Neurol 26:1193–1198

Attwell D, Laughlin SB (2001) An energy budget for signaling in the grey matter of the brain. J Cereb Blood Flow Metab 21:1133–1145

Auer RN (1986) Progress review: hypoglycemic brain damage. Stroke 17:699–708

Baltan S (2006) Surviving anoxia: a tale of two white matter tracts. Crit Rev Neurobiol 18 95–103

Baltan S (2014) Excitotoxicity and mitochondrial dysfunction underlie age-dependent ischemic white matter injury. Adv Neurobiol 11:151–170

Baltan S, Besancon EF, Mbow B, Ye Z, Hamner MA, Ransom BR (2008) White matter vulnerability to ischemic injury increases with age because of enhanced excitotoxicity. J Neurosci 28:1479–1489

Black JA, Waxman SG (1988) The perinodal astrocyte. Glia 1:169–183

Brown AM, Baltan Tekkök S, Ransom BR (2003) Glycogen regulation and functional role in mouse white matter. J Physiol 549(2):501–512

Brown AM, Evans RD, Black J, Ransom BR (2012) Schwann cell glycogen selectively supports myelinated axon function. Ann Neurol 72:406–418
Cataldo AM, Broadwell RD (1986) Cytochemical identification of cerebral glycogen and glucose-6-phosphatase activity under normal and experimental conditions. II. Choroid plexus and ependymal epithelia, endothelia and pericytes. J Neurocytol 15:511–524
Chih C, Lipton P, Roberts EL (2001) Do active cerebral neurons really use lactate rather than glucose? Trends Neurosci 24:573–578
Clarke LE, Liddelow SA, Chakraborty C, Munch AE, Heiman M, Barres BA (2018) Normal aging induces A1-like astrocyte reactivity. Proc Natl Acad Sci U S A 115:E1896–E1905
Dringen R, Gebhardt R, Hamprecht B (1993) Glycogen in astrocytes: possible function as lactate supply for neighboring cells. Brain Res 623:208–214
Drulis-Fajdasz D, Gizak A, Wojtowicz T, Wisniewski JR, Rakus D (2018) Aging-associated changes in hippocampal glycogen metabolism in mice. Evidence for and against astrocyte-to-neuron lactate shuttle. Glia 66:1481–1495
Frier BM, Fisher BM (1999) Hypoglycemia in clinical diabetes. Wiley, New York
Funfschilling U, Supplie LM, Mahad D, Boretius S, Saab AS, Edgar J, Brinkmann BG, Kassmann CM, Tzvetanova ID, Mobius W, Diaz F, Meijer D, Suter U, Hamprecht B, Sereda MW, Moraes CT, Frahm J, Goebbels S, Nave KA (2012) Glycolytic oligodendrocytes maintain myelin and long-term axonal integrity. Nature 485:517–521
Hajek I, Subbarao KV, Hertz L (1996) Acute and chronic effects of potassium and noradrenaline on Na+, K+-ATPase activity in cultured mouse neurons and astrocytes. Neurochem Int 28:335–342
Hamanaka G, Ohtomo R, Takase H, Lok J, Arai K (2018) Role of oligodendrocyte-neurovascular unit in white matter repair. Neurosci Lett 684:175–180
Harris JJ, Attwell D (2012) The energetics of CNS white matter. J Neurosci 32:356–371
Henn FA, Haljamae H, Hamberger A (1972) Glial cell function: active control of extracellular K+ concentration. Brain Res 43:437–443
Hertz L, Peng L, Dienel GA (2007) Energy metabolism in astrocytes: high rate of oxidative metabolism and spatiotemporal dependence on glycolysis/glycogenolysis. J Cereb Blood Flow Metab 27:219–249
Honegger P, Pardo B (1999) Separate neuronal and glial Na+, K+-ATPase isoforms regulate glucose utilization in response to membrane depolarization and elevated extracellular potassium. J Cereb Blood Flow Metab 19:1051–1059
Korol DL, Gold PE (1998) Glucose, memory, and aging. Am J Clin Nutr 67:764S–771S
Kuffler SW, Nicholis JG (1964) Glial cells in the central nervous system of the leech; their membrane potential and potassium content. Naunyn-Schmiedebergs Arch Exp Pathol Pharmakol 248:216–222
Kuffler SW, Nicholls JG (1966) The physiology of neuroglial cells. Ergeb Physiol Biol Chem Exp Pharmakol 57:1–90
Kuffler SW, Potter DD (1964) Glia in the leech central nervous system: physiological properties and neuron-glia relationship. J Neurophysiol 27:290–320
Kuffler SW, Nicholls JG, Orkand RK (1966) Physiological properties of glial cells in the central nervous system of amphibia. J Neurophysiol 29:768–787
Lee S, Leach MK, Redmond SA, Chong SY, Mellon SH, Tuck SJ, Feng ZQ, Corey JM, Chan JR (2012) A culture system to study oligodendrocyte myelination processes using engineered nanofibers. Nat Methods 9:917–922
Magistretti PJ, Pellerin L (1996) Cellular mechanisms of brain energy metabolism. Relevance to functional brain imaging and to neurodegenerative disorders. Ann N Y Acad Sci 777:380–387
Magistretti PJ, Sorg O, Naichen Y, Pellerin L, De Rham S, Martin JL (1994) Regulation of astrocyte energy metabolism by neurotransmitters. Ren Physiol Biochem 17:168–171
Maher F, Vannucci SJ, Simpson IA (1994) Glucose transporter proteins in brain. FASEB J 8:1003–1011
Moody DM, Bell MA, Challa VR (1990) Features of the cerebral vascular pattern that predict vulnerability to perfusion or oxygenation deficiency: an anatomic study. AJNR Am J Neuroradiol 11:431–439

Munzer JS, Daly SE, Jewell-Motz EA, Lingrel JB, Blostein R (1994) Tissue- and isoform-specific kinetic behavior of the Na,K-ATPase. J Biol Chem 269:16668–16676
Nave KA (2010a) Myelination and support of axonal integrity by glia. Nature 468:244–252
Nave KA (2010b) Myelination and the trophic support of long axons. Nat Rev Neurosci 11:275–283
Nave KA, Ehrenreich H (2014) Myelination and oligodendrocyte functions in psychiatric diseases. JAMA Psychiat 71:582–584
Nishizaki T, Yamauchi R, Tanimoto M, Okada Y (1988) Effects of temperature on the oxygen consumption in thin slices from different brain regions. Neurosci Lett 86:301–305
Pellerin L, Pellegri G, Bittar PG, Charnay Y, Bouras C, Martin JL, Stella N, Magistretti PJ (1998) Evidence supporting the existence of an activity-dependent astrocyte-neuron lactate shuttle. Dev Neurosci 20:291–299
Pellerin L, Bonvento G, Chatton JY, Pierre K, Magistretti PJ (2002) Role of neuron-glia interaction in the regulation of brain glucose utilization. Diabetes Nutr Metab 15:268–273. discussion 273
Pfeiffer-Guglielmi B, Fleckenstein B, Jung G, Hamprecht B (2003) Immunocytochemical localization of glycogen phosphorylase isozymes in rat nervous tissues by using isozyme-specific antibodies. J Neurochem 85:73–81
Phelps CH (1972) Barbiturate-induced glycogen accumulation in brain. An electron microscopic study. Brain Res 39:225–234
Ransom BR, Orkand RK (1996) Glial-neuronal interactions in non-synaptic areas of the brain: studies in the optic nerve. Trends Neurosci 19:352–358
Ros J, Pecinska N, Alessandri B, Landolt H, Fillenz M (2001) Lactate reduces glutamate-induced neurotoxicity in rat cortex. J Neurosci Res 66:790–794
Sanchez-Abarca LI, Tabernero A, Medina JM (2001) Oligodendrocytes use lactate as a source of energy and as a precursor of lipids. Glia 36:321–329
Simpson IA, Vannucci SJ, Maher F (1994) Glucose transporters in mammalian brain. Biochem Soc Trans 22:671–675
Soreq L, UK Brain Expression Consortium, North American Brain Expression Consortium, Rose J, Soreq E, Hardy J, Trabzuni D, Cookson MR, Smith C, Ryten M, Patani R, Ule J (2017) Major shifts in glial regional identity are a transcriptional Hallmark of human brain aging. Cell Rep 18:557–570
Souza DG, Bellaver B, Raupp GS, Souza DO, Quincozes-Santos A (2015) Astrocytes from adult Wistar rats aged in vitro show changes in glial functions. Neurochem Int 90:93–97
Stahon KE, Bastian C, Griffith S, Kidd GJ, Brunet S, Baltan S (2016) Age-related changes in axonal and mitochondrial ultrastructure and function in white matter. J Neurosci 36:9990–10001
Swanson RA, Choi DW (1993) Glial glycogen stores affect neuronal survival during glucose deprivation *in vitro*. J Cereb Blood Flow Metab 13:162–169
Tekkok SB, Brown AM, Ransom BR (2003) Axon function persists during anoxia in mammalian white matter. J Cereb Blood Flow Metab 23:1340–1347
Tekkok SB, Brown AM, Westenbroek R, Pellerin L, Ransom BR (2005) Transfer of glycogen-derived lactate from astrocytes to axons via specific monocarboxylate transporters supports mouse optic nerve activity. J Neurosci Res 81:644–652
Tekkok SB, Ye Z, Ransom BR (2007) Excitotoxic mechanisms of ischemic injury in myelinated white matter. J Cereb Blood Flow Metab 27:1540–1552
Tsacopoulos M, Veuthey AL, Saravelos SG, Perrottet P, Tsoupras G (1994) Glial cells transform glucose to alanine, which fuels the neurons in the honeybee retina. J Neurosci 14:1339–1351
Vannucci SJ, Maher F, Simpson IA (1997) Glucose transporter proteins in brain: delivery of glucose to neurons and glia. Glia 21:2–21
Wender R, Brown AM, Fern R, Swanson RA, Farrell K, Ransom BR (2000) Astrocytic glycogen influences axon function and survival during glucose deprivation in central white matter. J Neurosci 20:6804–6810
Zeppenfeld DM, Simon M, Haswell JD, D'abreo D, Murchison C, Quinn JF, Grafe MR, Woltjer RL, Kaye J, Iliff JJ (2017) Association of perivascular localization of aquaporin-4 with cognition and Alzheimer disease in aging brains. JAMA Neurol 74:91–99

Dysregulation of Glycogen Metabolism with Concomitant Spatial Memory Dysfunction in Type 2 Diabetes: Potential Beneficial Effects of Chronic Exercise

Mariko Soya, Subrina Jesmin, Takeru Shima, Takashi Matsui, and Hideaki Soya

Abstract Cognitive dysfunction is one of the comorbidities of diabetes mellitus, but hippocampus-dependent learning and memory, a component of cognitive function, shows particular decline in type 2 diabetes, suggesting an increased risk for dementia and Alzheimer's disease. Cognitive function is related to dysregulated glucose metabolism, which is the typical cause of type 2 diabetes; however, hippocampal glycogen and its metabolite lactate are also crucial for hippocampus-dependent memory function. Type 2 diabetes induced hippocampus-dependent learning and memory dysfunction can be improved by chronic exercise and this improvement may possibly mediate through an adaptation of the astrocyte-neuron lactate shuttle (ANLS). This chapter focuses on the dysregulation of hippocampal glycometabolism in type 2 diabetes examining both existing evidence as well as the potential underlying pathophysiological mechanism responsible for memory dysfunction in type 2 diabetes, and showing for the first time that chronic exercise could be an effective therapy for type-2-diabetes-induced hippocampal memory decline.

Keywords Type 2 diabetes mellitus · Hippocampus-dependent learning and memory · Hippocampal glycometabolism · Monocarboxylate transporter 2

M. Soya · S. Jesmin · T. Matsui · H. Soya (✉)
Sport Neuroscience Division, Advanced Research Initiative for Human High Performance (ARIHHP), Faculty of Health and Sport Sciences, University of Tsukuba, Tsukuba, Japan
e-mail: soya.hideaki.gt@u.tsukuba.ac.jp

T. Shima
Laboratory of Exercise Biochemistry and Neuroendocrinology, Faculty of Health and Sport Sciences, University of Tsukuba, Tsukuba, Japan

M. DiNuzzo, A. Schousboe (eds.), *Brain Glycogen Metabolism*, Advances in Neurobiology 23, https://doi.org/10.1007/978-3-030-27480-1_13

Abbreviations

ANLS	Astrocyte-neuron lactate shuttle
BDNF	Brain-derived neurotrophic factor
DAB	1,4-Dideoxy-1,4-imino-D-arabinitol
GLUT	Glucose transporter
GSK-3β	Phosphorylated-GS kinase-3β
LETO	Long-Evans Tokushima Otsuka
LTP	Long-term potentiation
MCT	Monocarboxylate transporters
MHPG	NA metabolites
NA	Noradrenaline
NMDA	N-methyl-D-aspartate
OLETF	Otsuka Long-Evans Tokushima Fatty
pCofilin	Phosphorylated-Cofilin
pCREB	Phosphorylated-cAMP-response-element-binding protein
PGK	Phosphoglycerate kinase
PTG	Protein targeting to glycogen
STZ	Streptozotocin
VIP	Vasoactive intestinal peptide
ZDF	Zucker diabetic fatty
ZL	Zucker lean
ZO	Zucker obese

1 Introduction

Diabetes mellitus is a metabolic disorder that is associated with a number of serious complications. The International Diabetes Federation reported 425 million persons as having diabetes mellitus in 2018, and this number is estimated to increase to 629 million by 2045. The World Health Organization (WHO) ranked diabetes mellitus among the top 10 causes of the 56.9 million deaths worldwide in 2016, indicating that diabetes mellitus is a common and serious disorder worldwide warranting global attention.

A complication of diabetes mellitus that has received considerable attention in recent years is hippocampal cognitive dysfunction, which is a recognized risk factor for dementia and Alzheimer disease (Heijer et al. 2003; Cukierman et al. 2005; Mccrimmon et al. 2012; Umegaki et al. 2013). Learning and memory deficits in cognitive components are more profound in patients with type 2 diabetes than in those with type 1 diabetes (Biessels et al. 2006; Biessels and Despa 2018; Mccrimmon et al. 2012), and this could be associated with hippocampal atrophy (Gold et al. 2007; Mccrimmon et al. 2012). The American College of Sports Medicine (ACSM) and the American Diabetes Association (ADA), based on a number of studies, recommend regular physical activity for improvement of diabetes and indicate that chronic exercise has health benefits that go beyond blood glucose control, weight control, and a reduction of the risk of well-known complications of type 2 diabetes although its effects are yet to be investigated for hippocampal memory dysfunction (Mu et al. 2001; Holloszy 2005; O'Gorman et al. 2006; Sriwijitkamol et al. 2007; Colberg et al. 2010; Lee et al. 2011; Jenkins et al. 2012)

At present, no consensus has been reached regarding the underlying mechanism of memory dysfunction in type 2 diabetes, including hippocampus-based memory decline, but dysregulation of glucose utilization in the brain (Sickmann et al. 2010), brain-derived neurotrophic factor (BDNF) (Stranahan et al. 2009), neuroinflammation (Whitmer 2007), and oxidative stress have all been associated with memory dysfunction in patients with type 2 diabetes (Whitmer 2007; Stranahan 2015). Interestingly, one metabolic adaptation shown in type 2 diabetes is elevated glycogen deposition in the heart (Bhattacharjee et al. 2006; Shearer et al. 2011). This suggests that a metabolic adaptation involving glycogen might also occur in the hippocampus, a crucial site for memory formation, and lead to memory dysfunction in type 2 diabetes as hippocampal glycogen-derived lactate in astrocytes and its transport into neurons plays a crucial role in hippocampus-dependent memory formation (Newman et al. 2011; Suzuki et al. 2011).

Chronic exercise, which positively affects hippocampus-dependent memory function in animals (van Praag et al. 1999; Brown et al. 2003; Liu et al. 2009, Creer et al. 2010), also increases hippocampal glycogen levels in normal rats (Matsui et al. 2012) and lactate transporter levels (monocarboxylate transporters; MCT) in type 1 diabetic rats (Aveseh et al. 2014), suggesting that chronic exercise could potentially ameliorate type-2-diabetes–related memory dysfunction through an adaptation of hippocampal glycometabolism which means glycogen metabolism (metabolism of sugars and other carbohydrate). In this chapter, we summarize the evidence for an association between brain glycometabolism dysregulation and cognitive decline in type 2 diabetes and the effects of chronic exercise on hippocampus-dependent memory dysfunction.

2 Cognitive Decline and Dysregulated Hippocampal Glycometabolism in Type 2 Diabetes

There are two types of diabetes mellitus: type 1 tends to develop in childhood and is characterized by a lack of insulin due to destruction of the βcells of the islet of Langerhans in the pancreas. By contrast, obesity and lack of physical activity are the crucial risk factors for type 2 diabetes, a disease condition with reduced insulin sensitivity and insulin resistance. Both type 1 and type 2 diabetes are associated with high blood glucose levels (hyperglycemia), which is the cause of numerous complications including cognitive dysfunction (Mccrimmon et al. 2012). Furthermore, insulin resistance is also a risk factor for cognitive dysfunction (Geroldi et al. 2005). In human studies, neuroimaging provides evidence that diabetes affects the brain, as adverse effects are evident in the cortical and subcortical regions, such as the medial temporal lobe (hippocampus and amygdala) (Heijer et al. 2003; Manschot et al. 2006), which are related to cognitive impairment associated with attention, executive function, information processing speed, and memory in patients with type 2 diabetes (Manschot et al. 2006). In these patients, the hippocampus may be the first site affected by type 2 diabetes (Gold et al. 2007). The neuropathological changes seen in the hippocampus with type 2 diabetes are also characteristic of Alzheimer's disease, but the same changes are not seen in patients diagnosed with dementia.

Conversely, in rodents, cognitive function tested using the T-maze and Morris water maze shows a decline in streptozotocin (STZ)-induced type 1 diabetic mice (Flood et al. 1990; Biessels et al. 1996; Kamal et al. 1999). In terms of alterations related to learning and memory at the cellular level, N-methyl-d-aspartate (NMDA)-dependent long-term potentiation (LTP) in the CA1 and NMDA-independent LTP in the CA3 fields of the hippocampus are also impaired in STZ-induced diabetic rats, and this impairment is correlated with the severity of hyperglycemia (Biessels et al. 1996; Chabot et al. 1997; Tekkök and Krnjevic 1999). Based on this research, type 2 diabetes could have a similar type of impact on cognitive function, particularly through its effects on the hippocampus. Much information is available regarding the effects of glucose metabolism on cognitive function, but the role of glycometabolism on cognitive deficits in type 2 diabetes is not yet clearly understood. Recent findings suggest that hippocampal glycogen plays an important role in memory processing (Newman et al. 2011; Suzuki et al. 2011), and that hippocampal glycometabolism could be altered/dysregulated in type 2 diabetes (Fig. 1).

2.1 Hippocampal Glycometabolism and Cognitive Function

The energy requirements of the brain are high, and brain energy metabolism changes depending on neuronal activity. Pioneering work on brain energy metabolism using the 2-deoxyglucose (2-DG) autoradiographic technique demonstrated that blood glucose is metabolized in the brain by neuronal activity, suggesting a metabolic coupling between neuronal activity and blood vessels (Newman et al. 2011). Furthermore, activation of glutamatergic neurons causes the release of glutamate into the intercellular space, and glutamate is a trigger that induces glycolysis in astrocytes (Magistretti and Pellerin 1999). The glutamate released from glutamatergic synapses is co-transported with Na^+ into astrocytes, which express the glutamate transporters GLAST and GLT1. This co-transport activates the astrocytic Na^+/K^+-ATPase, and glycolysis is subsequently stimulated in the astrocytes through activation of phosphoglycerate kinase (PGK) (Magistretti and Pellerin 1999). The lactate produced through glycolysis in the astrocytes is transported to the neurons via MCT (astrocyte: MCT1 and MCT4; neuron: MCT2) for use as an energy substrate, and this metabolic coupling between these two cell types has led to the proposal of a hypothetical 'astrocyte-neuron lactate shuttle' (ANLS) (Pellerin et al. 1998). In the

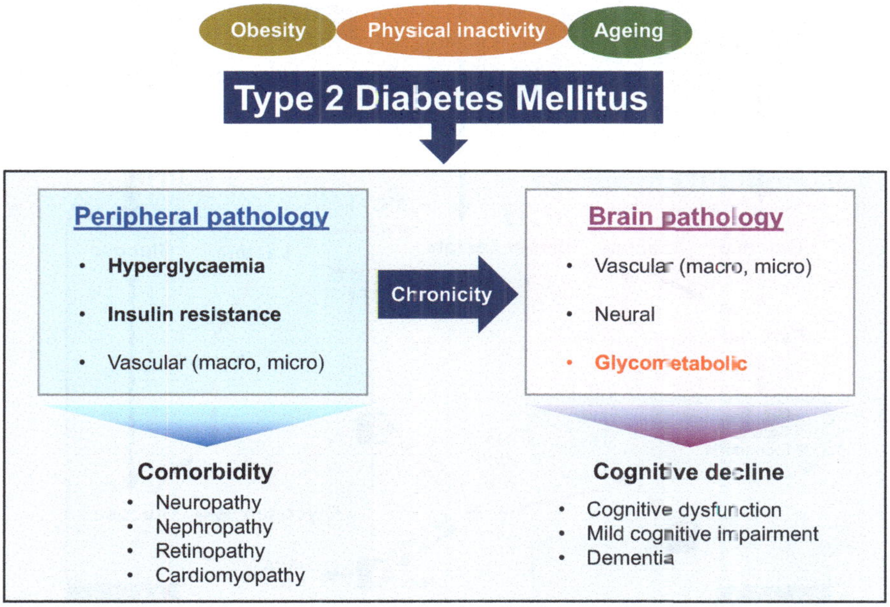

Fig. 1 Proposed pathophysiological mechanisms in type 2 diabetes. In contrast to type 1 diabetes, which is caused by insulin deficit due to the immune-mediated βcell destruction of pancreatic cells in the islets of Langerhans, type 2 diabetes occurs through insulin resistance and reduced insulin sensitivity that presents as hyperglycemia and is induced by genetic susceptibility, several environmental, lifestyle factors, such as obesity, physical inactivity, and ageing, leading to various comorbidities. The chronicity of these symptoms on the peripheral tissues affects the central nervous system and causes cognitive decline, which has been recently recognized as one of the comorbidities of diabetes. Brain glycometabolism is associated with cognitive function, particularly in the hippocampus, so dysregulated glycometabolism could be the underlying mechanism leading to the dysfunction of memory in type 2 diabetes

case of high or long-lasting neuronal activity, the glucose supply from the blood vessels to neurons becomes insufficient, thus, the contribution of glycogen stored exclusively in astrocytes is important for sustaining neuronal activity and maintaining a high rate of glycolysis (Swanson et al. 1992).

Astrocytic glycogen degradation (via glycogenolysis) is stimulated by noradrenaline (NA), vasoactive intestinal peptide (VIP), adenosine, and serotonin resulting in the release of lactate that is transported to neurons in the same manner as lactate derived from glycolysis (Magistretti et al. 1981; Hof et al. 1988; Sorg and Magistretti 1991, 1992; Dringen et al. 1993; Choi et al. 2012; Matsui et al. 2015) Interestingly, the tight coupling between neuronal activity and energy metabolism suggests that 'metabolic plasticity' underlies neuronal plasticity (Magistretti 2006; Choi et al. 2012). Indeed, metabolic adaptation occurs via glycogen metabolic processes under *in vitro* conditions. As mentioned above, neurotransmitters (namely NA, VIP, and adenosine) and neuromodulators induce glycogenolysis in a short time (seconds to minutes); however, NA has long-term effect in that it can stimulate a transcriptionally

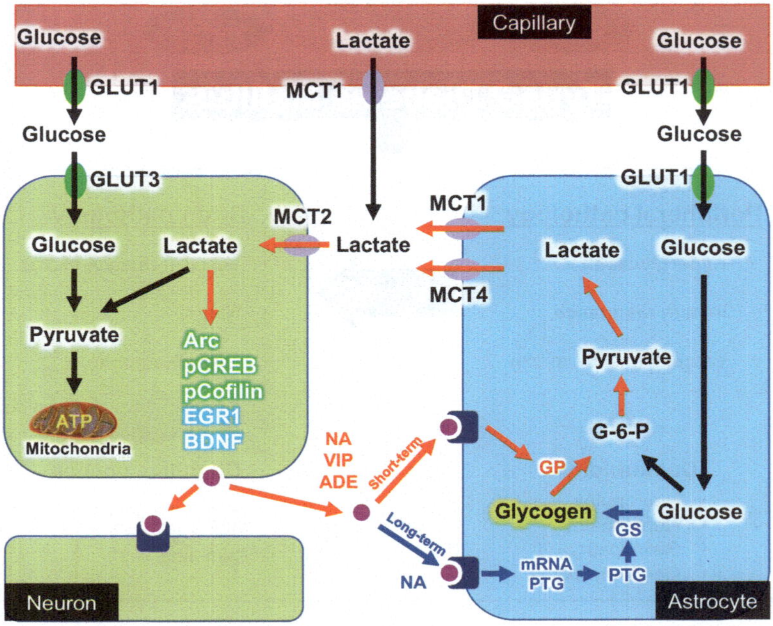

Fig. 2 Simplified illustration of the astrocyte-neuron lactate shuttle: Contribution of glycogen-derived lactate to cognitive function. During high or long-lasting neuronal activity, brain glycogen-derived lactate becomes an important energy source to satisfy neuronal energy demand. Glycogen degradation is induced by noradrenaline (NA), vasoactive intestinal peptide (VIP), and adenosine (ADE), resulting in lactate production. Particularly in the hippocampus, lactate is transported via monocarboxylate transporters (MCT) and serves both as an energy substrate and in memory processing through induction of protein expression related to neural plasticity, such as Arc, phosphorylated-cAMP-response-element-binding protein (pCREB), and phosphorylated-Cofilin (pCofilin). Early growth response protein 1 (EGR1) and brain-derived neurotrophic factor (BDNF) are also induced by lactate in *in vitro* experiments. These lactate effects are suppressed by blocking MCT2 or inhibiting glycogen degradation, suggesting that glycogen-derived lactate is crucial for memory processing and neuronal plasticity. NA induces both glycogenolysis as a short-term effect and glycogen synthesis as a long-term effect by inducing protein targeting to glycogen (PTG) and glycogen synthase (GS)

regulated action that results in glycogen synthesis (Allaman et al. 2000, 2003). The long-term effects of NA include NA-triggered cyclic-AMP production, which increases the expression of the transcriptional factor CCAAT/enhancer-binding protein (C/EBP), glycogen synthase, and protein targeting to glycogen (PTG) (Cardinaux and Magistretti 1996; Ruchti et al. 2016) (Fig. 2).

Recent studies have demonstrated that L-lactate is used as an energy substrate for neurons (Bélanger et al. 2011), and it also acts as a signaling molecule to induce the expression of neural plasticity related genes (i.e., activity-regulated cytoskeleton-associated protein [Arc], c-Fos, and Zif268) in cultured neurons (Yang et al. 2014). These are immediate early genes (IEGs), which are induced during both long-term memory and long-term plasticity (Alberini 2009; Bramham et al. 2010; Caroni et al.

2012). These effects of L-lactate are mediated by NMDA receptor activity in neurons and its downstream signaling Erk1/2 cascade, which means that L-lactate potentiates glutamatergic neuronal currents by ensuring an increase in intracellular calcium through NMDA receptor activation and changes in the cellular redox state, and that it acts as a neuromodulator to induce plasticity-related gene expression (Yang et al. 2014). Although most of these observations about ANLS were elucidated *in vitro* experiments using cortical cultured neurons or astrocytes, the most important current work on lactate effects associated with learning and memory in the hippocampus has shown that lactate derived from astrocytes is essential for long-term memory formation (Suzuki et al. 2011). Hippocampus-dependent long-term memory formation (measured using an inhibitory avoidance task) and LTP were inhibited when lactate production from astrocytes was blocked using DAB (1,4-dideoxy-1,4-imino-D-arabinitol). By contrast, co-injection of DAB and L-lactate rescued inhibited memory consolidation and LTP (Suzuki et al. 2011).

The underlying mechanisms by which glycogen-derived lactate induces memory formation could be associated with the induction of Arc, phosphorylated-cAMP-response-element-binding protein (pCREB), and phosphorylated-Cofilin (pCofilin), which are related to synaptic plasticity (Suzuki et al. 2011) (Fig. 2). A similar observation relating hippocampal glycogen to working memory has also been made with another behavioral paradigm (closed plus maze) (Newman et al. 2011). Little *in vivo* evidence supports a role for brain glycogen in memory function, but the blockade of glycogenolysis with DAB impaired a discrimination avoidance learning task in chicks (Gibbs et al. 2006). Similarly, the lack of brain glycogen generated by knockout of glycogen synthase specifically (GYS1) in the nervous system impaired memory in an operant conditioning task and LTP (Boury-Jamot et al. 2015).

2.2 *Dysregulated Hippocampal Glycometabolism and Cognitive Decline in Type 2 Diabetes*

A number of studies have shown that the central nervous system is affected by diabetes, and that diabetes-related cognitive decline is an independent risk factor for dementia and Alzheimer's disease (Biessels et al. 2006; Biessels and Despa 2018; Mccrimmon et al. 2012). These diseases are accompanied by morphological changes and metabolic alteration/dysregulation in the hippocampus and cortex, which are regions associated with cognitive function (Heijer et al. 2003; Sickmann et al. 2012; Biessels and Despa 2018). The hippocampus, in particular, shows atrophy in the initial stage of these neuro-degenerative disorders (Heijer et al. 2003; Gold et al. 2007). The energy supply to the brain depends strongly on glucose supply from blood vessels; thus, impairment of glucose metabolism by insulin deficiency, as commonly occurs in diabetes, could be one of the risk factors for impaired cognitive function. Moreover, hippocampal glycogen-derived lactate metabolism associated with cognitive function could also be affected by diabetes. However, little research has examined the relationship between hippocampal glycometabolism and cognitive decline/dysfunction in diabetes.

Previous studies examining the glycogen levels in the hippocampus and/or cortex in diabetic rats have demonstrated that STZ-induced type 1 diabetic rats had increases in glycogen levels in the cortex depending on the STZ treatment duration (Plaschke and Hoyer 1993). Conversely, in type 2 diabetes models, Zucker obese (ZO) rats had hyper-levels of hippocampal and cortical glycogen than did their counterpart Zucker lean (ZL) genetic control rats, whereas Zucker diabetic fatty (ZDF) rats had lower levels of cortical glycogen compared with control ZDF lean rats (Sickmann et al. 2010). One possible explanation for the conflicting results between these two studies is that the methodology used for brain tissue preparation was neither optimal nor standard/validated for glycogen measurement (Swanson et al. 1989; Dringen et al. 1993). Brain glycogenolysis degraded rapidly by ischemia; therefore, microwave irradiation can inactivate glycogenolysis enzymes needed for the accurate measurement of brain glycogen (Kong et al. 2002).

Our laboratory has endeavored to elucidate the glycometabolism pattern in type 2 diabetic rats by measuring glycogen and MCT levels in the hippocampus using microwave irradiation, with consideration to the issues mentioned above. For three reasons we used Otsuka Long-Evans Tokushima Fatty (OLETF) rats as our type 2 diabetes model: (a) OLETF rats exhibit hyperphagia; (b) they have a dysregulated sympathoadrenal response and show a decline in executive function (Suge et al. 2012); and (c) the onset of diabetes takes time, in agreement with human type 2 diabetes (Kawano et al. 1992). Type 2 diabetic rats exhibit hyper-glycogen levels and diminished MCT2 protein levels in the hippocampus when compared with their counterpart genetic control rats Long-Evans Tokushima Otsuka (LETO), and a negative relationship exists between the two components of the glycometabolism system in neurons (Shima et al. 2016a) (Fig. 3). Notably, the negative correlation exhibited in type 2 diabetic rats has been observed only in the hippocampus, and not in the hypothalamus and the cortex, even though these regions exhibit changes similar to those seen in the hippocampus, namely, glycogen (hypothalamus and cortex) and MCT2 (hypothalamus) increases, implying that the hippocampus may be the first region affected by dysregulation of glycometabolism induced by type 2 diabetes (Shima et al. 2016a). The significance of hyper-hippocampal glycogen in type 2 diabetic rats may indicate metabolic compensation for the decreased lactate utilization via MCT2 and this consequence may have linkage with ANSL although concrete evidence demands. Indeed, the same phenomenon was reported in the hearts of patients with type 2 diabetes (Bhattacharjee et al. 2006; Shearer et al. 2011). Notably, MCT1 and MCT4 are expressed in astrocytes and have a similar transport roles in astrocytes in all three brain regions (hippocampus, hypothalamus and cortex) as stated above (Shima et al. 2016a). Thus, collectively the findings observed in the study by Shima et al. at the progressive stage of type 2 diabetic hippocampus in concern to glycometabolism and the associated molecular mechanism warrant future studies to be conducted at pre diabetes or at the early stage of diabetes for the advancement of clinical improvement of diabetic complications.

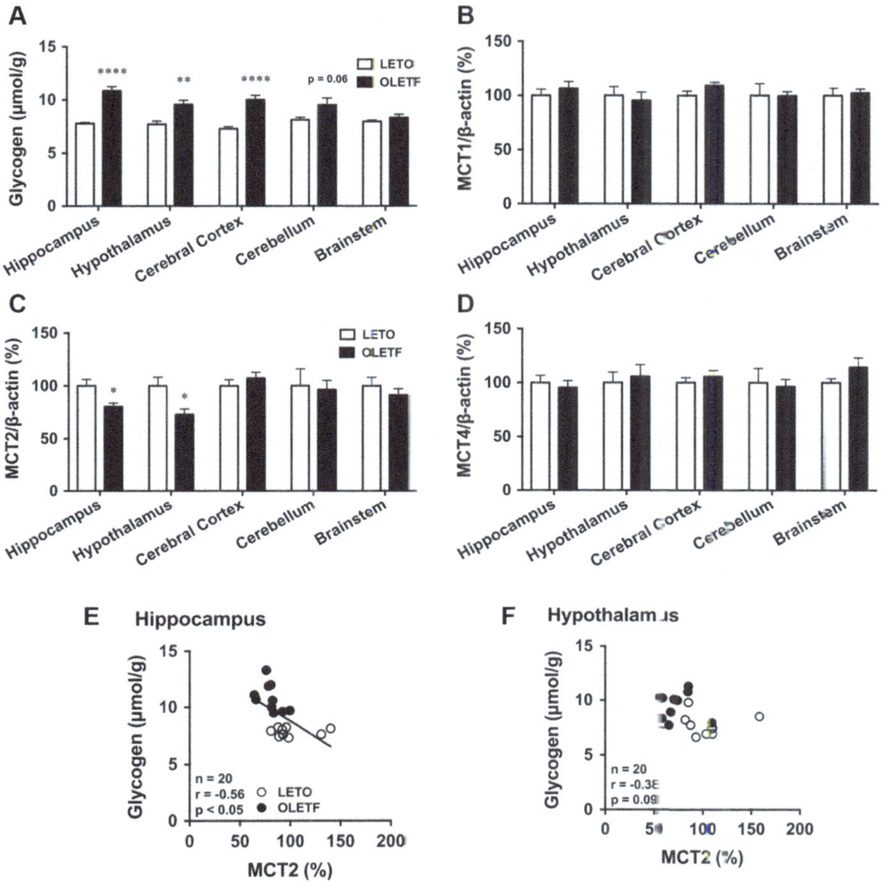

Fig. 3 Dysregulated hippocampal glycometabolism in a rat model of type 2 diabetes. Glycogen levels in the hippocampus, hypothalamus, and cerebellum are higher in the OLETF type 2 diabetes rat model than in control LETO rats. Conversely, MCT2 expression in the hippocampus and hypothalamus is lower in OLETF rats than in LETO rats, even though MCT1 and MCT4 levels show no significant differences between the LETO and OLETF rats. However, a significant correlation exists between MCT2 and glycogen levels, but only in the hippocampus suggesting that memory dysfunction in OLETF rats could be caused by dysregulated hippocampal glycometabolism. (**a**) Glycogen levels in the hippocampus, cortex, hypothalamus, cerebellum, and brainstem, (**b**) MCT1 levels, (**c**) MCT2 levels, (**d**) MCT4 levels, (**e**) Correlation between MCT2 and glycogen levels in the hippocampus, (**f**) Correlation between MCT2 and glycogen levels in the hypothalamus. (Referenced by Shima et al., *J Physiol Sci,* 2016)

3 Chronic Exercise Effects on Dysregulated Hippocampal Glycometabolism and Cognitive Decline in Type 2 Diabetes

Chronic exercise can normalize blood glucose level control by enhancing glucose uptake in the peripheral tissues in an insulin-independent manner and is an effective therapy for type 2 diabetes complications such as neuropathy (Winder et al. 1987; Mu et al. 2001; Balducci et al. 2006). Cognitive decline is one of the complications of type 2 diabetes (Gispen and Biessels 2000; Whitmer 2007), with hippocampal dysfunction being a particularly serious disability associated with the development of dementia and Alzheimer's disease (Pouwer et al. 2003; Witting et al. 2006; Ahtiluoto et al. 2010). Since a number of studies have demonstrated an improvement of cognitive function, such as hippocampus-dependent memory (van Praag et al. 1999; Liu et al. 2009; Creer et al. 2010; Bolz et al. 2015), with chronic exercise, this type of exercise could have the potential to affect hippocampal dysfunction in type 2 diabetes in a beneficial manner. Recently, our laboratory identified a chronic exercise regimen effective for hippocampal dysfunction in type 2 diabetes (Shima et al. 2016b); thus, the effects of acute and chronic exercise on hippocampal glycometabolism in normal and type 2 diabetic rats will be reviewed next.

3.1 Hippocampal Glycometabolism During Acute Exercise in Normal Rats

Exercise requires mobilization of every system of the body, including muscles and the brain (Vissing et al. 1996; Secher et al. 2008). Glucose molecules stored as glycogen play a crucial role in the immediate energy supply needed by exercising muscles to maintain muscular contraction because depletion of muscle glycogen during exercise causes a decline in endurance performance (Secher et al. 2008). Muscle glycogen decreases in an activity-dependent manner (Gollnick et al. 1974), and depletion of muscle and liver glycogen during hypoglycemia occurs when rats are subjected to prolonged exhaustive exercise (Winder et al. 1987). Exercise also affects the brain and increases neuronal activity (Saito and Soya 2004; Ohiwa et al. 2006, 2007, Soya et al. 2007a, b); furthermore, glucose is utilized as an energy source by neurons during exercise (Vissing et al. 1996). However, whether brain glycogen, and particularly hippocampal glycogen, decreases in a similar manner to that seen for muscle glycogen following prolonged exhaustive exercise remains an unanswered question.

In one previous experiment, glycogen levels in muscles and the liver were depleted after exhaustive exercise (treadmill running), which induced hypoglycemia (Matsui et al. 2011). Hippocampal glycogen also decreased with acute moderate exhaustive exercise (20 m/min, until exhaustion) (Matsui et al. 2011). Furthermore, levels of methoxyhydroxyphenylglycol (MHPG) and

5-Hydroxyindoleacetic acid (5-HIAA), which are the respective metabolites of NA and serotonin (5-TH), were increased in cerebral cortex after exercise, and their levels were negatively correlated with a decrease in cortical glycogen, suggesting that NA and 5-HT could be part of an underlying mechanism of exercise-induced brain glycogen degradation (Matsui et al. 2011).

Based on these results, next we will discuss exercise-induced hippocampal glycogen dynamics, especially in the recovery phase of glycometabolism. Muscle glycogen decreases with exhaustive exercise and then returns to above basal levels (supercompensation) 24 h after exercise (Bergström and Hultman 1966). Surprisingly, brain glycogen, including hippocampal glycogen, also exhibits supercompensation 6 h after exhaustive exercise (Matsui et al. 2012); furthermore, the rate of brain glycogen supercompensation depends on the rate of glycogen decrease during exercise, similar to the situation seen in muscles (Gaesser and Brooks 1980) and consistent with studies demonstrating brain glycogen supercompensation at 4–7 h after insulin-induced hypoglycaemia (Choi et al. 2003; Canada et al. 2011).

An adaptation to chronic exercise occurs, as indicated by glycogen supercompensation following exhaustive exercise (James and Kraegen 1984), as 4 weeks of chronic moderate exercise resulted in an increase of basal hippocampal and cortical glycogen levels in normal rats (Matsui et al. 2012). The underlying mechanisms of hippocampal glycogen synthesis is not clear, but exercise is reported to activate NA neurons (Kitaoka et al. 2010), and NA metabolites, such as MHPG, increase following exhaustive exercise and are negatively correlated with brain glycogen decreases (Matsui et al. 2011), suggesting that NA-induced glycogen synthesis might be involved (Allaman et al. 2000; Crosson et al. 2003; Ruchti et al. 2016) (Fig. 4).

3.2 *Hippocampal Glycometabolism with Acute Exercise in Type 2 Diabetes Rats*

A number of studies have shown that exercise is effective in restoring dysregulated glucose control in type 2 diabetes (Sigal et al. 2007; Ruchti et al. 2016); however, the most effective/beneficial exercise conditions and exercise effects on hippocampal glycometabolism remain unestablished. To address this, we have to use a relatively identical exercise intensity protocol for both non-diabetic rats and diabetic rats. Exercise increases blood lactate level which depends on the exercise duration and intensity; the exercise intensity at which blood lactate level begins to exponentially increase is called lactate threshold. Stress responses (plasma ACTH and adrenaline release) are also taken place at lactate threshold (Soya et al. 2007a), thus it is useful to determine the relatively same exercise intensity protocol based on the lactate threshold for the study rats to be compared. We used an acute moderate exercise protocol similar to a moderate exercise intensity based on the lactate threshold of type 2 diabetic rats (OLETF: 22.6 ± 0.3 m/min) and genetic control rats (LETO: 28.3 ± 1.8 m/min) (Shima et al. 2016b), and we then examined the effect of acute

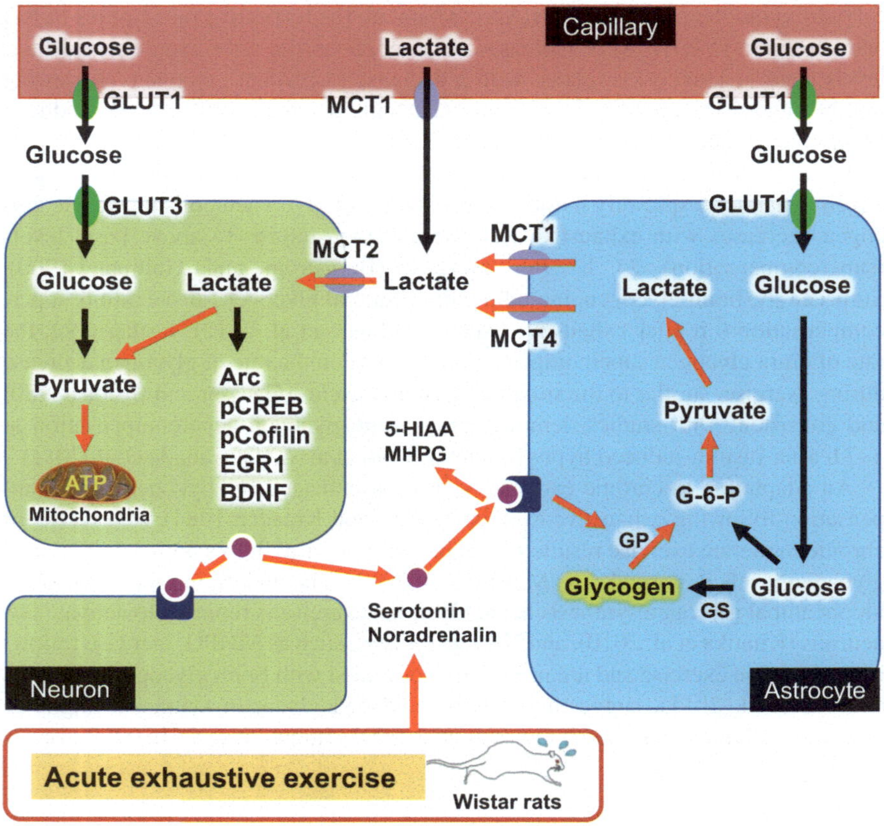

Fig. 4 Brain glycometabolism during acute exhaustive exercise. Brain glycogen levels decrease in the hippocampus, cortex, hypothalamus, cerebellum, and brainstem following acute exhaustive treadmill running exercise, while increases in 5-HIAA and MHPG, which are metabolites of serotonin (5-HT) and NA, respectively, occur after exercise. The decrease in glycogen levels in the cortex during exercise was negatively correlated with increased 5-HIAA and MHPG levels in the cortex, suggesting that 5-HT and NA may be involved in glycogenolysis during exercise. (Summarized figure based on Matsui *et al., J Physiol,* 2011)

moderate exercise on hippocampal glycometabolism (Shima et al. 2016b). We first noted a decrease in hippocampal glycogen following 30 minutes of acute moderate exercise in both OLETF and LETO rats that was positively correlated with the hippocampal lactate increase (Shima et al. 2016b). Diabetes patients exhibit hyper-glycogen in the heart (Bhattacharjee et al. 2006), and the OLETF rats exhibited hyper-glycogen in the hippocampus, which could be a result of adaptation to declining lactate utilization due to the downregulation of MCT2 (Shima et al. 2016a, b). These results suggest that hippocampal glycometabolism although dysregulated in type 2 diabetic rats is activated with acute moderate exercise and this activation with acute exercise is also observed in normal rats hippocampus.

3.3 Effects of Chronic Exercise on Type-2-Diabetes-Induced Dysregulation of Glycometabolism Accompanied with Cognitive Decline

Regular exercise is an effective therapy for type 2 diabetes because it leads to improvement of glycometabolism in peripheral organs (Mu et al. 2001; Holloszy 2005; O'Gorman et al. 2006; Sigal et al. 2007). In addition, chronic exercise not only enhances hippocampus-dependent memory but it also increases hippocampal glycogen levels in normal rats and MCT expression in both peripheral organs and brain tissues (hippocampus and cortex) in diabetes (van Praag et al. 1999; Juel et al. 2004; Liu et al. 2009; Creer et al. 2010; Nikooie et al. 2013; Aveseh et al. 2014; Bolz et al. 2015), suggesting that chronic exercise could be an appropriate exercise therapy for ameliorating hippocampal memory function. OLETF rats exhibited significant hippocampal-dependent memory dysfunction when compared to LETO rats, but, surprisingly, after four weeks of chronic moderate exercise, which is reported to increase hippocampal glycogen levels, memory function was also ameliorated in the OLETF rats (Shima et al. 2016b) (Fig. 5). Chronic exercise induced hyper-glycogen levels in the hippocampus of normal rats (Matsui et al. 2012) as well as in OLETF rats with concomitant normalization of downregulated MCT2 levels in neurons, increased glycogen synthase (GS) levels, and also ameliorated glucose transporter 1 (GLUT1) expression in astrocytes in the hippocampus without any significant alteration in the expression of GLUT3, MCT1 and MCT4 (Shima et al. 2016b). Interestingly, chronic exercise increased BDNF levels in the hippocampus of LETO rats, whereas no change in BDNF expression levels occurred in OLETF rats; furthermore, expression of other proteins associated with the downstream BDNF signaling, such as tyrosine receptor kinase B (TrkB), pCREB, and phosphorylated-GS kinase-3β(GSK-3β), were not changed by chronic exercise in either LETO or OLETF rats, implying that normalization of MCT2 levels and further increases in glycogen levels in the hippocampus could be an adaptation in response to chronic exercise that could contribute to the amelioration of hippocampus-dependent memory dysfunction in OLETF rats (Fig. 6).

A dysregulation of MCT2-mediated lactate uptake in neurons may be a novel etiology of memory dysfunction in type 2 diabetes, especially in the context of the hippocampus (Shima et al. 2016b). However, four weeks of chronic exercise did not improve well-known peripheral symptoms/abnormalities of type 2 diabetes, such as alterations in circulatory glucose, insulin and HbA1c levels, which means that more long-term exercise is needed to simultaneously ameliorate the pathology concerned not only with cognitive dysfunction but also with dysregulated glucose control at the peripheral level. Nevertheless, the amelioration of memory dysfunction induced with chronic (four weeks) exercise was exhibited earlier than other pathologies in the peripheral organs. Based on our preliminary findings, abnormal hippocampal glycometabolism and memory dysfunction are already evident at the pre-type2diabetes stage, so the nervous system may require more rapid adaptation to type 2 diabetes than do other organs. These findings suggest that conventional

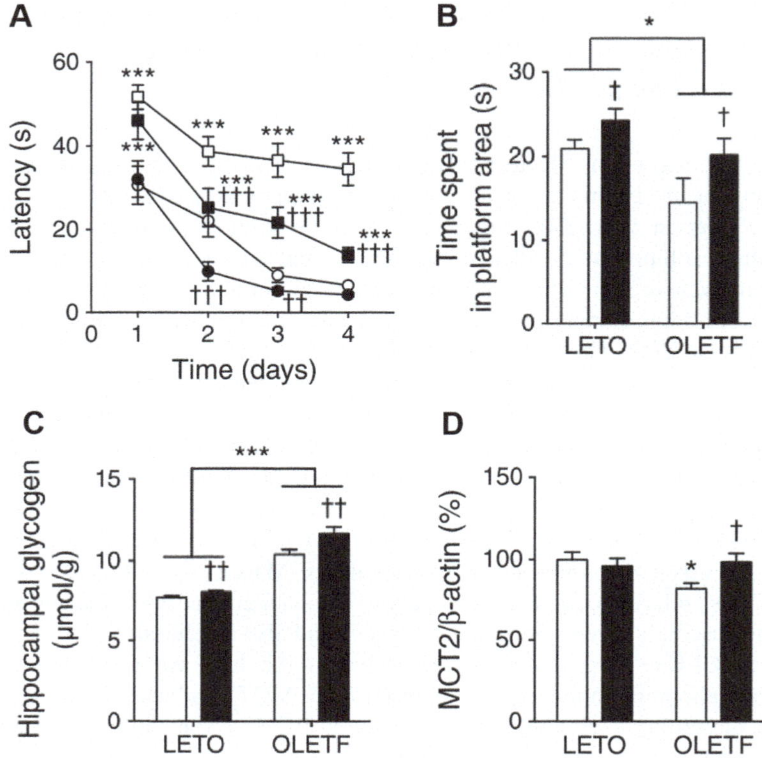

Fig. 5 The effects of chronic exercise on hippocampus-dependent memory and dysregulated hippocampal glycometabolism in type 2 diabetes. The escape latency during learning test of OLETF rats was shortened by 4 weeks of moderate exercise, the time spent in platform area during probe test was increased after 4 weeks of exercise. Furthermore, hippocampal glycogen levels were increased in exercised group of both LETO and OLETF rats, and MCT2 levels in the hippocampus were normalized in exercised OLETF rats group, suggesting that4 weeks of moderate exercise ameliorated hippocampus-dependent memory via improvement of MCT2-mediated lactate uptake into the neurons in OLETF rats (**a**) Escape latency, (**b**) Time spent in platform area during probe test, (**c**) Hippocampal glycogen levels, (**d**) Hippocampal MCT2 levels. Circles, LETO rats; squares, OLETF rats; white symbols and bars, sedentary rats; black symbols and bars, exercised rats, n = 6–8 rats per group. $^{*}p < 0.05$ and $^{***}p < 0.001$ vs LETO rats; $^{\dagger}p < 0.05$, $^{\dagger\dagger}p < 0.01$ and $^{\dagger\dagger\dagger}p < 0.001$ vs sedentary rats (Referenced by Shima *et al., Diabetologia,* 2016)

clinical diagnostic indices or criteria for type 2 diabetes, such as glucose and HbA1c levels, may not reflect the earliest organ involvement, which would be the brain with functional hippocampal impairment, as is evident in our unpublished observations and in other studies during the early phase of pre-type-2 diabetes (Soares et al. 2013). Thus, new diagnostic criteria might be necessary to prevent or delay the onset of type 2 diabetes with an urgent focus on the initiation and progression of brain memory dysfunction. Further, it is important to innovate an appropriate exercise regimen either from pre-diabetes or early diabetes, and even starting at advanced

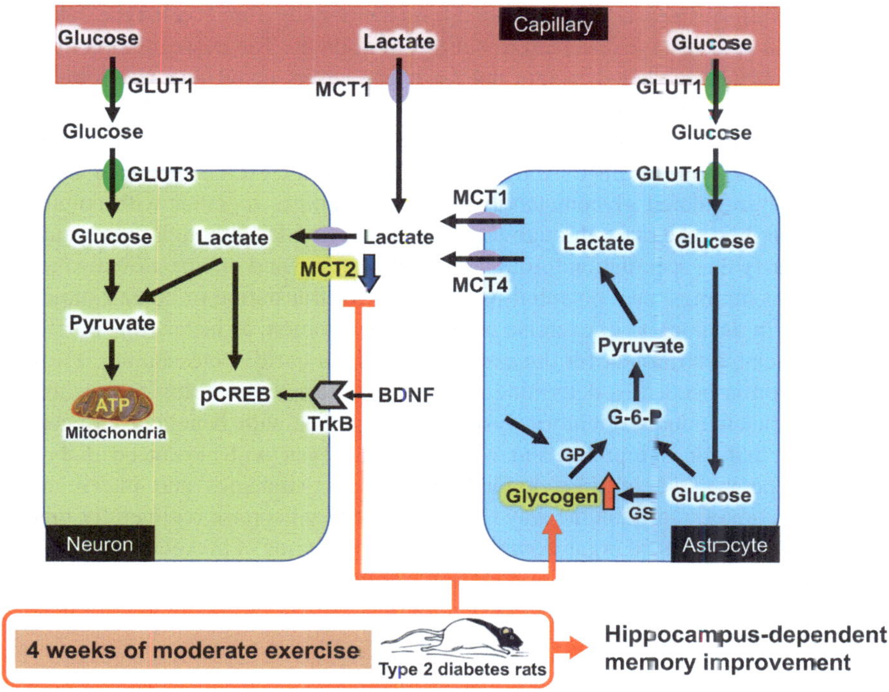

Fig. 6 Conceptual diagram of the chronic exercise effect on ameliorating hippocampus-dependent memory in type 2 diabetes. A 4-week regimen of moderate exercise improves memory dysfunction in type 2 diabetes via amelioration of hippocampal glycometabolism dysregulation. The dysregulated hippocampal glycometabolism and its relationship to lactate transport might be one of the possible etiologies of memory dysfunction in type 2 diabetes. (Summarized figure based on the Shima *et al., Diabetologia,* 2016)

stages, which would clinically improve peripheral diabetic or pre-diabetic symptoms with exerting a beneficial and effective positive role on memory dysfunction through in depth comprehensive future studies.

4 Concluding Remarks and Future Direction

The rapid increase in the number of patients with type 2 diabetes worldwide is a great global problem; hence, a strategy is needed apart from pharmacotherapy such as insulin treatment or other established anti-diabetic drugs, to treat or solve this problem. Cognitive decline is a particularly devastating complication of type 2 diabetes, as hippocampal dysfunction is a serious disability that is associated with dementia, Alzheimer disease, and depression (Pouwer et al. 2003; Ahtiluoto et al. 2010) ultimately causing a remarkable morbidity in social life. Although there are to date still very few studies focusing on this area, in this chapter we largely reviewed the ameliorating effects of chronic exercise on memory dysfunction in type 2 diabetes at the advanced stage observed in our recent findings. The dysregulation of hippocampal MCT2-mediated lactate uptake into neurons could be the etiology of hippocampus-dependent memory dysfunction in type 2 diabetes, and this is a completely novel addition to this field. Although we do not have any clear and concrete evidence linking our results to ANLS, we assume that the exercise induced amelioration of dysregulated glycometabolism in hippocampus together with cognitive impairment may take place through the adaptation of ANLS although future studies should clarify this speculation through extensive in vivo and in vitro investigations. The current findings have potential implications for translation to human subjects, especially in the context of exercise intensity and duration, as lactate threshold has been a crucial determinant for the exercise models in our diabetes studies. Further investigation is needed to determine the optimal exercise condition to ameliorate type-2-diabetes-induced memory dysfunction together with beneficial effects on peripheral diabetic symptoms and complications, even with advanced diabetes, from the perspective of lifestyle-based therapeutic strategies and intervention. Indeed, we are currently on the way to explore the best exercise regimen for diabetes and diabetes-induced organ complications both in terms of prevention and therapeutics through a multi-disciplinary and multi-faceted translational research approach.

References

Ahtiluoto S, Polvikoski T, Peltonen M, Solomon A, Tuomilehto J, Winblad B, Sulkava R, Kivipelto M (2010) Diabetes, Alzheimer disease, and vascular dementia: a population-based neuropathologic study. Neurology 75:1195–1202

Alberini CM (2009) Transcription factors in long-term memory and synaptic plasticity. Physiol Rev 89:121–145

Allaman I, Pellerin L, Magistretti PJ (2000) Protein targeting to glycogen mRNA expression is stimulated by noradrenaline in mouse cortical astrocytes. Glia 30:382–391

Allaman I, Pellerin L, Magistretti PJ (2003) Glucocorticoids modulate neurotransmitter-induced glycogen metabolism in cultured cortical astrocytes. J Neurochem 88:900–908

Aveseh M, Nikooie R, Sheibani V, Esmaeili-Mahani S (2014) Endurance training increases brain lactate uptake during hypoglycemia by up regulation of brain lactate transporters. Mol Cell Endocrinol 394:29–36

Balducci S, Iacobellis G, Parisi L, Di Biase N, Calandriello E, Leonetti F, Fallucca F (2006) Exercise training can modify the natural history of diabetic peripheral neuropathy. J Diabetes Complications 20:216–223

Bélanger M, Allaman I, Magistretti PJ (2011) Brain energy metabolism: focus on astrocyte-neuron metabolic cooperation. Cell Metab 14:724–738

Bergström J, Hultman E (1966) Muscle glycogen synthesis after exercise: an enhancing factor localized to the muscle cells in man. Nature 210:309–310

Bhattacharjee M, Venugopal B, Wong K, Goto Y, Bhattacharjee M (2006) Mitochondrial disorder, diabetes mellitus, and findings in three muscles, including the heart. Ultrastruct Pathol 30:481–487

Biessels GJ, Despa F (2018) Cognitive decline and dementia in diabetes mellitus: mechanisms and clinical implications. Nat Rev Endocrinol 14:591–604

Biessels G-J, Kamal A, Ramakers GM, Urban IJ, Spruijt BM, Erkelens DW, Gispen WH (1996) Place learning and hippocampal synaptic plasticity in streptozotocin-induced diabetic rats. Diabetes 45:1259–1266

Biessels GJ, Staekenborg S, Brunner E, Brayne C, Scheltens P (2006) Risk of dementia in diabetes mellitus: a systematic review. Lancet Neurol 5:64–74

Bolz L, Heigele S, Bischofberger J (2015) Running improves pattern separation during novel object recognition. Brain Plast 1:129–141

Boury-Jamot B, Carrad A, Martin J, Halfon O, Magistretti P, Boutrel B (2015) Disrupting astrocyte–neuron lactate transfer persistently reduces conditioned responses to cocaine. Mol Psychiatry 21:1–7

Bramham CR, Alme MN, Bittins M, Kuipers SD, Nair RR, Pai B, Panja D, Schubert M, Soule J, Tiron A, Wibrand K (2010) The arc of synaptic memory. Exp Brain Res 200:125–140

Brown J, Cooper-Kuhn CM, Kempermann G, Van Praag H, Winkler J, Gage FH, Kuhn HG (2003) Enriched environment and physical activity stimulate hippocampal but not olfactory bulb neurogenesis. Eur J Neurosci 17:2042–2046

Canada SE, Weaver SA, Sharpe SN, Pederson BA (2011) Brain glycogen supercompensation in the mouse after recovery from insulin-induced hypoglycemia. J Neurosci Res 89:585–591

Cardinaux JR, Magistretti PJ (1996) Vasoactive intestinal peptide, pituitary adenylate cyclase-activating peptide, and noradrenaline induce the transcription factors CCAAT/enhancer binding protein (C/EBP)-beta and C/EBP delta in mouse cortical astrocytes: involvement in cAMP-regulated glyco. J Neurosci 16:919–929

Caroni P, Donato F, Muller D (2012) Structural plasticity upon learning: regulation and functions. Nat Rev Neurosci 13:478–490

Chabot C, Massicotte G, Milot M, Trudeau F, Gagné J (1997) Impaired modulation of AMPA receptors by calcium-dependent processes in streptozotocin-induced diabetic rats. Brain Res 768:249–256

Choi IY, Seaquist ER, Gruetter R (2003) Effect of hypoglycemia on brain glycogen metabolism in vivo. J Neurosci Res 72:25–32

Choi HB, Gordon GRJ, Zhou N, Tai C, Rungta RL, Martinez J, Milner TA, Ryu JK, McLarnon JG, Tresguerres M, Levin LR, Buck J, MacVicar BA (2012) Metabolic communication between astrocytes and neurons via bicarbonate-responsive soluble adenylyl Cyclase. Neuron 75:1094–1104

Colberg SR, Sigal RJ, Fernhall B, Regensteiner JG, Blissmer BJ, Rubin RR, Chasan-Taber L, Albright AL, Braun B (2010) Exercise and type 2 diabetes: the American College of Sports Medicine and the American Diabetes Association: joint position statement. Diabetes Care 33:e147–e167

Creer DJ, Romberg C, Saksida LM, van Praag H, Bussey TJ (2010) Running enhances spatial pattern separation in mice. Proc Natl Acad Sci 107:2367–2372

Crosson SM, Khan A, Printen J, Pessin JE, Saltiel AR (2003) PTG gene deletion causes impaired glycogen synthesis and developmental insulin resistance. J Clin Invest 111:1423–1432

Cukierman T, Gerstein HC, Williamson JD (2005) Cognitive decline and dementia in diabetes - systematic overview of prospective observational studies. Diabetologia 48:2460–2469

Dringen R, Gebhardt R, Hamprecht B (1993) Glycogen in astrocytes: possible function as lactate supply for neighboring cells. Brain Res 623:208–214

Flood JF, Mooradian AD, Morley JE (1990) Characteristics of learning and memory in streptozocin-induced diabetic mice. Diabetes 39:1391–1398

Gaesser GA, Brooks GA (1980) Glycogen repletion following continuous and intermittent exercise to exhaustion. J Appl Physiol Resp Env Ex Physiol 49:722–728

Geroldi C, Frisoni GB, Paolisso G, Bandinelli S, Lamponi M, Abbatecola AM, Zanetti O, Guralnik JM, Ferrucci L (2005) Insulin resistance in cognitive impairment. Arch Neurol 62:1067–1072

Gibbs ME, O'Dowd BS, Hertz E, Hertz L (2006) Astrocytic energy metabolism consolidates memory in young chicks. Neuroscience 141:9–13

Gispen WH, Biessels GJ (2000) Cognition and synaptic plasticity in diabetes mellitus. Trends Neurosci 23:542–549

Gold SM, Dziobek I, Sweat V, Tirsi A, Rogers K, Bruehl H, Tsui W, Richardson S, Javier E, Convit A (2007) Hippocampal damage and memory impairments as possible early brain complications of type 2 diabetes. Diabetologia 50:711–719

Gollnick PD, Piehl K, Saltin B (1974) Selective glycogen depletion pattern in human muscle fibres after exercise of varying intensity and at varying pedalling rates. J Physiol 241:45–57

den Heijer T, Vermeer SE, van Dijk EJ, Prins ND, Koudstaal PJ, Hofman A, Breteler MMB (2003) Type 2 diabetes and atrophy of medial temporal lobe structures on brain MRI. Diabetologia 46:1604–1610

Hof PR, Pascale E, Magistretti PJ (1988) K+ at concentrations reached in the extracellular space during neuronal activity promotes a Ca2+–dependent glycogen hydrolysis in mouse cerebral cortex. J Neurosci 8:1922–1928

Holloszy JO (2005) Exercise-induced increase in muscle insulin sensitivity. J Appl Physiol 63110:338–343

James DE, Kraegen EW (1984) The effect of exercise training on glycogen, glycogen synthase and phosphorylase in muscle and liver. Eur J Appl Physiol Occup Physiol 52:276–281

Jenkins NT, Padilla J, Arce-Esquivel AA, Bayless DS, Martin JS, Leidy HJ, Booth FW, Rector RS, Laughlin MH (2012) Effects of endurance exercise training, metformin, and their combination on adipose tissue leptin and IL-10 secretion in OLETF rats. J Appl Physiol 113:1873–1883

Juel C, Holten MK, Dela F (2004) Effects of strength training on muscle lactate release and MCT1 and MCT4 content in healthy and type 2 diabetic humans. J Physiol 556:297–304

Kamal A, Biessels GJ, Urban IJGW (1999) Hippocampal synaptic plasticity in streptozotocin-diabetic rats: impairment of long-term potentiation and facilitation of long-term depression. Neuroscience 90:737–745

Kawano K, Hirashima T, Mori S, Saitoh Y, Kurosumi M, Natori T (1992) Spontaneous long-term hyperglycemic rat with diabetic complications. Otsuka long-Evans Tokushima fatty (OLETF) strain. Diabetes 41:1422–1428

Kitaoka R, Fujikawa T, Miyaki T, Matsumura S, Fushiki T, Inoue K (2010) Increased noradrenergic activity in the ventromedial hypothalamus during treadmill running in rats. J Nutr Sci Vitaminol (Tokyo) 56:185–190

Kong J, Shepel PN, Holden CP, Mackiewicz M, Pack AI, Geiger JD (2002) Brain glycogen decreases with increased periods of wakefulness: implications for homeostatic drive to sleep. J Neurosci 22:5581–5587

Lee H, Chang H, Park JY, Kim SY, Choi KM, Song W (2011) Exercise training improves basal blood glucose metabolism with no changes of cytosolic inhibitor κB kinase or c-Jun N-terminal kinase activation in skeletal muscle of Otsuka long-Evans Tokushima fatty rats. Exp Physiol 96:689–698

Liu YF, Chen HI, Wu CL, Kuo YM, Yu L, Huang AM, Sen WF, Chuang JI, Jen CJ (2009) Differential effects of treadmill running and wheel running on spatial or aversive learning and memory: roles of amygdalar brain-derived neurotrophic factor and synaptotagmin I. J Physiol 587:3221–3231

Magistretti PJ (2006) Neuron-glia metabolic coupling and plasticity. J Exp Biol 209:2304–2311

Magistretti PJ, Pellerin L (1999) Cellular mechanisms of brain energy metabolism and their relevance to functional brain imaging. Philos Trans R Soc Lond B Biol Sci 354:1155–1163

Magistretti PJ, Morrison JH, Shoemaker WJ, Sapin V, Bloom FE (1981) Vasoactive intestinal polypeptide induces glycogenolysis in mouse cortical slices: a possible regulatory mechanism for the local control of energy metabolism. Proc Natl Acad Sci U S A 78:6535–6539

Manschot SM, Brands AMA, van der Grond J, Kessels RPC, Algra A, Kappelle LJ, Biessels GJ (2006) Brain magnetic resonance imaging correlates of impaired cognition in patients with type 2 diabetes. Diabetes 55:1106–1113

Matsui T, Soya S, Okamoto M, Ichitani Y, Kawanaka K, Soya H (2011) Brain glycogen decreases during prolonged exercise. J Physiol 589:3383–3393

Matsui T, Ishikawa T, Ito H, Okamoto M, Inoue K, Lee MC, Fujikawa T, Ichitani Y, Kawanaka K, Soya H (2012) Brain glycogen supercompensation following exhaustive exercise. J Physiol 590:607–616

Matsui T, Soya S, Kawanaka K, Soya H (2015) Brain glycogen decreases during intense exercise without hypoglycemia: the possible involvement of serotonin. Neurochem Res 40:1333–1340

Mccrimmon RJ, Phd R, Mccrimmon RJ, Ryan CM, Frier BM (2012) Diabetes and cognitive dysfunction. Lancet 379:2291–2299

Mu J, Brozinick JT, Valladares O, Bucan M, Birnbaum MJ (2001) A role for AMP-activated protein kinase in contraction- and hypoxia-regulated glucose transport in skeletal muscle. Mol Cell 7:1085–1094

Newman LA, Korol DL, Gold PE (2011) Lactate produced by glycogenolysis in astrocytes regulates memory processing. PLoS One 6:e28427

Nikooie R, Rajabi H, Gharakhanlu R, Atabi F, Omidfar K, Aveseh M, Larijani B (2013) Exercise-induced changes of MCT1 in cardiac and skeletal muscles of diabetic rats induced by high-fat diet and STZ. J Physiol Biochem 69 865–877

O'Gorman DJ, Karlsson HKR, McQuaid S, Yousif O, Rahman Y, Gasparro D, Glund S, Chibalin AV, Zierath JR, Nolan JJ (2006) Exercise training increases insulin-stimulated glucose disposal and GLUT4 (SLC2A4) protein content in patients with type 2 diabetes. Diabetologia 49:2983–2992

Ohiwa N, Saito T, Chang H, Omori T, Fujikawa T, Asada T, Soya H (2006) Activation of A1 and A2 noradrenergic neurons in response to running in the rat. Neurosci Lett 395:46–50

Ohiwa N, Chang H, Saito T, Onaka T, Fujikawa T, Soya H (2007) Possible inhibitory role of prolactin-releasing peptide for ACTH release associated with running stress. Am J Physiol Regul Integr Comp Physiol 292:R497–R504

Pellerin L, Pellegri G, Bittar P, Charnay Y, Bouras C, Martin J, Stella N, Magistretti PJ (1998) Evidence supporting the existence of an activity dependent astrocyte-neuron lactate shuttle. Dev Neurosci 20:291–299

Plaschke K, Hoyer S (1993) Action of the diabetogenic drug streptozotocin on glycolytic and glycogenolytic metabolism in adult rat brain cortex and hippocampus. Int J Dev Neurosci 11:477–483

Pouwer F, Beekman ATF, Nijpels G, Dekker JM, Snoek FJ, Kostense PJ, Heine RJ, Deeg DJH (2003) Rates and risks for co-morbid depression in patients with type 2 diabetes mellitus: results from a community-based study. Diabetologia 46:892–898

Ruchti E, Roach PJ, DePaoli-Roach AA, Magistretti PJ, Allaman I (2016) Protein targeting to glycogen is a master regulator of glycogen synthesis in astrocytes. IBRO Rep 1:46–53

Saito T, Soya H (2004) Delineation of responsive AVP-containing neurons to running stress in the hypothalamus. Am J Physiol Regul Integr Comp Physiol 286:R484–R490

Secher NH, Seifert T, Van Lieshout JJ (2008) Cerebral blood flow and metabolism during exercise: implications for fatigue. J Appl Physiol 104:306–314

Shearer J, Ross KD, Hughey CC, Johnsen VL, Hittel DS, Severson DL (2011) Exercise training does not correct abnormal cardiac glycogen accumulation in the db/db mouse model of type 2 diabetes. Am J Physiol Endocrinol Metab 301:E31–E39

Shima T, Jesmin S, Matsui T, Soya M, Soya H (2016a) Differential effects of type 2 diabetes on brain glycometabolism in rats: focus on glycogen and monocarboxylate transporter 2. J Physiol Sci 68:69–75

Shima T, Takashi M, Jesmin S, Okamoto M, Soya M, Inoue K, Liu Y-F, Torres-Aleman I, McEwen BS, Soya H (2016b) Moderate exercise ameliorates dysregulated hippocampal glycometabolism and memory function in a rat model of type 2 diabetes. Diabetologia 60:597–606

Sickmann HM, Waagepetersen HS, Schousboe A, Benie AJ, Bouman SD (2010) Obesity and type 2 diabetes in rats are associated with altered brain glycogen and amino-acid homeostasis. J Cereb Blood Flow Metab 30:1527–1537

Sickmann HM, Waagepetersen HS, Schousboe A, Benie AJ, Bouman SD (2012) Brain glycogen and its role in supporting glutamate and GABA homeostasis in a type 2 diabetes rat model. Neurochem Int 60:267–275

Sigal RJ, Kenny GP, Boule NG, Wells GA, Prud D, Fortier M, Reid RD, Tulloch H, Coyle D, Phillips P, Jennings A, Jaffey J (2007) Effects of aerobic training, resistance training, or both on glycemic control in type 2 diabetes. Ann Intern Med 147:357–369

Soares E, Prediger RD, Nunes S, Castro AA, Viana SD, Lemos C, De Souza CM, Agostinho P, Cunha RA, Carvalho E, Fontes Ribeiro CA, Reis F, Pereira FC (2013) Spatial memory impairments in a prediabetic rat model. Neuroscience 250:565–577

Sorg O, Magistretti PJ (1991) Characterization of the glycogenolysis elicited by vasoactive intestinal peptide, noradrenaline and adenosine in primary cultures of mouse cerebral cortical astrocytes. Brain Res 563:227–233

Sorg O, Magistretti PJ (1992) Vasoactive intestinal peptide and noradrenaline exert long-term control on glycogen levels in astrocytes: blockade by protein synthesis inhibition. J Neurosci 12:4923–4931

Soya H, Mukai A, Deocaris CC, Ohiwa N, Chang H, Nishijima T, Fujikawa T, Togashi K, Saito T (2007a) Threshold-like pattern of neuronal activation in the hypothalamus during treadmill running: establishment of a minimum running stress (MRS) rat model. Neurosci Res 58:341–348

Soya H, Nakamura T, Deocaris CC, Kimpara A, Iimura M, Fujikawa T, Chang H, McEwen BS, Nishijima T (2007b) BDNF induction with mild exercise in the rat hippocampus. Biochem Biophys Res Commun 358:961–967

Sriwijitkamol A, Coletta DK, Wajcberg E, Balbontin GB, Reyna SM, Barrientes J, Eagan PA, Jenkinson CP, Cersosimo E, DeFronzo RA, Sakamoto K, Musi N (2007) Effect of acute exercise on AMPK signaling in skeletal muscle of subjects with type 2 diabetes: a time-course and dose-response study. Diabetes 56:836–848

Stranahan AM (2015) Models and mechanisms for hippocampal dysfunction in obesity and diabetes. Neuroscience 309:125–139

Stranahan AM, Lee K, Martin B, Maudsley S, Golden E, Cutler G, Mattson MP (2009) Voluntary exercise and caloric restriction enhance hippocampal dendritic spine density and BDNF levels in diabetic mice. Hippocampus 19:951–961

Suge R, Shimazu T, Hasegawa H, Inoue I, Hayashibe H, Nagasaka H, Araki N, Katayama S, Nomura M, Watanabe SI (2012) Cerebral antioxidant enzyme increase associated with learning deficit in type 2 diabetes rats. Brain Res 1481:97–106

Suzuki A, Stern SA, Bozdagi O, Huntley GW, Walker RH, Magistretti PJ, Alberini CM (2011) Astrocyte-neuron lactate transport is required for long-term memory formation. Cell 144:810–823

Swanson R, Sagar S, Sharp F (1989) Regional brain glycogen stores and metabolism during complete global ischaemia. Neurol Res 11:24–28

Swanson RA, Morton MM, Sagar SM, Sharp FR (1992) Sensory stimulation induces local cerebral glycogenolysis: Demonstration by autoradiography. Neuroscience 51:451–461

Tekkök S, Krnjevic K (1999) Diabetes mellitus preserves synaptic plasticity in hippocampal slices from middle-age rats. Neuroscience 91:185–191

Umegaki H, Hayashi T, Nomura H, Yanagawa M, Nonogaki Z, Nakshima H, Kuzuya M (2013) Cognitive dysfunction: an emerging concept of a new diabetic complication in the elderly. Geriatr Gerontol Int 13:28–34

van Praag H, Christie BR, Sejnowski TJ, Gage FH (1999) Running enhances neurogenesis, learning, and long-term potentiation in mice. Proc Natl Acad Sci U S A 96:13427–13431

Vissing J, Andersen M, Diemer NH (1996) Exercise-induced changes in local cerebral glucose utilization in the rat. J Cereb Blood Flow Metab 16:729–736

Whitmer RA (2007) Type 2 diabetes and risk of cognitive impairment and dementia. Curr Neurol Neurosci Rep 7:373–380

Winder WW, Yang HT, Jaussi AW, Hopkins CR (1987) Epinephrine, glucose, and lactate infusion in exercising adrenodemedullated rats. J Appl Physiol 62:1442–1447

Witting A, Chen L, Cudaback E, Straiker A, Walter L, Rickman B, Moller T, Brosnan C, Stella N (2006) Experimental autoimmune encephalomyelitis disrupts endocannabinoid-mediated neuroprotection. Proc Natl Acad Sci 103:6362–6367

Yang J, Ruchti E, Petit J-M, Jourdain P, Grenningloh G, Allaman I, Magistretti PJ (2014) Lactate promotes plasticity gene expression by potentiating NMDA signaling in neurons. Proc Natl Acad Sci U S A 111:12228–12233

Development of a Model to Test Whether Glycogenolysis Can Support Astrocytic Energy Demands of Na^+, K^+-ATPase and Glutamate-Glutamine Cycling, Sparing an Equivalent Amount of Glucose for Neurons

Douglas L. Rothman and Gerald A. Dienel

Abstract Recent studies of glycogen in brain have suggested a much more important role in brain energy metabolism and function than previously recognized, including findings of much higher than previously recognized concentrations, consumption at substantial rates compared with utilization of blood-borne glucose, and involvement in ion pumping and in neurotransmission and memory. However, it remains unclear how glycogenolysis is coupled to neuronal activity and provides support for neuronal as well as astroglial function. At present, quantitative aspects of glycogenolysis in brain functions are very difficult to assess due to its metabolic lability, heterogeneous distributions within and among cells, and extreme sensitivity to physiological stimuli. To begin to address this problem, the present study develops a model based on pathway fluxes, mass balance, and literature relevant to functions and turnover of pathways that intersect with glycogen mobilization. A series of equations is developed to describe the stoichiometric relationships between net glycogen consumption that is predominantly in astrocytes with the rate of the glutamate-glutamine cycle, rates of astrocytic and neuronal glycolytic and oxidative metabolism, and the energetics of sodium/potassium pumping in astrocytes and neurons during brain activation. Literature supporting the assumptions of the model is discussed in detail. The overall conclusion is that astrocyte glycogen metabolism is primarily coupled to neuronal function via fueling glycolytically pumping of Na^+ and K^+ and sparing

D. L. Rothman (✉)
Magnetic Resonance Research Center and Department of Radiology, Yale University, New Haven, CT, USA
e-mail: douglas.rothman@yale.edu

G. A. Dienel
Department of Neurology, University of Arkansas for Medical Sciences, Little Rock, AR, USA

Department of Cell Biology and Physiology, University of New Mexico, Albuquerque, NM, USA

M. DiNuzzo, A. Schousboe (eds.), *Brain Glycogen Metabolism*, Advances in Neurobiology 23, https://doi.org/10.1007/978-3-030-27480-1_14

glucose for neuronal oxidation, as opposed to previous proposals of coupling neurotransmission via glutamate transport, lactate shuttling, and neuronal oxidation of lactate.

Keywords Astrocyte · Brain · Brain activation · Glucose utilization · Glucose sparing · Glutamate-glutamine cycle · Glycogen · Glycogen shunt · Glycogen turnover · Lactate · Lactate shuttling · Oxygen consumption · Neuron · Neurotransmission · Potassium · Sodium/potassium ATPase

Abbreviations

$(A\text{-}V)_{substrate}$	Arteriovenous difference across the brain for the identified substrate
ANL shuttle	Astrocyte-neuron lactate shuttle
Asp	Aspartate
CBF	Cerebral blood flow rate
CF	Correction factor
CMR	Cerebral metabolic rate for substrate of interest = $CBF(A\text{-}V)_{substrate}$
CMR_{glc}	Cerebral metabolic rate for glucose = $CBF(A\text{-}V)_{glc}$
$CMR_{glc\text{-}nonox}$	Rate of non-oxidative metabolism of glucose
$CMR_{glc\text{-}nonox\text{-}A\text{-}b}$	Cerebral metabolic rate of nonoxidative metabolism of glucose in the astrocyte (A) under baseline conditions
$CMR_{glc\text{-}nonox\text{-}N\text{-}b}$	Cerebral metabolic rate of nonoxidative metabolism of glucose in the neuron (N) under baseline conditions.
$CMR_{glc\text{-}ox}$	Rate of oxidative metabolism of glucose
$CMR_{glc\text{-}ox\text{-}A\text{-}b}$	Cerebral metabolic rate of glucose oxidation in the astrocyte (A) under baseline conditions
$CMR_{glc\text{-}ox\text{-}N\text{-}b}$	Cerebral metabolic rate of glucose oxidation in the neuron (N) under baseline conditions.
$CMR_{glc\text{-}tot}$	Total rate of glucose utilization (i.e., at the hexokinase step)
$CMR_{glycogen}$	Cerebral metabolic rate for glycogen = Δ[glycogen]/time
CMR_{O2}	Cerebral metabolic rate for oxygen = $CBF(A\text{-}V)_{O2}$
CP-316,819	[R-R∗,S∗]-5-chloro-N-[2-hydroxy-3-(methoxymethylamino)-3-oxo-1-(phenylmethyl)propyl]-1H-indole-2-carboxamide
DAB	1,4-dideoxy-1,4-imino-d-arabinitol
DG	2-deoxy-D-glucose
DG-6-P	DG-6-phosphate
FDG	2-Fluoro-2-deoxy-D-glucose
FDG-6-P	FDG-6-phosphate
GABA	γ-Aminobutyric acid
GAD	Glutamate decarboxylase
Glc	Glucose

Glc-6-P	Glucose-6-phosphate
Gln	Glutamine
Glu	Glutamate
Lac	Lactate
LDH	Lactate dehydrogenase
MAS	Malate-aspartate shuttle
MCT	Monocarboxylic acid transporter; MCT1 and MCT4 isoforms are mainly astrocytic, whereas MCT2 is predominantly neuronal
MRS	Magnetic resonance spectroscopy
NAA	N-Acetylaspartate
NAL shuttle	Neuron-astrocyte lactate shuttle
OCI	Oxygen carbohydrate index = $CMR_{O2}/[CMR_{glc} + 0.5CMR_{lac} + CMR_{glycogen}] = (A\text{-}V)_{O2}/((A\text{-}V)_{glc} + 0.5(A\text{-}V)_{lac} + \Delta[\text{glycogen}])$, where lactate and [glycogen] are expressed in glucosyl units (2Lac = 1Glc)
OGI	Oxygen-glucose index = $CMR_{O2}/CMR_{glc} = (A\text{-}V)_{O2}/(A\text{-}V)_{glc}$ (CBF cancels out) This calculation assumes no other substrates are oxidized.
PAG	Phosphate-activated glutaminase
PPP	Pentose phosphate shunt pathway
TCA	Tricarboxylic acid
V_{ATP}	Rate of ATP production
V_{cycle}	Rate of the glutamate-glutamine cycle
$V_{cycle\text{-}b}$	Baseline rate of the glutamate glutamine cycle
V_{efflux}	Rate of lactate efflux
$V_{glu\text{-}ox\text{-}b}$	Baseline rate of astrocytic glutamate oxidation
$V_{PC\text{-}b}$	Baseline rate of pyruvate carboxylase
$V_{PDH\text{-}A\text{-}b}$	Baseline rate of pyruvate dehydrogenase in the neuron (N)
$V_{PDH\text{-}N\text{-}b}$	Baseline rate of pyruvate dehydrogenase in the astrocyte (A)
V_{shunt}	Rate of glycogenolysis = $CMR_{glycogen}$
ΔV_{cycle}	Incremental change in V_{cycle}

1 Introduction

1.1 Major Questions in Neuroenergetics

The last 20 years have seen substantial progress in understanding the energy requirements to support brain neuronal signaling. Among other discoveries, the major findings have been the measurement of the relationships between glutamate (Glu) and γ-aminobutyric acid (GABA) neurotransmitter cycling between neurons and astrocytes and cell-specific energetics and metabolic fluxes. Studies have found linear relationships between the glutamate-glutamine (Gln) and GABA-glutamine cycles, anaplerotic glutamine synthesis, and neuronal and glial glucose oxidation

(Yu et al. 2018; Lanz et al. 2013; Patel et al. 2010; Sonnay et al. 2016; Duarte et al. 2011). In the resting awake state almost 80% of total cerebral cortex energy consumption supports functional processes in neurons and glia. These experimental findings are consistent with energy budgets constructed from energy requirements determined experimentally from single cells (Yu et al. 2018; Harris et al. 2012; Howarth et al. 2012; Attwell and Laughlin 2001). Despite this progress several important questions remain unanswered regarding functional neuroenergetics including:

- **Why does the brain have an obligatory requirement for glucose consumption?** Many studies have shown that alternative substrates can be used by brain, but insulin-induced hypoglycemic coma in vivo is quickly reversed only by glucose (or maltose that is converted to glucose), not many other substrates including lactate, acetate, and β-hydroxybutyrate that are supplemental oxidative fuels in vivo (Clarke and Sokoloff 1999; Sokoloff 1960). Studies dating back to the 1960s have found that, even under conditions of extended fasting, brain glucose oxidation can only be partially replaced by fuels such as ketone bodies (Cahill Jr. 2006). More recently, it was shown that while the entire nonsignaling-coupled cerebral cortical energy requirements can be replaced by ketones, whereas only approximately 50% of the energetics used to support signaling can be replaced (Chowdhury et al. 2014) and metabolism of tracer amounts of β-[^{14}C]hydroxybutyrate does not rise during acoustic stimulation (Cruz et al. 2005). Similarly, studies in which lactate was infused or generated by exercise to produce high plasma levels have been unable to demonstrate complete suppression of brain glucose uptake in vivo (Quistorff et al. 2008; Smith et al. 2003).
- **Why is there a near-stoichiometric relationship between increments in glucose oxidation and glutamate/GABA neurotransmitter cycling?** As initially found by Sibson et al. (1998), there is close to a 1:1 molar relationship between increments in the GABA/glutamate/glutamine neurotransmitter cycle (V_{cycle}) and neuronal glucose oxidation. When the functional component of glial glucose oxidation is included, a 1:1 relationship is achieved (Yu et al. 2018; Lanz et al. 2013; Patel et al. 2010; Sonnay et al. 2016; Duarte et al. 2011). Figure 1a shows the slopes and intercepts of these relationships based upon recent meta-analyses. As discussed below, this relationship has been interpreted as implying the existence of a molecular

Fig. 1 (continued) pyruvate/lactate, with most of the lactate being shuttled to neurons as oxidative fuel. Numerical values are pathway fluxes in glucose equivalents normalized to $CMR_{glc\text{-}ox\text{-}N}$ = 1 µmol/g/min and the ATP yields from these fluxes. From Fig. 5 of (Hyder et al. 2006) © 2006, SAGE Publications with permission. **Abbreviations**: $CMR_{glc\text{-}ox\text{-}tot}$, total rate of cerebral glucose oxidation by all brain cells; $CMR_{glc\text{-}ox\text{-}N}$, rate of cerebral glucose oxidation in neurons; $CMR_{glc\text{-}ox\text{-}A}$, cerebral metabolic rate of glucose oxidation in astrocytes; V_{PC}, rate of pyruvate carboxylase in astrocytes; V_{cycle}, rate of glutamate-glutamine cycle; Glu, glutamate; Gln, glutamine; Glc, glucose; Pyr, pyruvate; Lac, lactate; OAA, oxaloacetate; GABA, γ-aminobutyric acid; Mal, malate, Suc, succinate; α-KG, α-ketoglutarate; Cit, citrate

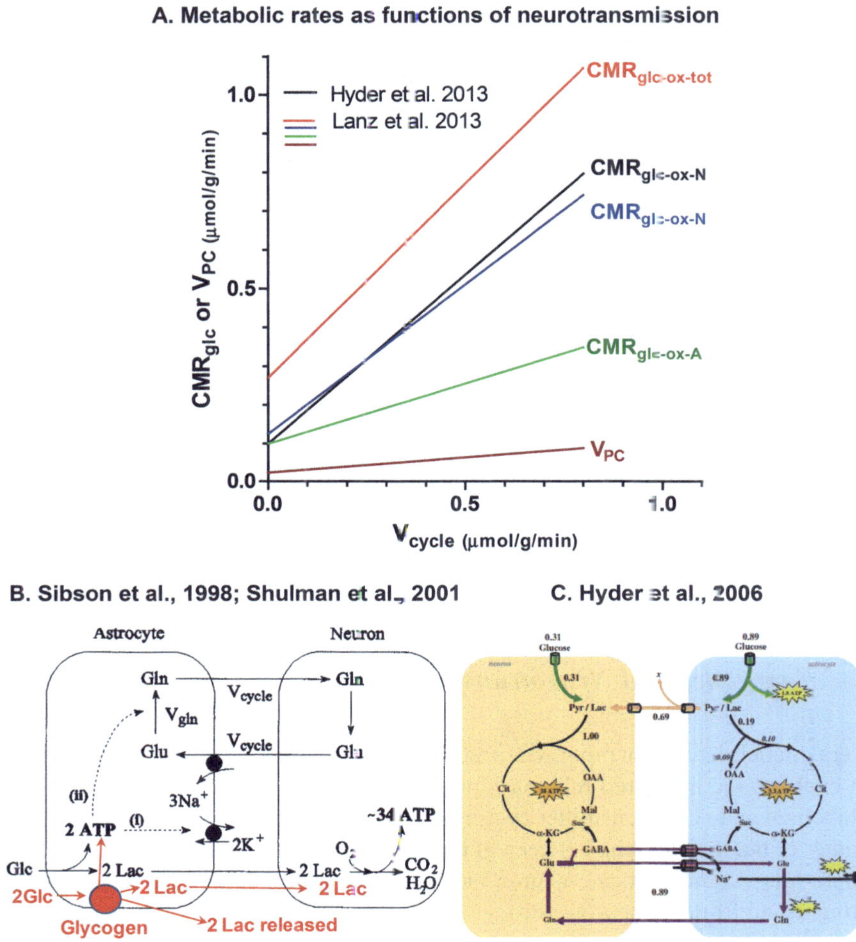

Fig. 1 Neuronal and glial metabolic fluxes and early models to explain the 1:1 relationship between neuronal glucose oxidation and glutamate-glutamine cycle rate. (**a**). Meta-analyses of the rates of glucose oxidation in neurons and astrocytes and pyruvate carboxylation in astrocytes determined by magnetic resonance spectroscopy (MRS) reveal linear relationships to the rate of the glutamate-glutamine cycle over a wide range of activities, from isoelectric to awake resting states. The y-axis intercept represents the metabolic rate at isoelectric EEG, which is interpreted as the energetics of nonsignaling brain activities. Increases in metabolic rate arise from functional signaling activities, with graded rates between isoelectricity and the awake resting state obtained by use of different anesthetics. Lines were plotted from the equation of the best-fit regression line in Fig. 2A of (Hyder et al. 2013) and from interpolated values calculated from the regression lines in Fig. 8 of (Lanz et al. 2013). (**b**) The black line drawing is Fig. 4 of (Sibson et al. 1998) © (1998) National Academy of Sciences, with permission of the authors. The red diagram added to the Sibson model is based on Fig. 1 of (Shulman et al. 2001a). Note that astrocyte-neuron lactate shuttling provides neuronal oxidative fuel in both models, but they differ in the source of lactate, either glucose or glycogen, and the stoichiometry. (**c**) Updated model in which glucose utilization is partitioned between astrocytes and neurons, with higher glycolytic activity in astrocytes to produce

mechanism coupling glycolytic ATP in the glia to the Na^+, K^+-ATPase activity associated with supporting glutamate transport into the astrocytes, with the associated transport of Na^+ ions into the cell requiring pumping out. However, no direct in vivo evidence has been found supporting this hypothesis.

- **Why is there an increase in nonoxidative glucose metabolism during intense sensory stimulation?** In the resting awake state, almost all of the ATP produced by the brain comes from glucose oxidation, with an oxygen glucose index ($OGI = CMR_{O2}/CMR_{glc}$, where CMR is cerebral metabolic rate for the subscripted substrate) ranging from the theoretical maximum of 6 to 5, depending on species and methodology (Hyder et al. 2006). Lear and Ackermann (1989) recognized that glycolysis is preferentially upregulated during stimulation and this increased flux causes a significant drop in the OGI and greater production of brain lactate (reviewed in Dienel 2019a, 2012; Dienel and Cruz 2016). Nevertheless, from an energetic standpoint, the extra ATP being generated during the activation by nonoxidative glycolysis remains a small fraction of the total ATP needs, suggesting that it is fueling specific processes. These specific processes and what cell types they occur in remain unidentified.

1.2 Early Mechanistic Models for Astrocyte-Neuron Energetics and Neurotransmission

Several models have been proposed to address the question of what is the mechanism that explains the measured relationships between neuronal and astrocytic glucose metabolism and neurotransmitter cycling. These models can be separated based on whether or not the coupling between glutamate/glutamine cycling and glucose consumption is via the astrocyte-neuron lactate shuttle (ANL shuttle), which is to be distinguished from the neuron-astrocyte lactate shuttle, NAL shuttle (Mangia et al. 2009, 2011; DiNuzzo et al. 2010a). In the ANL shuttle-based models, the glucose needed to fuel functional processes is taken up in astrocytes and converted to lactate. Once converted to lactate the majority is then transported out of the astrocyte to be taken up and oxidized by neurons. The ANL shuttle was proposed initially based on astrocyte cell culture studies to explain the elevation in lactate observed during sensory stimulation studies (Pellerin and Magistretti 1994). In these cell culture studies, it was found that added glutamate led to an increase in 2-deoxy-D-[^{3}H] glucose (DG) phosphorylation (Pellerin and Magistretti 1994). It was proposed that the nonoxidative glycolysis was due to the need for glycolytic ATP to fuel the Na^+,K^+-ATPase needed to pump the Na^+ co-transported with glutamate out of the astrocyte and to convert glutamate into glutamine, and that this process could explain in vivo nonoxidative glycolysis during sensory activation. In 1998 Sibson and coworkers initially measured the ~1:1 in vivo relationship between the glutamate/GABA/glutamine cycle and neuronal glucose oxidation. They further showed it was consistent with an ANL shuttle-based model, shown in Fig. 1b, in which astrocytic metabolism of glucose provided the glycolytic ATP needed to fuel the transport of glutamate into the astrocyte. The 1:1 stoichiometry arose because per glutamate molecule transported 1 ATP was needed to restore Na^+ and K^+ balance using the Na^+, K^+-ATPase and another ATP was needed to convert glutamate to glutamine. The use of glycolytic ATP was rationalized based on both of these events being rapid processes (Magistretti et al. 1999; Sibson et al. 1998).

Although the modified ANL shuttle model quantitatively predicted the in vivo stoichiometry between GABA/glutamate/glutamine neurotransmitter cycling and neuronal glucose oxidation, it provided no predictions of the amount that net non-oxidative glycolysis would increase as a function of GABA/glutamate/glutamine cycling. Shulman et al. (2001a) addressed this limitation by proposing that astrocytic glucose metabolism during intense activation was coupled to the glycogen shunt, a pathway originally described in exercising muscle to explain aerobic lactate generation (Shulman and Rothman 2001) and also proposed by Swanson (1992), which is shown schematically in the red text in Fig. 1b. In the brain, glycogen is almost entirely localized to astrocytes making it an ideal candidate for rapid lactate generation and transfer to neurons (Dringen et al. 1993). In the glycogen shunt, some of the astrocytic glucose uptake, instead of directly entering glycolysis, is first stored as glycogen. The glycogen is then broken down during intense bursts of activity to fuel the Na^+, K^+- ATPase and glutamine synthesis with glycolytic ATP. Due to the loss of 1 ATP in the synthesis of glycogen (relative to direct glycolysis), the net stoichiometry of the shunt would be 2 glucose per glutamate, with the lactate equivalents of 1 glucose transferred to the neuron for oxidation and the other equivalents being released into the blood. This stoichiometry was shown to agree with measurements during seizure. The shunt model was later modified by Hyder and coworkers (2006) to take into account astroglial glucose oxidation as shown in Fig. 1c.

Soon after the ANL shuttle was proposed it was criticized on a variety of grounds, including glutamate-evoked glycolysis not being a robust finding in many astrocyte cultures, it did not include glutamate oxidation upon its uptake, and the predicted stoichiometries of astrocytic lactate production and transfer did not match experimental results (reviewed in Dienel 2017, 2019a). To date, direct in vivo evidence of net transfer of lactate between astrocytes and neurons has not been demonstrated. In 2014, Patel and coworkers directly tested whether the increment in glucose uptake during seizure would be astrocytic at the rate relative to the GABA/glutamate/glutamine cycle predicted by the ANL shuttle-based glycogen shunt models shown in Fig. 1b, c (Patel et al. 2014). In the Patel study, the neuronal contribution was measured by extracting synaptosomes from animals infused with 2-fluoro-2-deoxy-D-glucose (FDG) and comparing the ratio of FDG-6-phosphate (FDG-6-P) to N-acetylaspartate (NAA) under control and seizure conditions. They found that neuronal glucose phosphorylation by hexokinase was proportional to total glucose phosphorylation from which they concluded that, within the accuracy of the study, there was little or no ANL shuttle activity.

As an alternative to ANL shuttle-based models, several models to explain mechanistic coupling of neuron-astrocyte energetics with neurotransmission have been proposed based on the localization of neuronal phosphate-activated glutaminase (PAG) in the inner membrane of mitochondria near presynaptic terminals. As a consequence of this localization, in order to transport a molecule of glutamate made from glutamine by PAG to the cytosol where it can be pumped into vesicles for use in neurotransmission, the malate-aspartate shuttle (MAS) needs to be used, as shown in Fig. 2a. The partial or pseudo-MAS requires an NADH from glycolysis which is subsequently transferred to the electron transport chain. The predicted stoichiometry of this mechanism is 1 glucose per 2 glutamates which is approximately half the experimentally-measured slope. However, as discussed in Sect. 2 the original proposals for this mechanism did not take into account that approximately half of neuronal glucose consumption is in the presynaptic terminals (Yu et al. 2018) where the majority of neurotransmitter glutamate will be synthesized, which changes the predicted stoichiometry to approximately 1:1. An additional potential coupling mechanism is due to pumping of glutamate into vesicles (Fig. 2b) via a glutamate-proton antiporter which requires an ATP per glutamate. This ATP appears to be preferentially provided by glycolysis (Ueda 2016) which also predicts an approximately 1:1 relationship when the distribution of glucose consumption between pre- and postsynaptic neuronal regions is taken into account (Sect. 2). These mechanisms are shown together in Fig. 2b. Although the proposed neuronal-based mechanisms (Fig. 2a, b) can explain the experimentally-measured neuronal stoichiometry between $CMR_{glc\text{-}ox\text{-}N}$ and V_{cycle} they do not explain the astroglial stoichiometry nor how neuronal and astroglial functional metabolism is coupled.

Fig. 2 (continued) glutamate that is loaded into synaptic vesicles. The pseudo-MAS model was based on studies in cultured neurons in which the NADH was derived from glycolytic metabolism of glucose. However, in brain in vivo, it is possible that metabolism of extracellular lactate could provide NADH for the pseudo-MAS. Modified from Fig. 5 of (Hertz and Chen 2017) ©2017 Hertz and Chen, open access with permission, that was derived from Fig. 3 of (Palaiologos et al. 1988) ©1988, John Wiley and Sons, with permission. (**b**) Energetics of conversion of glutamine into vesicular neurotransmitter glutamate. Under conditions when all glucose in the presynaptic terminal is oxidized, the 2 NADH produced by glycolytic metabolism of one glucose molecule can participate in the 'true' MAS and pseudo-MAS to support the glutamate-glutamine cycle via conversion of glutamine to glutamate and vesicular packaging of glutamate. From Fig. 4D of (Dienel 2019a), © 2019 American Physiological Society, with permission. **Abbreviations**: Glu, glutamate; $Glu_{vesicle}$, vesicular glutamate; Gln, glutamine; PAG, phosphate-activated glutaminase; Glc, glucose; Pyr, pyruvate; Pyr_{ox}, pyruvate oxidized; Lac, lactate; OAA, oxaloacetate; Asp, aspartate; ETC, electron transport chain; CMR_{O2}, cerebral metabolic rate of oxygen; mal, malate, Suc, succinate; αKG, α-ketoglutarate; GAP, glyceraldehyde-3-phosphate; GAPDH, GAP dehydrogenase; 1,3PG, 1,3-bisphosphoglycerate; 3PG, 3-phosphoglycerate; TCA, tricarboxylic acid

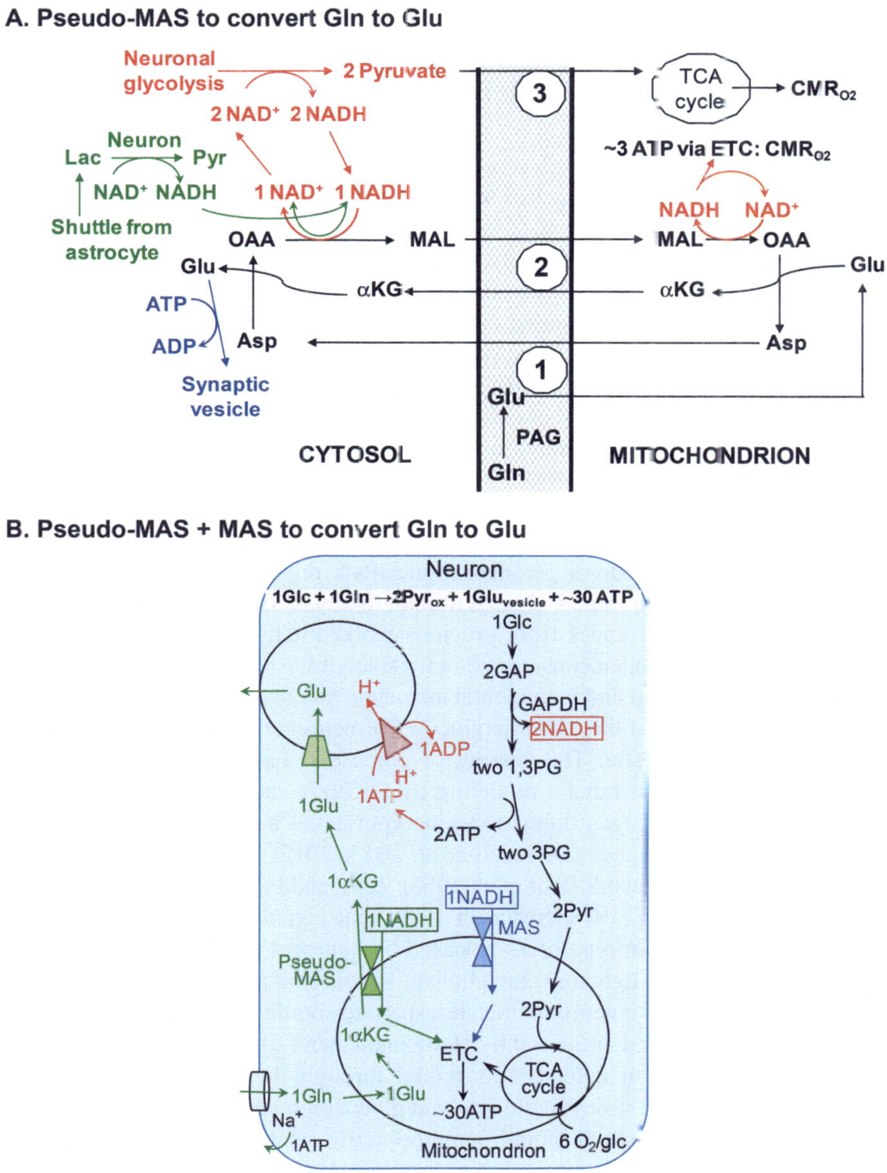

Fig. 2 Roles of the pseudo-malate-aspartate shuttle (MAS) and 'true'-MAS in Glu-Gln cycling. (**a**) Conversion of glutamate to glutamine requires the pseudo-MAS. The MAS transfers reducing equivalents from cytosol to mitochondria, enabling pyruvate to be taken up into mitochondria by its carrier (reaction 3) and oxidized. The pseudo-MAS differs from the 'true' MAS in that the glutamate is generated from the action of phosphate-activated glutaminase on glutamine derived from the glutamate-glutamine cycle that is taken up into neurons. The glutamate is transported into the matrix by the aspartate-glutamate carrier (aralar, reaction 1). Transamination produces α-ketoglutarate that is exported to the cytosol by the ketodicarboxylic acid carrier (reaction 2), where it is transaminated to produce

1.3 A New Stoichiometric Model for Roles of Glycogenolysis in Coupling of Astrocyte-Neuron Energetics with Neurotransmission in Activated Brain

Over the last two decades there has been increasing evidence for the importance of glycogen for brain functions, such as learning and memory (Hertz et al. 2013; Duran et al. 2013, 2019; Newman et al. 2011; Zhang et al. 2016; Suzuki et al. 2011). In addition, improvements in brain glycogen isolation and visualization (Cruz and Dienel 2002; Swanson 1992), as well as noninvasive magnetic resonance spectroscopic (MRS) methods have shown that brain glycogen concentrations are several-fold higher than previously reported (Öz et al. 2017; Öz et al. 2015; DiNuzzo 2013; Cruz and Dienel 2002; Oe et al. 2016; Kong et al. 2002), with proportional increases of the amount of glycogenolysis measured during intense neuronal activity (Dienel et al. 2002, 2007a). Given the important role of glycogen to provide rapid ATP without exhaustion of glucose levels during high energy demand in tissue such as muscle, it is surprising that glycogen is unable to indirectly play a similar role for neurons, which have 2 to 3 times higher energy requirements than astrocytes.

In this paper, we address the above questions regarding functional glucose metabolism by proposing a new glycogen shunt model that is shown in Fig. 3. There are three primary differences from previous work. First, as opposed to an ANL shuttle, astrocytic glycogen consumption effectively transfers glucose carbon to the neuron by reducing total and incremental astrocytic glucose uptake, thereby sparing an equivalent amount of blood-borne glucose for neuronal metabolism; there is no cell-cell substrate transfer. The sparing of glucose is based on the proposal by Swanson (1992) and metabolic modeling by DiNuzzo and coworkers (2010a, b, 2012). Second, the glycogen shunt model incorporates as a mechanism the proposal of DiNuzzo and colleagues (DiNuzzo et al. 2013, 2017; Mangia et al. 2013; Xu et al. 2013) and earlier work (Hof et al. 1988; Hertz and Peng 1992; Rosenthal and Sick 1992; Quach et al. 1978; Swanson 1992) that glycolytic ATP from glycogenolysis supports the pumping of K^+ released by neuronal activity, thereby linking astrocyte glycogen (and glucose) metabolism to neuronal activity. Third, while the relationship between the neuronal malate-aspartate shuttle and neuronal glutamate synthesis from glutamine proposed by Hertz and Chen (2017) (Fig. 2a) is incorporated into the model, it extends this model through the coupling mechanisms between neuronal and astrocytic glucose and glycogen metabolism.

From the glycogen shunt model, stoichiometric relationships can be derived between the rate of net glycogenolysis (V_{shunt}) and several fluxes including neuronal total (tot-N) and oxidative (ox-N) glucose consumption ($CMR_{glc\text{-}tot\text{-}N}$, $CMR_{glc\text{-}ox\text{-}N}$), the GABA/glutamate/glutamine cycle (V_{cycle}), and the relative neuronal and astrocytic functional neuroenergetic costs (as a function of V_{cycle}). We end the chapter by reviewing predictions of the model that can be tested experimentally.

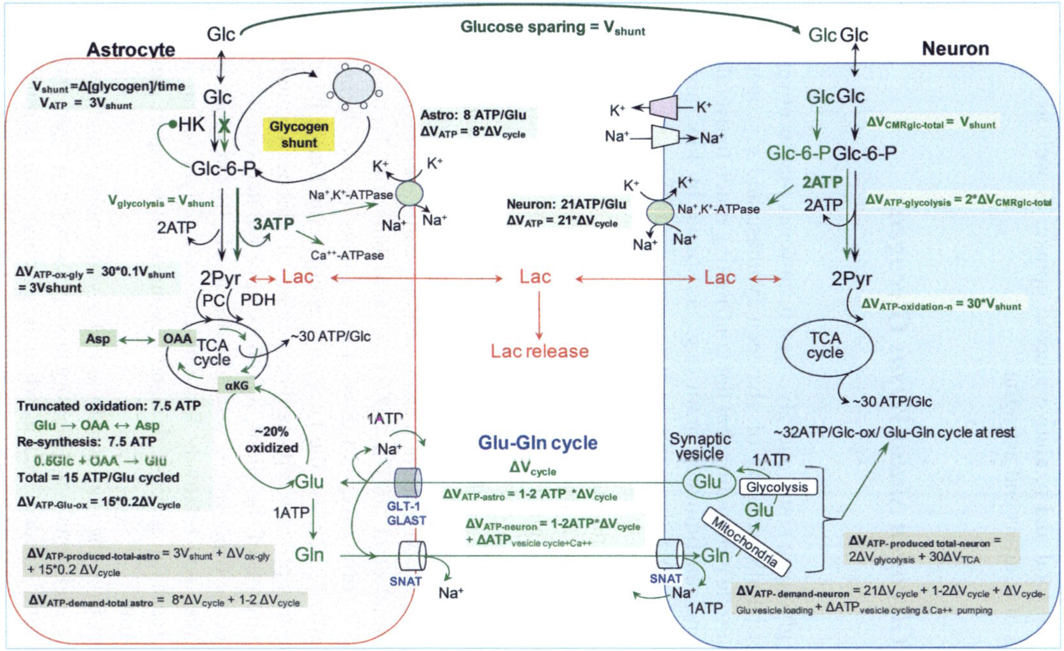

Fig. 3 The revised, updated glycogen shunt model. The glycogen shunt model shows the relationships between the rate of glycogenolysis during activation, glucose sparing for neurons, and metabolic pathway fluxes. In brief, during activation glycogenolysis in astrocytes replaces the incremental rise in astrocytic blood-borne glucose utilization, sparing an equivalent amount of glucose for neuronal glycolytic and oxidative metabolism. The model can be tested by evaluating the ability of glycogenolysis to satisfy the energetic demands of Na^+, K^+-ATPase and glutamate-glutamine cycling in astrocytes and neurons. The stoichiometric relationships derived from the model are given in Sect. 2 and the assumptions used in formulating the model and their justifications are discussed in detail in Sect. 3. The ATP requirements for Na^+, K^+-ATPase in neurons and astrocytes are taken from Fig. 2 of DiNuzzo et al. (2017). **Abbreviations**: Glu, glutamate; $Glu_{vesicle}$, vesicular glutamate; Gln, glutamine; Glc, glucose; Glc-6-P, glucose-6-phosphate; Pyr, pyruvate; Lac, lactate; OAA, oxaloacetate; Asp, aspartate; αKG, α-ketoglutarate; V_{shunt}, rate of glycogen utilization; V_{cycle}, rate of Glu-Gln cycle; $V_{glycolysis}$, glycolytic rate; GLT-1 and GLAST, glutamate transporters (co-transport of Na^+ with Glu), SNAT, solute neutral amino acid transporter (Na^+ linked Gln transporter); TCA, tricarboxylic acid; V_{TCA}, rate of TCA cycle; PC, pyruvate carboxylase; PDH, pyruvate dehydrogenase; V_{ATP}, rate of ATP production from the indicated sources; $V_{ATP-demand}$, rate of ATP consumption for the indicated cell type and processes; Δ, incremental change from baseline to activation of the indicated rate; V_{cycle}, rate of glutamate-glutamine cycle; Truncated Glu oxidation, entry of Glu into the TCA cycle at the αKG step and retention in the cycle or interconversion of OAA with Asp such that the 4-carbon backbone of Glu is used to re-synthesize Glu. See text for definitions of other fluxes

2 Description of the Glycogen Shunt Model

In this section, the glycogen shunt model is derived along with the predicted relationships between the rate of net glycogen breakdown (V_{shunt}), and the increments in Vcycle (ΔV_{cycle}), and neuronal total and oxidative glucose uptake, as well as the relative functional energetics of neurons and astroglia.

2.1 *Definition of the Glycogen Shunt Versus Glycogen Turnover*

The difference between the glycogen shunt and glycogen turnover is that turnover can be just due to low level parallel activity of glycogen synthase and glycogen phosphorylase involved in particle remodeling, but not contributing much to overall energetics. In liver and resting, unstimulated brain (e.g., as shown by low label incorporation and slow label release with no change in concentration Watanabe and Passonneau 1973), this is most likely what is happening. The caveat is that simply cycling of a glucose molecule into and out of glycogen costs 1 ATP, whereas net degradation of pre-formed glycogen yields 3 ATP per glucosyl unit compared with 2 ATP from glycolytic metabolism of glucose. Conversely, the cost of net synthesis can be paid during times of low energy demand and adequate glucose supply.

In the glycogen shunt there is also turnover, but net glycogen synthesis and net degradation are temporally separated such that there is net degradation or net synthesis of glycogen (i.e., substantial concentration changes along with label incorporation or label loss, e.g., Dienel et al. 2007a, 2002). A rapid burst of glycogen breakdown for glycolytic ATP (primarily) is followed by a longer period during which glycogen is re-synthesized. Because the necessary energy could have been supplied directly by glucose this process represents a ‘shunt’ of glucose through glycogen. The most likely purpose of the shunt, as others have pointed out, is to maintain homeostasis of glucose-6-phosphate and glucose levels. Brain glucose transport activity can be maintained relatively low due to having glycogen as a reserve for sudden large increases in glucose demand beyond what transport can supply at the onset of activation.

The glycogen shunt differs from the traditional view of glycogenolysis and glycogen synthesis in that in the traditional view the organ synthesizes glycogen and stores it for extended periods until needed, such as when the glycolytic flux is much higher than could be supplied from the blood (e.g., muscle glycogenolysis during a sprint). Alternatively, the glycogen shunt can work the other way in which a rapid burst of synthesis is followed by a longer period of slow breakdown such as in yeast where the role of glycogen in homeostasis of glucose and particularly glucose-6-phosphate is very clear.

Further discussion of these points is presented in the following papers (Shulman and Rothman 2001, 2015, 2017; Shulman et al. 2001a, b), while noting that the present report revises and updates the concept of the glycogen shunt in brain compared with in these earlier publications and the new model separates the glycogen shunt from involvement in the ANL shuttle model, as discussed above.

2.2 Explanation of the Model

Figure 3 shows pathway diagrams of the coupling between glucose metabolism and GABA/glutamate/glutamine cycling between neurons and astrocytes under nonactivated baseline and activated conditions. Because excitatory synapses are believed to account for the large majority of energy consumption during activation an excitatory neuron is diagrammed. However, the V_{cycle} flux in the relationships is the sum of the glutamate-glutamine and GABA-glutamine cycles and neuronal glucose metabolism, and neuronal energetics is the sum of GABAergic and glutamatergic neurons.

Based on meta-analyses of in vivo MRS studies (see Fig. 1a), the metabolic rates in the diagram can all be expressed as functions of the baseline (subscript b) rate of the GABA/glutamate/glutamine cycle ($V_{cycle-b}$) with intercepts at isoelectricity ($V_{cycle-b} = 0$) according to the following equations where CMR_{glc-ox} is the cerebral metabolic rate of glucose oxidation, $CMR_{glc-ox-N}$ is the cerebral metabolic rate of neuronal glucose oxidation, $CMR_{glc-ox-A}$ is the cerebral metabolic rate of astrocytic glucose oxidation, V_{PC} is the flux through pyruvate carboxylase in the astrocyte, and V_{glu-ox} is the flux through the astrocyte glutamate oxidation pathway. At steady state $V_{PC} = V_{glu-ox}$. The slopes and intercepts used in the equations are derived from studies primarily using ^{13}C-MRS of rat cerebral gray matter. However, as shown in Fig. 1a, results from humans are consistent with these relationships, as are recent results from other species including the tree shrew, rat, and mouse (Lanz et al. 2013; Sonnay et al. 2017b; Hyder et al. 2013). Equations (1)–(5) present the quantitative relationships of the above pathway fluxes to V_{cycle}, as shown in Fig. 1a.

$$CMR_{glc-ox} = 1.0\, V_{cycle} \tag{1}$$

$$CMR_{glc-ox-N-b} = 0.9\, V_{cycle} \tag{2}$$

$$CMR_{glc-ox-A-b} = V_{PC} + 0.5\, V_{PDH-A} = 0.3\, V_{cycle} \tag{3}$$

$$V_{PC} = 0.2\, V_{cycle} \tag{4}$$

$$V_{glu-ox} = 0.2\, V_{cycle} \tag{5}$$

Figure 3 also shows the incremental fluxes predicted by the model during activation when glycogenolysis (V_{shunt}) increases as a function of the increment in V_{cycle} (ΔV_{cycle}). Incremental neuronal glucose consumption is limited to the amount of glucose spared due to the astrocyte switching from glucose to glycogen consumption to produce glycolytic and oxidative ATP. The rate of V_{cycle} is determined by the rate of glucose oxidation in the presynaptic terminals via the pseudo-malate-aspartate shuttle mechanism (Hertz and Chen 2017), which is approximately ½ of total presynaptic neuronal glucose oxidation (Fig. 2). The rest of the neuronal glucose consumption is nonoxidative and fuels processes that use glycolytic ATP, such as glutamate pumping into synaptic vesicles (Ueda 2016). Note that, for simplicity, the GABA-glutamine cycle is not shown in the diagram but its contributions to V_{cycle} and $CMRg_{lc\text{-}ox\text{-}N}$ are included.

Based on the pathway diagrams in Fig. 3 and mass balance, the following steady state flux relationships in Eqs. (6)–(11) are derived for the incremental fluxes during activation. V_{shunt} is not given as an increment because under nonactivated conditions glycogen levels are constant so that there is no net flux, although turnover can occur, e.g. (Watanabe and Passonneau 1973; Dienel et al. 2002):

$$\Delta CMR_{glc-tot-N} = V_{shunt} \tag{6}$$

$$\Delta CMR_{glc-ox-N} = 0.5\,V_{shunt} \tag{7}$$

$$\Delta CMR_{glc-nonox-N} = 0.5\,V_{shunt} \tag{8}$$

$$\Delta V_{cycle} = 0.5\,V_{shunt} \tag{9}$$

$$\Delta V_{cycle} = \Delta CMR_{glc-ox-N} \tag{10}$$

$$\Delta CMR_{glc-tot-A} = -KV_{shunt} \tag{11}$$

The constant K in Eq. (11) ranges from 0 (in which case only incremental astrocytic glucose consumption is spared by V_{shunt}) to 1 (up until $V_{shunt} > CMR_{glc\text{-}A\text{-}b}$), in which case total incremental glucose consumption is reduced to 0.

As shown in these equations, the glycogen shunt model predicts several relationships between V_{shunt} and neuronal glucose metabolism and V_{cycle} that can be tested experimentally. Additional predictions regarding the coupling of glucose metabolism between neuronal and astrocytic energetics are derived in Sect. 2.3.

2.3 *Coupling Between Neuronal and Astrocytic Energetics in the Glycogen Shunt Model*

Figure 3 shows the predicted energetics of the glycogen shunt model relative to V_{shunt} and the increment in V_{cycle} (ΔV_{cycle}) and their coupling to Na^+ and K^+ pumping in the neurons and astrocytes. Based on the glycogen shunt model, the incremental ATP ($\Delta V_{ATP\text{-}A}$) generated by the astrocytes is given by Eq. (12) in which the first term is the glycolytic contribution from glycogenolysis and the second term is from glutamate oxidation and oxidation of lactate/pyruvate generated from glycogenolysis. Based on reports of no or little increase in anaplerosis during sensory stimulation or seizure (Sonnay et al. 2017a, b; Patel et al. 2004) partial glutamate oxidation as opposed to complete glutamate oxidation (which requires anaplerosis) is assumed.

$$\Delta V_{\text{ATP-A}} = 3\,\text{V}_{\text{shunt}} + 15\,\Delta V_{\text{glu}} \;=\; 3\,\text{V}_{\text{shunt}} + 15 * 0.1 * \text{V}_{\text{shunt}} = 4.5\,\text{V}_{\text{shunt}} \qquad (12)$$

Based on glucose sparing, the glycolytic and oxidative ATP generated by the neuron is also determined by the glycogen shunt with the first term from glycolysis and the second from oxidative phosphorylation derived from the glycolytic pathway and the TCA cycle.

$$\Delta V_{\text{ATP-N}} = \text{V}_{\text{shunt}} + 15\,\text{V}_{\text{shunt}} = 16\,\text{V}_{\text{shunt}} \qquad (13)$$

The ratio of the functional ATP consumption of astrocytes and neurons in the cerebral cortex is determined by the ratio of Eqs. (12) and (13). It is assumed in the glycogen shunt model that this ratio remains constant under conditions when the shunt is not active in order to maintain fidelity of synaptic transmission which can be tested by comparison with experimental measurements performed under nonactivated conditions. Due to the neuron shifting away from nonoxidative glycolysis under nonactivated conditions the ratio for comparison becomes.

$$\Delta V_{\text{ATP-N}} / \Delta V_{\text{ATP-A}} = 16 / 4.5 = 3.56 \qquad (14)$$

As with the glucose flux relationships (Eqs. 1–11), this assumption can be tested by comparison with experimental measurements of ATP production as a function of ΔV_{cycle}.

It is also assumed in the glycogen shunt model that all of the ATP generated by glycogenolysis is coupled to incremental Na^+ and K^+ pumping by the astrocytes to restore K^+ taken up by the astrocytes during activation back to the extracellular space; the ATP costs per glutamate cycled shown in Fig. 3 are from Fig. 2 of (DiNuzzo et al. 2017). Based upon this assumption and the calculation by DiNuzzo and colleagues regarding the fraction of total astroglial $V_{ATP\text{-}A}$ devoted to ion pumping the incremental Na^+ and K^+ fluxes are given by

$$\Delta V_{\text{K+-A}} = 2\left(\Delta V_{\text{ATPshunt-A}} + (2/3)\,\Delta V_{\text{ATP-glu-ox-A}}\right) = 8\,\text{V}\text{shunt} = 16\,\Delta V_{\text{cycle}} \qquad (15)$$

$$\Delta V_{\text{Na+-A}} = 3\,\Delta V_{\text{ATPshunt-A}} + (2/3)\,\Delta V_{\text{ATP-glu-ox-A}} = 12\,\text{V}_{\text{shunt}} = 24\,\Delta V_{\text{cycle}} \qquad (16)$$

Based on cellular measurements and energy budget calculations (Yu et al 2018) approximately 90% of incremental oxidative ATP generated by the neuron ($\Delta V_{ATP\text{-}ox\text{-}N}$) is used to fuel Na^+ and K^+ pumping ($\Delta V_{K+ \text{ or } Na+^-N}$).

$$\Delta V_{\text{K+-N}} = 0.9 * 2 * \Delta V_{\text{ATP-ox-N}} = 28.8\,\text{V}_{\text{shunt}} = 57.6\,\Delta V_{\text{cycle}} \qquad (17)$$

$$\Delta V_{Na+-N} = 0.9 * 3 * \Delta V_{ATP-ox-N} = 43.2\, V_{shunt} = 86.4\, \Delta V_{cycle} \quad (18)$$

2.4 Experimental and Theoretical Bases of the Shunt Model

In the following sections, the experimental and theoretical bases of the flux and energetic relationships (Eqs. 6–18) in the model are provided. Potential caveats and justifications are discussed in more detail in Sect. 3.

2.4.1 Glucose Sparing ($\Delta CMR_{glc\text{-}tot\text{-}N} = V_{shunt}$)

The relationships between V_{shunt} and changes in neuronal and astrocyte glucose consumption are derived from the concept of glucose sparing introduced by Swanson (1992) and formalized by DiNuzzo et al. (2012, 2010b, a). During activation, the use of glycogenolysis by the astrocyte reduces its incremental, and potentially total glucose consumption, relative to baseline, thereby sparing the equivalent amount of glucose for the neuron. In the glycogen shunt model, the extreme case of glucose sparing is assumed – in which the only incremental glucose available to the neuron is due to the glucose spared by glycogen consumption in the astrocytes. Although under fed conditions this limitation is unlikely to occur, in Sect. 3 we show that glucose transport can potentially limit brain neuroenergetics even under physiological fasting in the absence of the glycogen shunt. Although this limitation would not occur under nonactivated levels of activity, we predict that the relationship is maintained throughout the neuronal activity range in order to maintain a constant activity independent energetic cost of signaling per glutamate.

2.4.2 V_{cycle} Coupled to the Presynaptic Malate-Aspartate Shuttle and Glucose Oxidation ($\Delta CMR_{glc\text{-}ox\text{-}N} = \Delta V_{cycle}$)

As described in more detail in Sects. 3.6 and 3.7, based on cellular and enzyme localization studies, the maximum rate of glutamate synthesis from glutamine in the neuron is limited by the flux of the pseudo-MAS that removes newly-synthesized glutamate by PAG from the mitochondria (see Fig. 2a, b). The pseudo-MAS requires one reducing equivalent from glycolysis per glutamate removed. The pseudo-MAS model was developed from results of studies in cultured cerebellar glutamatergic neurons, and the NADH was derived from neuronal glycolysis. However, as shown in Fig. 2a, it is also glycogen-derived lactate shuttling in vivo that could potentially provide some NADH for the pseudo-MAS; this possibility is, however, considered highly unlikely and of very low magnitude, if any (see Sect. 3.8). When glucose is the oxidative substrate, the stoichiometry between glutamate synthesis from glutamine ($V_{glutamate}$) and neuronal glucose oxidation is 0.5 (Hertz and Chen 2017), and, with the further assumption of metabolic steady state, $V_{PAG-N} = V_{cycle}$:

$$2\,\Delta CMR_{glc-ox-N-presynaptic} = \Delta V_{PAG-N} = \Delta V_{cycle} = 0.5\,V_{shunt} \qquad (19)$$

The majority of PAG activity is believed to be localized on the presynaptic side of the neuron. Based on theoretical and experimental measurements (reviewed in Yu et al. 2018), neuronal glucose oxidation is approximately equal between the pre- and postsynaptic regions in the neuron. Substituting equation (19) into Eq. (9) (i.e., $\Delta V_{cycle} = 0.5\ V_{shunt}$) yields Eq. (10) (i.e., $\Delta V_{cycle} = \Delta CMR_{glc-ox\ N}$).

This relationship is in good agreement with the experimentally-measured relationships between the GABA/glutamate/glutamine cycle and neuronal glucose oxidation (Yu et al. 2018; Lanz et al. 2013) ranging from approximately 0.75 to 0.9 $\Delta V_{cycle.}$ Since the distribution between pre- and postsynaptic glucose oxidation is only approximately known we could choose one of these measured values but have avoided doing so in order that our calculations are independent of previous measurements of these fluxes in nonactivated states. We note that the meta-analyses published to date report the combined neurotransmitter cycling and glucose oxidation fluxes of glutamatergic and GABAergic neurons. A more precise testing of the glycogen shunt model would involve assessing the fluxes in each cell type separately.

As with the relation between V_{shunt} and neuronal glucose uptake and oxidation, we predict this relationship is maintained under non-shunt conditions in order to maintain the fidelity of neurotransmission, as it ensures a constant relationship between the number of glutamate molecules released and energy available from glucose oxidation to support ion pumping and other functional neuronal processes.

Note that the rate of mitochondrial glutamate production and ΔV_{cycle} could be lower than given in Eq. (20), since in the glycogen shunt model the relationship between neuronal and astrocytic functional glucose metabolism is primarily determined by K^+ and Na^+ flows across the cell membranes, as opposed to astrocyte glutamate uptake and lactate release, as in prior ANL shuttle models. Although the reason why the maximum amount of neurotransmitter glutamate is synthesized is not known, there is evidence that the number of glutamate molecules released per vesicle is required to ensure synaptic transmission fidelity with lower amounts leading to reductions in synaptic strength (Scimemi and Beato 2009).

2.4.3 Glycolytic ATP from Glycogenolysis Supports Astrocyte K^+ and Na^+ Pumping (ΔK^+-A = 2 $\Delta V_{ATP-shunt-A}$ = 6 V_{shunt})

This assumption in the glycogen shunt model is based upon a significant body of literature in cell and brain slice models as well as inhibitor studies in vivo (see discussion in Sect. 3.9). In addition, DiNuzzo et al. (2017) have estimated the K^+ and Na^+ fluxes across astrocyte membranes as a function of V_{cycle}, allowing direct comparison with the experimentally-measured rates of V_{shunt} and ΔV_{cycle}.

2.4.4 Constancy of the Ratios Between Neuronal and Astrocytic Glucose Metabolism and GABA/Glutamate/Glutamine Neurotransmitter Cycling Across the Full Range of Neuronal Activity

It is assumed in the glycogen shunt model that the metabolic relationships determined by the limiting case of the shunt model are maintained at all levels of neuronal activity. The basis of this assumption is that the membrane depolarizations associated with signaling events, associated with synaptic strength, depend on transmembrane ion flows that are directly coupled to energy metabolism via the Na^+/K^+ and other ATPases. These energy costs account for the very large majority of energy required to sustain brain function (Yu et al. 2018; Harris et al. 2012). Therefore, a change in the ratio between energy production from glucose metabolism and GABA/glutamate/glutamine cycling as a function of activity would mean that synaptic strength would be activity-dependent, impairing the fidelity of signaling.

2.4.5 Approximately-Equal Increment in Oxidative and Nonoxidative Neuronal Glucose Consumption During the Shunt ($\Delta CMR_{glc\text{-}ox\text{-}N} = \Delta CMR_{glc\text{-}nonox\text{-}N} = 0.5\ V_{shunt}$)

At present, the processes that lead to the experimentally-observed relationships between the increment in oxidative and nonoxidative glucose consumption during intense sensory activations are not known. However, a candidate process, which is incorporated into the model (Figs. 2b and 3), is the pumping of glutamate (and GABA) into presynaptic vesicles which requires 1 ATP per glutamate pumped (Ueda 2016). Evidence suggests that this ATP is preferentially derived from glycolysis (Ueda 2016). The stoichiometry of this process is given by:

$$\Delta V_{\text{Glu-vesicle}} = 2\,\Delta \text{CMR}_{\text{glc-nonox-N-presynaptic}} \qquad (20)$$

where $\Delta V_{\text{Glu-vesicle}}$ is the incremental rate of glutamate transport into vesicles, which at steady state is equal to ΔV_{cycle}.

However, since only approximately ½ of $\Delta CMR_{glc\text{-}ox\text{-}N}$ is presynaptic, Eq. (20) can be rewritten as

$$\Delta V_{\text{Glu-vesicle}} = \Delta \text{CMR}_{\text{glc-nonox}} = \Delta V_{\text{cycle}} \qquad (21)$$

Under conditions of lower activity where the shunt is not active there presumably is sufficient ATP from oxidative glucose metabolism and glycolysis to supply the energetic needs for this process. The remainder of nonoxidative neuronal glycolysis associated with intense activation would be due to yet unidentified processes requiring rapid ATP availability (e.g., synaptic vesicle recycling (Ashrafi et al. 2017; Ashrafi and Ryan 2017; Rangaraju et al. 2014) and energetics of postsynaptic spines that have few, if any, mitochondria (reviewed in Dienel 2019a).

To summarize, Eqs. (1)–(21) are based on assumptions derived from experimental measurements, mass balance, and flux analyses illustrated in Figs. 1–3. They represent a first approximation of the energetic relationships and can be refined in future studies in which the relationships are tested against data in which the rate of glycogenolysis, total glucose consumption, and total oxidative glucose consumption are measured under a variety of conditions.

3 Discussion of Evidence for and the Impact of the Glycogen Shunt Model: Justifications and Potential Caveats

The glycogen shunt model provides a formal, stoichiometric structure for quantitatively testing experimental data from both brain activation studies and measurements made under nonactivated conditions to investigate the role for astroglial glycogenolysis in determining the relationships between neuronal and astroglial glucose metabolism and function. Of particular importance will be the development of methods that allow real-time measurements of V_{shunt}, V_{cycle}, and pathways of cerebral glucose metabolism simultaneously under activated conditions in each subject.

3.1 The Glycogen Shunt Model Can Explain the Obligatory Need of the Brain for Glucose to Support Functional Processes and the Stoichiometry of Neuronal and Astroglial Glucose Metabolism with GABA/Glutamate/Glutamine Cycling

In this section, we discuss how the glycogen shunt can provide a quantitative explanation for the outstanding questions about the relationship between brain glucose metabolism and function summarized in the Introduction.

3.1.1 The Role of the Glycogen Shunt in the Obligatory Requirement of the Brain for Glucose as a Fuel During Nonactivated and Activated Conditions

Studies have shown that the brain requires glucose to support functional activity with the percentage increasing at higher levels of brain activity (Cremer et al. 1983, 1988; Hargreaves et al. 1986; Cunningham et al. 1986; Cunningham and Cremer 1981). Based on the analysis of experimental data with the shunt model, there are two major components of energetics that require glycolytic ATP from glucose (or glycogen in the astroglia). In the astroglia, the main process is the restoration of Na^+/K^+ and other ionic gradients via ATPases. In the neurons there is an increase in nonoxidative glucose consumption of ½ V_{shunt}, which we assign to supporting the pseudo-MAS and glycolytic ATP for glutamate and GABA transport into vesicles. These fluxes cannot be replaced by non-glucose oxidative fuel sources, such as lactate or ketones, and they help explain the preferential upregulation of glycolysis during activation (Dienel and Cruz 2016).

The need for rapid glycolytic ATP in the astroglia can partially explain the increase in lactate reported during activation (review of early studies by Korf and de Boer 1990; Korf 1996; other examples in Cruz et al. 2007; Fray et al. 1996; Dienel et al. 2002; Madsen et al. 1999) and under glycogen-depleted conditions some of the increase in nonoxidative glucose metabolism. However, based on the analysis there is also a requirement in the neuron for nonoxidative glycolysis during intense activation. The increase in nonoxidative glucose metabolism would also contribute to the reported increase in lactate levels in the above-cited studies. The findings of Dienel et al. (2002) that after 5 minutes the elevated lactate has the same ^{14}C fractional enrichment as glucose further supports an increase in neuronal nonoxidative glycolysis of blood-borne glucose. As described in Fig. 2 and discussed in more detail below, half of the measured neuronal nonoxidative glycolysis can be attributed to the need for glutamate pumping into vesicles, which has been shown experimentally to require ATP from glycolysis (Ueda 2016). The other processes that contribute to the requirement for neuronal glycolytic ATP at present are not known but could be related to the physical distance between mitochondria and postsynaptic spines and dendrites due to the small size of these structures. Energy budget models suggest that a considerable fraction of functional ATP consumption takes place in these structures in order to support the ion pumping associated with excitatory postsynaptic potentials. Even though, based on a ~1:1 split between neuronal oxidative and neuronal nonoxidative glycolysis, the glycolytic ATP contribution to this process will still be relatively small compared to ATP from glucose oxidation (unlike in the astroglia) it could play a critical role in making up for a short term energy gap at the initiation of the process.

Under nonactivated conditions fuel sources other than glucose can only replace 50% of functional neuronal glucose oxidation (Chowdhury et al. 2014; Quistorff et al. 2008; Smith et al. 2003). The remaining requirement for glucose can be explained by the continued need for glycolytic ATP and NADH production in order to maintain the GABA/glutamate/glutamine cycle through the pseudo-MAS and vesicle filling with glutamate and GABA (see Figs. 2 and 3 and sections below describing the mechanism). The stoichiometry of both of these processes is two glutamates per glucose undergoing oxidative glycolysis resulting in a minimum rate of glucose metabolism of $0.5{*}V_{cycle}$ as has been pointed out in several publications (see Hertz and Chen 2017).

3.1.2 The Glycogen Shunt and Glucose Sparing, *Not Glutamate Uptake into Astrocytes*, Determine the Relationship Between Astroglial and Neuronal Functional Energetics

As shown in Fig. 1 and Eqs. (1)–(5), over the last two decades experimental measurements, largely using MRS, have found a linear relationship between neuronal and astroglial glucose metabolism and GABA/glutamate/glutamine cycling. The linear relationship has close to 1:1 stoichiometry with total cerebral glucose oxidation (Fig. 1a) which led to earlier proposals, reviewed in the Introduction, that glutamate itself was responsible for the coupling via the need for glycolytic ATP in order to provide energy for its transport into the astroglia after its release from neurons (Fig. 1b). The glucose taken up by the astroglia to provide this ATP would subsequently be released as lactate and taken up by the neurons to provide oxidative ATP production.

In the revised glycogen shunt model (Fig. 3) proposed here, in contrast to glycolytic ATP requirements for astroglial glutamate transport being the coupling mechanism, the coupling is due to the requirement of glycolytic ATP from glycogenolysis to provide ATP to support close to the full incremental astroglial ion pumping requirements. The glucose spared by the use of glycogen instead of blood-borne glucose determines the maximum energy available to the neuron to support function. The proposal that glycogenolysis could help support neuron energetics via glucose sparing was originally put forth by Swanson (1992) and formalized with metabolic modeling by DiNuzzo, Mangia, and colleagues (DiNuzzo et al. 2010b, 2012), as was the proposal that total ion fluxes could couple neuronal and glial functional metabolism (DiNuzzo et al. 2013, 2017; Mangia et al. 2013).

When glycogen cannot be used, which physiologically could happen if glycogen stores are depleted, the astroglia replace it with glucose uptake and nonoxidative metabolism at approximately 3/2 times (i.e., to compensate for less ATP yield by glycolytic metabolism of glucose *vs.* glycogen) the rate of neuronal glucose consumption. The increased demand on brain glucose supplies would result in conditions where neuronal glucose oxidation was not sustainable. The neuron could then sacrifice nonoxidative glycolysis, which fuels processes mainly related to glutamate synthesis and vesicle packing which the cell could go without for an extended time due to the large reserve in the cytosolic glutamate and glutamate vesicle pools (Ottersen et al. 1990a, b, 1992), while protecting the much larger ATP yield from oxidative glycolysis which supports the ion pumping needed to sustain functional activity.

3.1.3 The Glycogen Shunt in Combination with the Pseudo-MAS Mechanism Determines the Stoichiometry Between Cerebral Glucose Metabolism and GABA/Glutamate/Glutamine Neurotransmitter Cycling

In the previous models of the glycogen shunt the coupling of the GABA/glutamate/glutamine cycle to glucose uptake was via astroglial glutamate uptake and glutamine synthesis (Hyder et al. 2006; Shulman et al. 2001a). Neuronal and astroglial glucose metabolism is then coupled via the ANL shuttle (Pellerin and Magistretti 1994). In the present model the coupling between neuronal and astroglial glucose metabolism is primarily through the relative ATP available under activated conditions for ion pumping. The relationship between V_{cycle} and $CMR_{glc\text{-}ox}$ is through the pseudo-MAS mechanism proposed by Hertz and Chen (2017) and described in Sects. 1 and 2, and discussed in more detail in Sects. 3.6 and 3.7, as opposed to direct astroglial coupling based on glutamate uptake (Pellerin and Magistretti 1994).

The pseudo-MAS mechanism predicts that the transport of a glutamate molecule derived from glutamine from the neuronal mitochondria to the cytosol requires an NADH reducing equivalent from *neuronal* glycolysis (see Sect. 3.8 that discusses why glycogen-derived lactate cannot support this process due to insufficient oxygen consumption to support oxidation of this lactate in neurons). The reducing equivalent via coupling to glucose oxidation and the electron transport chain drives the MAS with glutamate being transported out of the mitochondria for every NADH-derived reducing equivalent transported inward (see Fig. 2a, b). Based on the measured relationship between incremental neuronal glucose oxidation and GABA/glutamate/glutamine cycling (Eq. (2)), close to the maximum rate achievable by the pseudo-MAS mechanism is used. The high rate of synthesis is consistent with studies that have shown reducing cytosolic GABA and glutamate concentration results in impaired synaptic signaling (for a review see Scimemi and Beato 2009).

3.2 The Glycogen Shunt Helps Maintain Brain Glucose Homeostasis

The need for glucose sparing by the glycogen shunt appears inconsistent with the high amount of brain glucose delivery. Cerebral blood flow increases the same or more than glucose metabolism during sensory stimulation and seizure and as a result glucose delivery is always sufficient to meet the demands for both glycolytic and total ATP. However the brain also needs to maintain a low concentration of glucose relative to the blood as well as maintain homeostasis of glycolytic intermediates during large changes in glycolysis. As discussed in recent publications (Shulman and Rothman 2015, 2017), a key role of the glycogen shunt is to maintain low levels of intracellular glucose and glycolytic intermediates despite large increases in the glycolytic rate. By either glycogen synthesis taking up glucose or glycogenolysis releasing glucosyl units, glucose transport can be optimized to meet the average needs of the cell with the glycogen acting as a buffer which provides energy in the place of glucose during high activity or takes up excess glucose entering the brain during low metabolic activity or high postprandial plasma glucose levels (also see Swanson 1992). As a consequence, brain glucose transport activity can be reduced so that, under average activity, brain glucose concentrations are ~20% of those in the blood. The advantage of maintaining a low glucose level is due to the high reactivity of the glucose molecule as shown nonspecific glycation reactions being a major cause of cellular damage in diabetes (Giacco and Brownlee 2010; Rowan et al. 2018). Brain glucose levels in diabetic animal models rise from a control level of ~2 to ~7 μmol/g when the corresponding plasma levels are ~9 and 27 mM, respectively (Gandhi et al. 2010), indicating that brain glucose levels approaching 75–80% of the normal plasma level would cause diabetic complications.

Because of glycogen being predominantly localized to the astrocytes, its role as a buffer for providing energy to the neuron when overall glucose transport is inadequate is complicated relative to muscle and other tissues where glycogen is found in all cells. In the glycogen shunt model, the mechanism of energy provision is by glucose sparing with the astroglia obtaining their incremental, and under extreme conditions, possibly total energy needs largely from glycogenolysis. However, this mechanism assumes that there is an actual need for glucose sparing. In order to test this possibility, we calculated the maximum glucose transport under euglycemic and physiological hypoglycemic conditions in awake humans to assess the maximum increase in glucose consumption that can be supported with and without the glycogen shunt. In addition, we review studies that support the ability of the brain to sustain the increased electrical response to continuous activation is impaired under mild to moderate hypoglycemia.

3.2.1 Calculation of the Maximum Activation Sustainable in the Absence of the Glycogen Shunt

The transport parameters used for the calculation were taken from the ^{1}H MRS study of de Graaf et al. (2001), who separately measured these parameters for gray and white matter. These parameters were similar to those reported in other human studies (Gruetter et al. 1996, 1998; Seaquist et al. 2001; de Graaf et al. 2001). Parameters from human studies were used because they were measured in the awake state. The majority of studies on rodents have been performed under anesthesia. About 40% of the endothelial glucose transporter molecules are localized in the cytoplasmic space, providing a mechanism for upregulation of transport capacity across the blood-brain barrier (Farrell and Pardridge 1991). Due to glucose transport recruitment with increased activity (Hargreaves et al. 1986; Cremer et al. 1983; Cunningham and Cremer 1981), studies performed under conditions of low neuronal activity may underestimate transport capacity in the awake state.

The reversible two state Michaelis-Menten model was used for the calculations of glucose transport. Although a four state model more accurately describes the molecular mechanism of glucose transport, it has been shown that the two state model accurately reproduces the kinetics of glucose transport (Duarte and Gruetter 2012; Duarte et al. 2009; Simpson et al. 2007). Maximum glucose transport capacity was calculated using the approach of de Graaf et al. (2001) for calculating total glucose uptake capacity but without including hexokinase kinetics (equivalent to assuming K_m of hexokinase ~ 0) (Eq. 22). The parameters in the model are the volume of distribution of glucose (V_d), the half maximum transport constant (K_t), and the ratio of the maximum unidirectional rate of transport (V_{max}) over the rate of total glucose metabolism ($CMR_{glc\text{-}tot}$). For a given concentration of plasma glucose (G_o), the maximum net transport rate occurs when intracellular glucose (G_i) is at 0 concentration

$$G_i = -Vd\left[\left(\left(V_{max} / CMR_{glc-tot}\right)-1\right)G_o - K_t\right] / \left[\left(V_{max} / CMR_{glc-tot}\right)+1\right] \quad (22)$$

Setting G_i in Eq. (22) equal to 0 and rearranging yields the following expression

$$CMR_{glc-tot-max} = V_{max} / \left[\left(K_t / G_o\right)+1\right] \quad (23)$$

In the majority of MRS glucose transport studies, the ratio ($V_{max}/CMR_{glc\text{-}tot\text{-}b}$) (where the subscript b denotes baseline) has been measured as opposed to V_{max} directly. To incorporate this ratio, Eq. (23) can be rearranged into the following expression with F_{max} given by the fractional increase in $CMR_{glc\text{-}tot}$ above the baseline value.

$$\left(1+F_{max}\right) = \left(V_{max} / CMR_{glc-tot-b}\right) / [\left(K_t / G_{o)} + 1\right] \quad (24)$$

Using the transport constant values reported by de Graaf and coworkers for human gray matter ($K_t = 1.1$ mM, $V_{max}/CMR_{glc\text{-}tot\text{-}b} = 3.2$), the maximum sustainable fractional increase in glucose transport is $F_{max} = 0.88$ for a plasma glucose concentration under fed conditions of 5.5 mM (Hwang et al. 2019) and a value of $F_{max} = 0.58$ for a plasma glucose concentration of 3.5 mM which can occur under physiological conditions such as a 2 day fast (Rothman et al. 1991).

Under physiological hypoglycemia the maximum sustainable activation value is just above that reported from PET studies during sensory stimulation of 0.2 – 0.5 (see Table 1 in (Shulman et al. 2001b) during sensory stimulation.

However, in the absence of availability of glycogen, the astroglia need to increase their glucose consumption to meet the demand for glycolytic ATP. Due to the 2/3 lower efficiency of net ATP production per plasma glucose molecule versus glycogenolysis for a given increment in activity incremental glucose consumption would increase by

$$CMR_{glc-tot-no\,shunt} = CMR_{glc-tot\,shunt} + \left(3/2\right)V_{shunt} \quad (25)$$

From Eq. (25) it can be seen that in the absence of glycogenolysis the rate of glucose consumption during activation could exceed glucose transport capacity in human cerebral cortex. Furthermore, similar larger increases in functional glucose metabolism have been measured in animal models (e.g., Collins et al. 1987; Dienel et al. 2007b; Dienel et al. 2007a; Cruz et al. 2007; Cruz et al. 2005; Nakao et al. 2001). In addition, based on high resolution [^{14}C]DG autoradiography sub-regions within the sensory cortices (e.g., layer 4 and regions such as visual cortex columns) below the resolution of PET may have considerably higher increases in sustained glucose metabolism (Sokoloff 1981).

To summarize, glucose transport can potentially become limiting for brain metabolism during activation under hypoglycemic conditions. Although glucose transporter upregulation can occur after a lag, during the transient period at the start of an activation the switch of the astrocyte to glycogenolysis may spare sufficient glucose for the neuron to maintain functional activity. More work is required to establish the time courses and magnitudes during activation of glycogenolysis and glucose transporter mobilization at the blood brain barrier and in brain cells (also see Sect. 3.9.2).

3.2.2 Evidence that Intense Activation Cannot Be Sustained for Extended Periods Under Mild to Moderate Hypoglycemic Conditions in the Absence of Glycogen

It is well established in animal and human studies that sustained moderate hypoglycemia leads to an impairment of stimulation-induced increases in cortical activity, as measured through evoked potentials in humans and local field potentials and multi-unit activity in animal studies (Herzog et al. 2013; Amiel et al. 1991). Unfortunately, a direct measurement of the impact of blocking the glycogen shunt on the maximum supportable activation has not yet been performed. However, Swanson and coworkers (Suh et al. 2007) have shown that increasing brain glycogen concentration by 88% using a displaceable phosphorylase inhibitor (CP-316,819) increased the ability of the rats in the awake state to maintain brain electrical activity by 91 minutes under hypoglycemic conditions, consistent with the glycogen shunt sparing glucose consumption. Similar results have been reported by Gruetter, Öz, and coworkers when glycogen was super-compensated by pre-depletion and infusion of glucose and insulin (Duarte et al. 2017; Öz et al. 2012, 2017; Choi et al. 2003).

In addition to physiological stimulation, Blennow and coworkers looked at the impact of physiological hypoglycemia induced by 24 h of fasting, on the ability of rats to sustain seizure activity, as assessed by EEG (Blennow et al. 1977a, 1979). After the induction of seizure, both the increase in $CMR_{glc\text{-}ox}$ (measured by A-V difference for O_2) and electrical activity was the same in the fasted and fed groups. Consistent with glucose sparing by the glycogen shunt, after depletion of glycogen, the EEG of the fasted rats with plasma glucose levels below 3 mM ceased showing the burst and suppress pattern characteristic of bicuculline induced seizure (which was replaced by single spikes) and CMR_{O2} returned to pre-seizure values. Burst and suppress EEG activity could be restored with an injection of glucose but ceased after several minutes. Additional evidence for the importance of the glycogen shunt in sustaining seizure activity is that the level of lactate in fasted versus fed animals at 5 minutes differed by only approximately 1.5 μmol/g (~8.5 versus 7 μmol/g; see Fig. 4 in (Blennow et al. 1979)) despite the average brain glucose level during the first 5 minutes of seizure being approximately three-fold higher in fed rats (3 versus 1 μmol/g). If lactate were a major fuel that is rapidly and preferentially consumed compared with glucose during intense activity, its concentration should have been depleted, particularly in the fasted animals. Burst and suppress electrical activity in the fasted rats could be transiently restored by injection of glucose, underscoring the importance of glucose compared with lactate.

3.3 *Factors Determining the Relative Rates of Oxidative and Nonoxidative Neuronal Glucose Consumption*

Half of the measured increment of nonoxidative glucose consumption could be explained based on studies that have shown that the vacuolar H^+-ATPase that pumps protons into synaptic vesicles requires an estimated 1 ATP per 2 H^+ in order to maintain the proton gradient in vesicles; glycolytic ATP is preferentially used to fuel the proton pump, but the stoichiometry between glutamate pumping by VGLUT to load the vesicles and H^+, and Cl^- is not fully understood (Ueda 2016). The stoichiometric relation to nonoxidative glycolysis of this mechanism has been discussed in detail by Hertz and Chen (2017).

Hertz and Chen (2017) have also proposed that an additional need for nonoxidative glycolysis may be due to the need for redox equivalents to move glutamate out of the mitochondria using the pseudo-MAS (see Sects. 3.6 and 3.7, below). If these processes were separate, meaning that the ATP generated from glycolysis used to produce reducing equivalents was not involved in vesicle glutamate transport, they would explain the total measured nonoxidative glycolysis. However, given the close spatial proximity in the presynaptic terminals, complete separation is unlikely. In addition, the MAS mechanism can only operate transiently without an equal increase in the TCA cycle (and oxidative glucose and oxygen consumption) to remove the remove accumulating redox equivalents via the electron transport chain. As illustrated in Fig. 2a, b, if all of the NADH produced by glycolysis were solely used by the pseudo-MAS and the mitochondrial NADH were oxidized in the electron transport chain, pyruvate would accumulate in the cytoplasm. The pyruvate could not be converted to lactate because NADH is not available for the lactate dehydrogenase (LDH) reaction, and it would have to be oxidized via the TCA cycle or exported from the cell.

3.4 *Mechanism(s) of Coupling Between Neuronal Glutamatergic Activity and Glycogenolysis*

Relationships between glycogenolysis and neuronal activity are technically-difficult to study and are poorly understood. A number of reports using glycogen phosphorylase inhibitors or glycogen synthase knockout have implicated glycogen in neurotransmission and long-term potentiation in vitro and in vivo (Sickmann et al. 2009; Suzuki et al. 2011; Duran et al. 2013, 2019; Drulis-Fajdasz et al. 2015; Mozrzymas et al. 2011). However, the mechanisms by which glycogenolysis blockade affects neuronal functions remain to be established. A number of astrocytic processes are preferentially fueled by glycogen, and their disruption could secondarily affect neuronal activities, as discussed in detail elsewhere (Dienel 2019a, b; Dienel and Rothman 2019). Some possibilities include glucose supply deficits when glucose sparing is impaired, slower K^+ uptake from the synaptic cleft during high synaptic activity, abnormal astrocytic Ca^{++} homeostasis and gliotransmitter release to alter neuronal firing, reduced filopodial extensions from astrocytes to neurons to impair efficiency of K^+ and glutamate uptake from the synaptic cleft that may result in glutamate spillover and its interaction with presynaptic receptors to reduce firing. These and other possibilities remain to be evaluated experimentally.

Magistretti and colleagues reported that exposure of brain slices to glutamate does not stimulate glycogenolysis (Magistretti et al. 1981), whereas glycogenolysis is stimulated by K^+ (Quach et al. 1978) and it is linearly related to extracellular K^+ concentration over the range of 5–12 mM (Hof et al. 1988). Details of the signaling pathways involved in astrocytic K^+ uptake and glycogenolysis were worked out by Hertz and colleagues (Xu et al. 2013) when they followed up the hypothesis by Mangia, DiNuzzo and colleagues that K^+ uptake is fueled by glycogen (DiNuzzo et al. 2013, 2012, 2014; Mangia et al. 2013). Hertz and colleagues subsequently reviewed the details of linkage of Na^+,K^+-ATPase activity, astrocytic K^+ uptake, and glycogen mobilization (Hertz et al. 2015a, b, c). Clearly, ion fluxes associated with neurotransmission couple astrocytic glycogen mobilization with neuronal signaling.

Sickmann et al. (2009) evaluated effects of inhibition of glycogenolysis on glutamatergic neurotransmission in cerebellar astrocyte-neuron co-cultures, and concluded that glycogen-derived lactate shuttling to neurons is required to maintain excitatory neurotransmission. In brief, they depolarized the neurons and measured vesicular D-[^{3}H]aspartate (Asp) release from neurons and its reuptake mainly by astrocytes as representative of glutamate transmission. Inhibition of glycogenolysis with CP-316,819, but not with 1,4-dideoxy-1,4-imino-d-arabinitol (DAB), reduced release and re-uptake, and inclusion of 1 mM D-lactate to block L-lactate uptake into neurons further impaired release and reuptake.

Failure of DAB to impair neurotransmission seriously weakens the argument that lactate derived from glycogen is a necessary neuronal fuel. Also, D-lactate is likely to inhibit pyruvate uptake into mitochondria, reducing oxidative production of ATP and creating an energy crisis in both astrocytes and neurons that will stimulate glycolysis and lactate production (Dienel 2019a, b). Lactate released from astrocytes (their lactate transporters are not affected by 1 mM D-lactate) in higher amounts may interact with the neuronal lactate receptor HCAR1 to reduce excitatory and inhibitory firing (Bozzo et al. 2013) and interfere with neurotransmission. HCAR1 is mainly localized at pre- and postsynaptic sites of excitatory neurons (Lauritzen et al. 2014). In contrast, upregulation of glycolysis and lactate production in neurons will result in intracellular lactate accumulation due to the low K_m of MCT2 and its inhibition by D-lactate. The combination of CP-316,819 + D-lactate treatment during excitatory conditions deprives astrocytes of an important fuel and causes neuronal energy failure that will depress neurotransmission and glutamate uptake without lactate shuttling. With the popularity of the notion of astrocyte-neuron lactate shuttling, whether from glucose or glycogen, many authors interpret various findings as supportive of the idea without providing any direct experimental evidence. While there could conceivably be some glycogen-derived lactate shuttling, oxygen consumption is much too low to oxidize all carbohydrate consumed during activation (Dienel and Rothman 2019).

3.5 *Evidence for Glucose Sparing Versus the ANL Shuttle as the Mechanism by Which Astroglia Increase Neuronal Fuel Availability During Activation*

The hypothesis by Swanson (1992) and DiNuzzo and colleagues (2012; 2011; 2010a, b) that glycogenolysis spares glucose for neurons has not, to our knowledge, been tested in vitro or in vivo. This is an extremely difficult undertaking because it requires rigorous, quantitative procedures to measure the quantity of glycogen consumed, the amount of extracellular glucose metabolized by astrocytes, and the additional glucose consumed by neurons. Use of the glycogen shunt model developed herein can be used to test the ability of glycogen to satisfy the energy demands of Na^+-K^+ fluxes and V_{cycle} in astrocytes and to show that the equivalent amount of glucose consumed by neurons is sufficient for ATP expenditures for these processes in neurons, consistent with glucose sparing.

If, as proposed by DiNuzzo et al. (cited above), glycogenolysis increases the concentration of glucose-6-phosphate (Glc-6-P) in astrocytes to further inhibit hexokinase to an extent such that extracellular glucose is not further metabolized, the intracellular glucose concentration should rise. Indeed, when Prebil et al. (2011) treated cultured astrocytes with adrenaline or noradrenaline to trigger glycogenolysis, the cytosolic glucose level rose by 55%. This increase was prevented by DAB, consistent with reduced glucose utilization when glycogen is mobilized. The glucose is not produced by the action of Glc-6-phosphatase because cultured astrocytes do not release glucose to the culture medium when glycogen is degraded (Dringen et al. 1993). Although very low Glc-6-phosphatase activity is detectable in brain and brain cells, it does not convert Glc-6-P back to glucose in detectable amounts in vitro or in vivo (Gotoh et al. 2000; Dienel et al. 1988; Nelson et al. 1985, 1986). The limiting factor for phosphatase activity is extremely slow transport of Glc-6-P into the endoplasmic reticulum (compared with glycolytic rate) where it can be dephosphorylated by phosphatase (Dienel 2019c).

Fray et al. (1996) reported that extracellular lactate and glucose concentrations determined in microdialysates from awake rats are differentially regulated. Stimulating glycogenolysis with isoprenaline, a nonspecific β-adrenergic agonist, caused the glucose level to rise while lactate level was unchanged. Treatment with propranolol, a nonspecific β-adrenergic antagonist, blocked the rise in glucose level evoked by isoprenaline. A tail pinch caused both glucose and lactate levels to rise, and propranolol blocked the increase in glucose but not lactate. Failure of propranolol to abolish the rise in lactate level yet prevent the increase in glucose is consistent with the rise in glucose level secondary to astrocytes consuming less glucose when glycogen is mobilized. These findings also suggest that the lactate is not derived from glycogen. Instead, it comes from (perhaps neuronal) metabolism of blood-borne glucose. In both of these two examples, further work is required to evaluate the sources of glucose and lactate, but carefully-designed metabolic labeling studies can advance the field.

There is a fundamental conflict between the models in which glycogen spares glucose and shuttling of glycogen-derived lactate fuels activated neurons. In conjunction with the lactate shuttling concept, lactate has often been called the preferred brain fuel over glucose. Lactate can indeed be a supplemental brain fuel (and reduce glucose utilization by feedback mechanisms) when its plasma concentration rises during infusion or vigorous exercise. However, oxidation of available lactate does not substantially increase when present with glucose at normal concentrations in activated brain cells during intense activity. For example, cultured glutamatergic neurons do not increase lactate oxidation when stimulated in the presence of glucose plus lactate; instead, more glucose is oxidized (Bak et al. 2006, 2009, 2012). During seizures, glycogen is degraded, glycolysis is upregulated, and lactate accumulates to high levels in brain; both glucose and oxygen consumption rise substantially but OGI falls indicative of enhanced glycolysis (Chapman et al. 1977; Borgström et al. 1976). If lactate produced within the brain were the preferred oxidative fuel compared with glucose, OGI should be stable or increase above 6 if glycogen is oxidized (but not included in the OGI calculation). In starved rats subjected to seizures, the brain glucose level falls and lactate concentration rises to a level similar to that in fed, seizing rats (Blennow et al. 1977a, b). If lactate were an important neuronal fuel, especially when glucose supply is inadequate, lactate should be avidly consumed, substantially reducing or preventing its accumulation in brain. This is not, however, the observed experimental outcome.

3.6 *Evidence that the MAS Is Required for Neuronal Oxidation of Glucose*

Glycolytic metabolism of glucose consumes NAD^+ and produces NADH at the glyceraldehyde-3-phosphate dehydrogenase step. Because NAD^+ and NADH are present in brain in low concentrations (Medina et al. 1980), these compounds serve a catalytic role. The MAS regenerates cytoplasmic NAD^+ by transferring reducing equivalents from NADH in the cytosol to mitochondria so that pyruvate is available as an oxidative fuel. The MAS involves two shuttle systems across the mitochondrial membrane, one exchanges glutamate for aspartate and the other exchanges malate for α-ketoglutarate (as illustrated in Fig. 2a, except that the glutamate is from the cytoplasmic pool, not from glutamine as shown). The regulatory step is electrogenic influx of glutamate into the mitochondrial matrix along with a proton (LaNoue and Tischler 1974). NAD^+ can also be regenerated by the cytoplasmic LDH reaction to produce lactate that, when it diffuses from the cell, eliminates pyruvate as an oxidative fuel.

Many studies have shown that the MAS is essential for glucose oxidation in brain in vivo, in brain slices, and in isolated presynaptic endings (synaptosomes). In all three model systems, inhibition of the MAS substantially reduces glucose oxidation, increases the lactate/pyruvate ratio, stimulates lactate efflux, and reduces phosphocreatine (PCr) and ATP concentrations (Fitzpatrick et al. 1983; Kauppinen et al. 1987; Cheeseman and Clark 1988; McKenna et al. 1993). Similarly, respiration in cultured cortical neurons is reduced by half when the aspartate-glutamate transporter component of the MAS is knocked out (Llorente-Folch et al. 2013). Furthermore, L-lactate can rescue glutamate toxicity in wildtype neurons but not in those in which the MAS was knocked out (Llorente-Folch et al. 2016). Thus, MAS flux is a required pathway that is tightly integrated with brain and neuronal glucose oxidation.

Some authors (e.g., Schurr 2014) believe that lactate is the final product of glycolysis and that lactate is transported into mitochondria where it is oxidized by mitochondrial LDH to pyruvate that enters the TCA cycle. However, if lactate were the primary (or exclusive) neuronal oxidative substrate, it should have been able to support the rise in ATP demand caused by glutamate exposure, and it did not (Llorente-Folch et al. 2016). As noted above, partial inhibition of the MAS also led to glycolytic upregulation, lactate accumulation and release, and ATP deficits. These findings demonstrate that direct mitochondrial uptake and oxidation of lactate that accumulates when the MAS is impaired cannot support brain and neuronal energetics. Another strong argument against lactate being the major brain oxidative fuel is that LDH is long-considered to be a cytoplasmic marker in brain subcellular fractionation studies (Johnson and Whittaker 1963; Tamir et al. 1972). Analysis of the enzymatic composition of purified brain mitochondria demonstrated that <1% of the total LDH activity in the whole homogenate is recovered in mitochondria (Lai et al. 1977; Lai and Clark 1976). It is, therefore, not clear how such a small fraction of the total LDH can metabolize all of the lactate generated in the cytoplasm. The concept of lactate being the predominant oxidative fuel in brain is highly unlikely to be correct, and if it occurs, it must be a minor flux.

Ignoring the semantic argument whether pyruvate or lactate is the final product of glycolysis, the brain lactate/pyruvate ratio is about 10-14 to 1 (Veech et al. 1973; Siesjö 1978; Nordström 2010). This high ratio indicates that lactate rapidly equilibrates with pyruvate and raises the possibility lactate may be a dynamic buffer or reservoir for pyruvate, since α-ketoacids are chemically less stable than carboxylic acids. Lactate does serve as an overflow mechanism when glycolytic rate exceeds that of the MAS. The low K_m (0.7 mM) for the neuronal lactate transporter MCT2 (Bröer et al. 1999) restricts neuronal lactate influx and efflux when its concentration increases, and reduced lactate efflux during high glycolytic activity was proposed to help maintain a high intracellular substrate levels for neuronal oxidation (Chih and Roberts Jr 2003; Chih et al. 2001).

3.7 Involvement of the pseudo-MAS for Conversion of Glutamine to Transmitter Glutamate

Studies by the Hertz-Schousboe group elucidated mechanisms for conversion of glutamine to transmitter glutamate in cultured cerebellar glutamatergic granule neurons (reviewed in Schousboe et al. 2013). The phosphate-activated glutaminase (PAG) is thought to be located on the outer surface of the inner membrane of mitochondria, and the glutamate produced by PAG could either be released to the cytoplasm and pumped into synaptic vesicles, or it could enter the mitochondria, be metabolized and exit via the pseudo-MAS prior to vesicular loading (Fig. 2a). Their work demonstrated that inhibition of aspartate aminotransferase and the ketodicarboxylic acid carrier reduced transmitter glutamate release from cultured glutamatergic neurons (Palaiologos et al. 1988, 1989; Bak et al. 2008; Zielińska et al. 2004). Based on these and related findings, Hertz and Chen (2017) formulated the model illustrated in Fig. 2a to explain half of the stoichiometric relationship between glucose oxidation and glutamate cycling.

MRS studies revealed a linear 1:1 relationship between neuronal glucose oxidation and V_{cycle} (Fig. 1a), and involvement of the pseudo-MAS in transmitter glutamate production can explain half of this relationship, i.e., 0.5 glucose oxidized per glutamate cycled, and all of it when presynaptic glucose oxidation is taken as half of total neuronal oxidation. In this model, based on data obtained in cultured neurons, the NADH that participates in the pseudo-MAS must have been produced by neuronal glycolysis. Under resting conditions or under anesthesia when the oxygen-glucose index is close to 6 (i.e., nearly all of the glucose is oxidized), the 'true' MAS could oxidize the second NADH produced by glycolysis so that the presynaptic glucose oxidation:V_{cycle} stoichiometry is 1:1 (Fig. 2b). As long as the rates of the 'true' MAS and pseudo-MAS are matched, the 1:1 relationship could be maintained in presynaptic nerve endings even if there is disproportionate upregulation of glycolysis compared to oxidative metabolism. Other indentified mechanisms may also contribute to the 1:1 relationship.

3.8 Evidence Against Glycogen-Derived Lactate Shuttling Coupled with Neuronal Lactate Oxidation to Support the pseudo-MAS for Conversion of Glutamine to Transmitter Glutamate

An alternative in vivo mechanism to provide the NADH for the pseudo-MAS (instead of or supplementing neuronal glycolysis) is influx of extracellular lactate and its oxidation to pyruvate + NADH by LDH (Fig. 2a). In other words, glycogen-derived lactate shuttling from astrocytes to neurons may be coupled with the glutamate-glutamine cycle via the pseudo-MAS and serve as the mechanism linking glycogenolysis to glutamatergic neurotransmission. However, there is no compelling experimental evidence for this possibility. Shuttling has never been experimentally demonstrated to take place, and nearly all supportive evidence for lactate shuttle models is circumstantial and most is derived from immature cultured cells. In contrast, there is very strong in vivo evidence against lactate shuttling and use of lactate as major fuel during activation (reviewed in Dienel 2012, 2017, 2019a; Dienel and Cruz 2016).

In brief, the major arguments against lactate shuttling are as follows: Glutamate-evoked production of lactate is not a robust characteristic of cultured astrocytes, and many labs reported either no effect of glutamate treatment on astrocytic glycolysis or an inhibition of glycolysis (Dienel 2012; Dienel and Cruz 2004). Furthermore, the stoichiometry required by the glutamate-stimulated glycolysis shuttle model was not satisfied (Dienel 2017) in the original publication of the model by Pellerin and Magistretti (1994). The model has never been quantitatively validated and lactate shuttling has never been directly demonstrated. More important, lactate shuttling tightly coupled with its oxidation requires a parallel, quantitative rise in oxygen consumption (CMR_{O2}). However, measured increases in CMR_{O2} cannot support the oxidation of the carbohydrate consumed during activation, and when glycogen is included in the OGI calculation, the discordance increases substantially, as discussed in our companion article (Dienel and Rothman 2019). Metabolic studies have demonstrated that labeled products of ^{14}C labeled glucose are not quantitatively retained in activated brain tissue. This is important because label retention would occur if lactate were oxidized and the label incorporated into the large TCA cycle-derived amino acid pools. Detailed analysis of the basis for label loss identified lactate as the predominant, diffusible metabolite that carries label and carbon from the brain, explaining the fall in OGI, at least in large part. Upregulation of the pentose phosphate shunt pathway (PPP) during activation with recycling of the fructose-6-phosphate back into the PPP can also make a large contribution to the fall in OGI due to loss of labeled and unlabeled carbon one from blood-borne [1-^{14} or 1-^{13}C]glucose. Label is lost from carbon one in the first pass through the PPP and with extensive recycling, the resulting unlabeled fructose-6-P that that re-enters the PPP as glucose-6-P due to high isomerase activity will lose unlabeled CO_2 from carbon one, contributing to the fall in OGI to an extent varying with magnitude of recycling. The bottom line is that coupling of glycogen-derived lactate shuttling with neuronal transmitter glutamate synthesis via the pseudo-MAS is not a viable explanation for a critical role of glycogenolysis in the glutamate-glutamine cycle.

The following examples explain why it is highly unlikely for glycogen-derived lactate to be shuttled to neurons and metabolized as a major oxidative fuel using either the 'true' MAS to transfer the reducing equivalents produced by lactate dehydrogenase or to provide NADH for the pseudo-MAS to power the conversion of glutamine to glutamate and generate neurotransmitter glutamate.

- In the studies by Madsen et al. (1998, 1999), the respective metabolic rates during rest and activation for CMR_{O2} were 4.13 and 5.35 μmol/min/g and for CMR_{glc} were 0.75 and 1.08 μmol/min/g, giving incremental increases of 1.22 and

0.33 μmol/min/g, respectively. If all of this glucose were oxidized, it would consume 1.98 μmol O_2/min/g (6∗0.33), more than that of the oxygen actually utilized. Oxygen cannot support complete oxidation of the glucose taken up into brain, let alone the glycogen consumed.

- V_{shunt} in the same brains of the above Madsen studies calculated from net glycogen consumption in 6 min was 0.42 μmol/min/g (12.1-9.6 = 2.5 μmol/g; 2.5/6 = 0.42) (Cruz and Dienel 2002). Thus, the total incremental rate of carbohydrate (glucose plus glycogen) consumed is 0.33 + 0.42 = 0.75 μmol/g/min, the oxidation of which would require 4.5 μmol O_2/g/min. The incremental CMR_{O2} accounts for only 27% of the required amount to oxidize the carbohydrate, ignoring the oxygen consumed by transmitter glutamate oxidation. The fractions of each fuel oxidized in neurons and astrocytes remains to be established, but it is clear that most of the carbohydrate consumed cannot be oxidized.
- In the Madsen study, ΔV_{cycle} was calculated to be 0.2 μmol/g/min (on-going unpublished analysis of these data), which corresponds to 48% of V_{shunt} (0.2/0.42). From Eq. (7) of the glycogen shunt model, $\Delta CMR_{glc\text{-}ox\text{-}N} = 0.5\,V_{shunt} = 0.21$ μmol/g/min, the calculated neuronal rate of oxidation of glucose spared by glycogenolysis. $\Delta CMR_{glc\text{-}ox\text{-}N}$ would consume 1.26 μmol O_2/g/min, which is essentially equal to the measured incremental rise in CMR_{O2} of 1.22 μmol O_2/g/min. If all of the glycogen consumed were converted to lactate and shuttled to neurons for oxidation, it would require 2.52 μmol O_2/g/min (0.42∗6). In other words, after neuronal oxidation of the half of spared glucose there is no oxygen left over to oxidize lactate produced from glycogen in astrocytes.
- V_{cycle} consumes 1 NADH per glutamate inserted into the neurotransmitter pool from glutamine. If shuttled lactate provided all of this NADH, the lactate shuttle plus oxidation rate would be equal to the incremental rise in V_{cycle}, or 0.2 μmol/g/min (in hexose units). This would require 1.2 μmol O_2/g/min (6∗0.2), equal to all of the incremental rise in CMR_{O2}.This would mean that neurons could not have an incremental rise in glycolysis to satisfy the processes fueled by glycolytic ATP, e.g., vesicular glutamate packaging (Ueda 2016) and to oxidize the generated pyruvate to fuel other processes via ATP from the TCA cycle.
- Diaz-Garcia et al. (2017, 2018) clearly demonstrated that extracellular lactate derived from astrocytic glucose or glycogen was *not* consumed by hippocampal neurons during electrical stimulation or in whisker barrel cortex in vivo during whisker stimulation. This finding goes against the possibility that glycogen-derived lactate shuttling to provide NADH for the pseudo-MAS and conversion of glutamine to transmitter glutamate.
- The alternative mechanism proposed in the present study and supported by data from studies using a variety of experimental conditions strongly support the hypothesis that glycogenolysis fuels astrocytic energy demands related to Na^+,K^+-ATPase and Glu-Gln cycling, and glucose spared for neurons also accounts for the ATP required for these processes.

3.9 Evidence for Astrocytic and Neuronal Processes Requiring Glycolytic ATP

3.9.1 Astrocytic Processes Requiring Glycolytic ATP

The glycogen shunt model emphasizes the need for glycolytic ATP to support astrocytic processes that require a 'fast' supply of ATP, especially the Na^+,K^+-ATPase. On the other hand, oxidative ATP from glycogen-derived pyruvate oxidation and glutamate oxidation upon its uptake could fuel the much smaller, slower energy demands of the glutamate-glutamine cycle and larger, prolonged demands of other processes (Fig. 3). Preliminary unpublished calculations based on the glycogen shunt model are in excellent agreement with the conclusion that the 3 ATP produced per glucosyl unit by glycolytic metabolism of glycogen-derived Glc-6-P are sufficient to satisfy the energy demands of Na^+ and K^+ pumping. In addition, the excess ATP produced from glycogenolysis and oxidation can be used for Ca^{++} pumping and filopodial movements to and from active synapses to optimize clearance of K^+ and glutamate from the synaptic cleft during prolonged, intense activity. Thus, the glycogen shunt and glycolytic metabolism of glycogen can provide most of the ATP required for ion pumping during intense neuronal activity while sparing an equivalent amount of glucose for neurons. Further testing of available data sets is in progress.

Additional, indirect evidence supports the conclusion that glycogen is metabolized predominantly by glycolysis with release of lactate from brain. As noted above, the amount of oxygen consumed during activation in the Madsen et al. (1999) study cannot support the complete oxidation of the amount of glucose consumed, let along the glycogen metabolized. While partitioning of breakdown of these carbohydrates between the glycolytic and oxidative pathways requires additional experiments, an important clue is that glycogenolysis does not dilute the specific activity of brain lactate derived from blood-borne [6-^{14}C]glucose (Dienel and Cruz 2009) and the lactate equivalents of glycogen are not retained in the brain (Dienel et al. 2002; Cruz and Dienel 2002). If glycogen-derived lactate were retained in brain, the brain lactate level would have increased substantially, the lactate specific activity determined in homogenates would have been considerably diluted, and, if oxidized, oxygen consumption would have markedly increased. These finding indicate compartmentation of glycolytic metabolism of glucose and glycogen, with rapid release from brain of unlabeled lactate derived from glycogen (and potentially some oxidation of glycogen in the astrocyte which would further decrease astrocyte glucose phosphorylation). This phenomenon might be readily explained by predominant metabolism of glycogen in astrocytes and uptake and oxidation of blood-borne glucose by neurons, such that the labeled lactate recovered in brain during activation has neuronal origin, which is predicted by the glucose sparing component of the glycogen shunt model (Fig. 3).

At present we do not know the contributions of different molecular mechanisms that activate glycogenolysis in the astrocytes. Quach et al. (1978) showed that the glycogenolytic effect of noradrenaline in brain slices was evoked by a much lower concentration than the stimulation by cAMP accumulation (EC_{50} = 0.49 and 8.58 μM, respectively). In cultured astrocytes, the increases in cAMP level preceded the rise in intracellular glucose concentration after adrenergic stimulation (Prebil et al. 2011), and phasic increases in intracellular [Ca^{++}] mediated by α_1-adrenergic receptors were four-fold faster than cAMP/protein kinase A activation (Horvat et al. 2016). These findings are relevant to the sensory stimulation studies in which V_{shunt} can be determined (Dienel et al. 2002, 2007a; Madsen et al. 1999) because startling and alerting, a component of the sensory stimulation protocol, cause activation of the locus coeruleus noradrenergic system upon removal of the shelter in which the animal was sequestered prior to sensory stimulation. Hof et al. (1988) later showed that K^+-evoked glycogenolysis depends on influx of extracellular Ca^{++}. Hertz and colleagues using differentiated, cultured astrocytes revealed that activation of glycogenolysis is complex but starts with small increases in extracellular [K^+]. Through L-channel opening, the extracellular rise in [K^+] leads to elevation of intracellular [Ca^{++}], which then activates phosphorylase kinase. Phosphorylase kinase activates phosphorylase thereby stimulating glycogenolysis. Other pathways such as through cAMP play a minor role in activity coupled glycogenolysis (Hertz et al. 2015b, c; Xu et al. 2013, 2014). Thresholds for triggering glycogenolysis are most easily determined in cultured cells and brain slices, but must also be established in living brain.

The roles of glycogen in K^+ clearance from extracellular space also have important implications in brain disorders. Knockout of glycogen synthase causes mice to have greater susceptibility to hippocampal seizures during stimulation of Schaffer collaterals or treatment with kainate (Lopez-Ramos et al. 2015). Lafora disease is characterized by abnormal glycogen structure and severe epilepsy, ultimately leading to death (Gentry et al. 2018; DiNuzzo et al. 2014, 2015). These findings underscore the importance of improved understanding the functions of glycogen in normal brain function.

Not only does extracellular K^+ stimulate glycogenolysis, it can also have additional metabolic effects. For example, treatment of cultured astrocytes with high (>15 mM) concentrations of K^+ stimulates oxygen consumption in cultured astrocytes (Hertz and Kjeldsen 1985; Hertz et al. 1973), whereas physiological concentrations increase pyruvate carboxylase activity, with a 50% rise in the rate of CO_2 fixation when [K^+] was increased from 2 to 10 mM (Kaufman and Driscoll 1992). The muscle isoform of pyruvate kinase is predominant in brain and in cultured astrocytes (Farrar and Farrar 1995; Tolle et al. 1976; Terlecki 1989; Ngo and Ibsen 1989), and the muscle isoform has a requirement for and is activated by K^+ (Oria-Hernandez et al. 2005; Kachmar and Boyer 1953; Nowak and Suelter 1981). Thus, K^+ can stimulate anaplerosis and glycolysis to an extent that may vary with local [K^+].

3.9.2 Neuronal Processes Requiring Glycolytic ATP

Neurons are highly dependent on oxidative metabolism of glucose to satisfy their energy demands, with oxidation rates being directly proportional to V_{cycle} over a wide range (Fig. 1a). However, neurons also require glycolytic metabolism to fulfill critical processes. There is a long history of reports (often contradictory) indicating that glucose is required for normal neuronal function and that alternative substrates can only partially replace glucose (reviewed in Dienel 2019a). Many of these functions remain to be identified and quantified, but recent studies have demonstrated that upregulation in neuronal glycolytic metabolism of glucose is essential for synaptic function (reviewed by Ashrafi and Ryan 2017; Yellen 2018; Diaz-Garcia and Yellen 2018). Previous studies by Bachelard and colleagues also established that oxidative substrates cannot fully replace glucose to support evoked potentials (Cox and Bachelard 1982, 1988a, b; Cox et al. 1985; Bachelard et al. 1984). Subsequent critical and detailed reviews evaluated discordant results in which glucose replacement with lactate did or did not sustain neuronal functions; one reason was speed of brain slice cutting, with slower procedures enabling lactate to support evoked potentials (Chih and Roberts Jr 2003; Chih et al. 2001; Okada and Lipton 2007).

In addition, a significant finding is that the capacity for presynaptic glucose transport increases with higher activity. For example, treatment of cultured neurons with glutamate or N-methyl-D-aspartate (NMDA) increased plasma membrane localization of the glucose transporter GLUT3 (Ferreira et al. 2011; Weisova et al. 2009). When cultured neurons were stimulated during firing of 600 action potentials, an AMP-activated protein kinase-mediated mechanism triggered a ~three-fold rise in the expression of glucose transporter GLUT4 at the presynaptic membrane within about a minute (Ashrafi et al. 2017). Also, during memory consolidation in vivo, GLUT4 is upregulated, and memory is impaired when GLUT4 is inhibited (Pearson-Leary et al. 2018; Pearson-Leary and McNay 2016). Thus, presynaptic glucose transport capacity increases as a result of ATP-demanding activities, including cognitive processing. AMP-activated protein kinase is also necessary for the energetics of excitatory synaptic activation in cultured neurons, as well as regulation of expression of early intermediate genes that are involved in learning and memory; inhibition of hippocampal AMP kinase in vivo prior to footshock impairs long term avoidance memory when tested 24 h later (Didier et al. 2018; Marinangeli et al. 2018).

Not only does transport capability increase with neuronal activation, but glycolytic upregulation is necessary to generate ATP when energy demand rises. Endocytosis and recycling of synaptic vesicles, including vesicular acidification, were identified as a major energy cost to presynaptic terminals of cultured hippocampal cells (presumably astrocyte-neuron co-cultures), exceeding that of Na^+,K^+-ATPase (Rangaraju et al. 2014). In this study, blockade of glycolysis with deoxyglucose (but without an alternative substrate) or inhibition of oxidative phosphorylation with oligomycin severely impaired synaptic function. Inhibition of myosin light chain kinase reduced mobilization of the reserve pool of synaptic vesicles, but not the immediately releasable pool (Ryan 1999). The cost of reserve vesicle processing is not known.

Diaz-Garcia et al. (2017) provided compelling evidence that neurons upregulate glycolytic metabolism during electrical stimulation of hippocampal slices in vitro and during stimulation of whiskers to activate the barrel cortex in vivo. Additional evidence that glycolysis is essential for neurotransmission includes accumulation of glycolytic enzymes at presynaptic sites during metabolic stress (Jang et al 2016) and localization of glycolytic enzymes to synaptic vesicles (reviewed by Ashrafi and Ryan 2017; Ueda 2016).

The evidence above strongly supports an important role for non-oxidative glycolysis in neuronal metabolism during activation. Emphasis on the ANL shuttle in the literature has limited the study of these mechanisms. We hope that this paper and the studies cited above will help stimulate future work to evaluate essential roles of glucose in neuronal functions.

3.10 Neuronal Activity Requirements for the Glycogen Shunt to Be Used

Evaluation of the glycogen shunt model requires confining the analysis to the early phase of intense brain activation during which there is net consumption of glycogen. One of the limitations is that cycling of glucose-6-phosphate through glycogen cannot be measured with current technology, and the roles of continuous glycogenolysis and glucose sparing without net consumption under resting conditions cannot be determined. The likelihood of a threshold for net glycogenolysis is suggested by the study of Cruz and Dienel (2002) in which rats were given 30 seconds of sensory stimulation. Unlabeled glycogen levels were unchanged after the brief stimulation, whereas labeled glycogen was 16% lower (not quite reaching statistical significance, $p = 0.059$), suggesting net release of newly-synthesized outer tiers precedes consumption of unlabeled inner tiers.

Given the high amount of brain signaling activity in the resting awake state (which is several fold higher than the increments in signaling in even intense sensory activation) it is unclear why there appears to be a rapid transition to use of the glycogen shunt above this level of activity. Studies have shown that the total increment in signaling activity across the ensemble is proportional to the increment in glucose oxidation, consistent with the findings relative to V_{cycle} (Maandag et al. 2007) . However, the distribution of signaling frequency changes, with much of the increment being due to large percentage increases in the signaling frequency of a subset of neurons which have been called rapid signaling neurons (RSN, Maandag et al. 2007). The signaling frequencies of the RSN can be over five-fold higher than the overall ensemble average, and are likely associated with large release of K^+ which triggers glycogenolysis.

Even during resting awake and anesthetized conditions, there are still neurons that transiently signal at very high rates relative to the average resting awake value, although a smaller percentage. Evidence of continued activity of glycogen turnover related to the shunt, presumably in microdomains, is provided by measurements using ^{14}C and ^{13}C-labeled glucose of glycogen turnover (Watanabe and Passonneau 1973; Choi and Gruetter 2003; Choi et al. 1999). Recent analysis by DiNuzzo and coworkers (Öz et al. 2015, 2017; DiNuzzo 2013) suggest that the rate of turnover may be several fold higher than previously reported due to not taking into account the dynamics of glycogen particle synthesis and breakdown properly. Although outside of the scope of the present paper it would be of value to calculate from the turnover measurements what fraction of astroglia are involved in the glycogen shunt and compare the number to electrical recordings of the fraction of rapidly signaling neurons under those conditions.

4 Concluding Comments and Future Studies

A direct outcome of the glycogen shunt model development, its critical evaluation with respect to the literature, and its initial testing is that the glycogen shunt in combination with the pseudo-MAS governs the stoichiometry between glucose metabolism and GABA/glutamate/glutamine neurotransmitter cycling. Glutamate uptake into astrocytes linked to astrocyte-neuron lactate shuttling is *not the coupling mechanism*, and glycogen- and glucose-derived lactate are *not major oxidative fuels for neurons* although lactate can supplement glucose in both astrocytes and neurons.

Future fully-quantitative studies are needed to evaluate cellular and subcellular energetics in astrocytes and neurons and to identify functions served by ATP derived from glycolysis and oxidative metabolism. Among the key measurements that need to be performed to test and extend the model are the following:

- **Comparison between theoretical neuronal and astroglial glucose and glycogen metabolic fluxes and ΔV_{cycle} with experimental measurements during brain activation**. The flux relationships are tested against in vivo experimental data in which glycogenolysis, total and oxidative glucose uptake, and V_{cycle} are measured. In addition, the theoretical neuronal and astrocytic ATP production ($V_{ATP\text{-}N}$, $V_{ATP\text{-}A}$) relative to ΔV_{cycle} must be calculated and compared with experimental measurements.
- **Comparison of theoretical flux relationships between neuronal and astroglial energetics and ΔV_{cycle} and the values measured under nonactivated conditions**. In order to test whether the energetic relationships determined by the glycogen shunt are maintained at lower levels of brain activity the theoretical relationships between neuronal and astroglial ATP production and ΔV_{cycle} must be compared with the relationships determined by meta-analysis of experimental measurements done under different levels of brain activity from awake to isoelectricity, as shown in Fig. 1.
- **Comparison of the theoretical increment in glucose consumption when the shunt is blocked with experimental measurements**. In order to examine whether the glycogen shunt is able to spare glucose for neuronal consumption

the relationship between glucose consumption and ΔV_{cycle} when the shunt is blocked must be calculated and shown that it leads to an increase in incremental glucose requirements. This prediction is tested by comparing experimental measurements of the ratio between $CMR_{glc\text{-}tot}$ and $CMR_{glc\text{-}ox}$ with the shunt active with measurements performed when the shunt is either blocked by a glycogen phosphorylase inhibitor or through experimental conditions.

Along with methods development, studies are required to establish the thresholds and mechanisms for triggering glycogenolysis, the magnitudes, time courses, and durations of responses, the relationship between glucose sparing and upregulation of glucose transport capacities across the blood-brain barrier and plasma membranes of astrocytes and neurons, and the functions of glycolytic ATP in activated neurons.

Testing of the glycogen shunt model presented in this article has been initiated using the limited available literature in which sufficient quantitative measurements have been made. Initial results show excellent agreement between the quantitative predictions of the shunt model and the stoichiometric relationships of glycogenolysis rate with glutamate-glutamine cycle rate, fueling of astrocytic ion pumping by glycogen-derived ATP, and fueling of neuronal ion pumping and other neuronal energy demands by glycolytic and oxidative metabolism of the spared glucose. These results should stimulate development of new methods to quantify glycogenolysis in real time in living brain under different conditions in conjunction with the major pathway fluxes in astrocytes and neurons and rates of glutamatergic and GABA-ergic neurotransmission.

Acknowledgments *Sources of funding*: DLR: R01MH109159, R01 NS087568, R01 NS100106.

References

Amiel SA, Pottinger RC, Archibald HR, Chusney G, Cunnah DT, Prior FF, Gale EA (1991) Effect of antecedent glucose control on cerebral function during hypoglycemia. Diabetes Care 14(2):109–118. https://doi.org/10.2337/diacare.14.2.109

Ashrafi G, Ryan TA (2017) Glucose metabolism in nerve terminals. Curr Opin Neurobiol 45:156–161. https://doi.org/10.1016/j.conb.2017.03.007

Ashrafi G, Wu Z, Farrell RJ, Ryan TA (2017) GLUT4 Mobilization Supports Energetic Demands of Active Synapses. Neuron 93 (3):606-615.e603. doi:https://doi.org/10.1016/j.neuron.2016.12.020

Attwell D, Laughlin SB (2001) An energy budget for signaling in the grey matter of the brain. J Cereb Blood Flow Metab 21(10):1133–1145. https://doi.org/10.1097/00004647-200110000-00001

Bachelard HS, Cox DW, Drower J (1984) Sensitivity of guinea-pig hippocampal granule cell field potentials to hexoses in vitro: an effect on cell excitability? J Physiol 352(1):91–102. https://doi.org/10.1113/jphysiol.1984.sp015279

Bak LK, Obel LF, Walls AB, Schousboe A, Faek SA, Jajo FS, Waagepetersen HS (2012) Novel model of neuronal bioenergetics. Post-synaptic utilization of glucose but not lactate correlates positively with Ca2+ signaling in cultured mouse glutamatergic neurons. ASN Neuro 4(3):151–160. art:e00083. https://doi.org/10.1042/an20120004

Bak LK, Schousboe A, Sonnewald U, Waagepetersen HS (2006) Glucose is necessary to maintain neurotransmitter homeostasis during synaptic activity in cultured glutamatergic neurons. J Cereb Blood Flow Metab 26(10):1285–1297. https://doi.org/10.1038/sj.jcbfm.9600281

Bak LK, Walls AB, Schousboe A, Ring A, Sonnewald U, Waagepetersen HS (2009) Neuronal glucose but not lactate utilization is positively correlated with NMDA-induced neurotransmission and fluctuations in cytosolic Ca2+ levels. J Neurochem 109(Suppl 1):87–93. https://doi.org/10.1111/j.1471-4159.2009.05943.x

Bak LK, Ziemińska E, Waagepetersen HS, Schousboe A, Albrecht J (2008) Metabolism of [U-^{13}C]glutamine and [U-^{13}C]glutamate in isolated rat brain mitochondria suggests functional phosphate-activated glutaminase activity in matrix. Neurochem Res 33(2):273–278. https://doi.org/10.1007/s11064-007-9471-1

Blennow G, Folbergrova J, Nilsson B, Siesjo BK (1977a) Cerebral blood flow and oxygen utilization in rats during bicuculline-induced seizures. Effects of substrate depletion. Acta Neurol Scand Suppl 64:224–225

Blennow G, Folbergrova J, Nilsson B, Siesjo BK (1979) Effects of bicuculline-induced seizures on cerebral metabolism and circulation of rats rendered hypoglycemic by starvation. Ann Neurol 5(2):139–151. https://doi.org/10.1002/ana.410050207

Blennow G, Nilsson B, Siesjo BK (1977b) Sustained epileptic seizures complicated by hypoxia, arterial hypotension or hyperthermia: effects on celebral energy state. Acta Physiol Scand 100(1):126–128

Borgström L, Chapman AG, Siesjö BK (1976) Glucose consumption in the cerebral cortex of rat during bicuculline-induced status epilipticus. J Neurochem 27(4):971–973

Bozzo L, Puyal J, Chatton JY (2013) Lactate modulates the activity of primary cortical neurons through a receptor-mediated pathway. PLoS One 8(8):e71721. https://doi.org/10.1371/journal.pone.0071721

Bröer S, Bröer A, Schneider HP, Stegen C, Halestrap AP, Deitmer JW (1999) Characterization of the high-affinity monocarboxylate transporter MCT2 in Xenopus laevis oocytes. Biochem J 341:529–535

Cahill GF Jr (2006) Fuel metabolism in starvation. Annu Rev Nutr 26:1–22. https://doi.org/10.1146/annurev.nutr.26.061505.111258

Chapman AG, Meldrum BS, Siesjö BK (1977) Cerebral metabolic changes during prolonged epileptic seizures in rats. J Neurochem 28(5):1025–1035

Cheeseman AJ, Clark JB (1988) Influence of the malate-aspartate shuttle on oxidative metabolism in synaptosomes. J Neurochem 50(5):1559–1565

Chih CP, Lipton P, Roberts EL, Jr. (2001) Do active cerebral neurons really use lactate rather than glucose? Trends Neurosci 24 (10):573-578

Chih CP, Roberts EL Jr (2003) Energy substrates for neurons during neural activity: a critical review of the astrocyte-neuron lactate shuttle hypothesis. J Cereb Blood Flow Metab 23(11):1263–1281. https://doi.org/10.1097/01.wcb.0000081369.51727.6f

Choi IY, Gruetter R (2003) In vivo 13C NMR assessment of brain glycogen concentration and turnover in the awake rat. Neurochem Int 43(4-5):317–322

Choi IY, Seaquist ER, Gruetter R (2003) Effect of hypoglycemia on brain glycogen metabolism in vivo. J Neurosci Res 72(1):25–32. https://doi.org/10.1002/jnr.10574

Choi IY, Tkac I, Ugurbil K, Gruetter R (1999) Noninvasive measurements of [1-(13)C]glycogen concentrations and metabolism in rat brain in vivo. J Neurochem 73(3):1300–1308

Chowdhury GMI, Jiang L, Rothman DL, Behar KL (2014) The contribution of ketone bodies to basal and activity-dependent neuronal oxidation in vivo. J Cereb Blood Flow Metab 34(7):1233–1242. https://doi.org/10.1038/jcbfm.2014.77

Clarke DD, Sokoloff L (1999) Circulation and energy metabolism of the brain. In: Siegel GJ, Agranoff BW, Albers RW, Fisher SK, Uhler MD (eds) Basic Neurochemistry. Molecular, Cellular and Medical Aspects, 6th edn. Lippncott-Raven, Philadelphia, pp 637–669

Collins RC, McCandless DW, Wagman IL (1987) Cerebral glucose utilization: comparison of [^{14}C]deoxyglucose and [6-^{14}C]glucose quantitative autoradiography. J Neurochem 49(5):1564–1570

Cox DW, Bachelard HS (1982) Attenuation of evoked field potentials from dentate granule cells by low glucose, pyruvate + malate, and sodium fluoride. Brain Res 239(2):527–534

Cox DW, Bachelard HS (1988a) On the relationship between the excitability of dentate granule cell field potentials and their sensitivity to low glucose. Brain Res 440(1):195–198

Cox DW, Bachelard HS (1988b) Partial attenuation of dentate granule cell evoked activity by the alternative substrates, lactate and pyruvate: evidence for a postsynaptic action. Exp Brain Res 69(2):368–372

Cox DW, Drower J, Bachelard HS (1985) Effects of metabolic inhibitors on evoked activity and the energy state of hippocampal slices superfused in vitro. Exp Brain Res 57(3):464–470

Cremer JE, Cunningham VJ, Seville MP (1983) Relationships between extraction and metabolism of glucose, blood flow, and tissue blood volume in regions of rat brain. J Cereb Blood Flow Metab 3(3):291–302. https://doi.org/10.1038/jcbfm.1983.44

Cremer JE, Seville MP, Cunningham VJ (1988) Tracer 2-deoxyglucose kinetics in brain regions of rats given kainic acid. J Cereb Blood Flow Metab 8(2):244–253. https://doi.org/10.1038/jcbfm.1988.55

Cruz NF, Ball KK, Dienel GA (2007) Functional imaging of focal brain activation in conscious rats: impact of [(14)C]glucose metabolite spreading and release. J Neurosci Res 85(15):3254–3266. https://doi.org/10.1002/jnr.21193

Cruz NF, Dienel GA (2002) High glycogen levels in brains of rats with minimal environmental stimuli: implications for metabolic contributions of working astrocytes. J Cereb Blood Flow Metab 22(12):1476–1489. https://doi.org/10.1097/00004647-200212000-00008

Cruz NF, Lasater A, Zielke HR, Dienel GA (2005) Activation of astrocytes in brain of conscious rats during acoustic stimulation: acetate utilization in working brain. J Neurochem 92(4):934–947. https://doi.org/10.1111/j.1471-4159.2004.02935.x

Cunningham VJ, Cremer JE (1981) A method for the simultaneous estimation of regional rates of glucose influx and phosphorylation in rat brain using radiolabeled 2-deoxyglucose. Brain Res 221(2):319–330

Cunningham VJ, Hargreaves RJ, Pelling D, Moorhouse SR (1986) Regional blood-brain glucose transfer in the rat: a novel double-membrane kinetic analysis. J Cereb Blood Flow Metab 6(3):305–314. https://doi.org/10.1038/jcbfm.1986.53

de Graaf RA, Pan JW, Telang F, Lee JH, Brown P, Novotny EJ, Hetherington HP, Rothman DL (2001) Differentiation of glucose transport in human brain gray and white matter. J Cereb Blood Flow Metab 21(5):483–492. https://doi.org/10.1097/00004647-200105000-00002

Diaz-Garcia CM, Mongeon R, Lahmann C, Koveal D, Zucker H, Yellen G (2017) Neuronal stimulation triggers neuronal glycolysis and not lactate uptake. Cell Metab 26(2):361–374. e364. https://doi.org/10.1016/j.cmet.2017.06.021

Diaz-Garcia CM, Yellen G (2018) Neurons rely on glucose rather than astrocytic lactate during stimulation. J Neurosci Res 97(8):946–960. https://doi.org/10.1002/jnr.24374

Didier S, Sauve F, Domise M, Buee L, Marinangeli C, Vingtdeux V (2018) AMP-activated Protein Kinase Controls Immediate Early Genes Expression Following Synaptic Activation Through the PKA/CREB Pathway. Int J Mol Sci 19(12). https://doi.org/10.3390/ijms19123716

Dienel GA (2012) Brain lactate metabolism: the discoveries and the controversies. J Cereb Blood Flow Metab 32(7):1107–1138. https://doi.org/10.1038/jcbfm.2011.175

Dienel GA (2017) Lack of appropriate stoichiometry: Strong evidence against an energetically-important astrocyte-neuron lactate shuttle in brain. J Neurosci Res 95(11):2103–2125. https://doi.org/10.1002/jnr.24015

Dienel GA (2019a) Brain glucose metabolism: Integration of energetics with function. Physiol Rev 99(1):949–1045. https://doi.org/10.1152/physrev.00062.2017

Dienel GA (2019b) Does shuttling of glycogen-derived lactate from astrocytes to neurons take place during neurotransmission and memory consolidation? J Neuro Res 97(8):863–882. https://doi.org/10.1002/jnr.24387

Dienel GA (2019c) The "protected" glucose transport through the astrocytic endoplasmic reticulum is too slow to serve as a quantitatively-important highway for nutrient delivery. J Neuro Res 97(8):854–862. https://doi.org/10.1002/jnr.24432

Dienel GA, Ball KK, Cruz NF (2007a) A glycogen phosphorylase inhibitor selectively enhances local rates of glucose utilization in brain during sensory stimulation of conscious rats: implications for glycogen turnover. J Neurochem 102(2):466–478. https://doi.org/10.1111/j.1471-4159.2007.04595.x

Dienel GA, Cruz NF (2004) Nutrition during brain activation: does cell-to-cell lactate shuttling contribute significantly to sweet and sour food for thought? Neurochem Int 45(2-3):321–351. https://doi.org/10.1016/j.neuint.2003.10.011

Dienel GA, Cruz NF (2009) Exchange-mediated dilution of brain lactate specific activity: implications for the origin of glutamate dilution and the contributions of glutamine dilution and other pathways. J Neurochem 109(Suppl 1):30–37. https://doi.org/10.1111/j.1471-4159.2009.05859.x

Dienel GA, Cruz NF (2016) Aerobic glycolysis during brain activation: adrenergic regulation and influence of norepinephrine on astrocytic metabolism. J Neurochem 138(1):14–52. https://doi.org/10.1111/jnc.13630

Dienel GA, Nelson T, Cruz NF, Jay T, Crane AM, Sokoloff L (1988) Over-estimation of glucose-6-phosphatase activity in brain in vivo. Apparent difference in rates of [2-^{3}H]glucose and [U-^{14}C] glucose utilization is due to contamination of precursor pool with ^{14}C-labeled products and incomplete recovery of ^{14}C-labeled metabolites. J Biol Chem 263(36):19697–19708

Dienel GA, Rothman DL (2019) Glycogenolysis in cerebral cortex during sensory stimulation, acute hypoglycemia, and exercise: Impact on astrocytic energetics, aerobic glycolysis, and astrocyte-neuron interactions. Adv Neurobiol (23)

Dienel GA, Schmidt KC, Cruz NF (2007b) Astrocyte activation in vivo during graded photic stimulation. J Neurochem 103(4):1506–1522. https://doi.org/10.1111/j.1471-4159.2007.04859.x

Dienel GA, Wang RY, Cruz NF (2002) Generalized sensory stimulation of conscious rats increases labeling of oxidative pathways of glucose metabolism when the brain glucose-oxygen uptake ratio rises. J Cereb Blood Flow Metab 22(12):1490–1502. https://doi.org/10.1097/00004647-200212000-00009

DiNuzzo M (2013) Kinetic analysis of glycogen turnover: relevance to human brain ^{13}C-NMR spectroscopy. J Cereb Blood Flow Metab 33(10):1540–1548. https://doi.org/10.1038/jcbfm.2013.98

DiNuzzo M, Giove F, Maraviglia B, Mangia S (2017) Computational flux balance analysis predicts that stimulation of energy metabolism in astrocytes and their metabolic interactions with neurons depend on uptake of K^+ rather than glutamate. Neurochem Res 42(1):202–216. https://doi.org/10.1007/s11064-016-2048-0

DiNuzzo M, Mangia S, Maraviglia B, Giove F (2010a) Changes in glucose uptake rather than lactate shuttle take center stage in subserving neuroenergetics: evidence from mathematical modeling. J Cereb Blood Flow Metab 30(3):586–602. https://doi.org/10.1038/jcbfm.2009.232

DiNuzzo M, Mangia S, Maraviglia B, Giove F (2010b) Glycogenolysis in astrocytes supports blood-borne glucose channeling not glycogen-derived lactate shuttling to neurons: evidence from mathematical modeling. J Cereb Blood Flow Metab 30(12):1895–1904. https://doi.org/10.1038/jcbfm.2010.151

DiNuzzo M, Mangia S, Maraviglia B, Giove F (2012) The Role of Astrocytic Glycogen in Supporting the Energetics of Neuronal Activity. Neurochem Res 37:2432–2438. https://doi.org/10.1007/s11064-012-0802-5

DiNuzzo M, Mangia S, Maraviglia B, Giove F (2013) Regulatory mechanisms for glycogenolysis and K+ uptake in brain astrocytes. Neurochem Int 63(5):458–464. https://doi.org/10.1016/j.neuint.2013.08.004

DiNuzzo M, Mangia S, Maraviglia B, Giove F (2014) Physiological bases of the K^+ and the glutamate/GABA hypotheses of epilepsy. Epilepsy Res 108(6):995–1012. https://doi.org/10.1016/j.eplepsyres.2014.04.001

DiNuzzo M, Mangia S, Maraviglia B, Giove F (2015) Does abnormal glycogen structure contribute to increased susceptibility to seizures in epilepsy? Metab Brain Dis 30(1):307–316. https://doi.org/10.1007/s11011-014-9524-5

DiNuzzo M, Maraviglia B, Giove F (2011) Why does the brain (not) have glycogen? BioEssays 33(5):319–326. https://doi.org/10.1002/bies.201000151

Dringen R, Gebhardt R, Hamprecht B (1993) Glycogen in astrocytes: possible function as lactate supply for neighboring cells. Brain Res 623(2):208–214

Drulis-Fajdasz D, Wojtowicz T, Wawrzyniak M, Wlodarczyk J, Mozrzymas JW, Rakus D (2015) Involvement of cellular metabolism in age-related LTP modifications in rat hippocampal slices. Oncotarget 6(16):14065–14081. https://doi.org/10.18632/oncotarget.4188

Duarte JM, Lanz B, Gruetter R (2011) Compartmentalized Cerebral Metabolism of [1,6-(13)C] Glucose Determined by in vivo (13)C NMR Spectroscopy at 14.1 T. Front Neuroenerg 3(3). https://doi.org/10.3389/fnene.2011.00003

Duarte JM, Morgenthaler FD, Lei H, Poitry-Yamate C, Gruetter R (2009) Steady-state brain glucose transport kinetics re-evaluated with a four-state conformational model. Front Neuroenerg 1:6. https://doi.org/10.3389/neuro.14.006.2009

Duarte JMN, Gruetter R (2012) Characterization of cerebral glucose dynamics in vivo with a four-state conformational model of transport at the blood–brain barrier. J Neurochem 121(3):396–406. https://doi.org/10.1111/j.1471-4159.2012.07688.x

Duarte JMN, Morgenthaler FD, Gruetter R (2017) Glycogen Supercompensation in the Rat Brain After Acute Hypoglycemia is Independent of Glucose Levels During Recovery. Neurochem Res 42(6):1629–1635. https://doi.org/10.1007/s11064-017-2178-z

Duran J, Saez I, Gruart A, Guinovart JJ, Delgado-Garcia JM (2013) Impairment in long-term memory formation and learning-dependent synaptic plasticity in mice lacking glycogen synthase in the brain. J Cereb Blood Flow Metab 33(4):550–556. https://doi.org/10.1038/jcbfm.2012.200

Duran J, Gruart A, Varea O, López-Soldado I, Delgado-García JM, Guinovart JJ (2019) Lack of neuronal glycogen impairs memory formation and learning-dependent synaptic plasticity in mice. Front Cell Neurosci 13:374. https://doi.org/10.3389/fncel.2019.00374

Farrar G, Farrar WW (1995) Purification and properties of the pyruvate kinase isozyme M1 from the pig brain. Int J Biochem Cell Biol 27(11):1145–1151. https://doi.org/10.1016/1357-2725(95)00090-C

Farrell CL, Pardridge WM (1991) Blood-brain barrier glucose transporter is asymmetrically distributed on brain capillary endothelial lumenal and ablumenal membranes: an electron microscopic immunogold study. Proc Natl Acad Sci U S A 88(13):5779–5783

Ferreira JM, Burnett AL, Rameau GA (2011) Activity-dependent regulation of surface glucose transporter-3. J Neurosci 31(6):1991–1999. https://doi.org/10.1523/jneurosci.1850-09.2011

Fitzpatrick SM, Cooper AJL, Duffy TE (1983) Use of beta-methylene-D,L-aspartate to assess the role of aspartate aminotransferase in cerebral oxidative metabolism. J Neurochem 41(5):1370–1383

Fray AE, Forsyth RJ, Boutelle MG, Fillenz M (1996) The mechanisms controlling physiologically stimulated changes in rat brain glucose and lactate: a microdialysis study. J Physiol 496(1):49–57. https://doi.org/10.1113/jphysiol.1996.sp021664

Gandhi GK, Ball KK, Cruz NF, Dienel GA (2010) Hyperglycaemia and diabetes impair gap junctional communication among astrocytes. ASN Neuro 2(2):e00030. https://doi.org/10.1042/AN20090048

Gentry MS, Guinovart JJ, Minassian BA, Roach PJ, Serratosa J (2018) Lafora disease offers a unique window into neuronal glycogen. J Biol Chem 293(19):7117–7125. https://doi.org/10.1074/jbc.R117.803064

Giacco F, Brownlee M (2010) Oxidative stress and diabetic complications. Circ Res 107(9):1058–1070. https://doi.org/10.1161/circresaha.110.223545

Gotoh J, Itoh Y, Kuang TY, Cook M, Law MJ, Sokoloff L (2000) Negligible glucose-6-phosphatase activity in cultured astroglia. J Neurochem 74(4):1400–1408

Gruetter R, Novotny EJ, Boulware SD, Rothman DL, Shulman RG (1996) ^{1}H NMR studies of glucose transport in the human brain. J Cereb Blood Flow Metab 16(3) 427–438. https://doi.org/10.1097/00004647-199605000-00009

Gruetter R, Ugurbil K, Seaquist ER (1998) Steady-state cerebral glucose concentrations and transport in the human brain. J Neurochem 70(1):397–408

Hargreaves RJ, Planas AM, Cremer JE, Cunningham VJ (1986) Studies on the relationship between cerebral glucose transport and phosphorylation using 2-deoxyglucose. J Cereb Blood Flow Metab 6(6):708–716. https://doi.org/10.1038/jcbfm.1986.127

Harris JJ, Jolivet R, Attwell D (2012) Synaptic energy use and supply. Neuron 75(5):762–777. https://doi.org/10.1016/j.neuron.2012.08.019

Hertz L, Chen Y (2017) Integration between Glycolysis and Glutamate-Glutamine Cycle Flux May Explain Preferential Glycolytic Increase during Brain Activation, Requiring Glutamate. Front Integr Neurosci 11(18). https://doi.org/10.3389/fnint.2017.00018, https://doi.org/10.3389/fnint.2017.00018

Hertz L, Dittmann L, Mandel P (1973) K+ induced stimulation of oxygen uptake in cultured cerebral glial cells. Brain Res 60(2):517–520

Hertz L, Gerkau NJ, Xu J, Durry S, Song D, Rose CR, Peng L (2015a) Roles of astrocytic Na(+),K(+)-ATPase and glycogenolysis for K(+) homeostasis in mammalian brain. J Neurosci Res 93(7):1019–1030. https://doi.org/10.1002/jnr.23499

Hertz L, Kjeldsen CS (1985) Functional role of the potassium-induced stimulation of oxygen uptake in brain slices studied with cesium as a probe. J Neurosci Res 14(1):83–93. https://doi.org/10.1002/jnr.490140108

Hertz L, Peng L (1992) Energy metabolism at the cellular level of the CNS. Can J Physiol Pharmacol 70(Suppl):S145–S157

Hertz L, Song D, Xu J, Peng L, Gibbs ME (2015b) Role of the Astrocytic Na(+), K(+)-ATPase in K(+) Homeostasis in Brain: K(+) Uptake, Signaling Pathways and Substrate Utilization. Neurochem Res 40(12):2505–2516. https://doi.org/10.1007/s11064-014-1505-x

Hertz L, Xu J, Song D, Du T, Li B, Yan E, Peng L (2015c) Astrocytic glycogenolysis: mechanisms and functions. Metab Brain Dis 30(1):317–333. https://doi.org/10.1007/s11011-014-9536-1

Hertz L, Xu J, Song D, Du T, Yan E, Peng L (2013) Brain glycogenolysis, adrenoceptors, pyruvate carboxylase, Na(+),K(+)-ATPase and Marie E. Gibbs' pioneering learning studies. Front Integr Neurosci 7:20. https://doi.org/10.3389/fnint.2013.00020

Herzog RI, Jiang L, Herman P, Zhao C, Sanganahalli BG, Mason GF, Hyder F, Rothman DL, Sherwin RS, Behar KL (2013) Lactate preserves neuronal metabolism and function following antecedent recurrent hypoglycemia. J Clin Invest 123(5):1988–1998. https://doi.org/10.1172/jci65105

Hof PR, Pascale E, Magistretti PJ (1988) K^+ at concentrations reached in the extracellular space during neuronal activity promotes a Ca^{2+}-dependent glycogen hydrolysis in mouse cerebral cortex. J Neurosci 8(6):1922–1928

Horvat A, Zorec R, Vardjan N (2016) Adrenergic stimulation of single rat astrocytes results in distinct temporal changes in intracellular Ca(2+) and cAMP-dependent PKA responses. Cell Calcium 59(4):156–163. https://doi.org/10.1016/j.ceca.2016.01.002

Howarth C, Gleeson P, Attwell D (2012) Updated energy budgets for neural computation in the neocortex and cerebellum. J Cereb Blood Flow Metab 32(7):1222–1232. https://doi.org/10.1038/jcbfm.2012.35

Hwang JJ, Jiang L, Sanchez Rangel E, Fan X, Ding Y, Lam W, Leventhal J, Dai F, Rothman DL, Mason GF, Sherwin RS (2019) Glycemic Variability and Brain Glucose Levels in Type 1 Diabetes. Diabetes 68 (1):163-171. doi:https://doi.org/10.2337/db18-0722

Hyder F, Patel AB, Gjedde A, Rothman DL, Behar KL, Shulman RG (2006) Neuronal-glial glucose oxidation and glutamatergic-GABAergic function. J Cereb Blood Flow Metab 26(7):865–877. https://doi.org/10.1038/sj.jcbfm.9600263

Hyder F, Rothman DL, Bennett MR (2013) Cortical energy demands of signaling and nonsignaling components in brain are conserved across mammalian species and activity levels. Proc Natl Acad Sci U S A 110(9):3549–3554. https://doi.org/10.1073/pnas.1214912110

Jang S, Nelson Jessica C, Bend Eric G, Rodríguez-Laureano L, Tueros Felipe G, Cartagenova L, Underwood K, Jorgensen Erik M, Colón-Ramos Daniel A (2016) Glycolytic Enzymes Localize to Synapses under Energy Stress to Support Synaptic Function. Neuron 90(2):278–291. https://doi.org/10.1016/j.neuron.2016.03.011

Johnson MK, Whittaker VP (1963) Lactate dehydrogenase as a cytoplasmic marker in brain. Biochem J 88:404–409

Kachmar JF, Boyer PD (1953) Kinetic analysis of enzyme reactions: II. The potassium activation and calcium inhibition of pyruvic phosphoferase. J Biol Chem 200(2):669–682
Kaufman EE, Driscoll BF (1992) Carbon dioxide fixation in neuronal and astroglial cells in culture. J Neurochem 58(1):258–262
Kauppinen RA, Sihra TS, Nicholls DG (1987) Aminooxyacetic acid inhibits the malate-aspartate shuttle in isolated nerve terminals and prevents the mitochondria from utilizing glycolytic substrates. Biochim Biophys Acta 930(2):173–178
Kong J, Shepel PN, Holden CP, Mackiewicz M, Pack AI, Geiger JD (2002) Brain Glycogen Decreases with Increased Periods of Wakefulness: Implications for Homeostatic Drive to Sleep. J Neurosci 22(13):5581–5587
Korf J (1996) Intracerebral trafficking of lactate in vivo during stress, exercise, electroconvulsive shock and ischemia as studied with microdialysis. Dev Neurosci 18(5-6):405–414
Korf J, de Boer J (1990) Lactography as an approach to monitor glucose metabolism on-line in brain and muscle. Int J Biochem 22(12):1371–1378
Lai JC, Clark JB (1976) Preparation and properties of mitochondria derived from synaptosomes. Biochem J 154(2):423–432
Lai JCK, Walsh JM, Dennis SC, Clark JB (1977) Synaptic and non-synaptic mitochondria from rat brain: Isolation and characterization. J Neurochem 28(3):625–631. https://doi.org/10.1111/j.1471-4159.1977.tb10434.x
LaNoue KF, Tischler ME (1974) Electrogenic characteristics of the mitochondrial glutamate-aspartate antiporter. J Biol Chem 249(23):7522–7528
Lanz B, Gruetter R, Duarte JM (2013) Metabolic Flux and Compartmentation Analysis in the Brain In vivo. Front Endocrinol (Lausanne) 4:156. https://doi.org/10.3389/fendo.2013.00156
Lauritzen KH, Morland C, Puchades M, Holm-Hansen S, Hagelin EM, Lauritzen F, Attramadal H, Storm-Mathisen J, Gjedde A, Bergersen LH (2014) Lactate receptor sites link neurotransmission, neurovascular coupling, and brain energy metabolism. Cereb Cortex 24(10):2784–2795. https://doi.org/10.1093/cercor/bht136
Lear JL, Ackermann RF (1989) Why the deoxyglucose method has proven so useful in cerebral activation studies: the unappreciated prevalence of stimulation-induced glycolysis. J Cereb Blood Flow Metab 9(6):911–913. https://doi.org/10.1038/jcbfm.1989.128
Llorente-Folch I, Rueda CB, Amigo I, del Arco A, Saheki T, Pardo B, Satrustegui J (2013) Calcium-regulation of mitochondrial respiration maintains ATP homeostasis and requires ARALAR/AGC1-malate aspartate shuttle in intact cortical neurons. J Neurosci 33 (35):13957-13971, 13971a. doi:https://doi.org/10.1523/jneurosci.0929-13.2013
Llorente-Folch I, Rueda CB, Pérez-Liébana I, Satrústegui J, Pardo B (2016) l-Lactate-mediated neuroprotection against glutamate-induced excitotoxicity requires ARALAR/AGC1. J Neurosci 36(16):4443–4456. https://doi.org/10.1523/jneurosci.3691-15.2016
Lopez-Ramos JC, Duran J, Gruart A, Guinovart JJ, Delgado-Garcia JM (2015) Role of brain glycogen in the response to hypoxia and in susceptibility to epilepsy. Front Cell Neurosci 9:431. https://doi.org/10.3389/fncel.2015.00431
Maandag NJ, Coman D, Sanganahalli BG, Herman P, Smith AJ, Blumenfeld H, Shulman RG, Hyder F (2007) Energetics of neuronal signaling and fMRI activity. Proc Natl Acad Sci U S A 104(51):20546–20551. https://doi.org/10.1073/pnas.0709515104
Madsen PL, Linde R, Hasselbalch SG, Paulson OB, Lassen NA. (1998) Activation-induced resetting of cerebral oxygen and glucose uptake in the rat. J Cereb Blood Flow Metab 18(7): 742–748. https://doi.org/10.1097/00004647-199807000-00005
Madsen PL, Cruz NF, Sokoloff L, Dienel GA (1999) Cerebral oxygen/glucose ratio is low during sensory stimulation and rises above normal during recovery: excess glucose consumption during stimulation is not accounted for by lactate efflux from or accumulation in brain tissue. J Cereb Blood Flow Metab 19(4):393–400. https://doi.org/10.1097/00004647-199904000-00005
Magistretti PJ, Morrison JH, Shoemaker WJ, Sapin V, Bloom FE (1981) Vasoactive intestinal polypeptide induces glycogenolysis in mouse cortical slices: a possible regulatory mechanism for the local control of energy metabolism. Proc Natl Acad Sci U S A 78(10):6535–6539
Magistretti PJ, Pellerin L, Rothman DL, Shulman RG (1999) Energy on demand. Science 283(5401):496–497

Mangia S, DiNuzzo M, Giove F, Carruthers A, Simpson IA, Vannucci SJ (2011) Response to 'comment on recent modeling studies of astrocyte-neuron metabolic interactions': much ado about nothing. J Cereb Blood Flow Metab 31(6):1346–1353. https://doi.org/10.1038/jcbfm.2011.29
Mangia S, Giove F, DiNuzzo M (2013) K^+ homeostasis in the brain: a new role for glycogenolysis. Neurochem Res 38(3):470–471. https://doi.org/10.1007/s11064-012-0962-3
Mangia S, Simpson IA, Vannucci SJ, Carruthers A (2009) The in vivo neuron-to-astrocyte lactate shuttle in human brain: evidence from modeling of measured lactate levels during visual stimulation. J Neurochem 109(Suppl 1):55–62. https://doi.org/10.1111/j.1471-4159.2009.06003.x
Marinangeli C, Didier S, Ahmed T, Caillerez R, Domise M, Laloux C, Begard S, Carrier S, Colin M, Marchetti P, Ghesquiere B, Balschun D, Buee L, Kluza J, Vingtdeux V (2018) AMP-Activated Protein Kinase Is Essential for the Maintenance of Energy Levels during Synaptic Activation. iScience 9:1–13. https://doi.org/10.1016/j.isci.2018.10.006
McKenna MC, Tildon JT, Stevenson JH, Boatright R, Huang S (1993) Regulation of energy metabolism in synaptic terminals and cultured rat brain astrocytes: differences revealed using aminooxyacetate. Dev Neurosci 15(3-5):320–329
Medina MA, Deam AP, Stavinoha WB (1980) Inactivation of brain tissue by microwave fixation. In: Passonneau JV, Hawkins RA, Lust WD, Welsh FA (eds) Cerebral metabolism and neural function. Williams & Wilkins, Baltimore, pp 56–69
Mozrzymas J, Szczęsny T, Rakus D (2011) The effect of glycogen phosphorolysis on basal glutaminergic transmission. Biochem Biophys Res Commun 404(2):652–655. https://doi.org/10.1016/j.bbrc.2010.12.033
Nakao Y, Itoh Y, Kuang TY, Cook M, Jehle J, Sokoloff L (2001) Effects of anesthesia on functional activation of cerebral blood flow and metabolism. Proc Natl Acad Sci U S A 98(13):7593–7598. https://doi.org/10.1073/pnas.121179898
Nelson T, Lucignani G, Atlas S, Crane AM, Dienel GA, Sokoloff L (1985) Reexamination of glucose-6-phosphatase activity in the brain in vivo: no evidence for a futile cycle. Science 229(4708):60–62
Nelson T, Lucignani G, Goochee J, Crane AM, Sokoloff L (1986) Invalidity of criticisms of the deoxyglucose method based on alleged glucose-6-phosphatase activity in brain. J Neurochem 46(3):905–919
Newman LA, Korol DL, Gold PE (2011) Lactate produced by glycogenolysis in astrocytes regulates memory processing. PLoS One 6(12):e28427. https://doi.org/10.1371/journal.pone.0028427
Ngo JL, Ibsen KH (1989) Expression of pyruvate kinase in astrocytes induced to differentiate in vitro. J Neurosci Res 23(1):64–70. https://doi.org/10.1002/jnr.490230109
Nordström CH (2010) Cerebral energy metabolism and microdialysis in neurocritical care. Childs Nerv Syst 26(4):465–472. https://doi.org/10.1007/s00381-009-1035-z
Nowak T, Suelter C (1981) Pyruvate kinase: activation by and catalytic role of the monovalent and divalent cations. Mol Cell Biochem 35(2):65–75
Oe Y, Baba O, Ashida H, Nakamura KC, Hirase H (2016) Glycogen distribution in the microwave-fixed mouse brain reveals heterogeneous astrocytic patterns. Glia 64(9):1532–1545. https://doi.org/10.1002/glia.23020
Okada Y, Lipton P (2007) Glucose, Oxidative Energy Metabolism, and Neural Function in Brain Slices—Glycolysis Plays a Key Role in Neural Activity. In: Gibson GE, Dienel GA (eds) Brain Energetics. Integration of Molecular and Cellular Processes. Handbook of Neurochemistry and Molecular Neurobiology, 3rd edn. Springer, Berlin, pp 17–39. https://doi.org/10.1007/978-0-387-30411-3_2
Oria-Hernandez J, Cabrera N, Perez-Montfort R, Ramirez-Silva L (2005) Pyruvate kinase revisited: the activating effect of K+. J Biol Chem 280(45):37924–37929. https://doi.org/10.1074/jbc.M508490200
Ottersen OP, Laake JH, Storm-Mathisen J (1990a) Demonstration of a releasable pool of glutamate in cerebellar mossy and parallel fibre terminals by means of light and electron microscopic immunocytochemistry. Arch Ital Biol 128(2-4):111–125
Ottersen OP, Storm-Mathisen J, Bramham C, Torp R, Laake J, Gundersen V (1990b) A quantitative electron microscopic immunocytochemical study of the distribution and synaptic handling of glutamate in rat hippocampus. Prog Brain Res 83:99–114

Ottersen OP, Zhang N, Walberg F (1992) Metabolic compartmentation of glutamate and glutamine: morphological evidence obtained by quantitative immunocytochemistry in rat cerebellum. Neuroscience 46(3):519–534

Öz G, DiNuzzo M, Kumar A, Moheet A, Khowaja A, Kubisiak K, Eberly LE, Seaquist ER (2017) Cerebral glycogen in humans following acute and recurrent hypoglycemia: Implications on a role in hypoglycemia unawareness. J Cereb Blood Flow Metab 37(8):2883–2893. https://doi.org/10.1177/0271678x16678240

Öz G, DiNuzzo M, Kumar A, Moheet A, Seaquist ER (2015) Revisiting glycogen content in the human brain. Neurochem Res 40(12):2473–2481. https://doi.org/10.1007/s11064-015-1664-4

Öz G, Tesfaye N, Kumar A, Deelchand DK, Eberly LE, Seaquist ER (2012) Brain glycogen content and metabolism in subjects with type 1 diabetes and hypoglycemia unawareness. J Cereb Blood Flow Metab 32(2):256–263. https://doi.org/10.1038/jcbfm.2011.138

Palaiologos G, Hertz L, Schousboe A (1988) Evidence that aspartate aminotransferase activity and ketodicarboxylate carrier function are essential for biosynthesis of transmitter glutamate. J Neurochem 51(1):317–320

Palaiologos G, Hertz L, Schousboe A (1989) Role of aspartate aminotransferase and mitochondrial dicarboxylate transport for release of endogenously and exogenously supplied neurotransmitter in glutamatergic neurons. Neurochem Res 14(4):359–366

Patel AB, de Graaf RA, Mason GF, Kanamatsu T, Rothman DL, Shulman RG, Behar KL (2004) Glutamatergic neurotransmission and neuronal glucose oxidation are coupled during intense neuronal activation. J Cereb Blood Flow Metab 24(9):972–985. https://doi.org/10.1097/01.WCB.0000126234.16188.71

Patel AB, de Graaf RA, Rothman DL, Behar KL, Mason GF (2010) Evaluation of cerebral acetate transport and metabolic rates in the rat brain in vivo using 1H-[13C]-NMR. J Cereb Blood Flow Metab 30(6):1200–1213. https://doi.org/10.1038/jcbfm.2010.2

Patel AB, Lai JC, Chowdhury GM, Hyder F, Rothman DL, Shulman RG, Behar KL (2014) Direct evidence for activity-dependent glucose phosphorylation in neurons with implications for the astrocyte-to-neuron lactate shuttle. Proc Natl Acad Sci U S A 111(14):5385–5390. https://doi.org/10.1073/pnas.1403576111

Pearson-Leary J, Jahagirdar V, Sage J, McNay EC (2018) Insulin modulates hippocampally-mediated spatial working memory via glucose transporter-4. Behav Brain Res 338:32–39. https://doi.org/10.1016/j.bbr.2017.09.033

Pearson-Leary J, McNay EC (2016) Novel Roles for the Insulin-Regulated Glucose Transporter-4 in Hippocampally Dependent Memory. J Neurosci 36(47):11851–11864. https://doi.org/10.1523/jneurosci.1700-16.2016

Pellerin L, Magistretti PJ (1994) Glutamate uptake into astrocytes stimulates aerobic glycolysis: a mechanism coupling neuronal activity to glucose utilization. Proc Natl Acad Sci U S A 91(22):10625–10629

Prebil M, Vardjan N, Jensen J, Zorec R, Kreft M (2011) Dynamic monitoring of cytosolic glucose in single astrocytes. Glia 59(6):903–913. https://doi.org/10.1002/glia.21161

Quach TT, Rose C, Schwartz JC (1978) [3H]Glycogen hydrolysis in brain slices: responses to neurotransmitters and modulation of noradrenaline receptors. J Neurochem 30(6):1335–1341

Quistorff B, Secher NH, Van Lieshout JJ (2008) Lactate fuels the human brain during exercise. FASEB J 22(10):3443–3449. https://doi.org/10.1096/fj.08-106104

Rangaraju V, Calloway N, Ryan Timothy A (2014) Activity-Driven Local ATP Synthesis Is Required for Synaptic Function. Cell 156(4):825–835. https://doi.org/10.1016/j.cell.2013.12.042

Rosenthal M, Sick TJ (1992) Glycolytic and oxidative metabolic contributions to potassium ion transport in rat cerebral cortex. Can J Physiol Pharmacol 70(Suppl):S165–S169

Rothman DL, Magnusson I, Katz LD, Shulman RG, Shulman GI (1991) Quantitation of hepatic glycogenolysis and gluconeogenesis in fasting humans with 13C NMR. Science 254(5031):573–576

Rowan S, Bejarano E, Taylor A (2018) Mechanistic targeting of advanced glycation end-products in age-related diseases. Biochim Biophys Acta Mol basis Dis 1864(12):3631–3643. https://doi.org/10.1016/j.bbadis.2018.08.036

Ryan TA (1999) Inhibitors of myosin light chain kinase block synaptic vesicle pool mobilization during action potential firing. J Neurosci 19(4):1317–1323

Schousboe A, Bak LK, Waagepetersen HS (2013) Astrocytic control of biosynthesis and turnover of the neurotransmitters glutamate and GABA. Front Endocrinol (Lausanne) 4:102. https://doi.org/10.3389/fendo.2013.00102

Schurr A (2014) Cerebral glycolysis: a century of persistent misunderstanding and misconception. Front Neurosci 8:360. https://doi.org/10.3389/fnins.2014.00360

Scimemi A, Beato M (2009) Determining the neurotransmitter concentration profile at active synapses. Mol Neurobiol 40(3):289–306. https://doi.org/10.1007/s12035-009-8087-7

Seaquist ER, Damberg GS, Tkac I, Gruetter R (2001) The effect of insulin on in vivo cerebral glucose concentrations and rates of glucose transport/metabolism in humans. Diabetes 50(10):2203–2209

Shulman RG, Hyder F, Rothman DL (2001a) Cerebral energetics and the glycogen shunt: neurochemical basis of functional imaging. Proc Natl Acad Sci U S A 98(11):6417–6422. https://doi.org/10.1073/pnas.101129298

Shulman RG, Hyder F, Rothman DL (2001b) Lactate efflux and the neuroenergetic basis of brain function. NMR Biomed 14(7-8):389–396. https://doi.org/10.1002/nbm.741

Shulman RG, Rothman DL (2001) The "glycogen shunt" in exercising muscle: A role for glycogen in muscle energetics and fatigue. Proc Natl Acad Sci U S A 98(2):457–461. https://doi.org/10.1073/pnas.98.2.457

Shulman RG, Rothman DL (2015) Homeostasis and the glycogen shunt explains aerobic ethanol production in yeast. Proc Natl Acad Sci U S A 112(35):10902–10907. https://doi.org/10.1073/pnas.1510730112

Shulman RG, Rothman DL (2017) The Glycogen Shunt Maintains Glycolytic Homeostasis and the Warburg Effect in Cancer. Trends Cancer 3(11):761–767. https://doi.org/10.1016/j.trecan.2017.09.007

Sibson NR, Dhankhar A, Mason GF, Rothman DL, Behar KL, Shulman RG (1998) Stoichiometric coupling of brain glucose metabolism and glutamatergic neuronal activity. Proc Natl Acad Sci U S A 95(1):316–321

Sickmann HM, Walls AB, Schousboe A, Bouman SD, Waagepetersen HS (2009) Functional significance of brain glycogen in sustaining glutamatergic neurotransmission. J Neurochem 109(Suppl 1):80–86. https://doi.org/10.1111/j.1471-4159.2009.05915.x

Siesjö BK (1978) Brain Energy Metabolism. John Wiley & Sons, Chichester

Simpson IA, Carruthers A, Vannucci SJ (2007) Supply and demand in cerebral energy metabolism: the role of nutrient transporters. J Cereb Blood Flow Metab 27(11):1766–1791. https://doi.org/10.1038/sj.jcbfm.9600521

Smith D, Pernet A, Hallett WA, Bingham E, Marsden PK, Amiel SA (2003) Lactate: a preferred fuel for human brain metabolism in vivo. J Cereb Blood Flow Metab 23(6):658–664. https://doi.org/10.1097/01.wcb.0000063991.19746.11

Sokoloff L (1960) Metabolism of the central nervous system in vivo. In: Field J, Magoun HW, Hall VE (eds) Handbook of physiology —neurophysiology, vol 3. American Physiological Society. Washington, DC, pp 1843–1864

Sokoloff L (1981) Localization of functional activity in the central nervous system by measurement of glucose utilization with radioactive deoxyglucose. J Cereb Blood Flow Metab 1(1):7–36. https://doi.org/10.1038/jcbfm.1981.4

Sonnay S, Duarte JMN, Just N, Gruetter R (2016) Compartmentalised energy metabolism supporting glutamatergic neurotransmission in response to increased activity in the rat cerebral cortex: A ^{13}C MRS study in vivo at 14.1 T. J Cereb Blood Flow Metab 36(5):928–940. https://doi.org/10.1177/0271678x16629482

Sonnay S, Duarte JMN, Just N, Gruetter R (2017a) Energy metabolism in the rat cortex under thiopental anaesthesia measured in vivo by ^{13}C MRS. J Neurosci Res 95(11):2297–2306. https://doi.org/10.1002/jnr.24032

Sonnay S, Poirot J, Just N, Clerc A-C, Gruetter R, Rainer G, Duarte JMN (2017b) Astrocytic and neuronal oxidative metabolism are coupled to the rate of glutamate–glutamine cycle in the tree shrew visual cortex. Glia 66(3):477–491. https://doi.org/10.1002/glia.23259

Suh SW, Bergher JP, Anderson CM, Treadway JL, Fosgerau K, Swanson RA (2007) Astrocyte glycogen sustains neuronal activity during hypoglycemia: studies with the glycogen phosphorylase inhibitor CP-316,819 ([R-R*,S*]-5-chloro-N-[2-hydroxy-3-(methoxymethylamino)-3-oxo-1-(phenylmethyl)pro pyl]-1H-indole-2-carboxamide). J Pharmacol Exp Ther 321(1):45–50. https://doi.org/10.1124/jpet.106.115550

Suzuki A, Stern SA, Bozdagi O, Huntley GW, Walker RH, Magistretti PJ, Alberini CM (2011) Astrocyte-neuron lactate transport is required for long-term memory formation. Cell 144(5):810–823. https://doi.org/10.1016/j.cell.2011.02.018

Swanson RA (1992) Physiologic coupling of glial glycogen metabolism to neuronal activity in brain. Canadian J Physiol Pharmacol 70(Suppl):S138–S144

Swanson RA, Morton MM, Sagar SM, Sharp FR (1992) Sensory stimulation induces local cerebral glycogenolysis: demonstration by autoradiography. Neurosci 51(2):451–461

Tamir H, Kaufman H, Rapport MM (1972) Subcellular distribution of pyruvate kinase (EC 2.7.1.40) in cerebral cortex. J Neurochem 19(7):1759–1768. https://doi.org/10.1111/j.1471-4159.1972.tb06220.x

Terlecki G (1989) Purification and properties of pyruvate kinase type M1 from bovine brain. Int J Biochem 21(9):1053–1060

Tolle SW, Dyson RD, Newburgh RW, Cardenas JM (1976) Pyruvate kinase isozymes in neurons, glia, neuroblastoma, and glioblastoma. J Neurochem 27(6):1355–1360

Ueda T (2016) Vesicular glutamate uptake. Adv Neurobiol 13:173–221. https://doi.org/10.1007/978-3-319-45096-4_7

Veech RL, Harris RL, Veloso D, Veech EH (1973) Freeze-blowing: a new technique for the study of brain in vivo. J Neurochem 20(1):183–188

Watanabe H, Passonneau JV (1973) Factors affecting the turnover of cerebral glycogen and limit dextrin in vivo. J Neurochem 20(6):1543–1554

Weisova P, Concannon CG, Devocelle M, Prehn JH, Ward MW (2009) Regulation of glucose transporter 3 surface expression by the AMP-activated protein kinase mediates tolerance to glutamate excitation in neurons. J Neurosci 29(9):2997–3008. https://doi.org/10.1523/jneurosci.0354-09.2009

Xu J, Song D, Bai Q, Cai L, Hertz L, Peng L (2014) Basic mechanism leading to stimulation of glycogenolysis by isoproterenol, EGF, elevated extracellular K+ concentrations or GABA. Neurochem Res 39(4):661–667. https://doi.org/10.1007/s11064-014-1244-z

Xu J, Song D, Xue Z, Gu L, Hertz L, Peng L (2013) Requirement of glycogenolysis for uptake of increased extracellular K+ in astrocytes: potential implications for K+ homeostasis and glycogen usage in brain. Neurochem Res 38(3):472–485. https://doi.org/10.1007/s11064-012-0938-3

Yellen G (2018) Fueling thought: Management of glycolysis and oxidative phosphorylation in neuronal metabolism. J Cell Biol 217(7):2235–2246. https://doi.org/10.1083/jcb.201803152

Yu Y, Herman P, Rothman DL, Agarwal D, Hyder F (2018) Evaluating the gray and white matter energy budgets of human brain function. J Cereb Blood Flow Metab 38(8):0339–1353. https://doi.org/10.1177/0271678x17708691

Zhang Y, Xue Y, Meng S, Luo Y, Liang J, Li J, Ai S, Sun C, Shen H, Zhu W, Wu P, Lu L, Shi J (2016) Inhibition of lactate transport erases drug memory and prevents drug relapse. Biol Psychiatry 79(11):928–929. https://doi.org/10.1016/j.biopsych.2015.07.007

Ziemińska E, Hilgier W, Waagepetersen HS, Hertz L, Sonnewald U, Schousboe A, Albrecht J (2004) Analysis of glutamine accumulation in rat brain mitochondria in the presence of a glutamine uptake inhibitor, histidine, reveals glutamine pools with a distinct access to deamidation. Neurochem Res 29(11):2121–2123. https://doi.org/10.1007/s11064-004-6885-x

Index

M. DiNuzzo, A. Schousboe (eds.), *Brain Glycogen Metabolism*, Advances in Neurobiology 23, https://doi.org/10.1007/978-3-030-27480-1

L

M

N

The manufacturer's authorised representative in the EU is Springer Nature Customer Service Centre GmbH, Europaplatz 3, 69115 Heidelberg, Germany. If you have any concerns regarding our products, please contact ProductSafety@springernature.com

Printed and bound by CPI Group (UK) Ltd, Croydon, CR0 4YY

25/09/2026

02222851-0008

Springer Polar Sciences

Springer Polar Sciences is an interdisciplinary book series that is dedicated to research in the Arctic, sub-Arctic regions, and the Antarctic. In recent years, the polar regions have received increased scientific and public interest. Both the Arctic and Antarctic have been recognized as key regions in the regulation of the global climate, and polar ecosystems have been identified to be particularly susceptible to the ongoing environmental changes. Consequently, the international efforts in polar research have been enhanced considerably, and a wealth of new findings is being produced at a growing rate by the international community of polar researchers.

Springer Polar Sciences aims to present a broad platform that will include state-of-the-art research, bringing together both science and humanities to facilitate an exchange of knowledge between the various polar science communities. The Series offers an outlet to publish contributions, monographs, edited works, conference proceedings, etc. Topics and perspectives will be broad and will include, but not be limited to: climate change impacts, environmental change, polar ecology, governance, health, economics, indigenous populations, tourism and resource extraction activities. Books published in the series will appeal to scientists, students, polar researchers and policy makers.

More information about this series at https://link.springer.com/bookseries/15180

Robert P. Wheelersburg

The Security of Iceland and the Arctic 2030

A Recommendation for Increased Geopolitical Stability

Springer

Robert P. Wheelersburg
Department of Sociology and Anthropology (retired)
Elizabethtown College
Elizabethtown, PA, USA

ISSN 2510-0475 ISSN 2510-0483 (electronic)
Springer Polar Sciences
ISBN 978-3-030-89947-9 ISBN 978-3-030-89948-6 (eBook)
https://doi.org/10.1007/978-3-030-89948-6

This Springer imprint is published by the registered company Springer Nature Switzerland AG
The registered company address is: Gewerbestrasse 11, 6330 Cham, Switzerland

[Joshua, a computer with artificial intelligence, after learning from playing several simulated games of Global Thermal Nuclear Warfare that there can be no winner states]. "Greetings Professor Falken. A strange game. The only winning move is not to play. How about a nice game of chess?" Lasker and Parkes (1983). WARGAMES

This book is dedicated to Arnór Sigurjónsson, former Director of the Security and Defence Department of the Ministry for Foreign Affairs of Iceland. Besides helping me with virtually every phase of researching and writing this book during the past five years, Arnór is also a brother in arms who I met on my first assignment in Iceland during 1990. From his training with the Norwegian Army, to commanding the Viking Squad, to being the former Icelandic representative to NATO, Colonel Sigurjónsson has spent over half a century working tirelessly to improve Iceland's security. Although his assistance on this book has been invaluable, especially by ensuring accuracy in my writing, the ideas presented here are my own. Takk fyrir Arnór.

This book is dedicated to the thousands of dependents of Icelandic Defense Force (IDF) personnel who served at the Keflavik NATO Base from 1951 to 2006. While their spouses and parents protected the North Atlantic by flying aircraft around the North Pole hunting Soviet bombers and boomers (i.e., missile submarines), wives/husbands and children lived on a wind-swept, sub-Arctic peninsula, enclosed behind wire and Icelandic customs checkpoints.

With few exceptions, over the 60 years that the NATO base flourished, the relationship between the American soldiers and Icelanders was never a close one. The Icelandic government made those relationships difficult due to the incredible trauma the country suffered during WWII, now referred to as "The Situation." During the 1940s, when there were 50,000 American soldiers in Iceland, girls as young as 14 attended USO dances and private parties with soldiers. They drank alcohol at a time when their country still had a prohibition on beer. Iceland had no dating culture until the Americans came. They brought customs like paying for dinner and movies, taking women to dances, holding doors and treating women respectfully, and giving their dates presents such as nylons, cigarettes, chocolate, and of course liquor. Consequently, after a few years of former country girls moving to the capital region and behaving shamelessly (according to many Icelanders), the government appointed the country's first Icelandic policewoman to control the situation. After spying on girls with especially bad reputations, she forcibly banished many of them to outlying farms or institutionalized them for cavorting with the GIs (many of whom were only 18 or 19 themselves). After the NATO base was built, neither the Icelandic government nor the country's leading citizens wanted a repeat of The Situation and took steps to prevent it from happening. Those steps included restricting the times and places that service members and their families could go off base. Eventually, the base was completely sealed off from the rest of Iceland, with both military personnel and their dependents required to pass through customs and have permission to stay overnight in Iceland.

Yet despite the difficult living conditions, over several generations the little American community at the Keflavik base saw children born, classes taught, sports played, and excursions taken to the exotic foreign country outside the fence. As much as anyone, IDF dependents helped the world win the Cold War without firing a shot by providing American pilots, air crew, and ground personnel with a stable family and community. That stability gave service members a real reason to live instead of an ideal to die for like a political system. Frankly their sacrifices are neither remembered nor valued either by Icelanders or the world today. But their contributions to peace over nearly six decades formed the winning move not to play the real game of Global Thermal Nuclear Warfare in the Arctic. Live long and prosper!

Foreword to Robert P. Wheelersburg, The Security of Iceland and the Arctic 2030: A Recommendation for North Atlantic Geostrategic Stability

The Arctic region provides a template for ways in which global security challenges are evolving in the twenty-first century. The recession of the polar ice cap provoked by climate change is transforming an area once considered remote into a focal point of geostrategic competition. A contest for control over sea lines of communication and once inaccessible caches of strategic resources has emerged, with considerable relevance for great power competitors (the USA, Russia, China) as well as regional states. Traditional national security concerns such as territorial defense now overlap with new threats generated by environmental disintegration and its social and economic consequences.

The Security of Iceland and the Arctic 2030 addresses the implications of these changes for Iceland, a small but strategically significant Arctic state. Any attention accorded to the security concerns of small states is of course to be welcomed. Robert Wheelersburg's treatment embeds the dilemmas of Icelandic security within larger trends that are transforming the regional and global environments. Traditionally considered a bastion of peace and stability, a NATO member state but without indigenous armed forces, Iceland is now confronted by emerging *sub-strategic* security threats that it is not fully prepared to manage. The special relevance of Wheelersburg's text is the detailed attention devoted to identifying these challenges, and the hard-minded effort to specify policy initiatives capable of addressing them. The text addresses the *context* of Arctic security, but also the *substance* of security policy. In the longer term how Arctic states including Iceland adapt to the changing strategic environment will determine whether the desired goal of enduring geostrategic stability might in fact be achieved.

Robert Wheelersburg's impressive competencies (firsthand knowledge of Iceland and the larger Arctic region, cultural sensitivity, military professionalism, scholarly specializations in global and regional security affairs) make him uniquely qualified to address these issues. Arctic security has become and will remain a critical part of the global security agenda. The nuanced analysis presented here takes up the consequences of these changes for a small but vital regional state and recommends ways in which policy priorities might be adjusted to address them. It represents an important contribution to the expanding literature on Arctic security in all its dimensions.

Distinguished Fellow, U.S. Army War College, Carlisle, PA, USA
22 September 2021

R. Craig Nation

Preface

> They were conquerors, and for that you want only brute force--nothing to boast of, when you have it, since your strength is just an accident arising from the weakness of others.
>
> They grabbed what they could get for the sake of what was to be got.
>
> It was just robbery with violence, aggravated murder on a great scale, and men going at it blind –
>
> As is very proper for those who tackle a darkness.
>
> Joseph Conrad, *Heart of Darkness*

Heart of Darkness

In 1979, I watched with fascination the film *Apocalypse Now*. The most shocking scene occurred as U.S. Army counterintelligence officer Captain Benjamin L. Willard (Marin Sheen) received his orders to "terminate with extreme prejudice" fellow American, Colonel Walter E. Kurtz (Marlin Brando), for using "unsound" methods to kill the enemy during the Vietnam War (Milius and Coppola 1979).

A decade later, I completed U.S. Army intelligence officer school and reported for duty as assistant counterintelligence section chief with the 258th Military Intelligence Detachment in Richmond, Virginia.

Fortunately for me, since watching the film, President Gerald Ford had issued Executive Order 11905 with its Section 5(g) prohibiting political assassination by any agency of the U.S. Government (Ford 1976). Presidents Jimmy Carter (1978) and Ronald Reagan (1981) reaffirmed that prohibition also through executive orders. Thus, despite my assignment as a counterintelligence officer, I assumed I would never be ordered to assassinate anyone, let alone kill Americans in the course of my duties. Fat chance.

During my 25-year military career, I spent much of my time killing Americans albeit imaginary ones. Whether using pen and paper or digital media, my notional targets were Americans nonetheless. As an exercise writer and red gamer (i.e., enemy force player) for organizations ranging from First U.S. Army, the North Atlantic Treaty Organization (NATO), the Icelandic Defense Force (IDF), and the Federal Emergency Management Agency (FEMA), I created scenarios that killed imaginary people to help train combat units and response agencies. If I didn't do my duty, I rationalized, real Americans might die. I also assumed the role of an enemy commander or international terrorist leader, competent in my craft, consumed with hatred of America and the West, and ruthless in the application of my methods. Army, joint service, and international military and civilian schools, along with domestic and foreign exercises, provided the training and practice for me to sharpen the skills necessary to accomplish my mission.

Yet one thing in my preparation was missing. To create and conduct realistic scenarios, I needed to develop a "heart of darkness" that would allow me to do an imaginary assassin's job without mercy. I attained my dark heart by studying Okinawan Karate for 7 years, earning a black belt in what is a ruthless killing art. Whether using open hands or weapons, that karate style is so deadly that sparring with protective equipment does not provide an equivalent experience to real fighting. True practitioners do not bother with sparring.

Early on I made a vow that, despite my duties, I would try my best never to hurt a living human being either intentionally or unintentionally. I adopted the personal philosophy articulated by Master Kan to Caine in the pilot movie for the T.V. show *Kung Fu* (1972), "Avoid, rather than check. Check, rather than hurt. Hurt, rather than maim. Maim, rather than kill. For all life is precious, nor can any be replaced." At another point in the film, Caine asked "What is the best way to deal with force, Master"? Master Teh responded, "As we prize peace and quiet above victory, there is a simple and preferred method ... run away" (Spielman and Friedlander 1972).

In the early twenty-first century, we live in a terrifying world where there is often no chance to run away. Thus, I used my dark heart to create the frightening scenarios contained in this book to show the successful application of my recommended Icelandic security force. Of all the material contained herein, I believe those scenarios will disturb Icelanders the most, especially my artistic and creative friends. As members of a culture with no military, no tradition of armed service, no monuments to war or its fighters, and no changing of the palace guard, many Icelanders will find it hard to understand why I undertook this work with such fervor.

I know from my interactions with Icelanders that many do not accept the need for a security force because they believe there is no potential threat to them requiring such a capability. I think that view is wrong. Still, it is a view many Icelanders hold without compromise, which I accept. Consequently, rather than ask for their understanding, I ask only for their forgiveness. I undertook this work with their interests at heart. I did so with no thought of profit or benefit to myself. Thus, I present this work with no reservations, regardless of how others receive it.

Much of the research for this project was conducted while I was a National Science Foundation-Fulbright Arctic Scholar and affiliated researcher with the Centre for Arctic Policy Studies at the University of Iceland. While there, I also taught a graduate public policy course entitled Images of the Arctic: Power Dynamics in Everyday Life where I created some of the material in this book. Later, I began teaching graduate courses for my home institution's new master's program in public policy and supervising theses on Icelandic modern culture. These theses included participant observation by a female Muslim student working with asylum seekers in Iceland during Ramadan and a female student examining the relationship between traditional pagan sorcery and modern Icelandic women's movements.

The entrance into public policy scholarship allowed me to join together my two, formerly disparate, careers as an academic and an Army officer. Instead of merely studying a topic, I desired to create a recommendation that could be evaluated by, and perhaps assist, Icelandic policy makers in meeting their future security needs.

Making a recommendation for an Icelandic security force is an extension of my decade of commitment to the protection of the country's citizens as a member of the Icelandic Defense Force during the 1990s. I believe there is a need to have a security force that protects the country's territorial integrity and its internal security. To me, that need is critical enough for Icelanders to accommodate new security structures such as the one presented here. As I document in Chap. 5, Icelandic cultural history is filled with the creation of such ad-hoc security forces, whether they were a Viking raiding party or a deputized mob.

The changing circumstances that make it critical for Iceland to create a security force is global climate change and the future Arctic human dynamics it will create. Predictions are that the melting ice and the exposure of new transportation routes, emerging territories, and accessible natural resources will draw evil state and non-state actors seeking murder and mayhem in the circumpolar region. There is no other option than to have Iceland or some other entity create a security force prepared to defeat such threats as they arise there.

New structures and methods should be developed as part of Icelandic security sector reform (SSR) to face those emerging challenges coming from the Arctic. Without SSR, Icelanders will go forward unprotected into a wide-open Arctic where nations, corporations, criminals, and terrorists operate freely within the country's sovereign territory. The lack of a consensus-built, appropriate security force condemns the country's citizens to being powerless in the face of those threats, while having others dictate their futures (including choosing to harm them).

Yet, if Iceland will not address its own sub-strategic security needs, NATO must do it for them, taking away the country's self-determination. The stakes are too great for the other Arctic Council Members to ignore and they must take action to prevent catastrophe before it happens. The recommendation presented herein could provide for Iceland's future security that has local ownership, supported by both elected officials and their constituencies, and not imposed by outsiders. Additionally, the proposed security force is designed to reduce the need for foreigners to be stationed or to conduct military operations on Icelandic soil for its protection. Icelanders taking responsibility for their own sovereignty and internal safety represents a compromise with political parties who believe there should be no military forces in the world, let alone Iceland.

As U.S. Navy Commander Mike "Viper" Metcalf told incoming student pilots in the 1986 film *Top Gun*, "Gentlemen [and ladies], this school is about combat. There are no points for second place" (Cash and Epps 1986) (Fig. 1).

Fig. 1 U.S. Army Captain Wheelersburg, exchange officer with the Norrland Dragoon Regiment (K-4 Swedish Ranger) Battalion, in 1994 at the Arctic Combat School, Umeå, Sweden. I am firing an M-2 (Ma Deuce) Browning .50 caliber machine gun with laser sights mounted on a Swedish BV1 armored personnel carrier at a Soviet-made T-34 tank. After firing several dozen rounds, the sergeant said, "Jag tror det räcker" [I think that is enough]. I asked, "Träffade jag målet?" [Did I hit the target], to which the instructor replied "Jajamensan" [You bet!]. (Photo by unknown Swedish Army officer)

Elizabethtown, PA, USA Robert P. Wheelersburg

References

Carter, Jimmy. 1978. *Executive Order 12036 United States Foreign Intelligence Activities*. Section 2 Restrictions on Intelligence Activities, 2-305 Prohibition on Assassination. January 24, 1978. Washington, DC: Federation of American Scientists. Accessed September 14, 2020, https://fas.org/irp/offdocs/eo/eo-12036.htm.

Cash, Jim, and Jack Epps Jr. 1986. *Top Gun*. Hollywood: Paramount Pictures.

Conrad, Joseph. 2007. *Heart of Darkness*. Claremont: Coyote Canyon Press.

Ford, Gerald R. 1976. *Executive Order 11905: United States Foreign Intelligence Activities*. Section 5. Restrictions on Intelligence Activities, (g) Prohibition of Assassination. February

18, 1976. Washington, DC: Federation of American Scientists. Accessed September 14, 2020, https://fas.org/irp/offdocs/eo11905.htm#SEC.%205.

Milius, John, and Francis Coppola. 1979. *Apocalypse Now*. Hollywood: United Artists.

Reagan, Ronald. 1981. *Executive Order 12333: United States Intelligence Activities. Part 2.11 Prohibition on Assassination*. December 4, 1981. Washington, DC: Federation of American Scientists. Accessed September 14, 2020, https://fas.org/irp/offdocs/eo12333.htm.

Spielman, Ed, and Howard Friedlander. 1972. *Kung Fu*. Pilot episode. Hollywood: Warner Brothers Television.

Acknowledgments

Dreh dich nich um,
Der Kommissar geht um,
Er hat die Kraft und wir sind klein und dumm,
Und dieser Frust macht uns stumm.

[Don't turn around,
The Commissar's in town,
He's got the power and you're so weak,
And your frustration will not let you speak]
Robert Ponger and Falco, *Der Kommissar*

This research project was conducted during a 5-year period from mid-2015 through 2020. I wrote the manuscript in 2020 and early 2021 while isolated during the pandemic. The initial research and core of the manuscript was developed while I was a U.S. Fulbright-National Science Foundation Arctic Scholar at the University of Iceland's Centre for Arctic Policy Studies during fall semester 2015. Thanks to Pia and Margret for their support. Some of the concepts and hypotheses in the work were tested on students in a course I taught during that semester, along with Kristinn Schram, Ph.D., who is an associate professor of anthropology at the University of Iceland. Thanks to Belinda Theriault, Icelandic Fulbright Commission Director who continues to support my work in the country and my alumni connection to Fulbright. Anna Kertulla, former manager of the National Science Foundation's Arctic Social Science program, has funded and otherwise facilitating my work in the Arctic for nearly two decades. A special thanks to Margaret Deignan of the Springer Polar Science Series who encouraged me to begin, and more importantly, to complete this project. Thank you to Kara Hunsberger who edited the manuscript. Many of the sources cited in the work also involved personal contacts with the scholars, and I very much appreciate the time and effort they took in helping me understand their work. A special shoutout to my friends at Prikið Kaffihús. Your creativity and enthusiasm for life taught me much about modern Icelandic culture. You also allowed me to discuss with you my experiences in the country, helping my understanding of Iceland evolve from my first time there in 1990 to my most recent visit in 2021. Finally, thank you very much to my wife Susan who was my traveling partner to Iceland during many of my trips there, and for providing a relief from my work on this project. In a world populated by unscrupulous and powerful leaders, not to mention their less powerful but no less dangerous counterparts like criminals and terrorists, Susan teaches children to express themselves creatively using color, form, line, shape, space, and texture. It is those children who give us hope for the future.

Contents

Chapter 1
Introduction and Setting the Stage

Homo homini lupus *[Man is wolf to man].*
Titus Maccius Plautus (c. 212 B.C.E), Asinaria

Abstract This chapter discusses the current security situation in the North Atlantic. Historical circumstances in the Arctic are presented beginning with World War II, continuing through the Cold War, and projecting into the future (i.e., 2030). The predicted Arctic peace dividend following the fall of the Soviet Union is examined, along with how that peace dividend in the region failed to materialize. In the chapter I also define security, hard versus soft, and introduce a new level, which I termed sub-strategic security. Sub-strategic threats include non-state sources such as terrorism, crime, smuggling, and human trafficking, along with state actions like conflicts over natural resources or limited territorial incursion. I conclude that Iceland is currently unable to defeat such threats because it has no security force above the police or the coast guard. Finally, the chapter outlines the book's remaining structure and implications.

1.1 Introduction to North Atlantic Security

> Iceland faces a security dilemma: how to keep the High North an area of low tensions as Russia builds up its military presence and China's Arctic ambitions grow. [...] While remaining the only Ally without an armed forces and Ministry of Defence, Iceland is investing in new defence and security capabilities [... since ... c]rucially, Iceland serves as the Alliance's outpost in the North Atlantic (NATO 2019a).

The quote above following an official NATO visit to Iceland in May 2019 seemed incongruous with the predicted post-Cold War geopolitical situation in the North Atlantic. After the fall of the Soviet Union in 1991, both scholars and government officials proclaimed the Arctic security situation peaceful. Most trumpeted a unique opportunity to use international cooperation to resolve disputes through the Arctic Council formed in 1996. For example,

R. P. Wheelersburg, *The Security of Iceland and the Arctic 2030*, Springer Polar Sciences, https://doi.org/10.1007/978-3-030-89948-6_1

> Starting with Mikhail Gorbachev's "Arctic zone of peace" speech in October, 1987, the region became a target of opportunity for those interested in promoting various forms of international and transnational cooperation (Young 2011: xxii).

More recently, some scholars began referring to the anticipated Arctic peaceful cooperation as "exceptionalism", while other scholars sought to redefine what security meant in the Arctic (Gjørv and Hodgson 2019; Pic and Lasserre 2019). Politicians jumped on the peace bandwagon, and statements ensued from that perspective, such as the one from Norwegian Defence Minister Søreide that "[c] ompared to most quarters of the world, the Arctic is a region of stability, respect for international law, and well-functioning multilateral institutions" (Søreide 2016: 49).

The area was considered so secure that many believed North Atlantic demilitarization would continue well into the future (Wezeman 2016). Perhaps the best example was the U.S. government closing the Keflavik base in 2006, having previously removed its permanent contingent of Navy submarine-patrol aircraft and Air Force bomber-interceptor planes. Initially, the short-lived Icelandic Defence Agency assumed control of the base in 2008 until the agency was abolished by the Leftist Government in 2010. The Icelandic Coast Guard took over base operations, and it looked like Arctic demilitarization was progressing as predicted, although some analysts maintained the Arctic never demilitarized (e.g., Keupp 2015). Given these events, we might ask how the Arctic got back to its former high tempo of strategic operations three decades after the end of the Cold War? The answer lies in the out-of-area military activities of would-be superpowers like the Russian invasion of Crimea and the Chinese take-over of the Spratly Islands, which illustrated that such states will engage in unilateral actions without consideration of their global political implications.

While these actions precipitated a response by NATO members, it is the predicted melting of the Arctic ice that will expose new, resource-rich territories and allow rapid transit of the Transpolar Sea Route (TSR) (Humpert and Raspotnik 2012; Stephenson 2013) that is leading to the massive military buildup in the Arctic. The emerging territories and TSR could lead to competing sovereignty and resource claims in the Arctic, along with creating possible conflicts about transportation access such as the decades-long dispute between the U.S. and Canada over the Northwest Passage (Shea 2019). New territorial disputes are already occurring including the one over Lomonosov Ridge claimed by Canada, Denmark and Russia (Arctic Monitor 2016). Competition could change quickly to conflict, producing skirmishes between nation-states like the Cod Wars did between Iceland and Great Britain from the 1950s to the mid-1970s (GuĐmundsson 2006). While international bodies like the Arctic Council might reduce tensions in the region, the problem is that unilateral actions by state and non-state actors could determine the requirements for future Arctic security.

One major problem with solving conflicts in the circumpolar region is that the Arctic Council is forbidden from discussing military issues because of its founding document (Arctic Council 2016). After the Crimea invasion by Russia in 2014, some Arctic Council members called for establishing an Arctic security forum (Stephen

2016), but that never took place. Along with generating strategic conflicts, anticipated future events could increase security challenges below the strategic level for countries in the region. An increased number of people moving through the North Atlantic along the TSR could also create security threats for Iceland, including terrorism, international crime, and human trafficking (Byers 2009).

There are several terms used to describe security in the Arctic. The highest level is strategic (often called "hard") security practiced by states and non-state actors (e.g., NATO) using conventional military forces. Specifically, strategic security means actual warfare between the armed forces of states to defend their sovereign territory, to acquire or protect natural resources, or to advance their ideology (e.g., democracy versus communism) including religious beliefs (e.g., Christianity versus Islam). On the other hand, security below the strategic level (often called "soft") is practiced by states and non-state actors (e.g., Interpol) using unconventional forces and civilian agencies such as the police or border patrol. In this book, I will concentrate on sub-strategic security (defined in the next paragraph) to distinguish between threats requiring military action and threats that require capabilities above the police but below a military.

For the purpose of my recommendation, I use the term "Arctic sub-strategic security" to refer to the capability of a country in the region to counter threats to maintain its sovereignty and protect its citizens, whether alone or as part of an alliance (Gjørv and Hodgson 2019). As defined by Huebert (2011), sovereignty has four characteristics: a government firmly establishing state borders, other states accepting those borders, a national population living within the state's borders, and a government making decisions about its state's internal and foreign affairs.

Using these characteristics of sovereignty, my definition of Arctic sub-strategic security differs somewhat from that presented in the 2019 Arctic Yearbook. As articulated by the editors

> ... [Arctic] security may be redefined across the globe as environmental issues and local challenges become more pertinent vis à vis traditional state security issues across the globe. It can also show how environmental and societal challenges are increasingly becoming 'traditional' security issues in and of themselves (Heininen et al. 2019: 7).

The editors' definition also included such soft security aspects as indigenous home rule, egalitarian economic development, and environmental protection. My recommendation for Arctic sub-strategic security addresses only the editors' "local challenges", which requires a state to establish a security force to defend territory/resources and maintain internal public safety against external (e.g., terrorism) and internal (e.g., organized crime) threats.

Sub-strategic security includes defeating threats from internal forces (e.g., violent ethnic groups like white supremacists), external threats such as limited border disputes (i.e., not involving the entire country), and international efforts to combat organized crime, terrorism, smuggling, and human trafficking (Fatić 2002). States respond to sub-strategic threats using police forces to fight crime, border guards to control immigration and human trafficking, and coast guards to protect maritime territory and resources. They also typically have a security force like the Alaskan National Guard or a paramilitary unit such as the Royal Canadian Mounted Police to assist in defeating threats that overcome those front-line civilian agencies. Iceland currently has the appropriate security protection for some sub-strategic security threats through their police department and coast guard. I designed the security force proposed in this book to defeat potential future sub-strategic security threats in Iceland should they arise without depending upon the armed forces of another country for an initial response.

The key to Iceland's sub-strategic security preparedness is that it must address threats below those sufficient to trigger a strategic response involving militaries conducting conventional warfare. Specifically, while adapting the sub-security structure recommended herein, Iceland could ask for help from other states without invoking NATO Article 5. For example, either in preparation for or as a response to a sub-strategic security threat, Iceland could request support from Canada or Norway individually but not appeal to NATO formally that would invoke the required mutual protection response. Even if a foreign state were the aggressor, Iceland meeting a sub-strategic security threat without foreign assistance would not escalate the conflict to the strategic level wherein NATO and another major power engage in conventional combat. Thus, Iceland adopting its own security force could actually improve the chance for peace rather than threatening it.

Using minimal requests for assistance from other states on an informal basis during peacetime, rather than through an alliance or bilateral treaty, is provided for by the United Nations policy on regional war response. A national government threatened by a second state could request that a third state serve as an executive agent to lead an UN-sanctioned, informal coalition (or merely the third state alone) to intervene in the conflict to maintain peace. Although the UN would have to endorse the action, if the Security Council were deadlocked or if one of the permanent members was an aggressor in the potential conflict and vetoed the endorsement, the General Assembly could provide the necessary authorization. Such an informal coalition could bypass complex bureaucracy like elected-officials debating the action to provide a rapid, tailored, and limited response to the threat of incursion or violence by another country (Fitz-Gerald 2003).

Therefore, there are three advantages to the recommendation presented in this book. First, the proposed security force would allow Iceland to deter or defeat sub-strategic threats within its own territory using its own assets. Consequently, in combination with an informal Iceland-requested coalition, the proposed security force could meet both Iceland's security policy and the requirements for UN-sanctioned peacekeeping, reducing the need to permanently station foreign military troops in the country. Perhaps most importantly, by using its own assets, along with an informal coalition to meet sub-strategic security threats, Iceland would not need to invoke NATO Article 5, limiting the operation to a regional-level and preventing escalation of the conflict to a strategic (and possibly nuclear) war.

1.2 Past Is Prologue – Militarization of the Arctic

For most of the latter half of the twentieth century, the North Atlantic was the stage for high-tempo military operations between NATO and the Warsaw Pact (i.e., the U.S.S.R.), representing one of the greatest strategic threats to peace during the Cold War. Although the Allies had been in the North Atlantic since 1940 with the occupation of Iceland (NATO n.d.-a), it was in the decades following World War II that Russia finally achieved Peter the Great's dream of becoming a strategic power. The immense buildup of the post-war Soviet Navy helped make it a legitimate superpower, as strong navies had contributed previously to creating the British and American "empires" in the eighteenth and nineteenth centuries. Using its navy, Russia projected its military might to far corners of the world, no longer constrained by the physical need to transport troops and equipment overland (Sollie 1988). Strategic bombers also provided a long-range capability that projected Russian military might through the North Atlantic into North America.

Much of the Russian naval presence in the West centered on the North Atlantic Arctic, allowing the country to pursue a unilateral security policy in the region. Russia backed its newly-created maritime power with ballistic missile submarines that roamed the North Atlantic depths. As the only ice-free ports in Russia's western Arctic, Murmansk and other submarine bases on the Kola Peninsula sent "boomers" (i.e., missile boats) and "hunters" (i.e., attack subs) through the GIUK (Greenland, Iceland, United Kingdom) Gap. Overhead, Soviet bombers transited the North Atlantic in the sky rather than below the surface. In response to the Russian strategic threat, for over 50 years NATO aircraft stationed at Keflavik Naval Air Station hunted Soviet subs and bombers in the North Atlantic during the Cold War (NATO n.d.-a).

There were also many soft-security threats to the Arctic related to the militarization of the region following World War II. Those threats included environmental pollution caused by both NATO and the Warsaw Pact, resulting from nuclear accidents and atomic waste disposal. One example was the U.S. Air Force's B-52 crash at Thule Air Base, Greenland in 1968, when a Stratofortress carrying four nuclear weapons crash-landed on the ice after a cabin fire caused the crew to bail out. The accident ruptured and dispersed the nuclear payload, leading to severe localized radioactive contamination. Although the U.S. and Denmark conducted an emergency cleanup, including disposing of contaminated snow on American soil, one weapon was lost and the area was radioactive for decades. Human damage occurred as well through health problems and loss of indigenous peoples' traditional resource-exploitation areas. The loss of Greenlandic Inuit's indigenous territory and village home also occurred earlier when the U.S. Air Force built Thule Air Base in 1953 (Brøsted and Fægteborg 1985; Ackrén 2019).

Russians had their own nuclear accidents that threatened the Arctic Ocean's environmental security, especially in the North Atlantic. Those accidents reached their apex in 2000 with the sinking of the *Kursk*, an Oscar-II class nuclear-powered missile boat, in the Barents Sea. With both atomic warheads and nuclear reactors onboard, the sub's sinking and subsequent explosions created an environmental catastrophe. While the largest incident, the *Kursk* was only one of eleven Russian submarine accidents that involved sinking and/or fires in the Arctic since 2000, including one in 2019 that killed fourteen crew members (RFE/RL 2019).

In addition, the Russian government admitted that from the 1960s through the 1990s it scuttled four nuclear submarines in the Arctic Ocean with reactors and weapons onboard. Russia also dumped radioactive contamination from other parts of the Soviet Union near Novaya Zemlya Island (Sullivan 1992). Poorly dismantled or improperly maintained reactors from rusting hulks in the Murmansk submarine bases also leaked radioactive contamination into the Barents Sea. Combined with fallout from Russian (and American) above-ground nuclear weapons tests in the 1950s and 1960s, the 1986 Chernobyl disaster nearly made the North Atlantic waters a nuclear dead zone (Amundsen et al. 2001).

Yet there were lasting benefits of Arctic militarization resulting from World War II and the Cold War. One of the largest positives was the building of infrastructure projects, especially international airports in a region without rail or road networks and few sea ports. British construction of the Reykjavík (now Domestic) Airport in 1941 allowed Iceland's first non-military international flight in 1945 (Jóhannsdóttir 2020). Building the Keflavik NATO airfield, which became Leif Erikson International Airport in the 1950s, changed Iceland by making it a leader in global travel during the 2000s. International airfield building improved other parts of the North Atlantic, including Søndre Strømfjord in Greenland, Gander in Newfoundland, Kirkenes in Norway (built by the Luftwaffe), and Iqaluit in Nunavut. The Russians also built Arctic airbases like Rogachevo (Novaya Zemlya Island) and Greem Bell (Graham Bell Island).

Addionally, during WWII the Allies completed transportation infrastructure projects throughout Iceland such as improving roads and bridges on what would become Route 1 (Þjóðvegur 1 or the "Ring Road"). The British and the Americans constructed ports and harbors in Iceland that are still in use today and built barracks later used as low-income civilian housing into the 1970s. Combined with demographic changes the new construction required, such as much of the population moving to the capital region for employment opportunities, Allied operations helped make Iceland the modern, independent country it is today (Office of Strategic Services 1942).

Following a relatively peaceful 20 years after the fall of the Soviet Union, during the past decade Russia and NATO have increased strategic operations in the Arctic to a point approaching Cold War levels. Starting in 2008, NATO and other partner-nation (e.g., Finland and Sweden) aircraft from Keflavik are once again hunting boomers and bombers through anti-submarine patrols and air-policing missions (NATO n.d.-b). The 2019 U.S. defense budget included $19 million for improving Keflavik hangers to house the P-8 Poseidon anti-submarine patrol aircraft (Hafstað 2020), which are much larger than the older P-3 Orion (Mcleary 2016).

On the other side of the Arctic Ocean, Russia is rebuilding and reoccupying some of its abandoned northern bases and creating new military units to staff them (Garamone 2019). These facilities and units include airbases and radar stations, anti-aircraft missile and rocket-artillery units, an Arctic brigade, and modern T-80BVM tanks under the Joint Strategic Command (Aliyev 2019).

With the recent increase in operational activities on both sides of the NATO-Russian divide, the Cold War is back, and the North Atlantic Arctic is "ground zero" for military exercises and potential challenges between the opposing forces. An example of increased military activity in the North Atlantic was Trident Junction 2018, the largest NATO exercise since 2015, that took place in Norway to simulate Arctic warfare conditions (NATO 2018). Twenty thousand American troops participated, including U.S. Marines training in Iceland to conduct military operations in the Arctic before heading to Norway for the main exercise (Friedl 2018). In July 2020, NATO conducted Operation Dynamic Mongoose with British Navy and Icelandic Coast Guard ships patrolling the North Atlantic (Icelandic Coast Guard Facebook 2020).

Russia is also conducting exercises and taking hostile actions against NATO in the North Atlantic, including their first ever amphibious assault exercise in the Arctic during 2018 (Ahlenss 2019). The following year, ten Russian submarines left their Arctic bases to intimidate NATO shipping in the largest such action since the end of the Cold War (Grove and Marson 2020). In April 2020, the Russians conducted a high-altitude paratrooper drop on Franz Josef Land followed by a three-day ground combat exercise (Brewster 2020). After similar penetrations in 2016 and 2019, Russian bombers entered NATO airspace near Iceland without a flight plan or active radar on July 3, 2020 and were escorted back to international airspace by Italian F-35 fighters from Keflavik (Askham 2020). During August 2020, the Russian Navy "attacked" American fishing vessels in the Bering Sea within the Alaskan Exclusive Economic Zone by broadcasting warnings that Russian submarines were firing missiles (Baker 2020).

1.3 Iceland's Internal Security Posture

Iceland is not a fully contributing member of NATO, which leads to two security problems in the circumpolar region. First, Iceland pays its required contribution to NATO. The renegotiated cost share of NATO members for the calendar year 2020 had Iceland paying 0.0597% to the alliance (NATO 2020). Yet because it has no military, only one other country, Montenegro, pays a lower NATO cost share than Iceland. That discrepancy requires larger countries to pay much more than their share, leading to insufficient funding in other areas (NATO 2019b). Second, Iceland's lack of a viable sub-strategic security apparatus following the 2006 U.-S. withdrawal, led the Icelandic Government to proclaim that a security vacuum existed in the North Atlantic (Ingimundarson 2011). Therefore, by creating and funding a sub-strategic security force, Iceland could contribute to increased North Atlantic geopolitical stability by paying its share of the NATO commitment and by strengthening the regional defence posture.

Iceland's Coast Guard and Police are well trained and dedicated to addressing sub-strategic threats in the country, to a degree. Yet an emergency involving multiple events or overwhelming force could exceed the ability of those agencies to defeat threats. For example, the homegrown terrorism by Anders Behring Breivik on July 22, 2011 provided a case study of a "two-front" attack that overwhelmed the Norwegian police. After Breivik detonated an explosion in Oslo that caused police to concentrate their efforts on the capital, the killer moved to an island where he systematically gunned down young people while police on the shore waited for a helicopter. After nearly 2 h, a police anti-terrorism unit arrived to arrest Breivik but not until the two attacks killed almost eighty people and wounded three hundred (Hemmingby and Bjørgo 2018).

Without the Norwegian anti-terrorism security force, Breivik's death toll could have been much higher. One could anticipate a similar tragedy occurring in Iceland if there were no security force prepared, equipped, and able to respond to multiple emergencies. In fact, many states have altered their militaries' capabilities to perform police functions for that very reason (Mittleman 2010), although that presents its own problems as discussed below. Since Iceland has no military, it should create a security force to perform those types of multiple event missions.

An actual example of how multiple events could overwhelm the Icelandic police occurred on June 26, 2018. The first event was the Secret Solstice music festival in Reykjavík that drew 15,000 young people (Kyzer 2019). A few miles away from the festival, thousands (including me) watched the Icelandic national soccer team play Croatia in the World Cup after the city set up large screens in public parks and stadiums throughout the Capital Region (Culpepper and Allen 2018; Fig. 1.1). On the same day, approximately 8000 tourists were transiting through the Leif Erikson International Airport only 50 kilometers from Reykjavík (Icelandic Tourist Board 2020). Also, on June 26, 2018, 180 kilometers north of Reykjavík, the cruise ship Costa Mediterranea with 2600 passengers aboard docked at Grundarfjörður (CruiseTT 2018).

Thus, on a single day, within a 200-kilometer radius from the center of Reykjavík, approximately 35,000 people were visiting or attending events that required a police presence, which is equivalent to 10% of Iceland's population. That figure does not include the approximately 7000 tourists travelling the highways or walking around the capital area that day but not attending the events (Icelandic Tourist Board 2020).

Icelandic Police from Reykjavík and the surrounding areas provided security; however, there were only 730 active and reserve police officers in the whole country. The number on duty within that 200-kilometer radius must have been less than half of the total (Overseas Advisory Council 2019). A large-scale emergency like a terrorist attack occurring at the same time could have overwhelmed the police. To help defeat a sub-strategic security threat, and to provide other capabilities above the police (e.g., anti-air defence), I make a recommendation for Iceland's security structure in the year 2030. By that year, experts predict that the ice-free Arctic Ocean will allow anyone with a sea-worthy boat or flyable aircraft to land in Iceland or operate within its territorial waters to challenge the country's sovereignty and perhaps harm its citizens.

Fig. 1.1 Icelanders and tourists watching the Iceland – Croatia World Cup Soccer Match in Ingolfstorg Square, Reykjavik, June 26, 2018. (Photograph by Gretar Ivarsson, used by permission)

1.4 Recommendation for Icelandic Security

This book represents a departure from many others on Arctic security because it actually makes a policy recommendation, which most scholarly literature on the topic does not. For example, perhaps the first comprehensive book published on Arctic security, *Arctic security in an age of climate change*, edited by James Kraska (2011), presented fifteen essays on that topic without a single policy recommendation despite being sponsored by the U.S. Naval War College. The most recent compendium on Arctic security, the 2019 Arctic Yearbook (Heininen et al. 2019) contained over thirty scholarly articles, briefing notes, and commentaries, yet it presented exactly three policy recommendations. The first by Exner-Pirot (2019: 73–74) called for "neutralization" or the elimination of all warfare in the Arctic. Exner-Pirot made a second recommendation, to reestablish the chiefs of staff meetings for Arctic nations. The third recommendation by Depledge et al. (2019: 85) proposed to create an Arctic military code of conduct of "'acceptable' and 'legitimate' military practice in the Arctic". With the exception of the chiefs of staff meetings, which actually happened twice in the past (Strader 2012), the other recommendations have virtually no chance of being accepted.

My hope is that the security force recommended in this book is never needed by Iceland. Unfortunately, it is hard to predict the future of the Arctic. My primary goal is to spur discussions about how to create a security structure that is both effective and acceptable so that the country can be protected should the need ever arise.

Indeed, critics will note that, as it has for over a decade, the Global Peace Index ranked Iceland the safest country in the world (Institute for Economics and Peace 2020a). That could mean, of course, that Iceland does not need a security force. Yet in 2010 the GPI ranked Norway the fifth safest country in the world, only 0.110 points below that year's number two country, Iceland (Institute for Economics and Peace 2020b). Six months into that year, Breivik executed the Norwegian home-grown terrorism described above, which was the deadliest terrorist attack ever in the Nordic countries.

My recommendation presents a concrete recommendation for Icelandic territorial integrity and internal safety that could serve as a starting point for discussions about future security. The recommendation is based upon the country's current security structure where the Icelandic Justice Ministry controls the intelligence, police, and other security assets like the coast guard. Although a Security and Defence Directorate exists within the Ministry of Foreign Affairs, its responsibilities lie largely sponsoring NATO military operations at the strategic level, especially the air policing mission, coordinating exercises like Northern Viking, and supporting other international missions such as training air traffic controllers in Afghanistan. Given Iceland's cultural history and current security situation, the recommendation is for the Icelandic Parliament to create a national security force using the country's existing resources. That security force recommended here consists primarily of its world-class search-and-rescue teams, supplemented by trained reserve personnel.

To place that recommendation in context, the book is organized in the following way. In Chap. 2, I review scholarship on globalism, combining it with current events that test globalism's assumptions in relation to Iceland's sub-strategic security. In Chap. 3, I examine policy statements by the Arctic Council Members to understand how their security forces are poised to handle sub-strategic threats in the circumpolar region. In Chap. 4, I recommend an Icelandic security force's organization and capabilities, including rank structures, troop levels, equipment, and unit locations. In Chap. 5, after hypothesizing the North Atlantic geopolitical situation in 2030, I present four scenarios to illustrate successful deployment of that security force against various threats. In Chap. 6, I summarize Iceland's conflict history and outline the country's current attitudes toward security, using that information to predict whether the public will accept the recommendation presented here.

References

Ackrén, Marie. 2019. From Bilateral to Trilateral Agreement: The Case of Thule Air Base. In *Redefining Arctic Security, Arctic Yearbook 2019*, ed. Lassi Heininen, Heather Exner-Pirot, and Joël Plouffee, 33–42. Akureyri: Arctic Portal. Accessed July 11, 2020, https://issuu.com/arcticportal/docs/ay2019.

Ahlenss, Ellen A. 2019. Incompatible Futures: Frontier Nostalgia and Southern Discourses of the Arctic. In *Redefining Arctic Security, Arctic Yearbook 2019*, ed. Lassi Heininen, Heather Exner-Pirot, and Joël Plouffee, 231–249. Akureyri: Arctic Portal. Accessed July 11, 2020, https://issuu.com/arcticportal/docs/ay2019.

Aliyev, Nurlan. 2019. Russia's Military Capabilities in the Arctic. *Diplomaatia*, No. 190/191, June 28 2019, pp. 1–22. Tallinn: International Centre for Defence and Security. Accessed October 28, 2020, https://icds.ee/en/russias-military-capabilities-in-the-arctic/#:~:text=Based%20on%20the%20Northern%20Fleet,operations%20from%20sea%20and%20air.

Amundsen, Ingar, Bjørn Lind, Ole Reistad, Knut Gussgaard, Mikhail Losipe, and Morten Sickel. 2001. *The Kursk Accident*. Norwegian Radiation Protection Authority 2001: 5. Østerås: Statens strålevern. Accessed July 2, 2020, https://www.dsa.no/publication/straalevernrapport-2001-5-the-kursk-accident.pdf.

Arctic Council. 2016. *Declaration on the Establishment of the Arctic Council (The Ottawa Declaration)*, p. 2, first footnote. Accessed October 29, 2020, https://oaarchive.arctic-council.org/bitstream/handle/11374/94/EDOCS-1200-v4-All_Arctic_Council_Declarations_1996-2017_Searchable.PDF?sequence=7&isAllowed=y.

Arctic Monitor. 2016. Arctic Territorial Claims. September 15, 2016. Accessed July 17, 2020, https://thearcticmonitor.wordpress.com/tag/territorial-claim/.

Askham, Poppy. 2020. Russian Long-Range Bombers Spotted Within NATO Airspace Near Iceland. *The Reykjavík Grapevine*, July 6, 2020. Accessed July 9, 2020, https://grapevine.is/news/2020/07/06/russian-long-range-bombers-spotted-within-nato-airspace-near-iceland/.

Baker, Mike. 2020. 'Are We Getting Invaded?' U.S. Boats Faced Russian Aggression Near Alaska. *The New York Times*, November 12, 2020. Accessed November 13, 2020, https://www.nytimes.com/2020/11/12/us/russia-military-alaska-arctic-fishing.html?searchResultPosition=8.

Brewster, Murray. 2020. Russian Arctic military draws awe and concern. *Eye on the Arctic*, Monday, May 11, 2020. Accessed July 17, 2020, https://www.rcinet.ca/eye-on-the-arctic/2020/05/11/russian-arctic-military-exercise-draws-awe-and-concern/.

Brøsted, Jens, and Mads Fægteborg. 1985. Expulsion of the Great People. When U.S. Air Force Came to Thule. An Analysis of Colonial Myth and Actual Incidents. In *Native Power. The Quest for Autonomy and Nationhood of Indigenous Peoples*, ed. Jens Brøsted, Jens Dahl, Andrew Gray, Hans Christian Gulløv, Georg Henriksen, Jørgen Brøchner Jørgensen, and Inge Kleivan, 213–238. Bergen: Universitetsforlaget AS.

Byers, Michael. 2009. *Who Owns the Arctic? Understanding Sovereignty Disputes in the North*. Vancouver: Douglas and McIntrye.

CruiseTT (Time Table). 2018. *Port of Akureyri Cruise Ship Timetable for June 2018*. M.CruiseTT.com. Accessed July 17, 2020, http://m.cruisett.com/s/Costa_Cruise_Lines/Costa_Mediterranea/June_2018/.

Culpepper, Chuck, and Scott Allen. 2018. Croatia Wins Again, Bringing Iceland's World Cup Appearance to an End. *Washington Post*, June 26, 2018. Accessed July 11, 2020, https://www.washingtonpost.com/news/soccer-insider/wp/2018/06/26/iceland-vs-croatia-2018-world-cup/.

Depledge, Duncan, Mathieu Boulégue, Andrew Foxall, and Dmitriy Tulupov. 2019. Why We Need to Talk About Military Activity in the Arctic: Towards an Arctic Military Code of Conduct. In *Redefining Arctic Security, Arctic Yearbook 2019*, ed. Lassi Heininen, Heather Exner-Pirot, and Justin Barnes, 83–86. Akureyri: Arctic Portal. Accessed July 11, 2020, https://issuu.com/arcticportal/docs/ay2019.

Exner-Pirot, Heather. 2019. Between Militarization and Disarmament: Constructing Peace in the Arctic. In *Redefining Arctic Security, Arctic Yearbook 2019*, ed. Lassi Heininen, Heather Exner-Pirot, and Justin Barnes, 72–74. Akureyri: Arctic Portal. Accessed July 11, 2020, https://issuu.com/arcticportal/docs/ay2019.

Fatić, Aleksandar. 2002. Conventional and Unconventional – 'Hard' and 'Soft' Security: The Distinction. *Seer: Journal for Labour and Social Affairs in Eastern Europe* 5(3): 93–98. October 2002. Accessed July 20, 2020, https://www.jstor.org/stable/43292068.

Fitz-Gerald, Ann M. 2003. Multinational Land Force Interoperability: Meeting the Challenge of Different Cultural Backgrounds in Chapter VI Peace Support Operations. *The Journal of Conflict Studies* 23(1): 60–85. Winter 2003. Accessed July 30, 2020, https://journals.lib.unb.ca/index.php/jcs/article/view/351/548.

Friedl, Megan. 2018. *U.S. Joins NATO's Trident Junction Exercise*. U.S. Department of Defense, October 18, 2018. Accessed July 18, 2020, https://www.defense.gov/Explore/News/Article/Article/1666272/us-joins-natos-trident-juncture-exercise/.

Garamone, Jim. 2019. Changing Environment Means Changing Arctic Strategy. *U.S. Department of Defense News*, June 14, 2019. Accessed July 17, 2020, https://www.defense.gov/Explore/News/Article/Article/1865978/changing-environment-means-changing-arctic-strategy/.

Gjørv, Gunhild Hoogensen, and Karen K. Hodgson. 2019. "Arctic Exceptionalism" or "Comprehensive Security"? Understanding Security in the Arctic. In *Redefining Arctic Security, Arctic Yearbook 2019*, ed. Lassi Heininen, Heather Enxer-Pirot, and Justin Barnes, 218–230. Akureyri, Iceland: Arctic Portal. Accessed July 11, 2020, https://issuu.com/arcticportal/docs/ay2019.

Grove, Thomas, and James Marson. 2020. Russian Submarines Test NATO in Icy North Atlantic. *Wall Street Journal*, July 2, 2020. Accessed July 17, 2020, https://www.wsj.com/articles/russian-submarines-test-nato-in-icy-north-atlantic-11593682201.

GuÐmundsson, GuÐmundur J. 2006. The Cod and the Cold War. *Scandinavian Journal of History* 31(2): 97–118. 22 August 2006. Accessed July 2, 2020, https://www.tandfonline.com/doi/abs/10.1080/03468750600604184.

Hafstað, Vala. 2020. *US Navy to Begin Construction at Kef Airport*. Iceland Monitor, Politics and Society. Friday, April 17, 2020. Accessed October 22, 2020, https://icelandmonitor.mbl.is/news/politics_and_society/2020/04/17/us_navy_to_begin_construction_at_kef_airport/.

Heininen, Lassi, Heather Exner-Pirot, and Justin Barnes, eds. 2019. *Redefining Arctic Security, Arctic Yearbook 2019*. Akureyri: Arctic Portal. Accessed July 11, 2020, https://issuu.com/arcticportal/docs/ay2019.

Hemmingby, Cato, and Tore Bjørgo. 2018. Terrorist Target Selection: The Case of Anders Behring Breivik. *Perspectives on Terrorism* 12(6): 164–176. Terrorism Research Initiative and the Institute of Security and Global Affairs, University of Leiden. Accessed July 13, 2020, https://www.universiteitleiden.nl/binaries/content/assets/customsites/perspectives-on-terrorism/2018/issue-6/a11-hemmingby-bjorgo.pdf.

Huebert, Rob. 2011. Canada and the Newly Emerging International Arctic Security Regime. In *Arctic Security in an Age of Climatic Change*, ed. James Kraska, 193–217. New York: Cambridge University Press. Accessed July 21, 2020, https://doi.org/10.1017/CBO9780511994784.

Humpert, Malte, and Andreas Raspotnik. 2012. The Future of Shipping Along the Transpolar Sea Route. In *Arctic Yearbook 2012 (1)*, ed. Lassi Heininen, Heather Exner-Pirot, and Joël Plouffee, 281–307. Akureyri: Arctic Portal. Accessed July 11, 2020, https://web.archive.org/web/20160121203700/http://www.arcticyearbook.com/images/Articles_2012/Humpert_and_Raspotnik.pdf.

Icelandic Coast Guard. 2020. *Icelandic Coast Guard Facebook*. Accessed July 9, 2020, https://www.facebook.com/permalink.php?story_fbid=3490581087630033&id=377056962315810.

Icelandic Tourist Board. 2020. *Tourism in Iceland in Figures*. Reykjavík. Accessed July 17, 2020, https://www.ferdamalastofa.is/en/recearch-and-statistics/tourism-in-iceland-in-figures.

Ingimundarson, Valur. 2011. Territorial Discourses and Identity Politics. Iceland's Role in the Arctic. In *Arctic Security in an Age of Climatic Change*, ed. James Kraska, 174–189. New York: Cambridge University Press. Accessed July 21, 2020, https://doi.org/10.1017/CBO9780511994784.

Institute for Economics and Peace. 2020a. *Global Peace Index 2020. Measuring Peace in a Complex World*. Sydney: Institute for Economics and Peace. Accessed July 27, 2020, https://www.visionofhumanity.org/wp-content/uploads/2020/10/GPI_2020_web.pdf.

———. 2020b. *Global Peace Index 2010. 2010 Methodology, Results & Findings*. Sydney: Institute for Economics and Peace. Accessed July 27, 2020, https://www.peacewomen.org/sites/default/files/wps.iep_gpi2010_methodologyresultsfindings.2010_0.pdf.

Jóhannsdóttir, Hildur Margrét. 2020. 75 Years Since the First Flight From Iceland [in Icelandic]. *RUV*, July 11, 2020. Accessed July 16, 2020, https://www.icelandreview.com/travel/seventy-five-years-since-first-international-flight-from-iceland/.

Keupp, Marcus Matthias. 2015. Arctic Security, Sovereignty, and Rights of Utilization: Implications for the Northern Sea Route. In *The Northern Sea Route*, ed. Marcus Matthias Keupp, 21–38. Wiesbaden: Springer Gabler. Accessed July 13, 2020, https //doi.org/10.1007/978-3-658-04081-9_2.

Kraska, James. ed. 2011. *Arctic Security in an Age of Climatic Change*. New York: Cambridge University Press. Accessed July 21, 2020, https://doi.org/10.1017/CBO9780511994784.

Kyzer, Larissa. 2019. Secret Solstice May Relocate Amidst Nonpayment Complaints. *Iceland Review*, April 12, 2019. Accessed July 11, 2020, https://www icelandreview.com/news/secret-solstice-may-relocate-amidst-nonpayment-complaints/.

Mcleary, Paul. 2016. Pentagon Looking at Iceland Hub for Its Spy Planes. The Cable. *Foreign Policy*. February 11, 2016. Accessed July 11, 2020, https://foreignpolicy.com/2016/02/11/pentagon-looking-at-iceland-hub-for-its-spy-planes/.

Mittleman, James H. 2010. *Hyperconflict: Globalism and Insecurity*. Palo Alto: Stanford University Press.

NATO. 2018. Trident Juncture 2018. *Media Resources*, October 25, 2018. Accessed July 18, 2020, https://www.nato.int/cps/en/natohq/news_158620.htm.

———. 2019a. *Iceland's Role in Transatlantic Security Growing*. 14 May 2019. Brussels. Accessed July 1, 2020, https://www.nato-pa.int/news/icelands-role-transatlantic-security-growing.

———. 2019b. *Financial Statements (Abridged version) for the year ended 31 December 2018*. NATO Parliamentary Assembly, June 2019. Brussels. Accessed July 1, 2020, https://www.nato-pa.int/download-file?filename=sites/default/files/2019-06/116%20FIN%2019%20E%20fin%20-%20Audited%20Financial%20Statements%202018.pdf.

———. 2020. *Funding NATO*. 5 May 2020. Brussels. Accessed July 1, 2020, https://www.nato.int/cps/en/natohq/topics_67655.htm.

———. n.d.-a. *Origins*. Brussels. Accessed July 2, 2020, https://www.nato.int/cps/en/SID-21A87C2D-BE06D3F1/natolive/declassified_162083.htm?selectedLocale=en.

———. n.d.-b. *Iceland's Peacetime Preparedness Needs*. Brussels. Accessed July 18, 2020, https://shape.nato.int/icelands-peacetime-preparedness-needs.

Office of Strategic Services (OSS). 1942. *Iceland. Part 1*. Accessed July 9, 2020, https://www.youtube.com/watch?v=TiyML1_0CBU.

Overseas Advisory Council. 2019. *Iceland 2019 Crime & Safety Report. Overseas Advisory Council (OSAC), March, 8, 2019*. Country Security Report. Diplomatic Security Service. Accessed November 13, 2020, https://www.osac.gov/Country/Iceland/Content/Detail/Report/2d8cdd85-eb54-4ec7-828a-15f4aeb18469.

Pic, Pauline, and Frédéric Lasserre. 2019. What is 'Arctic' about 'Arctic security'? In *Redefining Arctic Security, Arctic Yearbook 2019*, ed. Lassi Heininen, Heather Exner-Pirot, and Justin Barnes, 1–17. Akureyri: Arctic Portal. Accessed July 11, 2020, https://issuu.com/arcticportal/docs/ay2019.

RFE/RL (Radio Free Europe/Radio Liberty). 2019. *Major Russian Submarine Accidents Since 2000, July 2, 2019*. Washington, DC: U.S. Agency for Global Media. Accessed July 18, 2020, https://www.rferl.org/a/major-russian-submarine-accidents-since-2000/30033592.html.

Shea, Neil. 2019. Scenes from the new Cold War unfolding at the top of the world. *National Geographic, Environment*, May 8, 2019. Accessed July 2, 2020, https://www.nationalgeographic.com/environment/2018/10/new-cold-war-brews-as-arctic-ice-melts/#close.

Sollie, Finn. 1988. The Soviet Union in Northern Waters – Implications for Resources and Security. In *The Soviet Union and Northern Waters*, ed. Clive Archer, 13–44. London: Routledge.

Søreide, Ine Eriksen. 2016. *NATO and the North Atlantic: Revitalizing Collective Defense and the Maritime Domain. Prism, Volume 6, No. 2, July 12, 2016, Features, NATO in Context: Geopolitics and the Problem of Russian Power*. Washington, DC: Center for Complex Operations. Accessed July 1, 2020, https://cco.ndu.edu/PRISM/PRISM-Volume-6-no-2/Article/835074/nato-and-the-north-atlantic-revitalizing-collective-defense-and-the-maritime-do/.

Stephen, Kathrin. 2016. *An Arctic Security Forum? Please, no! Commentary, Defense and Security Law and Governance, May 26, 2016*. Washington, DC: The Arctic Institute. Accessed October 29, 2020, https://www.thearcticinstitute.org/arctic-security-forum-please-dont/?cn-reloaded=1.

Stephenson, Scott R. 2013. New Trans-Arctic shipping routes navigable by midcentury. *Proceedings of the National Academy of Sciences of the United States of America (26 March 2013)* 110 (13): 1191–1195. Accessed July 13, 2020, https://www.ncbi.nlm.nih.gov/pmc/articles/PMC3612651/.

Strader, Olin. 2012. *Arctic Chiefs of Defence Staff Conference: An Opportunity to Formalize Arctic Security*. Washington, DC: The Arctic Institute, Center for Circumpolar Security Studies.

Sullivan, Walter. 1992. Soviet Nuclear Dumps Disclosed. *The New York Times*, November 24, 1992, p. 9. Accessed July 2, 2020, https://www.nytimes.com/1992/11/24/science/soviet-nuclear-dumps-disclosed.html.

Wezeman, Siemon T. 2016. *Military capabilities in the Arctic: A New Cold War in the High North?* Stockholm International Peace Research Institute (SIPRI) Background Paper, October 2016. Accessed July 2, 2020, https://www.sipri.org/sites/default/files/Military-capabilities-in-the-Arctic.pdf.

Young, Oran R. 2011. Forward. Arctic Futures: The Politics of Transformation. In *Arctic Security in an Age of Climatic Change*, ed. James Kraska, xxi–xxvii. New York: Cambridge University Press. Accessed July 21, 2020, https://doi.org/10.1017/CBO9780511994784.

Chapter 2
Globalism Failed – Iceland Alone in an Insecure Arctic

Bad dreamer, what's your name?
Looks like we're riding on the same train
Looks as though there'll be more pain
There's gonna' be a showdown
And it's raining, all over the world
Jeff Lynne (Electric Light Orchestra), (1973), Showdown

Abstract Following a brief summary of how forms of globalism provided strategic-level protection for Iceland since the mid-thirteenth century, this chapter outlines the academic perspective of globalism in its three basic forms. The discussion continues with why the preceding theoretical model of the interaction between states (i.e., international relations) was a methodologically flawed and incomplete construct because it failed to consider non-state actors. In addition, I provide the rationale and some examples illustrating why scholars using various forms of globalism failed in their predictions that the state will cease being responsible for its own national security. Specific examples of states acting on their own argue against the premise of globalism, such as Russia engaging in aggressive behavior in the Arctic and China conducting extensive intellectual property theft. The recent Covid-19 pandemic illustrates how globalism has failed, resulting in states creating their own responses such as closing borders and practicing "vaccine nationalism". Finally, I discuss whether globalism and its premise of the decreased need for state-level national security applies to Iceland, or whether the country should provide its own sub-strategic security based upon its recent successful emergency responses.

2.1 Introduction to Globalism and Iceland

One of the most used terms today in discussions about relations between international actors is globalism, which refers to the world's interconnectedness through economic, cultural, ideological, and communication behaviors. To some, globalism arises from the work of Neo-Marxists like Immanuel Wallerstein, whose world-

R. P. Wheelersburg, *The Security of Iceland and the Arctic 2030*, Springer Polar Sciences, https://doi.org/10.1007/978-3-030-89948-6_2

systems perspective emphasizes levels of analysis above the nation-state to understand how capitalism (evolving out of feudalism) became a world economy that shaped people's lives. As extended by scholars like Eric Wolf (1982), that perspective involves the effects of the European-transmitted (through force) transnational capitalist system on local cultural production and trade. A global society arose influenced by laws, treaties, customs, and governments during the development of peripheral areas by central governments. Production and trade involved alterations in traditional economic behaviors like land and resource use, settlement patterns, social structures, and cultural characteristics such as religion and language, as influenced by such national activities as taxation (Wallerstein 2004).

As larger entities evolved (e.g., kingdoms, empires, alliances), they incorporated smaller units into their security spheres. Iceland allowed its strategic security to be provided by other countries for three quarters of a millennium. Following decades of civil war, the Icelandic chieftains submitted to the control of the Norwegian King Haakon IV and his descendants from 1262 until 1380, when the Kalmar Union arose and Iceland came under Danish control. Denmark provided strategic security for Iceland until 1944, although arguably it lost the ability to influence events in Iceland after the Germans occupied Denmark in May 1940. That occupation was followed the next month by the British "invasion" of Iceland. One year later, 50,000 American troops occupied the country, helping Iceland declare its independence from Denmark in 1944. The American military left in 1947, though it returned in 1951 when Iceland became a founding member of the newly created NATO. The Americans withdrew from Iceland in 2006 by abandoning the Keflavik Naval Air Station (by that time an Air Force base). Without a military of its own, Iceland needs NATO forces to provide strategic-level security to this day.

Despite falling under the protection of other countries for 750 years, the strategic security umbrella did not prevent Iceland from suffering sub-strategic harm such as religious assassinations, murder and kidnapping by foreign pirates, plagues, natural disasters, famine, accused witch murders, forced dissolution of parliament, invasions by foreign powers, violent disputes over resources, insurrection by communists, and economic collapse. Thus, globalism and its earlier forms (e.g., colonialism) provided the strategic-level protection that Iceland needed to survive as a state, but it did not prevent sub-strategic threats from doing great harm to the country. That failure suggests that globalism might not help Iceland protect its sovereignty and its citizens in the future, if as predicted sub-strategic security threats in the Arctic increase with the retreating ice.

Today Iceland is almost totally dependent upon globalism for its strategic security. Its 2016 national security policy states that Iceland's national defence depends upon global organization (UN), multi-national alliances (NATO), regional entities (Nordic Defence Cooperation), and bilateral treaties such as with the U.S. (Ministry for Foreign Affairs n.d.).

> The alliance with the North Atlantic Treaty Organization (NATO) and the Bilateral Defence Agreement of 1951 between Iceland and the United States remain fundamental pillars of Iceland's security and defence. Iceland is a country without a military and has emphasized a

comprehensive and multilateral approach in security affairs and as a member of key organizations, such as the United Nations, NATO and the OSCE (Organization for Security and Co-operation in Europe).

Iceland's future security needs will be addressed in the chapters below. This chapter's next section presents a comprehensive discussion about the inability of globalism as a theoretical construct to contribute to security policy formation and the failure of global forces to control world events as states remain the preeminent actors (but not the only actors) in security activities.

2.2 Globalism and the Nation-State

Following the end of the Cold War, globalists predicted the death of the state as regional and global organizations began to provide collective security, assuming the primary role of national governments for defence (Holsti 1996). The decline in the autonomy of the state appeared especially likely in the Arctic where cooperative institutions such as the Arctic Council, alliances especially NATO, and non-state actors like the European Union (EU) influenced national policies in the region (Galimullin and Matveenko 2019). Examples of cooperative policies involving Iceland included the Arctic Council's Search-and-Rescue Agreement and the EU's Dublin Regulation concerning asylum seekers (Ástvaldsson 2019). Additionally, the decline of traditional powers in the circumpolar region, along with the growth of Arctic autonomous areas like Greenland and Nunavut, reduced the security effectiveness of states by diffusing previous power relationships. Non-state actors including multi-national corporations also began to engage in activities once conducted by states such as resource exploration and extraction.

After the collapse of the Soviet Union in 1991, globalists (e.g., James 2006) predicted a reduced need for state-level national security when it appeared that countries' defense spending decreased, universal conscription in states like Sweden ended, and the number of conflicts between nations plummeted. Events that took place over the past decade, however, illustrated how globalists incorrectly predicted the end of the state and failed to anticipate the nationalistic responses to security threats. The conclusion derived from that failed prediction suggests that instead of depending upon the process of globalism to meet sub-strategic security threats, Iceland should use its existing capabilities and policy framework to create its own national security capability to defeat them.

In security studies, international relations (IR) preceded globalism as a theoretical framework to understand the relations between states. However, in my view, globalism represents a more comprehensive approach and therefore has replaced IR as the primary concept to understand the dealings between international actors. Beginning in the 1990s globalization began to dominate scholarship over international relations because it included interactions between entities other than the state, which was the primary unit in IR analyses (Buzan and Little 2001).

Also, IR represented a failed approach toward understanding how international actors cooperate to ensure collective national security for two additional reasons. Despite voluminous publication on the topic, attempts to reach a valid operational definition of the perspective's methodology have been unsuccessful, as has attaining consensus on what constitutes an effective research program to apply the theory to specific cases (Thies 2008). In addition, IR as a theoretical perspective is by its very nature speculative because there is no opportunity to test any of its suppositions due to the methodological shortcoming just described (Waltz 2003).

These three weaknesses meant that discussions about IR hypotheses devolved into meaningless and unsolvable operational problems that resolved nothing related to the practical application of the theoretical framework. While successful as an academic discipline to attract students, according to some scholars, "IR has generated almost no significant public debate outside the academia ... [with] ... 'big names' in IR theory ... virtually unknown outside the discipline" (Buzan and Little 2001: 21). I also do not address alliance theory, a subordinate concept of IR because it relates to larger, strategic-level security relationships between states like those within NATO (Snyder 1990). Instead, I place the focus on globalism and its failure to acknowledge that the state's ability to provide its own sub-strategic security has primacy over every other entity.

2.3 Types of Globalism and Their Predicted Outcomes

While there are variations of globalism, I consider three basic types as described by Ripsman and Paul (2010) among others, which I term economic/commercial, democratic peace/political, and universal/cultural. While the three types differ in how they apply to national security, all of them predict that global-level security policies, processes, and apparatuses eventually would overcome the need for states to maintain their own national security capabilities.

Economic globalists maintain that the flow of people, goods, and services freely across international borders contributes to the decline of the state's importance in providing security. Also termed market globalism, economic globalists believe there was a direct relationship between the advantages of states becoming trading partners and the disadvantages of being enemies (Steger and James 2010: x). Warfare decreases a state's ability to trade by degrading its infrastructure, destroying its production capabilities (e.g., factories), and draining available financial and human resources. Rather than attacking a potential foe over something static like territory, states prefer to engage in mutually beneficial economic partnerships, minimizing the need for national security. Thus, a state's incentive to eschew warfare results from its need to participate in a viable global economy concentrating its resources on exchange activities (Ripsman and Paul 2010). Europe shows the importance of transnational economic cooperation in reducing warfare since after a thousand years of destructive wars, the creation of the European Community (replaced by the EU) led to mostly peaceful relations for the last half century.

A second type is democratic peace globalism, also called “political” globalism or “justice globalism” (Steger and James 2010: ix). That position assumes that a rational state understands how the advantages of providing democratic reforms to its citizens leads to lower global condemnation for its repressive policies. The spread of democratic values like free speech and gender equality reduces conflict both within and between states, precluding the need for national armed forces to attain security (Ripsman and Paul 2010). An international entity that controls a state's border, such as the European Union's Schengen Area, provides international security protection to its member states through coherent, shared, and humane immigration policies and procedures.

The third type of globalism involves the creation and proliferation of a global popular culture, sometimes referred to as “McWorld” (Barber 1992). While the global cultural position also involves acceptance of democratic norms, through shared interests in music, film, literature, etc. a global culture arises making people more similar across national boundaries. For example, people in different countries who enjoy the same popular culture are more likely to emphasize their common interests, thereby reducing nationalism. The very processes of globalization, especially communication capabilities such as the Internet and social media platforms that are available to most world citizens, facilitate the creation of a global culture (Ripsman and Paul 2010). As with the other two types of globalism, the emergence of a global popular culture reduces the importance of the state in national security because people begin accepting a global form of government as a consequence.

2.4 Globalism – Theory Versus Reality

Globalism achieved its zenith in Arctic politics during the halcyon days after the Soviet Union's fall in 1991 and the emergence of a lone superpower. The following two decades of relative calm caused President George W. Bush and Prime Minister Margaret Thatcher to proclaim a “peace dividend” (Hoagl 1990). Politicians and scholars predicted that the peace dividend would lead to more productive use of resources than warfare and create bonds across national borders to provide regional or global solutions to security challenges. Not surprisingly, that period coincided with the growth of globalism as an academic discipline. To globalists the evidence was clear; countries' military expenditures shrank, international conflicts especially interstate warfare decreased, and the traditional functions of individual states diminished. Thus, the state's functions were reduced to the point that many considered them unnecessary (Mittleman 2000).

In the extreme globalist view, future allegiances of individual citizens lie above the state, directed instead toward regional entities like the European Union or world organizations such as the United Nations (Carpio 2019). Yet Ripsman and Paul (2010) found that during the height of globalism and the supposed death of the state, there was little evidence that nations reduced their traditional security activities such as funding strong militaries. The authors concluded that far from being dead, the state retained the primary role in providing its citizens with the security necessary to participate freely in a global economy, political system, and culture.

There was other evidence that assumptions underlying globalism were not fully supportable. A study by Sarkees et al. (2003) concluded that since the Napoleonic Era began in 1803, the risk of global conflict has remained relatively constant. In fact, the last two centuries saw an increase in wars of all types when using an expanded definition that included inter-state wars, extra-state wars involving both dependent and independent non-state actors (e.g., Iranian militias in Syria [Hezbolla] and ISIS, respectively), and intra-state (i.e., civil) wars. Although their study was done in 2003, the number of low-level conflicts increased during the ensuing decade. Around 2010, low-level conflicts and the number of battle deaths actually began to increase, peaking around 2014 (Strand et al. 2019: Figure 2). Thus, instead of a peace dividend following the breakup of the Soviet Union, in fact there has been a marked increase in wars of varying types through the last decades of the Cold War and into the first quarter of the twenty-first century. Given the evidence, apparently global-level security efforts have not reduced warfare, and states retain their unique national security needs.

A former champion of globalism James Mittleman (2000), more recently altered his earlier position about global influences on the state's national security functions. Mittleman abandoned the absolute aspects of globalism, due to the variety and nature of the threats that arose beyond armed conflict between states. Actions of non-state actors, or sometimes inequities within states, created a variety of threats to states' national security including environmental degradation, disease, international crime, and multi-national terrorism. These "post-national" threats emanated from both military and non-military sources and involved activities such as bombings, kidnapping, and cybercrime (Mittleman 2010: 159). Post-national threats like terrorist attacks that alliances cannot respond to quickly enough require a state-level apparatus combining policing and border security (Ripsman and Paul 2010; Wheelersburg 2021).

Events that took place in 2020 provided evidence that regional security structures were foundering as individual states undertook efforts to retain control over their own peoples, terrorists, and economies during the pandemic.

Starting with economic globalism, relations between China and its trading partners illustrated how economic connectedness across borders was ultimately subordinated to Chinese national interests. Despite having most-favored nation status that allowed many European and American economies to pump trillions into China's manufacturing, not to mention China owning another trillion of their debts, the Chinese engaged in intellectual property piracy including stealing commercial videos and music on an unprecedented scale. That theft of popular culture pales in comparison to Chinese industrial espionage, which according to some scholars, represented "the greatest transfer of wealth in history" (Blair and Alexander 2017).

To conduct its industrial espionage program, China created a vast spy network according to FBI director Christopher Wray (2020). Using a variety of techniques such as planting intelligence agents, corrupting trusted American scientists through its "Thousand Talents Plan" (Department of Justice 2020), and requiring exchange students to engage in passive collection, the Chinese do not care who they offend with their spying. As a consequence, America is taking steps to prevent and punish Chinese theft and espionage, such as imposing trade sanctions and passing new legislation to protect taxpayer-funded research. In July 2020, the U.S. closed the Chinese consulate in Houston after criminal charges were filed against hackers trying to steal pharmaceutical companies' Covid-19 vaccine research (BBC 2020). Thus, instead of achieving more peaceful relations through global economic interconnectedness, it is those economic connections that are creating the potential for conflict between China and America.

The democratic values position assumes states will subscribe to the spread of democratic norms across the world for their own benefit by upholding human rights. Yet during the past decade authoritarian regimes continued repressing their citizens. These same regimes also engaged in military adventurism. In Russia, Vladimir Putin solidified his position as a dictator for life by arresting and assassinating political opponents and pushing legislation to remain in power using undemocratic means (Kantchev 2020). In addition, Russia's dictator did not care about being condemned by the world community for his illegal annexation of Crimea (Pifer 2020). Islamic repressive states such as Iran and Saudi Arabia engage in killings and mass arrests of opponents and journalists to quell dissent (Human Rights Watch 2020), not to mention the former committing atrocities in the Syrian War and the latter doing the same in Yemen (Decker 2017; Human Rights Watch 2019).

North Korea, perhaps the most repressive state on the planet, continues to torture its own people by suppressing their human rights and allowing them to starve to death by the millions. Undeterred by global sanctions, the country's rogue despot maintains an aggressive stance by conducting nuclear weapons testing and firing missiles into international waters (Reuters 2017). With the world distracted by the Covid-19 Pandemic, China implemented a new security policy against Hong Kong. By eliminating human rights through squashing protests and arresting anti-government politicians, China is receiving criticism and risking further trade sanctions (Bradsher 2020). China's seizure of atolls in the South Pacific has created regional conflict, risking a new Cold War with the U.S. Thus, whether the ideology is Slavic dictatorship, Muslim theocracy, or Asian communism, authoritarian states reject human rights and engage in repression and hegemony without concern for global pushback.

The third globalism type involved the spread of a universal popular culture around the world, overwhelming national identities and local traditions, making people in separate states more alike than different. Of all the global hypotheses, the McWorld type is perhaps the most naïve, assuming that because someone accepts a piece of technology or particular genre of music, they will accept democratic norms. It is a bit harder to illustrate the failure of the popular culture position because, while the spread of music and film is obvious, whether global culture carries with it some democratic norms is more tenuous to determine. To see the specious nature of the popular culture globalist position, one can only imagine the North Korean people being given human rights because Kim Jong-un is a fan of K-Pop and likes to dance Gangnam Style.

Incredible as it may seem, I wrote the previous sentence in January 2021. Six months later, Kim Jong Un banned K-Pop music, South Korean movies, fashion, hair styles, and slang because those cultural behaviors foster counter-revolutionary beliefs and threaten the North Korean way of life (Yeung and Seo 2021). Thus, instead of cultural globalism leading to improved human rights under dictators, in North Korea the opposite is true. Rather than allowing his people some amusement to detract from the mass starvation and suffering under his repressive regime, Jong Un took that amusement away to keep his people on the right path toward oblivion. What an appropriate example to illustrate the failure of the universal popular culture type of globalism!

There are other examples of the failure of the global culture position. Iceland demonstrates how its people accept universal popular culture but reject the values that accompany it. Based upon my 30 years of ethnographic research in the country, it is obvious that Iceland is dominated by American culture, which has become global culture to a large extent. Yet Icelanders reject values such as militarism and racism that they perceive to be inherent in American society. In terms of racism, despite consuming American Hip Hop or Rap music, many Icelanders reject completely American values they say are hypocritical; i.e., spouting equality publicly but discriminating privately. The Black Lives Matter (BLM) protests in the U.S. being "attacked" by police confirmed to Icelanders just how corrupt America culture really is. A supportive BLM demonstration took place in Reykjavík on June 3, 2020 (Ćirić 2020) and a long-time Icelandic performing artist friend of mine told me on a visit to Reykjavik in June 2021 that she "felt sorry for me because I lived in such a racist country" (her identity withheld).

Iceland refuses to accept other American values despite welcoming its popular culture. The country's English newspaper, *The Reykjavík Grapevine*, derided the American government for treating the pandemic response as if it was a war (Grettisson 2020). In the latest American values debacle in Iceland, the Trump-appointed Ambassador, Jeffrey Ross Gunter, asked to carry a handgun and upgrade his security detail with armed guards after retweeting Trump referring to Covid-19 as the "China Virus". Icelanders universally condemned such nationalistic rhetoric publicly, and Gunter's subsequent request for a concealed weapon permit further insulted Icelanders who despise handguns and violent crime they produce (Ingvarsson 2020).

2.5 Globalism Dies Due to a Virus

Covid-19 presented a paradox for even the strongest globalist. On the one hand, the pandemic illustrated that global communication and transportation are real. Spread through rapid travel across international borders by infected people and goods, the virus demonstrated that within days, a disease originating in a local area moved around the world unchecked. On the other hand, the response to Covid-19 became nationalistic almost immediately with many governments shutting their borders and providing emergency transportation to get their citizens home from abroad. A new phrase "vaccine nationalism" (Karp 2020) was coined to describe states' monopolization of Covid-19 vaccine that reached the extreme in countries like Italy, where pharmaceutical executives were arrested for planning to ship vaccine out of the country.

The entire world suffers today during Covid-19 because a global response has not materialized, despite world leaders such as Bill Gates stating that the disease will not be conquered without one (Cockburn 2020). Instead, the most powerful leader in the world rejected a global response thusly during the height of the pandemic

> Globalists. You know what a globalist is? They want the globe to do well, but they don't care about us. Now we want everybody to do well. But we have to take care of America first. . . . And you know what? Other countries say their country first. Why wouldn't they do that? But we didn't do that. We had a bunch of globalists; they didn't know what the hell they were doing (Trump 2020).

With the exception of Sweden, the Nordic countries were among the first to close borders, led by Denmark but followed quickly by Iceland. Internal actions such as lockdowns, testing, and social distancing (including wearing masks) followed, leaving individual countries to respond essentially alone. While the country's response to the pandemic is ongoing, Iceland has been relatively successful in preventing deaths with only 30 recorded as of late July, 2021 (Reuters 2021; Kolbert 2020). Conversely, both international governmental and non-governmental organizations such as the World Health Organization, the European Union, and NATO have been impotent in the face of a major security challenge presented by the pandemic.

Of particular note, the United States, which provided leadership to its European allies throughout the twentieth century, has not only lost that leadership position through inaction (Mead 2020), but the bumbling federal response and Trump's failure to acknowledge the pandemic made America a global laughingstock. In fact, the U.S. response moved from the national to state level based upon the Federalist principles. Trump went a step further by concentrating the country's support to states depending upon whether the governor was Democrat or Republican (Steib 2020). While the vaccine roll-out initially precipitated a positive response by Americans, the current resistance to getting the vaccine or even observing protocols like masking is occurring along political lines with many Trump backers and conservative Republicans refusing to protect themselves or the society as a whole (Enten 2021).

Using Trump's own words, with its one third of the global deaths from Covid-19, its overwhelmed and inadequate health care system, and its inability to provide citizens with adequate testing, the U.S. became a "shithole country" (Kenci 2019). A Norwegian university, from the country Trump used as an example of acceptable immigrants (Aizenman 2018), warned its students to return home from overseas "especially if you are staying in a country with poorly developed health services and infrastructure and/or collective infrastructure, for example the USA" (Langlois 2020). In addition, the European Community closed its borders to American citizens because of its inability to control the pandemic (Thiessen 2020). Finally, in a startling Wall Street Journal Op-Ed, political economist Nicholas Eberstadt (2020) suggested that Covid-19 could not only destroy a stable world economy, but that "the pandemic could mark the end of the postwar stage of globalization" [i.e., Pax Americana].

The process occurring over the past decade whereby states again began to provide for their own security, accelerated rapidly in a few months with the Covid-10 pandemic. Thus, in making my recommendation for Icelandic national security, I reject the position that globalism creates a collective security apparatus that eliminates the need for individual states to take care of their own national security. Iceland is alone when confronted with sub-strategic security threats such as terrorism, and the country must develop a capability to defeat them. There is evidence that the country can meet sub-strategic security threats alone. In addition to its successful Covid-19 response, Iceland recently defeated national security threats alone in two areas – natural disasters including volcanoes and economic collapse with the failure of its banking system.

An example of Iceland's successful response to a natural disaster occurred during the 2010 Eyjafjallajökull eruption, which threatened local Icelandic residents through lava flows, glacial outwashes, and ash fallout. The volcanic threat was met by the Icelandic Police and its subordinate Department of Civil Protection and Emergency Management, the Icelandic Red Cross, and national search-and-rescue teams (ICE-SAR). All of these organizations were well-trained, experienced, and communicated their intentions well, so that most local residents were able to evacuate the area safely. The successful operation resulted from shared cultural values including trust in government, subordination of individual needs for collective good, and historical knowledge of the dangers inherent in volcanic eruptions (Bird and Gísladóttir 2018).

Just as with the pandemic and volcanoes, Iceland dealt with its 2008 financial collapse alone (Boyes 2009). Initially, Iceland sought help from the United States, which declined to intervene despite paying billions of dollars to defend Iceland militarily for 60 years. Iceland next applied for EU membership, although that was withdrawn by the incoming government in 2013, possibly over Brussel's inability to guarantee Icelandic fishing quotas (The Guardian 2015). Although it did receive assistance from the International Monetary Fund, Iceland worked its own way out of debt by fostering massive tourism (helped initially by the Eyjafjallajökull eruption) that brought almost two million visitors to the country in 2018 (Icelandic Tourist Board 2019). The currently active volcano, Geldingadalir, is helping end the country's pandemic-caused economic problems by generating global interest that is bringing tourism back to Iceland.

In sum, my view about Icelandic security rejects the position that globalism has minimized the need for individual states to manage their own national security interests. Based upon its proven ability to defeat national security threats, I recommend that Iceland develop its own sub-strategic security force to assist the police and coast guard in protecting the country and its people.

References

Aizenman, Nurith. 2018. Trump Wishes We Had More Immigrants from Norway. Turns Out We Once Did. Goats and Soda, Immigrants. *National Public Radio*, January 12, 2018. Accessed October 20, 2020, https://www.npr.org/sections/goatsandsoda/2018/01/12/577673191/trump-wishes-we-had-more-immigrants-from-norway-turns-out-we-once-did.

Ástvaldsson, Jóhann Páll. 2019. Asylum Seeker Deportations. *Iceland Review*, August 2, 2019. Accessed May 3, 2020, https://www.icelandreview.com/society/in-focus-asylum-seeker-deportations/.

Barber, Benjamin R. 1992. Jihad vs. McWorld: How Globalism and Tribalism Are Reshaping the World. *The Atlantic* 269(3): 53–65, March 1992. Accessed June 5, 2020, https://www.theatlantic.com/magazine/archive/1992/03/jihad-vs-mcworld/303882/.

Bird, Deanne K., and Guðrún Gísladóttir. 2018. Responding to Volcanic Eruptions in Iceland: from the Small to the Catastrophic. *Palgrave Communications* 4: 151. Accessed June 23, 2020, https://doi.org/10.1057/s41599-018-0205-6.

Blair, Dennis C., and Keith Alexander. 2017. China's Intellectual Property Theft Must Stop. *Opinion. The New York Times*, August 15, 2017. Accessed June 2, 2020, https://www.nytimes.com/2017/08/15/opinion/china-us-intellectual-property-trump.html.

Boyes, Roger. 2009. *Meltdown Iceland. Lessons on the World Financial Crises from a Small Bankrupt Island*. New York: Bloomsbury, USA. Accessed June 23, 2020, https://www.economist.com/media/pdf/meltdown-iceland-boyes-e.pdf.

Bradsher, Keith. 2020. China Approves Plan to Rein in Hong Kong, Defying Worldwide Outcry. *The New York Times*, May 20, 2020. Accessed June 3, 2020, https://www.nytimes.com/2020/05/28/world/asia/china-hong-kong-crackdown.html.

British Broadcasting Corporation (BBC). 2020. *Chinese Consulate in Houston Ordered to Close by US. July 23, 2020*. Accessed July 28, 2020, https://www.bbc.com/news/world-us-canada-53497193#:~:text=The%20US%20has%20ordered%20China,was%20%22stealing%22%20intellectual%20property.

Buzan, Barry, and Richard Little. 2001. Why International Relations Has Failed as an Intellectual Project and What to Do About It. *Millennium: Journal of International Studies* 30 (1): 19–39. Accessed June 5, 2020, https://journals.sagepub.com/doi/abs/10.1177/03058298010300010401.

Carpio, Alvin. 2019. The Answer to Nationalist Fervour Isn't Less Globalism. It's more. *World Economic Forum Annual Meeting*, January 16, 2019. Retrieved June 4, 2020, https://www.weforum.org/agenda/2019/01/the-answer-to-nationalist-fervour-isnt-less-globalism-its-more/.

Ćirić, Jelena. 2020. Black Lives Matter Gathering Held in Reykjavík Tomorrow. *Iceland Review*, June 2, 2020. Accessed July 12, 2020, https://www.icelandreview.com/society/black-lives-matter-gathering-held-in-reykjavik-tomorrow/.

Cockburn, Harry. 2020. Coronavirus Pandemic Needs Global Response, Urges Bill Gates. *Independent*, Sunday, April 12, 2020. Accessed October 29, 2020, https://www.independent.co.uk/news/world/americas/coronavirus-bill-gates-global-response-pandemic-equipment-g20-countries-a9461391.html.

Decker, Benjamin T. 2017. Who Will Talk About Iran's War Crimes in Syria? *The Tower Magazine* 45, January 2017. Accessed June 2, 2020, http://www.thetower.org/article/who-will-talk-about-iran-war-crimes-in-syria/.

Department of Justice. 2020. *Harvard University Professor and Two Chinese Nationals Charged in Three Separate China Related Cases*, January 28, 2020. Washington DC: The United States Department of Justice. Accessed June 19, 2020, https://www.justice.gov/opa/pr/harvard-university-professor-and-two-chinese-nationals-charged-three-separate-china-related.

Eberstadt, Nicholas. 2020. Can Pax Americana Survive the Coronavirus? *Opinion, The New York Times*, June 19, 2020. Accessed June 19, 2020, https://www.wsj.com/articles/can-pax-americana-survive-the-coronavirus-11592586385?mod=hp_opin_pos_2.

Enten, Harry. 2021. The 2021 Vaccination Map Looks Like the 2020 Election Map. *CNN News*, May 23, 2021. Accessed June 22, 2021, https://www.cnn.com/2021/05/23/politics/vaccine-map-2020-election-analysis/index.html.

Galimullin, Eduard, and Yuri Matveenko. 2019. The Ongoing Formation of Russia's Arctic Policy: A New Stage? In *Redefining Arctic Security, Arctic Yearbook 2019*, ed. Lassi Heininen, Heather Exner-Pirot, and Joël Plouffee, 305–321. Akureyri: Arctic Portal. Accessed July 11, 2020, https://issuu.com/arcticportal/docs/ay2019.

Grettisson, Valur. 2020. Editorial: The Coronavirus Is Not War. *The Reykjavík Grapevine*, May 20, 2020. Accessed June 3, 2020, https://grapevine.is/mag/editorial/2020/05/29/editorial-the-coronavirus-is-not-war/.

Hoagl, Jim. 1990. The Cost of the 'Peace Dividend'. *The Washington Post*, April 12, 1990. Accessed June 24, 2020, https://www.washingtonpost.com/archive/opinions/1990/04/12/the-cost-of-the-peace-dividend/3e1e4dc0-352d-4787-bf73-19da7ee7f45e/.

Holsti, Kalevi J. 1996. *The State, War, and the State of War*. Cambridge: Cambridge University Press.

Human Rights Watch 2019. World Report 2019. New York. Accessed June 22, 2021, https://www.hrw.org/worldreport/2019

Human Rights Watch 2020. World Report 2020. New York. Accessed June 22, 2021, https://www.hrw.org/worldreport/2020#

Icelandic Tourist Board. 2019. *Tourism in Iceland in Figures – January 2019*. Reykjavík. Accessed June 20, 2020, https://www.ferdamalastofa.is/static/files/ferdamalastofa/tolur_utgafur/january-2019.pdf.

Ingvarsson, Ásgeir. 2020. US Ambassador to Iceland Causes a Stir. *Monitor Iceland*, July 27, 2020. Accessed July 28, 2020, https://icelandmonitor.mbl.is/news/news/2020/07/27/us_ambassador_to_iceland_causes_a_stir/.

James, Paul. 2006. *Globalism, Nationalism, Tribalism: Bringing Theory Back In*. London: Sage. Accessed June 4, 2020, https://www.academia.edu/1642214/Globalism_Nationalism_Tribalism_Bringing_Theory_Back_In_2006.

Kantchev, Georgi. 2020. Putin's Bid to Extend Rule Is Approved by Russia's Parliament. *The Wall Street Journal*, March 11, 2020. Accessed June 2, 2020, https://www.wsj.com/articles/putins-bid-to-extend-rule-is-approved-by-russias-parliament-11583932096.

Karp, Paul. 2020. Former WHO Board Member Warns World Against Coronavirus 'Vaccine Nationalism'. *The Guardian*, May 18, 2020. Accessed June 15, 2020, https://www.theguardian.com/world/2020/may/18/former-who-board-member-warns-world-against-coronavirus-vaccine-nationalism.

Kendi, Ibram X. 2019. The Day *Shithole* Entered the Presidential Lexicon. *The Atlantic*, January 13, 2019. Accessed June 25, 2020, https://www.theatlantic.com/politics/archive/2019/01/shithole-countries/580054/.

Kolbert, Elizabeth. 2020. How Iceland Beat the Coronavirus. *The New Yorker*, June 1, 2020. Accessed June 24, 2020, https://www.newyorker.com/magazine/2020/06/08/how-iceland-beat-the-coronavirus.

Langlois, Shawn. 2020. Norwegian University Urges Students to Return Home from the U.S., Cities 'Poorly Developed Health Services. *MarketWatch, The Margin*, March 15, 2020. Accessed October 23, 2020, https://www.marketwatch.com/story/norwegian-university-urges-students-to-return-home-from-the-us-cites-poorly-developed-health-services-2020-03-15.

Mead, Walter Russell. 2020. The Imperial Presidency Will Outlast Trump. Opinion. *The Wall Street Journal*, June 22, 2020. Accessed June 23, 2020, https://www.wsj.com/articles/the-imperial-presidency-will-outlast-trump-11592865672?mod=hp_opin_pos_3.

Ministry for Foreign Affairs. n.d. *National Security*. Reykjavík: The Government of Iceland. Accessed July 20, 2020, https://www.government.is/topics/foreign-affairs/national-security/.

Mittleman, James H. 2000. *The Globalization Syndrome: Transformation and Resistance*. Princeton: Princeton University Press.

———. 2010. *Hyperconflict: Globalism and Insecurity*. Palo Alto: Stanford University Press.

Pifer, Steven. 2020. Order from Chaos. Crimea: Six Years After Illegal Annexation. *Brookings Institution*, Tuesday, March 17, 2020. Accessed June 2, 2020, https://www.brookings.edu/blog/order-from-chaos/2020/03/17/crimea-six-years-after-illegal-annexation/.

Reuters. 2017. North Korea nuclear test prompts global condemnation. *Reuters World News*, September 3, 2017. Accessed June 3, 2020, https://www.reuters.com/article/us-northkorea-missiles-reaction/north-korean-nuclear-test-prompts-global-condemnation-idUSKCN1BE0L4.

———. 2021. Covid-19 Tracker. Reuters.com, July 30, 2021. Accessed July 30, 2021, https://graphics.reuters.com/world-coronavirus-tracker-and-maps/countries-and-territories/iceland/.

Ripsman, Norrin M., and T.V. Paul. 2010. *Globalization and the National Security State*. New York: Oxford University Press.

Sarkees, Meredith Reid, Frank Whelon Wayman, and J. David Singer. 2003. Inter-State, Intra-State, and Extra-State Wars: A Comprehensive Look at Their Distribution over Time, 1816–1997. *International Studies Quarterly* 47(1): 49–70, March 2003. Accessed June 14, 2020, https://doi.org/10.1111/1468-2478.4701003.

Snyder, Glenn H. 1990. Alliance Theory: A neorealist first cut. *Journal of International Affairs, Spring/Summer 1990* 44 (1): 103–123.

Steger, Manfred B., and Paul James. 2010. Introduction: Ideologies of Globalism. In *Globalism and Culture, Volume IV: Ideologies of Globalism*, ed. Paul James and Manfred B. Steger, viii–xxxi. London: Sage. Accessed June 4, 2020, https://www.researchgate.net/publication/270900922_Globalization_and_Culture_Vol_4_Ideologies_of_Globalism.

Steib, Matt. 2020. Trump's Disregard for Blue States Is at the Heart of His Shoddy COVID Response. *New York Magazine. Intelligencer, Coronavirus*, July 31, 2020. Accessed October 25, 2020, https://nymag.com/intelligencer/2020/07/trumps-war-on-blue-states-is-worse-than-previously-thought.html.

Strand, Håvard, Siri Aas Rustad, Henrik Urdal, and Håvard Mokleiv Nygård. 2019. *Trends in Armed Conflict, 1946–2018*. PRIO, Conflict Trends. Oslo: The Peace Research Institute. Accessed July 9, 2020, https://www.prio.org/Publications/Publication/?x=11349.

The Guardian. 2015. Iceland Drops EU Membership Bid. *The Guardian*, March 12, 2015. Accessed June 20, 2020, https://www.theguardian.com/world/2015/mar/12/iceland-drops-european-union-membership-bid.

Thies, Cameron G. 2008. Book Review, Progress in International Relations Theory: Appraising the Field. Edited by Colin Elman and Mirium Fendius Elman, MIT Press, 2003. *The Journal of Politics* 70(1): 285–286, January 2008. Accessed June 21, 2020, https://www.journals.uchicago.edu/doi/pdfplus/10.1017/S002238160708022X.

Thiessen, Tamara 2020 Europe Travel Ban, Here Are 5 Things Americans Must Know. *Forbes*. July 1, 2020. Accessed July 10, 2010, https://www.forbes.com/sites/tamarathiessen/2020/07/01/us-europe-eu-travel-ban-5-things-americans-must-know/?sh=53a625fc6c4c.

Trump, Donald. 2020. *Remarks by President Trump at Owens & Minor, Inc.* Allentown/Washington, DC: Distribution Center/The White House. Accessed June 26, 2020, https://www.whitehouse.gov/briefings-statements/remarks-president-trump-owens-minor-inc-distribution-center-allentown-pa/.

Wallerstein, Immanuel. 2004. *World-Systems Analysis: An Introduction*. Durham: Duke University Press.

Waltz, Kenneth N. 2003. Foreword. Thoughts about Assaying Theories. In *Progress in International Relations Theory. Appraising the Field*, BCSIA Studies in International Security, ed. Colin Elman and Miriam Fendius, vii–xii. Cambridge, MA: MIT Press.

Wheelersburg, Robert P. 2021. North Atlantic Small State Security 2025: The West Nordic Security Zone. In *Small States and the New Security Environment*, The World of Small States, Vol. 7, Chapter 7, ed. Anne-Marie Brady and Baldur Thorhallsson, 87–101. Basel: Springer Nature.

Wolf, Eric. 1982. *Europe and the People Without History*. Oakland: University of California Press.

Wray, Christopher. 2020. *Responding Effectively to the Chinese Economic Espionage*. Department of Justice China Initiative Conference, Center for Strategic and International Studies. Washington, DC, February 6, 2020. Accessed June 2, 2020, https://www.fbi.gov/news/speeches/responding-effectively-to-the-chinese-economic-espionage-threat.

Yeung, Jessie, and Yoonjung Seo. 2021. Why North Korea Is So Afraid of K-Pop. *CNN, World, Asia*. July 23, 2021. Accessed July 25, 2021, https://www.cnn.com/2021/07/23/asia/north-korea-culture-war-kpop-intl-hnk-dst/index.html.

Chapter 3
Arctic States' Comparative Sub-Strategic Security Policies

And Caesar's spirit, ranging for revenge
With Ate by his side come hot from hell
Shall in these confines with a monarch's voice
Cry 'havoc,' and let slip the dogs of war.
William Shakespeare, Julius Caesar

Abstract This chapter summarizes the current security policies and operational capabilities for each of the Arctic Eight countries. While those policies usually address the strategic level, to formulate the recommendation contained in this book, I concentrated on sub-strategic security policies, potential threats, and capabilities countries possess to meet those threats. For each country, I review the official publications containing Arctic security policy and highlight any recent changes to that policy. In addition, I examine how the countries intend to monitor their territories to identify suspicious activities using surveillance and reconnaissance. Next, I use official websites from security agencies and military units to describe the organization, personnel, and equipment they would use to defeat sub-strategic security threats like terrorism. The comparative analysis indicates that Iceland alone among the Arctic Eight is ill-prepared to deal with sub-strategic security threats that would overwhelm the police or coast guard. Based upon that analysis, I reach the conclusion that to improve its preparedness, Iceland should develop a security force. The recommended security force could be capable of meeting emergencies by assisting the police and coast guard, and in combination with other nations' security forces.

3.1 Introduction to Arctic Security

James Kraska's edited volume *Arctic Security in an Age of Climate Change* "[was] the first book to examine Arctic defence policy and military security from the perspective of all eight Arctic states" (2011: i). In the book's forward, Oran Young (2011) suggested that, although the Arctic was an important part of global

R. P. Wheelersburg, *The Security of Iceland and the Arctic 2030*, Springer Polar Sciences, https://doi.org/10.1007/978-3-030-89948-6_3

security politics, it attained that status not for what was taking place in the region, but because of its location. Control of the North Atlantic Arctic was necessary to protect a primary corridor for armed forces to transport their strategic weapons systems like ballistic missile submarines, long-range bombers, and troop/equipment ships transiting between Europe and North America.

Alyson Belies, former British ambassador to Finland, said as much during a class session we co-taught at the University of Iceland in 2015, declaring "Arctic military operations are based upon activities outside of the region". That situation began 70 years ago when the Allies occupied Iceland to prevent the Germans from using the country as a base to attack North America or to stop allied forces from sailing to Europe. For five decades starting in 1951, NATO forces operated at a high tempo from Keflavik to defend the Greenland-Iceland-United Kingdom (GIUK) Gap from Warsaw Pact forces. Those efforts were not intended to improve the security of Nuuk, Rovaniemi, or Murmansk; instead, they related to what was happening in Moscow and Washington. After the breakup of the Soviet Union, the anticipated melting ice in the Arctic Ocean caused the NATO security focus to change. As Oran Young (2011) explained, the attention of world leaders has only recently been drawn to the Arctic due to the emergence of new territories, abundant natural resources, and rapid transport corridors located there.

The North Atlantic is arguably the most politically diverse part of the Arctic, which results in a variety of strategic policies and defence objectives that influences all northern residents, indigenous, traditional (i.e., settler), and recent. The North Atlantic includes large states such as Russia, Canada, the United Kingdom, and the United States. It also contains small states with high Arctic territories including Denmark (Greenland), Iceland (i.e., its exclusive economic zone extends into the Arctic Ocean), and Norway (Svalbard). Lastly, the North Atlantic consists of the land-locked states of Finland and Sweden, which possess territories within the Arctic Circle and also encroach on the North Atlantic through aviation and maritime assets.

Military alliances add complexity to the region, as five of the Arctic states are members of NATO and three are neutral (if one includes Russia), although two of those neutral states participate in NATO's Enhanced Opportunity Partners (EOP) program. Since the Arctic Council is prevented from taking up military issues by its founding document, the Ottawa Declaration (Arctic Council 1996), and given the fact that seven of the eight Arctic nations participate in the Iceland air-policing mission, NATO and Russia control security in the North Atlantic. Without the ability to discuss military activities in the circumpolar region, the Arctic Council is relegated to addressing soft-security issues like environmental protection, economic equity, and indigenous rights (Greaves and Pomerants 2017).

As nations begin to compete with each other for pre-existing and emerging Arctic landscapes, especially the natural resources they contain, there is a possibility that competition can develop into conflict involving large segments of the circumpolar region. Preventing conflicts in the North Atlantic should be the focus of Arctic policy, which needs to be developed cooperatively and stated clearly through recognized treaty efforts. Yet given the inability of the Arctic Council to address military issues, the cooperative process is not working as states develop unilateral security policies that can diverge from each other. The exception is that, to counter Russia in the region, the other Arctic Eight countries seem to be tightening their security relationships with each other. In addition to strategic-level activities, states bordering the Arctic Ocean are concentrating their efforts on defending their sovereignty and protecting their citizens from a variety of threats (Wezeman 2016).

Another level of security operation that adds complexity to the North Atlantic Arctic is sub-strategic security threats, which I defined in Chap. 1. Despite the remote and seemingly unassailable nature of the Arctic merely a century ago, as the ice melts the Transpolar Sea Route (Bennett 2019) could provide access to the region for a variety of actors (both state and non-state) from within and from outside of the Arctic. The emergence of what Canada refers to as "ungoverned" or "borderless" territories (Government of Canada 2019: 14, 51) means that Arctic states must guard their sovereignty and protect their citizens from other Arctic Council nations or rogue states, such as China, seeking a foothold in the circumpolar region. Under international law, ungoverned areas are called *terra nullis*, meaning territory over which no state has legally exercised sovereignty. The appearance of ungoverned areas could increase competition as more unclaimed lands and resources like natural gas deposits become exposed (Nicol et al. 2015).

Non-state actors such as corporations (looking to exploit the natural resources regardless of impact), terrorists, and international criminal syndicates could increase Arctic states' sub-strategic security demands. Non-state actors do not operate in the same manner as states, for example they do not require methods like diplomacy to resolve conflict. Such a dynamic changes the nature of security operations by requiring strategic-level assets like militaries to assist civil authorities during emergencies including search-and-rescue operations, natural disasters, and other threatening situations (Government of Canada 2019).

In order to support the book's recommendation for Icelandic security, I will examine each Arctic Council member's security policy as it relates directly to North Atlantic sub-strategic threats. That examination deals especially with territorial defence (Mendershausen 1980) and citizens' protection following the opening of the Transpolar Sea Route (TSR) and other access avenues into the region. The topics addressed in each section include the country's goal for Arctic security, the specific threats it anticipates, how it intends to monitor its respective territories, and what actions it would take to defeat threats when encountered. The resulting comparative framework will provide context for my Icelandic security force recommendation in Chap. 4.

3.2 Arctic States' Sub-strategic Security Policies

3.2.1 Canada's Arctic Security Policy

World War II made Canada a strategic military power, allowing it to build one of the world's largest armies, navies, and air forces, with over a million troops under arms (Veterans Affairs Canada 2020). Yet it was decided that military troops were not enough to ensure complete defence and security for the Canadian Arctic; rather, a combination of mobile force and frontier scouts was necessary. This resulted in a security force involving both regular army paratrooper units called the Mobile Striking Force and a type of reserve home guard named the Canadian Rangers. The Canadian Rangers are local residents (mostly First Nation members, Inuit; i.e., indigenous peoples) able to transit the countryside, survive its hardships, provide reconnaissance, and provide their own protection if necessary (Coates et al. 2008).

Today, Canada bases its Northern Strategy on exercising sovereignty and protecting its citizens living in the Arctic (Wezeman 2016). Its 2017 policy statement insists that all Canadian Arctic defence depends upon protecting the nation's territory (Government of Canada 2017). Given competing claims over territory in the Arctic (including the Northwest Passage dispute with the U.S.), Canada seeks international acknowledgment of its claim to the extended continental shelf containing the Lomonosov Ridge along with other disputed territories like Hans Island. Canada's strategic operations to maintain security in the circumpolar region outside its borders are performed through bilateral treaties with the U.S., the North American Aerospace Defense Command (NORAD), and NATO (NATO n.d.).

Canada created a multi-layered defence network during World War II as illustrated in *49th Parallel* (Pressburger 1941), a powerful British propaganda film designed to influence American entry into the war by illustrating Canada's vulnerability. In the film's plot, the Royal Air Force sinks a German U-Boat in Hudson Bay, leaving only members of a shore party who try to reach neutral America using escape and evasion. They raid a local Hudson Bay Trading Post, killing its staff and massacring Inuit men, women, and children, with the goal of escaping in a float plane. As the plane takes off, an Inuit hunter shoots and kills one of the Germans. The plane runs out of fuel and crashes, so the Nazi leader decides his men will walk across Canada to Vancouver, where they can catch a freighter to Japan. On their journey, the Germans kill a travelling salesman and steal his car. They drive to Winnipeg, but the group is down to two when the Royal Canadian Mounted Police (RCMP) captures a third crew member at a festival. On their way to freedom, they walk across the Rocky Mountains through a national park where a tourist/writer exposes them as Nazis. The leader finds himself alone when his final crew member is taken prisoner. He continues, hopping a train crossing the border at Niagara Falls, but he is finally trapped in one of the boxcars by an AWOL Canadian soldier.

Remarkably, the film predicted the country's future sub-security apparatus. Today, Canada monitors its interior and defends against any threat to its citizens with overlapping levels of surveillance including Canadian Rangers, the RCMP, private citizens, and the military.

The key to understanding sub-strategic security threats for the Commonwealth member nation is contained in its 2019 policy, *Statement on Canada's Arctic Foreign Policy*. The policy led to increased government monitoring and security capability in Canada's Arctic territory consisting of Nunavut Province and the Northwest and Yukon Territories that extend to the North Pole. The area comprises 40% of Canada's land mass, 75% of its coastline, and includes 36,000 islands. The large, remote area presents a challenge to monitor given that it is without roads or railroads and is sparsely populated by mostly indigenous peoples (Government of Canada 2019).

Potential sub-strategic security threats to Canada's Arctic were summarized in its 2017 defence policy.

> States and non-state actors are increasingly pursuing their agendas using hybrid methods in the "grey zone" that exists just below the threshold of armed conflict. Hybrid methods involve the coordinated application of diplomatic, informational, cyber, military and economic instruments to achieve strategic or operational objectives. . . . The potential of terrorist groups to spread into ungoverned spaces and exploit information technology to form alliances and far-flung trans-regional networks also poses a security challenge (Government of Canada 2017: 53).

Those potential threats differ slightly from the country's earlier policy that named international crime syndicates and smuggling people and drugs as top challenges. Current Canadian sub-strategic security depends upon a strong and flexible military to assist in defeating the threat posed by international terrorists and violent extremists who could target ungoverned spaces in the Arctic. Terrorists, criminals, and others who might use violence to achieve their missions would be countered by the civil authorities, law enforcement, and the armed forces. In addition, the Canadian military would assist with natural disasters, multiple simultaneous domestic emergencies, or complex search-and-rescue situations caused by increased movement into and around the Arctic (Government of Canada 2019).

How capable is Canada to conduct sub-strategic security operations in its Arctic regions? In the past 10 years, the country has begun improving its personnel, training, and equipment for Arctic security operations. In terms of surveillance, Canada is working with NORAD to improve monitoring of the northern area through new space-based assets. In addition, there are plans to acquire sophisticated unmanned aerial vehicles (i.e., drones) to increase the military's ability to investigate suspicious activities in remote areas. At the local level, efforts are underway to improve the relationship with First Nation, Inuit, and local groups by including them in decision making (Kikkert and Lackenbauer 2019), which will further connect Canada's rural areas with urban ones [for a critique of these efforts see Salter 2019].

In addition, the Canadian Rangers have been increased in size by nearly 25% and are now fully integrated into the armed forces by providing southern-based troops with Arctic training. The Rangers have also been outfitted with new rifles, replacing their outdated weapons. To supplement a 100-member Reserve Arctic unit, the regular Army will form a battalion (500 soldiers) of Arctic troops to improve operational capabilities. Arctic security transport capabilities are also being increased, as the government plans to purchase new all-terrain vehicles, snowmobiles, and semi-amphibious personnel carriers. Canada will acquire new weapons systems, especially portable ground-based air-defence and anti-armor missile systems so that they can be moved to areas where they are needed. In terms of aviation transport, Canada's C-130J Super Hercules and rotary-wing aircraft (i.e., helicopters) are capable of using improvised/temporary landing strips, creating the flexibility to address rapidly developing sub-strategic emergencies (Wezeman 2016; Government of Canada 2019).

3.2.2 *Denmark's Arctic Security Policy*

Formally called the Kingdom of Denmark, the Danish core area (e.g., Jutland) along with the Faroe Islands and Greenland form a major North Atlantic power. At the beginning of the 2010s, the country's strategy was based upon exploiting the so-called peace dividend. The strategy had the "goal of preventing conflicts, avoiding the militarization of the Arctic, and actively helping to preserve the Arctic as a region characterized by trust, cooperation and mutually beneficial partnerships" (Ministry of Foreign Affairs of Denmark 2011: 9). Missions for the Danish military consisted primarily of defending The Realm, maintaining sovereignty by showing the flag, conducting search-and-rescue operations, protecting the environment, and providing Defense Support to Civil Authorities (Jørgensen 2016).

That goal changed substantially due to Russian activity in the region, which is explained in the country's current Arctic defence policy. The policy specifies that the Danish armed forces have a primary mission of providing security to The Kingdom's territories in the North Atlantic. The military will accomplish that mission by establishing and protecting the nation's sovereignty and by conducting surveillance over the circumpolar region (Danish Ministry of Defence 2019).

Sovereignty is a consistent theme in the most recent Danish security policy, which is not surprising given current territorial disputes including Hans Island and the Lomonosov Ridge. In its attempt to gain a foothold in the Arctic as part of its Silk Road initiative, China attempted to build a factory, road, airfield, and a mine on an abandoned navy base in Greenland. Under pressure from the U.S. Department of Defense, the Chinese proposal was refused due to their security implications (Matzen 2017). Armed forces are also ordered to show the flag, support the civilian population in remote areas, and to protect Danish shipping in its maritime zones. After 2025, Denmark is not restricting its threat assessment to the summer only, believing that the ice-free period will extend into the winter as well (Defence Command 2013).

Perceived threats to Danish Arctic territory include terrorism and organized crime. The most recent security policy directs the armed forces to support the police in countering those threats by developing a light infantry battalion known as the Arctic Response Force (ARF). Although there would not be a separate and permanent ARF, completing the training for the ARF would improve overall Arctic capabilities for all military units because they have trained Arctic soldiers. The ARF would be formed from various units at particular times to participate in exercises or to assist civilian authorities during emergencies in Greenland (Danish Military Command 2020). In addition, the ARF would support other missions in international Arctic territories by assisting civil authorities in disaster response by conducting early-stage reconnaissance, engaging in difficult search-and-rescue operations, providing immediate medical support, supplying fuel/electricity to first responders, and reestablishing communications after a disaster (Defence Command 2013).

The current Arctic policy requires Danish special operations forces (SOF) such as the Frogmen Corps (*Frømandskorpset*; the equivalent of U.S. Navy Seals) stationed in Greenland to participate in search-and-rescue (S&R) operations and other emergencies. Along with Danish Rangers (*Jaeger*) and the Sirius Sled Patrol in Greenland, Danish SOF could assist with surveillance requirements and provide a quick response to emergencies. Under the control of the Arctic Command in Nuuk, the Arctic-capable forces and special operations forces could assist civil efforts to prevent terrorism and international crime. They could also enforce exclusive economic zones and fishing grounds to preserve Danish sovereignty over natural resources as well as territory (Danish Ministry of Defence 2019).

Given the importance of maritime enforcement of The Kingdom's North Atlantic holdings, the Danish Navy is improving its Arctic capabilities. For decades, the Danish Navy has been sending older, Thetis-class patrol frigates to Greenland from Reykjavík to resupply military bases and research stations. These older frigates are being replaced by new patrol vessels like the HDMS Ejnar Mikkelsen (P-571) and patrol frigates including Hvidbjørnen (F-360), improving both range and firepower in northern waters. Surface vessels work with long-range reconnaissance jet aircraft like the Challenger (CL-604), along with new Sea Hawk (MH-60R) helicopters that can land on Danish frigates and corvettes for more localized operations. The armed forces can use satellites and are expanding their use of unmanned aerial vehicles (UAV) for surveillance and reconnaissance missions. The navy has a small base in the south of Greenland at Kangilinnguit to direct local operations if the need arises (Wezeman 2016).

Exercises help Denmark's military to enforce sovereignty in the Arctic. Operation Nanook 2010, for example, joined together Danish Arctic patrol vessels with Canadian Coast Guard cutters to practice SAR, sail around Greenland, and transit the Northwest Passage (Government of Canada 2010). In 2015, the Danish Air Force conducted NATO's air-policing mission in Keflavik, engaging in practice landings at the secondary Icelandic airports of Akureyri and Egilstaðir (Iceland Review 2015). Northern Falcon 2019, conducted with the Ukrainian logistics command, transported fuel from Thule Airbase to Station Nord, the Danish military and science station in northeast Greenland (Ministry of Foreign Affairs of Denmark 2019).

It is important to note that, while Denmark continues to control Greenland's defence, as independence proceeds, the Greenlandic Home Rule government insists that Denmark subordinate security concerns to economic development. Called de-secularization, the Greenlandic Home Rule government prioritizes the use of the armed forces for civil missions that help support the local economy. For example, while the Danish military currently leads search-and-rescue operations and patrols fishery zones, training locals to perform those missions could help the fledgling Greenlandic government develop its own security capability. De-secularization also means the military should help with youth employment, and steps have been taken to support that goal. For example, young Greenlanders may now take their conscript tests and serve at home without traveling to Denmark (Ministry of Foreign Affairs of Denmark 2018). Use of the American-controlled Thule Airbase to support Greenland's economic development also has entered policy discussions, especially after recent service contracts went to American firms instead of Greenlandic ones (McGwin 2017). However, Denmark indicated it may retain control over defence and foreign affairs, which could hinder Greenland's independence but ultimately allow the Kingdom to continue its Arctic Council membership (Rasmussen 2019).

3.2.3 Finland's Arctic Security Policy

Finland's security policy, unlike those of some other Nordic states such as Sweden and Iceland, derives from the government's self-identity of the entire country as an Arctic nation. Finland considers itself part of Western Europe, influenced, as other nations have been, by Russia's recent military activities (Finnish Prime Minister 2016). Although the country has no specific Arctic security policy, Finnish defence forces base their operational effectiveness on an extensive cold-weather capability acquired through Arctic training and experience. Defence forces are prepared to counter sub-strategic security threats such as organized crime and terrorism, both internally and by cooperating with neighboring states like Norway and Sweden in the European circumpolar region. Finland participates in the Nordic Defence Cooperation (NORDEFCO), the Barents Euro-Arctic Council (BEAC), and the UK-led Joint Expeditionary Force (Finnish Prime Minister 2013; Kuusela 2020). To defeat sub-strategic security threats, Finland uses the border guard, coast guard, home guard, and police, along with organizations in private industry and non-governmental agencies like the Red Cross.

As part of its defence mission, Finland maintains that surveillance is critical for identifying sub-strategic security threats. Finland uses its border guard as a part of the overall defence plan to monitor and secure its territorial integrity, working with the military to plan, train, and exercise its commands. The Finnish Border Guard and Coast Guard are part of the Ministry of the Interior. They patrol the country's external borders, provide police and customs-control duties, and if necessary, are ready to assist with search-and-rescue operations. The border guard also includes a Special Intervention Unit that can defeat threats at the border and supplement the Finnish police's response in non-border areas (Finnish Border Guard 2020).

Of course, the Finnish Army is the main part of the country's defence. Constructed primarily of reservists and conscripts, the army maintains a small cadre of full-time officers and senior non-commissioned officers to train, organize, and lead the part-time soldiers. The army's tasks are to defend the country's territory, support civil authorities during natural and human-made emergencies, train with international security organizations, and selectively provide troops and equipment for international crisis management (Finnish Army 2020).

Although official policy claims that there are no specialized Arctic units in the military, in fact the Jaeger [Ranger] Brigade maintains the expertise and training to operate in the far north. It also provides ground-based air defence, trains other units including the border guards in Arctic combat, and conducts research on Arctic warfare. The Ranger Brigade hosts and participates in Arctic exercises such as

Lappi 20 that took place on February 24–28, 2020. Lappi 20 brought together Lapland's provincial authorities, police, northern border guards, regional hospital staff, and search-and-rescue personnel to train and practice security tasks and to gain experience in working with multiple entities under a unified command (Finnish Army 2020).

Finnish sub-strategic security improvements include upgrading its navy to perform patrol functions in the Baltic Sea and more northern waters. The Squadron 2020 program to modernize the navy includes acquiring four Pohjanmaa-class corvettes that can accomplish more sub-strategic security missions than frigates or larger ships because they can operate in littoral (i.e., shallow) waters (Finnish Ministry of Defence 2020). The corvettes, which will also perform icebreaking duties, can be used for smuggling interdiction or other sub-strategic security missions (O'Dwyer 2020).

Since the inhabitants of the area surrounding the Baltic Sea on both sides of the border are mostly Swedish speakers, the Finnish Navy's Nyland Brigade consists mainly of Swedish-speaking conscripts. The brigade, which could be described as a combination of marines, naval infantry, and coastal rangers, is trained to provide support to other branches of the Finnish armed forces or the border guard using a rapid response that requires amphibious capabilities. The unit also trains in crisis management using English so that it can participate in international operations as an amphibious unit (Finnish Defence Command 2020). The Nyland Brigade will soon have the Jehu-class landing craft (i.e., the M18 AMC-armored modular craft) that carries 24 troops and is armed with modern crew-served machine guns (Marine Alutech 2020).

Yet due to border changes related to World War II, Finland lost access to its Barents Sea ports when it gradually gave up its Petsamo District to the Soviet Union. Russia took part of the Rybachy Peninsula on the Barents Sea from Finland through the 1941 Treat of Paris, while the former port of Liinahamari, situated 100 kilometers east of Kirkenes, Norway was transferred with the Moscow Armistice in 1944. The former Finnish district is now known as Pechengsky Municipal District, part of Murmansk Oblast (province) with the industrial city Nikel being the capital (Holtsmark 1993). Without territory in the high Arctic, Finland's capability to respond quickly to emergencies in the circumpolar region is limited.

Despite that limitation, Finland is taking steps to gain access to circumpolar shipping lanes as the Arctic Ocean opens. Shipping lanes including the TSR and the Northern Sea Route (NSR; controlled by Russia) could provide Finnish commercial vessels with rapid access to Pacific markets. There are proposals to build a railroad link from either the Finnish Baltic port of Oulu to the Norwegian port at Kirkenes or from Rovaniemi to the Russian port of Murmansk. The future rail link, called the Arctic Ocean Railway, is encountering practical obstacles, like high cost versus low shipping volume, and environmental barriers, especially the destruction of reindeer grazing lands (Standish 2018). Should the Arctic Ocean Railway come to fruition, the Finnish navy could increase its activities in the circumpolar region to protect its shipping.

3.2.4 Iceland's Arctic Security Policy

Early in the 2010s Iceland believed that "safeguarding broadly defined security interests in the Arctic region [must occur] through civilian means [while] working against any kind of militarization of the Arctic" (Icelandic Parliament 2011: 2). That demilitarization did not occur, as shown by the increasing operational tempo of the Keflavik base in the past decade. The military friction between NATO and Russia represents a strategic-level security threat. In addition, the Cod Wars of the mid-twentieth century between Iceland and Britain demonstrated that peaceful cooperation has not always been possible, even between allies, in the wake of natural resource competition (Storey 1992; United Nations 2011). Future conflicts over resources and territory in the Arctic could also involve non-state actors such as corporations.

The 2016 Parliamentary Resolution establishing a National Security Policy for Iceland stated that the country's security goal is to "ensure Iceland's independence, the sovereignty and integrity of its borders, the security of its citizens and the protection of its system of government and the fundamental structures of society". While the government recognizes that Iceland "has neither the capacity nor the desire to maintain an army", the policy gives particular consideration to Iceland's security interests in the Arctic. Along with maintaining its membership in NATO, the security policy requires monitoring the ability of the 1951 bilateral U.S. defence agreement to meet strategic threats and other security challenges. The policy also requires Iceland to cooperate with the other Nordic countries on security and defence. The resolution orders appropriate ministries "to ensure that Iceland has the defence structures, equipment, capacity and expert knowledge needed to meet the challenges it faces in the field of security and defence". It directs governmental ministries to create policies that take into account other threats requiring full attention, such as terrorism and organized crime (Icelandic Parliament 2016: 1–2).

Importantly, Iceland's 2016 security policy established the National Security Council. Members of the council include the prime minister, the foreign minister, and the justice minister, along with the three permanent secretaries of those ministries. The council also includes two members of parliament, one from a ruling coalition party and one from an opposition party. Operationally, the council includes the Director General of the Coast Guard, the National Police Commissioner, and a member of the Icelandic Search and Rescue Association. The national security policy designates the foreign minister as the government's lead authority on defence issues, and the policy defines the role of government agencies in security activities (Icelandic Parliament 2016).

In 2009, an interdisciplinary committee formed by the foreign ministry produced a risk assessment for Iceland. The committee report stated that, while the risk for direct terrorism was low in Iceland, it was more probable that the country could be used as staging area for launching attacks on another country (Icelandic Ministry of Foreign Affairs Ad-Hoc Committee 2009). As an example, in 1976, Croatian terrorists landed a hijacked TWA airliner at Keflavik airport, where they faced unarmed Icelandic police. The incident eventually led to the creation of the country's special weapons unit, the Viking Squad, which in 2013 became the first ever police unit in Iceland to kill someone with a weapon (Balestrieri 2019). Increased immigration from the Middle East could lead to homegrown terrorism by Islamic extremists or nationalistic anti-immigrant groups. That dynamic could result in attacks within Iceland, just as it increased such risks in Sweden (Swedish Security Service 2019).

The 2009 committee also anticipated that organized crime was a growing threat in Iceland due to drug trafficking, violent crime, money laundering, and human trafficking. These crimes could be committed by Icelanders as well as by international gangs with the committee claiming that "[t]he police must be supplied with adequate means, equipment and resources to fight international organized crime" (Icelandic Ministry of Foreign Affairs Ad-Hoc Committee 2009: 15). Organized crime is present in Iceland today, indicated by higher levels of hard drug usage compared to the other Nordic countries, which is controlled by criminal gangs. In the past 5 years, the number of opioid deaths and the use rate places Iceland on par with the U.S (Ehrat 2018). A study of Reykjavík's wastewater, along with statistics for driving under the influence and drug seizures, revealed that Iceland had a higher amphetamine-use level than other Nordic countries (Löve et al. 2018). All of these indicators suggested that the drug trade run by organized crime is rampant in Iceland, representing a serious threat to its sub-strategic security.

Another threat named in the 2009 report was human trafficking. The Icelandic Red Cross report on human trafficking called the country both a transit station and a final destination for enslaved women, especially in prostitution. Human trafficking also involved men and women forced to work in Iceland's service and construction industries. Most of the female sex industry workers came from Africa, Asia, Eastern Europe, Russia, and South America, while undocumented male workers were brought from Eastern Europe and the Baltic Countries (U.S. Department of State 2010). Iceland remained in Tier 2 during 2017, meaning that the government did not meet the Trafficking Victims Protection Act minimum standards. Those TVPA minimum standards included whether a country's government prohibited sex trafficking (especially of children) and severely punished those who do while making serious efforts to eliminate the illicit practice (United States Code 2021; U.-S. Department of State 2017a). Yet the trafficking continues amid a construction boom, as a recent tragedy in Reykjavík exposed when three Polish workers died as a fire swept through an overcrowded flophouse for such workers (Askham 2020).

How prepared is Iceland to meet possible sub-strategic security threats with its own resources? The two organizations that are responsible for Icelandic security are the police and the coast guard, both of which come under the Minister of Justice. A report by the U.S. Department of State's Diplomatic Security Service Overseas Security Advisory Council concluded that the Icelandic police is an effective, professional force that suffers from low staffing and funding. In 2019 there were 730 active and reserve police officers, most of whom performed their duties unarmed, although there are provisions for allowing cops to have weapons in their patrol vehicles (Overseas Security Advisory Council 2019).

Their special weapons tactical unit, colloquially called the Viking Squad, possesses 150 Norwegian-gifted Heckler and Koch MP5 submachine guns (Björnsson 2014). The Viking Squad is divided into five sections: an Explosive Ordinance Disposal (EOD) section, a seaborne section, a sniper section, an intelligence section, and an airborne unit. The Viking Squad assumed Iceland's paramilitary role in overseas peacekeeping operations after the government de-armed Iceland's Crisis Response Unit (ICRU), which continues to participate in NATO peacekeeping missions as civilian experts under the defence directorate (Balestrieri 2019; Icelandic Ministry of Foreign Affairs n.d.).

The Icelandic Coast Guard (ICG) is fully trained, equipped, and dedicated to surveil and defend Iceland's territorial waters. The ICG also conducts other operations such as patrolling fishing grounds, ensuring environmental protection, and participating in maritime search-and-rescue operations. The coast guard consists of 250 officers and enlisted personnel. Operating four surface ships, the ICG's newest ship is a Kongsberg's UT 512L patrol vessel (Kongsberg 2019). The ICG also has two Ægir offshore patrol vessels and a Baldur research ship. ICG aviation assets include a Bombardier Dash 8, one Aerospatiale Super Puma AS-332L1, and two Airbus Super Puma H225s. In addition, the coast guard is responsible for the Icelandic Air Defence System and the Combat Readiness Center at Keflavik. The service also has an EOD unit (Icelandic Coast Guard n.d.). The coast guard has a certain level of interoperability with other security or emergency organizations in the country. Recently, the ICG and Viking Squad conducted a real-life combined operation to land on an underway fishing vessel at sea and arrest two murder suspects (Balestrieri 2019).

In order to bolster the country's security, on April 20, 2021 Iceland joined the UK-led Joint Expeditionary Force (JEF). The force includes the Nordic countries, the Baltic countries, the Netherlands, and the United Kingdom. While it contributes no military forces to the JEF, according to Icelandic Foreign Minister Gudlaugur Thór Thódarson, Iceland joined to create an entity capable of "deeper regional consultation and cooperation on security and defence, as well as possible coordinated civil emergency and humanitarian assistance" (Wallace 2021). Arnór Sigurjónsson, director of the Icelandic Security and Defence Directorate, maintained that Russia's invasion of Crimea changed the North Atlantic security dynamic. It caused the U.K to reengage in the Arctic with sub-hunting P-8s and the aircraft carrier the HMS Queen Elizabeth. The U.K. Ministry of Defence asked Iceland to join the JEF in order to give it access to the country's land, air, and sea for security purposes. The benefit is a political alignment on regional security issues, including information sharing and crisis management (Sigurjónsson 2021).

Yet Iceland lacks the ability to engage ground-based, sub-strategic security threats on a larger scale than what the police or the coast guard can handle (Paumgarten 2015). Even a small assault force with trained troops, armored vehicles, attack helicopters, or amphibious assault vessels could quickly overwhelm the police. As described in Chap. 1, terrorists or criminals attacking multiple targets or taking large numbers of hostages could also surpass the combined police and coast guard capability to respond. While Covid-19 has decreased tourism to Iceland during

2020 and early 2021, once the pandemic is over, the tourist industry expects the prior level of tourism to resume (Iceland Monitor 2019). Therefore, as recommended in Chap. 4, to defeat sub-strategic security threats, Iceland should develop a security force that possess the personnel, equipment, and training to respond effectively and rapidly to them.

3.2.5 Norway's Arctic Security Policy

Norway represents a unique case, positioned on the border of a large country with an extensive military presence in the North Atlantic, but part of an alliance that stands against its larger neighbor. Norway is a member of NATO with a strong military, but it must maintain a special relationship to Russia because of its geopolitical location. This dual position shows the narrow path the country must tread to maintain peace, given its precarious security relationships. In addition, Norway now recognizes that its security requirements extend to the North Pole, increasing the territory it must defend in an emergency.

Norway's 2006 security policy was focused more on North Norway than on the Arctic. The policy stressed Norwegian sovereignty, stating its purpose was

> [T]o ensure a *presence* in the High North both through maintaining settlements and through being present in the areas under Norway's jurisdiction including Svalbard, our sea areas, coastline, and settlements and towns on land (Norwegian Ministry of Foreign Affairs 2006: 2).

The tension caused by Russia's 2014 seizure of Crimea led to negative sanctions by NATO member states, which is perhaps a preview of the future as Russia struggles to secure natural resource access in the North Atlantic. That tension caused Norway to make some difficult policy decisions that will affect its relationship with Russia into the next decade. The country's 2017 security plan was designed to ensure that Norway is capable of defending itself against any aggression by another country (Norwegian Ministry of Foreign Affairs 2017).

Norwegian defence planning consists of three elements that would increase its Arctic strategic readiness. First, Norway is expanding the scope and locations for U.S. Marines to conduct winter combat training over the next 5 years, making them the first foreign troops to be based in Norway since the Second WWII (Fouche 2018). Second, the recent acquisition of 40 F-35 Lightning IIs says quite clearly "do not take transiting Norwegian territory for granted" (Norwegian Ministry of Defence 2020). Third, and perhaps most telling, is the plan to hold several large-scale NATO exercises in Norway over several years that began with Trident Juncture 2018. That exercise involved 40,000 participants, 130 aircraft, and 70 naval ships from 30 nations with training taking place in Sweden, Iceland, and Norway (NATO 2018). Predictably, Russia has protested these developments, calling them aggressive actions that illustrate how the expansion of NATO threatens their sovereignty (Watling 2018).

Threats to Norway could come from a variety of sources, both natural and human. A 2016 security assessment produced a threat matrix of 20 potential risks to the country. Eight of those possible threats were natural disasters such as earthquakes and a long-term volcanic eruption in Iceland. Seven were human-made or technological disasters like a nuclear accident or a ship collision within an urban area. Finally, four potential risks were from state or non-state actors including strategic engagements, terrorism, and cyberattacks. Interestingly enough, the matrix also included a pandemic, which of course has come to pass (Norwegian Government Security and Service Organization 2016: 19, Figure 4.8).

The Norwegian Police Security Service (the country's domestic counterintelligence agency) believes strategic spying against the government, parliament, and military is a primary threat to the nation's security. Along with spies from foreign intelligence services stationed in Norway or operating from third countries to recruit native Norwegians, the PST believes that terrorists digitally mapping critical infrastructure as a prelude to sabotage represents a considerable threat (Police Security Service 2020). Terrorism includes both domestic and foreign actors, with the Breivik incident described in Chap. 1 an example of the former and Islamic terrorism from ISIS or another group an example of the latter.

Other threats to Norway include international crime syndicates smuggling humans, drugs, or weapons; piracy, and cyberattacks (Norwegian Ministry of Foreign Affairs 2015; Norwegian Government Security and Service Organization 2016: 19). It seems unusual that Norway considers ocean piracy a major sub-strategic security threat, although the logic is clear when one considers the country's global maritime activities. With nearly 1600 Norwegian-flagged merchant marine vessels traveling the world's oceans in 2017 (CIA 2017), piracy is a concern. During 2009, one of its Nansen-class combat ships exchanged fire with a pirate vessel while patrolling in the Gulf of Aden (Royal House of Norway 2013). In addition, the numerous Norwegian oil rigs in the North Atlantic could be easy targets for pirates if not for such military capabilities in the country's arsenal as the Coastal Rangers.

Norway believes a major problem with both threat analysis and detection is that much of the planning and preparation for sub-strategic attacks takes place in the digital realm, using social media and other encrypted communications (Police Security Service 2020). In addition, potential sub-strategic combatants use "hybrid warfare", which utilizes both conventional and covert assets to conduct an attack (Norwegian Ministry of Foreign Affairs 2015: 10). Hybrid warfare can combine both military and non-military means, such as special forces operating without uniforms (or in fake uniforms), or involve state-sponsored terrorism. The country's concern with domestic terrorism (regardless of the source) as the number one sub-strategic security threat seems appropriate considering the current situation. To defeat that threat, the country's Norwegian Police Security Service, Intelligence Service, and military intelligence operations established the Joint Counter Terrorism Center in 2014 (U.S. Department of State 2018).

In 2020, the Ministry of Defence published its new security policy for the country, which ensures the protection of "Norwegian sovereignty, territorial integrity, democratic institutions and freedom of action against political, military and other pressure" (Norwegian Ministry of Defence 2020: 4). Norway must be able to defend itself alone or with its allies, either other NATO members or those countries with whom they have a bilateral agreement. With so many interests and activities in northern Norway and the circumpolar region, the Arctic must be a primary location for the military and security assets. Consequently, the Norwegian armed forces maintain a substantial presence in the Arctic, including recently moving the headquarters for the Defence Ministry, the army, and the coast guard north of the Arctic Circle (Wezeman 2016).

Norway is building Arctic readiness throughout the armed forces and civil entities (including private business) as part of its "Total Defence" concept. All national-level resources can become involved in civil-military cooperation (CIMIC) activities, with civil authorities and private entities involved in contingency planning, crisis management, and emergency response (Norwegian Ministry of Defence 2020: 16). Within Total Defence planning, by definition the Norwegian military must participate in the response to a sub-strategic security threat by providing assistance to civil authorities. For example, during the Covid-19 Pandemic, the Home Guard established additional check points at EU's Schengen Border to prevent infected people from entering the country (Norwegian Government Security and Service Organization 2016: 19).

The Norwegian Army is increasing its Arctic capability by adding personnel and equipment to create Brigade North with four maneuver battalions, making it the Army's largest unit. It is adding firepower with new main battle tanks and new mobile air defence systems. As part of the army, the Home Guard is developing a more rapid response by forward-stationing equipment and ammunition. The 17th Guard District based in Finnmark will add new anti-armor missiles (Norwegian Ministry of Defence 2019).

The navy is modernizing its Nansen-class frigates and upgrading its six ULA-class submarines to operate effectively under Arctic conditions. Along with patrolling, submarines can transport special operations troops to emergencies. Coast guard replacement of its existing Nordkapp-class off-shore vessels with Jan Mayen-class Arctic ships will provide additional territorial and border protection and strengthen search-and-rescue capabilities. Special operations forces are purchasing new helicopters for added mobility. Coastal Rangers will train for counterterrorism and to support other emergencies along with their primary mission of intelligence, reconnaissance, and surveillance (Norwegian Ministry of Defence 2019).

To achieve an increased level of readiness and to support an expanded reserve, Norway will continue the conscription of both males and females (Wezeman 2016; Norwegian Ministry of Defence 2019; Norwegian Ministry of Defence 2020).

3.2.6 *Russia's Arctic Security Policy*

Russian Arctic policy calls for the protection of its Arctic territories containing important natural resources, especially petroleum. The policy also mandates developing and protecting the Northern Sea Route (NSR) as it increases in importance for Arctic shipping (Galimullin and Matveenko 2019). Exactly what Russian Arctic territory consists of is uncertain. The 2014 annexation of Crimea is not the only example of Russian hegemony. Russia annexed the Arctic region of Petsamo from Finland in 1944, acquiring nearly 20,000 square miles of land, of which 5,000 square miles was located on the Arctic Ocean (Kostanick 1951). Since Russia currently claims both the Mendeleev and Lomonosov Ridges under the Arctic Ocean, the annexation of other territories in the North Atlantic could occur.

Russia's 2008 Arctic strategy, updated in 2016, laid out the federal Arctic policy through 2020, illustrating the country's intent to pursue an independent and aggressive course of action in the region. Russia constructed its national strategy to maintain the country's security and sovereignty in the Arctic despite the region's difficult climate, sparse settlement, and long distance from industrial areas. Russian Arctic policy seeks to (1) increase the country's natural resources, (2) defend the state's territories and boundaries, (3) limit the effects of global climate change and increased exploitation of resources, (4) connect the north with other parts of Russia through information technology, and (5) ensure that other countries enter into and honor treaties respecting Russian sovereign territory and access to natural resources (Burgess 2010).

In addition, Russian Arctic policy stresses sub-strategic security threats that could arise as a result of the melting ice. With both military and security troops (especially uniformed members of the Russian Federal Security Service [FSB]), Russia will control its borders by strictly enforcing entry points to the Federation including newly exposed terrain features such as river mouths and shorelines along the NSR. Russia will also develop Arctic border facilities and improve the staffing, preparation, and equipment for border troops to fight the emerging problems of "maritime terrorism, contraband [smuggling], illegal migration, and the protection of sea-based resources" (Rusnak and Berman 2008: 102).

In 2013, Putin approved Russia's Arctic national security policy designed to guide the country's defence strategy through 2020. By that time, the Russian military, security, and border forces should have attained the readiness level necessary to identify and defeat all threats to the Federation. Readiness includes the ability for its forces to defend Russia's sovereign territory in the Arctic, including not only the borders but also the exclusive economic zone that extends to the continental shelf. The policy calls for increasing the number of troops operating in the Arctic and improving the training and support for those based in the region and those capable of operating within it but stationed outside the circumpolar area (Devyatkin 2018). The Russian military is the prime mover in the enforcement of the country's Arctic strategy, becoming especially active after nearly 25 years of relative dormancy (Winger and Petursson 2016).

Russia is placing renewed emphasis on Arctic defence by building or reoccupying bases, developing new weapons systems, creating new units, and reconfiguring existing units. These changes include: "a new Arctic command, four new Arctic brigade combat teams (approximately 6,000 troops), 14 operational airfields [in the region], 16 deep water ports, and 40 icebreakers with an additional 11 in development" (Gramer 2017). New sensors placed on Arctic islands like Franz Josef Land provide early warning of approaches through the NSR. The FSB will increase the number of uniformed border guards operating in the Arctic, although they are not part of the military (Mandraud 2014). New Russian Long-Range Aviation (LRA) forces contain bombers with ranges of up to 10,000 kilometers, specifically to reach into the Arctic from more southern bases. To manage increased Arctic military power, one of the five new joint strategic commands (OSKs) the Russians created was the OSK North, whose Arctic boundary leaves no doubt of its purpose. OSK North will project Russian military power into the circumpolar region to coincide with opening of the NSR (Defense Intelligence Agency 2017).

Following a period when Russia seemed to truly seek cooperation with other Arctic states, relations in the region and with NATO countries generally began to deteriorate due to Russian activity in Crimea and Syria. Additionally, Russian maritime policy, including improvements to the country's northern fleet and command elements, is targeted on the North Atlantic due to the NATO presence there. These developments, along with the too-oft mentioned flag planting on the Arctic Ocean floor in 2007, clearly illustrated Putin's desire to enforce Russia's will in the Arctic through military power (Klimenko 2020).

Russia's latest Arctic policy document, "Basic Principles of Russian Federation State Policy in the Arctic to 2035", continues most objectives of its earlier policy (Klimenko 2020; Krutikov 2020). The document presents the first overt statement about Russia's national security in the Arctic, and the change related to security indicates the increasing tension and military activity in the region between Russia and NATO. The new Arctic policy's number one goal is ensuring Russia's sovereignty and territorial integrity through enforcement operations by the armed forces, coast guard, and border guard on land, on the sea, and in the air. While expanding the economic impacts of the NSR, Russia intends to monitor and regulate foreign military vessels transiting the route through increased air/maritime patrols (Klimenko 2020). Opposition to Russian Arctic actions, especially those involving territorial disputes with other states, will be considered a security threat. An example might be the 1920 Svalbard Treaty, which Norway interprets currently as permitting its exclusive economic zone including the surrounding fishing beds (Holtsmark 1993; Klimenko 2020).

As the country's policy states, terrorism is a primary threat to Russian sub-strategic security in the Arctic. Importantly, Russia considers any attack on its territory to be a domestic issue, whether in the Arctic or elsewhere. That means Russia will send special operations troops (SOF) anywhere the government considers its own interests without regard to the implications of using them in either another country or international space. Russia viewed Crimea as its own territory and responded appropriately, including using its SOF wearing Ukrainian nationalist uniforms to assassinate anti-government demonstrators opposing nationalistic forces (Galeotti 2020).

The Russians utilize their SOF whether in a counterterrorism or a counter-insurgency role, such as in the Syrian War, which is providing them with real combat experience. Whether it's the army's Spetsnaz or the FSB's Alpha Squad, SOF act directly under Putin's standing authority, not requiring approval from the armed forces command to attack any threat (Sukhankin 2019). SOF have also participated in African anti-piracy operations and are engaged in Venezuela (Galeotti 2020; Sukhankin 2019). Thus, Russian SOF are well-trained, equipped, motived, and prepared to destroy any threat to their Arctic territories, passages, or resources.

Based on their recent actions elsewhere, there is no doubt that Russia will meet any circumpolar threats on its own territory in a similar fashion. I can only assume that, if Russia should engage in any Arctic sub-strategic offensive operations on another state's sovereign territory or within a disputed area, their response will be aggressive and final (Ilyushina and Smith-Spark 2020). After the utopian view some analysts held about how successfully the Arctic Council cooperated early on, it is chilling to realize that recent Russian military actions have brought a level of conflict not seen in the North Atlantic since the Cold War.

> All of Russia's Arctic neighbors have introduced economic sanctions and restrictive measures against Russia since 2014. Military cooperation has also been suspended for an unknown period. Despite the fact that tensions have stemmed from events outside the Arctic, the Arctic states have re-evaluated their cooperation with Russia in the region (Smith 2016: iv).

3.2.7 Sweden's Arctic Security Policy

Sweden now claims itself Arctic while recognizing that the country has no territory bordering on the Arctic Ocean. The authors of the 2011 policy document *Sweden's strategy for the Arctic region* spent much time justifying the country's membership in the Arctic Eight (it was the chair of the Arctic Council at the time). The government asserted Sweden's Arctic identity was based on historical and cultural grounds including being home to an indigenous Arctic people, sponsoring explorations into the circumpolar region, and conducting research activities on Spitsbergen and in their own Arctic territory (Swedish Ministry for Foreign Affairs 2011).

Sweden attempted to take advantage of the peace dividend following the end of the Cold War when it deactivated military units hundreds of years old and ended its mandatory conscription in 2010. That dividend lasted only a few years as Russian military actions in Georgia, Ukraine, and Syria caused the kingdom to begin rebuilding its security apparatus. Sweden's view was that it had not kept up with the Russian military growth. Consequently, the country had to strengthen EU collective security, expand the partnership with NATO, and honor the Nordic security agreement. Although Sweden is not subject to NATO's Article 5, Swedish policy is that it will respond to any attack or other disaster involving EU members as well as Iceland and Norway, expecting those countries to reciprocate (Swedish Defence Ministry 2020).

The 2020 Swedish security policy is built on the kingdom's 2017 total defence concept (Swedish Defence Ministry 2017). The concept maintained that Sweden's government, infrastructure, and economy must be prepared to operate for 3 months should an attack disrupt the country. The kingdom must also be able to function for the same amount of time should conflict reduce the ability of Europe or of a bordering country to operate. The policy anticipates a security threat that could use either military or hybrid activities to disrupt Sweden or its neighbors. To defeat the threat, Sweden must fight and win on its own territory using every means at its disposal from the military, security forces, and civil infrastructure (Swedish Defence Ministry 2020).

Importantly, Sweden believes it does not have the organization, units, equipment, personnel, or training to effectively defeat a strategic wartime threat. The report estimated that it would take until at least 2026 to have the capability to defeat such threats. The policy called for acquiring new equipment, reinstituting conscription, and developing the logistics and command structure needed for a modern military. It also recommended the reorganization of some existing forces, the activation of certain decommissioned units, and the creation of new ones. Many of these proposed changes involved conventional forces, such as a mechanized infantry brigade in Norrland, which could play a role in Sweden's defence of its Arctic territory. Importantly for sub-strategic security threats, the plan proposed that Sweden establish territorial units to assist in guarding and protecting key elements of the total defence including important infrastructure assets (Swedish Defence Ministry 2020).

In order to defeat sub-strategic security threats, the government recommended strengthening the capabilities of the Home Guard and SOF that could provide Defense Support to Civilian Authorities. Under certain circumstances, component commands like the army or the navy, which also have SOF, are placed under the Swedish Special Operations Command that was created in 2011. Through advanced training and equipment, Swedish special forces can operate domestically or within international coalitions, providing reconnaissance and human intelligence. SOF can perform hostage rescue, eliminate high-value targets, and provide military assistance to civilian governments (Swedish Defence Ministry 2020).

Other parts of Sweden's Arctic sub-strategic security are more difficult to identify because, like Finland, there are no units dedicated solely to the circumpolar region. Recently the defence minister noted that the military should establish border patrol units able to operate in sub-Arctic conditions. Those patrol units could work with Finland and Norway to secure borders or to prevent incursions. In addition, customs units, which are police entities under the Ministry of Finance, can function as border guards (Nilsen 2020). Since Sweden is an EU member surrounded almost exclusively by other EU states, it follows Schengen Border rules to control immigration and cross-border traffic. The exception is Norway that participates in the Nordic security agreement and the Nordic Defence Cooperation (Swedish Defence Ministry 2020).

Sweden also has a Civil Contingencies Agency that helps plan and execute emergency management under the Ministry of Health and Social Affairs (Swedish Civil Contingencies Agency 2020). Within the police authority, Sweden has a branch of the National Operations Department called the National Task Force. Its mission is to operate as a tactical unit during hostage and terrorism situations and to raise the public's terrorism awareness (Polisen 2020).

The Swedish State Security Service (SAPO) produces an annual terrorist threat assessment for the country. The latest document lists Islamic and domestic right-wing extremists as the top terrorism threats, with future attacks on targets in Sweden possibly carried out by lone perpetrators. Propaganda produced and disseminated by such organizations as ISIS combines with personal characteristics including mental illness, triggering factors like deaths of their family or friends in conflict zones, and personal association with violence-tolerating radical groups to induce terrorist activities. Right-wing extremists in Sweden are motivated by cultural-based hatred of foreigners who do not learn or accept traditional values even after several generations. Right-wing extremists utilize social media to convey their messages of hate, often spurred on by politicians elected using anti-foreigner platforms. Whether attacks are directed at the country's citizens or infrastructure, or they are carried out against other nationalities in the country, the terrorist threat in Sweden is real (NTC 2020).

Sweden presents a unique case for Arctic sub-strategic security analysis. Since early in the Cold War, the kingdom has planned for a situation where either the government falls or a large section of the country (probably the Arctic) is occupied by foreign troops during a limited war. Whether Sweden's neighbors would provide assistance during such times is uncertain, but the country possesses a public sector that can implement a response in its Arctic territory. It is uncertain whether the military or security forces will have the capability to defeat threats that may arise in the future Arctic, especially when Sweden admitted that currently they would be unable to do so.

3.2.8 United States' Arctic Security Policy

In reality, there is no U.S. Arctic policy as of this writing because the federal government has not emphasized the region for nearly a decade. Perhaps, the most glaring example of that disinterest in the Arctic was the Trump administration's recent diversion of funds earmarked for a coast guard icebreaker to build the border wall with Mexico (Wyland 2019). Despite government agencies like the air force having their own Arctic policies, the U.S. armed forces' focus on Middle Eastern conflicts and upon heightened Asian tensions has made America vulnerable to sub-strategic security threats in the North Atlantic.

The lack of emphasis on U.S. Arctic policy was illustrated perhaps best in President Trump's appointments to the Arctic Council. All of the Arctic Eight countries appointed ambassador-level officials (including flag officers like generals or admirals) to represent them on the Arctic Council. During the Obama Administration, Coast Guard Admiral Pape held the council's U.S. seat. However, Trump nominated low-level foreign service officers to be the U.S. Arctic representative (U.S. Department of State 2017b). His most recent Arctic Council appointment, Meredith Rubin, was a career foreign service officer with little Arctic experience. Despite being assigned to the U.S. Embassy in Reykjavík for 3 years in the early 2010s, in her own words she "did not know a lot about the Arctic Council" (Arctic Council 2020). From a security perspective, using career foreign service officers below the ambassador rank instead of flag officers severely damages U.S. Arctic security preparedness by giving the position less clout.

American Arctic security policy was traditionally dominated by strategic-level missions like early warning, missile defence, long-range bombers, and missile submarines (Bergh 2012). Yet Arctic sub-strategic security may be increasing in importance for the U.S., at least to senior military officers, as evidenced by Air Force General O'Shaughnessy in his testimony to congress on March 11, 2020. General O'Shaughnessy, then commander of the U.S. Northern Command with responsibility for the American homeland and the Arctic, argued against the Department of Defense's (DoD) current opinion that the Arctic is low risk. He insisted that the Arctic had a high conflict potential and therefore should be elevated to a high security priority (Burke 2020).

The 2020 DoD Arctic strategy is defensive in nature, stating that U.S. priorities are to "defend the homeland, compete when necessary to maintain favorable regional balances of power, and ensure common domains remain free and open". While the overwhelming majority of DoD-level Arctic operations involve strategic objectives and assets, one of the strategy's goals is to increase the ability of joint forces to perform Defense Support of Civil Authorities (DSCA). DSCA activities include working with state and local/tribal officials during emergencies such as "mass casualty incidents and natural disasters" (U.S. Department of Defense 2019: 6–7, 12).

The U.S. Air Force (USAF) has the greatest American military presence in the circumpolar region, controlling 80% of DoD's fixed assets in Alaska, Canada, and Greenland. While most Air Force and Space Force facilities, units, and activities concentrate on strategic missions like satellite communication, missile telemetry, early warning, and bomber intercepts, some could be used for sub-strategic protection and response. These include the New York Air National Guard's ski-capable LC-130s, pararescue airmen, and security forces located at Thule Air Base and Raven Camp in Greenland. The Air Force is ready to provide support to joint force, interagency, and civil activities in the Arctic, since it is the force with the most rapid capability to provide humanitarian assistance, disaster response, and personnel recovery (U.S. Air Force 2020).

U.S. Coast Guard operations in the Arctic involve both military and civilian aspects, which is unique in the American armed forces. In addition to its wartime role that can be transferred to the navy for military operations, the coast guard also has maritime interdiction authority over civilian vessels involved in illegal fishing or smuggling. That role in the Arctic has existed since the U.S. purchased Alaska from

Russia in 1867, when "Coast Guard Revenue Cutters first sailed to Alaska to establish U.S. sovereignty" (U.S. Coast Guard 2019: 4).

Although it is severely outnumbered by the Russians in the number of icebreakers it has, the service is making up for that shortfall by being innovative and adaptive. For example, instead of building an icebreaker fleet, the U.S. Coast Guard is developing a polar security cutter program to produce vessels that can perform combat missions as well as other tasks normally performed by icebreakers (Woody 2020). Those steps will be insufficient to improve security in the region, however, if the following comment by a former U.S. State Department official is true.

> The U.S. Coast Guard that is the lead agency with responsibilities for protecting the Arctic, continues to rely on outdated capabilities and thinly resourced budgets [that] equate to a seasonal presence (July-October) in the American Arctic. Should an incident occur [there, to be successful it must happen] during this season and preferably near a pre-positioned U.S. maritime asset [navy as well]. Years of underinvestment now leaves the United States ill-prepared as other nations prioritize the region as one of future geostrategic value (Conley 2019).

The U.S. Army released its Arctic policy during 2021 (U.S. Army 2021) with the grandiose title, "United States Army Regaining Arctic Dominance". As I detail in Chap. 4, neither the American Army nor the Marines has ever held dominance over ground combat in the Arctic. In fact, both ground forces have been incompetent in Arctic warfare dating back to World War II. The American inability to fight and win in the Arctic was described in the late 1940s thusly.

> A seasoned former Finnish army officer . . . in 1948 . . . observed, 'American infantry winter warfare ability is such that unless a miracle happens, the Russian Ski Divisions will cut these infantry divisions to pieces without mercy' (U.S. Army 2021: 13).

The situation remains the same today and the Army's new Arctic policy essentially is department-level (i.e., ministry-level) nonsense that makes political rather than operational statements about the readiness of its units to conduct Arctic combat. The policy is both descriptive and future-looking, but it lacks the proof that the Army could engage in extensive and long-term combat in the Arctic successfully. Problems defending the Arctic include the region being part of three combat commands that diffuses the leadership chain, the lack of training throughout units for cold weather environments, and the inability of the Army to create and maintain dedicated units for Arctic operations. With units stationed in the Arctic undergoing rotational deployments to the Middle East over the past 20 years, soldiers and the leaders stationed in Alaska are more prepared for desert warfare than Arctic combat.

To improve, the Army would need to use joint forces to operate in the Arctic, involving not only Active Component troops but also Army Reserve and National Guard units. Although it is not outlined specifically, the Army policy contends that Arctic units should incorporate indigenous peoples' knowledge about survivability and mobility in the circumpolar region. One way the Army will use indigenous knowledge is by standing up Alaska Scouts (similar to Canadian Rangers) as part of the Alaskan Army National Guard (U.S. Army 2021).

The policy states that the first step toward the Army "regaining" dominance will be to station a multi-domain task force (MDTF) in Alaska that is trained and equipped to fight and win in the Arctic. The U. S. Army's Alaska Command controls several assets that are dedicated to defeating threats in the circumpolar region. These units include the 4th Airborne Infantry Brigade, the only airborne infantry brigade in the Pacific Theater that could provide a rapid military response to an Arctic crisis. Alaska Command also includes the 1st Attack Reconnaissance Battalion (nicknamed the Arctic Attack Battalion) that uses AH-64 Apache and support helicopters. While the army maintains that these units are stationed there to defend the Arctic, those forces could be transferred quickly to the operational control of the U.S. Pacific Command for an Asian crisis (U.S. Army 2020).

That is not the official position of the Alaska Commander who claims that the service branch has made a significant "pivot . . . to train and operate more in Alaska to rebuild skills". That pivot, according to Major General Andrysiak, consists of bringing troops fighting in the Middle East back to Alaska, where they can train for cold-weather fighting. It also consists of making investments in Alaskan facilities and developing an Army-specific Arctic strategy. The general also claimed there was a demonstrably increased interest in the Arctic because the Secretaries of the Army, Navy, and the Air Force visited Alaska during 2020 (Bye 2020: 1).

The U.S. Navy and its amphibious force the Marines are truly global assets. The Navy is reestablishing its presence in the North Atlantic by restarting anti-submarine patrols out of Keflavik, conducting submarine hunting and ballistic missile operations, and patrolling in the Barents Sea with major combat ships. Some analysts characterize that high tempo as "The U.S. Navy is back in the Arctic Circle after being gone for 30 years" (Larter 2020).

The U.S. Marines have been stationed in Norway for 4 years, training and participating in exercises like Trident Junction, although their rotational periods will change from 6 months to shorter periods depending on the type of exercise being completed (Harkins 2020). Marines also trained in Iceland and other cold weather areas to a lesser degree. Marines have a new uniform for cold weather; however, whether that uniform will be available in sufficient numbers if the force needs to rapidly respond to emergencies is uncertain. Marines have the potential to participate in sub-strategic security missions throughout the Arctic. The problem with the Marines, as I have written elsewhere (Wheelersburg 2021) is that they are not always available for a rapid deployment (i.e., within 24 hours) for threats like terrorist attacks in the circumpolar region. For that reason, the Navy Department, including the Marines, does not seem suited for Arctic sub-strategic security missions when speed is of the essence.

In addition to the armed forces and coast guard, Alaskan sub-strategic security threats can be met at the federal level by the U.S. Customs and Border Protection (CBP). CBP maintains a Special Operations Group (SOG) to respond rapidly to tactical or emergency response missions (U.S. Customs and Border Protection 2020). Based upon mission requirements provided by the CBP's intelligence unit, the SOG can respond to terrorist or humanitarian emergencies. Below the federal level, sub-strategic security threats can be attacked by the Alaskan State Police's

Special Emergency Response Team (SERT). SERT can accomplish missions like hostage rescue in the northern wilderness with its own aviation or water transport, and its own self-sufficient logistical support (Hall 2008).

3.3 Summary

For two decades, demilitarization characterized the Arctic (Wezeman 2016), perhaps symbolized most by the sudden abandonment of the NATO base at Keflavik in 2006. In fact, I believe the U.S. should have closed the base at least a decade earlier following the breakup of the Soviet Union and the subsequent disappearance of Russian submarines and bombers from the North Atlantic. During the 1990s there was essentially no threat to the region from Russia, and much of what the Keflavik Base did had nothing to do with strategic security operations. Rather, the U.-S. provided important infrastructure construction and financial support (including employment) to the surrounding community.

One important accomplishment of the Keflavik NATO Base operations that occurred during the post-Cold War period was an initial test of the search-and-rescue agreement developed by the Arctic Council and signed in 2011 (Arctic Council 2011). Ambassador Benedikt Ásgeirsson twice served as director of the defence and security directorate within the Ministry of Foreign Affairs during the 1990s. During his time as director, Keflavik hosted NATO Partnership for Peace activities such as the 1997 Cooperative Safeguard disaster assistance exercise, in which I participated. Ásgeirsson believes those exercises, which included Russian and former Warsaw Pact troops laid the groundwork for the search-and-rescue agreement that was one of the Arctic Council's first accomplishments (Ásgeirsson 2015).

Before its closing, with no requirement to fly anti-submarine patrols, in 2002 the Navy turned Keflavik over to the U.S. Air Force command and left. Gradually, even the Air Force bomber intercept missions from Keflavik ended because the Global War on Terror in Afghanistan and Iraq drew their fighter aircraft away. Without the jets, the Keflavik base's helicopters were no longer needed to rescue downed NATO pilots and crews. Instead, those helicopters spent several years plucking Icelandic commercial fishermen from distressed vessels in the North Atlantic. For example, during early 2006, HH-60 helicopters rescued Icelandic fishermen in two separate incidents (Eisele 2006). The base continued to deactivate with schools and medical/dental facilities closing, dependent families leaving, and military units departing until Secretary of Defense Rumsfeld closed the American portion of the base in 2006 (Lachowski 2007).

The short-lived Icelandic Defence Agency took over operation of the base after the Americans left, followed by the Icelandic Coast Guard that assumed control over Keflavik when the agency was disbanded. Since 2008, however, NATO and its Partnership for Peace nations such as Finland and Sweden began conducting several annual air-policing missions against Russian bombers in the Icelandic Air Defence Identification Zone (Vandiver 2020). In addition, the 2018 U.S. defence budget included $17 million to improve Keflavik hangers to house the Navy's P-8 Poseidon anti-submarine patrol aircraft, which is much larger than the older P-3 Orion (Cisneros 2020). With the recent increase in strategic activities on both sides of the NATO–Russian conflict, the Cold War is back in the North Atlantic, leaving the Arctic Council an impotent observer.

Along with the increase in strategic activities in the North Atlantic, global climate change is presenting opportunities for sub-strategic security threats to pursue their goals in the Arctic. The melting ice in the Arctic Ocean will open up unimproved landing strips and natural harbors in Iceland that can be used by those engaging in criminal activities, terrorism, territorial incursions, and natural resource theft. At present, Iceland is left to its own devices when it comes to providing an immediate and effective response to such sub-strategic security threats.

While the simplistic conclusion is that Iceland is unprepared because it lacks a military, the comparative analysis presented above indicates that Iceland is insufficiently prepared to defeat future sub-strategic security threats in the following areas:

- Iceland lacks a reconnaissance unit to monitor remote areas for limited border incursion by another country (although ICE-SAR could readily assume that mission).
- Iceland has no border guard with sufficient weaponry to respond to organized criminals or terrorists attempting to enter the country using remote ports or airfields.
- Iceland has no fixed-wing aircraft (e.g., C-130J) capable of transporting personnel and equipment rapidly to a site in sufficient numbers to respond to a large-scale emergency.
- Iceland has no portable anti-armor or anti-air weapons to defeat a sub-strategic assault by an enemy combatant attacking with armored vehicles or aircraft.
- Icelandic security forces cannot contribute substantial Defense Support to Civilian Authorities (DSCA) in an emergency.
- Iceland has no Arctic ground-based security capabilities to work with other nations' forces responding to threats traversing the TSR into the North Atlantic.

The proposed Icelandic security force recommended in the next chapter could eliminate all of those deficiencies without creating a military. If adopted, the proposed security force could increase the country's ability to counter future sub-strategic security threats effectively, both by deterring enemies before they attack and by being able to defeat them once they do. In addition to reducing the country's vulnerability, the proposed security force could help meet Iceland's military contribution to NATO through both direct funding and personnel. Iceland being able to defend itself against sub-strategic security threats could fill the vacuum that currently exists in the North Atlantic. Rather than secretly purchasing weapons, arming police, or sending resources outside of the country to participate in NATO combat operations, none of which seem acceptable to the public, an Icelandic security force prepared to meet sub-strategic security threats in its own territory would indicate to Icelanders, allies, and foes alike that the country is secure – but on its own terms.

The next three chapters present, in order, a recommendation for a possible Icelandic security force, four scenarios that illustrate successful employment of that security force against different threats, and a discussion of how that proposed security force could be accepted by Icelanders while conforming to the country's existing policies, long history, and anti-militaristic culture.

References

Arctic Council. 1996. *Declaration on the Establishment of the Arctic Council*. Joint Communique of the Governments of the Arctic Countries on the Establishment of the Arctic Council. September 16, 1996. Ottawa. Retrieved July 8, 2020, https://www.maritime-executive.com/editorials/the-arctic-shipping-route-no-one-s-talking-about.

———. 2011. *Agreement on Cooperation on Aeronautical and Maritime Search and Rescue in the Arctic*. Tromsø: Arctic Council. Accessed November 10, 2020, https://oaarchive.arctic-council.org/handle/11374/531.

———. 2020. *Interview with Meredith Rubin, The United States' Senior Arctic Official, December 2, 2020*. Accessed October 23, 2020, https://arctic-council.org/en/news/interview-with-meredith-rubin-the-united-states-senior-arctic-official/.

Ásgeirsson, Benedikt. 2015. Personal Interview, November 11, 2015. Icelandic Ministry of Foreign Affairs, Reykjavík.

Askham, Poppy. 2020. Growing Calls for Reform Following House Fire That Killed Three Polish Workers. *The Reykjavík Grapevine*, June 30, 2020. Accessed August 30, 2020, https://grapevine.is/news/2020/06/30/growing-calls-for-reform-following-house-fire-that-killed-three-polish-workers/.

Balestrieri, Steve. 2019. Fire and Ice: Iceland's Elite, "Viking Squad" Swat Team. *SOFREP*, February 12, 2019. Accessed September 1, 2020, https://sofrep.com/specialoperations/fire-and-ice-icelands-elite-viking-squad-swat-team/.

Bennett, Mia. 2019. The Arctic Shipping Route No One's Talking About. *The Maritime Executive*, May 8, 2019. Retrieved July 8, 2020 https://www.maritime-executive.com/editorials/the-arctic-shipping-route-no-one-s-talking-about.

Bergh, Kristofer. 2012. *The Arctic Policies of Canada and the United States: Domestic Motives and International Context*. Stockholm International Peace Research Institute (SIPRI) Insights on Peace and Security, No. 2012/1, July 2012. Accessed October 29, 2020, https://www.sipri.org/publications/2012/sipri-insights-peace-and-security/arctic-policies-canada-and-united-states-domestic-motives-and-international-context.

Björnsson, Anna Margrét. 2014. The Gun Plot Thickens? *Iceland Monitor*, October 24, 2014. Accessed July 2, 2020, https://icelandmonitor.mbl.is/news/news/2014/10/24/the_gun_plot_thickens/.

Burgess, Philip. 2010. *Foundations of the Russian Federation's State Policy in the Arctic Until 2020 and Beyond*. Translation [Original in Russian] December 1, 2010. Akureyri: Arctic Portal. Accessed November 1, 2020, https://icr.arcticportal.org/index.php?option=com_content&%20view=article&%20id=1791%3.

Burke, Ryan. 2020. *Great-Power Competition in the "Snow of Far-Off Northern Lands": Why We Need a New Approach to Arctic Security*. April 8, 2020. West Point: Modern War Institute. Accessed October 29, 2020, https://mwi.usma.edu/great-power-competition-snow-far-off-northern-lands-need-new-approach-arctic-security/.

Bye, Hilde-Gunn. 2020. Increased Interest in the Arctic: "The U.S. Army Has Made a Significant Pivot. *High North News*, October 22, 2020. Accessed November 10, 2020, https://www.highnorthnews.com/en/increased-interest-arctic-us-army-has-made-significant-pivot.

CIA. 2017. *Central Intelligence Agency, World Factbook. Country Comparison, Marchant Marine*. Washington, DC. Accessed September 14, 2020, https://www.cia.gov/library/publications/the-world-factbook/rankorder/2108rank.html.

Cisneros, Mila. 2020. Air Force Awards Multiple Contracts for Airfield Construction at NAS Keflavik. *U.S. Air Force News*. September 24, 3030. Accessed September 26, 2020, https://www.af.mil/News/Article-Display/Article/2359356/air-force-awards-multiple-contracts-for-airfield-construction-at-nas-keflavik/.

Coates, Ken S., Greg Poetzer, William R. Morrion, and P. Whitney. 2008. *Arctic Front: Defending Canada in the Far North*. Toronto: T. Allen Publishers.

Conley, Heather A. 2019. *The Implications of U.S. Policy Stagnation Toward the Arctic Region*. Washington, DC: Center for Strategic & International Studies. Accessed November 10, 2020, https://www.csis.org/analysis/implications-us-policy-stagnation-toward-arctic-region.

Danish Military Command. 2020. Personal communication. September 2, 2020.

Danish Ministry of Defence. 2019. *Agreement for Danish Defence 2018–2023*. Accessed August 21, 2020, https://fmn.dk/en/topics/agreements-and-economi/agreement-for-danish-defence-2018%2D%2D-2023/.

Defence Command. 2013. *Report on the Designation of an Arctic Contingency* [original in Danish] Copenhagen.

Defense Intelligence Agency. 2017. *Russia Military Power*. Washington, DC: U.S. Defense Intelligence Agency. Accessed November 22, 2020, https://www.dia.mil/Portals/27/Documents/News/Military%20Power%20Publications/Russia%20Military%20Power%20Report%202017.pdf.

Devyatkin, Pavel. 2018. *Russia's Arctic Strategy: Military and Security (Part II)*. Washington, DC: The Arctic Institute, February 13, 2018. Accessed September 22, 2020, https://www.thearcticinstitute.org/russias-arctic-military-and-security-part-two/.

Ehrat, Noemi. 2018. The Icelandic Opioid Crises: Iceland Approaching U.S. Numbers in Drug-Related Deaths. *The Reykjavík Grapevine*, August 27, 2018. Accessed August 30, 2020, https://grapevine.is/news/2018/08/27/the-icelandic-opioid-crisis-iceland-approaching-u-s-numbers-in-drug-related-deaths/.

Eisele, Travis D. 2006. *Airmen Rescue Icelandic Fisherman. U.S. Air Forces in Europe and Air Forces Africa*. March 20, 2006. Accessed November 1, 2020, https://www.usafe.af.mil/News/Article-Display/Article/256469/airmen-rescue-icelandic-fisherman/.

Finnish Army. 2020. *The Finnish Army*. Accessed August 26, 2020, https://maavoimat.fi/en/frontpage.

Finnish Border Guard. 2020. *Raja [Border] 100*. Accessed August 26, 2020, https://www.raja.fi/en.

Finnish Defence Command. 2020. *The Nyland Brigade Produces Coastal Jæger Troops*. Helsinki: Finnish Defence Command. Accessed October 28, 2020, https://merivoimat.fi/en/nyland-brigade.

Finnish Ministry of Defence. 2020. *Defence Policy*. Accessed August 25, 2020, https://www.defmin.fi/en#c4fcb92c.

Finnish Prime Minister. 2013. *Finland's Strategy for the Arctic Region 2013*. Government Resolution on 23 August 2013. Prime Minister's Office Publications 16/2013. Helsinki: Edita Prima. Accessed August 24, 2020, https://vnk.fi/documents/10616/334509/Arktinen+strategia+2013+en.pdf/6b6fb723-40ec-4c17-b286-5b5910fbecf4/Arktinen+strategia+2013+en.pdf.

———. 2016. *Government Report on Finnish Foreign and Security Policy*. Helsinki: Prime Minister's Office Publications, 9/2016. Accessed November 3, 2020, https://valtioneuvosto.fi/documents/10616/1986338/VNKJ092016+en.pdf/b33c3703-29f4-4cce-a910-b05e32b676b9.

Fouche, Gwladys. 2018. Norway to Invite More U.S. Marines, for Longer and Closer to Russia. *Reuters*, June 12, 2018. Accessed September 20, 2020, https://www.reuters.com/article/us-norway-us-russia/norway-to-invite-more-u-s-marines-for-longer-and-closer-to-russia-idUSKBN1J8149.

Galeotti, Mark. 2020. *Spetsnaz: Operational Intelligence, Political Warfare, and Battlefield Role*. Security Insights, Russia, February 2020, Number 046. Garmisch-Partenkirchen: George C. Marshall European Center for Security Studies. Accessed November 24, 2020, https://www.marshallcenter.org/en/publications/security-insights/spetsnaz-operational-intelligence-political-warfare-and-battlefield-role-0.

Galimullin, Eduard, and Yuri Matveenko. 2019. The Ongoing Formation of Russia's Arctic Policy: A New Stage? In *Redefining Arctic Security, Arctic Yearbook 2019*, ed. Lassi Heininen, Heather Exner-Pirot, and Joël Plouffee, 305–321. Akureyri: Arctic Portal. Retrieved on July 11, 2020, https://issuu.com/arcticportal/docs/ay2019.

Government of Canada. 2010. *Canada Command – OP Nanook 2010. National Defense and the Canadian Forces*. Accessed December 2, 2020, https://web.archive.org/web/20101130083856/http://www.canadacom.forces.gc.ca/daily/archive-nanook10-eng.asp#050810.

———. 2017. *Strong, Secure, Engaged. Canada's Defence Policy*. Ottawa: Ministry of Defence. Accessed July 7, 2020 https://www.canada.ca/en/department-national-defence/corporate/reports-publications/canada-defence-policy.html.

———. 2019. *Statement on Canada's Arctic Foreign Policy. Exercising Sovereignty and Promoting Canada's Northern Strategy Abroad*. Ottawa. Retrieved July 8, 2020, https://www.international.gc.ca/world-monde/assets/pdfs/canada_arctic_foreign_policy-eng.pdf.

Gramer, Robbie. 2017. Here's What Russia's Military Build-Up in the Arctic Looks Like. *Foreign Policy*, January 25, 2017. Accessed September 23, 2020, https://foreignpolicy.com/2017/01/25/heres-what-russias-military-build-up-in-the-arctic-looks-like-trump-oil-military-high-north-infographic-map/.

Greaves, Wilfrid, and Daniel Pomerants. 2017. 'Soft Securitization': Unconventional Security Issues and the Arctic Council. *Politik*, Nummer 3, Årgang 20: 2017, Theme: Arctic International Relations in a Widened Security Perspective", guest edited by Marc Jacobsen and Victoria Herrmann, 31–46.

Hall, Jeff. 2008. Alaska Special Emergency Response Team. *Tactical Life*, May 1, 2008. Accessed November 9, 2020, https://www.tactical-life.com/firearms/alaska-special-emergency-response-team/.

Harkins, Gina. 2020. Major Changes Coming for Marines' Norway Deployments. Military.com, August 7, 2020. Accessed November 18, https://www.military.com/daily-news/2020/08/07/major-changes-coming-marines-norway-deployments.html#:~:text=The%20Marine%20Corps%20will%20overhaul,months%20for%20nearly%20four%20years.

Holtsmark, Sven G. 1993. *A Soviet Grab for the High North? USSR, Svalbard, and Northern Norway 1920–1953*, Defense Studies, 7/1993. Oslo: Institute for Defence Studies. Accessed October 22, 2020, https://core.ac.uk/download/pdf/52082845.pdf.

Iceland Monitor. 2019. Record Number of Cruise Ships to Iceland This Year. *Iceland Monitor*, January 22, 2019. Accessed August 30, 2020, https://icelandmonitor.mbl.is/news/nature_and_travel/2019/01/22/record_number_of_cruise_ships_to_iceland_this_year/#:~:text=On%20July%202019%20a%20historic,some%20345%20metres%20in%20length.

Iceland Review. 2015. Four F-16 Fighter Jets Coming. *Iceland Review*, August 28, 2015. Accessed August 23, 2020, https://www.icelandreview.com/news/four-f-16-fighter-jets-coming/.

Icelandic Coast Guard. n.d. *The Icelandic Coast Guard "Always Prepared"*. Accessed September 2, 2020, https://www.lhg.is/media/LHG80/Landhelgisgasla_Islands_enska2_.pdf.

Icelandic Ministry for Foreign Affairs. n.d. *Iceland Crises Response Unit*. Reykjavík: Government of Iceland. Accessed September 2, 2020, https://www.government.is/topics/foreign-affairs/icru/.

Icelandic Ministry for Foreign Affairs Ad-hoc Committee. 2009. *A Risk Assessment for Iceland: Global, Societal, and Military Factors*. English Summary of the Findings of an Interdisciplinary Commission Appointed by the Icelandic Foreign Ministry. Reykjavík. Accessed August 30, 2020, https://www.stjornarradid.is/media/utanrikisraduneyti-media/media/Skyrslur/A_Risk_Assessment_for_Iceland_-_English_Summary.pdf.

Icelandic Parliament. 2011. *A Parliamentary Resolution on Iceland's Arctic Policy*. Althingi 139th legislative session, March 28, 2011. Reykjavík. Accessed August 29, 2020, https://www.government.is/media/utanrikisraduneyti-media/media/nordurlandaskrifstofa/A-Parliamentary-Resolution-on-ICE-Arctic-Policy-approved-by-Althingi.pdf.

———. 2016. *Parliamentary Resolution Establishing a National Security Policy for Iceland (no. 26/145; April 13, 2016)*. Accessed July 20, 2020, https://www.government.is/

library/01-Ministries/Prime-Ministrers-Office/Parl-Res-Nat-Sec-Council-25-1-5-2016-tr-br-prot-SKIL-310817-.pdf.

Ilyushina, Mary, and Laura Smith-Spark. 2020. Putin Says If Russia Wanted to Kill Opposition Leader Navalny, It Would Have 'Finished' the Job. *CNN*, December 17, 2020. Accessed November 22, 2020, https://www.cnn.com/2020/12/17/europe/putin-annual-press-conference-intl/index.html.

Jørgensen, Kim J. 2016. *Denmark's Military Capacities in the Arctic Region. Defence Committee 2015–16*. FOU Alm.del Bilag 138. Offentligt. Copenhagen. Accessed August 21 2020, https://www.ft.dk/samling/20151/almdel/fou/bilag/138/1642107.pdf.

Kikkert, Peter, and P. Whitney Lackenbauer. 2019. Canada's Arctic and Northern Policy Framework: A Roadmap for the Future? In *Redefining Arctic Security. Arctic Yearbook 2019*, ed. Lassi Heininen, Heather Exner-Pirot, and Joël Plouffee, 322–331. Akureyri Arctic Portal. Retrieved on July 11, 2020, https://issuu.com/arcticportal/docs/ay2019.

Klimenko, Ekaterina. 2020. Russia's New Arctic Policy Document Signals Continuity Rather than Change. *SIPRI*, 6 April 2020. Accessed September 15, 2020, https://www.sipri.org/commentary/essay/2020/russias-new-arctic-policy-document-signals-continuity-rather-change.

Kongsberg. 2019. *Coastguard Vessel Design – UT 512 Coastguard*. Accessed September 2, 2020, https://www.kongsberg.com/maritime/products/ship-design/other-specialised-vessel-designs/coastguard-vessel-designs/coastguard-vessel-design%2D%2D-ut-512-coastguard/.

Kostanick, Huey Louis. 1951. Soviet Territorial Annexations in Eastern Europe. *Yearbook of the Association of Pacific Coast Geographers* 13: 14–18. University of Hawaii Press.

Kraska, James, ed. 2011. *Arctic Security in an Age of Climate Change*. Cambridge: Cambridge University Press.

Krutikov, Alexander V. 2020. *Russia's Ten Priorities in the Arctic*. Rovaniemi: University of the Arctic. Accessed September 15, 2020, https://www.uarctic.org/shared-voices/shared-voices-magazine-2020/russia-s-ten-priorities-in-the-arctic/.

Kuusela, Janne. 2020. A View from Finland: Security and Defence in the Arctic. *Defence News. Frozen Pathways*, May 12, 2020. Accessed August 24, 2020. https://www.defencenews.com/smr/frozen-pathways/2020/05/11/a-view-from-finland-security-and-defence-in-the-arctic/.

Lachowski, Zdzislaw. 2007. *Foreign Military Bases in Eurasia*, SIPRI Policy Paper No. 18. Stockholm: Stockholm International Peace Research Institute Accessed September 22, 2020, https://www.files.ethz.ch/isn/31819/Policypaper18.pdf.

Larter, David B. 2020. The US Navy returns to an increasingly militarized Arctic. *Defence News*, May 12, 2020. Accessed November 12, 2020, https://www.defensenews.com/naval/2020/05/11/the-us-navy-returns-to-an-increasingly-militarized-arctic/.

Löve, Arndís Sue Ching, Jose Antonio Baz-Lomba, Malcom J. Reid, Aino Kankaanpää, Teemu Gunnar, Maria Dam, Kristín Ólafsdóttir, and Kevin V. Thomas. 2018. Analysis of Stimulant Drugs in the Wastewater of Five Nordic Capitals. *Science of the Total Environment* 627: 1039–1047, 15 June 2018.

Mandraud, Isabelle. 2014. Russia Prepares for Ice-Cold War with Show of Military Force in the Arctic. *The Guardian*, October 21, 2014. Accessed September 22, 2020, https://www.theguardian.com/world/2014/oct/21/russia-arctic-military-oil-gas-putin.

Marine Alutech. 2020. *Combat Support Service Vessel*. Watercat M18 AMC. Accessed October 29, 2020, http://www.marinealutech.com/boats/landing_crafts/watercat_m18_amc.

Matzen, Erik. 2017. Denmark Spurned Chinese Offer for Greenland Base Over Security: Sources. *Reuters, World News*, April 6, 2017. Accessed August 23, 2020. https://www.reuters.com/article/us-denmark-china-greenland-base/denmark-spurned-chinese-offer-for-greenland-base-over-security-sources-idUSKBN1782EE.

McGwin, Kevin. 2017. For the Second Time a Local Firm Has Lost a US Defence Contract in Greenland – Despite a Lower Bid. *Arctic Today*. Accessed August 23, 2020, https://www.arctictoday.com/for-the-second-time-a-local-firm-has-lost-a-us-defence-contract-in-greenland-despite-a-lower-bid/.

Mendershavsen, Horst. 1980. *Reflections on Territorial Defence*. Santa Monica: Rand Corporation. Accessed August 25, 2020, https://www.rand.org/pubs/notes/N1265.html#:~:text=A%20territorial%20defence%20posture%20is,different%20in%20type%20and%20composition.

Ministry of Foreign Affairs of Denmark. 2011. *Denmark, Greenland, and the Faroe Islands. Kingdom of Denmark. Strategy for the Arctic 2011–2020*. Copenhagen: Rosendahls-Shultz grafisk a/s. Accessed August 21, 2020, https://um.dk/en/foreign-policy/the-arctic.

———. 2018. *The Danish Government Foreign and Security Policy Strategy 2019–2020*. Copenhagen. Accessed August 21, 2020, https://www.fmn.dk/globalassets/fmn/dokumenter/nyheder/engelske/-2018-11-15-foreign-and-security-policy-strategy-2019-2020-.pdf.

———. 2019. *Ukrainian-Danish Fuel Transportation Operation "Northern Falcon 2019" Has Started*. Copenhagen. Accessed December 2, 2020, https://ukraine.um.dk/en/about-us/news/newsdisplaypage/?newsid=4f2accb6-2531-4fd6-a59f-9df5cc94c9c0.

NATO. 2018. Trident Junction 2018. Brussels. Accessed October 23, 2020, https://www.nato.int/cps/en/natohq/157833.htm.

———. n.d. *Origins*. Brussels: My Country and NATO. Canada and NATO. Accessed October 22, 2020, https://www.nato.int/cps/en/natohq/declassified_161511.htm?selectedLocale=en.

Nicol, Heather, Barret Weber, Joshua Barkan, Philip Steinberg, Jeremy Tasch, and Hannes Gerhardt. 2015. *Contesting the Arctic: Politics and Imaginaries in the Circumpolar North*. New York: I. B. Tauris.

Nilsen, Thomas. 2020. Sweden to Re-establish Northern Regiment, Fears Armed Attack. *The Barents Observer*, October 17, 2020. Accessed November 18, 2020, https://thebarentsobserver.com/en/security/2020/10/sweden-re-establish-northern-regiment-fears-armed-attack.

Norwegian Government Security and Service Organization. 2016. *Interaction for Security. Protection of Basic Social Functions in a Changing Time*. NOU 2016: 19. Oslo. Accessed September 8, 2020, https://www.regjeringen.no/contentassets/03960058f3f94fbe9d290593bee22c1a/no/pdfs/nou201620160019000dddpdfs.pdf.

Norwegian Ministry of Defence. 2019. *Future acquisitions for the Norwegian Defence Sector 2019–2026*, March 2019. Oslo: Ministry of Defence. Accessed September 13, 2020, https://www.regjeringen.no/contentassets/7635ac0d48d44fc180fac57f58be7518/faf-2019-2026-english---final.pdf.

———. 2020. *The Defence of Norway. Capability and Readiness. Long Term Defence Plan 2020*. Oslo: Ministry of Defence. Accessed September 11, 2020, https://www.regjeringen.no/contentassets/3a2d2a3cfb694aa3ab4c6cb5649448d4/long-term-defence-plan-norway-2020---english-summary.pdf.

Norwegian Ministry of Foreign Affairs. 2006. *The Norwegian Government's High North Strategy*. Oslo/Tromsø: Norwegian Ministry of Foreign Affairs. Accessed September 11, 2020, https://www.regjeringen.no/globalassets/upload/ud/vedlegg/strategien.pdf.

———. 2015. *Global Security Challenges in Norway's Foreign Policy. Terrorism, Organized Crime, Piracy and Cyber Threats. Meld. St. 37 (2014–2015) Report to the Storting (White Paper) Summary*. Oslo: Norwegian Ministry of Foreign Affairs. Accessed September 6, 2020, https://www.regjeringen.no/contentassets/bdf4bd40d57d4dc79409de87419a2217/en-gb/pdfs/stm201420150037000engpdfs.pdf.

———. 2017. *Setting the Course for Norwegian Foreign and Security Policy. Meld. St. 36 (2016–2017) Report to the Storting (White Paper)*. Oslo: Norwegian Ministry of Foreign Affairs. Accessed September 5, 2020, https://www.regjeringen.no/contentassets/0688496c2b764f029955cc6e2f27799c/en-gb/pdfs/stm201620170036000engpdfs.pdf.

NTC. 2020. *The Terrorist Threat to Sweden 2020*. Stockholm: National Centre for Terrorist Threat Assessment. Accessed November 2, 2020, https://www.sakerhetspolisen.se/download/18.f2735ce171767402ba3eb/1600433792019/NCT-one-year-assessment-2020.pdf.

O'Dwyer, Gerard. 2020. Finland to bolster Navy's surface fleet with new ships, more missiles. *Defence News*, January 13, 2020. Accessed October 29, 2020, https://www.defencenews.com/

digital-show-dailies/surface-navy-association/2020/01/13/fin and-to-bolster-navys-surface-fleet-with-new-ships-more-missiles/.

Overseas Security Advisory Council. 2019. *Iceland 2019 Crime & Safety Report*. Overseas Security Advisory Council (OSAC). March, 8, 2019. Country Security Report, U.S. Diplomatic Security Service. Accessed November 13, 2020, https://www.osac.gov/Country/Iceland/Content/Detail/Report/2d8cdd85-eb54-4ec7-828a-15f4aeb18469.

Paumgarten, Nick. 2015. Life Is Rescues. Looking for Trouble with a National Team of Emergency-Response Volunteers. *The New Yorker*, Letter from Iceland, November 9, 2015. Accessed December 2, 2020, https://www.newyorker.com/magazine/2015/11/09/life-is-rescues.

Police Security Service. 2020. *National Threat Assessment 2020* [original in Norwegian]. Oslo: PST. Accessed September 7, 2020, https://www.pst.no/alle-artikler/trusselvurderinger/nasjonal-trusselvurdering-2020/.

Polisen. 2020. *The Swedish Police Authority*. Accessed November 1, 2020, https://polisen.se/en/the-swedish-police/the-swedish-police-authority/.

Pressburger, Emeric. 1941. *49th Parallel*. London: Ortus Film. Accessed July 20, 2020, https://www.youtube.com/watch?v=AtEtMUXjuKw.

Rasmussen, Rasmus Kjærgaard. 2019. The Desecuritization of Greenland's Security? How the Greenlandic Self-Government Envision [sic] Post-independence National Defence and Security Policy. In *Redefining Arctic Security, Arctic Yearbook 2019*, ed. Lassi Heininen, Heather Exner-Pirot, and Joël Plouffee, 287–304. Akureyri: Arctic Portal. Retrieved on July 11, 2020, https://issuu.com/arcticportal/docs/ay2019.

Royal House of Norway. 2013. *Visited Operation Ocean Shield*. October 23, 2013. Oslo. Accessed September 22, 2020, https://www.royalcourt.no/nyhet.html?tid=118458.

Rusnak, Maxim, and Ilan Berman, translators. 2008. *Russia's New Arctic Strategy. The Foundations of Russian Federation Policy in the Arctic 2020 and Beyond*. Accessed June 15, 2020, http://www.arctic.or.kr/files/pdf/m4/rusia_eng.pdf.

Salter, Mark B. 2019. Arctic Security, Territory, Population: Canadian Sovereignty and the International. *International Political Sociology* 13(4): 358–374, December 2019. Accessed August 20, 2020, https://academic.oup.com/ips/article-abstract/13/4/358/5568066?redirectedFrom=fulltext.

Sigurjónsson, Arnór. 2021. *Personal interview at the Icelandic Ministry for Foreign Affairs*, June 10, 2021.

Smith, Dan. 2016. Preface. In *Russia's Arctic Security Policy. Still Quiet in the High North?* SIPRI Policy Paper No. 45, ed. Ekaterina Klimenko, iv. Stockholm: Stockholm International Peace Research Institute. Accessed November 2, 2020, https://www.sipri.org/sites/default/files/SIPRIPP45.pdf.

Standish, Reid. 2018. *Finland's race for Arctic riches*. Politico, March 28, 2018. Brussels: Politico. Accessed October 29, 2020, https://www.politico.eu/contact-us/

Storey, Norman. 1992. *What Price Cod? A Tugmaster's View of the Cod Wars*. Beverley: Hulton Press.

Sukhankin, Sergey. 2019. Russian Special Operations Forces: Image Versus Substance. *Eurasia Daily Monitor* 16(43), March 27, 2019. Accessed October 28 2020, https://jamestown.org/program/russian-special-operations-forces-image-versus-substance/

Swedish Civil Contingencies Agency. 2020. *Swedish Civil Contingencies Agency*. Accessed November 8, 2020, https://www.msb.se/en/.

Swedish Defence Ministry. 2017. Resilience Focus of the Total Defence and the Design of the Civil Defence 2012–2025 [original in Swedish]. *Ds* 2017: 66, 20 December 2017. Stockholm. Accessed October 29, 2020. https://www.regeringen.se/rattsliga-dokument/departementsserien-och-promemorior/2017/12/ds-201766/.

———. 2020. *The Swedish Defence Commission's White Book on Sweden's Security Policy and the Development of the Military Defence 2021–2025. Stockholm*. Accessed October 29, 2020, https://www.government.se/4ada4f/globalassets/government/dokument/forsvarsdepartementet/forsvarsberedningen/defence-commissions-white-book-english-summary.pdf.

Swedish Ministry for Foreign Affairs. 2011. *Sweden's Strategy for the Arctic Region*. Stockholm: Ministry for Foreign Affairs, Department for Easter Europe and Central Asia, Arctic Secretariat. Accessed November 22, 2020, https://www.government.se/4ab1ed/contentassets/85de9103bbbe4373b55eddd7f71608da/swedens-strategy-for-the-arctic-region.

Swedish Security Service. 2019. *Security Service Yearbook 2019*. Stockholm. Accessed October 28, 2020, https://www.sakerhetspolisen.se/en/swedish-security-service.html.

U.S. Air Force. 2020. *The Department of the Air Force Arctic Strategy*. July 21, 2020. Washington, DC. Accessed November 14, 2020, https://www.af.mil/Portals/1/documents/2020SAF/July/ArcticStrategy.pdf.

U.S. Army. 2020. *U.S. Army Alaska. America's Arctic Warriors*. Accessed November 4, 2020, https://home.army.mil/alaska/index.php/USARAK-units.

———. 2021. *Regaining Arctic Dominance. The U.S. Army in the Arctic*. Headquarters, Department of the Army, January 19, 2021, Chief of Staff Paper #3. Washington, DC. Accessed May 23, 2021, https://api.army.mil/e2/c/downloads/2021/03/15/9944046e/regaining-arctic-dominance-us-army-in-the-arctic-19-january-2021-unclassified.pdf.

U.S. Coast Guard. 2019. *United States Coast Guard Arctic Strategic Outlook*. April 2019. Washington, DC. Accessed November 14, 2020, https://www.uscg.mil/Portals/0/Images/arctic/Arctic_Strategy_Book_APR_2019.pdf.

U.S. Customs and Border Protection. 2020. *Border Patrol Special Operations Group (SOG)*. Accessed November 20, 2020, https://www.cbp.gov/sites/default/files/assets/documents/2018-Aug/border-patrol-special-operations-group.pdf.

U.S. Department of Defense. 2019. *Report to Congress. Department of Defense Arctic Strategy*. June 2019. Washington, DC. Accessed November 12, 2020, https://media.defense.gov/2019/Jun/06/2002141657/-1/-1/1/2019-DOD-ARCTIC-STRATEGY.PDF.

U.S. Department of State. 2010. *Trafficking in Persons Report 2010*. Washington, DC. Accessed May 4, 2020, https://2009-2017.state.gov/j/tip/rls/tiprpt/2010//index.htm.

———. 2017a. *Trafficking in Persons Report, June 2017*. Washington, DC. Accessed August 28, 2020, https://www.state.gov/reports/2017-trafficking-in-persons-report/.

———. 2017b. *Key State Department Arctic Officials*. Accessed November 26, 2017, https://2009-2017.state.gov/e/oes/ocns/opa/arc/ao/index.htm

———. 2018. *Country Reports on Terrorism 2017 – Norway*, 19 September 2018. Accessed September 12, 2020, https://www.refworld.org/docid/5bcf1f8d13.html.

United Nations. 2011. *Iceland*. Food and Agriculture Organization of the United Nations, C1053, Iceland, 21 April 2011. Rome, Italy. Accessed August 29, 2020, http://www.fao.org/3/i2099e/i2099e01.pdf.

United States Code. 2021. *Title 22, Chapter 78, Section 7106*. Accessed June 22, 2021, https://uscode.house.gov/view.xhtml?req=(title:22%20section:7106%20edition:prelim).

Vandiver, John. 2020. F-15s Out of Lakenheath Patrol the Skies Above Iceland. *Stars and Stripes*, October 13, 2020, Accessed November 20, 2020, https://www.stripes.com/news/f-15s-out-of-lakenheath-patrol-the-skies-above-iceland-1.648384#:~:text=The%20deployment%20is%20part%20of,Iceland%2C%20March%2016%2C%202020.

Veterans Affairs Canada. 2020. *Canada's Role in the Second World War*. Ottawa. March 9, 2020. Retrieved July 8, 2020, https://www.veterans.gc.ca/eng/remembrance/history/second-world-war.

Wallace, Ben. 2021. *Iceland Becomes 10th Nation to Join UK-led Joint Expeditionary Force*. Gov. UK, News. London. Accessed June 20, 2021, https://www.gov.uk/government/news/iceland-becomes-10th-nation-to-join-uk-led-joint-expeditionary-force.

Watling, Jack. 2018. *NATO's Trident Junction 2018 Exercise: Political Theatre with a Purpose*. RUSI, 20 November 2018. London: Royal United Services Institute. Accessed June 22, 2021, https://rusi.org/explore-our-research/publications/commentary/nato%E2%80%99s-trident-juncture-2018-exercise-political-theatre-purpose.

Wezeman, Siemon T. 2016. *Military Capabilities in the Arctic: A New Cold War in the High North?* Stockholm: Stockholm International Peace Research Institute (SIPRI), October 2016. Accessed August 8, 2020, https://www.sipri.org/publications/2016/sipri-background-papers/military-capabilities-arctic.

Wheelersburg, Robert P. 2021. North Atlantic Small State Security 2025: The West Nordic Security Zone. In *Small States and the new Security Environment, The World of Small States*, ed. Anne-Marie Brady and Baldur Thorhallsson, 87–101. Cham: Springer.

Winger, Gregory, and Gustav Petursson. 2016. Iceland's Cold War Legacy Reappraised. *Foreign Affairs*, February 24, 2016. Accessed September 20, 2020, https://www.foreignaffairs.com/articles/united-states/2016-02-24/return-keflavik-station?fbclid=IwAR1GE-QXEICENg52MtSLuyZt2FxK3oNTA-teKiPD8lYFxP-C9-K9Byzc364.

Woody, Christopher. 2020. As US Tries to Close 'Icebreaker Gap' with Russia, Its Only Working Icebreaker Is Making a Rare Trip North. *Business Insider*, November 9, 2020. Accessed June 22, https://www.businessinsider.com/us-worried-about-icebreaker-gap-with-russia-in-arctic-2020-11.

Wyland, Scott. 2019. Congress Restores Funding for Coast Guard's First New Heavy Icebreaker in 40 Years. *Stars and Stripes*, February 19, 2019. Accessed November 2, 2020, https://www.stripes.com/news/congress-restores-funding-for-coast-guard-s-first-new-heavy-icebreaker-in-40-years-1.569253.

Young, Oran. 2011. Forward. Arctic Futures: The Politics of Transformation. In *Arctic Security in an Age of Climate Change*, ed. J. Kraska, xxi–xxvii. Cambridge University Press.

Chapter 4
Icelandic Security Sector Reform and Proposed Security Force

They called upon Thor, who came immediately,
and the next thing to happen was that the hammer Mjölnir
was in the air.
[Thus], Thor put an end to the giant's life in Jötunheim.
He struck the first blow in such a way that the giant's skull
broke into small pieces.
Snorri Sturluson, The Prose Edda

Abstract This chapter recommends that an Icelandic security force be created to meet threats arising from Arctic climate change creating new transportation routes into the North Atlantic, along with exposing new territories and natural resources. Following an analysis of European Security Sector Reform (SSR) in Germany and Georgia, a comparison between Iceland and Costa Rica illustrates the need for SSR in the former. Using the Swedish Rangers and Canadian Rangers as models, the proposed Icelandic security source could include cadre (i.e., full-time members), reservists, and volunteers from Icelandic Search-and-Rescue (ICE-SAR). Based upon a cavalry regiment whose mission is to conduct reconnaissance while providing for its own self-defence, the security force could be structured as a battalion with two companies. Each company could include a Headquarters Recon Patrol, a Heavy Recon Patrol, a Weapons Recon Patrol, and a Light Recon Patrol, the last of which could be composed of volunteers from ICE-SAR. The security force could receive its equipment and training from U.S. and NATO security assistance programs, and by meeting the country's NATO recommended defence expenditure (2% of GDP). The proposal also recommends that Iceland purchase a C-130J Super Hercules aircraft to improve its security preparedness, which could also be used by the coast guard and police for patrolling and in emergencies. The proposed security force is necessary to provide an effective and rapid response to emergencies. Historical evidence reveals that the U.S. is unable to perform sub-strategic security in Iceland due to its inability to mount a rapid response against threats like terrorism or to conduct effective operations in cold-weather environments. Thus, in order not to depend on unreliable outside help, Iceland should create its own security force

R. P. Wheelersburg, *The Security of Iceland and the Arctic 2030*, Springer Polar Sciences, https://doi.org/10.1007/978-3-030-89948-6_4

4.1 Introduction to Security Sector Reform

Creating an Icelandic security force can be considered part of European Security Sector Reform (SSR) related to new threats that cannot be defeated through the country's existing security apparatus. Although it is not a member of the European Union (EU), Iceland subscribes to a number of EU policies. These policies include the Schengen Area agreement for border control and the Dublin Regulation for asylum seekers, both of which involve potential security challenges for Iceland. Therefore, I will use EU SSR to illustrate the changes Iceland should undergo to strengthen its ability to protect itself by using its own current and future innovative capabilities, personnel, and resources.

Europe has been a focal point for SSR since the 1990s due to the evolution of new sub-strategic security threats after the fall of the Soviet Union. The EU integrated SSR into its 2005 European Security and Defence Policy after dozens of conventional conflicts broke out during the late twentieth and early twenty-first centuries. These conflicts arose between Russia and former Soviet Republics like Georgia and Ukraine, between previous Yugoslavian provinces such as Kosovo and Slovenia, and from insurgencies in Chechnya and the Caucasus (Dursun-Özkanca and Vandemoortele 2012).

SSR happens through the evolution of security routines (i.e., practices) when they fail to effectively solve new challenges created by social or cultural change. When that failure occurs, it creates a need to develop innovative solutions to those existing routines in order to successfully meet those new challenges.

> An innovation can originate from the puzzles or anomalies that relate to existing routines particularly as surrounding circumstances change and the routine becomes less beneficial to the organization. Furthermore, innovation often involves new combinations of elements of existing routines (Piotukh and Wilson 2009:8).

Without innovating existing routines or adding novel ones in the face of new challenges, the security sector could become unable to perform its functions. According to the Geneva Centre for the Democratic Control of Armed Forces (DCAF),

> [A]n ineffective . . . security sector cannot credibly perform its missions in national defence, law enforcement or public assistance, and itself poses a threat to both the state and the population. . . . SSR provides for national defence while contributing to international peace and security. An ineffective . . . security sector poses a threat to other states, either by harbouring potentially hostile criminal or political elements or by succumbing itself to violent conflict (DCAF 2015:4).

Despite the need to improve national and regional stability in a country, SSR is more effective when it is not imposed from above by another state, a union, an alliance, or a global-level entity. SSR should be a participatory activity with a country's elected officials and their constituents involved at all stages of the process. The lack of such local ownership is one of the primary barriers to the successful implementation of SSR programs (Dursun-Özkanca and Vandemoortele 2012). While it is not necessary to have a referendum on every step of the SSR process, transparency and communication with the electorate by the decision makers is critical for the public's acceptance of new security organizations and activities. That approval is especially necessary if SSR requires a reallocation of public funds from other areas that could lead to public sacrifices, for example, transferring funds from necessary social services to security activities. Without approval based upon trust and the public's agreement with decisions to engage in and complete SSR, the process could fail, leaving the country less protected than before (DCAF 2015).

4.2 Examples of SSR: Germany, Georgia, and Costa Rica

There are several European countries that have undergone SSR over the past several decades, and there also are countries that have only recently begun the process. Germany is an example of a country undergoing long-term SSR. After the Nazi regime failed, the country was split into two nations, both of which underwent radical SSR to rebuild their security apparatuses. Over the last 50 years, The Federal Republic of Germany underwent SSR as a country after it was the victim of domestic terrorism against Germans, foreign terrorist attacks within Germany targeting third-country nationals, and foreign terrorist attacks directed against Germans in other countries.

Before starting SSR in the mid-1970s, Germany had a security force capable of defeating some sub-strategic security threats, which was the border guard or Bundesgrenzschutz (BGS). The BGS conducted security patrols within a 30-kilometer (19-mile) border zone to prevent limited territorial incursions and smuggling. It also performed internal security missions such as protecting airports and foreign embassies. The BGS had capabilities above the police consisting of its own armored vehicles, anti-armor missiles, heavy machine guns, and aviation assets (Shears 1970). Yet the inability of the German police or BGS to defeat all sub-strategic security threats was revealed at the Munich Olympics of 1972, when Israeli coaches and athletes were murdered by Palestinian terrorists. That event led to the creation of an anti-terrorism/hostage rescue unit named Grenzschutzgruppe 9 (GSG-9). GSG-9 was originally part of the BGS but became a special unit of the Federal police. An example illustrating the effectiveness of Germany's SSR occurred in 1977. After a Lufthansa flight was hijacked by Palestinian terrorists, GSG-9 rescued most of the passengers in Somali using a small-scale raid (Katz 2005).

After the large-scale terrorist attacks against the U.S. on September 11, 2001, Germany and other European countries faced new challenges requiring an innovative response. Germany initiated additional SSR measures including developing new policies and laws that allowed monitoring of religious and extremist groups. Other laws and capabilities facilitated increased communication between various German and other nations' intelligence agencies to identify and track terrorist and criminal groups. Germany also implemented EU regulations against money laundering. In addition to strengthening the capabilities of GSG-9, those measures allowed

Germany to effectively counter a variety of new sub-strategic security threats (Miko and Froehlich 2004).

An example of a country that recently underwent SSR is Georgia. Georgia's SSR started in the mid-1990s as a former Soviet Republic that needed to develop its own national security policy and organization. Georgia created and implemented its first national military strategy in 2005; however, the 2008 Russian-Georgian War led to the need for additional SSR. In 2014 Georgia revised its national military strategy to address new challenges that arose. These challenges included Russians occupying parts of the country and committing terrorist (or insurgent) activities against the Georgian Government. Both of these threatened the country's sovereignty and regional security. Georgia's current National Military Strategy is based upon democratic control of the armed forces, peaceful conflict resolution, transparency, and support to civilian authorities during natural and manmade disasters (Ministry of Defence of Georgia 2014).

While there were many actors involved in the development of Georgia's SSR, they can be categorized into three groups. The first group included new government entities needed to manage security. One of the most critical Georgian reforms was establishing a national security council (NSC) in 1996, which consisted of the Georgian Prime Minister, the Speaker of the Parliament, and the Ministers of Foreign Affairs, Defence, Interior, and Finance. As constituted, the NSC advises the president on both domestic and foreign security issues (Simon 2014).

Security providers like the police and the armed forces comprised the second group of Georgia's SSR actors. As part of the Soviet Union, Georgia was forced to participate in Warsaw Pact military operations. After gaining independence, Georgia reformed the police, making it more professional, merit-based, and subject to administrative control. In the ministry of defence, SSR included building new units, staffing them with professional personnel, developing individual and unit training, acquiring new equipment, and providing civilian oversight. The government also eliminated militia forces that threatened national security objectives with their own political agendas. Those actions allowed Georgia to better align with NATO and its other international allies. In 1991 the Georgian Defence Forces (GDF) were created, which included an army, navy, air force, national guard, and special operations forces (SOF). GDF missions involved defending the country's sovereignty and protecting its territorial integrity. Georgia also formed a territorial defence force and an army reserve to assist the civilian authorities in an emergency by guarding critical infrastructure and evacuating civilians (Simon 2014).

The third group of actors was made up of entities that involved themselves in the political process to help or hinder SSR for various reasons. In Georgia, the Civil Council on Defence and Security (CCDS) was started in 2005 by a number of private entities such as the Caucasian Institute for Peace, Democracy, and Development (Simon 2014). The CCDS "is a non-governmental, non-partisan, not-for-profit organization which contributes to democratic reforms in Georgia's security and defence sector and supports the country's democratic transition and Euro-Atlantic integration" (Civil Council on Defence and Security 2020).

There are also actors who were opposed to Georgia's SSR efforts. The Georgian Parliament voted to join NATO in 2006, followed by a 2008 referendum when an overwhelming majority approved the action. Subsequently, following Russian protests, several conflicts with Russian troops and their supporters on Georgian soil created a barrier toward the country joining NATO. Separatists from the Georgian regions of Abkhazia and South Ossetia along the Russian border threatened to split permanently from Georgia if the country became a NATO member (Lowe 2008). The main problem for these separatists about the country joining NATO was whether Article 5 should involve all of Georgia's territory since there were existing cross-border conflicts that did not require intervention by the entire alliance (Simon 2014).

Yet Georgian security policy continued to be based primarily on eventually joining NATO. The 2016 Substantial NATO-Georgia Package provided contributions from 20 member countries to enable further SSR, such as constructing new training facilities and conducting multinational exercises (NATO n.d.). Overall, the evolution of EU-based SSR improved Georgia's ability to provide for its own defence by strengthening the police and security forces, gaining the population's trust of agencies exercising civilian control over security forces, and involving the public in important security decisions.

While there are similarities between Georgia and Iceland, there are many differences that may make the comparison of limited use. Since 1940, Iceland has been at the center of Allied and NATO defence efforts in the North Atlantic both during World War II and throughout the Cold War. Georgia is still aspiring to join NATO while Iceland was a founding member. Georgia has its own armed forces, but Iceland is a non-military nation. In addition, Georgia has a population nearly ten times that of Iceland in only three-quarters the amount of territory (Suny 2020).

A non-military nation outside Europe such as Costa Rica may make a better comparison for Iceland to consider when deciding to implement the appropriate type and scope of SSR. Both Iceland and Costa Rica are described as some of the safest and most democratic countries in the world. Although its membership in NATO makes Iceland technically a militarily-aligned country, Iceland is considered a non-military country because it has no armed forces. Iceland relies on its NATO membership and the 1951 Bilateral Treaty with the U.S. (reaffirmed in 2016 by joint declaration) for much of its security (U.S. Department of State 2017). As a non-military country, Costa Rica also depends on an alliance and a bilateral treaty to protect the country from external threats. Costa Rica is part of the Inter-American Treaty of Reciprocal Assistance (Rio Treaty) and has a close security relationship with the U.S. (Dursun-Özkanca 2017).

In typical non-military countries, police forces are responsible for both internal (e.g., law enforcement) and external security missions (e.g., border guards/customs). A lack of armed forces could cause police forces to be overwhelmed when threat scenarios change and require different responses outside of existing police capabilities. Militarizing the police is not a good option since it could reduce the population's confidence when the police force oversteps its responsibilities leading to repression, corruption, and alienation (Dursun-Özkanca 2017). Small-scale

Fig. 4.1 Political cartoon of the Icelandic Police carrying machine guns during the 2017 Independence Day Parade in Reykjavík. (Illustration by Gunnar Karlsson, used with permission)

militarization has happened recently in Iceland when the police obtained machine guns from Norway and began openly carrying them at public events like the Secret Solstice Festival and the annual Colour Run (Þórsson 2017). Icelanders were almost universally opposed to further armed displays by the police at public events (Fig. 4.1).

Yet Costa Rica faces many of the same challenges as other non-military countries with only the police to accomplish internal security. Organized crime, especially drug cartels loaded with cash, are a threat to both Costa Rica's security sectors and the country as a whole. There are potential problems for Costa Rica's police to fight such threats including: "insufficient size of the police force, lack of resources, lack of proper training, assignment of external security responsibilities to the police, and organizational fragmentation" (Dursun-Özkanca 2017: 18). Those same problems are also inherent in Iceland's police force, leading to the need for the country to conduct SSR to prevent it from being overwhelmed by threats or overstepping its responsibilities (Overseas Security Advisory Council 2019).

4.3 Iceland and Security Sector Reform

This book's recommendation for Icelandic SSR results from recent changes involving the collapse of the Soviet Union, the end of the Cold War, and rise of global warming. When the decrease of Russia's military presence in the North Atlantic Arctic brought an end to the balance of power, peace was maintained in the region for three decades by Pax Americana as the U.S. remained the world's only superpower (Nye 1990). While that peaceful period strengthened Iceland's strategic-level capabilities, at the same time Iceland's sub-strategic security decreased with the abandonment of the Keflavik base in 2006. Another change that could occur in the future Arctic (i.e., 2030) is how state and non-state actors respond to emerging territories, increased accessibility of natural resources, and the availability of new transport corridors.

The first step in conducting Iceland's SSR requires evaluating the effectiveness of the police and coast guard, which are responsible for national security. Those agencies were well suited to accomplish the country's past security challenges, especially when the country came under the protection of the U.S. military stationed at the Keflavik base. Yet emerging challenges such as territorial and resource claims by Arctic and rogue states, along with non-state actors like terrorists and organized criminals, require the creation of new security structures through innovation. The innovation recommended in this case is the creation of an Icelandic security force that can meet those new potential threats.

As stated above, some militarization has occurred recently within the Icelandic police, resulting in "mission creep". Mission creep is the increase in the variety and number of responsibilities of a security agency that impacts both the effectiveness of the police and the public's perception of it. For example, the 1976 terrorism incident at Leif Erikson International Airport mentioned in Chap. 3 led to the Icelandic police assuming a counter-terrorism role and capability. That new mission resulted in two potential problems: first, that the police might not be able to defeat a terrorist attack, and second, that the public disapproves of the police being armed at public events. Thus, in order to prevent additional mission creep, which could actually reduce the ability of the Icelandic Police to accomplish their primary mission of public safety, SSR should take place in the country to develop a new sub-strategic security organization.

In order for SSR to be successful in Iceland, the primary requirement of local ownership must be met. As described previously, SSR requires transparency between elected decision makers and their constituents. That is especially important in Iceland where the failure of past governments to include the public in the country's security decisions caused dissent, protest, and even violence, all of which occurred during the anti-NATO riots of March 30, 1949. Those riots will be discussed in Chap. 6, but essentially, they were caused by secret negotiations between the Icelandic government and the U.S. and the reluctance to have foreign troops in the country.

The Atom Station, a fictional story written by Halldór Laxness, portrayed the lack of transparency Icelandic security decisions and negotiations. Published in 1948, the book takes place in post-WWII Reykjavík where Laxness predicted the actions of a few senior government officials who conducted secret negotiations with the U.-S. government. The novel's fictional decisions preceded the country actually joining NATO 1 year later, permitting the Keflavik base, and stationing American troops on its soil. *The Atom Station*, the only book in Iceland ever to sell out on the first day of printing, revealed the topic's importance to the public (Laxness 1982). Membership in NATO remains controversial in Iceland and is a possible source of resistance to future SSR unless the public is included in security decisions.

The SSR principles of participatory and transparent decisions, combined with historical events, influenced an address by former U.S. Ambassador to Iceland Robert C. Barber given to the Icelandic association *Varðberg* [Organization for Western Cooperation and International Affairs], on November 19, 2015. In his talk "Iceland, the United States, and North Atlantic and European Security", Barber declared that

> The state of the world today presents many challenges to the rule of law, to peace and stability, and to happiness, which we all deserve and strive to achieve. That is the harsh reality staring us right in our faces, and it means tough decisions lie ahead for the U.S., for NATO, for the Nordics, and for Iceland. I would like to urge all Icelanders to participate in a national discussion about security. Have the conversation amongst yourselves about Iceland's role and responsibilities in its own security and in North Atlantic and European security more broadly. How will Iceland as a whole bear the costs, economic and otherwise, of its collective defense commitments, in a changed and ever-changing security environment, especially if the costs fall disproportionately on particular geographic areas or sectors of Icelandic society? What will Icelanders accept, and what will you not accept? Have the conversation. I would further urge you to take collective responsibility as citizens of this country to have the conversation. Do not, through lack of participation make just a few people within your government responsible for these decisions (Barber 2015).

One of the situations calling for such conversations between the policy makers and the public occurs when a new security policy is created, which Iceland did in 2016 when it created its new national security policy. That policy established a novel relationship between "security providers" like the police and coast guard and "security management and oversight bodies" including the justice ministry, the ministry of foreign affairs, and the parliament (DCAF 2015: 6). Specifically, Iceland's 2016 security policy created the National Security Council (NSC) to provide oversight to the country's security agencies (Icelandic Parliament 2016). The NSC operates under and is directed by the 2016 National Security Council Act (Prime Minister 2016).

Yet the 2016 security policy did not create a new security provider such as a border guard. According to the basic principles established by the United Nations, the evolution of Iceland's SSR requires a new form of security provider to fully realize the aims of the security policy (United Nations and the Rule of Law n.d.). Since the policy required the NSC to review it every 5 years, it is possible that the upcoming review in 2022 could conclude that such a security structure is necessary to fully implement the policy (Icelandic Parliament 2016).

That review requirement means that within the time frame of this recommendation, the national security council should have reviewed Iceland's policy twice, in 2022 and 2027 (or within a reasonable time period). That time period provides ample opportunity for the NSC, working through the appropriate ministries, to develop a plan for a new security provider and to communicate that plan to the public by 2025. That schedule would provide enough time for the public to provide feedback through their elected representatives and/or a referendum before building a security force by the end of the decade. Thus, either a new policy or a revised one could be instituted by 2030, allowing time for my recommendation to be considered.

While there may be a new national security policy in place by 2030, my recommendation complies with the current one. Specifically, the proposed security force could "ensure Iceland's independence and sovereignty, territorial integrity, the safety of its citizens, and the protection of its governmental system and social infrastructure". A security force rather than a national guard or other military unit recognizes that Iceland "has neither the resources nor the desire to maintain an army". With a future orientation, the Parliamentary "policy extends to … military risks [and] give[s] particular consideration to Iceland's … security interests in the Arctic" while maintaining its membership in NATO, monitoring the ability of the 1951 bilateral defence agreement to meet "military threats", and strengthening "further cooperation among the Nordic countries on security and defence". The resolution orders appropriate ministries to "ensure that Iceland has the defence infrastructure, equipment, capacity, and expertise needed both to respond to the challenges facing the country in connection with security and defence and to honour its international commitments". It directs that "government policies take into account other threats that require full attention, such as terrorism [and] organized crime" (Icelandic Parliament 2016).

Iceland has civil entities that support and those that oppose any form of security above the police, including the bilateral treaty with the U.S. and membership in NATO (Simon 2014). On the supportive side is *Varðberg*. Started in 1961, *Varðberg* has as its primary purpose to host lectures and discussions about security and defence in the North Atlantic and to support NATO activities taking place in Iceland (*Varðberg* 2020). In the past, *Varðberg* leadership called for a new security force using Icelandic volunteers. In 1995, Björn Bjarnason, Iceland's Minister of Culture and Education, and later chairman of Varðberg, called for such a force during a lecture to a conference on security issues in the North Atlantic Region.

> A small number of Icelanders or between 500 – 1000 people in uniform on either voluntary or full-time basis who could train other reserves would not create any difficulties … More importantly the creation of some type of Icelandic National Guard or Home Guard Units either independently or alongside American forces based at Keflavik would allow the Icelandic authorities the flexibility that does not exist today to enhance security at key objectives around the country without calling for reinforcements from the United States in situations that fall short of national military emergency and could also be used to augment existing civil defense and natural relief efforts if required. … what I am specifically talking about is the participation of Icelanders in the ground defense of their country by creating a small nucleus of officers and non-commissioned officers that would be responsible for

certain security tasks as well as training voluntary or conscript reserves across the nation (Bjarnason 1995: 4).

My recommendation for an Icelandic security force builds on Bjarnason's earlier model and expands on it to create a specific table of organization, staffing, and equipment (TO&E).

On the resistance side, organizations like the Icelandic Peace and the Anti-War Movements hold protests and marches to fight against the country's attempts at security alliances, including demonstrations against NATO exercises (Ćirić 2019). Such activities began when Iceland joined NATO and continued into the late twentieth century with marches between Reykjavík and the Keflavik Base (Markham 1982). Yet there is evidence that such resistance to Iceland's membership in NATO has not continued on a large scale into the twenty-first century. For example, the 2018 NATO exercise conducted at the base and Marine testing of cold-weather gear in the country's interior generated no protests or marches. In fact, although Marines and sailors drank Reykjavík out of beer after the exercises, the locals appreciated the American service members who "were well-behaved, polite and friendly, and caused no problems at all" (English 2018).

An even greater barrier to SSR is the presence of anti-military parties in government, including both members of parliament and ministers. The current prime minister, Katrín Jakobsdóttir, is the leader of the Left-Greens, the country's second-largest political party. The Left-Greens oppose any military activities and is anti-NATO (Veal 2019). In 2018, Prime Minister Jakobsdóttir refused to accompany other ministers on a tour of the American aircraft carrier U.S.S. Harry Truman while the ship was off the Icelandic coast (Kyzer 2018). In fact, Iceland represents a unique case in that political parties that have anti-military positions hold seats in parliament and, at times, national power (i.e., prime minister). It also stands alone in being the only NATO member not to contribute to its own defence and is second to last among member nations in making its contribution to NATO. Frankly, given such positions and actions, it was incredibly hypocritical of Jakobsdóttir to stand facing the electronic monolith of the NATO banner with the other member leaders at the NATO Summit in Brussels on June 14, 2021.

An important requirement that would help establish credibility for Iceland's SSR is that the government should decide who will provide the advice and expertise to helping institute such reforms. Although the country has many connections to the EU, it may not be culturally acceptable for Iceland to invite the union to initiate SSR. The same is true of NATO, which may also not be culturally acceptable to the Icelandic population due to some people's long-standing opposition to the alliance. Choosing Denmark or the U.K. to assist in SSR is probably not a good fit, since each of those countries generated negative feelings during their occupation periods of Iceland or later conflicts with the country (e.g., Cod Wars).

The 2011 EKOS Research Associates survey provided an answer to who Iceland should turn for such advice and expertise by asking the questions; "Preferred Partner in Dealing with Arctic Issues" and "Which of the following countries would you be most comfortable with (your country) dealing with on Arctic issues?" The responses from the Icelanders surveyed in order were Scandinavia, Canada, and the United States (EKOS Research Associates 2011: 36). Therefore, I would recommend that either Sweden (since it is not yet a member of NATO) or Canada provide SSR support to Iceland.

4.4 Swedish and Canadian Rangers as Models for an Icelandic Security Force

The Icelandic security force needs to be effective enough with its capabilities and training to deter any form of aggression, whether it comes from state or non-state actors. At the same time, to prevent developing a security force that could be involved in military adventurism outside of Iceland, it must be defensive in nature. A defensive security force has neither the offensive weapons nor the operational capabilities to attack an aggressor outside of its national territory. With only domestic capabilities, the primary goals for the Icelandic security force could be to deter potential threats from attacking the country through actions taken within its own territory. "The aim of *deterrence* is to convince potential attackers not to attack because the consequences could be unacceptably costly to them, including the failure to gain their objectives (Sharp 1990: 3)."

If deterrence fails, an effective security force must be armed and trained appropriately to defeat a variety of sub-strategic security threats. Whether between states or involving non-state actors, most armed conflict is fought with small arms and light weapons, which are standard terms in arms-control language. Small arms consist of individual weapons from handguns to assault rifles while light weapons include crew-served machine guns and portable anti-tank/anti-aircraft missiles. Therefore, Iceland does not require heavy weapons like tanks or helicopter gunships to defend against everything from a terrorist attack to a light infantry assault (United Nations 2013).

The circumstances for such limited conflict are even greater in the Arctic where poor ground transportation exists throughout most of the circumpolar region. That factor is especially true in the summer when the top soil layers above the permafrost melt, preventing tracked vehicles like main battle tanks and self-propelled artillery from traveling across unpaved areas. That means, in order to counter sub-strategic threats in the Arctic, the Icelandic security force needs to match its opponents' use of small arms and light weapons and not rely on heavy weapons. It also needs to be able to transport itself throughout the Icelandic countryside, which the nation's search-and-rescue teams do on a daily basis.

To meet the country's sub-strategic security threats, I recommend an Icelandic security force that combines the capabilities and operations of both the Swedish Rangers and the Canadian Rangers. The Swedish Rangers is a military unit made up of reserves (i.e., conscripts) while the Canadian Rangers contains only volunteers. The proposed Icelandic security force could include both types of personnel, conscripts and volunteers. Both of these forces utilize small-arms for their missions. The Swedish Rangers, unlike the Canadian Rangers, also use light weapons against threats. Thus, to accomplish a wider variety of missions, the security force recommendation suggests the volunteers use only small arms while the reserves could also use light weapons after more extensive training.

Depending upon the mission, the Swedish Rangers are equipped with individual weapons like rifles, along with crew-served light machine guns, portable anti-tank missiles, and portable anti-aircraft missiles. The Swedish Ranger-influenced portion of the Icelandic recommendation uses light weapons, along with limited numbers of wheeled armored personnel carriers and all-terrain vehicles. The Swedish Rangers also have the capability for small-scale airborne responses (i.e., parachuting) or air assaults (i.e., aircraft landing to disembark troops) that I included in the Icelandic security force recommendations (Swedish Army Chief of Staff 1990).

Due to the knowledge, skills, and experience needed to perform effectively as assault troops, I used the Swedish Rangers as a model for the Heavy Recon Patrol and the Weapons Recon Patrol of the Icelandic security force, both of which have the mission to assault threats. Swedish Rangers are trained as soldiers, including both conscripts and professional full-time cadre, who receive both individual skills education and unit training. In my proposal, each member of the Heavy Recon Patrol and the Weapons Recon Patrol would be a reservist who completes his/her individual military occupation specialty (MOS) training in Sweden, Canada, or other country.

The Canadian Rangers are the model for the Light Recon Patrol portion of my recommended security force, which could be made up of civilian volunteers recruited from local Icelandic search-and-rescue teams (ICE-SAR). The Canadian Rangers consists of rural volunteers (mostly indigenous peoples) who are trained and equipped to conduct reconnaissance and provide for their own self-defence. The Canadian Rangers conduct sovereignty patrols, assist the civilian governments with search-and-rescue tasks, and give other aid during emergencies. They can also provide for their own defence, but are not capable of assaulting threats when encountered. Using knowledge of local terrain, their own mobility, and survivability experience in the Arctic, the Canadian Rangers support troops who have the weapons needed to assault aggressors.

Canadian Rangers are divided into five patrol groups spread throughout the country's northernmost region, with each patrol group containing squads at the village level (Government of Canada 2017). That dispersed pattern could easily be incorporated into the current ICE-SAR structure, which has teams of approximately 100 members spread throughout the countryside. Following training, the Icelandic volunteers could be activated by the NSC or foreign minister in an emergency, allowing a quick response by the Light Recon Patrol to perform reconnaissance and gather intelligence on a threat operating within Icelandic territory. They would also be in a position to receive, brief, and transport assault troops when they arrive at the threat location.

There is some criticism of the Canadian Government for not providing more military-type training to its rangers. Unless they have served in the army, Canadian

Rangers are not viable soldiers because their training consists only of learning mobility and survivability in the Arctic (Smol 2020). They use their own vehicles such as snowmobiles and all-terrain vehicles (ATVs) to transport themselves, for which they are reimbursed. Once they arrive at a location requiring defensive fires for self-protection, they use skills learned while hunting. Each Canadian Ranger is issued a government rifle, which has recently been upgraded to a Finnish-designed Tikka T3 CTR bolt-action rifle (Tikka 2020). However, Canadian Rangers do not receive infantry training such as operating crew-served weapons or gathering intelligence, which is a weakness.

An alternate model for the Icelandic Light Recon Patrol is the volunteer security force in Greenland, known as the Sirius Dog Sled Patrol, which consists of 12 patrolmen divided into 6 dogsled teams under the operational control of the Danish Arctic Command in Nuuk. Each patrol is assigned to a geographically dispersed logistical network of depots containing security equipment and supplies within the operational area, which is Greenland National Park on the island's northernmost part. The patrols are specially trained for Arctic conditions and can operate independently for long periods. Patrols are active year-round, although patrolling with dogsleds takes place only in the winter. In summer they patrol on foot, drive ATVs, use boats, or fly charter aircraft. The Danish Navy provides long-distance transportation for the teams to other parts of the island depending upon the mission (Robinson 2016).

The use of indigenous and traditional groups in Arctic countries for their own sub-strategic security needs is critical for stability in the circumpolar region. These groups include the Alaskan National Guard (soon to include the newly-created Alaskan Rangers), Canadian Rangers, Danish Sirius Dog Patrol, and to a certain extent the Icelandic search-and-rescue teams. Other countries use either military or security force troops in the Arctic such as the Norwegian Rangers, Swedish Rangers, Finnish Border Rangers, and the Russian Border Service. Each of these security forces is its country's primary means of reconnaissance in their respective Arctic regions, capable of transporting and surviving in the remote areas and providing limited security functions backed up by police or armed forces.

Although fictional, the recent Icelandic television series on Netflix *Katla* (Kjartasnnon and Kirmákur 2021) illustrated how important ICE-SAR is for recon, intelligence, and support for the police during a national emergency. In the series, the year-long ash eruption of the active volcano Katla caused the civil defence office to evacuate a fallout hazard zone near *Vík* í Mýrdal, allowing only those with a permit to enter the zone. The search-and-rescue personnel navigated freely throughout the hazard zone with their all-terrain and amphibious vehicles, even providing transportation support to the police. Although I will not reveal the plot, it becomes apparent during the series that the ICE-SAR team also needed to carry self-defence weapons to prevent the police from being the only armed entity within the hazard zone. While the series was fictional, it accurately depicted the reality of an Icelandic emergency during which only ICE-SAR could operate by using its own transportation, terrain familiarity, and survivability skills.

4.5 Recommendation for an Icelandic Security Force

I present the table of organization and equipment (TO&E) for a recommended Icelandic security force in Fig. 4.2. While my intent is to create a flexible model for use by Icelandic decision makers, presenting a concrete structure has two advantages. First, the model's characteristics allow for discussion and debate of a specific structure. Second, the structure allows me to illustrate how the security force would be deployed within various scenarios as depicted in Chap. 5.

The recommended Icelandic security force could use the regimental system of cavalry units, whose primary mission is to provide reconnaissance, surveillance (i.e., reconnaissance behind enemy lines), and security. Given the size of the recommended security force, I have designated it as a battalion. The battalion contains two companies, each with four recon patrols (Headquarters, Light, Heavy, and Weapons). Creating a battalion with two subordinate companies could be accomplished with 400 personnel, although the exact size and structure of the force should be designed by military advisors and the Icelandic Government. The battalion commander, company commander or executive officer, and either an officer or non-commissioned officer in each recon patrol could be full-time cadre. The remainder of the force could consist of either reservists or volunteers based on their patrol assignment. During peacetime, civilian contractors could oversee the

ICELANDIC SECURITY FORCE TABLE OF ORGANIZATION AND EQUIPMENT

BATTALION COMMANDER (OF-5 or equivalent) 400 personnel

Two security companies each with the following Table of Organization and Equipment

COMPANY COMMANDER (OF-3 or equivalent) 200 personnel
9 officers
191 enlisted

HQ Recon Patrol	Light Recon Patrol (ICE-SAR)	Heavy Recon Patrol	Weapons Recon Patrol
-command group	-recon squads x 4	-recon squads x 4	-anti-armor squad FGM 148 x2
-signal	-airborne/air assault	-8 Italian Puma APC (or similar)	-air defense artillery squad (RBS x 2)
-intelligence	-foot/all-terrain vehicles		-machine gun squad M-240 x2
-medical	-UAV surveillance squad		-mortar squad M224 60 mm x2
20 personnel	60 personnel	60 personnel	60 personnel
3 officers	2 officers	2 officers	2 officers
17 enlisted	58 enlisted	58 enlisted	58 enlisted

Rank structure using Icelandic Crises Unit designations. ICE-SAR ranks convert to security force structure upon activation.

OFFICERS	ENLISTED
Battalion Commander – Ofursti	Battalion Sergeant Major - Flokkstjóri 1
Battalion Executive Officer – Undirofursti	Company First Sergeant - Flokkstjóri 2
Company Commander – Majór	Recon Patrol Corporal - Korporáll
Company Executive Officer – Höfuðsmaður	Recon Patrol Trooper - Óbreyttur
Recon Patrol Leader – Liðsforingi	
Recon Patrol Assistant Leader – Undirliðsforingi	

Fig. 4.2 Recommended Icelandic security force table of organization and equipment

administration of battalion facilities as well as provide vehicle maintenance, both of which could be located at the Keflavik base.

To facilitate command and control during emergencies, I incorporated Iceland's former Crisis Response Unit's rank structure. While the ICE-SAR has its own hierarchical structure, to maintain discipline in emergencies, I recommend that the search-and-rescue volunteers assume security force ranks when activated. My structure includes a colonel as the battalion commander (Ofursti), a lieutenant colonel as executive officer (Undirofursti) and a sergeant major (Flokkstjóri 1).

The mission of each company's Headquarters (HQ) Recon Patrol could be to provide command and control of the subordinate three recon patrols, including organizing training and exercises while on regular duty and providing leadership during emergencies. Each of the two HQ Recon Patrols would contain a command group including the company commander (Majór), a company executive officer (Hófuðsmaður), and a company first sergeant (Flikkstóri 2). The command group could also contain specialists including two troopers who are signal (i.e., communication) specialists, two troopers who are intelligence specialists, and two troopers who are medical specialists. The HQ Recon Patrols could be stationed at the Keflavik base.

The mission of the Light Recon Patrol could be to participate in training and exercises during regular duty and to provide reconnaissance and surveillance in emergencies. The patrols could conduct their monitoring missions on foot, using ATVs, or with unmanned aerial vehicles (UAV). Each of the Light Recon Patrols could include a patrol leader (Liðsforingi) and an assistant patrol leader (Undirliðsforingi). There could be four recon squads in every patrol, each of which could be led by a corporal (Korporáll) and consist of a medic and 12 troopers (Óbreyttur). The Light Recon Patrol could have two troopers as a UAV squad. The eight Light Recon Patrol squads could be made up of volunteers stationed with existing ICE-SAR teams in towns with airports throughout the country. Those towns could include Akureyri, Egilsstaðir, Höfn, Husavik, Ísafjörður, Reykjavík, Vik, and Vopnafjörður. Upon activation, the teams could report to their patrol headquarters in Keflavik, to coast guard units operating locally, or to the district Icelandic Police to assist with emergency tasks as assigned, while remaining under the command of their security force leadership.

Members of ICE-SAR teams could volunteer to join a Light Recon Patrol. ICE-SAR members already have the knowledge and experience to navigate and survive in remote and harsh Icelandic conditions. Light Recon Patrol volunteers could receive a government-issued rifle and ammunition, along with whatever uniform items the government deems appropriate. To perform their security missions, Icelandic Light Patrol members could receive additional training in marksmanship, reconnaissance, communication, camouflage, concealment, cover, fire, and movement. Also, while Light Recon Patrols could utilize their own transportation during missions, on a volunteer basis their members could complete airborne or air-assault training with a foreign military, given permission by the NSC on a case-by-case basis.

Fig. 4.3 Four Swedish Rangers armed with AK-5 automatic rifles on skis wearing Arctic camouflage aligned in an "L-shaped" ambush formation, November 15, 1994, K-4 Ranger Base, Arvidsjaur, Norrbotten, Sweden. (Photo by U.S. Army Captain Robert Wheelersburg)

The mission of the Heavy Recon Patrol is to complete training and exercises during regular duty, along with providing mechanized reconnaissance, surveillance and assault troops during emergencies. Each Heavy Recon Patrol could use the same rank structure as the Light Recon Patrol; however, the troopers could be reservists. In addition, the troopers' MOS training could be tailored to attain their squad's required tactical capabilities, such as mechanized infantry or armored personnel carrier driver. The Heavy Recon Patrols could be stationed with their equipment at the Keflavik NATO Base. The patrols could either drive themselves to their required locations using their wheeled armored vehicles, or they could be transported by cargo aircraft (discussed below). At their duty site, Heavy Recon Patrol squads could link up with local Light Recon Patrol squads to receive intelligence, terrain orientation, transportation, and logistical support before continuing the mission. In actual conflict, the Heavy Recon Patrol could conduct small-scale assaults against a threat using such tactics as ambushes (Fig. 4.3).

The mission of the Weapons Recon Patrol is to form an assault team with a Heavy Recon Patrol squad and provide it with fire support including portable anti-armor missiles, anti-aircraft missiles, mortars, and light machine guns. Each Weapons Recon Patrol could consist of the same rank structure as the Light Recon Patrol, except that the troopers could be reservists. In addition, the troopers' MOSs would be tailored to their squad's tactical capability and weapons, such as anti-tank missile, anti-aircraft missile, mortar, or machine gun. The Weapons Recon Patrols could be stationed with their equipment at the Keflavik NATO Base. Since they are dismounted, the weapons squads could be transported to their required locations by Light Recon Patrol vehicles, local ICE-SAR teams, or cargo aircraft. At their duty site Weapons Recon Patrol squads could link up with Heavy Recon Patrol squads to plan fire support.

Reservists could complete individual infantry training in programs either hosted by a foreign nation or held in Iceland using foreign instructors. Specialists (e.g., signal or medical) could complete an MOS school abroad. In terms of unit training, one patrol from each company could participate in a training assembly one weekend a quarter. Once a year, the entire company could engage in simultaneous drills at Keflavik or other designated locations in the countryside. In alternate years, the battalion could train together for two weeks each summer, including conducting live fire exercises. To train as a unit in survivability and mobility in a cold-weather environment, the battalion could conduct an exercise during one weekend every 2 years during winter at the battalion, company, or patrol level.

4.6 Security Force Management and Operational Principles

The mission of the recommended security force is to provide a short-term, limited response to sub-strategic security threats beyond the capabilities of the Icelandic Police, including the Viking Squad and/or the coast guard. The security force would not be a viable military unit capable of sustained combat operations against a determined enemy presenting overwhelming odds. Instead, its limited missions could include conducting reconnaissance or surveillance, maintaining or restoring territorial integrity in cooperation with other Icelandic units (i.e., police or coast guard), and performing a delaying action to allow time for a NATO or a Nordic security force to arrive. Above all, the purpose of the Icelandic Security Force is to deter aggressive action by state or non-state actors through training and exercises that communicate the forces' capabilities and intentions clearly to potential threats.

The Icelandic Security Force could be assigned to the Ministry of Foreign Affair's Directorate of Security and Defence for administration and training during regular duty. When the security force is activated during an emergency, it could be managed by the directorate in consultation with the justice ministry. During larger-scale emergencies, the security force could be transferred to the Ministry of Justice and placed under the Icelandic Coast Guard (ICG) for operational control (OPCON). The ICG has a higher rank structure and a larger national security mission than the security force to attain Iceland's national defence objectives as outlined in the 2016 national security policy. That transfer is similar to that of the U.S. Coast Guard (USCG), which during peacetime comes under the authority of the Department of Homeland Security. In time of declared war, the USCG is transferred to the Defense Department and placed under the operational control of the U.S. Navy for OPCON (Dolbow and Howe 2017).

Above all, the proposed security force would create no responsibility issues between ministries or their subordinate agencies, as it should be an operational

ICELAND'S CURRENT SECURITY/DEFENSE MANAGEMENT

National Security Council

Justice Ministry
Icelandic Police
-intelligence, policing
-border control, immigration
-terrorism

Almannavarnadeild
-civil defence operations
-natural/human disaster exercises
-mass movement planning
-command center ICESAR

Icelandic Coast Guard
-fisheries, environmental, search and rescue
-defence tasks at Keflavik base (requested by Foreign Minister)

Foreign Ministry
Security and Defence Directorate
-NATO
-exercises/operations
-air policing
-defence facilities
-infrastructure (Keflavik)
-APODs, SPODs*
-reinforcement
-defence plans
-defence budget
-host nation support
Security Force (proposed)
-DSCA#/terrorism/crime

* = Airport of embarkation, sea port of embarkation
# = Defence Support to Civilian Authorities

Source: Arnór Sigurjónsson, Director General, Directorate of Security and Defence, June 9, 2021.

Fig. 4.4 Proposed security force management structure

entity without supervisory capability over other agencies or programs. Falling directly under the Icelandic Foreign Ministry's Security and Defence Directorate places the proposed security force under civilian control, with the ability to activate and command resting with the national security council. Thus, the security force would conform to the emergency management structure outlined in Iceland's existing security policy by providing defence support to civil authorities (DSCA; Fig. 4.4).

To prevent mission creep, the security force could be restricted by law from participating in training or emergency operations outside the country except for individual MOS education as described above. Cadre or reserve security force troops observing Northern Viking or other exercises in Iceland could be permitted only on a case-by-case basis with express approval of the foreign minister after NSC review. The security force could be restricted by law from participating in NATO or foreign military exercises in Iceland except as members of ICE-SAR or if requested by civil authorities (e.g., an accident in a remote area), and then only with the approval of the foreign minister after NSC review.

A second but important part of this recommendation is that Iceland could seek to have NATO provide the equipment needed for the security force including armored personnel carriers, anti-armor/aircraft weapons, and a C-130J Super Hercules cargo aircraft. The equipment could be supplied by the U.S. Defense Security Cooperation Agency (DSCA), whose mission is to support American security cooperation "in its efforts to train, educate, advise, and equip foreign partners" (DSCA 2020). Iceland could also request equipment and training from NATO through its Partnership Interoperability Initiative (PII). One of the main PII goals is to have partners who are "able to operate together with NATO forces according to NATO standards, rules, procedures and using similar equipment" (NATO 2020). The equipment could be surplus or reconditioned, with training provided by the SSR advisor country or other NSC-approved nation.

Iceland could benefit tremendously from having its own C-130J, especially the coast guard, due to its extended range and time-on-station capabilities compared to helicopters or jet aircraft. The C-130J could complete peacetime missions such as patrolling the country's Exclusive Economic Zone, dropping supplies or flotation devices (e.g., rafts) to stranded mariners, or assisting with emergency evacuations. With the decrease in commercial domestic flights using turboprop aircraft over the past few years, Iceland currently has a substantial number of unemployed younger pilots who could cross-train in the C-130J as part of a reserve assignment.

C-130J cargo planes have a long history of use in the Arctic region, and American and Danish militaries have operated them in Greenland and in Antarctica for decades. The C-130J has as its primary strength the ability to deliver a large volume of materials and personnel in remote areas. A C-130J can transport approximately 80 troops with individual equipment or two armored personnel carriers and their crews. The aircraft can land at most airports in Iceland and, in certain circumstances, directly on the ice using skis. The New York Air National Guard, which flies the missions to Greenland and Antarctica using either landing gear or skis, could provide orientation and training to Icelandic pilots and crew. At the same time, Iceland already has circumpolar aviation experience through Icelandair's flights to Antarctica. At least four times, Icelandair has flown to Antarctica to retrieve Norwegian scientists who are ending several months of research on the continent. While the company uses large commercial jet aircraft to perform the missions, the institutional management, flight operations, and maintenance experience in the Antarctic would allow the country to operate a C-130J under the country's subarctic conditions (Icelandair 2021).

Perhaps the most important management question is, how will Iceland pay for its recommended security force? Funding for the proposed security force could be provided by Iceland meeting NATO's military expenditure requirement of 2% of Gross Domestic Product (GDP), of which 20% should go toward equipment. In 2020, Iceland contributed nothing for military expenditures - the only country in NATO that failed to do so. That year, Iceland's GDP of nearly $25 billion meant that it could spend $500 million on security to meet its recommended NATO contribution of 2%, and $100,000,000 on equipment. Given that amount of funding, Iceland could support the entire proposed security force, finance training and new equipment, and fund the ICE-SAR that currently receives no government money (Forces Net 2021; Trading Economics 2021).

The projected cost of the security force is difficult to determine, but it is possible to use the cost of a light infantry battalion in the U.S. Army National Guard as an estimate since most personnel are not full time. The U.S. Congressional Budget Office (CBO) is responsible for providing the Legislative Branch with analysis on budgets and expenditures related to funding the Executive Branch's (i.e., president) activities. The CBO estimate for the cost of a National Guard light infantry battalion is approximately $25,000,000 per year, although I extrapolated that figure from the cost of a larger Army brigade (CBO 2021). If you add in the projected cost of a

C-130J (i.e., flyaway cost), that figure comes to $100,000,000, which could be spread over the lifetime of the C-130J (with an average age of 25 years the cost of the C-130J would be $4,000,000 per year). Thus, the funds required to develop and operate the proposed security force would be less than 20% of the recommended NATO military expenditure of $500,000,000 per year. Clearly, Iceland could and should fund its own security force if it intends to remain a member of NATO.

4.7 U.S. Inability to Survive and Fight in the Arctic

> Snow and cold can cause difficulties that must be overcome. But remember, that your difficulties are also your enemy's. He who best overcomes them has the upper hand. *The winter is a friend, if you are a more skillful winter soldier than your opposition* [italics in original] (Swedish Army Chief of Staff 1968: 7).

As a strategic power, American armed forces must be capable of being deployed quickly to various regions of the globe without equipment and training dedicated to those regions (e.g., desert). That is a serious problem, especially in the Arctic. Lack of American ground-defence effectiveness in Icelandic weather conditions goes back to the initial occupation troops during the Second World War. The first American troops to relieve British forces in 1941 were the Marines, who stayed until 1942, when the U.S. Army took over Iceland's defence. Army troops consisted mostly of draftees, who at the time, were forbidden by U.S. law to leave the Western Hemisphere. Marines were an all-volunteer force and under no such restrictions. Therefore, to meet the Icelandic Government's request to have U.S. troops occupy the country, a Marine unit from California already sailing to the Caribbean was diverted to assume the Iceland mission.

While in Charleston, South Carolina, the Marines were supplied with cold-weather gear described as "the darndest collection of winter clothing ever assembled; as there were bits and pieces of everything". After arriving in Iceland as winter approached, the Marines were unprepared for the coming cold. They used most of their time winterizing their camps, which prevented them from conducting military training. The Marines, who were trained and equipped as warm-weather amphibious assault troops, spent the winter "buttoned up … huddled around the stoves in the Nissen huts. … [asking] were they to be left forgotten in the wrong ocean?" (Clifford 1970). Even after the Army replaced the Marines, the troops were unprepared to survive in the cold weather, let alone to fight in it.

One can surmise what could happen to troops mobilized for extended ground combat in Iceland using the U.S. Aleutian Islands Campaign during WWII as an example. The initial attack in 1943 was against the 500 Japanese troops occupying Attu Island. The assault was conducted by the American 7th Infantry Division, which was untrained and poorly equipped for the frigid Arctic weather. Although they retook Attu Island, the Americans suffered more non-battle casualties attributed to weather, disease, and vehicle accidents than losses in combat. The inability to resupply U.S. troops caused them to endure starvation making them less battle worthy (NBC 1953; American Battle Monuments Commission 2018).

Kiska was the second island attacked by the Allies. These troops included Canadian and U.S. Special Forces trained and equipped for Arctic combat. While the Japanese had withdrawn from Kiska by the time the Americans retook the island, even with the prepared troops and no enemy, the Americans still suffered half of the casualties that occurred on Attu Island. That fact indicates that American troops were unprepared to navigate and survive in an Arctic setting, let alone conduct successful combat operations. Retaking both Aleutian Islands produced not only substantial non-battle casualties, but the logistical efforts to transport, resupply, evacuate wounded, and reestablish control were overwhelming. While the main benefit from the fighting there was improved home-front morale by removing Japanese forces from their only occupation of American soil during the war, the territory was not strategically important in the war effort enough to lose nearly two thousand troops (History.com 2020). One can hypothesize from the Aleutian Islands Campaign that using American ground troops to fight in the Arctic, whether prepared or not, could overwhelm logistics and drain resources from the host country beyond the value of the security they provide.

During the Korean War, the cold weather created challenges to both personnel and operations, which were difficult to overcome. Due to the extreme cold and snowfall during the winter of 1951, U.S. and U.N. forces were subjected to temperatures well below zero. The cold froze weapons and vehicles and caused a substantial number of non-combat injuries due to frostbite. Mobility was restricted due to the cold as well the snow and ice that blocked mountain passes, causing main supply routes at lower altitudes to be clogged with excess vehicle traffic. Ultimately, the cold reduced the combat-effectiveness of Allied forces, resulting in extreme losses due to both combat and non-combat injuries.

The reason for the American inability to operate in the severe winter weather was related to the lack of preparation of the troops that the U.S. sent to fight in Korea. Instead of sending properly trained and equipped cold-weather units, commanders in Tokyo (i.e., General Macarthur) sent occupation forces from Japan. Those occupation troops spent most of their time maintaining their dress uniforms and practicing close order drill (i.e., ceremonial marching) for the numerous parades in which they participated (Woodhouse 2011). They were totally unprepared for combat, especially in extreme cold-weather conditions.

In addition, Marines from the East Coast of the U.S. were sent to Korea without proper training or equipment, with virtually no cold-weather uniforms (Owen 1997). The result was near disaster for American troops, especially the poorly equipped Marines, during the cold-weather fighting that took place from November 26th through December 13th. Perhaps even worse, the Marines were issued cold-weather uniforms when they were already fighting on the frontline under Arctic conditions. With no time to be trained in the use of these uniforms, the troops sustained a large number of cold injuries, especially frostbite to the feet. The battle resulted in

> America's darkest hour in the Korean War: the harrowing retreat from North Korea's Chosin Reservoir, where American forces were surrounded, vastly outnumbered and facing mass slaughter in brutally cold mountains near the Chinese border (Janssen 2019).

Throughout its operational period, lack of cold-weather equipment and capabilities severely limited the ground defence of the Keflavik base. While a small contingent of Marine and Air Force security, along with Navy and Air Force personnel not trained primarily for ground combat, was supposedly capable of defending the base against ground attacks, the reality was very different. The 1994 Viking Thunder base-defence exercise demonstrated that inability.

> Exercises like the recent Viking Thunder, which lasted three days last week, always have a tendency to disrupt the normal flow of life. Not that life is all that normal in Iceland, but somehow the cold, the snow and the infamous Icelandic wind are a bit easier to deal with inside the shelter of a warm building rather than outside. … ground defence personnel not only had to hold off surprise enemy attacks, but also during the interim periods they had to contend with the rigors of life in the field. Although space heaters were used in bunkers where security forces held their guard, life was far from luxurious. "It took me an hour and a half just to heat my Spaghettios," commented one Air Force patrolman (Bongloanni 1994).

I participated in three Northern Vikings during the 1990s that were held in the summer. Units from the 187th Separate Infantry Brigade (Army Reserve) participated in those exercises, both to provide base defence and to practice patrolling in Akureyri and Egilsstaðir that contained military-capable airports. While accompanying those units as a NATO civil-military-cooperation officer (CIMIC), I observed that the American units were not properly equipped with the uniforms, supplies, or transportation to engage in sustained operations in Iceland, even in the summer (Fig. 4.5). The troops were not physically or mentally prepared for the rigors of cold-weather combat, and few (if any) of the officers or non-commissioned officers had completed cold-weather training. If the exercises had involved real emergencies, the American troops would have failed to accomplish their mission after they suffered substantial losses from non-combat injuries.

The inability of American troops to operate in cold weather is only one reason to develop an Icelandic security force. Another reason is that American or other NATO troops might not be able to provide a timely response. When meeting threats that require an immediate response (i.e., under 24 h), Iceland is essentially alone because larger countries like Canada and the U.S. will take a long time to arrive. For example, at the invitation of the Icelandic Foreign Ministry's Security and Defence Directorate, I observed Trident Junction '18 Iceland first hand. The exercise involved terrorists taking over the Icelandic Coast Guard Crisis Response Center (CRC) at the Keflavik base.

Based on my observations of the exercise and my previous military experience, I believe that the responding U.S. Marines failed to defeat the threat for four reasons. First, the Marine landing ship, the U.S.S. Iwo Jima (LHD-7) was delayed for a day due to strategic deployment issues. Second, despite excellent weather conditions for amphibious operations, the Marine commander with the task force decided to use OV-22 Osprey tilt-wing aircraft and CH-53E Super Sea Stallion helicopters to land the Marines (Fig. 4.6). Third, the decision to use aircraft instead of amphibious

Fig. 4.5 187th Separate Infantry Brigade airmobile troops disembark a U.S. army CH-47 Chinook Helicopter at Akureyri Airport During Northern Viking '95. (Photograph by U.S. Army Captain Robert Wheelersburg)

vehicles left the assault troops without organic transportation, requiring the Marines to "attack" the terrorists using tourist buses. Fourth, with no cold-weather gear, the Marines would have had difficulty surviving the night on the Reykjanes Peninsula without sustaining cold injuries. Instead, as part of the exercise, the next day the Marines went on an overnight march wearing their newly issued cold-weather uniforms. Yet given the past history of Marine and Army deployments to Iceland and elsewhere in the Arctic, the U.S. government could send troops into the region without cold-weather equipment or training, jeopardizing their ability to defeat the threat (Wheelersburg 2021).

Icelanders comprising a security force, especially one using search-and-rescue volunteers, would experience no such cold-weather issues. The survivability and mobility of ICE-SAR teams even in the Icelandic winter would allow them to counter a threat during an emergency, while their enemies would be hampered by weather constraints.

It is indeed an irony that the most dangerous weapon Icelanders have to defeat sub-strategic security threats is that their country is essentially Jötunheim, the Ice World of Norse mythology brought to the country by the original Viking settlers. In their Arctic-like country today, those descendants of Vikings are the ice giants of Jötunheim who would survive and fight enemies unable to operate in Iceland.

Fig. 4.6 Icelandic Police Viking Squad member guarding U.S. Marine OV-22 Osprey landing at Keflavik NATO base during Trident Junction '18, October 17, 2018. (Photograph by Robert Wheelersburg)

4.8 Summary

The possibility of new sub-strategic security threats like terrorism and organized crime arising in Iceland due to ease of transit into the North Atlantic because of melting ice means that, by 2030, the country should create its own security force to deter and if necessary, defeat those threats. Considering the country's current 2016 security policy, existing bilateral and alliance security partnerships, and assets like ICE-SAR and the Keflavik base, Iceland already has the support network to develop and support that force. Tailored to accomplish domestic missions by meeting threats to Iceland's sovereignty, natural resources, and people, the security force could increase the stability of the country and the region by deterring threats. If necessary, the recommended security force could have the capability of defeating those threats as well.

The cost-effectiveness of the proposed security force is readily apparent. Continuing Iceland's peaceful existence into the future has benefits beyond virtually any cost, monetary or otherwise. Engaging in Security Sector Reform presents other advantages to Iceland aside from having a more secure existence. If some form of the recommendation presented here is adopted, it would reduce the need for foreign troops to help defend the country, which many Icelanders do not want. The country's self-determination in a rapidly changing Arctic could ultimately remove the need for occupying forces, whether Danish, British, or American, allowing the country to finally gain its independence from sheltering by dominant states.

The training and part-time employment opportunities provided to young Icelandic men and women who join the security force could provide a buffer for some rural dwellers in financially hard times. Another benefit would be that the use of ICE-SAR volunteers mostly from the countryside could unite the rural areas with urban ones since the reservists would probably reside on the Reykjanes Peninsula (nearly 2/3 of the population does). In addition, the country receiving and operating a C-130J cargo aircraft as part of the coast guard and security force could have tremendous benefits for Iceland's security and civilian agencies. It seems that creating the Icelandic Security Force could be a win-win situation for the country with no potential barriers to its acceptance – except one.

Would the Icelandic public accept such a security force given the country's history, culture, and attitudes? That question will be discussed in Chap. 6. The lack of a military and the virtual exclusion of conflict from the country over many years might make it seem that such a step would be unnecessary. But given the predicted sub-strategic security threats that global warming is bringing to the circumpolar region, could Iceland afford to be caught unprepared if it faces a severe emergency? I believe strongly that the answer is no.

References

American Battle Monuments Commission. 2018. *Remembering the Aleutian Campaign of World War II*. May 11, 2018. Accessed, May 2, 2020, https://www.abmc.gov/news-events/news/remembering-aleutian-campaign-world-war-ii.

Barber, Robert C. 2015. *Iceland, the United States, and North Atlantic and European Security*. Address to Varðberg, Reykjavík, Iceland, November 19, 2015.

Bjarnason, Björn. 1995. *Icelandic Security and Transatlantic Affairs*. Security and Environmental Issues in the North Atlantic Region Conference in Reykjavík and Keflavik September 7–9, 1995.

Bongloanni, Carlos. 1994. Viking Thunder: Practice makes perfect. *The White Falcon* 53(8): 6–7. February 25. Accessed November 2, 2020, https://timarit.is/page/6097906?iabr=on#page/n5/mode/2up.

CBO. 2021. *The U.S. Military's Force Structure: A Primer, 2021 Update*. Washington, DC: Congressional Budget Office. Accessed June 23, 2021, https://www.cbo.gov/system/files/2021-05/57088-Force-Structure-Primer.pdf

Ćirić, Jelena. 2019. We All Protest! *Iceland Review*, July 15, 2019. Reykjavík. Accessed November 20, 2020, https://www.icelandreview.com/society/we-all-protest/.

Civil Council on Defence and Security. 2020. *About the Civil Council on Defence and Security*. Tbilisi. Accessed November 19, 2020, http://civilcouncil.org/about.

Clifford, Kenneth J. 1970. *The United States Marines in Iceland, 1941–1942. Marine Corps Historical Reference Pamphlet*. Historical Division, Headquarters, U.S. Marine Corps. Washington, DC. Accessed December 2, 2020, https://www.marines.mil/Portals/1/

Publications/The%20United%20States%20Marines%20in%20Iceland,%201941-1942%20%20PCN%2019000412300.pdf.

DCAF. 2015. *Security Sector Reform*. SSR Backgrounder Series, Geneva Centre for the Democratic Control of Armed Forces. Accessed November 20, 2020, https://www.dcaf.ch/sites/default/files/publications/documents/DCAF_BG_2_Security%20Sector%20Reform.pdf.

Dolbow, Jim, and Jim Howe. 2017. Shift the Coast Guard to DoD. *U.S. Naval Institute, Proceedings* 143/2/1,368, February 2017. Annapolis: U.S. Naval Institute. Accessed November 21, 2020, https://www.usni.org/magazines/proceedings/2017/february/shift-coast-guard-dod.

DSCA. 2020. *Defense Security Cooperation Agency. Mission, Vision, Values*. Washington, DC: U.S. Department of Defense. Accessed December 12, 2020, https://www.dsca.mil/.

Dursun-Özkanca, Oya. 2017. Pitfalls of Police Reform in Costa Rica: Insights into Security Sector Reform in Non-military Countries. *Peacebuilding*, 2017. Accessed November 12, 2020, https://doi.org/10.1080/21647259.2017.1293350.

Dursun-Özkanca, Oya, and Antoine Vandemoortele. 2012. The European Union and Security Sector Reform: Current Practices and Challenges of Implementation. *European Security* 21 (1): 139–160.

EKOS Research Associates. 2011. *Rethinking the Top of the World: Arctic Security Public Opinion Survey*. Final report, January 2011. Ottawa: EKOS Research Associates. Accessed November 20, 2020, https://munkschool.utoronto.ca/wp-content/uploads/2012/07/ArcticCouncil_GordonFoundation_2011.pdf.

English, Martin. 2018. Iceland Bars Run Out of Beer Supplying Thirsty US Sailors, Marines. *Stars and Stripes*, October 29, 2018. Accessed December 13, 2020, https://www.stripes.com/news/iceland-bars-run-out-of-beer-supplying-thirsty-us-sailors-marines-1.554176.

Forces Net. 2021. NATO: Which Countries Pay Their Share on Defence? *Forces Net, NATO*, March 18, 2021. Buckinghamshire. Accessed May 1, 2021, https://www.forces.net/news/world/nato-which-countries-pay-their-share-defence.

Government of Canada. 2017. *Strong, Secure, Engaged*. Canada's Defence Policy. Ottawa: Ministry of Defence. Accessed July 7, 2020 https://www.canada.ca/en/department-national-defence/corporate/reports-publications/canada-defence-policy.html.

History.com. 2020. Battle of the Aleutian Islands. History.com Editors. New York: A&E Television Networks. Accessed June 1, 2021, https://www.history.com/topics/world-war-ii/battle-of-the-aleutian-islands.

Icelandair. 2021. *Icelandair Flies to Antarctica: Putting the Ice Back in Icelandair*. 3/01/21. Accessed June 22, 2021, https://www.icelandair.com/blog/icelandair-flies-to-antarctica/.

Icelandic Parliament. 2016. *Parliamentary resolution on a national security policy for Iceland*. 145th Legislative Session, 2015-2016, Parliamentary document 1166-327th matter, No. 16/145. Accessed November 2, 2020, https://www.government.is/library/01-Ministries/Prime-Ministrers-Office/Parl-Res-Nat-Sec-Council-25-145-2016-tr-br-prot-SKIL-310817-.pdf.

Janssen, Volker. 2019. *The Most Harrowing Battle of the Korean War. At the Chosin Reservoir, Subzero Temperatures Were the Enemy as Communists: Frozen Bodies Were Used as Sandbags. History*. Accessed December 7, 2020, https://www.history.com/news/korean-war-chosin-reservoir-veterans-stories.

Katz, Samuel M. 2005. *Global Counterstrike. International Counterterrorism*. Minneapolis: Lerner Publications.

Kjartasnnon, Sigurjón, and Baltasar Kirmákur. 2021. *Katla*. Reykjavik: RVK Studios.

Kyzer, Larissa. 2018. Prime Minister Turns Down US Naval Invitation. *Iceland Review*, September 21, 2018. Accessed November 25, 2020, https://www.icelandreview.com/politics/prime-minister-turns-naval-invitation/.

Laxness, Halldór. 1982. *The Atom Station*. New York: Second Chance Press.

Lowe, Christian. 2008. Russia Warns Against Georgia NATO Membership. *Reuters, World News*, March 11, 2008. Accessed November 22, 2020, https://www.reuters.com/article/us-russia-nato-georgia/russia-warns-against-georgia-nato-membership-idUSL1185178620080311.

Markham, James M. 1982. Iceland's Elves Are Enlisted in Anti-NATO Effort. *New York Times*, March 30, 1982. Accessed December 15, 2020, https://www.nytimes.com/1982/03/30/world/iceland-s-elves-are-enlisted-in-anti-nato-effort.html.

Miko, Francis T., and Christian Froehlich. 2004. *Germanys Role in Fighting Terrorism: Implications for U.S. Policy. Congressional Research Service Report for Congress*. Washington, DC: U.S. Congress. Accessed November 24, 2020, https://fas.org/irp/crs/RL32710.pdf.

Ministry of Defence of Georgia. 2014. *National Military Strategy*. Tbilisi. Accessed September 21, 2020, https://mod.gov.ge/uploads/2018/pdf/NMS-ENG.pdf.

NATO. 2020. *Partnership Interoperability Initiative*. Brussels. Accessed November 21, 2020, https://www.nato.int/cps/en/natohq/topics_132726.htm.

———. n.d. *Substantial NATO-Georgia Package (SNGP)*. North Atlantic Treaty Organization Factsheet. Brussels. Accessed November 20, 2020, https://www.nato.int/nato_static_fl2014/assets/pdf/pdf_2016_06/20160614_1606-georgia-sngp-factsheet-en.pdf.

NBC. 1953. *Victory at Sea. The Magnetic North 401582203*. War from Murmansk to Alaska, Episode 11, January 18, 1953. Universal City: NBC Television Network. Accessed June 19, 2021, https://www.youtube.com/watch?v=RNK8DyqbduU.

Nye, Joseph S. 1990. The Changing Nature of World Power. *Political Science Quarterly* 105 (2, Summer, 1990): 177–192.

Overseas Security Advisory Council. 2019. *Iceland 2019 Crime & Safety Report*. Overseas Security Advisory Council (OSAC), March, 8, 2019. Country Security Report, U.S. Diplomatic Security Service. Accessed November 13, 2020, https://www.osac.gov/Country/Iceland/Content/Detail/Report/2d8cdd85-eb54-4ec7-828a-15f4aeb184e9.

Owen, Joseph R. 1997. *Colder than Hell. A Marine Rifle Company at Chosin Reservoir*. New York: Ivy Books.

Piotukh, Volha, and Peter Wilson. 2009. *Security Sector Evolution. Understanding and Influencing How Security Institutions Change*. Accessed November 20, 2020, https://issat.dcaf.ch/Learn/Resource-Library2/Policy-and-Research-Papers/Security-Sector-Evolution-Understanding-and-influencing-how-security-institutions-change.

Prime Minister. 2016. *National Security Council Act No. 98, 20, September 2016*. Reykjavík: Prime Minister's Office. Accessed November 2, 2020, https://www.government.is/publications/legislation/lex/2018/01/19/National-Security-Council-Act-No.-98-20-September-2016/.

Robinson, Denver David. 2016. The World's Most Unusual Military Unit. Why Members of an Elite Danish Force Patrol the Rooftop of the World ... by Dog Sled. *The Christian Science Monitor*, June 22, 2016. Accessed November 21, 2020, https://www.csmonitor.com/World/2016/0622/The-world-s-most-unusual-military-unit.

Sharp, Gene. 1990. *Civilian-Based Defense. A Post-Military Weapons System*. Albert Einstein Institution. Princeton University Press. Accessed November 22, 2020, https://www.aeinstein.org/wp-content/uploads/2013/09/Civilian-Based-Defense-English.pdf.

Shears, David. 1970. *The Ugly Frontier*. London: Chatto & Windus.

Simon, Greg. 2014. Security Sector Reform and Georgia: The European Union's Challenge in the Southern Caucasus. In *The European Union as an Actor in Security Sector Reform*, ed. Oya Dursun-Özkanca, 134–155. New York: Routledge.

Smol, Robert. 2020. *A Canadian Rangers Reset Would Help Armed Forces Keep Pace with a Changing North*. Canadian Broadcasting Corporation (CBC), Opinion, October 20, 2020. Accessed November 20, 2020, https://www.cbc.ca/news/opinion/opinion-canadian-rangers-arctic-sovreignty-1.5763215.

Sturluson, Snorri. 2006. *The Prose Edda: Norse Mythology* (Penguin Classics). Translated by Jesse L. Byock, 74. London: Penguin Press.

Suny, Ronald G. 2020. *Georgia*. Britannica, December 28, 2020. Chicago. Accessed October 22, 2020, https://www.britannica.com/place/Georgia/Independent-Georgia.

Swedish Army Chief of Staff. 1968. *Winter Soldier* [original in Swedish]. Stockholm: Swedish Defence Ministry.

———. 1990. *Ranger Company, Norrland's Ranger Squadron* [original in Swedish]. Stockholm/Borås: Swedish Defence Ministry/Libergraf.

Þórsson, Elías. 2017. Police with Guns at All Future Public Gatherings in Iceland. *The Reykjavík Grapevine*, June 13, 2017. Accessed November 12, 2020, https://grapevine.is/news/2017/06/13/police-with-guns-at-all-future-public-gatherings-in-iceland/.

Tikka. 2020. *T3X Arctic*. The Rifle That Canadian Rangers Trust. Tikka Arms, Rikimäki. Accessed November 22, 2020, https://www.tikka.fi/rifles/tikka-t3x/t3x-arctic.

Trading Economics. 2021. *Iceland GDP*. New York. Accessed May 1, 2021, https://tradingeconomics.com/iceland/gdp.

U.S. Department of State. 2017. *Agreement Between the United States of America and Iceland. Effected by Exchange of Notes at Reykjavík October 13 and 17, 2017 with Attachments and Minutes*. Accessed November 22, 2020, https://www.state.gov/wp-content/uploads/2019/02/17-1017-Iceland-Defense-Coop-Notes.pdf.

United Nations. 2013. *International Instruments to Enable States to Identify and Trace, in a Timely and Reliable Manner, Illicit Small Arms and Light Weapon*. Vienna: United Nations Office on Drugs and Crime, February 25, 2013. Accessed August 7, 2020, http://www.unodc.org/documents/organized-crime/Firearms/ITI.pdf.

United Nations and the Rule of Law. n.d. *Security Sector Reform (SSR)*. New York: Office of the Secretary General. Accessed November 21, 2020, https://peacekeeping.un.org/en/security-sector-reform.

Varðberg. 2020. *Organization for Western Cooperation and International Affairs*. Reykjavík. Accessed November 22, 2020, http://vardberg.is/upphafid/.

Veal, Lowana. 2019. Concerns About Military Build-up in Iceland. *IDN-InDepthNews*, September 14, 2019. New York: National Press Syndicate. Accessed November 23, 2020, https://indepthnews.net/index.php/armaments/nuclear-weapons/2972-concerns-about-military-build-up-in-iceland.

Wheelersburg, Robert P. 2021. North Atlantic Small State Security 2025: The West Nordic Security Zone. In *Small States and the New Security Environment, The World of Small States*, ed. Anne-Marie Brady and Baldur Thorhallsson, 87–101. Cham: Springer.

Woodhouse, Dale B. 2011. *Operational Lessons Learned in the Korean War*. Fort Leavenworth: School of Advanced Military Studies, United States Army Command and General Staff College. Accessed July 26, 2020, https://apps.dtic.mil/dtic/tr/fulltext/u2/a557709.pdf.

Chapter 5
Icelandic Security Force Scenarios 2030

Some died by the Glenside, some died near a stranger,
And wise men have told us their cause was a failure.
But they stood by old Ireland and never feared danger,
We may have brave men, but we'll never have better
Glory O, Glory O, to the bold Fenian men.
Peadar Kearney, Down by the Glenside *(Irish rebel song)*

Abstract This chapter presents four scenarios involving fictional state and non-state actors threating Icelandic territory and people to illustrate successful use of the proposed security force. I begin the section on scenario development by examining appropriate conflicts for the security force missions, referred to as military operations other than war (MOOTW) or irregular warfare (IW). Rather than be prepared for conventional war, the proposed Icelandic Security Force is structured to provide assistance to civil authorities and other security forces including the coast guard and police during emergencies. Principles for exercise design and scenario development according to U.S. Army doctrine are presented, as are important concepts such as operational environment and operational themes used to create the four scenarios. The scenarios follow a fictional geopolitical situation in 2030 after the polar ice cap has melted, opening new transport corridors, exposing new territories, and permitting access to natural resources that presented opportunities for states, terrorists, and criminals to compete for in the Arctic. The first scenario consists of a cruise-ship hijacking in Akureyri. The second scenario involves an attempt by a state to intrude into the country near Þorlákshöfn following a powerful volcano. The third scenario concerns a non-Arctic state attempting to occupy abandoned Grímsey Island. The fourth scenario presents a human tracking operation by international criminal syndicates in Seyðisfjörður and Egilsstaðir.

R. P. Wheelersburg, *The Security of Iceland and the Arctic 2030*, Springer Polar Sciences, https://doi.org/10.1007/978-3-030-89948-6_5

5.1 Introduction to Operations Other Than War/Irregular Conflicts

The term I introduced in Chap. 1, sub-strategic security, is often referred to in other ways such as military operations other than war (MOOTW) or irregular warfare (IW). The concepts from those two types of conflict create the appropriate situations necessary to develop scenarios illustrating the capabilities of the recommended security force. Yet developing accurate MOOTW/IW scenarios is more difficult than those created for conventional warfare because historical examples of the latter are much more plentiful. Therefore, before presenting the four scenarios, it is important to review the basic concepts of MOOTW/IW to understand how they influenced the scenarios' development.

After the fall of the Soviet Union and the First Gulf War, both of which took place in 1991, signaled an end to America focusing primarily on strategic-level, land-based conventional warfare, it was no surprise that the manual for U.S. MOOTW was published in 1995. The Chairman of the Joint Chiefs of Staff stressed MOOTW's importance in the manual's forward.

> While we have historically focused on warfighting, our military profession is increasingly changing its focus to a complex array of military operations – other than war. . .Participation in MOOTW is critical in the changing international security environment. Although the goals and end states may not be crystal clear, you should spare no effort in planning and executing MOOTW. . . our Nation and its independence may depend upon your knowledge [of it] (Shalikashvili 1995: forward).

Preparing conventional war fighters to adopt new methods required a radical shift in conceptional framework. The difference between conventional war and MOOTW is fundamental to planning and conducting military operations since the former involves "large-scale, sustained combat operations to achieve national objectives" while the latter places its "focus on deterring war and promoting peace" (Joint Chiefs of Staff 1995: vii). Some aspects of conventional war fighting did not fit into the MOOTW framework; for example, field artillery batteries or tank brigades have little use in a natural disaster response. Thus, U.S. defence doctrine had to conduct Security Sector Reform by adapting existing force structures and evolving new ones to accomplish novel MOOTW missions.

Moving from land-based conventional warfare in the defence of Western Europe to an MOOTW doctrine also required a tremendous change in the structure and operations of American and NATO forces. The last fight for NATO Cold Warriors, who spent decades preparing to repel Warsaw Pact armored divisions in places like West Germany's Fulda Gap, occurred in a Middle-Eastern desert. During the First Gulf War (Desert Storm) from August 1990 through February 1991, coalition forces faced Iraqi tank divisions that were equipped with Soviet weapons. Resulting in perhaps as many as 50,000 enemy deaths, the coalition ground attack was preceded by a massive air campaign that reduced Saddam Hussein's feared tank divisions to rubble piled high with dead Iraqi soldiers (Holland 2018). While American warfighters were conducting the Gulf War Campaign, planners at the Pentagon prepared to downsize the U.S. military (especially the army in Europe) to make it more flexible and able to address multiple threats throughout the world. For the past three decades, the American Army and Marines, and by proxy other NATO ground forces, have spent much of their efforts planning for and executing MCOTW.

The influence on the nature and capabilities of armed forces under each type of security operation is obvious. For example, one of the most important methods for deterring conflict by using MOOTW is for a security force to provide support to civilian authorities during an emergency (i.e., crisis response) such as evacuation or recovery operations in a natural disaster. Crisis response prevents emergencies from turning into security failures as threats could exploit the confusion arising from an event like a natural disaster to attack. MOOTW may also provide support to civil authorities during rapidly unfolding hostile threat situations through missions such as counterterrorism or hostage rescue. Reserve forces are especially important in domestic MOOTW due to their familiarity with local circumstances and their ability to interact with the civilian population using language and cultural knowledge (Joint Chiefs of Staff 1995).

At all levels, those engaging in MOOTW must understand that their actions involve political issues that must be resolved to establish legitimacy with the public. MOOTW gains public approval when it deters aggression by demonstrating that the national command authority will use force to protect the country's territory and people. During peacetime operations, a security force providing MOOTW, such as humanitarian assistance, facilitates peaceful solutions to emergencies, which will also increase a population's confidence in that security force. Security forces based upon MOOTW must respond rapidly, regardless of the nature of the crisis, and they must work in a coordinated fashion with civil authorities and agencies like the police and emergency management. Lastly, effective security forces capable of conducting successful MOOTW missions, either unilaterally or within multi-national alliances, contribute to their country's national security and strengthen regional stability (Joint Chiefs of Staff 1995).

A more recent interpretation of MOOTW is irregular warfare (IW), which also constitutes a sub-strategic security threat to a country's sovereignty and public safety. Some analysts describe IW thusly.

> The twenty-first-century security environment is characterized by an abundance of violent nonstate actors who do not adhere to the principles of regular warfare. This equally applies to the rise of hybrid warfare in which state and nonstate actors alike preferably operate below the threshold of overt war and revert to irregular activities instead. Consequently, the capability to take on missions outside the conventional comfort zone should be a quintessential part of any modern military's tool box (Kitzen 2020: 2).

Examples of irregular conflict include guerrilla warfare, terrorism, and organized crime. We can also count limited clashes between state-level aggressors such as limited border incursions or territorial disputes over natural resources (e.g., the Cod Wars) as irregular warfare. One of the problems security forces have in dealing with sophisticated IW is that the most advanced threats have the capability to transition to a conventional engagement rapidly, albeit on a smaller scale than standing militaries.

That means a security force engaged in an irregular warfare must make that transition just as quickly as the threat or risk defeat (Kitzen 2020).

IW often is part of an asymmetrical conflict between a state-level military force and guerillas or terrorists. A difference between guerillas and terrorists is that guerrillas prefer not to attack civilians because it risks alienating the society in which they must operate. Terrorists, on the other hand, want to induce shock and fear among civilians who are often the targets of their attacks. While IW actors may not be able to influence natives to engage in violence against their own countries, selective targeting of individuals allows terrorists to shape the political arena in their favor on issues like immigration (Kitzen 2020). For example, inspiring right-wing, anti-immigration sentiment in Norway is what motivated Ander Breivik's home-grown terrorist attacks against left-wing youth groups. Such operations are called "false flag" wherein a state or non-state threat acts as if they were conducting IW as the agent of or to benefit another country.

Irregular forces do not expose themselves by attacking non-military targets, which permits them to hide among the civilian population awaiting the opportunity to attack security forces (Kitzen 2020). Understanding that tactic is critical in Iceland where there are astronomical opportunities for terrorists, criminals, or saboteurs to blend in with the country's civilian population. In 2018, nearly ten million passengers transited Keflavik airport, 250,000 cruise-ship passengers docked in either Reykjavík or Akureyri, and 20,000 passengers came by ferry from Denmark to Seyðisfjörður. Thus, in a single year, visitors to Iceland outnumbered the resident population by a factor of 30, not counting the unknown number of people carried by private ships or aircraft that arrived in the country during the same period. Those numbers could increase substantially after the Arctic Ocean ice cap melts, with a greater potential for illegal visitors to enter Iceland using dozens of unimproved landing strips or natural harbors located throughout the country.

Each of these two terms, MOOTW and IW, includes a word that is not relevant to the proposed Icelandic Security Force; military in the first term and warfare in the second. Thus, for our purposes I will refer to the two types of security activities as Operations Other Than War (OOTW) and irregular conflict (IC), which present a more accurate description of the proposed Icelandic security force's mission capabilities. The basic principles remain the same despite using modified terms to describe the situational considerations.

Based upon these concepts, to generate the following scenarios I made five assumptions about how the proposed security force could be prepared to conduct OOTW/IC missions. Those assumptions were that

- a security force's fundamental purpose is to deter aggression by state or non-state actors,
- a security force's support to civil authorities in a crisis increases its public acceptance,
- a security force's missions include counterterrorism and fighting organized crime,
- a security force must anticipate that enemies may infiltrate the civilian population,

- a security force should be prepared to use limited conventional tactics to defeat threats.

5.2 Exercise Design and Scenario Development

An effective exercise scenario must include reasonable courses of action for friendly units and the opposing force (OPFOR) consistent with the selected operational environment (OE). In this case, the friendly units are the proposed Icelandic Security Force (ISF), the Icelandic Police (IP), Icelandic Coast Guard (ICG), and the Icelandic Search and Rescue (ICE-SAR) teams. OPFOR in the scenarios include terrorists, an Arctic state, a non-Arctic state, and criminal organizations.

The operational environment (OE) is a critical consideration for creating scenarios because it is "a composite of the conditions, circumstances, and influences that affect the employment of capabilities and bear on the decisions of the commander" (Department of Defense 2010: 254). For example, the OE in Syria today, where vast areas are not controlled by a government and lawlessness reigns, allows for militias and terrorist groups to occupy territory and operate openly. Conversely, in Europe's OE, irregular threats are often intermingled with the civilian population and must operate covertly.

The operational environment contained in the following scenarios is based upon my predicted geopolitical relationships in 2030 following substantial reduction of the ice cap in the Arctic Ocean. Conditions in the OE included Iceland's subarctic terrain and weather, the nature of the country's surrounding territorial waters, and the opening of the Transpolar-Sea Route (TSR). I also used several possible future circumstances when determining the OE such as a decrease in the price of oil with the emergence of new Arctic petroleum fields. According to current practices, I also considered the influences of possible future political, military, economic, and social variables influencing the OE to build realistic scenarios (Department of the Army 2010).

Within my hypothesized operational environment, the security force's capabilities and commanders' decisions are the result of operational themes. "The operational theme is the character of the dominant major operation being conducted at any time within a land force commander's area of operations" (Department of the Army 2010: 2–4).

There are five operational themes, but the ones used here are OOTW/IC, peace operations, and limited intervention. Since the Area of Operations (AO) for these scenarios is the country of Iceland and its surrounding waters, we can assume that the country's security forces have an intimate knowledge of the AO and possess the capability to survive and maneuver in it. We can also make the assumption that OPFOR comprised of out-of-area states, terrorists, and criminals do not have such knowledge or capability for operating in the Icelandic AO, although other Arctic states may be somewhat knowledgeable and capable in generic subarctic conditions

I based these scenarios on the possible future operational environment assessment (OEA) for Iceland and the circumpolar region as taken from various threat assessments compiled by Arctic Council members presented in Chap. 3. These combined threat assessments predicted four different potential threats to Iceland's sovereignty and national security that could follow global climate change in the Arctic. Each of these potential threats is used to create the operational environment and the opposing forces for one of the scenarios.

First, terrorists seeking to influence the public either in their home regions or globally may use the relative openness and ease of access to Arctic territories to commit atrocities. The terrorists may target the residents of a country, visitors from another country that is the main target (e.g., tourists from a European country while in the Arctic), or random attacks to achieve wealth, freedom (i.e., prison release), or political sympathy.

Second, military confrontations could increase as Arctic nations position themselves to control portions of the Arctic Ocean and the North Atlantic. Those conflicts may include a state's efforts to extend its own security forces or to degrade the security forces of potential adversaries in the region. One could expect limited border incursions to occur during such confrontations.

Third, out-of-area states seeking to gain a foothold in the Arctic may attempt to claim newly emerged landscapes either through legal or aggressive means. That attempt could involve challenges to a country's sovereignty over land that is unoccupied or abandoned. Out-of-area states may intrude on a nation's Exclusive Economic Zone to exploit newly-accessible resources.

Fourth, exponential growth in travel through the Arctic created by tourism, cruise ships, transportation carriers, research organizations, and exploration companies could allow international criminal organizations to infiltrate countries to engage in illegal activities. The opportunities for smuggling drugs, weapons, and hazardous material (e.g., plutonium) may increase due to that essentially unmanageable population movement through Iceland, with human trafficking representing one of the most substantial threats to the region.

To illustrate the potential capabilities of the recommended Icelandic Security Force to provide support to civil agencies in an emergency, I organize the security force assets to work with the IP, ICG, and ICE-SAR to deter or defeat the threat. In addition, I use the legal framework, especially the 2016 Icelandic National Security Policy and governmental organization (i.e., responsible ministries), to activate, deploy, and demobilize the security force.

The scenarios portray the effective use of the recommended Icelandic security force in emergencies. By convention, scenarios are merely illustrative and not subject to criticism, which detracts from the main goal of the exercise to portray the activities of friendly units for analysis.

Each of the scenarios involves the security force utilizing activities from the U.S. Army's tactical task list to demonstrate its ability to deter or defeat a threat using force (Department of the Army 2015). While the tactical tasks also include other appropriate operations such as reconnaissance, to illustrate the potential for the security force to engage in hostile action against an aggressor, I base the four scenarios on conflict-laden tactical tasks.

Scenario one is an assault against terrorists in Akureyri who have hijacked a cruise ship and taken the passengers hostage. By definition, "an assault is an attack that destroys an enemy force through firepower and the physical occupation and/or destruction of his position" (Department of the Army 2010: B-1).

Scenario two is an ambush against an Arctic country's forces that attempted to invade Þorlákshöfn following a devastating earthquake. By definition, "an ambush is a surprise attack from a concealed position, used against moving targets to render the enemy combat ineffective" (Department of the Army 2010: B-3).

Scenario three is a defence from a battle position against a non-Arctic state preparing to illegally occupy Grímsey Island. By definition, "a defence from a battle position is oriented on the most likely enemy objective [where] defenders will take all actions necessary to prevent enemy penetration" (Department of the Army 2010: B-6).

Scenario four is a fire and maneuver engagement against a criminal syndicate attempting human trafficking in Seyðisfjörður. By definition, "fire and maneuver is the way small units move with a support element directing suppressive fire against the enemy while the action element moves to secure the objective" (Department of the Army 2010: B-14).

The scenarios are presented at various locations throughout Iceland in chronological order to differentiate them in space and time; however, any of them could occur at any place or date as neither the locations nor the times are required for the depicted events (Fig. 5.1).

The fictional amphibious landing at Þorlákshöfn following a massive volcanic eruption on the Reykjanes Peninsula in scenario two was created 2 years prior to the U.S. Marine exercise there that was supposed to include an amphibious landing. As described in Chap. 4, despite excellent conditions for that landing to take place, the task force commander decided to use aircraft to deliver the Marines to engage the threat at a site at Keflavik. Thus, the Marine exercise provided a test for whether a large ally like the U.S. could provide Iceland with sub-strategic security in a rapid and self-sufficient manner–a test which it failed. In addition, despite not having a volcanic eruption on the Reykjanes Peninsula for 800 years, the scenario was developed 4 years prior to the eruption occurring there as this manuscript goes to print.

5.3 A Possible North Atlantic Geopolitical Situation in 2030

Frustrated with its ongoing observer status on the Arctic Council, in 2018 China issued a new Arctic Policy with four goals. The first goal was building a "Polar Silk Road" along the Transpolar Sea Route and facilitating "its enterprises to participate in the infrastructure construction for these routes and conduct commercial trial voyages" (Gao 2018: 1). The second goal was to increase China's natural resource

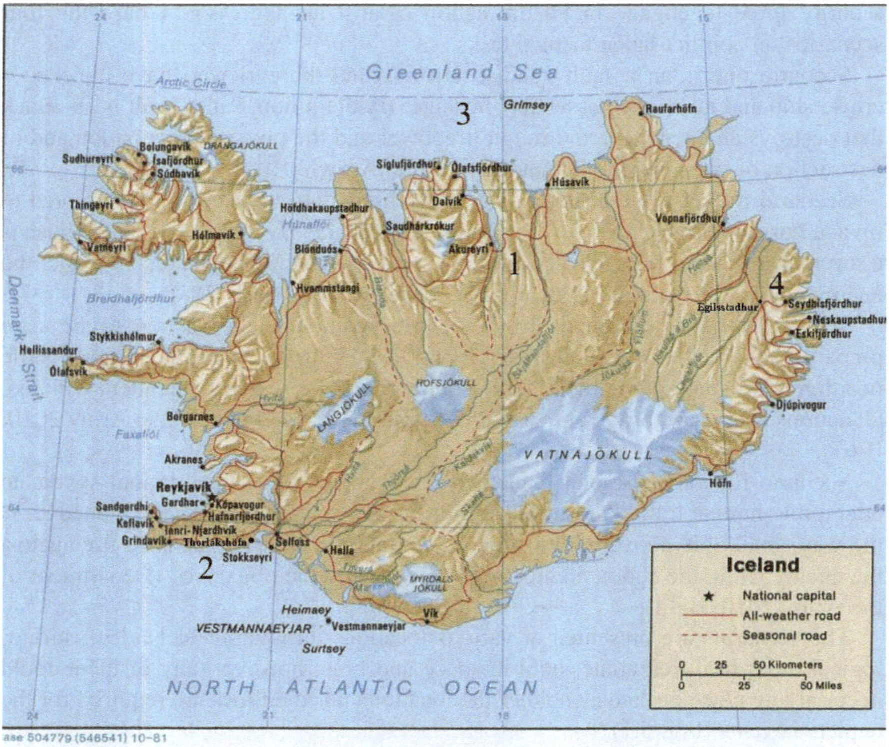

Fig. 5.1 Map of Iceland with scenario locations. (Modified from U.S. Central Intelligence Agency Map of Iceland 1981)

exploration in the circumpolar region especially for oil and gas. The third goal was for Chinese fishing vessels to exploit waters throughout the Arctic with unrestricted access. The fourth goal was to increase Chinese tourism in the Arctic to establish the country's claims over emerging or unoccupied territory in the region, as it has done in the disputed Paracel Islands in the South Pacific (Wang 2021).

After losing its territory in the Middle East in 2023, The Islamic State in Iraq and Syria (ISIS) vowed to take the fight to Europe and the U.S. until countries in those regions supported the establishment of a Caliphate. Using disgruntled immigrants and planted terrorist cells among asylum seekers, ISIS carried out daily bombings in Great Britain, Northern Europe, and the United States. In 2025 when Muslim residents reached 10% of the population in the Nordic countries (20% in Sweden), ISIS implemented a plan to engage in terrorism throughout the region until the Nordics recognized ISIS for UN membership.

In 2026, Russia invaded and annexed Ida-Viru County, Estonia "to repatriate the Russian peoples held captive there" (Putin speech to the Duma). Led by Bulgaria, Turkey, and Georgia, a majority of NATO members voted against invoking Article 5 for the Estonian crisis. Emboldened by NATO's lack of action, in 2027 a company of Russia's 76th Guards Airborne Division parachuted at night onto Sweden's Gotland Island. With the Russian Government claiming it was a navigational error, the airborne troops left the next morning after ground-truthing satellite imagery of Swedish defense facilities on that key approach to the Baltic Sea. After oil prices dropped below $20 a barrel, Russia declared bankruptcy and appealed for help to the International Monetary Fund (IMF). The IMF denied the request because of increased Russian military activities.

During early 2028, the United Nations Office of Drugs and Crime (UNODC) identified Iceland as a leading haven for international criminal syndicates involved in trafficking humans, drugs, and weapons. The Icelandic Coast Guard and Police gradually became overwhelmed with the volume and sophistication of the traffickers and were unable to stem the tide. European intelligence agencies warned of the increasing threat and began intense monitoring of aircraft and ocean vessels owned by known and suspected criminal syndicates travelling to and from Iceland. The U.S. National Security Agency also began to track suspected travel for illegal purposes in the North Atlantic Arctic.

After Syrian dictator Bashar al-Assad was assassinated in 2029, Iranian militia groups blamed the U.S. and vowed to strike in the heart of America. Later that year, the U.S. Secret Service determined that a dirty nuclear device smuggled into Chicago by Hezbollah originated with the Pakistani Inter-Services Intelligence Agency and Yemeni-based Houthi rebels who supplied the terrorists with radioactive materials, the explosive device to detonate them, funding, and logistics. In response, the U.S. opened a two-front incursion into southern Pakistan and northwestern Yemen to destroy terrorist bases there.

With the U.S. otherwise engaged and NATO countries no longer bound absolutely by Article 5, in 2029 the Nordic countries strengthened their security alliance. Members of the Nordic Defence Cooperation (NORDEFCO) signed a treaty guaranteeing mutual security support during a crisis. The main provision of the Nordic Security Alliance was that a member nation needed a security force in order for the other states to respond to emergencies in that member's country. To meet that requirement, the Icelandic NSC formed a security force in January 2030.

I created the following scenarios from events depicted in this hypothetical geopolitical situation.

5.4 Icelandic Security Force Scenarios 2030

5.4.1 Scenario 1 – Cruise Ship Hijacking, Akureyri, North Iceland

At 1430 on June 5, 2030, the Icelandic Coast Guard received a distress call from a cruise ship enroute to Qaqortoq, Greenland carrying 5000 passengers while transiting the Transpolar Sea Route (TSR). Terrorists hidden among the passengers aboard the ship forced the captain to make the emergency call and subsequently contacted a terrorist cell in deep cover at a safe house located in northern Iceland. Icelandic Coast Guard command authorized an emergency port of call at Akureyri to fix its severe mechanical problems. The newly commissioned Coast Guard Patrol Vessel Gryla, trailed the apparently disabled cruise ship into Akureryi fjord and launched a helicopter to conduct aerial surveillance of the ship as it approached the dock.

After the onshore terrorists disabled security on the dock, at 1700 on June 5, 2030, the onboard terrorists removed nearly 1000 Europeans from the ship, forcing the captain to sail away immediately with the remaining passengers. The hostages were herded into a fish processing complex in the harbor, and the terrorists blew up several warehouses nearby, using the debris and heavy equipment on the dock to fortify their position. They demanded $10 billion in gold and safe passage immediately and began executing a prisoner every 15 min until those demands were met. The district Icelandic Police (IP) responded, but the terrorists were armed with automatic weapons and hand grenades. The police took heavy casualties while fighting to free the passengers and could not penetrate the enemy stronghold. The IP District Commissioner ordered the remaining officers to the Akureyri airport, holding it for reinforcements (Fig. 5.2).

The crisis action team of the National Security Council (Prime Minister, Justice Minister, and Foreign Minister) declared a state of emergency at 1745 on June 5, 2030. It authorized direct action against the terrorists and ordered the Icelandic Security Force to activate and transfer its command to the Justice Ministry. After notification of the emergency by the Foreign Minister, the Security and Defence Directorate ordered the security force commander to report to the Icelandic Coast Guard Chief of Operations at the Keflavik Base. The NSC appointed the Icelandic Coast Guard's Deputy Chief of the Defence Division as incident commander (IC), who flew to Akureyri and established a command post at police headquarters.

After receiving orders from the justice minister, at 1900 on June 5, 2030, the security force battalion commander formed a task force. The task force consisted of a mechanized squad from the Heavy Recon Patrol, along with a machine gun squad and an unmanned aerial vehicle (UAV) squad from the Weapons Recon Patrol. The task force with one of its armored personnel carriers (APC) boarded the Coast Guard's C-130J cargo plane, along with members of the Viking Squad.

The task force took off and flew north at 2030 on June 5, 2030. One hour later, the C-130 landed at Akureyri airport, and the task force disembarked. Following surveillance of the objective by the UAV squad, the Heavy Recon Squad drove from the airport to the harbor in its APC. The machine gun squad remained at the airport to support the police and established two tactical positions near the terminal.

At 2145 on June 5, 2030, the security force battalion commander in Keflavik activated a squad from the Light Recon Patrol based in Akureyri, made up of ICE-SAR personnel, which mustered at the British World War II hanger south of the city. The recon squad drove in their all-terrain vehicles to the airport terminal, where they made contact with the machine-gun squad, and joined the two tactical positions armed with their rifles.

Fig. 5.2 Terrorists hijack a cruise ship in Akureyri Harbor. (Illustration by Elin Flóki Storeide)

At 2215 on June 5, 2030, a terrorist suicide team from the harbor assaulted the airport terminal using a stolen truck while firing automatic weapons and throwing grenades. Although the security force lost one tactical position in the assault the second was still combat effective, and the airport remained in Icelandic hands.

At 2230 on June 5, 2030, using its APC, the mechanized squad assaulted the terrorist's fortified position, dispatching the enemy combatants after a 15-min struggle. The accompanying Viking Squad members arrested the remaining terrorists while the local police took up positions at the harbor to begin crowd control and to process surviving hostages.

Nearly three dozen hostages and five troops were killed during the conflict; however, over 950 hostages survived unharmed, and the residents of Akureyri remained safe.

5.4.2 Scenario 2 – Arctic-State Limited Territorial Incursion, Þorlákshöfn, South Iceland

At 0730 on December 29, 2031, an earthquake near Grindavík, followed by an accompanying large lava-forming eruption, closed the main road from the Leif Eriksson International Airport in Keflavik to Reykjavík (Fig. 5.3). With the fear that a second eruption might occur soon, at 1000 on December 30, 2031, the Icelandic Department of Civil Protection and Emergency Management ordered an evacuation of a 40-square-kilometer hazard zone from Grindavík to Þorlákshöfn. The evacuation displaced some 16,000 people, which left the region abandoned.

At 1100 on December 31, 2031, an Arctic Council Member informed the Icelandic Ministry of Foreign Affairs that under the 2011 Arctic Search and Rescue Agreement, it would land an amphibious force on the beaches to the east of Þorlákshöfn. The foreign landing force's mission was to secure the town's ferry port in preparation for a second landing of troops to conduct humanitarian operations to aid Icelandic civil authorities during the emergency (Fig. 5.4).

At 1200 on December 31, 2031, the Icelandic Prime Minister (PM) warned the foreign nation's ambassador that under its national security policy, violating the territorial integrity of Iceland was considered an act of war. Subsequently, the PM

Fig. 5.3 Active lava flow from Fagradalsfjall Volcano near Grindavík in early June, 2021. (Photograph by Robert Wheelersburg)

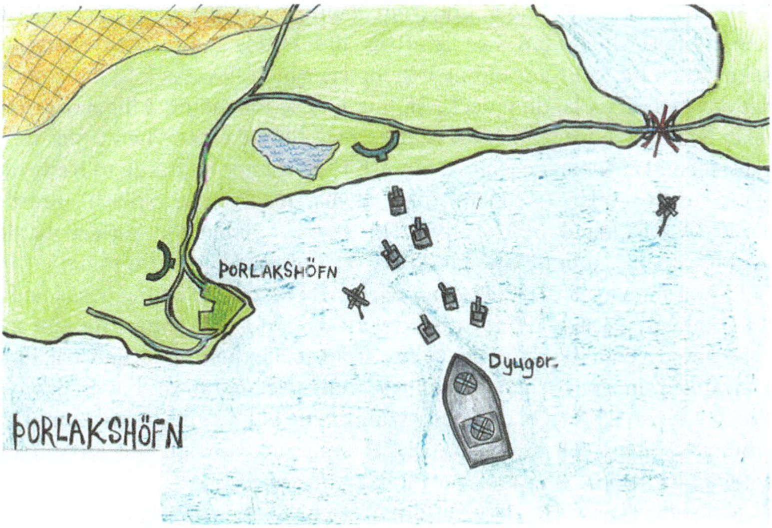

Fig. 5.4 Arctic-state limited territorial incursion east of Þorlákshöfn. (Illustration by Elin Flóki Storeide)

declared an emergency and the foreign minister activated the Icelandic Security Force.

At 1430 on December 31, 2031, the security force battalion commander formed a task force composed of two squads from a Heavy Recon Patrol along with a squad of air-defence artillery and a squad of mortars from a Weapons Recon Patrol. At midnight on December 31, 2031, the Icelandic task force drove its armored personnel carriers and ICE-SAR all-terrain vehicles from the Keflavik Base to Þorlákshön, where it took up two defensive positions east of the town. Each position had an APC with its machine gun, an air-defence team, and a mortar team.

At 0930 on January 1, 2032, an enemy amphibious flotilla appeared off the southeastern coast east of Grindavík, near the location of the U.S. Naval Radio Transmission Facility (NRTF-N62). NRTF-N62 provided communication between NATO forces in the North Atlantic and their commands in North America and Europe. If NRTF-N62 were lost, NATO naval forces in the Arctic Ocean would be out of transmission range of headquarters, limiting their effectiveness to monitor and counter aggressive actions in the North Atlantic.

At 1100 on January 1, 2032, two Dyugon-class landing craft sailed toward Þorlákshöfn, 60 kilometers east of Gríndavik, one carrying a platoon of armored naval infantry and the other a naval tank platoon. One landing craft disembarked four amphibious BTR-80 APCs with naval infantry squads in the littoral water east of the town. The other landing craft, carrying three T-80 main battle tanks, anchored just outside the town's Westman Islands' ferry dock. An Ivan Rogov-class landing ship carrying four Ka-52 K Alligator assault helicopters moved into position to support the landing's first wave. The Rogov launched two enemy assault helicopters that flew above the amphibious BTRs as they approached the beach at Eyrarbakki.

The helicopters came under fire from members of the Icelandic task force's air-defense team, which fired Swedish RBS-20 missiles at the approaching aircraft. A missile struck the tail rotor of one helicopter, which flew back to the amphibious ship but crashed on its landing deck, fouling it. The remaining Alligator dispatched the Icelandic air-defense team with cannon fire. As the enemy helicopter turned to attack the second air-defence team, a Finnish F-22 Raptor returning to Keflavik after air-policing duty under "weapons free" status, received a "splash" order from NORDEFCO Command in Helsinki. The Raptor joined the surface battle and exploded the enemy aircraft with an AIM-260 Vulture air-to-air missile.

At 1145 on January 1, 2032, the four enemy BTRs landed on the beach and drove inland to highway 34, where they turned west toward Þorlákshöfn. As it neared the town, the enemy convoy came under fire from Icelandic mortar and machine gun teams arranged in an L-shaped ambush formation north of the highway. The crossfire set three BTRs ablaze, destroying them before the troops could escape the burning vehicles. The remaining BTR continued toward the ferry dock and disembarked its troops, who took positions on the waterfront.

The Icelandic Heavy Recon team drove its APC to the ferry dock in the town, dismounted its troops, and assaulted the enemy position. After a half-hour fire fight with casualties on both sides, at 1500 on January 1, 2032, the Icelandic task force took the remaining enemy troops prisoner and set up secure positions on the dock.

At 1530 on January 1, 2032, the Icelandic Coast Guard cutter Týr engaged the Ivan Rogov landing ship with 40 mm naval cannon fire. After sustaining a hit on its bow, the Rogov ship maneuvered westward to avoid the incoming projectiles. At 1715 on January 1, 2032, the enemy naval flotilla disappeared into the Atlantic Ocean. The security force stood down, while the police reestablished control in the town.

At 0430 on January 2, 2032, airborne troops from the 1st Canadian Parachute Battalion jumped from two C-130J Hercules that took off from Gander, Newfoundland, landing on Þorlákshöfn's ferry dock to relieve the Icelandic Security Force, which returned to the Keflavik base and demobilized.

5.4.3 Scenario 3 – Non-Arctic State Occupies Grimsey Island, North Iceland

In an effort to retain sovereignty over Grímsey Island, on August 5, 2032, the Icelandic Government took measures to ensure the island remained populated to maintain its territorial integrity through occupation by its citizens. These measures included raising the fishing quota for the surrounding waters, establishing a low-cost grocery and supply delivery service, and subsidizing home heating for the island's permanent residents.

Cruise ship visits bringing nearly one hundred thousand tourists to the island each year helped support the economy; however, many residents still had only part-time seasonal employment. Lack of jobs did not allow young people to return to the island after completing schooling on the mainland. With no children left on Grímsey Island, the elementary school in operation since 1904 closed at the end of the academic year.

In late August 2032, several older lifetime residents of the island died or became ill and required long-term institutional care in Akureyri or Reykjavik. On September 15, 2032, the island's last ten residents held a meeting and voted to move to the mainland, using government funds to help them relocate. By the end of 2032 the island was deserted except for occasional visits by scientists and tourists.

On July 1, 2033, the Icelandic staff at the Joint Aurora Observatory (JAO) departed for a month-long vacation were replaced by ten foreign scientists from an Arctic Council Observer country. On July 3, 2033, the JAO began intercepting signals from NATO and commercial ships reporting navigational and sea ice conditions in the Arctic Ocean. North American Aerospace Defense Command (NORAD) monitored the JAO's subsequent retransmissions of the information to foreign ocean-going vessels in the Bering Strait preparing to enter the Transpolar Sea Route (TSR) on their way to the North Atlantic.

On July 15, 2033, the Arctic Council received notification from a Pacific nation that it regarded Kolbeinsey Island north of Grímsey as flotsam, or abandoned property (*terra nullis*) awash at sea. As such, according to international maritime law, Kolbeinsey was subject to a territorial claim by a third party because Iceland had relinquished sovereignty over the island. The Pacific nation also informed the Icelandic Foreign Minister of its intention to occupy neighboring Grímsey Island. The foreign power would use Grímsey as a staging area for naval engineers and support troops while they were constructing an artificial island around Kolbeinsey containing a landing strip, harbor, and air-defense weapons.

At 0400 on July 17, 2033, a U.S. satellite identified a naval task force exiting the TSR sailing toward northern Iceland, consisting of a Type 052D Kunming-class guided missile destroyer, a Yuting-II-class landing ship with a company of naval engineers, and four dredging ships.

At 1000 on July 17, 2033, upon activation by Iceland's National Security Council, two mechanized squads with their APCs from the Heavy Recon Patrol loaded onto a C-130J for transport to Grímsey Island. After landing 2 h later, the patrol's squads deployed in tactical positions at the airport using their APCs and other abandoned vehicles on the island. Orders from the Coast Guard incident commander to the security force on Grímsey Island were to defend the island from a simple battle position (Fig. 5.5).

At 1300 on July 17, 2033, the Icelandic Prime Minister called the Chairman of the NORDEFCO Command in Stockholm and requested additional support. Three hours later, the Danish Naval vessel HDMS Lauge Koch (P-572), on routine patrol southeast of Greenland, steamed toward the aggressor's flotilla and fired a warning shot from its 76 mm naval cannon over the bow of one of the landing ships.

At 1315 on July 17, 2033, two Norwegian Navy CB-90N fast assault craft carrying 40 Norwegian Coastal Rangers transiting from Jan Mayen to a training site in Greenland were diverted to Grímsey Island. At 2000 on July 17, 2033, under orders from NATO Joint Arctic Command in Nuuk, Greenland, the Norwegian

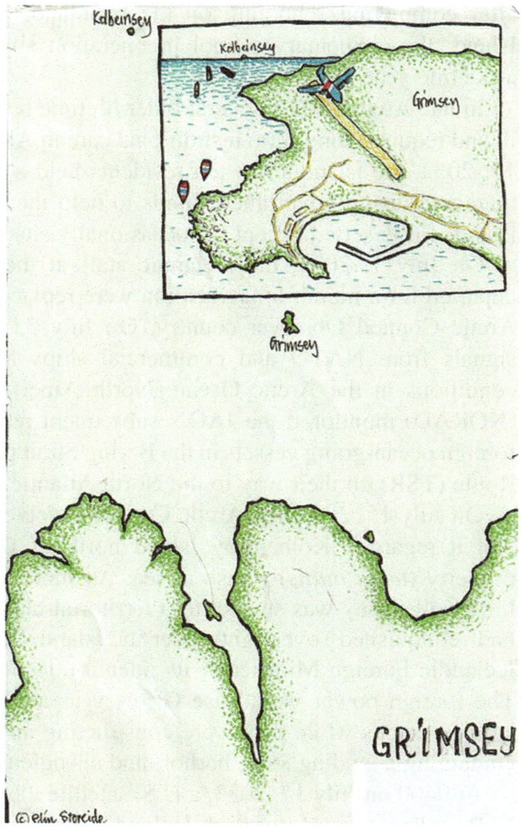

Fig. 5.5 Non-Arctic state tries to occupy Grímsey Island. (Illustration by Elin Flóki Storeide)

Coastal Rangers disembarked on the island and took up positions to support the Icelandic Security Force.

At 0300 the next day, the Icelandic Prime Minister informed the foreign power's embassy in Reykjavík that any attempt to land on Grímsey Island would be considered an act of war. The invasion would be repelled using all the tactical capabilities of the assembled multi-national security assets. At 0600 on July 18, 2033, the aggressor-nation's satellite reconnaissance confirmed Grímsey Island's defence posture, and its naval task force turned northward and began transiting the Transpolar Sea Route back to the Pacific Ocean.

The Danish ship and the Norwegian Rangers resumed their missions in the North Atlantic, leaving the Icelandic Security Force in control of Grímsey Island. At the same time, the police in Akureyri arrested the foreign staff at the Joint Arctic Observatory on suspicion of espionage.

5.4.4 *Scenario 4 – Human Trafficking in Seyðisfjörður and Egilsstaðir, East Iceland*

With three million tourists arriving annually, in 2034 Iceland became a major transshipment point for human trafficking. Europol's human intelligence (HUMINT) service learned that criminal syndicates planned to begin transshipping sex slaves from Eastern Europe through Iceland to other parts of the world.

On May 21, 2034, Europol tracked a 120-meter superyacht as it sailed from Durrës Harbor, Albania through the Strait of Gibraltar and into the Atlantic Ocean. According to its manifest, the craft was bound for the ferry port at Seyðisfjörður carrying a crew of twelve, listing its purpose as a "tourist charter to Iceland". HUMINT sources reported that prior to sailing, two panel vans delivered ten young females to yacht personnel who took them aboard the ship. Europol notified the Icelandic Coast Guard, which moved the patrol vessel Þór to the east coast to block the yacht once it was in port at Seyðisfjörður.

At 1000 on May 23, 2034, the U.S. National Security Agency (NSA) intercepted voice communications from a known drug cartel's Gulfstream V leaving Barranquilla, Columbia bound for Egilsstaðir. NSA notified the Icelandic Prime Minister through the U.S. Northern Command about the suspicious flight.

The aircraft had six pilots and two crew members. The flight plan listed a refueling stop and change of pilots at Newark International Airport, New Jersey, with an ongoing flight to Gander, Newfoundland. Upon landing at Egilsstaðir airport at 1400 on May 24, 2034, the flight plan designated an on-ground time of 2 h to board passengers and refuel. The Gulfstream would fly directly to Culiacán, Mexico after a stopover in Gander to change pilots and refuel.

At 0600 on May 24, 2034, members of the Viking Squad reported to the NATO base at Keflavik. There, they met up with two Heavy Weapons Patrol squads from the security force with their armored personnel carriers (APCs). Both the police and security force squads boarded the Icelandic Coast Guard's C-130 bound for Egilsstaðir Airport, where it landed at noon. Viking Squad members established positions around the airport in anticipation of the Gulfstream's arrival. The security force squads driving in their APCs were met by the local ICE-SAR team who guided them to the airport (Fig. 5.6).

At 1000 on May 24, 2034, the Icelandic National Security Council activated the Light Recon Patrol squad in Seyðisfjörður. Local ICE-SAR teams transported the recon squad to the Seyðisfjörður ferry dock where it took up positions to support the police. Four hours later, the yacht docked in the town, and the Coast Guard vessel Þór moved into position to prevent the yacht from escaping. Police officers went aboard to inspect the cargo. The captain insisted that he could not disturb the charter guests and refused to bring them on deck.

The police aboard the ship called for backup, but as soon as additional officers arrived, automatic weapon fire from the yacht's helipad deck killed or wounded several of them. Immediately, the Light Recon Patrol squad around Seyðisfjörður engaged the enemy with rifle fire. As it assaulted the yacht's gangplank, the squad

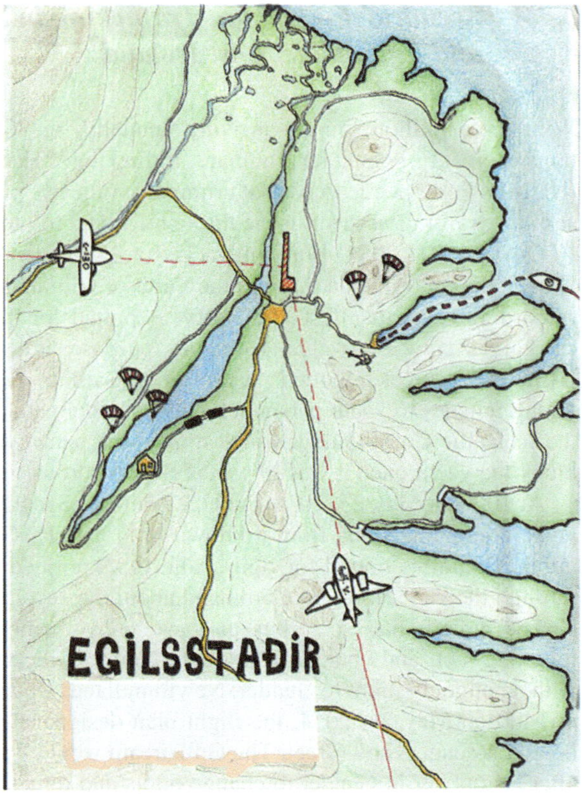

Fig. 5.6 Human trafficking in Seyðisfjörður and Egilsstaðir. (Illustration by Elin Flóki Storeide)

dispatched several of the criminals while the others surrendered. At 1500 on May 24, 2034, the squad secured the hostages on the yacht.

At 1415 on May 24, 2034, the Gulfstream landed at Egilsstaðir and disembarked personnel. The Viking Squad approached the plane but came under automatic weapons fire from the aircraft.

After fierce fighting at the airport, at 1430 on May 24, 2034, the surviving criminals boarded rental vans and headed south on Highway 1 toward the yacht. After seeing the yacht captured, the criminals turned the vans onto Route 931 to escape. The drivers pulled into the Hallormsstað Artisan and Home Economics Boarding School, entered the dormitory, and took the students and staff hostage. After demanding safe passage out of the country, the criminals murdered the headmistress to demonstrate their resolve.

In the meantime, an airborne squad from the local Light Recon Patrol boarded the Icelandic C-130J at Egilsstaðir Airport. The aircraft took off and flew at 500 feet above the school. The recon squad jumped out of the cargo ramp at the rear of the plane, landing near the hostages, where they took up support positions using their rifles.

The security force APC drivers followed the vans to the school and dismounted their troops. The Icelandic airborne team provided supporting fire while the dismounted troopers maneuvered to the dormitory and captured the criminals. At 1500 on May 34, 2034, the hostages were freed and the Icelandic security squad took control of them until the police arrived. At 1515 on May 24, 2034, police secured the ferry dock, airport, and school. One hour later the Viking Squad and security force members flew to Keflavik on the C-130J and demobilized.

5.5 Summary

This chapter presented four different fictional scenarios during which the recommended Icelandic Security Force encountered various hypothetical threats to the country's territorial integrity and public safety. The scenarios were created using standard exercise practices, concepts, and terms according to U.S. Army doctrine. The operational environments in which the scenarios took place were based upon predicted effects from global climate change during the 2030s when the Arctic could be more accessible than in the past.

In the scenarios, state and non-state actors involved in territorial incursions, terrorist attacks, or criminal activities were deterred or defeated by the capabilities of the recommended security force. The scenarios were not important in and of themselves. Rather, they served merely to illustrate possible ways to deploy and utilize the security force with the proposed table of organization and equipment. Using both reserves and volunteers, the security force, as activated and controlled by the relevant government agencies and according to Iceland's current security policy, was organized into teams that were able to deter or defeat the threat. During the scenarios, the security force provided support to the Iceland Coast Guard and Police. In two of the scenarios, other NATO and NORDEFCO nations provided assistance to Iceland; however, most of the successful responses to the threats came from the Icelandic Security Force.

The scenarios themselves are disturbing regardless of the fact that they are fictional. My hope is that the terrorism, crime, and assaults on sovereignty they portray never occur in Iceland. At the same time, Iceland is low-hanging fruit, ripe for the picking by foreign troops, terrorists, or criminals anxious to exploit the country's openness and lack of security preparedness. Some aspects of the scenarios are already happening in Iceland, including human trafficking. More threats may follow according to informed predictions.

Despite being dark, the scenarios were intended to stimulate discussion about the types of threats Iceland may face in the future and what sort of security apparatus would be needed to defeat them. Rather than focus on the scenarios, I encourage policy makers and academics who intend to contribute to the policy process to discuss whether the security force as proposed is appropriate to meet such threats.

A positive outcome of that discussion would be that my recommended security force, while requiring changes to the final product, provides a model for the need to protect Iceland's territory and people in the future Arctic operational environment. Even if only the existing ICE-SAR teams are incorporated into a reconnaissance organization as volunteers during future emergencies (a small step from the fictional portrayal in *Katla*), I will have achieved my goal.

I have been successful with scenario design before; unfortunately, I predicted events that actually occurred. For example, during late spring 2001, as an Army emergency preparedness liaison officer assigned to the U.S. Federal Emergency Management Agency, I wrote an exercise to train troops in the Washington, D.C. area to respond to a terrorist attack. The scenario I created called for a delivery truck bomb to explode at the Pentagon and the table-top exercise took place in August 2001 involving the troops that would respond to such an emergency. Those same troops responded 1 month later to a real terrorist event when a hijacked commercial aircraft crashed into the Pentagon.

I feel now as I felt in 2001; that is, I would rather be wrong than right about Iceland's security threats in 2030. Yet just as before with the 9/11 hijackers, in 2030 I will have no control over whether my predicted scenarios, or the need for Iceland to develop its own security force, come to pass. If any of them do occur in some form, I can only hope that I will have contributed to Iceland's national security by helping increase its preparedness.

References

Department of Defense. 2010. *Department of Defense Dictionary of Military and Associated Terms*. Joint Publication 1-02, November 8, 2009. Washington, DC: Joint Chiefs of Staff. Accessed May 10, 2020, https://www.acqnotes.com/Attachments/Joint%20Publication%201-02.pdf.

Department of the Army. 2010. *Exercise Design*. TC 7-101. Headquarters, Department of the Army, November 2010. Washington, DC: Army Publishing Directorate. Accessed December 2, 2020, https://armypubs.army.mil/epubs/DR_pubs/DR_a/pdf/web/tc7_101.pdf.

———. 2015. *The Army Universal Task List*. ADRP 1-03. Army Doctrine Reference Publication No. 1-03. Headquarters, Department of the Army, Washington, DC, October 2, 2015. Accessed November 8, 2020, https://armypubs.army.mil/epubs/DR_pubs/DR_a/pdf/web/adrp1_03.pdf.

Gao, Charlotte. 2018. China Issues Its Arctic Policy. *The Diplomat*, January 26, 2018. Accessed September 14, 2020, https://thediplomat.com/2018/01/china-issues-its-arctic-policy/.

Holland, Brynn. 2018. Operation Desert Storm. *History*, August 22, 2018. Accessed November 15, 2020, https://www.history.com/news/history-vault-operation-desert-storm.

Joint Chiefs of Staff. 1995. *Joint Doctrine for Military Operations Other Than War*. Joint Pub 3-07, June 16, 1995. Washington, DC: Joint Chiefs of Staff. Accessed December 10, 2020, https://www.bits.de/NRANEU/others/jp-doctrine/jp3_07.pdf.

Kearney, Paedar. n.d. *Down by the Glenside (Bold Fenian Men)*. Dublin: Unknown Publisher.
Kitzen, Martijn. 2020. Operations in Irregular Warfare. In *Handbook of Military Sciences*, ed. A. Sookermany, 1–21. Cham: Springer. Accessed December 2, 2020. https://doi.org/10.1007/978-3-030-02866-4_81-1.
Shalikashvili, John M. 1995. Forward. In *Joint Doctrine for Military Operations Other Than War*. Joint Pub 3-07, June 16, 1995. Washington, DC: Joint Chiefs of Staff. Accessed December 10, 2020, https://www.bits.de/NRANEU/others/jp-doctrine/jp3_07.pdf.
Wang, Serenitie. 2021. Contentious Cruising: China's Paracel Islands Tourism Is About More Than Travel. *CNN Travel*, January 24, 2021. Accessed January 25 2021, https://www.cnn.com/travel/article/south-china-sea-paracel-islands-cruise-tourism-intl-hnk/index.html.

Chapter 6
Iceland's Conflict History and Attitudes Toward Security

Is it better to out-monster the monster or to be quietly devoured?
Beware that, when fighting monsters, you yourself do not become a monster . . .
for when you gaze long into the abyss. The abyss gazes also into you.

Friedrich W. Nietzsche

Ísland úr NATO og herinn burt (*Iceland out of NATO and the Army out of Iceland).*
Icelandic political slogan chanted during 1960s' anti-NATO marches

Abstract This chapter discusses the Icelandic public's support for a sub-strategic security force using the country's conflict history and modern attitudes toward defence. In chronological order, Iceland's conflict history over the past millennium included territorial combat, feuding, civil war, religious violence, "witch" murders, piracy and human trafficking, political insurrection, global warfare, "war bride" persecution, and violent protests. These conflicts were addressed by Icelandic and/or foreign security forces, demonstrating that the country has a long tradition of using them. Previous surveys measuring security attitudes revealed Icelanders are not uniform in their views toward security. Substantial numbers of respondents are confused about the country's existing security relationships. Nearly half of Icelanders believe their country neutral despite being a member of NATO and having security treaties/agreements with the U.S. and the Nordic countries. Icelanders think their greatest security challenges are natural disasters and financial collapse but rank organized crime and terrorism as possible threats. Although Icelanders support immigrants, nearly half believe that terrorism and crime could increase because of Muslim immigration (refugees/asylum seekers), with one quarter wanting to reduce immigration. Taken as a whole, Iceland's conflict history and current security attitudes could allow the public to support the sub-strategic security force recommended in this book.

R. P. Wheelersburg, *The Security of Iceland and the Arctic 2030*, Springer Polar Sciences, https://doi.org/10.1007/978-3-030-89948-6_6

6.1 Iceland 2030 – Organians in a Dark World

I understand the Icelandic public's view about national security by studying the plot of the *Star Trek* episode titled "Errand of Mercy" (Coon 1967). After their recently declared war on the Klingons, the Federation ordered the U.S.S. Enterprise to the planet Organia on the border of the Klingon Empire to prevent it from being conquered. After transporting down to the planet's surface, Captain Kirk and Science Officer Spock encountered a primitive agricultural society apparently without a government. Seeing a castle on the hillside above the settlement, Mr. Spock told the captain that such a structure required a feudal system with a king, nobles, warriors, and peasants, which would be impossible at the level of society they were observing.

Kirk and Spock tried to convince the Organians that the Federation was benevolent and would allow them to remain free, compared to the dictatorial Klingons who would enslave them. However, the planet's informal council assured the Enterprise crew that they were in no danger. After the Klingons apparently slaughtered hostages to end resistance to their invasion, conflict ensued with the two alien contingents attacking each other. As the Enterprise and Klingon space ship prepared to do battle in orbit, the Organians used tremendous power to make the ships' weapons systems so hot that they could not be touched by their human or Klingon operators. The hosts then demonstrated their power by turning into diamond shapes of pure energy and disappearing. Spock's interpretation was that the Organians had evolved to the point that they no longer needed physical shape and only assumed it when more primitive alien visitors required a conventional reference point. Spock likened humans' evolutionary relationship to the Organians as the amoeba was to humans. Therefore, as the Organians insisted all along, they were so advanced that they could not be harmed by the primitive weapons of their attackers (Coon 1967).

Importantly for my purposes here, "Errand of Mercy" was the first appearance of the aliens known as Klingons. Before they were rehabilitated and joined the Federation in the second television series, "Star Trek: The Next Generation" (Fontana and Roddenberry 1987), the Klingons were deadly enemies of humans, Vulcans, and virtually every other intelligent species they encountered. Living in a feudal and dictatorial society, the militaristic Klingons were the antithesis of the democratic and free Federation. Instead, the Klingons were ruthless killers who destroyed anyone who refused to submit to their will. Their fictional warrior code was modelled after Japanese Bushido to the point that, like *Kamikaze*, a Klingon suicide that resulted in the death of an enemy was considered the highest form of bravery.

While Icelanders are not as evolved as Organians, many Icelanders believe that they do not require protection against sub-strategic threats, foreign or domestic because they have evolved beyond the need for weapons and protection. In reality, because Icelanders are not made up of pure energy, it does not matter what they believe, since there are Klingons in the world today who would destroy them brutally to accomplish their goals. Whether it be authoritarian states like Russia and China or non-state actors such as terrorists and criminals, if there is some reason for a malevolent power to harm Iceland and its citizens, it will.

6.2 Threats to Iceland 2030

Frankly, the naïve belief by many Icelanders that their people are basically good and that evil forces can bring no harm to their country is more dangerous than any other view against strengthening their national security. Icelanders are not isolated from the threats of terrorism, organized crime, or states intending them harm, all of which are prevalent in the world today. As bloodthirsty Mexican drug lords illustrated during the past decade and a half, they will kill anyone who stands in their way. Since 2006, at least 80,000 people have gone missing in Mexico, ostensibly victims of the drug cartels. The pace of the disappearances surpasses even those in Latin and South America during the Cold War, when repressive regimes with U.S. support for their anti-Communism kidnapped and murdered perhaps 100,000 of their leftist opponents over several decades (Sheridan 2020).

Organized crime syndicates have also permeated global society, expanding out from their Sicilian origins to take control of much of the illegal activity in Europe. While they became the darlings of film and television, the Italian Mafia had tremendous power that involved murder, kidnappings, theft, and political corruption stretching into all parts of the world. One of their most powerful crime families remains an evil presence in over 30 countries, responsible for most of the cocaine smuggling in Europe, and has infiltrated other criminal and legitimate businesses there. The problem is that, unlike Hollywood portrayals that show the Mafia only fighting amongst themselves, the unfeeling killers eliminate anyone who gets in their way, intentionally or not. The high-profile murders of Sicilian prosecutors in 1992, along with their families and police escorts, illustrated the criminals' ruthlessness (Donadio 2020).

There is some indication that portions of the Icelandic government understand the need to develop the ability to counter those evil entities who will do their country harm. The Security and Defence Directorate within the Ministry of Foreign Affairs has realized that possibility for a long time, developing the Crisis Response Unit and supporting the NATO and the U.S. bilateral treaties. Although it is not yet law, in late 2020 the Minister of Justice, Áslaug Arna Sigurbjörnsdóttir announced she was proposing a bill to allow foreign police officers to conduct investigations in Iceland. More importantly, foreign police would be permitted to carry weapons during the course of their duties there. The bill is predicted to pass despite initial questioning by Pirate Party MPs. Should the proposal become law, foreign police such as the U.S. Federal Bureau of Investigation (FBI) or Interpol could, under certain circumstances, enter Iceland with weapons, although they would be subject to regulation by the responsible ministry (Fontaine 2020b).

That step represents a potential change in Icelandic political opinion about security, especially given recent events that precipitated negative public reaction. In 2020, following criticism about former U.S. Ambassador Jeffrey Ross Gunter, a Trump appointee and sycophant, retweeting a Trump Tweet that referred to Covid-19 as "the invisible China virus", the ambassador requested armed security guards. Embassies in Iceland do not have armed security guards, and the request to have such escorts was strongly condemned, as was his permit application to carry a sidearm. While there are nearly 70,000 registered firearms in Iceland (and probably another 10,000 unregistered weapons), the process to own a gun in the country is complex, especially for handguns (Kyzer 2020).

The conclusion to be drawn from both the growth of threats and the possibility of changing attitudes on security in Iceland suggests that, while there will be resistance to my recommended security force, there may also be support. As former Icelandic Ambassador to several European countries and the U.S., Einar Benediktsson asked after the Americans left Keflavik, "Of course, Russia was not going to attack Iceland, but what about the other threats? Could a small band of terrorists take over and hold hostage a full-fledged NATO member country?" (Benediktsson 2011). As this book goes to press, that scenario has occurred in Afghanistan where a formal government with a well-equipped and trained military estimated at 300,000 soldiers, supported by the U.S. airpower, collapsed in weeks to disorganized bands of terrorist rabble with approximately 60,000 fighters driving old trucks or walking (Beale 2021).

The overall purpose of this recommendation, as stated above, is to help start the discussion about the country's sub-strategic security between both groups, supporters of the concept and those who oppose it. In order to help start that discussion, I need to examine how Icelandic attitudes toward security developed, including the belief that the country is somehow impervious to attack like the fictional Organians had.

My analysis begins with Iceland's history of conflict. The substantial amount of conflict and violence occurring there over time may surprise those without knowledge of the country's history. Iceland's dark side includes territorial battles, blood feuds, generational combat over political control of the country, religious assassination including persecution of witches, piracy and mass kidnappings, world war, and global nuclear confrontation. Thus, it is important to examine each of the conflict periods briefly to understand how they contribute to Icelandic security attitudes today. The analysis suggests that my recommended security force conforms to those attitudes so that the public could accept it.

We must also review the history of foreign powers taking over the nation's external security needs. Larger states providing national security for smaller states, often those without a military, is referred to as sheltering (Thorhallsson 2018). Even with that foreign sheltering against external threat, which did not always work to protect Iceland from such threats as pirates and kidnappers, Iceland internally was at times a hotbed of violent strife that sheltering could not prevent. Thus, the country has not been a place of peaceful relations that many Icelanders (and tourists) believe today. In fact, Iceland's history of conflict has been as violent and destructive as the history of many other states, although the country has avoided being the location of conventional warfare.

6.3 Iceland's History of Conflict

6.3.1 Conflict During the Age of Settlement

According to the sagas, the initial settlers to Iceland dispersed themselves sufficiently enough that they were able to avoid warfare despite many of them claiming large territories. The relative peace during the Age of Settlement (874–930) led to the creation of a homogeneous and egalitarian society of warrior-priest chieftains called goðar. Without a king or even a superior chieftain, each of the 36 goði contributed to the defence of Iceland using his followers (i.e., farmers and other warriors) as militias according to Nordic tradition. During the latter part of the Age of Settlement conflicts began to occur between the initial settlers and later arrivals who also wanted land (Byock 1990; Karlsson 2000).

Yet the process that happened in the Scandinavian countries where fighting between petty kings created an absolutist monarchy did not take place in Iceland due to three societal characteristics. One, unlike the Viking core area in Denmark, Norway, and Sweden, Iceland did not develop a nobility, petty kingdoms, or a territorial-based defense system that required armies. Two, Icelandic chieftains had no financial means to raise or sustain militaries, since their only source of wealth was land and limited taxation. Three, the chieftains' support was due to allegiance not territory, and many regions contained farmers whose allegiances were intermixed, leaving no consolidated territory to defend. Together, those three characteristics prevented the Icelandic settlers from creating armies and escalating conflicts to the point of societal-level warfare (Byock 1990, 2003).

Toward the end of the Age of Settlement, Iceland developed assemblies (Things) to resolve disputes by creating local and national laws, along with establishing courts at the local level to enforce them. At that time, one of the main sources of armed conflict in Iceland was feuding between families based upon vendettas (Karlsson 2000). As happened in many other cultures, the society developed ways to limit actual bloodshed to either the individuals involved or their immediate families (i.e., brothers and sons). When those failed, social change led to the revision of the judicial system during the Commonwealth Period (930–1262), when more formal means to limit conflict developed (Byock 2003).

Thus, Icelandic values acquired two cultural traits concerning security during the Age of Settlement. The first was that armed conflict was a matter between families based on personal feuding not territorial warfare. The second was that Icelandic society was founded on the rule of law not the power of might. Although elements of these traits pervade Icelandic attitudes toward defense today, neither would survive intact into the Icelandic Commonwealth.

6.3.2 Conflict During the Icelandic Commonwealth and the Age of Sturlungs

The initial decades of the Icelandic Commonwealth saw the division of the country into quarters, each with a certain number of goðar and the total number fixed at 39. Several ways to prevent feuding or warfare arose including the creation of quarter courts. Quarter courts allowed both the accused and the aggrieved to accept the legal decisions because the change of venue provided an impartial process, which could not be guaranteed when trying someone in his own court (Byock 2003). In addition, a person accused of a crime or transgression could plead guilty, allowing the aggrieved party to determine a settlement. Lastly, the ability to settle disputes between individual freemen prevented the potentially warring factions from engaging in combat, leading to a peaceful solution. The revised legal system during the tenth century was designed to "end fighting by peaceful settlement and [by creating] the machinery to arrange such resolutions". Further, by the start of the eleventh century, Icelandic law forbade the duel, "probably because it embodied outdated values incompatible with the system of negotiation and compromise which by then had become firmly entrenched" (Byock 1990: 66, 69).

Iceland entered a period of relative peace once again. That peace would be short-lived as the struggle for power among the goðar during the late twelfth and thirteenth centuries led to full-scale battles involving thousands of troops and even naval warfare. When the Icelandic Commonwealth ended, during the Age of the Sturlungs (1220–1260) the goðar system collapsed as six families gained power by combining chieftaincies. The large chieftains or storgoðar formed a new social rank similar to lower-level nobility such as a *jarl* [earl] in the Norse core kingdoms. Many of the chieftainships were acquired by marriage or through inheritance, which was not controlled with rigidly enforced generational transmission rules like in formal monarchies. Eventually, civil war broke out, which in the absence of a centralized government meant conflict between warring factions seeking to take control of the entire country (Gersovitz and Kriger 2013). Iceland's armies consisted of storgoðar, their families, and any others who could be recruited or forced to fight for their chieftains (Sveinsson 1953; Byock 1990; Karlsson 2000).

During the mid-thirteenth century, three major battles took place within Iceland between these armies. The Battle of Örlygsstaðir was fought on August 21, 1238 in northern Iceland between the Sturlungs family and the alliance of the Ásbirningar and Haukdaeler families. Probably equal to European-style warfare at the time, the battle involved nearly 3000 troops and fortresses, although there were few battle deaths (Karlsson 2000; Saga Museum 2020). The second was the naval battle Flóabardagi fought with Viking long-boats in 1244. In 1246, the "bloodiest battle" in Icelandic history occurred at Haugsnes in northern Iceland. The clash was again between the Sturlungs and the Ásbirningar families, the latter of whom was defeated with the loss of their chieftain, ending his attempt to seize power (Saga Museum 2020; Historic Places in Northwest Iceland n.d.).

Although individual goði security was subordinated to collective security by the establishment and acceptance of the rule of law, civil/internal warfare continued into the late thirteenth century as ruling families gained larger fiefdoms through bloody conflict (Byock 1990; Short 2010). In order to eliminate fighting, in 1262 leading families including the Sturlungs requested and received protection from King Haakon IV. The Norwegian Crown ruled Iceland until the founding of the Kalmar Union in 1380, which brought Norway, Sweden (including Finland), Denmark, and Iceland under one flag (Karlsson 2000; Lindow 2001). When the absolutist King of Sweden, Gustav Vasa, dissolved the Kalmar Union in 1523, Denmark was left with control over Norway, the Faroe Islands, Greenland, and of course Iceland (Gustafsson 2006).

Thus, two different values contributing to the modern Icelandic attitudes towards security were acquired during the Icelandic Commonwealth. First, internal strife to gain political control was ineffective and costly, leading only to more bloodshed and civil war. Second, and perhaps more importantly, Icelanders learned to accept foreign control over their country's security, a practice that would continue for nearly 750 years.

6.3.3 *Conflict During the Reformation and the Witch Trials*

The next conflict period involved religious violence including the assassination of religious leaders, officials, and Danish citizens. The conversion to Christianity occurred in relative peace as the Icelandic AlÞingi voted to accept baptism around the year 1000, following an epiphany by the pagan Lawspeaker, Þorgeir Þorkelsson. The conversion was easier to accept perhaps because there was a period wherein the old ways continued in tandem with Christian practices. In addition, Icelanders may have seen similarities between the two religions. For example, the pagan religion consisted of a polytheistic structure of Odin, Thor, and Loki, with Freyja and other female goddesses also having a prominent role. Christianity (actually Catholicism) had a similar structure, with God in the three forms of Father, Son, and Holy Spirit. Also, Mary was essentially God-like, immaculately conceived, giving birth to a god, Jesus, then taken into heaven without death. Thus, the Vikings may have adopted Christianity without concern for the loss of their pagan religion (Sawyer and Sawyer 1993).

The acceptance of Catholicism gradually changed to resistance as the religion began to permeate all aspects of life. In the Nordic countries, the Catholic Church became not just a religious entity, but it structured the political system and government. In countries like Sweden, Christian parishes assumed primary administrative roles such as tax collection and keeping records of births and deaths. While still under Danish control, Iceland remained Catholic, with the country divided into two Bishoprics. These were Hólar in the North and Skálholt in the southwest near modern-day Reykjavík, which were held by Icelandic bishops until the Reformation (Buck 1957; Karlsson 2000). Perhaps the most important aspect of Catholicism in Iceland was that it brought literacy to the country that allowed poet priests like Snorri Sturluson and his pupils to write down the orally-transmitted sagas of the pagan

Vikings; perhaps the only time in history that the victors recorded the history of the losers. Catholicism changed the country forever as Iceland would not be the same today without its saga literature.

Religious violence in Iceland began as the Reformation reached the country during the mid-sixteenth century. Upon his coronation in 1536, King Christian III established the Danish Lutheran Church and decreed that all Danish subjects must convert to Protestantism, except in Iceland. The two Icelandic Catholic bishops wielded both secular and ecclesiastic power in the country while Denmark was absorbed in the civil war to dissolve the Kalmar Union. The new Danish Royal Governor to Iceland, Klaus von Mervits, was commanded by the king in 1539 to advance the Reformation and to seize any Catholic Church property. Governor von Mervits did just that when his sheriff took over the monastery on Viðey Island in Reykjavík Harbor. In 1541, Danish soldiers took Biship Ögmundur hostage, sending him to Denmark where he died the next year. The other Icelandic Bishop, Jón Arason, resisted and in 1550 was beheaded along with his two sons (Karlsson 2000). Ultimately, the loss of the Catholic Church and conversion to Protestantism tightened Denmark's grip over Iceland (Buck 1957).

Half a century later, witch trials began in Iceland. As the witch persecution spread throughout Europe beginning in the mid-1400s, executions approached genocide levels. For example, some scholars estimate that between 60,000 and 100,000 "witches" were executed, while other estimates are much lower (Levack 2006). The Nordic countries did not escape the hysteria and between 1634 and 1779 over 500 witch trials took place in the court at Göta, Sweden alone (Hagen 2013; van Gent and van Gent 2008). Denmark conducted witch trials and seemed unrepentant about the murders that took place in the name of religion, numbering 500 executions between 1609 and 1687 on Jylland alone (Ankarloo and Henningsen 1987).

The Icelandic witch trials lasted from 1554 until 1719, ultimately resulting in at least 25 murders of accused witches. Iceland executed mostly male witches, 23 men compared to two women (Karlsson 2000), which was different than elsewhere where mainly women were killed. Eventually, the homogeneity of the Icelandic people eliminated religious violence. Pagans were not persecuted since the witch trials did not focus on those who practiced seiðr, non-Christian sorcerers whose supernatural powers derived from nature and preceded Christianity (McCoy 2019). Conversely, witchcraft was based on the Christian belief that Satan was the dark power behind witches and to engage in it was to make a pact with the devil.

Icelanders acquired another cultural value during the period of religious conflict lasting from the mid-1500s to the mid-1700s. That value was religious tolerance and acceptance of non-Christian belief systems. Religious tolerance extends today to Muslim immigrants who are relatively new to the country but who have three mosques there. In addition, pagan religious practitioners including seiðrs (sorcerers) and followers of the Norse Pantheon like members of the Ásatrú Society are not only recognized by the state and respected by the public, but a new pagan temple is being built in downtown Reykjavík. One of my Icelandic friends has a father who is a leading member of the society. According to the father, many members resist building the temple because they do not believe paganism is a religion. About one quarter of Iceland's population today is either not registered with any faith or is part of pagan societies (Fontaine 2020a).

6.3.4 Conflict During Piracy and Human Trafficking

The next stage of Icelandic conflict involved piracy and human trafficking. The stage was set by Danish Crown policies that protected only trade goods moving to and from Iceland during the Monopoly Period (1602–1786). Although there were Danish warships in the region, they played virtually no role in protecting the Icelandic people.

> There has been repeated evidence of how utterly defenceless Iceland was in the centuries. In the monopoly period the Crown usually sent a warship to protect the trading ships on their way to Iceland, and sometimes these warships stayed in the waters around the country during the summer, trying to keep foreign fishing vessels away from the coast, until the trading ships were ready to return to Denmark. Sometimes the same ship was commanded to protect both Iceland and the Faroe Islands, countries separated by hundreds of nautical miles (Karlsson 2000: 143).

During the seventeenth century, whaling and cod fishing brought seafarers from many nationalities into the North Atlantic. In 1615, three Basque whaling ships were lost in Icelandic waters during a storm, with nearly eighty crewmembers surviving. The sailors washed ashore in the West Fjords and began stealing food and other goods from local farmers to survive. After the sheriff wrote a decree allowing residents to kill the "Spaniards" as they were mistakenly called, he and some farmers used ancient halberds supplied by Denmark to massacre dozens of the Basque sailors (Miglio 2015).

A decade later, Barbary Pirate expeditions to Iceland resulted in the kidnapping of nearly five hundred men, women, and children, who were taken to Africa and sold into slavery. The attacks occurred mostly along the southern coast and in the Eastfjords (Karlsson 2000). Some of the attacks were led by a Dutch privateer captain named Murad Reis. Reis was captured and taken to the Barbary Pirate region where he converted to Islam, which meant he was accepted by the local authorities who ran the slave trade. As an admiral, Reis approved attacks by other pirates and sanctioned corsair expeditions to the North Atlantic. He led two expeditions to Iceland himself in 1627 and 1631. On the first raid, Reis took a single ship into Grindavík harbor, where he encountered a Danish merchantman. With the number of ships sailing around Iceland flying various flags, no one seemed to notice the pirate vessel. Reis' crew quickly subdued the Danish vessel, taking its crew hostage (Hreinsson and Nichols 2020).

Grindavík was one of the few natural harbors on the southern coast of Iceland. A Danish trading post with warehouses and unimproved docks on the shore for unloading fish from local vessels, Grindavík was not really a town. Like much of the Icelandic countryside at the time, it consisted of a few widely dispersed farmsteads. The area was defenseless, with no real Danish contingent or concentration of Icelandic chieftains to challenge threats. With his 30 men, Murad Reis attacked the farms, killing anyone who resisted. He took the survivors as hostages, especially women and young girls, to be sold as slaves in North Africa. Stealing anything of value, the pirates burned the remaining buildings (Karlsson 2000; Hreinsson and Nichols 2020).

Fig. 6.1 Christians being sold in the Algerian slave market 1684. (Illustration by Jan Luyken, used with permission)

Next, the pirates sailed northwest toward the Danish Royal Governor's residence at Bessastaðir near today's Hafnarfjörður. Governor Holger Rosencrantz received word about the earlier raid and built a hasty fortification made of cut sod. The Danish Governor had cannon manned by among others, the famed Icelandic cannoneer Jón Ólafsson, along with two merchantmen armed with cannons in the harbor. The artillery, along with support from visiting chieftains from the north and their retinues, allowed the governor to engage the pirates. The ad-hoc Icelandic security force exchanged fire with the pirate fleet, but there were no serious casualties or vessels sunk on either side. In the ensuing confusion, one of Admiral Reis' ships ran aground. Unfortunately, the governor did not continue to fire on the pirates, and the tide floated their vessel free. The ships sailed away with their hostages toward the Eastfjords, where Reis took over a hundred additional captives (Karlsson 2000; Hreinsson and Nichols 2020).

From the east, Admiral Reis sailed toward the Westman Islands after joining up with another pirate vessel. When they landed their pirates at Heimaey, however, it was a different, bloodier battle than at the governor's residence. Murder, rape, and kidnapping virtually destroyed the small community of Heimaey. With no defence, since the Danish fort on the island was abandoned at the time, the pirates spent days killing over 40 people including women and children. They took 250 captives for the Algerian slave market (Fig. 6.1), looted the area, and set fire to the buildings

Fig. 6.2 Drawing of the Sixteenth Century Danish Fort Skansinn, Heimaey Island, Iceland. (Illustration by Susan Wheelersburg, used with permission)

including farms, Danish warehouses, and the village church. Only a few of the captives escaped or bought their freedom from the slave owners (Karlsson 2000; Hreinsson and Nichols 2020).

Paradoxically, the Danish King had built a fort on Heimaey Island in 1586 (Fig. 6.2). The fort, called "Skansinn", had a militia and cannons to protect the Danish stores and keep the English ships and crews from raiding them. After the pirate raid of the island in 1627, Danish royal troops manned the fort once again. In 1808, the fort's garrison was unable to fight against a British military fleet returning to England after bombarding Copenhagen and destroying Danish and Norwegian ships (Karlsson 2000; Welcome to the Vestmannaeyjar n.d.). Despite the death and destruction on Haimaey, the Danes abandoned Skansinn in the nineteenth century, leaving the Westman Islands defenseless again.

During the mid-nineteenth century, Westman Islands Sheriff Andreas August von Kohl stood up a security force (or militia) using some funding from the Danish King. Von Kohl had been a captain in the Danish Army and was well prepared to recruit and train a militia. He recruited one hundred members (along with a training unit for boys) and formed it into 60 armed infantrymen, consisting of four squads of 15 militiamen each. The militia was armed with rifles, uniforms, and other equipment. The Westman Islands militia had as its primary mission to repel attacks by foreigners, with a secondary mission to maintain internal order. To accomplish its missions, the militia coordinated with the police on the island, establishing the precedent that Icelandic militias provide military support to civil authorities (Sands 2022). The force drilled at Skansinn until von Kohl's death in 1860, when Pétur Bjarnasen became sheriff. Bjarnasen commanded the garrison until his death later that decade, when the militia disbanded and the fort was essentially abandoned for any military purpose. The British occupation forces used Skansinn as an army post on the Westman Islands during World War II (Welcome to the Vestmannaeyjar n.d.). Iceland would not have another fortification with the exception of the Keflavik NATO base, but even that facility was not protection against ground attack.

Icelanders learned from the pirate attacks that no other country is ultimately responsible for their sub-strategic security. Attempts by Denmark to provide external security ultimately failed for several reasons, including lack of political will or resources as the kingdom became entangled in costly wars elsewhere. The distances between Denmark and Iceland eventually required Iceland to provide for its own security by creating a fledgling security force. Unfortunately, the lack of sheltering from another country, whether Norway, Denmark, or the U.S. led to a cultural value that militaries are bad things, not worthy of Icelandic values.

6.3.5 Crown Sheriffs, Icelandic Police, and Provisional Security Forces

One of the benefits of being part of a European kingdom was having the services of a royal administration referred to as "The Crown". Those services were lost in 1918 when Iceland became independent. While there were negative characteristics of the Danish administration of Iceland, a positive one that led to an increase in security was the presence of royal sheriffs. Sheriffs have a long history in the Nordic countries, and in rural areas of the kingdoms they often represented the sole royal authority by keeping peace, making arrests, maintaining jails, collecting taxes, and of course providing security against insurrection.

In Iceland, Danish sheriffs provided internal and some external security well into the eighteenth century. When people were detained, the sheriffs would either keep them in local jails or send them to Copenhagen, both of which were very costly. The Crown denied a request by the sheriffs to merely hang convicts, and instead prisoners were put to work building the first prison in Reykjavík, which was completed in 1764. Despite having a true lockup, convicts escaped periodically and caused violence and crime until they were apprehended again. Clearly another level of security was required to protect the population.

Along with acting as night watchmen by patrolling the country's first industrial site in Reykjavík, Icelandic contractors were given the task of keeping order in the city using archaic spiked maces. Along with a mace, the watchmen each carried a lantern and an hour glass, singing out the hours throughout the night to show that they were on patrol. In 1791, a major criminal escaped from the prison, and the private citizens hired their own watchman. After the Royal Governor Erichsen stated that Reykjavík was without any security, in 1803 a Danish bailiff and two Danish soldiers were appointed as Iceland's first official policemen. Other Danish appointees to the fledgling police force followed until 1859, when the last Dane left service and only Icelanders became police officers (Hauptmann 2013). Since then, the police have formed Iceland's only ground-based security force, evolving over time to include such capabilities as explosive ordinance disposal and tactical operations.

The next potential security threat to the country came as Europe was drawn into World War I. During the war, Iceland did not face external threats since their ruling country Denmark was officially neutral. Around 1200 Icelanders fought in the war, mostly those who had emigrated to Canada or their descendants. Icelanders fought on the Allied side along the Western Front, and nearly 150 were killed in battle or died of the Spanish Flu shortly thereafter. Some Icelanders did not look kindly on their fellow citizens fighting in the war, and members of Parliament tried to get those who did sanctioned. A few others spoke positively about their heroism and dedication to their adopted country (Guðmundsson 2014).

The end of World War I, although it had minimal impact on Iceland itself, brought a new element into the country that would lead to conflict – the Communist Party. The 1918 Danish-Icelandic Act of Union granted the country limited independence, but it also removed Iceland from colonial status and eliminated the political parties that formed under it. In 1918 Icelandic university students who studied in Europe brought Communist ideas and values home to fill the vacuum, forming a Young Communist League, and creating new political parties. In 1921 Icelandic Communists were not officially affiliated with the Communist International's Second Congress; however, they were represented (Olmsted 1958; Gissurarson 2011). That growth in Iceland's Communist movement caused some consternation locally. The October Revolution in Russia involved such a violent overthrow of the Tsarist Government and total destruction of the existing society that it scared many other European governments about the Communist threat.

While attending the Third Communist International Congress in Moscow during 2021, Ólufur Friðriksson, editor of *The People's Paper*, met a young Russian orphan boy whom he decided to bring back to Iceland. Apparently under the premise of the boy having a communicable disease, again only a few years removed from the tragic Spanish Flu, the government tried to send the boy back to Russia. Friðriksson refused to allow the boy to be removed, organized a "para-military force" of nearly one hundred armed men, barricading himself in his Reykjavík house near the current city hall. The government organized a "reserve force of between four and five hundred men drawn from the ranks of citizens and athletes [led by the captain of the country's only gunboat]; and the city was in a state of war" (Friðriksson 2014: 200). The provisional security force assaulted Friðriksson's house on November 23, captured the boy, and sent him back to Russia (Gissurarson 2011). Once again, Iceland garnered a security force to defend itself from a perceived sub-strategic security threat.

This is how a leading Icelandic newspaper described the event on November 24, 2021.

> [s]ince the violence was perpetrated against the police of this town by Ólufur Friðriksson (ÓF), there has been no other discussed army in the town. The event was so unique and unprecedented in the history of the country. And it was no secret that he could precede even more serious events if the government did not take the reins firmly and decisively. The townspeople saw this, and so did the government. And in silence they worked to carry out what had failed that day, and also to detain those who had been violent in the face of the police force. The leader of that group was a former Danish commander, an Icelandic man, Johann Jonsson, who had been captain of a rescue ship in the Westman Islands, where he had been appointed chief of police, and Axel Tulinius, former sheriff. Before noon, the police

and their assistants went up to the house demanding to be admitted, but this was refused. The door was then broken and an easy passage into the house. It took a very short time until Ol Friðriksson and two or three others came out and were immediately taken to prison. Most were kept free, but there were a few placed in irons who committed some form of resistance or violence. The house was emptied after a short while, and about 20 people will have been taken out of it and placed in custody. OF's wife and the foreign boy were taken to a sanatorium. All of this was done with the utmost calm and without agitation. Those who may have had some urge to riot will have seen that the security was so tight that it was hopeless to start such a thing. And the reason why, among other things, what this was good is that no injuries, beatings or bloodshed would take place at the arrest of these many men, as could well have happened. It is safe to say that the general public thinks that the army has carried out a town clean-up. The air was a military mix. More serious riots than what happened the other day could erupt at any moment. But now the ignition, the initiators are arrested. Then first there is hope that it will be possible to protect the peace of our society and not least of this community (Morgunblaþiþ 1921).

Thus, the period from the mid-18th to the mid-twentieth century produced several aspects of the Icelandic value system toward security that remain today. The loss of Danish sheriffs led to the creation of a police force that continues to provide the country's internal security. However, the conflict history during this period also revealed that occasionally, Iceland had to resort to a volunteer security force to meet threats that exceeded the capabilities of the police. That practice continued through the 1940s after the American Army left the country, as violent anti-NATO riots threatened the democratic process in Iceland.

6.3.6 Iceland as "Nonbelligerent Ally" in a Global Conflict

In order to understand the evolution of Icelandic attitudes towards security, we have to move forward in time to the country's involvement in World War II. WW II brought changes to Iceland that made it a modern nation, including the Allies building infrastructure like the Reykjavík Airport and Highway 1 (Ring Road). The war also ended the financial depression that had impoverished the country. In 1939, Iceland was one of the poorest countries in Europe, and by 1945 it was one of the richest per-capita countries there. The improved economy led to a wage spiral and inflation, causing mass migration from the countryside to the capitol region (Ingimundarson 2011). Ultimately the war made Iceland the country it is today.

World War II also helped the country create its emergency preparedness system. During the war, there was an expectation that the Germans might bomb Iceland. The Ministry of Justice, working with the National Police Commissioner, established Air Defense Committees throughout the country. These committees designated civil defence districts, each with a governor and shelters for air raids. The governors appointed bunker guards, couriers, and scouts to find people displaced by air raids. The civil defence had a rank structure and uniforms, including helmets donated from the U.S., U.K., and Sweden that indicated their positions (FBI n.d.). Although the country was not bombed except for ships at sea, that civil defence preparedness carried over to today's excellent emergency system in the country.

World War II also gave Iceland its formal independence. Denmark's reign over Iceland continued nominally until 1944, when the Icelandic Parliament declared independence from Denmark's government in absentia. One factor contributing to Iceland making that declaration was the 50,000 American troops occupying the country, along with America being the first country to formally recognize Iceland's independence. Although the U.S. left Iceland following the signing of the bilateral Keflavik Agreement in 1946, within 3 years plans were made to join NATO, bring the Americans back, and build the Keflavik base (Ingimundarson 2011).

Unfortunately, the Second World War also brought a terrible cost to many young Icelandic women and girls who ended up being persecuted by the fellow countrymen for engaging in romantic relationships with foreign troops stationed there. When the Americans replaced the British, first with Marines and then with Army soldiers, 50,000 mostly single young males with money in their pockets were placed in close proximity to Icelandic females in and around Reykjavík. That caused many problems, not the least of which was jealousy by the Icelandic males who sometimes persecuted their countrywomen (Benjamin 2016).

The same situation took place in other countries the U.S. occupied during the war, especially the United Kingdom and its Commonwealth countries of Australia and New Zealand. However, there was one difference between Iceland and the United Kingdom – in the U.K. and Commonwealth, the young men were mostly away fighting in the war. Nearly 400,000 of them did not come back. Thus, when the American soldiers entered Great Britain, Australia, and New Zealand, the men were gone and there was no issue of jealousy. Yet that unrestricted access to native women created another problem. Tens of thousands of war babies were born out of wedlock, often to American soldiers with a wife in the States or to single soldiers who were moving on to fight in Europe or Asia. The American government took steps to discourage marriage between the local women and soldiers, such as requiring permission to get married from lower-level commanders, who could deny permission without reason. Other barriers such as the U.S. not providing the locations of accused fathers who had gone to duty elsewhere or were killed in battle also limited the number of marriages (Ollove 1991).

Called "The Situation" (*Ástandið*), the war caused problems at a societal level because Icelandic women often preferred the Americans to their own men for various reasons. The Americans had different dating customs, including actually dating instead of merely "hooking up" in the barnyard or hay loft. Americans gave Icelandic girls presents like candy, cigarettes, nylons, and alcohol. Since many Icelanders came from the countryside where there was no bars or restaurants, unlike native males, American soldiers also paid for drinks, meals, and movies and invited them to dances at the USO or on the bases. With parents' permission, Icelandic girls as young as 14 could attend those dances.

That led to a second difference between The Situation in Iceland and the situation in the U.K. As described by Ingimundarson (2011: 50), "during WWII, the Icelandic government had imposed tight controls on young Icelandic girls who fraternized with soldiers, even to the point of incarcerating them (between 30 and 40 girls were treated this way)". Those tight controls meant slut shaming, investigations, and actual incarceration for girls said to have bad moral character or mental problems because they consorted with soldiers (Benjamin 2016).

Women having children out of wedlock with occupying foreign soldiers always has been a universal phenomenon of wars between nations regardless of the time period, culture, or country involved. More than a century of U.S. troops being spread over Asia, Europe, and America has produced tens of thousands of war babies who did not receive assistance from either the father or his government. That fact is an overlooked part of the gendered discrimination contributing to the tragedy of warfare, especially on a global scale.

As an example, Norway just attempted to correct its discrimination against women and war babies of German soldiers. As with France, Holland, and other European countries, local women who consorted with Germans in Norway were outcasts, often marked by having their heads shaved, being paraded through the streets, and ending up financially stranded with a child.

> In a national reckoning with its past, the Norwegian government has offered an official apology to women – and their offspring – who were ostracized, stigmatized and in some cases deported because of their relationships with German soldiers during World War II (Magra 2018).

The situation was of course a bit different with countries occupied by German troops, some of whom killed the men they were displacing.

Still, Icelanders (at least the government and the country's men) considered these women who were dating American soldiers to be traitors because they were taking away the emerging national identity from the newly independent country. Traditionally, Icelandic women were considered to be bearers of the country's traditions and culture (Benjamin 2016).

One of the most striking items related to WWII at the Icelandic National Archives I reviewed is the Reykjavík Police blotter (i.e., log book) dedicated to paternity cases. The blotter contained the names of dozens of American soldiers, their female accusers, and the abandoned children. Although some were claimed by the fathers and moved to America, many children's names would forever be "Hansson" (his son) or "Hermansson" (soldier's son). Even more disturbing was the way many of the girls and women were treated by their government, in particular by Iceland's first policewoman.

Jóhanna Knudsen was a nurse and the first Icelandic female police officer, appointed in 1941. The 1933 Police Act did not lead to a national police force until 1977, and Knudsen was appointed as a Reykjavík city police officer while the country was still under Danish rule. Another item I reviewed in the Icelandic National Archives was a manual requested in 1941 by the Ministry of Justice from the New York City Police Department about appropriate uniform wear and behavior of female officers. Despite that guidance, Knudsen was unregulated by the authorities to the point that she essentially operated like an Icelandic Gestapo officer: authorizing illegal searches, denying appeals, relying on anecdotal information, and assigning summary imprisonment. Although many of the offenders were banished to relatives in the countryside, Knudsen confined some women at a place called Kleppjárnsreykir (Ómarsdóttir 2015).

> Per the legislation, police were to follow up on any leads about children and teenagers consorting with soldiers and pass them onto child protection services. In spite of this, Jóhanna refrained from passing on any of the information she gleaned, opting to investigate and interrogate suspects on her own. She would then take the girls to court, and push for maximum sentences, while the laws, as they were written, plainly suggested that the first step was to be a warning to the young women and their families. In just two months in 1941, Jóhanna had gathered the names of some 500 girls that were under suspicion. By the end of her tenure, she had recorded some 800 names in her journals, spearheading the most pervasive espionage operations on Icelandic citizens' private lives to date (Benjamin 2016).

Thus, although no actual combat took place there, during World War II Icelanders learned both the advantages and the high cost of having foreign troops stationed on their soil. As part of the positives, American and British troops constructed a tremendous amount of the nation's infrastructure, provided substantial employment, and ensured Iceland's independence. On the negative side, interactions between young soldiers and Icelandic girls/women had consequences even when those relationships ended in marriage. Although the exact number is unknown, "between the years 1942 and 1952, about one million American soldiers married foreign women from 50 different countries" (Wilt 2018). Iceland had its share too, with 332 women marrying foreign soldiers and leaving the country as war brides (Gunnarsson 2012). The ambivalence about foreign troops in Iceland as protectors led many to consider them a necessary evil, albeit one who deserves scorn and to be exploited (e.g., exorbitant salaries for natives on their bases for performing menial tasks like washing dishes and mowing lawns).

6.3.7 The Atom Station, *NATO, and Violent Protests*

Toward the end of the war, the U.S. wanted to maintain some of its forces in Iceland and proposed to its government that they retain three bases: one for sea planes in Reykjavík, one at Keflavik airport for patrolling, and one for a naval base at Hvalfjörður. Resistance to that proposal mainly came from Iceland's anti-militarism groups, but some of it came from the country's communist movement, headed nominally by Halldór Laxness. In 1945, the Soviet Union became Iceland's largest trading partner, and many in government did not want to jeopardize that relationship. Yet several events caused support for the Soviet Communists to fade. In 1948 the Soviets stopped trading with Iceland, doing damage to its fishing industry. The U.S. helped make up for the loss of financial support by giving Iceland grants and loans for their fishing industry and infrastructure. Later that year, there was a communist coup in Czechoslovakia that led to the Nordic governments worrying about the Soviets and their own countries. With the Americans gone and only 120 police officers as a security force, some in the Icelandic government believed that the Soviets might invade and take over the Reykjavík and Keflavik airfields for strategic reasons (Whitehead 1998).

Three major events that shaped Iceland's future security occurred shortly after WWII. The first was the signing of the Keflavik Treaty in 1946, which allowed the Americans to keep some equipment and weapons in the country even though troops were denied permission to stay. The second event took place in 1949 when Iceland became a founding member of NATO, allowing the alliance to build the Keflavik base. The third event was signing the Bilateral Defence Treaty with the U.S. in 1951. Shortly afterward, the Icelandic government fell, and a permanent schism was created between those who supported NATO and those who did not (Hardarson 1985).

Within a few years after leaving Iceland, the Americans returned to Keflavik and stayed for over half a century. That security relationship came at a cost, with NATO membership and base construction in Iceland making the country a nuclear target. This led many to conclude the base was not worth the cost, ultimately culminating in violent anti-NATO riots in 1949 (NATO n.d.). Iceland becoming a nuclear target, along with secret deals to conduct base negotiations (which actually occurred), was the theme of Laxness' *Atom Station*, which included the following quote.

> He asked whether I did not read the newspapers, but I laughed and said that I was from the north. Then he showed me an article in the evening paper, which said that a request had been received from one of the Great Powers that Iceland should sell, lend or give it her capital city, Reykjavík, otherwise named Smoky Bay, or some other bay equally suitable for attack or defense in an atomic war (Laxness 1982: 32).

It seems incongruous today that peaceful Iceland was the scene of an insurrection when its citizens rioted against the government's decision to join NATO and allowed the U.S. to build the Keflavik base. Iceland was split in two on the NATO question. While other countries had mixed views about joining NATO, Iceland was the only country to have a riot in its capital city over the issue. The relations between the pro- and anti-NATO factions became so heated that Prime Minister Stefánsson (Social Democratic Party) and Foreign Minister Benediktsson (Independence Party) each wanted to stand up a militia in case there was violence. The Independence Party actually did create an unarmed militia in 1949, but it was too small to have an impact on the violence that actually took place (Ingimundarson 2011).

Finally, the Icelandic police had to use tear gas to separate the two factions, leaving images of violence that stunned the country and the world (Figs. 6.3 and 6.4). During the parliamentary vote for NATO membership, protestors threw stones, broke windows, and assaulted each other. Three policemen and a dozen civilians were injured seriously and twenty insurrectionists were given prison sentences (NATO n.d.; Karlsson 2000). It would be 60 years before police again used tear gas on protestors rioting in front of the Althing.

Communists were involved in the anti-NATO movement, with Laxness as a figurehead for it. Laxness continued to deepen his Communist ties, even if only for show. In 1950 the Soviet-Icelandic Friendship Association was created with Laxness as president. That same year the Icelandic Peace Committee started and

Fig. 6.3 Fighting breaks out between anti- and pro-NATO supporters, and police. The windows of the House of the Althing have been smashed. March 30th 1949. Arnaldur Grétarsson & family, https://creativecommons.org/licenses/by-sa/3.0/deed.en

Laxness became a member of the Communist World Peace Council. An opinion poll in 1955 conducted in Iceland for the U.S. government, showed that the Soviet Union was more popular among the public than the U.S. Eleven percent of the respondents indicated they were Communists, a higher number than the other Nordic countries. In 1955 Laxness won the Nobel Prize for Literature, becoming the only Icelander to do so, which garnered him additional prestige and influence. In 1956 Iceland became the only western parliament with a Communist member. When the Progressive and Labor parties withdrew their support for U.S. forces in Iceland, the coalition government stated publicly that it would close the military bases. It seemed as if the leftists won and that the U.S. military would soon leave (Gissaurason 2011).

Although his book accurately depicted what was going on in secret between the U.S. and Icelandic governments, others say that Laxness stole the story used in The Atom Station from an earlier Czech book (Gissurarson 2005). In addition, Laxness was not a good communist because he liked his creature comforts too much. When touring his house (i.e., manor), I saw that he lived a rich life with luxuries such as a swimming pool, expensive furniture, and elaborate decorations. He also owned a 1969 Mark 2 Jaguar, the first Jaguar in Iceland. Laxness dressed like a wealthy man, travelled and stayed in the finest hotels, and often did not pay his workers. He cheated on his wife (Gissurarson 2015). Ultimately the nominal leader of the anti-NATO movement was a disingenuous capitalist who supported his own

Fig. 6.4 Depiction of the anti-NATO riots 1949. (Illustration by Þrándur Þórarinsson, used with permission)

high-standard of living, merely pretending he was taking the anti-militaristic, moral high ground.

As part of the negotiations over the bilateral treaty, the Icelandic government promised to develop a security force or home guard with an air contingent that would soon replace the American troops. Yet the anti-militarism sentiment in the country prevented that promise from being fulfilled, so the government created an Icelandic Defence Force (IDF) composed of the Americans. The IDF included a U.S. Army infantry battalion with artillery support, an Air Force interceptor squadron, and a Navy anti-submarine patrol squadron. The American contingent also contained an airbase squadron to maintain the Keflavik airfield and technicians to operate the radar stations along Iceland's coast. In May 1951, 200 U.S. troops arrived in the country, a number that grew to 3500 by 1953 (Whitehead 1998; Roehner 2009). A similar organization and troop level would continue for over 40 years, although a Reserve or a National Guard brigade, rather than an active duty one, eventually formed the ground defense portion of the IDF.

Iceland's anti-militarists continued to oppose the Americans throughout the 1950s. Just when it seemed that the Icelandic leftists would prevail in removing the NATO forces from the country, the Russians once again made serious missteps that caused communist popularity in Iceland to plummet. In 1956, the Hungarian Revolution scared the Icelandic government as had the earlier Czech conflict. The resulting fear of possible Soviet hegemony in Iceland led to the ruling coalition persuading the Americans to remain at Keflavik (Whitehead 1998).

The Cold War was on with Iceland at its center in the volatile and active North Atlantic. Throughout the 1950s, 60s, and the early 70s, the conflicts in Korea and Vietnam provided a relief valve so that a strategic-level war between NATO and the Warsaw Pact never took place. While the military tempo was high throughout the earlier part of the Cold War, in 1979 it got hotter when the Soviets invaded Afghanistan. With the U.S. backing anti-Soviet forces in Afghanistan, withdrawing from the SALT II Treaty, withholding grain and technology, and boycotting the 1980 Moscow Summer Olympics, the Cold War was in full force (Gaddis 2005).

The NATO Alliance survived several challenges over the years. Perhaps the largest challenge started in 1952, when Iceland became entangled in a sub-strategic security conflict with a NATO ally that would last until 1976. The so-called "Cod Wars" began as Iceland terminated its fishing treaty with Great Britain to begin the process of extending its exclusive economic zone (EEZ). Following the Parliamentary Act of 1948, called the Law on the Scientific Protection of Fishing Grounds on the Shelf, Iceland declared itself able to determine its own exclusive fishing waters. The country extended the EEZ in 1952 to four miles, in 1958 to 12 miles, in 1972 to 50 miles, and in 1975 to 200 miles (Karlsson 2000). Each of these limits put pressure on Great Britain to prevent its fishing trawlers from entering the newly created no-fishing zones that ultimately led to physical conflict between the two countries.

During the first Cod War from 1952 to 1956, there was no direct confrontation within the 4-mile limit, but Great Britain embargoed Icelandic fish, which hampered Iceland's attempt to improve its economy. From 1958 to 1961, British warships repeatedly tried to intimidate the Icelandic Coast Guard (ICG) as it attempted to enforce a 12-mile EEZ. When the 50-mile limit was instituted in 1972 and the 200-mile limit in 1975, open conflict broke out ICG vessels began cutting British trawler fishing nets. Coast guard cutters also rammed Royal Navy frigates to move them off so they could go after the trawlers. During the final confrontation, Iceland broke off diplomatic relations with Great Britain and threatened to leave NATO itself. In the end, Iceland actually won the Cod Wars as Great Britain relented to Iceland's 200-mile limit, and diplomatic relations were restored between the two NATO allies (Guðmundsson 2006). The Cod Wars illustrated that, while Iceland had no real defence on land, it could take care of itself at sea by enforcing its EEZ with the coast guard.

Yet the NATO base created problems from its start in the 1950s. NATO indicated to Iceland that, once Keflavik was completed, it would need a small security force of about 200 troops to protect it against sabotage. The coalition government insisted that such a force should consist of Icelanders not foreigners. Since Iceland lacked a military culture, not to mention the weapons or training to constitute a competent security force, the Americans said that the force should be from the U.S. or another NATO country. In addition, there was no contemporary support in parliament or among the public for an Icelandic security force. Ultimately, the U.S. was considered a safe choice for the force until Iceland could create its own national security policy and security force after the Americans left in a few years (Ingimundarson 2011).

The Americans returned but things were different this time. After the circumstances that led to The Situation during World War II, along with fears of a black market developing, the Icelandic government ensured that the Keflavik base would be nothing more than an armed camp filled with hired guns (although it was the U.S. paying for its own guns) for the country's protection. The newly formed U.S. Department of Defense made many concessions, creating a hardship duty assignment that was one of the most extreme for American troops and their dependents in any country. Initially, troops were not allowed to leave the Keflavik base, but that prohibition was eventually removed. Still, there was an effort to restrict the U.S. troops stationed in the country to those who were married to prevent young, single, well-paid soldiers from courting Icelandic women. In addition, although I have never found a primary source document stating the Icelandic government's position officially, secondary sources indicated that African-American troops were not to be stationed at the NATO base (Johnson 1971). If Iceland demanded racism against foreign troops for its own defence, then that demand was despicable.

Eventually, in a gutless capitulation similar to not allowing Armed Forces Radio to play Christmas music for the troops in Kuwait preparing to invade Iraq during the First Gulf War, the Americans gave way to the Icelandic demands for restricting the troops' activities to the base. In 1954, the U.S. completely fenced in the base, putting guard posts at the few gates. In addition, Iceland stationed customs posts at the gates as if those coming or going were crossing an international border. Service men and their families leaving the base were tightly controlled, passing through Icelandic Customs with the limitations of having only one opened and one unopened pack of cigarettes, one six pack of beer or bottle of wine, enough food for a picnic for each person in the vehicle, and receipts for such items as cameras or radios that they had to show upon leaving and reentering the base.

Additionally, the U.S. sailors and airmen could not be off the base between the hours of 10:00 p.m. and 6:00 a.m. except on Wednesdays, the night when Icelandic restaurants were not permitted to serve alcohol. While there were other duty stations in Iceland, such as radar facilities and the naval fuel depot in Hvalfjördur, those were mostly staffed by Icelanders or civilian contractors. As Ingimundarson (2011: 50) pointed out, "The timing of the Icelandic action was motivated by the mounting opposition to the US military presence, but patriarchal ideas – the desire to protect

Icelandic women from foreigners – provided that rationale". The restrictions and the high cost of living outside the wire was enough to make experiencing Iceland for military personnel and their dependents prohibitive, preventing Americans learning about the Icelanders, and vice versa. One of the most ridiculous acts by the Icelandic Government was preventing the Armed Forces Television from broadcasting outside the base; which remained a source of contention for many Icelanders my age who suddenly lost the ability to watch cartoons, movies, and their favorite TV shows (Lederer 1989). That censorship was not undertaken in a vacuum as at the time Icelandic government owned TV stations (the only kind in the country) could not broadcast on Thursdays so that the people could participate in traditional activities like knitting circles or music groups.

The result was a caged ally living on a barren and windswept outpost who had little contact with the host nation. As proof, one only has to read the words of a local Icelander who visited the base frequently with her grandfather who was employed there.

> When U.S. troops finally evacuated in September 2006, I was primarily concerned with the inevitable end to Wendy's runs with my grandfather, and wondering what he would do, now that his job (only ever explained to me as "lightbulb purchaser for the U.S. Army") was gone.
>
> The segregation of the base's inhabitants from the surrounding community was tangible – not in the least due to the barbed-wire fence separating the two. While some Icelanders worked for the army [sic] as non-military personnel – such as my grandfather – and were thus somewhat incorporated into daily life on the base, the perception that this little piece of land on the Reykjanes peninsula wasn't truly a part of Iceland, continued to dominate. Vague memories I have from visiting the base as a small child in the nineties and early naughts, are shrouded in mystery and intrigue, and feel more at home with my recollections of Saturday morning cartoon favorites [broadcast on the Armed Forces Network until censured by the Icelandic government], like *Little Nemo: Adventures in Slumberland.*
>
> Unintelligible writing on road signs, foreign voices everywhere, and Wendy's Frostys – paired with the threat that I better behave impeccably because Americans were not to be messed with and wouldn't see my young age as any excuse (and I'd seen enough episodes of Law and Order at Grandma's house to believe it) – made each trip feel like an adventure as scary and exhilarating as any rollercoaster. Especially interesting were the children I'd sometimes see; on playgrounds by the very Icelandic-looking apartment complexes, eating out at Wendy's with their military parents, kicking a soccer ball around a school yard – so similar to my friends and neighbors, but yet somehow so different.
>
> So, seeing the video [below] distributed to members of the U.S. Navy about to be stationed in Iceland was absolutely fascinating – being able to get even just a quick peek at how the people I found so entrancing and fantastical, saw *me* and *my side* of the barbed-wire fence (Kaaber 2015). https://www.youtube.com/watch?v=0HAT6O6gZws

Kaaber's piece accurately depicted the plight of the Americans willing to be shot down or vaporized protecting Iceland and upholding its NATO commitment. What makes her story even more compelling is that at the same time they were defending Iceland, Americans were subject to sexism, racism, ethnocentrism and other negative "isms" (e.g., isolationism through physical and cultural means). Kaaber's essay documented the poor treatment of Americans during 60 years of being held ransom by the Icelandic government for the "privilege" of serving as its mercenaries. Laxness's book should have been titled "The Atom Station, Inc." given the billions of dollars the U.S. paid to Iceland for infrastructure improvements and employment over the 60 years at the NATO base, while the people defending it had no real chance to experience their host nation except on short trips. Many Americans were happy to serve in Iceland; however, compared to other duty stations where they did or could serve like Germany or England, Iceland was a hardship post especially for junior enlisted who had no family with them.

After the fall of the Soviet Union in 1991, the Americans gradually withdrew strategic forces (e.g., anti-submarine patrols and bomber intercepts). Subsequently Iceland became a security vacuum in the North Atlantic as various organizations from the Keflavik Base departed to join the fight in the Global War on Terrorism in the Middle East. By the time the base closed in 2006, Keflavik provided no strategic assets to the NATO alliance.

During the American use of the Keflavik base, Icelanders learned to control foreign troops on their soil and to benefit financially from the country's bilateral security relationship with the U.S. Despite the political difficulties experienced during Keflavik's operational period, Iceland once again allowed a strategic power to take care of its security, putting off its own responsibility to protect the country's territory and citizens. When the Americans left unexpectedly, many Icelanders both inside and outside of government questioned how the country could protect itself. That question remains unanswered, especially if future threats appear that cannot be handled by the police or coast guard.

Since 2006, there were two homegrown violent incidents that resulted in the Icelandic police being overwhelmed to the point that cops were injured, despite using chemical irritants for the first time since the 1949 Anti-NATO riots. The first of these incidents followed the bank failure in late 2008, when Icelandic criminal bankers and corrupt politicians hid their money in illegal off-shore accounts, causing the collapse of the country's economy. During a five-day period involving thousands of protestors outside the Icelandic Parliament (Alþingi), late on January 22, 2009 some protestors began to throw stones and other objects at the building and police. After two police officers were hospitalized, a smaller group continued the violent protest into the early morning, causing the police to use tear-gas on the assembled crowd to disperse it (McLaughlin 2009; The New York Times 2009; D'Amico 2015).

The second incident took place in 2019 when protestors gathered outside parliament to protest the deportation and treatment of asylum seekers (Figs. 6.5 and 6.6). Both Icelanders and asylum seekers carried banners and signs while occupying the square across from the Althing. The protestors demanded that mass deportations of asylum seekers be stopped, that cases be reviewed individually, and that asylum seekers be given equal access to health care and improved living conditions. The protest started peacefully, but after some protestors began to erect a tent in the square, police moved in and suppressed the crowd with batons and pepper spray, with some being arrested (Iceland Monitor 2019).

Fig. 6.5 Violent protests at the Alþingi over asylum-seeker treatment 2019. (Photograph by Ómar Sverrisson, used by permission)

Fig. 6.6 Police use teargas to clear protestors from Austurvöllur Square across from the Icelandic parliament during protests against the treatment of asylum seekers in 2019. (Photograph by Ómar Sverrisson, used by permission)

These two incidents illustrated that the political situation in Iceland has changed somewhat, creating a climate where distrust of the government is sometimes met with violent clashes between protestors and the police. There could come a time in the near future, unfortunately, when those clashes involve more than tear gas or pepper spray, as armed protesters use Iceland's free speech to cover criminal or terrorist attempts to overthrow the government. Such an incident occurred in the U.S. on January 6, 2021 when armed insurgents stormed the capital in an attempt to overthrow the democratically elected government to keep a would-be dictator in power. If such an insurrection can happen in the U.S., it can happen anywhere. Thus, along with the evolving circumstances in the circumpolar region due to climate change that could lead to external threats, Iceland's potential internal conflicts could escalate to the point that there is a need for a domestic security force.

6.4 Modern Icelandic Attitudes Toward Security

Many of my younger Icelandic friends love the global culture phenomenon, *Star Wars*, which has as its premise a Just War philosophy of militarism. That philosophy meant that the Rebel Alliance was justified morally in fighting the Empire using insurrection and guerilla tactics after a repressive emperor seized control and eliminated the democratically elected galactic senate. Many Icelanders have no problem cheering for Hans Solo killing aliens with a blaster or Luke Skywalker slicing up imperial soldiers with his light saber, but the idea of an Icelandic police officer firing a weapon to prevent terrorists from slaughtering women and children is reprehensible. Other Icelanders recognize the importance of providing the country's own sub-strategic security and support efforts by the to do foreign and justice ministries to do so, revealing the schism in the public's view of security.

There have been two opinion surveys that illustrated the schism. In 1984, the Icelandic Commission on Security and International Affairs asked respondents ($n = 1003$) about their attitudes concerning Iceland's NATO membership, about the Keflavik Base, about whether Iceland should receive payment for the base, and about whether Iceland needed military defence. The results indicated that, at the time, there was strong support for NATO membership with 53% approving of it with only 13% disapproving (Hardarson 1985). Importantly, one third (34%) had no opinion. Most respondents who did not express an opinion described themselves as having no interest in politics. The survey also suggested that younger people were more opposed to NATO membership than older voters; however, most younger respondents had no opinion. Lastly, respondents with only a high-school education were more likely to have no opinion than those with a college degree (42% to 10% respectively) (Hardarson 1985: Tables 1, 3, 4, and 5).

The 1984 survey indicated that 54% of respondents approved of the Keflavik base, 30% disapproved, and 15% had no opinion. There was a distinction between approval of NATO membership and the Keflavik base, with more respondents opposed to the base than to NATO. Those answering that they were in favor of NATO were more likely to approve of the Keflavik Base (76%). Attitudes toward the

Keflavik base varied by age, with younger voters showing greater opposition. In addition, 63% of respondents believed that Iceland should receive payment for the Keflavik base, while 29% disagreed (Hardarson 1985: Tables 8 9, 10, and 16).

The 1984 survey asked whether Iceland needed military defence. Of those answering (n = 318), 45% indicated that they agreed Iceland needed defence while 43% said they disagreed. The remaining 12% thought that the need for defence was dependent upon the circumstances at the time. In his analysis, Hardarson (1985) found that most of those opposed to the Keflavik base also believed Iceland did not need military defence. Only 16% of those who approved of the Keflavik base thought Iceland did not need military defence.

Although those findings seemed like contradictions, Hardarson provided three possible reasons for them. The first reason was respondent error by some who did not connect the two questions. The second reason was that some Icelanders may have thought that the Keflavik base was important for reasons other than defence, such as employment or financial gain for the country. The third possible reason was survey error, because those answering "no" may have thought that the question asked whether Iceland should have a military. Thus, the 1984 survey suggested that "a majority of Icelandic voters supported the two strands of Icelandic foreign policy the past two decades – membership in NATO and the Keflavik base" (Hardarson 1985: 315).

A more recent survey revealed some of the same characteristics in Icelandic attitudes toward NATO and defence. In late 2016, the Social Science Research Institute at the University of Iceland conducted a survey on Icelanders' attitudes toward foreign affairs and security. The survey, which included four questions about the security of Iceland and Icelanders, had a much larger sample size (n = 4848) than the 1984 survey.

One question asked whether Iceland was neutral in military affairs. Amazingly, 44% of the respondents thought that Iceland was neutral, a belief held by more than half (52%) of those between the ages of 18 and 29. Importantly, the people who answered neutral were not from the anti-military parties like the Social Democratic Alliance or the Pirates. Instead, the coalition government at the time, made up of two long-term established parties, the Progressive and the Independence Parties, had the most members who believed Iceland neutral (54% and 52%, respectively). There was no observed relationship between respondents' attitudes and education level. Analysts thought that there may have been a problem with the wording of the question, in that respondents may have assumed the absence of a military in Iceland meant the country is non-aligned. The data also suggested that the general public was more concerned with sub-strategic security threats rather than conventional ones like an invasion (Ómarsdóttir 2018).

Another question in the 2016 survey asked respondents (n = 3607) about the greatest threats Iceland faced as a country. Survey takers were asked to rank eleven different threats in order of potential harm to Iceland's security. Not surprisingly given the recent emergencies during the years prior to the survey, especially the 2008 financial crisis and the 2010 Eyjafjallajökul eruption, financial and natural disasters were deemed greater threats than military attack. One quarter of respondents ranked natural disasters as the number one threat, while environmental problems were ranked number two. In third place was a financial crisis (20.3%), with organized crime (11.2%) in fourth place, and terrorism in fifth place (5.8%). Interestingly, given today's events, migration (2.8%) and epidemics (2.4%) ranked near the bottom, while only 1.8% of respondents thought that armed conflict represented a threat to Iceland (Ómarsdóttir 2018).

A third question asked individuals (n = 3197) to rank those same eleven potential threats in order of the personal danger each represented to them and their families. With that change from the country to the personal level, financial crises moved up to number one (37.5%) and natural disaster was number two (18.7%). Violence, organized crime, terrorism, epidemics, and armed conflict all ranked much lower with less than 6.1% each (Ómarsdóttir 2018).

The final question related to security asked respondents (n = 3645) to rank the greatest contributions to Iceland's security. The highest number of respondents (41%) considered peaceful relations with neighboring countries as making the greatest contribution to the country's security. Interestingly enough, aside from the country's small size (17%), both Iceland's NATO membership (17%) and its non-military status (16%) were considered the next important contributing factors. Lastly, 14% of respondents believed that Iceland's cooperation with Western countries outside of NATO contributed to its security (Ómarsdóttir 2018).

Combined, the two surveys indicated a pattern related to the recommendation for Icelandic security contained in this book. The 1984 survey showed that few Icelanders disapproved of NATO and the Keflavik base during the height of the Cold War, with nearly equal numbers believing that the country needed a defence and that it did not. Three decades later, in 2016, Icelanders were uncertain about their security relationships, with nearly half of the respondents believing that the country was neutral. Interestingly, over half of the members of traditional Iceland political parties thought the country neutral. After threats from natural disasters and financial collapse, respondents ranked organized crime and terrorism much lower threats both to the country and to them personally. The more recent survey also illustrated that Icelanders believed their peaceful relations with neighbors was the most important aspect of their security, followed by the country's size, NATO membership, non-military status, and cooperation with non-NATO countries.

As with much of Europe, some Icelanders view increasing immigration as a challenge to security. A 2020 study conducted by researchers at the University of Akureyri sampled opinions concerning immigrants. The study asked questions about immigrants learning Icelandic, voting, and being invited to social events. The survey also included a question about whether immigrants had a positive impact on their municipalities. The vast majority of respondents believed that immigrants should learn Icelandic (86% [n = 3585]). Forty-one percent believed that immigrants should accept the customs and values of their host country. Three quarters of the respondents (n = 3427) agreed that immigrants had a positive economic impact on their communities (Murdock 2020: 36, Fig. 1; 40, Fig. 7; 41, Fig. 9). From the survey responses, it seems clear that Icelanders welcome immigrants, believing they should be assimilated by learning Icelandic language and culture. Overwhelmingly, Icelanders thought that immigrants improved their communities.

Yet a 2019 study provided contradictory evidence about Icelandic perceptions of whether Muslim immigrants represented a security threat to the country. Just under half (44%) of the respondents (n = 1575) answered that Muslim immigrants increased the chance of terrorism occurring in Iceland. In addition, 27% believed that Iceland should reduce the number of Islamic refugees. Just over 30% of the surveyed respondents answered that immigrants will ultimately increase the crime rate in Iceland. The results showed that, while Icelanders are accepting of immigrants and the contributions they can make to their new country, new arrivals who are culturally different (e.g., Middle Eastern) than those from Europe or the Nordic countries are not viewed in as positive a light (Valdimarsdóttir and Jónsdóttir 2020: 223, Fig. 2; 230, Figs. 5 & 6).

The two surveys about immigrants revealed a distinction in Icelandic attitudes about security. Icelanders were overwhelmingly positive about the influence new members of society had on their economy and culture. They also wanted immigrants to learn Icelandic language and culture. Yet there was a substantial amount of distrust toward Muslim immigrants among respondents who believed that they could support or conduct terrorism in Iceland. Given recent events in Europe and the Nordic countries, there is every possibility that future threats could arise from the mutual mistrust between Muslim immigrants and right-wing nationalists. The threat can be diminished by the natives and newcomers coming to understand each other, especially through learning about the other's culture and language. One thing is certain; potential threats to Iceland could arise from processes beyond global climate change in the Arctic. Therefore, just like the external threats arising from changes in the circumpolar region, internal threats from other sources also contribute to the need for a security force.

6.5 Icelandic Sub-strategic Security Force

The question remains, what is the best means for Iceland to ensure that it can meet future sub-strategic security challenges effectively? Can Icelanders merely assume that there is no real threat to their peaceful existence and therefore no additional security preparedness is necessary? Is the solution to rely on others to provide for their security by stationing foreign troops on their soil? Or, as is suggested herein, is the answer that Iceland create a sub-strategic security force to support the coast guard, police, civil defence, and other civilian authorities during emergencies involving attacks by a malevolent actor?

To reach the answer, one must consider the fact that Iceland had a violent history over the past millennium, including both internal and external conflicts, during which Icelandic and/or outside security forces protected the country (or failed to do so). Thus, security preparedness has been a part of Iceland's past just as it should be part of its future. Relying on others for sub-strategic security is not the answer because foreign troops may not respond effectively (i.e., operating in Iceland's climate and terrain). In addition, having essentially foreign mercenaries as guarantors of the country's security means that Iceland is a militaristic society-just not one in which Icelanders have the courage to provide for their own security.

As stated previously, European-based Security Sector Reform (SSR) provides the answer. Iceland has the security management, including policies and government entities, effective security providers to meet most (but not all) threats, and non-government supporters to engage in that SSR. All that is missing is a bridge between the nation's security providers (e.g., police and coast guard) and foreign providers (e.g., NATO, the U.S.). That bridge can be built by Iceland creating its own sub-strategic security force.

Iceland has had some missteps in trying to develop its own defence capabilities through SSR, both at the management and provider levels. In terms of security management, the government previously created an Icelandic Defence Agency (IDA) with Parliamentary Defence Act no. 34/2008. The act authorized the IDA to administer all defence activities in Icelandic territory, to maintain relations with foreign states, and to cooperate with military authorities and defence organizations (e.g., NATO) for such activities as hosting exercises. The act's main goal was to ensure that democratic principles and entities controlled defence planning and operations. Also, the act separated governmental management of civilian security agencies, such as the police, from those agencies involving the country's defence (Ministry for Foreign Affairs n.d.-a).

As is often the case in Iceland, given its volatile political system and seemingly constant coalition shuffling, the IDA was always controversial due to the anti-militaristic attitudes held by some of the public and some political parties. The defence agency lost favor when a new coalition government was formed. The Parliamentary Defence Act no. 98/2010 eliminated the IDA and reverted its defence-oriented activities to the national police or the coast guard. The Ministry of Foreign Affairs maintained a Security and Defence Directorate that performed security functions like coordinating exercises and scheduling the air-policing missions (Icelandic Parliament 2010).

The second misstep in Iceland's recent defence activity was the creation of the Crisis Response Unit (CRU) as a uniformed and armed force that operated outside of the country. The CRU was contained within the Ministry of Foreign Affairs, and had the mission of conducting various peacekeeping activities during NATO or U.N. operations. CRU personnel were expert civilian employees such as medical staff, police, air traffic controllers, and logistic specialists. They were sent to Afghanistan, Palestine, Africa, and the Balkans (Ministry for Foreign Affairs n.d.-b). In October 2004, however, the Crisis Response Unit became involved in a controversy that would change the nature of the organization.

While serving as peacekeepers in Kabul, Afghanistan, members of the CRU were outfitted in military uniforms and issued personal arms including pistols, assault rifles, and submachine guns. A group of CRU members were in a market in Kabul on a personal errand when a suicide bomber targeted them and exploded a device nearby. Several of the CRU members were wounded, along with an Afghan girl and an American translator who were both killed. News footage sent back to Reykjavík showed the CRU members in uniform and carrying weapons, which outraged the anti-militaristic elements in the public and some political parties. The Ministry of Foreign Affairs contended that the CRU personnel were armed only for personal protection but maintained their civilian status (Lyall 2004). That policy of arming the CRU lasted until 2008 when CRU personnel changed to civilian clothes and operated abroad without weapons.

While Iceland's previous attempts at creating security managers and providers were unsuccessful, there is no reason not to try again. If you could travel back in time to the seventeenth century and talk to the hundreds of Icelanders taken away by pirates as slaves, I am sure most would confirm the need for such a force. Yet despite what some more vocal elements of the society maintain, Icelandic attitudes toward defence and security do not prevent the country from creating its own security force.

6.6 Summary and Conclusion

Are Icelanders truly Organians and unable to be harmed because they are so highly evolved that they do not need security protection? I hope so; however, I've never seen any of them turn into a gleaming pure energy diamond (although I believe I know one or two who might be able to do so). Therefore, for the time being, Icelanders should be prepared to protect their country and themselves against sub-strategic security threats that could arise during the next decade. In this book, I provided a recommendation for a security force with Icelanders as security managers, security providers, and security supporters. Although the country has used such security forces several times in the past, albeit on a provisional and limited basis, there has been some opposition to Iceland providing its own security.

That opposition was based partially on attitudes against a military. Yet there were several reasons why those attitudes were hypocritical. First, although anti-military movements and political parties were very vocal on several occasions against NATO and bilateral defence agreements, when the time came to act, Iceland remained in NATO, protected by foreign troops. Second, recent surveys of Icelandic attitudes toward the country's security revealed no clear condemnation of NATO, the Keflavik base, or bilateral treaties. In fact, the results were quite the opposite as those surveyed showed substantial support for those issues. Third, the intellectuals (symbolized by Laxness) who purported to be communist or leftist and therefore unable to support a relationship to the West for defence purposes, were often pretenders. In reality, these intellectuals took the best that democracy and capitalism had to offer for their own personal comfort and gain. In addition, many of those intellectuals hid behind hired guns carried by foreign troops while at the same time criticizing those who protected their security. As articulated so well in the movie *A Few Good Men*, Jack Nicholson's character addresses those who depend upon the blood and sacrifice of others to protect them, whom they then condemn for that protection.

> I have neither the time nor the inclination to explain myself to [those] who rise and sleep under the blanket of the very freedom I provide and then question the manner in which I

> provide it. I would rather that you just said "thank you" and went on your way. Otherwise, I suggest you pick up a weapon and stand a post (Sorkin 1992).

Regardless of the obvious hypocrisy by some elitist Icelanders, there are valid concerns within the general public and political parties involving the creation of an Icelandic security force.

The recommendation presented here addresses those concerns by using my understanding and knowledge of today's Arctic situation, along with Iceland's cultural value system and conflict history, both acquired during my four decades of working in the Arctic and Iceland. The first solution is that the proposed security force would operate only in Icelandic territory. While the security force could be part of the Directorate of Security and Defence under the Ministry of Foreign Affairs for administration and training, actual missions could be conducted under the Ministry of Justice in support of civilian agencies like the police and coast guard. Security support to civilian authorities during emergencies including natural disasters as well as threats by state and non-state actors, could be provided by reservists and civilian volunteers only. In addition, adopting my recommendation could restrict the security force from participating in exercises involving foreign armed forces, unless that participation is approved by the National Security Council (or foreign minister in a rapid reaction scenario) under highly controlled circumstances.

Iceland's future security threats could come from both external and internal sources. As I write these words, it has been less than a month since a bunch of domestic terrorists and morons took over the U.S. Capital and beat/killed policemen, spurred on by a traitorous, gangster president. The images shocked the world, but the incident graphically illustrated how Ancient Rome fell with half a million legionnaires in the field and one of the strongest forms of government in history – from within. Over time, insurrections and small border conflicts weakened the state and allowed barbarians to overrun and destroy the powerful Roman Empire.

On January 6, 2020, the U.S. had 11 nuclear-powered aircraft carriers, each of which could destroy many countries around the globe, 18 ballistic missile submarines and an unknown number of missile silos capable of ending human life on the planet, along with almost 2,500,000 troops under arms. Yet for several hours, a few thousand punks wielding sticks and stun devices, nearly overthrew the United States government at its most critical point of practicing democracy-certification of the election that facilitated the peaceful transfer of presidential power. That sequence of events following the presidential election in America showed yet another example of how Iceland could be lulled into believing that they are safe and do not require security, since it is provided by NATO and the U.S.

Today, the Icelandic Police and Coast Guard are capable of providing most of the country's sub-strategic security needs. To meet more serious threats, the recommended security force is necessary to make Iceland and the North Atlantic Arctic more secure in the future. At the strategic level, the risks in the Arctic are increasing due to the military adventurism being displayed in the circumpolar region. For example, in February 2021 the U.S. deployed B-1 Lancer Strategic Bombers in Norway for the first time to patrol the Arctic (Starr 2021).

At the sub-strategic security level, past events in Iceland suggested that bad things can come from a more open Arctic after the ice melts. Criminals involved in piracy and human trafficking, terrorists seeking political gain, and state-level actors claiming territory and natural resources near Iceland represent substantial threats. The potential for future internal conflicts arising in Iceland, also a prominent feature of its history, seems a distinct possibility as well. With two violent clashes occurring during the past decade that overwhelmed the police, along with a belief by Icelanders that the situation could get worse, a security force seems necessary to protect the country.

As this manuscript is being completed, an organized-crime murder took place in Reykjavík. On February 13, 2021, an Albanian man was shot to death on the street outside his home, with the police arresting a foreign national. With evidence that the murder was a gang-style execution, the Icelandic Police's Viking Squad raided a safehouse in Selfoss, a town 40 miles (60 km) from Reykjavík. During the raid, the police arrested three men, an Icelander and two foreign nationals. Europol is assisting in the investigation (meaning international involvement) and their appearance in court involved security by multiple police and Viking Squad vehicles and personnel just like mob trials in the U.S. and Europe. In all, nine males were arrested, eight of them foreign nationals (Ćirić 2021). Thus, Iceland enters the world of criminal mob violence, with the public learning that those criminals play for keeps.

The Minister of Justice, Áslaug Árna Sigurbjörnsdóttir in an interview with Icelandic state news RÚV on February 26, 2021, revealed that there were 15 organized criminal gangs operating in Iceland today. The multi-national syndicates in Iceland function both within the country and abroad, and like most mob organizations, they engage in both illegal and legal activities. To combat that threat, the government is adding nearly ISK3,500,000 ($30,000) to the country's law enforcement budget, stepped up domestic police activities, and expended international police cooperation. In the minister's words, "It is important that the Icelandic police – and in fact the Icelandic legal system – have the capacity, ability, and knowledge to deal with the complex tasks we face in the hardening world of organized crime" (Þórisdóttir 2021). Aside from these steps is there anything else Iceland can do to combat this growing threat to their security? Yes, they can accept and implement my recommendation for an Icelandic security force!

As part of a European Security Sector Reform analysis, my recommendation acknowledges that Iceland has the policies/laws, management, and public support to create such a security force. Luckily, some Icelanders feel the same way. Given a substantial effort by the government to seek the public's true feelings toward the idea and involving the people in creating a security force, I believe that many more Icelanders would also support it. Unfortunately, it might take a major attack on the country to indicate the necessity of having such a force. Until they are the evolutionary equivalent of Organians, Icelanders should provide for their own sub-strategic security. I hope the ideas presented in this book present a model for discussing that possibility.

References

Ankarloo, Bengt, and Gust Henningsen, eds. 1987. *The Witches' Europe 1400–1700. Historical and Anthropological Studies* [Original in Swedish]. Skrifter. Bind 13. Stockholm: Nerenius & Santérus.

Beale, Jonathan. 2021. *Afghanistan: How the Taliban Gained Ground So Quickly*. BBC, News, Asia. London. Accessed August 15, 2021, https://www.bbc.com/news/world-asia-58187410.

Benediktsson, Einar. 2011. *At Crossroads: Iceland's Defence and Security Relations, 1940–2011. Strategic Studies Institute, August 18, 2011*. Carlisle: U.S. Army War College Press. Accessed December 22, 2020, https://web.archive.org/web/20180316151727/http://ssi.armywarcollege.edu/index.cfm/articles/Icelands-Defense-and-Security-Relations-1940-2011/2011/8/18.

Benjamin, Gabriel. 2016. Slut-Shaming and State-Sponsored Persecution in "Situation" Era Iceland. *The Reykjavík Grapevine*, March 4, 2016. Accessed November 24, 2020, https://grapevine.is/mag/feature/2016/03/04/slut-shaming-and-state-sponsored-persecution-in-situation-era-iceland/.

Buck, Thomas. 1957. An Icelandic Martyr: Jón Arason. *Studies: An Irish Quarterly Review* 46 (182): 212–222.

Byock, Jesse. 1990. *Medieval Iceland: Society, Sagas, and Power*. Berkley: University of California Press.

Byock, Jesse L. 2003. Feuding in Viking Iceland's Great Village. In *Conflict in Medieval Europe: Changing Perspectives on Society and Culture*, ed. Warren C. Brown and Piotr Górecki, 229–241. Farnham: Ashgate Publishing. Accessed December 2, 2020, http://www.viking.ucla.edu/publications/articles/feuding_viking_age_iceland_byock_vengeance.pdf.

Ćirić, Jelena. 2021. Seven Now in Custody Due to Reykjavík Murder. *Iceland Review*, February 18, 2021. Accessed February 19, 2021, https://www.icelandreview.com/society/three-more-arrests-in-connection-with-reykjavik-murder/.

Coon, Gene L. 1967. *Errand of Mercy. Star Trek, Season 1, Episode 26, March 23, 1967*. Hollywood: Desilu Productions.

D'Amico, Guiliano. 2015. The Whole World Is One *Atom Station*: Laxness, the Cold War, Postcolonialism, and the Economic Crisis in Iceland. *Scandinavian Studies* 87 (4): 457–488.

Donadio, Rachel. 2020. Mob Justice. *The Atlantic*, Dispatches, October 2020, 22–25.

FBI. n.d. *Emergency Preparedness in Iceland, the War Years*. Forgotten Brotherhood of Iceland (FBI). Brotherhood of the FBI. Accessed December 2, 2020, http://www.ww2museum.is/emergency-preparedness-in-iceland-war-years.

Fontaine, Andie Sophia. 2020a. Pagans Are the Largest Non-Christian Faith in Iceland. *The Reykjavík Grapevine*, May 22, 2020. Accessed November 22, 2020, https://grapevine.is/news/2020/05/22/pagans-are-the-largest-non-christian-faith-in-iceland/.

———. 2020b. New Bill Would Allow Police from Other Countries to Operate in Iceland. *The Reykjavík Grapevine*, December 18, 2020. Accessed December 20, 2020, https://grapevine.is/news/2020/12/18/new-bill-would-allow-police-from-other-countries-to-operate-in-iceland/.

Fontana, D.C., and Gene Roddenberry. 1987. *Encounter at Farpoint. Star Trek: The Next Generation, Season 1, Episode 1, September 28, 1987*. Hollywood: Paramount Television.

Friðriksson, Guðjón. 2014. *Reykjavík Walks*. Reykjavík: Bókautgáfank Hildur.

Gaddis, John Lewis. 2005. *The Cold War: A New History*. London: Penguin Press.

Gersovitz, Mark, and Norma Kriger. 2013. *What Is a Civil War? A Critical Review of Its Definition and (Econometric) Consequences*. The World Bank Research Observer, May 13, 2013. Accessed November 12, 2020, https://openknowledge.worldbank.org/bitstream/handle/10986/21428/wbro_28_2_159.pdf;jsessionid=D8A16E73171DE00FEE3E7A441ED0751D?sequence=1.

Gissurarson, Hannes H. 2005. *Halldor K. Laxness as the 20th Century Intellectual*. Paper Presented at the Mont Perelin Society Regional Meeting, Iceland, 21–24 August 2005.

———. 2011. *The Icelandic Communist Movement, 1918–1998*. Reykjavík: Almenna bókafélagið.

———. 2015. *Personal Interview*, November 4, 2015. University of Iceland.

Guðmundsson, Guðmundur J. 2006. The Cod and the Cold War. *Scandinavian Journal of History* 31(2): 97–118. Accessed November 19, 2020, http://rse.hi.is/wp-content/uploads/2017/11/Gudmundsson.CodColdWar.SJH_pdf.

Guðmundsson, Helgi Hrafn. 2014. *Icelandic Blood Has Been Shed.* About 1,200 Icelanders Fought in World War I. The Reykjavík Grapevine, June 26, 2014. Accessed December 22, 2020, https://grapevine.is/mag/articles/2014/06/26/icelandic-blood-has-been-shed/.

Gunnarsson, Valur. 2012. Love in the Time of War. *The Reykjavík Grapevine*, August 31, 2012. Accessed December 22, 2020, https://grapevine.is/mag/feature/2012/08/31/love-in-the-time-of-war/.

Gustafsson, Harald. 2006. A State That Failed? *Scandinavian Journal of History* 32 (3): 205–220.

Hagen, Rune Blix. 2013. Witchcraft Criminality and Witchcraft Research in the Nordic Countries In *The Oxford Handbook of Witchcraft in Early Modern Europe and Colonial America.* ed. Brian P. Levack. Oxford University Press.

Hardarson, Ólafur Th. 1985. *Icelandic Security and Foreign Policy: The Public Attitude*. Cooperation and Conflict XX (1985), Research Note.

Hauptmann, Katharina. 2013. The Night-Watchmen. The Beginning of the Icelandic Police. *Wall Street International Magazine, Economy & Politics*, September , 2013 Accessed December 22, 2020, https://wsimag.com/economy-and-politics/4947-the-night-watchmen.

Historic Places in Northwest Iceland. n.d. *Haugsnes.* The Official Travel Guide for Northwest Iceland. https://web.archive.org/web/20110722021902; http://www.northwest.is/1sogulegirstadir2.asp.

Hreinsson, Karl Smári, and Adam Nichols. 2020. *Northern Captives. The Story of the Barbary Corsair Raid on Grindavík in 1627*. Keflavik: Saga Akademia.

Iceland Monitor. 2019. *Asylum Seeker Protest Ends in Violence*. Iceland Monitor, Politics and Society, Tuesday, March 12, 2019. Accessed December 20, 2020, https://icelandmonitor.mbl.is/news/politics_and_society/2019/03/12/asylum_seeker_protest_ends_in_violence/.

Icelandair. 2021. *Icelandair Flies to Antarctica: Putting the Ice Back in Icelandair*. Reykjavik. Accessed March 3, 2021, https://www.icelandair.com/blog/icelandair-flies-to-antarctica/.

Icelandic Parliament. 2010. *Act Amending the Defence Act, no. 34/2008.* Parliamentary Act, no. 98, June 28, 2010. Accessed December 5, 2020, https://www.althingi.is/altext/stjt/2010.098.html.

Ingimundarson, Valur. 2011. *The Rebellious Ally: Iceland, the United States, and the Politics of Empire 1945–2006*. Dordrecht: Republic of Letters.

Johnson, Thomas A. 1971. Bias Is Charged on G.I.'s in Iceland. *The New York Times*, November 17, 1971.

Kaaber, Emilía S. Ólafsdóttir. 2015. In Retrospect (ESOK). Views. *Iceland Review*, July 20, 2015.

Karlsson, Gunnar. 2000. *The History of Iceland*. Minneapolis: University of Minnesota Press.

Kyzer, Larissa. 2020. US Ambassador to Iceland Wants to Carry a Gun. *Iceland Review*, July 26, 2020. Accessed November 18, 2020, https://www.icelandreview.com/politics/us-ambassador-to-iceland-wants-to-carry-a-gun/.

Laxness, Halldór. 1982. *The Atom Station*. Sag Harbor/New York: Second Chance Press.

Lederer, Edith M. 1989. Fiercely Independent Iceland Plays Reluctant Host to American Forces. *The Los Angeles Times*, July 2, 1989. Accessed June 22, 2021, https://www.latimes.com/archives/la-xpm-1989-07-02-mn-4723-story.html.

Levack, Brian P. 2006. *The Witch Hunt in Early Modern Europe*. London: Pearson Longman.

Lindow, John. 2001. *Norse Mythology. A Guide to the Gods, Heroes, Rituals, and Beliefs.* New York: Oxford University Press.

Lyall, Sarah. 2004. Disquiet in Iceland That Its Peacekeepers Dress for War. *New York Times, Reykjavík Journal*, December 20, 2004. Accessed November 9, 2020, https://www.nytimes.com/2004/12/20/world/europe/disquiet-in-iceland-that-its-peacekeepers-dress-for-war.html.

Magra, Iliana. 2018. Norway Apologizes, 70 Years Later, to Women Who Had Relationships with WWII Germans. *New York Times*, October 19, 2018. Accessed November 29, 2020, https://www.nytimes.com/2018/10/19/world/europe/norway-lebensborn-german-girls.html.

McCoy, Daniel. 2019. *Norse Mythology for Smart People*. Seidr. Accessed June 22, 2021, https://norse-mythology.org/concepts/seidr/

McLaughlin, Kim. 2009. *Protestors Demand Iceland Government Quit Now*. Reuters, Latest Crises, January 24, 2009. Accessed December 22, 2020, https://www.reuters.com/article/idUSLO529629.

Miglio, Viola G. 2015. Chapter 3: A True Account of the Shipwreck of the Spaniards and Their Slaying by Jón Guðmundsson the Learned. In *Basque Whale Hunting in the North Atlantic*, ed. Xabier Lruzo and Viola G. Miglio, 81–108. Santa Barbara: Barandiaran Basque Studies Series & Strandagaldur.

Ministry for Foreign Affairs. n.d.-a. *National Security*. Government of Iceland, Reykjavík. Accessed December 22, 2020, https://www.government.is/topics/foreign-affairs/national-security/.

———. n.d.-b. *Iceland Crisis Response Unit (ICRU)*. Government of Iceland, Ministry for Foreign Affairs. Accessed December 22, 2020, https://www.government.is/topics/foreign-affairs/icru/.

Morgunblaþiþ. 1921. *The Law in Force* [Original in Icelandic]. Morgunblaþiþ, 9th year, 20th issue, page 1, Thursday, November 24, 1921. Timarit.is. Accessed December 21, 2020, https://timarit.is/page/1207062#page/n0/mode/2up.

Murdock, Elke. 2020. Perception of Immigration to Iceland: The Host Country Perspective. In *A Society of Diversity. Local Communication and Immigrants in Iceland*, ed. Meckl Markus and Gunnþórsdóttir Hermína, 35–44 [in Icelandic] Akureyri University. Accessed January 14, 2021, https://inclusivesocietiesunak.files.wordpress.com/2021/01/samfelfjolbr_stakar_corrected-20.01-1.pdf.

Museum, Saga. 2020. *The Battle at Orlygsstadir*. Reykjavík: Saga Museum. https://www.sagamuseum.is/overview/#the-battle-at-orlygsstadir.

NATO. n.d. *North Atlantic Treaty Organization*. Brussels: Iceland and NATO. Accessed January 17, 2021, https://www.nato.int/cps/en/SID-21A87C2D-BE06D3F1/natolive/declassified_162083.htm?selectedLocale=en.

New York Times. 2009. Protests Rattle Iceland's Government. *The New York Times*, Europe, February 2, 2009.

Ollove, Michael. 1991. *Finding of Wartime Fathers. Children of GIs from World War II Get U.S. Help at Last*. The Baltimore Sun, January 1, 1991. Accessed December 23, 2020, https://www.baltimoresun.com/news/bs-xpm-1991-01-01-1991001094-story.html.

Olmsted, Mary S. 1958. Communism in Iceland. *Foreign Affairs* 36 (2): 340–347.

Ómarsdóttir, Alma. 2015. *The Girls at Kleppjárnsreykir* [Original in Icelandic]. Reykjavík: RÚV (Ríkisútvarpið), Iceland's Public Service Broadcasting Network.

Ómarsdóttir, Silja Bára. 2018. Sýn Íslendinga á utanríkis- og öryggismál [Icelanders Views on Foreign Affairs and Security]. *Icelandic Review of Politics & Administration (Stjórnmál & Stjórnsýsla)* 1 (2).

Roehner, Bertrand M. 2009. *Relations Between Allied Forces and the Population of Iceland*. Institute for Theoretical and High Energy Physics, University of Paris 6, Working Paper. Accessed December 22, 2020, https://www.lpthe.jussieu.fr/~roehner/oci.pdf.

Sands, Frank Walter. 2022. The Westman Islands Army. How a Danish Captain Transformed Heimaey. *Iceland Review*, 3: 78–81. Reykjavik.

Sawyer, Birgit, and Peter Sawyer. 1993. *From Conversion to Reformation Circa 800–1500*. Minneapolis: University of Minnesota Press.

Sheridan, Mary Beth. 2020. Losing Control. The Search for the Disappeared Points to Mexico's Darkest Secrets. *The Washington Post*, December 3, 2020. Accessed December 4, 2020, https://www.washingtonpost.com/graphics/2020/world/mexico-losing-control/mexico-disappeared-drug-war/.

Short, William R. 2010. *Icelanders in the Viking Age. The People of the Sagas*. Jefferson: McFarland & Company.

Sorkin, Aaron. 1992. *A Few Good Men*. Culver City: Columbia Pictures.

Starr, Barbara. 2021. *US Deploying B-1 Bombers to Norway to Send a Message to Russia*. CNN, February 8, 2021. Atlanta. Accessed February 8, 2021, https://www.cnn.com/2021/02/08/politics/us-b-1-bombers-norway/index.html.

Sveinsson, Einar. 1953. *The Age of the Sturlungs: Icelandic Civilization in the Thirteenth Century*. Ithaca: Cornell University Press.

Thorhallsson, Baldur. 2018. A Small State in World Politics: Iceland's Search for Shelter. In *Icelandic Review of Politics & Administration (Stjórnmál & Stjórnsýsla)*, ed Guðmundur Heiðar Frímannsson, 14 (1): 61–82. Special Issue on Power and Democracy in Iceland.

Þórisdóttir, Anna Lilja. 2021 *The Minister of Justice Announces Action Against Criminal Groups* [Original in Icelandic] RÚV, February 26, 2021. Accessed February 26, 2021, https://www.ruv.is/frett/2021/02/26/domsmalaradherra-bodar-adgerdir-gegn-glaepahopum?fbclid=IwAR30 6hT9wFzKJ1HoVJS-ScIgQDvlo8AJzO4YcNrTpm0rdA_-5Xz2hEpLq3c.

Valdimarsdóttir, Margrét, and Guðbjörg Andrea Jónsdóttir. 2020. Attitudes Towards Refugees and Muslim Immigrants in Iceland: The Perceived Link to Terrorism. *Icelandic Review of Politics and Administration (Stjórnmál & stjórnsýsla)* 16 (2): 217–242. Accessed January 14, 2021, http://www.irpa.is/article/view/a.2020.16.2.7/pdf.

van Gent, Jacqueline, and Jacqueline van Gent. 2008. *Magic, Body, and the Self in Eighteenth-Century Sweden*. Leiden: Brill.

Welcome to the Vestmannaeyiar. n.d. *Skansinn*. Accessed December 2, 2020, https://visitvestmannaeyjar.is/en/skansinn/.

Wheelersburg, Robert P. 2021. North Atlantic Small State Security 2025: The West Nordic Security Zone. In *Small States and the new Security Environment, the World of Small States*, ed. Anne-Marie Brady and Baldur Thorhallsson, 87–101. Cham: Springer.

Whitehead, Thor. 1998. *The Ally Who Came in from the Cold. A Survey of Icelandic Foreign Policy 1946–1956. Centre for International Studies*. Reykjavík: University of Iceland Press.

Wilt, Brenda J. 2018. *War Brides. America in WWII*. Harrisburg: 310 Publishing. Accessed December 22, 2020, http://www.americainwwii.com/articles/war-brides/.

The manufacturer's authorised representative in the EU is Springer Nature Customer Service Centre GmbH, Europaplatz 3, 69115 Heidelberg, Germany. If you have any concerns regarding our products, please contact ProductSafety@springernature.com

Printed and bound by CPI Group (UK) Ltd, Croydon, CR0 4YY
30/09/2026
02225465-0016